MOST OF THE WORLD

MOST OF THE WORLD

THE PEOPLES OF

Africa, Latin America, and the East

TODAY

EDITED BY RALPH LINTON

New York : COLUMBIA UNIVERSITY PRESS

1949

This symposium and its publication were made possible by funds granted by the Viking Fund, Inc., a foundation created and endowed at the instance of A. L. Wenner-Gren for scientific, educational, and charitable purposes. The Viking Fund, Inc., is not, however, the author or publisher of this publication, and is not to be understood as approving, by virtue of its grant, any of the statements made, or views expressed, therein.

7,2
L

PUBLISHED IN GREAT BRITAIN, CANADA, AND INDIA
BY GEOFFREY CUMBERLEGE, OXFORD UNIVERSITY PRESS,
LONDON, TORONTO, AND BOMBAY

First printing January, 1949
Second printing May, 1949

To A. L. WENNER-GREN

PATRON OF ART AND SCIENCE
FOUNDER OF THE VIKING FUND

CONTENTS

CONTENTS

MAPS

MOST OF THE WORLD

Ralph Linton

INTRODUCTION

It is well-nigh impossible for the people of any period to tell what contemporary events are going to make an enduring mark on history, and when they guess, they usually guess wrong. If Americans were asked what had been the most important event of the twentieth century, most of them would certainly answer that it was the release of atomic energy. This is natural enough for the members of a society like our own, with its devotion to mechanical improvements and laboratory sciences. The atom bomb and the peacetime applications which we dimly perceive behind it mark the culmination of our drive for control over the forces of nature. The prospect of flying the Atlantic on a pound of fuel is the veritable apotheosis of the Machine Age.

Spectacular as these developments are, it seems highly improbable that they mark any turning point in history. Barring the universal catastrophe of an atomic war, there is no reason to believe that the new atomic discoveries will bring about any changes in our patterns of living which would not have been brought about by the continuing development of our pre-atomic technology. A full-scale atomic war would be less a turning point in history than its rounding off, a neat conclusion to the far from creditable record of our species; but the chances of such a debacle will diminish as our understanding of its full consequences increases. No nation was ever deterred from using a weapon because it was too terrible or too inhuman, but a fair number of weapons have been laid aside because they were too dangerous. It will scarcely pay a nation to wipe out an antagonist at the cost of spending the next few thousand years in an atmosphere supercharged with fission products.

Turning from such doubtful musings, it seems likely that the historians of a few millennia hence will be less impressed than we are with our machines and laboratories. By that time mankind will have grown used to the control of nature which we now find so novel and thrilling and will be able to see the various steps in the development of this control in their proper perspective. One may hazard a guess that the nineteenth century will be the one called "The Century of Technological Invention." The twentieth will be known not as "The Century of Atomic Energy" but as either "The Century of Social and Political Invention" or "The Beginning of the Dark Ages." To be sure "Dark Ages" already has a particular meaning for Europeans, but it is improbable that the world-minded historians of the future will apply this term to a slight regression in the stream of European culture which was more than counterbalanced by the artistic accomplishment of the Tang dynasty and the vivid intellectual life of the Caliphate. If the twentieth is not the century of successful social and political invention, the oncoming Dark Age will be lit by no such beacons.

Whether we like it or not, *one world* is today a functional reality, and the unification has gone far enough so that the peoples of the world must stand or fall together. Like so many other significant developments of our time, this world unification has emerged without plan or intention. It has been an accidental by-product of the technological and commercial developments on which the attention of the West has been focused. These developments brought needs for new materials and, under capitalism, for new markets. They also made possible the conquest and domination of territories which would supply these needs. More recently, such developments as the airplane and radio have played their part in drawing tighter the web which commerce and industry had woven. The new ideologies which disturb us and the current attempts to develop some sort of world political organization are simply recognitions of the *fait accompli*, belated attempts to bring political and social forms into some sort of adjustment with current reality.

That human life has changed more in the last hundred and fifty years than in the previous five thousand is so obvious that it scarcely needs to be said, yet it is hard for most of us to realize how revolutionary the changes have been. Only a century and a half ago, most of the world's inhabitants lived very much as their ancestors had lived since the close of the Stone Age. In fact in many parts of the world this Age had not closed. Each little local group could and did produce its own necessities and traded its scanty surplus only for luxury objects or novelties. Here and there a group found itself dependent on its neighbors for some necessity such as iron or salt, but the quantities needed were small.

This narrow economic horizon was matched by an even narrower intellectual one. The average man's world ended very near the limits of his village's fields. The curious habits and ideas of people outside that world came to him as amusing bits of folklore, sources of self-congratulation. They had no bearing on his everyday life. Meeting a stranger was not only an event, but one fraught with peril, since, except for pilgrims and peddlers, honest people stayed at home. Religion centered in the local deity or the ancestral tomb, justice in the village elders gathered at the threshing floor in the cool of the evening, and government, where it existed, was represented by the tax collector, a pest regarded on a par with locusts or drought.

Today, thanks to the development of modern means of transport and communication, this age-long mold of human life has been broken. The docile villagers of Shangri La can tune in on Radio Moscow. Even the New Guinea pigmy turns from the cultivation of his vertical yam patch to wave a ragged O.D. jacket at passing planes, wondering why this magic gesture no longer brings chocolate from heaven. Even when the members of different societies cannot see or hear each other, they can still feel each other plucking at the strands which bind the modern world together. A new tariff in the United States deprives South Sea Islanders of badly needed matches and fishhooks, and the lagging production of English cotton mills causes political unrest in Nigeria. Most of the world has become so interdependent that a collapse of the present network of trade and communication would be a calamity on a par with the Black Death.

It is highly ironic that at the very time when the technical possibilities for communication seem to be rising to a crescendo, the lagging development of our social and political forms should threaten the whole structure. The liberation of India marks the end of a period of political unification unique in world history. From the end of the eighteenth century up until World War I, most of the planet was under unified control. After the early period of colonial wars, Europe settled down to a long peace based on the balance of power. In spite of quarrels within the continent, the European nations in their dealings with the rest of the world functioned practically as a confederacy. Such feeble members of the European family as Portugal and Holland were allowed to keep the colonial empires won by their vigorous ancestors and even Belgium was given the Congo. The few parts of the world which had not been brought under European flags during the earlier period of triumphant piracy were divided into spheres of influence which had all the advantages of colonial possessions without the expenses of colonial administration. Strikingly similar systems of government and exploitation were imposed wherever the Europeans came, and in spite of the varying flags and languages the *pax Europa* was as unified as the *pax Romana* had been.

Within the shadow of this peace there sprouted the beginnings of a real world society, an organic growth as accidental as the conditions which made it possible. Never before in the world's history had individuals enjoyed an equal freedom of either spacial or social movement. I still remember vividly the first time that I obtained a passport, in 1912. The passport office was in a slightly remodeled brownstone dwelling and there was only one bored clerk in attendance. When I stated my business he looked puzzled and asked: "Where are you going? Turkey?" The document, when I received it, looked like a high school diploma and was clearly designed to impress illiterate minor officials. Prior to World War I it was possible to travel almost anywhere on earth with no better credentials than a checkbook. The priests of the great deity Commerce wandered with the freedom of the religious mendicants of earlier times, and, like these mendicants, they were bound together by a sort of informal freemasonry. International settlements were scattered from Kabul to Cuzco, little groups of expatriates who always damned the locality and talked sentimentally of home, but who were never happy when they went home. Membership in such groups depended only on finances and foreignness. The local native was pretty sure to be excluded from the local club, but anyone above a certain degree of affluence and sophistication was welcome if he came from outside.

The freemasonry of commerce was as nothing compared to that

of science. No budding research worker felt that his training was complete unless he had studied in at least one country beside his own, and in later years he treasured friendships with fellow workers of all nationalities. In general, the scientist was more consciously international in his outlook than the businessman and was inclined to deprecate patriotism. He saw science as a sort of universal religion which would bring the salvation of all mankind and was eager to pool his findings with those of other scientists. Actually, ideas were passed about with complete indifference to national lines. The latest discovery of an Indian physicist would be tested in Germany and America only weeks after it was announced.

During this period, it even seemed that the world might be moving toward a common basic culture modeled on that of Europe. The prestige of the European was enormous and the advantages to be derived from taking over his science and technology were quite obvious. While those members of non-European groups who derived their advantage from their position in old-style native society might fight Europeanization, most native peoples were ready to copy white ways uncritically and as completely as they could. Their attitudes toward their European rulers were ambivalent; hatred mixed with admiration and resentment of current social discrimination mitigated by the expectation that when they learned to behave like whites they would be treated like whites. In many parts of the European colonial empire, the French territories for example, the promise of the last was direct and explicit. The discovery by the native elite that they were not treated like Europeans even when formally admitted to citizenship has probably caused more bitterness than all the commercial exploitation to which native groups have been subjected.

It would be wrong to picture this period as a golden age. There was plenty of injustice and exploitation under the *pax Europa* and this was by no means limited to the colonies. At the very time when the wealth of the world was being siphoned off into Europe, most European labor lived and worked under conditions which any modern American would consider intolerable. Native peoples often existed, as they still do, at a level of deprivation which the same American would consider not so much intolerable as incredible. However, with all its shortcomings, this period represented many genuine gains over anything that had gone before. The possibility of one world as well as its desirability was widely recognized for the first time and mankind was given a glimpse of what such a unified humanity might accomplish.

The *pax Europa* ended in August, 1914, with the breaking of the united front which Europeans had presented to the colonial peoples for almost a century. Even though no native peoples won their independence during the thirty years that followed, the ferment of revolt spread steadily through the colonial fabric. In retrospect, it seems that the Europeans showed a curious lack of realism in dealing with their subjects. The European hegemony had been built on a combination of military power and technological superiority. Native mercenaries were always relied on heavily in colonial wars, but now they were used against other Europeans, thus giving them a superior training in modern mechanized warfare. This might not have been of too great importance if the science and technology required to implement a modern fighting force had been kept a European monopoly but, with few exceptions, Europeans have provided their subjects with opportunities for modern education and have allowed mechanization in the colonies at the whim of private capital. In effect, they have done their best to place in the hands of the subject peoples the very weapons used to keep them in subjection. One is reminded of the quaint duels of the now extinct Tasmanians in which one man hit another on the head with a club, then handed the club to him and waited for the return blow.

The renewed clash in Europe and the battle with Japan have given the subject peoples their chances, and the last semblance of European hegemony is disappearing as the Asiatic colonies break away one by one and the various "spheres of influence" reject the policies of their former guardians. Although some of the feebler or less-advanced groups may be kept in colonial status for a few generations longer, it is obvious that the colonial system is on the way out. The significance of this for the present European and North American economic systems is obvious. With the last checks on mechanization removed, it will be only a question of time until most of the regions which now provide Europe with markets and sources of raw material will become competitors. What Japan did can be done by most of the other long-civilized oriental groups.

Behind this threat to the present economic system there lies another threat which is at least equally serious. During the period of the European hegemony, the world seemed to be moving toward cultural unity under the influence of European prestige. The possibility of such unity has now retreated into the remote future. The debacle in Europe has thrown the shortcomings of Western civilization into glaring relief, and the rest of the world is becoming more and more selective in its cultural borrowings. While Western technology and science are in demand everywhere, many of the former subjects do not want the rest of Western civilization. There is a widespread resurgence of cultural as well as political nationalism, the revival of certain of the old ways becoming symbolic of the new freedom. The next fifty years will certainly witness a whole series of attempts to synthesize modern science and technology with the Chinese, Indonesian, Indian, and Islamic civilizations. That such syntheses can be successful is shown by the pre-war situation in Japan where modern science and Emperor worship were able to flourish side by side.

To the stresses set up by increasing economic competition there will be added, therefore, those resulting from the clash of cultures. Increasing ease of transportation and communication are quite as likely to intensify these stresses as to diminish them. The frequently repeated statement that if you really know the members of another society you will like them is wishful thinking at its worst. There are many societies whose attitudes and values are so antithetical that the better their members know each other, the less they will like each other. Any Latins who were given an opportunity to get well acquainted with Nazi Germans can give testimony on this point. Moreover, frequent contacts between members of different societies have in themselves potentialities for trouble. It is difficult to insult either individuals or groups at long range, distressingly easy to do so face-to-face. Ego injuries rankle even more than physical ones and a few unfortunate episodes may produce lasting hostility toward all members of a foreign group.

Engrossed as we are in the present struggle with Russia, most Americans tend to ignore the nations which are now winning their freedom from political or economic domination. These nations do not belong on either side of the iron curtain. They have experienced

Western European domination at its worst and as the earlier idealistic aspects of Communism dissolve into too familiar patterns for
Russian world conquest, their enthusiasm for the new religion has
waned. They view the possibility of an armed clash between the
two foci of power with the pleased resignation of a man who antici-
pates that one of his enemies will kill the other and get hung for it.
Capitalism and Communism, Democracy and Totalitarianism are
equally foreign to these nations, and any apparent conversion to any
of them will be only a transitory episode in the slow unfolding of
their own deeply rooted civilizations.

It is hard for Americans, reared in the traditions of European
ethnocentrism, to appreciate the importance of these emergent
powers. In population and natural resources they represent most
of the world and they are moving toward technological equality
with the West at a startling rate. The purpose of this book is to
give an accurate picture of the conditions which exist in most of
the world today in the hope that this may assist in the formation
of public opinion and may provide a basis of sound knowledge for
future planning. The task which confronts us now is that of trying
to reconstitute one world on the basis of collaboration rather than
domination. We must devise techniques to conserve the advantages
of the former European hegemony as far as possible while rectify-
ing its injustices. Any realist must recognize that the chances of
accomplishing this are far from good at present. However, we
must find what consolation we can in the knowledge that if we fail
there will be others to try again, and again, until world unity is
achieved.

Howard A. Meyerhoff

NATURAL RESOURCES IN MOST
OF THE WORLD

SINCE COLONIAL EXPANSION started in the fifteenth century, the earth has belonged to—or, at least, has been ruled by—peoples comprising less than a quarter of its population and, in their original European homelands, occupying a scant 6 percent of its surface. Some of these peoples went into the business of colonization early, and the effects of their energy have left indelible impressions in every part of the globe. The Portuguese, the Spaniards, the French, the British, and the Dutch fought for possessions in the Indies, the Americas, Africa, Asia, and the Pacific, while the Russians contented themselves with a less obtrusive program of contiguous territorial acquisition, which ultimately placed more than 8,000,000 square miles of Eurasia within their political control. Late comers had to be content with what was left. King Leopold II of Belgium acquired the Congo as a private property, turning it over to the Belgian nation only under the pressure of world reaction to a ruthless program of exploitation. Italy and Germany, belatedly organized into national units, took what they could in Africa, by violence when other means failed. The United States fought and bought its way to the Pacific and to Bering Strait and then, administering a final push to tottering Spain, moved into the West Indies and the Western Pacific. Japan made the most of China's and Russia's military weakness and of Germany's defeat in 1918 to enlarge its spheres of direct control and economic penetration.

During five centuries of territorial partition, national motifs and international attitudes underwent slow but drastic changes. All of the exploration was economically motivated, but the early acquisitions of territory were aimed to supply the luxury markets of aristocratic Europe, and to bolster national treasuries with newly mined gold and silver. The Church was also concerned with the salvation of the heathen, but religion had an uphill struggle to counteract the forced labor and slavery which became synonymous with colonization. This pattern changed as the effects of the industrial revolution spread, yet even the British, who were quick to grasp the significance of customers in the colonies, were slow to comprehend the meaning of industrial raw materials. As late as 1763, British statesmen attached as much importance to a small sugar- and rum-producing island in the West Indies as to the whole of Canada.

Perspective sharpened in the nineteenth century, when modern industrial trends began to fashion the course of empire. Lacking sources of energy to power a navy and a merchant marine, Spain watched a huge colonial empire crumble. Most of her territorial losses were suffered at the hands of her own nationals. The Spaniards who settled Middle and South America were willing enough to endure despotic rule so long as it gave a measure of military protection and individual liberty. When, however, the mother country failed to provide either, resentment against maladministration fanned the fires of revolution and prompted the uprisings which ultimately freed the Spanish-American colonies. In Brazil, Portugal contended with a similar reaction; and, much earlier, England suffered her only colonial setback when she misjudged the temper of her American colonists. It was learned that white could not domineer white, and the world's colonial empires shrank to the dimensions of the lands occupied by the so-called "subject races," and to lands too thinly populated to maintain an independent existence. Yet, of the nations which thus acquired independence, only the United States has reached and surpassed the political and economic stature of the mother country. The others, though partially freed of colonial psychology, have remained shackled by colonial economies. Now, more and more of these countries are seeking or asserting economic independence, which implies the ability to pay, each its own way, in the world's markets, and to do it on the basis of the produce of the land, the products of the mines, and the productivity of native labor. It is pertinent, therefore, to inquire into the resource factor in these independent nations which are shedding the last vestiges of colonial dependence.

Asia and Africa were well partitioned when Germany and Italy entered the race, and among the older colonial nations aggression had gone out of style. With no more worlds worth conquering, the countries of northwestern Europe could afford to exhibit pious indignation, not to mention some diplomatic anxiety, regarding the belated ambitions of two neighbors who were feeling their military strength. But in their concern over the shifting sands of Africa, they missed the import of the Russian-Japanese War: an Asiatic nation had demonstrated that the white race was not invincible. For some time the defeat of the Russians seemed little more than a passing military incident. Japan entered the councils of the white nations to an increasing degree, backed the winners in the first World War, and was appropriately rewarded with seemingly inconsequential German holdings in the Pacific and a firmer footing on the Asiatic mainland. Meanwhile the conviction that the white man was vulnerable and dispensable gained momentum in Japan and spread throughout the East. Inadvertently the United States helped spread it, for there were many who were more willing to believe that Philippine independence was a confession of military impotence than the altruistic realization of an ideal. India grew more restive; Burma, French Indo-China, and the Netherlands Indies exhibited unmistakable signs of discontent. The war did not allay, but enhanced racial and national consciousness, and now the colonial empires in Asia are falling apart. Only Africa offers the appearance of uneasy peace.

With political disruption imminent, with white men ready and anxious to lay their Asiatic burdens down, it is appropriate to examine and assess the continent and also to extend the survey to Africa, where independence is more remote, but where native resources may hereafter be more important to the mother countries which will suffer serious losses in Asia. At a time when rehabilitation is a major concern, a knowledge of what the rest of the world has for its own subsistence and for the recuperation of Europe is crucial, though unfortunately it rarely reaches those alleged statesmen who attach more significance to an historical but otherwise meaningless boundary than they do to the integrity of natural geographic regions upon which the welfare of entire populations depends.

North Africa and the Near East

The study may begin with North Africa and the Near East, for this region is contiguous to Europe and will, for geographic reasons, be strongly influenced by European economy and politics. How strong that influence is likely to be may be judged from two statistics: the region has an area approximating 5,300,000 square miles or 10 percent of the earth's land surface, whereas it supports only 3.8 percent of the earth's people (80,000,000); and if the dense population of the Lower Nile valley is excluded, the remaining area contains but 3 percent. Historically, this region holds some of the oldest civilizations known; hence the ratio of area to population is a direct measure of the land's utility.

The explanation is, of course, on everyone's tongue: except for a narrow strip along the Mediterranean and for mountain slopes of limited extent, the region is a desert which receives less than twenty inches of rainfall annually, and 90 percent of it gets less than ten inches. In any latitude precipitation as low as this produces semiarid to desert conditions, but in latitudes 20° to 35° north, where high temperatures and prevailing high atmospheric pressures increase the evaporation rate to an incredible figure, twenty inches or less are insufficient to nourish any but the hardiest desert vegetation or to keep streams fed with water enough for intermittent flow. Nearly all of Africa and at least 75 percent of the Near East lie within these latitudinal limits.

In North Africa desert conditions are alleviated locally or regionally by three influences, which provide the only habitable areas of any considerable size. First is the Mediterranean climate, which is characterized by hot, dry summers and cool, rainy winters. The winter period of rain becomes progressively shorter southward, and the north coast of North Africa lies so far south that it barely gets within the zone of winter precipitation. In northern Morocco, coastal Algeria, and northern Tunisia, the winter rainfall suffices for those subtropical or "Mediterranean" crops that require abundant water for seeding and growth but can mature in drought. Except for the bulge of the Barca Plateau in Cyrenaica, the Libyan coastline lies too far south to receive much winter rain, but locally there are springs and seeps that augment the meager precipitation and provide sites for settlement and modest agricultural activities.

A second element which has brought climatic benefits to Morocco, northern Algeria, and part of Tunisia is the Atlas Range, which rises high enough to receive heavier rainful and to enjoy a lower average temperature. In combination these factors reduce the evaporation rate and make larger volumes of water available for agriculture and drainage. Although the alpine topography restricts full use of the area which enjoys these meteorological advantages, the latter are reflected in the distribution and density of population in the North African colonies.

	Area (sq. mi.)	Population	Density
Morocco (Spanish)	11,000	825,000	77
Morocco (French)	155,000	6,700,000	43
Algeria (all)	850,000	8,000,000	9
Algeria (coastal and mountain belt)	120,000	7,500,000	62.5
Tunisia (all)	48,300	2,812,000	57
Tunisia (north)	30,000	2,400,000	80
Libya	679,000	900,000	1.3
Rio de Oro	111,000	25,000	0.2

If population statistics were available for smaller geographic units, the effects of precipitation could be made startling, and the result would more truly represent the facts. The tillable acreage would be reduced to a very small figure, and 90 percent of the population would be found in limited sections of each country, with densities closer to 200 per square mile in the habitable stretches. The desert population would be found in clusters around oases, with unpeopled wastes between. Nonetheless, the figures given above are useful in one important respect: they reveal that the region at its best supports but eighty people per square mile and at its worst can support no life at all.

Archaeology and geology indicate that extreme desert conditions have descended on North Africa during the past few thousand years, and that in earlier times it supported large urban, agricultural and pastoral populations in districts which are uninhabitable now. What led to the relentless change is immaterial. It has occurred, and it has rendered 1,450,000 square miles of country in Rio de Oro,

Morocco, Algeria, and Libya useless, and even larger areas to the south and east in French West Africa, French Equatorial Africa, Egypt and the Anglo-Egyptian Sudan. Only 300,000 square miles contain adequate water and soil resources for habitation and development, and even this figure must be trimmed to eliminate rocky alpine wastes and semidesert reaches to which water cannot be brought. Thanks again to the distribution of rainfall, the coastal belt is most densely settled, but agricultural and pastoral populations make extensive use of the uplands lying between the coast and the mountains, as well as of the intermontane valleys. Of the latter the largest is the "Plateau of the Shotts" or lakes—a linear depression between the Little Atlas Mountains and the Saharan Atlas. This depression attains its broadest development in Algeria, but it extends across northern Tunisia to the Gulf of Tunis.

On the flanks of the Little Atlas Mountains, and more particularly near the Mediterranean Coast, agriculture thrives. Algeria, which is politically integrated with France, has been somewhat more intensively cultivated than Morocco and Tunisia, but the northern slopes and valleys of all these countries comprise an agricultural region of considerable importance. Inevitably the typical Mediterranean crops feature in the coastal sections, and olives, grapes, and citrus fruits are important elements in trade with continental Europe, particularly France. Somewhat more surprising, however, is the magnitude of grain production. As many as 8,000,000 acres have been utilized for raising barley, and only slightly less for winter wheat. Corn, or maize, has achieved some importance in Morocco, and there is also a modest yield of oats in Algeria and Tunisia. With an earlier season than France and Spain, these countries have, to some extent, anticipated seasonal food requirements in the mother countries and have tended to restrict commercial agricultural production to crops of this kind.

Historically the native peoples have been more seriously concerned with pastoral pursuits than with sedentary agriculture, and animals still play a significant role in the economy. The sheep population is conservatively figured at 10,000,000; goats at 6,000,000; cattle at 1,500,000; and beasts of burden at 1,000,000. Animal husbandry is the mainstay of the desert tribes, but in and out of the desert, it is chiefly concerned with subsistence and plays a negligible part in overseas or trans-Mediterranean trade.

Modern industry has made few inroads in these traditionally non-industrial lands. The small deposits of low-grade coal in Morocco and Algeria have never yielded more than 200,000 metric tons, and both countries have produced minute quantities of oil. With so small an allotment of energy resources, other minerals, when mined at all, are shipped to Europe for processing. Deposits of magnetic iron ore extend throughout the length of the Atlas Mountains, and typical pre-war production figures merely suggest the potentialities of the region as a source of iron ore.

	Metric Tons
Algeria	3,000,000
Spanish Morocco	1,400,000
Tunisia	945,000
French Morocco	265,000
Total	5,610,000

Reliable estimates of reserves are not available, but they are larger than the production indicates. Even so, as a region, North Africa ranks ninth or tenth among the iron-producing districts of the world.

Other metals are present. Tunisia has annually mined 25 to 30 thousand metric tons of lead for several decades: Algeria, approximately 12,000. These two countries likewise mine zinc, but for this metal the production figures are reversed—about 32,000 tons for Algeria with 10,000 for Tunisia. Silver production spurted just before the war in the French colonies, but the yield of 430,000 fine ounces made an insignificant contribution to the world total. Relatively more important in a meager world supply were 1,000 tons of antimony, chiefly from Algeria. The list of metals is not quite endless, but it includes modest amounts of manganese, molybdenum, mercury, copper, tin, and gold. It reveals the rich variety of minerals which the Atlas Range contains, but is not indicative of great mineral wealth. This part of Africa has been carefully explored and, though the mineral possibilities are neither completely known nor fully developed, it is safe to conclude that the mountain region in the north is not a storehouse of fabulous metallic wealth. Knowledge of the desert to the south is insufficient to warrant any conclusion.

In the nonmetallic field, latent possibilities were being rapidly realized before the war, and developments will undoubtedly continue as quickly as economic conditions become stabilized. North Africa contains the second largest reserve of phosphate rock in the world, but pre-war production was just beginning to reflect its significance.

	Metric Tons
French Morocco	1,500,000
Algeria	600,000
Tunisia	1,900,000
Total	4,000,000

When it is recalled that the United States rarely extracts more than 4,000,000 tons per year, the importance of these deposits to Europe's agriculture and chemical industries is obvious. Salt flats and the waters of salt lakes provide larger reserves of sodium chloride, gypsum, and other salines than are currently in demand. Again the list could be extended to include sulphur (from pyrite), fluorite, barite, and other nonmetallic products, but it is already long enough to show that North Africa was not badly treated when Nature passed out mineral supplies.

Closely linked with Spain, France, and Italy across the Mediterranean, these countries suffer a major handicap, not only in the absence of fuels within their own borders, but also in a deficiency of the energy resources in the mother countries. France has achieved a measure of industrialization through hydroelectrification and the importation of coal and oil. Her heavy industry is situated in the northeast upon low-grade iron ores which would profit by beneficiation with high-grade magnetite from North Africa. Unfortunately, no location could be worse in so far as use of Algerian iron and manganese is concerned. Although Algeria, Morocco, and Tunisia may supplement the mineral needs of the mother country, the lack of fuel and the accidents of geography offer minimum encouragement for the full development of their mineral resources. In Spanish Morocco, what little progress has been made in the mining field has been made in spite of the mother country, which, for lack of coal, must export her own mineral raw materials. From any standpoint Libya is virtually a total loss, and of its 679,000 square miles a scant 3 percent is habitable. Even its sheep population (about 900,000) is no greater than its human population; and goats, which have the reputation of living where nothing else can, number only 725,000. As an Italian colony it was a financial liability. It absorbed few Italian colonists, and it contributed almost nothing to Italian economy.

Egypt would be little better off than Libya were it not the beneficiary of a third element which operates to ameliorate desert conditions. It receives the waters of the Nile, which gathers them from the rain-drenched uplands of equatorial Africa and delivers them to the Mediterranean via the Anglo-Egyptian Sudan and Egypt. A thin, broken strip of Mediterranean coastline is settled, but most Egyptians live on the delta or on the constricted alluvial plain in the valley, depending on its methodical seasonal overflow for water and the enrichment of their hard-worked lands or upon the more dependable technique of scientifically controlled irrigation. In Egypt the valley is 700 miles long, flaring at the delta, which begins at Cairo. Its area is less than 20,000 square miles, but so intensively is it utilized that the rural population has an average density of 700 per square mile and more than 1,500 on the delta. Yet in the desert, beyond the reach of floods, canals, and distributaries, the density drops to the low Libyan average. Egypt is the Nile.

Egypt's most important contribution to the world's industrial economy is cotton, which constitutes approximately 80 percent of her exports and gives her foreign trade second place among the political divisions of Africa. Normally the annual yield is slightly under 2,000,000 bales, more than 90 percent of which is exported. Since the boll weevil claimed American Sea Island cotton, the Nile valley has become the principal source of long staple cotton. The limits of production have, however, been reached and can be increased only at the expense of essential food crops. There is little, if any, more land accessible to irrigation, and the Nile is already carrying the maximum irrigation load. The rapidly increasing population has steadily reduced the food crop surplus and, through subdivision of farm holdings, has impaired the efficiency of land use. In consequence, the Nile, with its large wheat crop, is still the granary of Egypt, but no longer the "granary of the East." Indeed, within two decades, Egyptians may be confronted with the choice of reducing the cotton crop to maintain essential food production, or of becoming partially dependent upon foreign sources for basic food requirements. Commercial cotton culture has been greatly expanded in the Anglo-Egyptian Sudan, where there are still additional possibilities for greater production.

Mineral raw materials are not abundant in Egypt and the Anglo-Egyptian Sudan. Geologically the belt facing the Red Sea is an extension or outpost of the Near East geological province, and as such its oil possibilities have attracted increasing attention. In 1937 four major oil companies started exploration aimed to increase production, and the program continues, following temporary suspension during the war. Since 1939, production has doubled, and in 1944 it reached 9,431,000 barrels. Although this is but a drop in the world's petroleum bucket, it is strategically located and it is of crucial importance in a country that lacks coal. Exploration proceeds in the desert west and east of the Nile and in the Sinai Peninsula, and some optimism may be felt regarding further increases, despite the fact that recent discoveries have been limited to deeper horizons in old pools rather than new fields or new pools in old fields.

Mining acquires little significance in Egyptian economy, and none at all in the Anglo-Egyptian Sudan. From the standpoint of volume and value, the nonmetallic minerals far outrank the metals. At peak production 700,000 tons of gypsum and 550,000 tons of phosphate rock have been extracted in a single year, and 440,000 tons of salt have entered the export market. Small tonnages of feldspar, talc, and soapstone have also been exported. The ruins surrounding an exhausted copper mine in the upland between the Nile and the Red Sea have provided archaeologists with clear but otherwise unrecorded evidence of ancient mining activity; but the copper was gone before history could record the operation. In a short list of metals currently produced, copper is absent, and the only one of any significance is manganese. Granting that the desert has not been systematically examined for mineral deposits, there is reason to believe that Egypt and the Anglo-Egyptian Sudan will never make important contributions to the world's metal supply.

Although Doughty, Musil, and Lawrence have shown that the English language already contains a sufficient number of synonyms for the word *desolation* adequately to describe the Arabian peninsula, it is safe to predict that the picturesque vocabularies of geologists, exploration geophysicists, and drillers will bring new adjectives and word combinations into the descriptive legend and literature of the region, as they try to recover the 5,000,000,000 barrels of oil it is believed to contain, and as they search for the additional ten to fifteen billion barrels which may be there. Approximately 1,000,000 square miles in extent, Arabia is two-thirds as large as India, but it contains only 2 percent of India's population. Possibly the most remarkable fact is that 8,000,000 people find it possible to live in so barren a waste. Except in the marginal highlands extending from Mecca south-southeastward through Yemen to Aden, no part of the huge peninsula receives as much as ten inches of rainfall; and at latitude 13° to 32° north, precipitation of this order means desert conditions of the most rigorous character. Much of the country rises above the 2,000-foot contour, and the elevation makes day temperatures tolerable and night temperatures cool to cold; but it takes rain, rather than temperature, to furnish a livelihood even for a small population. It is inevitable, therefore, that more than half the population should live in the meagerly watered southwestern highlands; and, also, that a substantial fraction of the country's financial support should come from devout Moslem pilgrims who visit the Prophet's birthplace and tomb. While it lasts, oil may change all that, but it cannot alter the permanent inadequacies of the land as a means of support.

Comparable aridity obtains to the north in Transjordan and Iraq, but in Iraq climatic conditions are ameliorated by the Tigris-Euphrates River system, which, like the Nile, heads in more amply watered mountains—the mountains of Kurdistan. The Murat, a major tributary of the Euphrates, is fed by the snows of Mount Ararat, a 16,916-foot peak situated at the corner where Turkey, Iran, and Armenia meet. Unlike Egypt, however, Iraq has done little to utilize the waters of its rivers, for only 8 percent of the alluvial lowland is irrigated and under cultivation. Even the ancient Chaldeans, Assyrians, and Babylonians did better than that long before the Christian era began; and unless Iraq is content to rest its future and its reputation on being the world's chief source of dates, it may profitably devote the income from its oil to the development of the alluvial lowland. Its possibilities are not unlike those of the Imperial Valley of California; and, under good political and economic management, the Tigris-Euphrates valley can temper its reputation as one of the hottest places on earth (in summer), by becoming one of the most fertile. Foreign capital is developing the country's other major resource—its oil; but much as the oil companies may contribute to the progress of Iraq, they are not likely to display interest or concern in assuring its agricultural future.

The Mediterranean countries of the Near East—Palestine, Syria and Turkey—enjoy a heavier rainfall than the lands beyond the Jordan, and this fact is reflected in denser populations and more widely dispersed agricultural and pastoral activities. Palestine momentarily occupies the spotlight of public attention as the unfortunate victim of emotional prejudices which are completely blind to the inherent limitations of the country itself. Palestine is approximately the size of Maryland, and it contains only 300,000 fewer people. The rural population is definitely larger than it is in Maryland, but the country and its hinterland lack the rich resources that make the city of Baltimore one of the busiest manufacturing centers and ports of the United States. The people of Palestine are dependent upon a land that receives a scant twenty inches of rainfall, and which has only the limited hydroelectric energy supplied by the Jordan River and the commercial salts in the Dead Sea to broaden the agricultural-pastoral base upon which the economy of the country is founded. Upon such stubborn facts as these the destinies and the welfare of people depend, and it would be a long step forward if the emotion which dominates the Palestinian problem were tempered by the realistic consideration of resources—the means of livelihood for the people who are there and for those who wish to come.

In Syria and Lebanon there is little to augment the agricultural and pastoral pursuits of their 3,000,000 people, most of whom live near the Mediterranean. The coast is far enough north to benefit from moisture-laden westerlies, which bring nearly thirty inches of rain and provide ample winter precipitation for wheat, barley, and the usual Mediterranean tree crops. Mineral deposits, which enrich the near-by island of Cyprus and the mountains of Turkey, and reservoirs of oil, which curve S-like around the Persian Gulf and the southern Caspian basin, by-pass Syria; its heavier rainfall and its topography likewise deprive it of the saline wealth which the more intense aridity and the Jordan-Dead Sea lowland have given Palestine. Even the forests of Lebanon are so depleted as to be of historic interest as a resource which abetted the Phoenicians rather than the modern inhabitants. It is the destiny of the latter, by judicious use of the more generous water supply that geographic location has given them, to raise crops and animals in excess of domestic needs, and to help in this small way to feed less favored Mediterranean neighbors.

Turkey, with 300,000 square miles of country and a population of 19,000,000, owes its critical importance in international affairs almost entirely to its geographic position athwart the land route from Europe to southern Asia, and the land and water routes from the Black Sea ports of its gigantic neighbor, Russia, to the Mediterranean. Although such a position has advantages, its drawbacks are formidable. Turkey has learned, however, that the mountainous fringes and dissected intermontane plateau of Asia Minor, or Anatolia, are nearly impregnable; and, despite a fanatical attachment to Istanbul, it has concentrated its program of internal development in this stronghold. There it has a living for its rapidly growing population. In contrast with the arid Moslem states to the south it has, at worst, a semiarid climate. The land is good, if somewhat rough, for crops and animals; and if it rises to, or above, the altitudinal limits of economic use in the east, the elevation insures a perennial water supply in streams that reach the Black, Caspian, and Mediterranean seas and the Persian Gulf.

Notwithstanding a more vigorous and liberal government than Turkey has ever had before, there is little to mar the archaic rusticity of the Turkish landscape. Local irrigation is practiced chiefly where it was instituted generations ago; negligible use is made of the considerable hydroelectric potential; known mineral deposits lie neglected or are worked by leisurely and outmoded methods. The mineral resources are not fabulous, and none of them is large or

significant enough to be a serious temptation to the industrialized
nations of Europe. But there are more poorly endowed countries
that would like the chance to double or treble the coal production
of 2,700,000 tons; or the iron output of 143,000 tons; or the copper
yield of 8,000 tons; not to mention the small quantities of silver,
manganese, molybdenum, mercury, lead, zinc, sulphur and asbestos.
War demand and war prices did push chromite production up to
213,000 tons, giving Turkey a precarious hold on second place
among the world's producers. There is, however, neither the domestic demand for, nor the domestic interest in, a national industrial
economy, nor is it possible to detect a native incentive to increase
the volume of foreign trade. The latter stands roughly at $150,000,-
000; and, among exports, tobacco tops the list!

Iran and Afghanistan are the easternmost outposts of the desert.
Together they comprise just under 900,000 square miles of formidable mountains and rugged, arid intermontane uplands. Iran, with
15,000,000 inhabitants, has frontage on the Persian Gulf and the
Gulf of Oman, as well as on the Caspian Sea. Landlocked Afghanistan, with a population in excess of 12,000,000, is bisected by lofty
outliers of the Hindu Kush Mountains, which function as an effective barrier to transportation and to political and economic unification. Actually there is little to unify in either country, for the
primitive agriculture and grazing are on a sectionally and regionally
self-sustaining basis, which creates little domestic commerce, and
practically nothing except Persian rugs for foreign trade. Until the
discovery of oil in Iran, the two countries served nominally as buffer states between Russia and British holdings and interests in
southern Asia. The fact was that the forbidding terrane was buffer
enough, and at no time have the governments of these countries
wielded a particle of influence in the balance of power in Asia. For
Iran, however, the development of petroleum resources changed
all that, and the alignment of powers in the second World War converted her temporarily into a vital corridor for lend-lease goods. A
strangely useless national railroad suddenly became an all-important
transportation link with an ally that was almost completely isolated. Iran thus drifted into the main current of world affairs,
and while her oil fields are being drained, she is destined to stay
there.

This brief survey of North African, Near Eastern, and Middle Eastern geography indicates that this great region has a limited but clearly defined role in world affairs. It is desert, and the driest countries—Libya and Arabia—can barely support a single human being on a square mile of land, except in such better watered spots as the Barca Plateau of Cyrenaica and the uplands of Yemen. The best watered—Morocco, Algeria, Tunisia, Egypt, and Turkey—are as densely populated as agricultural and pastoral opportunities will permit. Of them, French Morocco, Algeria, and Tunisia, as parts of the French Colonial Empire, have expanded their economies somewhat by the partial utilization of mineral reserves, notably phosphates and iron ore. Egypt is benefiting from the exploitation of her oil. Turkey, on the other hand, displays little interest in mineral wealth which, though imperfectly known, is of much greater importance than current production suggests. Turkey, Syria, Iran, Afghanistan, and even Iraq, with its neglected agricultural possibilities, are actually supporting denser rural populations than the intermontane region of the United States, which has a comparable climate, modern and mechanized facilities, and as much industrialization as mineral resources and regional demand warrant.

In the Nile Valley, a river, irrigation, improved sanitation and a crop that brings cash in the world market have conspired to bring the problem of overpopulation to Egypt, which is now demanding complete freedom to expand southward into the Sudan. Insistence on "the unification of the Nile" makes no sense except as the effort of a harrassed government to relieve the pressure of an unwieldly population. In Palestine, Zionism is moving—inadvertently, perhaps—toward the same problem. There the heavier rainfall of the Mediterranean slope and the Jordan River favor more intensive agriculture than is possible in many other sections of the Near and Middle East; and an energetic and resourceful immigrant population is making the most of this situation, as well as of the few saline mineral resources which the Dead Sea contains. But these assets are not enough for limitless expansion, and before the population goes much beyond its present density of 150 to 200 per square mile, the potentialities of Palestine should be resurveyed. At the moment it looks as if religious and racial zeal threatens to exceed nature's dowry in the Holy Land.

North Africa, the Near East, and the Middle East are the dismembered fragments of ancient empires in which there was too little cohesion for unity in any field, with the single exception of religion. Even the common problem of position—of being in the middle and often in the way—has failed to bring these peoples together for concerted action, or for common defense. Those countries lying close to the predaceous nations of western Europe lost their independence; the others have kept theirs, not through internal strength or a united front against aggressors, but simply because the region was thought to contain too little to repay the cost of conquest and administration.

Then came oil. It started innocently enough, for gas and oil seeps had been known for centuries, and it occasioned no excitement, even when the Russians started to exploit the oil pools on the slopes of the Caucasus Mountains. It was not long, however, before Russia exhibited a proprietary interest in Iran. In 1872, she forced the cancelation of a concession which would have given Baron Julius von Reuter, a naturalized Briton, broad and inclusive mining rights in Persia; but in 1901, she failed to block a more significant transaction. In that year a New Zealander named D'Arcy obtained an exclusive monopoly on petroleum and petroleum products in all of Persia, except the five northern provinces. The organization of the Anglo-Persian (now Anglo-Iranian) Oil Company and, later (1914), of the Turkish Petroleum Company for the exploitation of oil in what is now Iraq, and subsequently the granting of concessions in Bahrein (1922) and Saudi Arabia (1933) to American interests, were events that set the pattern for bitter diplomatic war between the British and the Americans, who resented the British policy of exclusion, particularly in mandated territory. Meanwhile Russia, under new management, thwarted the efforts of American companies to establish oil rights in northern Iran; and that the none too subtle hand of the U.S.S.R. is still at work was evident in the recent Azerbaijan incident, whereby Communist sympathizers tried to split off the northern provinces of Iran.

Evidently the stakes are high when the British were willing to risk amicable relations with the United States during the twenties; when Communists foment revolution and abet secession; when a

Cabinet officer hazards the wrath of the Senate; and when oil companies spend hundreds of millions of dollars in wells, tank "farms," housing, refineries, harbor facilities, and pipe lines through deserts where upkeep against the elements and unsympathetic nomads will be a major problem. The stakes are a ten- to twenty-year supply of oil. Put more dramatically, the Middle East contains the largest proven oil reserve in the world.

Curving around the northern rim of the Persian Gulf from Iran, through Iraq, to Bahrein Island and into Saudi Arabia—perhaps even farther to the oil fields along the Red Sea in Egypt—is a belt of petroliferous rocks which exceed even the most optimistic oil man's dream of heaven. That the climate, the terrane, and working conditions in general bear a closer resemblance to hell does not alter his concept, for here and around the margins of the southern Caspian basin he believes there are approximately 25 billion barrels of oil, with probable reserves of 30 billion more. He is even willing to guess that there could be 80 to 90 billion, but this figure is scarcely more than wishful speculation. True, the Russians at one time announced that incomplete prospecting had located a reserve of 61 billion barrels in the U.S.S.R.,[1] but the claim lacked substantiation and, in any case, a goodly fraction of the estimate involved oil within the Caspian-Middle East region. The Russian estimate has been discounted, though there is no disposition to underrate the quantity and importance of Soviet reserves. In the United States the supply of proven petroleum is being depleted more rapidly than new drilling can prove additional resources. Oil experts are confident of recovering 20 billion barrels, but a supply of this magnitude is little more than a source of worry in a country which produces and consumes one and a half billion barrels yearly. For this reason the Middle East looks good to American oil men and, latterly, to the American Government. It is even more attractive to other governments, which see an opportunity to equalize the production under their political or economic control with that of the United States.

The extent to which development of Middle Eastern oil fields has proceeded may be gauged by 1945 production statistics:

[1] I. M. Goubkin, "World Petroleum Supplies," International Geological Congress, *Report of the XVII Session, 1937*, I (Moscow, 1939), 177-188.

	Barrels
Bahrein	7,309,000
Saudi Arabia	22,211,000
Iraq	31,000,000
Iran	120,200,000
Total	180,720,000

If Egyptian production of nearly 10,000,000 barrels is added, the total is still somewhat less than the output of the Caucasian fields in the U.S.S.R., which are also a part of this petroliferous province. Analysis of the statistics reveals a temporary wartime slump in production, except in Saudi Arabia. In Bahrein and Iraq there has been little more than bare recovery from the slump, whereas in Iran and Saudi Arabia there have been phenomenal increases in annual yield, with still larger ones to come. Even so, the entire region, including the Caucasus, is producing at a rate of only 1.5 percent of allegedly proven resources, in comparison with 7.5 percent in the United States. As producers hit their strides in the effort to keep output abreast of world demand, the course of economic events is as clear as the trend of political events is obscure. For some years to come the Middle Eastern picture will be painted in oil.

India

In respect to latitude, Arabia and India lie side by side. Arabia extends from 13° to 32° north; India, from 6° to 36° north. India, with an area of 1,575,000 square miles, is 50 percent larger than Arabia, but its population of 385,000,000 is nearly 5,000 percent greater. The westernmost section of Baluchistan is only three degrees east of easternmost Arabia, and in these nearest points, the climates of the two are similar. East of the Indus, however, precipitation increases, exceeding eighty inches in the Brahmaputra valley along the Burma boundary. A combination of meteorological features accounts for the climatic change. Chief of them are the higher degree of saturation of the trade winds which blow across the Deccan Plateau, the monsoon winds which blow seasonably in and out of Central Asia; and the Himalaya Mountains, which bar passage of the moisture, if not of the north-bound winds.

With minor modifications related primarily to relief and elevation, a map illustrating the density of population in India would coincide with a hydrographic-rainfall map showing the distribution of water. The lower Indus valley, lying in a desert, supports a population in excess of 100 people per square mile because, like the Nile, the Indus redistributes the run-off from the western slopes of the Himalaya, Trans-Himalaya, and Karakorum ranges. Crowded as India is, the adjustment of population to water supply and relief is delicate; for the land sustains the bulk of the population, and the response of the earth to water is direct and immediate in a tropical and subtropical country. The soil is not rich, and it is seriously overtaxed. Only the rapid rate of rock-weathering, the seasonal overflow of streams, and the irrigation of nearly one-quarter of the cropped land keep production as high as it is.

The price of human fertility is hard manual labor, even to maintain a low level of subsistence. The undernourishment and poverty of millions of India's inhabitants have been so effectively publicized that comparatively few people think of their country as one of the most productive, agriculturally, in the world. Yet it is just that. The rice crop of 110 billion pounds is a very close second to China's. Only the U.S.S.R. and the United States raise more wheat, and India ranks second as a producer of millet, fourth as a producer of barley. There are three times as many cattle (over 200 million) as in the United States, but the Hindu religion makes them a severely limited economic asset. She raises enough sheep for a small export trade in wool, and, though a poor second, she ranks next to the United States as a raiser and exporter of cotton. Linseed is another important cash crop. Most surprising is the fact that India produces twice as much sugar as any other country and nearly one-third of the world total, yet she imports a small quantity to meet the domestic demand. With Ceylon, she raises more tea than the rest of the world combined, and her tobacco crop in normal years is larger than that of the United States.

With so large a population and so much of it impoverished, India might logically be expected to make full use of available agricultural land, but this is far from the fact. Surveys indicate that, in the British provinces, 42 percent of the land is arable, whereas only 34 percent is actually under cultivation. Of the total acreage so used, four-fifths are utilized for food crops, the balance for commercial crops. Despite the need for high yields, the acreage returns are generally low, partly because the soil has been overworked, and poor seed used; but the principal causes for low returns are the small individual land holdings and the inefficient methods and equipment which are employed. There is little mechanization, for in a country where human life and labor come cheap, there is bound to be a wasteful expenditure of physical effort. Integration of farms, the application of soil restoratives, the introduction of new methods and modern machinery would, without question, increase the yield per acre and the gross food supply substantially, but only a little progress has been made in any of these directions.

In spite of the domestic turmoil that is retarding progress, the Indian Government has tackled one approach to the problem—namely, the expansion of acreage by means of improved and new irrigation projects. Linked with studies now in progress is the harnessing of undeveloped water power for rural and industrial electrification. Such steps are important, but the acreage which lies within reach of reclamation is small as compared with the land that nature waters. The artificial introduction of water invariably raises land values and crop costs, which must be studied closely in relation to India's economy to determine whether the investment is yet warranted. More urgent than reclamation is the proper and efficient use of the much larger acreage which needs no irrigation, but which has suffered from long use, and which must serve in the future, as it has in the past, as the mainstay of India's agriculture. Some steps have been taken to assure improved yields, but better seeds, crop rotation, fertilization to replace the ingredients which a tropical climate and constant cropping so quickly remove, and a program of farm integration are essential parts of any plan which aims to improve India's lot.

At the present time crop acreage figures at less than one acre per capita. In the Ganges valley there are locally as many as 2,000 people per square mile trying to earn a living from farming. Although more land may relieve the situation, relief will at best be temporary in a country where the population is increasing so rapidly. Greater productivity per acre and greater productivity per worker are imperative. The former calls for intensive practical education, and the latter will require integration of farms and the displacement of farmers—a procedure which any political group will hesitate to undertake unless there is remunerative work for the displaced population to do. In a country hampered by the caste system, by religious taboos, and by a high percentage of illiteracy, education and displacement are slow processes. A dictatorship might accomplish the latter, as the Communists effected "collectivization" of farms by liquidating the so-called kulaks; but it is to be hoped that the arbitrary methods of Soviet Russia will not be duplicated, and that more humane inducements can be devised by a progressive and enlightened government to consolidate farm acreage and to manage it efficiently for greater production.

In India's predominantly agricultural economy, individual incomes from all sources are extremely low, notwithstanding the fact that a high percentage of the farming is devoted to cash crops for the domestic market—for example, grain and sugar—and for foreign trade (cotton, linseed, tea). Manufacturing and the extractive industries, though far from negligible, furnish a small fraction of the national income. Yet, in a country with so tremendous a reservoir of manpower, the development of natural resources for use in native industrial plants is an obvious step to increase employment, raise living standards, and achieve a greater measure of economic independence, as well as government revenue from taxes. Although the war did much to accelerate industrial activity, there is still rather low productive capacity in the manufacture of essential goods. With independence a reality, the most imperative need at the moment is expansion of industrial capacity and employment. This will require capital, equipment, and raw materials, and it must be asked whether these basic needs are available.

There is, of course, a great deal of private wealth in India, although only a fraction of it has been industrially utilized. Now it is urgently needed, for public funds are committed to an ambitious and equally important program of irrigation, hydroelectrification, transportation, education, and other public works that will require a huge investment and heavy borrowing. Private capital is necessary to carry the staggering burden of investment, and it should properly concern itself with those projects in which a reasonable return may be anticipated. Indian industry is better prepared now than ever before to make machines for production, but initially much of the equipment for modern industrial plants must be purchased abroad. Technicians to erect and repair the machines are just as essential, but capital, technicians, and machines can be obtained only if India achieves a greater measure of internal stability than she enjoys at the present time. Assuming an amicable settlement of the political structure in independent India, natural resources become the crucial factor in the development of the country. Are they adequate for the minimum task of industrialization which must be undertaken? Only a qualified answer can be given to this question, but the outlook is by no means dismal.

Leaving aside such essential occupations as the processing of food, agricultural resources for industrial use can be made adequate without impairing food production, provided steps are taken to obtain increased acreage yields at a lower expenditure of manpower. A cotton crop of 4,000,000 bales, half of which was exported before the war, should easily sustain a greatly expanded cotton textile industry; silk, wool, and hide production could be brought up to domestic requirements without serious difficulty. Greater industrial use could be made of the forests, although the best stands of timber are situated in Burma, which is excluded from the newly constituted Indian states. And if the Indian market were not large enough, or prosperous enough, to absorb all the manufactures that modern establishments can turn out, the dense populations of near-by countries should satisfy the most ambitious industrialist.

As a solid foundation for industry, India is well endowed with coal. The two most important regions which contain coal of good industrial grade are the northeastern coal province, extending from the northeastern corner of the Central Provinces across Orissa and Bihar, south of the Ganges, into western Bengal; and the central, or Godavari, coal province, reaching in discontinuous deposits from the northwestern boundary of the Central Provinces southeastward through eastern Hyderabad to Madras. Local deposits occur elsewhere, but are poor in quality, small in quantity. The Geological Survey of India estimates that these two coal regions contain reserves of sixteen to seventeen billion tons. Mining is actively carried on, particularly in the northeast, where the proximity of iron ore and the requirements of the large metropolitan center of Calcutta have stimulated the demand. Although the quantity mined (28–

30,000,000 tons) is scarcely 5 percent of production in the United States, India currently ranks sixth and normally ranks seventh in coal output.

Nature thoughtfully placed medium to high-grade hematite iron ores in Orissa, hardly twenty-five miles from the nearest coal deposit. No more happy circumstance could be devised for the economic smelting of iron ore and the manufacture of iron and steel. Ore is being extracted from deposits which contain reserves of more than three billion tons, at a rate approximating 3,000,000 tons a year. Pig iron production ranges from 1,750,000 to 2,000,000 tons annually, giving the country a rank of eighth among the nations of the world. With ample reserves of coking coal and iron ore, production can be stepped up virtually to any level by expanding mining operations and increasing furnace and mill capacity. As if this were not enough, India has annually supplied the industrial world with 600,000 to 1,000,000 tons of manganese—far more than the modest requirements of her own steel industry. Other ferrous alloys include chromite in moderate quantity and tungsten in very small amounts.

Bauxite is quite widely dispersed, but output has been small. In 1942, India started producing metallic aluminum and, according to latest statistical reports, was placing 1,600 tons on the market. This is little if measured against the performance of many other nations, but it heralds a start in what may be an important adjunct to modernization. Among the other metals, copper, silver, and gold are the only ones in production. Copper output is small and the reserves appear to be limited, and silver production is sporadic; but the 300,-000 fine ounces of gold constitute a helpful contribution to the Indian economy. Deficiencies are made up, in part, in Burma, where lead, zinc, tin, and nickel deposits are being exploited. Nonmetallic minerals have not been mined to any great extent, except for ilmenite and mica, both of which have featured in India's foreign trade. Salt has been the basis of an industry and a controversy between the Hindus and the British government, and India's production is large. But the salt mined or obtained from the sea is for domestic rather than industrial use. With few exceptions, noteworthy among them being sulphur and potash, one need merely name a nonmetallic mineral and India has it, though not always in proven economic quantity: barite, magnesium, fluorspar, feldspar, gypsum, lime, and phosphate rock have all appeared in statistical tables of mineral output. Some of them could be mined in larger quantities if the domestic demand should warrant it.

India was treated badly with respect to one important need—namely, oil. For several years production has remained fairly steady at a paltry two to two and one-quarter million barrels—a quantity far from adequate to satisfy the demand in a great and growing state. Even with the seven to eight million barrels from Burmese pools, there is not enough, nor is there a reasonable prospect of increasing the reserves or the annual yield to satisfactory proportions. Approximately one-eighth of the petroleum comes from the Punjab in northwestern India; the balance, from Assam near the Burmese border. Some of the local pools have already been exhausted, but the possibilities of extending both fields are good. Even the proverbial optimism of the oil profession does not, however, anticipate a great future for Indian oil.

Notwithstanding this unfortunate deficiency, the country is not badly off with respect to energy resources. The coal supply is ample for all needs, and if there is reason to complain about its uneven geographic distribution, India has no more to complain about than the United States, Canada, and the U.S.S.R., and much less than Argentina, Spain, Italy, and other countries in which coal is extremely scarce. In addition to solid fuels, India and its Himalayan mountain fringe, part of which lies beyond the political boundaries of the country but within economic transmission distance of Indian cities, have a water-power potential of 28 to 30 million horsepower. Only a minute fraction of this impressive amount has been developed—in fact, less than enough to meet existing demand. This estimate, moreover, is based on the minimum continuous potential at low-water stages of the rivers; hence, with judicious use of steam plants powered by coal and with integration of steam and hydroelectric plants, a higher water-power recovery could be effected, with steam generation making up deficiencies in the dry season. Even with regard to oil, India lies within easy and economical reach of the refineries of the Persian Gulf, which draw crude oil from Saudi Arabia, Bahrein, Iraq, and Iran; and not too far from those in the East Indies. In the absence of domestic reserves, she could not be more strategically situated. For heavy and light industry and for transportation, the country is far better off than most, and her future will be circumscribed only to the extent that her people fail to realize its full possibilities.

Failure is by no means a remote or idle threat. Irreconcilable religious differences have already split India into Pakistan and the Dominion of India, or Hindustan, with separate political administrations. The split has not only partitioned the land but also the mineral resources, leaving most of the important ones except oil, and roughly 90 percent of the total, in Hindustan. Both dominions have suffered losses of reserves which can be neutralized if the differences end with religion and are not pushed into the economic field. If political boundaries become economic barriers, there will be pointless waste of money in reorienting the transportation system and in relocating and building new industrial sites, while old ones will presumably be left to decay. The energy, money, and time can be spent with greater profit in improving the lot and the opportunities of Moslems and Hindus alike by means of education, public works, and expansion of existing industrial facilities, wherever located. Any other procedure will arrest progress to such an extent that India—Moslem and Hindu—will be regarded as a poor financial risk. New capital will be difficult to get; and without capital, industrial expansion will take place slowly or not at all. Without industrial expansion, it will be slow work to improve the standard of living, and the vicious circle will be complete. Fortunately there are Indians with vision who are aware of this potent threat of social and economic stagnation and deterioration. It may be hoped that they can wield sufficient influence in time to stop the threatened disintegration short with political partition, and that they will inject reason and immutable geographic and geologic facts into a sane solution of social-economic problems. If they succeed, India has an assured future.[2]

Southeast Asia and Indonesia

With India's change of status, Burma is now part of Southeast Asia politically, as well as geographically and racially, sharing the

[2] For a discussion of mineral resources in relation to partition, see Charles H.

peninsula that lies between the Bay of Bengal and the South China
Sea with Siam, French Indo-China, and British Malaya. This entire
region lies within the tropics, except the northern part of Burma,
which extends nearly five degrees into the temperate zone. Save
for the highest mountain ridges along its border, however, the monsoon winds and rains make all the habitable sections of the country
distinctly tropical, regardless of latitude. In the Mandalay section
of the Irrawaddy valley precipitation drops slightly below forty
inches, but everywhere else in Burma and its eastern and southern
neighbors there is a rainfall in excess of sixty inches. In fact, most
of the Malay Peninsula is a soaking, tropical rain forest.

Whatever geographic handicaps may beset this part of the world,
a deficiency of rainfall is not one of them, but topography is. At
its eastern end the Himalaya Mountain System breaks into a bewildering number of linear ranges which fan outward. One group
swings southward around the Indian massif into Southeastern Asia,
giving it the tortured topography that is one of the outstanding characteristics of the region. One range, the Arakan Yoma, forms the
all but impassable boundary between Assam and Bengal, India, and
Upper Burma. With a divide which rarely drops below 6,000 feet,
and with individual peaks rising above 9,000 feet, the names applied
to the northern sections of this chain are modest understatements:
the Patkai Hills, Naga Hills, Lushai Hills. To the south this mountain element is partially submerged, forming the Andaman and
Nicobar islands, and to all appearances it continues in the Mentawai
Islands, or possibly in the Barisan Chain of Sumatra. Lower ranges
furrow the interior of Burma, but another high and persistent mass
cuts up the Shan States, providing a natural boundary and barrier
to Siam, and forming the 1,000-mile-long Malay peninsula. Somewhat similar north-south elevations transect Siam, but they are
lower, and it is not nearly so difficult to travel across Siam from the
Bangkok basin in the west to the Upper Mekong basin in the east,
as it is to travel by land from Bangkok to Rangoon, or from Mandalay to Calcutta. In French Indo-China the Annamitic Cordillera
almost fills Laos and Annam, but narrows southward in an open
hook that partially encloses the alluvial plain of the Mekong in

Behre, Jr., "India's Mineral Wealth and Political Future," *Foreign Affairs,* XXII
(1943), 78–93.

Cambodia and Cochin China. Other branches of the shattered Himalayas leave Tongking with little flat land, except the valley and delta
of the Songka (Sungkoi or Red River).

Formidable physical barriers have hindered transportation and
unification of this region, which embraces 800,000 square miles; and,
to a considerable degree, they have exerted a controlling influence
on settlement, growth and spread of population, and economic development. The gross population is 61–62,000,000, distributed as
follows:

	Area	Population	Density
Burma	261,610	17,000,000	69
Malay States	51,382	5,500,000	106
Siam	198,247	15,000,000	73
French Indo-China	286,119	24,000,000	83
Total	797,358	61,500,000	

More precise analysis by districts would show that, with local exceptions, population density conforms with the rule; it is densest in
the fertile lowlands, and it thins in more rugged terrane. Especially
is this true in Burma, where the northern mountains tie directly into
the Himalayas, and where inaccessibility became a by-word, as the
world learned about the Burma Road and Japanese infiltration
through an indefensible wilderness.

In combination the mountains and rainfall have shaped the destinies of these countries by creating fertile alluvial and delta plains
which are intensively cropped; by providing land that yields better
values in forest products than in crops; and by introducing a varied
assortment of metals, some of which have become so important to
the Western nations than their military and civilian operations were
partially impaired by the Japanese conquest of the area. No less
important is the moderate population density, which is great enough
to supply lumbermen in the teakwood forests of Burma, workers
for the rubber plantations of Malaya, miners for the mineral deposits,
and farmers for the crops; but which is not so great as to consume
all of the food raised. This is a region of surpluses—food surpluses
for overcrowded neighbors, and mineral and forest surpluses for the
nations of the West.

For overpopulous and famine-ridden China and India, all of these
countries except British Malaya have been a godsend. Burma, with

a normal rice crop of seven and one-half million tons, has raised
enough to satisfy her own needs, to make up the deficit in India, and
still to export one and one-half million tons. French Indo-China exports nearly two million of its six-million-ton crop, and Siam, with
five million tons, has one and eight-tenths million for needy neighbors. Only British Malaya appears not to meet domestic requirements and, statistically, to rank as the world's largest importer. This
situation is extraordinary enough to call for brief comment. The
Malay States nominally have the densest population among the countries comprising Southeastern Asia, but over 10 percent of the people live and work in Singapore and its environs. The balance are
thinly scattered over a mountainous, rain-drenched and fever-ridden
jungle, gathering latex, working tin deposits, raising rice, or picking
coconuts. Nearly 40 percent are foreign born and have come to this
part of the world to earn a living in agricultural, mining, commercial, and manufacturing establishments and not to conquer the jungle
as the native populations of the other countries have done. British
Malaya is, therefore, a small producer of rice, but the fact is, little
of the rice imported is consumed. Singapore is one of the largest
commercial centers in the East, and it buys to sell. A substantial fraction of its foreign trade ($600–650,000,000, annually) represents
nothing more than trading activity in the products of near-by countries, and its rice exchange has long been the largest in the world.

The list of other food crops raised in these countries is long, but
most of them are consumed locally. Tea, condiments, spices, and
herbs feature in the export trade, and French Indo-China ships a
considerable quantity and variety of foods to France. Of greater
importance, however, are such forest products as rubber, copra, and
wood. British Malaya leads in rubber production, with a normal
yearly output of 400,000 tons. Siam has a 10.5 million-ton coconut
crop and a sizable export trade in copra. All but French Indo-China
have a large timber cut, with teakwood featuring in Burma and Siam.

In the rich assortment of minerals which the mountains of these
countries contain, tin is outstanding. As food consumers forcefully
learned during the war, Southeast Asia and Indonesia comprise the
most important of three regions endowed with commercial supplies
of tin. Except in Burma, exports have been rigidly restricted, chiefly
through British control of smelting technique and facilities, with

near-disastrous results when the Japanese seized these countries.
The full capacity of the mines is unknown, but in 1940 and 1941,
years of minimum restriction, 153,600 and 152,000 long tons, respectively, were extracted. Production was distributed as follows:

	1940	1941
Malay States	85,384	78,000
Netherlands East Indies	43,193	51,000
Siam	17,447	16,250
Burma	5,500	5,500 (estimated)
Indo-China	2,098	1,430
	153,622	152,180

Virtually all of the ore was smelted in the Straits Settlements and
Netherlands Indies, for it was not until 1940 that smelting rights
were wrung from the reluctant British-Dutch International Tin
Committee, and facilities were put into operation in the United
States. It was 1945 before these smelters developed sufficient capacity to handle Bolivian production of 40,000 tons. Until that time
Bolivian ore had gone to Britain; and the United States, with no
visible reserves of ore and a sporadic production ranging from zero
to 56 tons, was completely at the mercy of a tightly controlled cartel, which held all but 10 percent of the world's smelting and refining facilities. Naturally there is some concern in this country regarding the future of tin in Southeast Asia and Indonesia. Like the
rubber cartel, the so-called International Tin Committee has always
been more deeply concerned with profits than with far-sighted international policies, but its influence on the future of the metal is
only part of the problem. With the British grip on areas of production loosened, and the Dutch grip definitely broken, the native peoples may have something to say about policies and production, for
it is native labor that works the mines. In this situation it is not unreasonable to insist that the world's chief consumer, the United
States, take a hand in curbing one of the more sinister monopolies.

Even without tin, Southeast Asia is an important mineral region,
the potentialities of which are only partly developed. A table showing current production will indicate which of the economic minerals are present, as well as the distribution of mining activities
among the four nations. It will be noted that, apart from tin, tung-

sten is the only other metal which is mined in important amounts, when measured against world production; but figures for some of the other metals are not to be lightly dismissed. Unlisted are gem stones, particularly rubies, sapphires, and jade, which Burma produces in quantity and quality unsurpassed elsewhere. Yet the entire assemblage does not have the makings of a regional economy. Although the geology of Siam is not so favorable for mineralization as the geology of the three colonial countries, it is evident that the latter have been more intensively explored than the one independent nation, whose people are not industrially inclined and have no obligations to a "mother" country. There is an evident deficiency in energy resources. Burma's oil might be adequate for the limited demand which currently exists, but it is heavily mortgaged. East Indian oil, however, is handy, and the Middle East should be able to take care of any demands which the East Indies can not meet. The coal reserves are meager and have limited industrial utility. From the standpoint of industry Indo-China would be better off with a good grade of coking coal than with one of the earth's few reserves of anthracite, and Malaya's bituminous reserves are small. There is more iron ore available than is being mined, but there is little that can be done with it. Copper deposits are so few and so small that a copper mining industry has been unable to maintain itself. Mining operations have thus been concentrated upon a few nonmetals which are needed locally, and upon those metals which bring a good price in distant markets.

An awakening of latent industrial ambitions in the peoples of Southeast Asia appears remote. The immigrant population of Singapore has brought both trading and industrial life to the Straits Settlements, but elsewhere there is little or none. In 1938, there were 83,000 workers in industrial establishments in Burma, and apparently the numbers were so low in Siam and French Indo-China as to make a statistical count superfluous. These countries contain water-power sites from which 12,000,000 horsepower or more could be generated; yet very little of it is harnessed, and there is no program for expansion. None of this is written in criticism. There is no reason why all the world should go modern and industrial—indeed, it may be good that one section of it should continue the even tenor of its rustic ways, particularly when there is no urgent prob-

Mineral Production in Southeast Asia

Mineral	Burma	Siam	French Indo-China	Malay States	World Production	
Gold	1,200	12,700	8,000	40,000	40,000,000	ounces
Silver	6,800,000	..	1,700	3,500	267,000,000	ounces
Lead	80,000	..	10	..	1,700,000	tons
Zinc	68,000	..	6,000	..	1,600,000	tons
Iron	2,700	..	134,000	2,000,000	200,000,000	tons
Manganese	..	..	2,400	32,000	5,200,000	tons
Nickel	1,200	..	..	..	120,000	tons
Cobalt	240	..	..	..	3,700	tons
Tungsten	8,000	400	650	1,350	40,000	tons
Bauxite	..	..	3,500	140,000	14,000,000	tons
Antimony	84	..	83		40,000	tons
Tin	5,500	17,447	2,098	85,000	240,000	tons
Titanium	..	..	..	11,000	320,000	tons
Coal						
Bituminous	..	..	..	794,000	600,000,000	tons
Anthracite	..	..	2,500,000	..	110,000,000	tons
Oil	7,700,000	..	..	..	2,500,000,000	barrels
Phosphate rock		..	35,600	..	11,500,000	tons
Salt	40,000	150,000	210,000	..	27,500,000	tons
Cement	..	..	300,000	..	90,000,000	tons

lem to be solved, such as the employment of underprivileged or unemployed millions. Two important functions seem to define the
future of the region, as they have the past; this section of Asia can
supplement the food deficiencies of India and China, and it can
supply the Western nations with critical raw materials—notably tin,
rubber, and tungsten.

The function of the East Indies is not so easily defined, if only
because there are so many unexpected contrasts among islands
which, by any process of reasoning, should be alike. Geologically
they comprise an arcuate, partially submerged, pair of parallel mountain ranges fronting the Sunda Deep on the south, with northward
extensions from the ends of the arc that apex in the Philippine Islands and Formosa. The islands, numbering many thousands, which
constitute the East Indies, straddle the equator from latitude 7° north
to latitude 11° south, and they stretch from longitude 95° to 155°
east—more than 4,000 miles. Rainfall is heavy in the west, but
diminishes eastward, though it nowhere drops below the humid
category. In Sumatra, Java, and Borneo precipitation is heavy
enough for the requirements of a tropical rain forest, and although
the forest exists, notably in Borneo, it has no chance on Java, where
42,000,000 people have crowded it out. A tabulation of the major
islands with their areas and populations will bring out a few of the
contrasts.

	Area	Population	Average Density
Netherlands Indies	733,790	70,000,000	95
Sumatra	182,860	8,700,000	47
Borneo	206,810	2,500,000	12
Java	51,030	50,200,000	1,000
Celebes and New Guinea	293,000	8,600,000	30
British-Australian Indies	256,726	1,911,000	9
Brunei	2,226	36,000	16
Sarawak	42,000	500,000	12
British North Borneo	31,000	300,000	10
New Guinea	91,000	775,000	8
Papua	90,500	300,000	3
Portuguese-Timor	7,330	480,000	66

With almost a million square miles of land surface and 72,500,000 inhabitants, the East Indies are more nearly comparable with Southeast Asia than one might initially think. Like the latter, the Indies are ribbed with high, untamed mountains, with peaks rising above 13,000 feet in North Borneo and above 12,000 in Sumatra, Java, Lombok, and Papua. Still in the formative stage, the mountains present the hazards of earthquakes and volcanic eruptions, while the sea at times brings disastrous typhoons and, more rarely, *tsunamis* or "tidal" waves. Like aprons protecting the mountain chains, piedmont plains rise from the sea to the alpine slopes and edge into the mountain fronts in alluvium-filled valleys. The plains offer ideal sites for sugar, rubber, and rice plantations, though the cultivation of rice and rubber does not stop with the mountain borders. In crowded Java the native rain forest has given way to a shrewdly conceived system of terraces which conserve the soil and the rains for the growing of rice; but, in islands where thinner populations exert less pressure on the land the forest still has its way. In the east, where precipitation is lower, savanna lands are a more usable substitute for the forest.

There is a tendency to confuse the history and development of the East Indies with the phenomenal growth of Java. This island became one of the richest colonial possessions in the world in the course of a century, primarily because the Dutch recognized the superior abilities of its native population and so centered their colonial administration there. The growth of economic opportunities led to an extraordinary increase in population, from an estimated 4,500,000 in 1845, to 41,700,000 in the census of 1930, and an estimated 50,000,000 at the time of Japanese occupation. Java, with an area about the size of Alabama (population approximately 3,000,000) tucks nearly 1,000 people into every square mile of land. As is always the case in thickly settled regions, there is delicate adjustment of density to topography, soil fertility, and economic opportunity, but there is little of the island that is not needed to accommodate so many Javanese. The low density in other Dutch islands reflects smaller initial populations, a lower degree of intelligence and energy among the people, and a slower program of economic exploitation. The British and Australian holdings contain more natural wealth than human wealth, though only in Sarawak has there been any serious effort to develop both. In Timor, where the Portuguese found a denser population with a mental capacity more nearly equal to that of the Javanese, there is a direct reflection of the contrasts in colonial administration.

It was the genius of the Dutch to recognize the economic value of Java's human resources and to exploit them—benignly, to be sure, but so firmly that, when Japanese invaders came, it was a question whether the Javanese loved the Japanese less or disliked their Dutch masters more. The situation was aggravated by waxing sentiment in favor of independence, which was especially strong at the time in consequence of American action in the Philippines. How the current movement toward independence will affect the economy of Java and the other islands is unknown, but just from an historical standpoint the Dutch have chalked up an achievement of humane but efficient utilization of human resources that is unmatched anywhere in the annals of colonial expansion. As such, it merits brief analysis.

The first need for so large a population is food, and the Dutch have made the raising of food obligatory among the Javanese farmers—there must be an acceptable ratio of food-crop to cash-crop production, however lucrative and attractive cash crops may appear at any given time. As a result, over 80 percent of the farm-plantation acreage is utilized for rice and corn. The yield is nearly 7,000,000 tons of rice and 2,200,000 tons of corn; yet, at times, this must be augmented by imports of 400,000 tons of rice to meet requirements. Other foods are imported, but generally they comprise not more than 20 percent of total imports, which average $250,-000,000 a year. Java is not quite self-sufficient, but it misses self-sufficiency by a relatively modest margin.

Cash crops are grown on nearly all of the remaining acreage. Featuring in the export trade are, in order of dollar value, rubber, vegetable oils and fats, tea, sugar, tobacco, drugs and spices, fibers, and coffee. Although British Malaya and the British East Indies outproduce the Netherlands Indies in rubber, one suspects that the production ratios established by the British-Dutch monopoly have more to do with the statistics than the capacity of the Dutch plantations. In sugar production Java ranks sixth (third in cane sugar), but is second only to Cuba among exporters. And the Netherlands

Indies have acquired a virtual monopoly on cinchona, as well as a firm grip on the vegetable oil, spice, and drug trade. The position of Java and the other Dutch islands in respect to rubber, sugar, and cinchona has been established within the past half-century. These products have two characteristics in common: They require heavy precipitation and perennial tropical to subtropical temperatures; and their cultivation and harvesting require a great deal of manual labor. Java has both climate and manpower, and it is one of the few places in the world where the two occur in combination. It is the only tropical country where the full economic value of the combination has been realized.

From this realization has evolved the rubber monopoly, which has become one of three important elements in the economy of Southeast Asia and Indonesia. It started in the last quarter of the nineteenth century, when the latex-yielding plant *Hevea brasiliensis* was successfully transplanted from the jungles of the Amazon basin to the Far Eastern colonies of the British and Dutch. Although the British were content to plant it and let it grow with native vegetation, the Dutch conceived the idea of planting it alone and of keeping competitive vegetation down. Such a method of cultivation requires plenty of labor, and the Dutch had it. In 1910, the first important commercial shipment was made from this region. Despite the rapidly rising demand for rubber, cheap East Indian and Malaysian labor quickly put the Brazilian industry out of business so completely that Ford's millions were unable to revive it in the twenties, and American money and ingenuity met with no better success during the war. In Brazil, *Hevea brasiliensis* now has merely the status of a botanical specimen.

With the demise of Brazil's rubber industry, British and Dutch producers were not slow in recognizing that they had a monopoly on their hands. As in the case of tin, they worked it to the limit of its possibilities, again chiefly at the expense of their largest customer, the United States. Careful control of plantings and rigid restrictions on latex-gathering, rubber production and exports kept the price steady—and high—regardless of economic conditions. For a time during the boom years the supply was kept so low that American users could not meet their requirements. Official protests led to slight relaxation of quota controls but to no fundamental change in policy. In 1937, 1938, and 1939, the producers paid little heed to mounting tension in the Far East; and despite sharply increased quotas in 1940 and virtually unrestricted output in 1941, it was too late to anticipate military and industrial needs over the prolonged period of Japanese occupation. Ceylon was the only rubber-producing area that remained free, and it ordinarily accounts for 7 percent of the world's output. A comparable Dutch monopoly of cinchona had the same result, but in this case there is no proof of carefully planned restriction.

The story of synthetic rubber has no place in this exposition of natural resources, but it may seriously affect the economy of the Netherlands Indies and British Malaya. Use has demonstrated that it can substitute for natural crude rubber, whether or not it is quite as good as the real thing or quite as cheap. The monopoly is definitely broken, but the commercialization of synthetic processes of manufacture may prove to be more than a belated instance of poetic justice. With the cost of manpower rising and the unit cost of synthesis dropping, an industry that for many years provided the Netherlands Indies with 20 to 25 percent of their export trade may be wrecked. With it will go the livelihood of many Javanese, whose energies must be diverted to other occupations—if only to making up the deficiency in domestic food production.

Of less vital concern to the native population but of far-reaching importance to the oil-poor nations on the western rim of the Pacific is the petroleum industry of the Netherlands Indies, Sarawak, and Brunei. Before the Japanese came, 62,000,000 barrels of oil were being obtained from wells in Java, Sumatra, and Dutch Borneo; and exploration and development had stopped just short of production in New Guinea. Another 7,000,000 barrels were being obtained in the British sections of Borneo—Sarawak and Brunei. So effective was Dutch destruction of wells and refineries that only 9,000,000 barrels were obtained during the first year of Japanese occupation, and it took two years to build production up to 40,000,000 (1944). In the British holdings, production was halved temporarily but was very nearly normal in 1944. Indonesia has been in the oil business for many years, and despite the setback occasioned by the Japanese conquest, there is every prospect of long-continued activity. The reserves are limited, and in recognition of this fact there has been no feverish overdevelopment that will lead to premature exhaustion. Like everything else the Dutch have controlled, oil exploitation has proceeded in a sane, businesslike manner, and only the uncertainties of the internal situation will retard rehabilitation of the industry to its pre-war status. The reserves, estimated conservatively at 3,000,000,000 barrels, are crucial in the Western Pacific, for there is little production anywhere else in this part of the world. China, Formosa, Japan, and Sakhalin have a combined output of less than 10,000,000 barrels; Australia and New Zealand have practically none. Obviously, the market is greater than the visible supply.

Aside from a small textile industry, tin smelting, the ramifications of oil refining, and the preparation of rubber for overseas shipment, there is little manufacturing in Indonesia, even in Java, where more than 6,000 small manufacturing establishments are listed. Natural gas and oil furnish much of the heat and power required, but coal is a supplementary source of fuel. The small reserves are of medium quality and have rarely been called upon for more than 200,000 tons in any one year. The hydroelectric potential is considerable in islands with so heavy a rainfall and such high relief, and it has not been overlooked. In Java nearly all the available energy is utilized, and hydroelectrification has made some progress in Sumatra, though there is much more untapped energy there. New Guinea with 5,-000,000 horsepower, Borneo with more than 2,000,000, and Celebes with more than 1,000,000 have as yet put little of it to use.

Like the adjacent mainland and for the same geologic reasons, the islands of the archipelago contain many minerals. Dutch territory has been rather thoroughly prospected; British territory, less so. Deposits of bauxite, copper, gold, silver, platinum, iron, manganese, nickel, tungsten, asphalt, sulphur, phosphate rock, and salt are recorded, in addition to the tin, petroleum, and coal, which have already been discussed. Recovery of gold and silver has been taken seriously, and the Netherlands Indies have reported 69,000 fine ounces of the former and 1,500,000 ounces of silver in a good year. Sarawak has a small silver production and approximately 17,000 ounces of gold. Copper and iron are known but not mined, but there was some pre-war recovery of manganese (12,000 tons), nickel (2,-200 tons), bauxite (300,000 tons), phosphate rock (34,000 tons), and salt (390,000 tons).

It would take an inspired and perspicacious prophet to foretell the future of Southeast Asia and the East Indies, but the geography and geology clearly delimit the possibilities. The startling fact that one island of 51,000 square miles at latitude 6°–8° south is supporting 50,000,000 people is a tribute to Javanese industry and Dutch managerial skill, but it is not a measure of what may be anticipated in other parts of the region. The other islands and sizable sections of the mainland are under no pressure, and they can absorb the overflow from Java and China for some time to come. They can also expand their function of raising food surpluses for the crowded Asiatic mainland. Temperamentally this is all the native populations are inclined to do, but under the tutelage of Western nations they have done much more: they have taken over the rubber industry, and they have commercialized sugar, copra, cinchona, tea, and teakwood production, without deviating from their traditional agricultural bent. Mining, oil recovery, and mass-manufacturing are, however, distinctly foreign elements in their life; and a nation like Siam, which has been left more to its own devices than the other countries, has shown no disposition to adopt Western industrialism. There is no reason to believe that it will, at least in the immediate future.

In Java, however, population growth has exceeded the potentialities of the land; and, whether the Dutch remain in partial or full control, or are ejected altogether, there must be industrialization or emigration, or both. It looks as if there may be some emulation of the Filipinos, who, in fifty years association with Americans, have adopted and adapted many features of the American outlook. The economies of the two peoples run parallel, although the 16,000,000 Filipinos who inhabit the 114,400 square land miles in the archipelago can not produce as much as 50,000,000 Javanese. Yet they do well, as comparative statistics indicate.

Product	Java [a]	Philippine Islands [a]
Rice	7,000,000	2,000,000
Sugar	1,500,000	1,100,000
Coconuts	1,300,000,000 (nuts)	1,800,000,000 (nuts)
Corn	2,200,000	550,000
Tobacco	45,000	40,000
Abaca	..	180,000
Kapok	23,000	..

[a] Export in tons, unless otherwise designated.

Product	Java	Philippine Islands
Rubber	400,000	1,000
Cinchona	13,600	..
Gum and Resins	..	49
Logs, sawn timber	10,400,000 (bd. ft.)	250,000,000 (bd. ft.)
Oil	62,000,000 (bbls.)	..
Coal	2,000,000	47,000
Gold	89,000 (oz.)	1,000,000 (oz.)
Silver	1,500,000 (oz.)	1,350,000 (oz.)
Copper	100	9,000
Iron	..	1,200,000
Manganese	12,000	58,000
Chromite	..	300,000
Nickel	2,200	..
Bauxite	300,000	..
Tin	51,000	..

Value of overseas trade: Java, $650,000,000 (13.00 per capita); Philippines, $250,000,000 (18.62 per capita).

The oil, tin, and rubber of Java account for a substantial fraction of the difference in the export figures for the two countries, but part of the difference is due to the consolidation of production statistics for all the Netherlands Indies. In the table Java has been given credit for some commodities raised in Sumatra, Celebes, and Borneo. Otherwise, comparison of the two economies is direct, and although the Filipinos may be better prepared for independence, as political entities the two countries may be expected to pursue parallel and analogous lines of economic activity and to become embroiled, as competitors always do, when they seek outlets for their products in the same world markets. In a competitive race the advantage which the Filipinos have through their commercial ties with the United States is offset by Java's possession of tin and oil. Whether rubber is an asset or a worrisome liability depends upon the course which the synthetic rubber industry will take.

China

For the earth's most populous nation and one of its largest, China has exerted rather little influence in world affairs but has, instead, been a perennial victim of them. The trouble seems to lie with the amorphous mass of Chinese people, who have become so obsessed with the difficult task of subsisting that they have no background, no interest, and no time for comprehension and action on domestic issues, much less on the affairs of the outside world. This preoccupation on the part of the public, incompetence—or worse— among public officials at the provincial level, and consequent weakness and ignorance at higher levels create, in combination, an overcast through which one can catch only an occasional and partial glimpse of the true situation beneath. Under the circumstances it is little short of presumptuous to attempt an appraisal of China's resources, yet it must be done. A country that contains 20 percent or more of the earth's population, and 6.5 percent of its land surface, cannot be ignored or dealt with sketchily, even though its people have always exhibited carelessness and nonchalance with regard to facts and figures.

Problems beset the geographer from the start of his study, for it is difficult to assign precise geographic and political boundaries, and virtually impossible to put down even an approximate figure for China's population. If one agrees to accept the territorial claims of the Nanking government, he may incur wrath in some quarters, but he will deal with the maximum area—4,314,000 square miles—which can be included under the name of China as a geographic designation. For the purpose of the present survey, this arbitrary procedure is the most practical and will be adopted. In this sense, China is a polyglot country, with even greater variety in its geographic regions than among its peoples. Starting at longitude 74° east in disputed Sinkiang, it touches longitude 135° east in controversial Manchuria. In dubious Mongolia, China reaches latitude 52° north, and in recently regained Hainan it extends to the subtropical latitude of 18° north. Although it fails to reach subarctic latitudes, it attains subarctic altitudes in the Plateau of Tibet.

In such a large sector of the earth there can be no simple climatic pattern and no geographic unity, and political unity could be achieved only by measures beyond the purse and the power of the central government. Mass education, easy communication, and accessibility through modern transportation might weld the population into a national unit in spite of the diversity of agricultural pursuits and economic interests. But there is no popular system of education; 200,000 telephones are not even adequate for the transaction of business; fifty radio stations reach only 500,000 radio sets; 70,000 miles of improved highways, 7,000 of railway lines, and 12,000 of regularly flown airlines, supplemented by temperamental and hazardous river navigation, are weak ties in so big a country. Until they are greatly strengthened, China will be of little use to herself, and will constitute a major ideological risk to the rest of the world. Chinese geography illumines, but does not solve the problem.

Southeastern China is well watered, but precipitation decreases gradually northward along the coast to little more than twenty inches in Manchuria. The declining evaporation rate partially compensates for the lower rainfall, but the lower population density in Manchuria (approximately 90 per square mile) reflects Manchu history and political turmoil as much as the drier and more rigorous climate. Westward, in the interior, precipitation drops more rapidly to arid and desert conditions in the Gobi Desert of Mongolia and Tarim Basin of Sinkiang. In the latter, drainage from the encircling mountains is ample to support a population much larger than the 2,700,000 who now live there. These mountains, notably the Tien Shan Range, are also reputed to contain mineral deposits which have aroused the interest of Soviet Russia to the point of making earnest efforts to detach this province from the territory under Nanking rule. As large as Alaska (estimates of the area range from 550,000 to 634,000 square miles), and with a climate not unlike that of Utah, Arizona, and New Mexico, Sinkiang could—and should—occupy a significant place in Chinese economy. As matters stand, it is little more than a source of political worry. Located almost in the geographic center of Asia, it is one of the most isolated spots left on this shrinking planet. Although separated from the Tadzhik and Kirghiz Autonomous Soviet Socialist Republics by the formidable Tien Shan Mountains, its most accessible outlet is the railroad at Andizhan, across the range in the U.S.S.R. With no equally facile connections with, or counterinfluence from, east China, it is not surprising to find Soviet interest and penetration so strong in Sinkiang.

Mongolia differs from Sinkiang in only three respects: it is larger (1,368,000 square miles); it has only one-third the population (875,-

ooo); and it lies much nearer the seat of government. Proximity has
meant something to Inner Mongolia, which is served by a few pass-
able roads and by a single railroad line; but for Outer Mongolia it is
as easy to cross the Soviet border to the north to obtain access to
transportation, as it is to travel south or southeast within the confines
of China. The fact is, Urga is scarcely half as far from the Trans-
Siberian Railway at Lake Baikal as it is from the rail facilities at
Kalgan or Sui Yuan. Water resources are even more limited in Mongolia than in Sinkiang, for the mountain ranges are lower and far
apart, and most of the drainage of Tibet is so oriented as to miss
Mongolia completely. The thin population is nomadic and pastoral,
for rather little agricultural use is made of the small amount of water
that makes its way down from the mountains into the undrained
basins of the Gobi. A few metals have been reported, notably gold
and copper, in the Altai and other ranges, and as much as 60,000
ounces of gold have been washed from placer gravels in the years
before political conditions became unsettled. Without transportation and with none in prospect, there is little inducement to mine
other metals or to attempt to raise crops on irrigated lands. Although the country is so dry and cold that settlement will be severely
restricted,[3] the examples of other deserts indicate that a region as
large as Mongolia, though incapable of supporting very many of
China's millions, can unquestionably provide a living for several
times as many people as are there now. But its remoteness and its
tremendous empty spaces, which are too vast and worthless to tie
in by means of an expensive transportation system, are immutable
factors circumscribing its potentialities and offering an eloquent
commentary on the worthlessness of the heart of the Heartland—
the crux of the Geopolitik which inspired Hitler, and which, from
all appearances, is guiding Soviet policy at the present time.

If Tibet serves no other function, it provides China with a magnificent southern boundary. Although such a perfect natural barrier might have been more helpful along the northern border of
the country, there is little likelihood of territorial conflict between
India and China. The Tibetan Plateau is 700 miles broad north of

[3] See George B. Cressey, *China's Geographic Foundation* (New York, 1934), pp.
268-271.

Assam and Bihar; and even though it narrows westward, it is still an impenetrable wall between the Punjab and the Tarim Basin. In length it exceeds 1,600 miles. New provinces have been chipped off its more habitable fringes, but as a geomorphic unit it embraces approximately 450,000 square miles. Estimates of population range from 1,500,000 to 4,000,000, practically all of it restricted to the valleys—the only areas which drop far enough below the snowline to be habitable. The comparatively low precipitation, decreasing so markedly to the north that the snowline rises more than 4,000 feet poleward, is merely of scientific interest, for most of the valleys are watered by melting snow in the short "summer" season. During a substantial part of the year everything is frozen tight. Topography, soil, and climate conspire against agricultural activity, and the population depends upon the restricted grazing lands for its meager living. Yak and goats are practically the only animals that can survive in such a regime, and they are moved upward from one sparse pasture to another during the warmer weather and down again as the Tibetan winter closes in. There is no place for horizontal expansion, and the valleys, though long, already contain as many people as the economy will permit. Even communication between one valley and the next is impossible in a highland with such stupendous relief and treacherous weather. Geographic exploration is far from complete, and geologic exploration has scarcely been begun. Economic minerals, if they exist, are unknown. There is nothing in the entire region upon which further development can be predicated.

Quite clearly from this survey of western China, the Chinese are in eastern China—420,000,000 of them in approximately 1,550,000 square miles. Although it must be a phenomenal land that has given subsistence to so many for so long, analysis shows that, with the possible exception of the loess, it is exactly like similarly located regions in other parts of the world; and that frugal management and substandard objectives have enabled it to do more than any comparable area has done. The use of the land changes with latitude. In the south, rice is the main subsistence crop, whereas in the north it is wheat. Millet, corn, kaoliang or sorghum, and barley are important accessory crops; and no less important in an essentially vegetarian diet are potatoes and, of course, soy beans. Annual production statistics are nonexistent, but from an analytical study of available government surveys, Cressey [4] gives the following figures, which will serve as a reasonably reliable index of Chinese agriculture.

Product	Pounds	
Rice	130,000,000,000	
Wheat	56,500,000,000	(940,000,000 bu.)
Kaoliang	31,000,000,000	
Millet	29,000,000,000	(500,000,000 bu.)
Corn	20,000,000,000	(360,000,000 bu.)
Barley	17,000,000,000	(300,000,000 bu.)
Potatoes, sweet	36,000,000,000	(600,000,000 bu.)
Potatoes, white	5,400,000,000	(90,000,000 bu.)
Soy beans	30,000,000,000	(500,000,000 bu.)
Sugar cane	6,500,000,000	(3,250,000 tons)

The figures are large, but they should not be surprising. What is surprising is the fact that, by dint of intensive cultivation and soil management, the Chinese farmer is getting a higher yield per acre for many of his crops than the world average. For example, the average for rice is 32 bushels, whereas the Chinese are obtaining more than fifty. As for total volume, China vies with India for first place in rice production, though fewer of her people use it as a principal item of food. Her wheat crop rates her with the U.S.S.R. and United States. She leads in the yields of sweet potatoes, kaoliang, soy beans, peanuts, millet, and barley. As in India, sheer numbers of people necessarily make China a productive agricultural nation. Cultivation reaches its peak of intensity on the delta plain of the Yangtze-Kiang in the Shanghai district, but it is nearly as intense on the delta lands northward to the Yellow River and beyond to Tientsin. The high percentage of land use can be traced up the flood plains of every valley, reaching noteworthy intensities in the Hankow basin and in the Chunking section of the Yangtze River system. Nearly all the land in the extensive interstream loess region is in agricultural use, but the intensity of farming operations decreases westward as distance from transportation increases, and as relief heightens toward the mountain ranges which twist southward and southeastward from

[4] Cressey, *op. cit.*, pp. 90–103.

the east end of the Himalayan-Tibetan highlands. Thanks to ample rainfall, warm temperatures, and rapid rock-weathering, even the plateaus and mountains of Yunnan and Kweichow contain an average of 160 people in every square mile, with the usual variations and adjustments to the local flood plains, mountain slopes, and canyons which dissect the region.

For many generations the Chinese have been "moving over" to make room for the annual increase in population, but only at continued sacrifices in the seriously substandard level of living. With so large a population in China Proper, and no place to go except, possibly, to Manchuria, the fact that this is the world's greatest food-producing region loses its meaning, for it is, at the same time, its most impoverished land. In a richer country deficiencies in food production would be met by imports, but there is not enough purchasing power to bring much additional food into China. Even if it were imported, distribution would present insurmountable problems. The transportation facilities are not adequate for movements of food within the country. Because of this situation each section has developed a degree of precarious self-sufficiency; hence, in the event of flood, drought, or other agricultural calamity, there are no surpluses which can be moved into the stricken areas. And if there were, the means of local distribution are unsatisfactory.

At the present time and for some time to come, there is no prospect of improving the agricultural situation. Without spectacular expansion and modernization of transportation there is no point in increasing productive acreage by irrigating the piedmont sections in arid Sinkiang and Mongolia. Until there is education to teach people to use the marginal subhumid to semiarid lands of Inner Mongolia intelligently, it is better that they remain unused, for misuse will entail an ultimate loss of more serious proportions than nonuse. Without greater river control, serious crop losses from floods will continue in China Proper, bringing famine and death to the afflicted areas. In brief, there is no quick relief for the food and nutritional plight of China, and conditions are such that progress should first be sought in other economic fields, with the possible exception of measures that will further reduce flood risks in the larger and most densely settled deltas and floodplains. Before these other fields are considered, however, the economic roles of Formosa and Manchuria should be reviewed.

Formosa, with an area of 13,857 square miles, is approximately as large as Massachusetts, Connecticut, and Rhode Island. It has a population of 5,350,000, mostly rural, as compared with a predominantly urban population of 6,500,000 in the three southern states of New England. In no other respects are these widely separated areas comparable, because Formosa lies upon the Tropic of Cancer, with equable, subtropical temperatures at most elevations, and a copious, though not heavy, rainfall. Topographically it is rugged, its mountain backbone rising above 10,000 feet and at least one prominent peak ascending to 12,000 feet. The topography limits and concentrates the population and agricultural activity at low and moderate elevations, where rice, sweet potatoes, sugar cane, bananas, and tea are the major crops. Following its acquisition by the Japanese, Formosa became a source of food supply for needy Japan. Of exports valued at $125–150,000,000, 85 percent went to Japan, and 75 percent consisted of rice, sugar, bananas, and canned pineapple. Whether this surplus will now be routed to China is doubtful, for the Chinese have little money for purchase and few manufactured products for trade. Moreover, the Chinese government has not endeared itself to Formosans by imposing on them an inept and tyrannical postwar administration. If the Chinese want cooperation from one of the few sections of their reconstituted nation that has surplus food, they must do differently and better than they have done since August, 1945.

Manchuria's 44,000,000 people live in a country which is slightly smaller than the Province of Quebec but fifteen times as populous. Situated roughly six degrees of latitude farther south than Quebec, it has a comparable winter climate with slightly higher mean temperatures along the south coast, but summer temperatures are nearly ten degrees warmer, and there is less rain. The severity of the winter weather made it an unpopular assignment among Japanese government officials, and willing colonists were difficult to find. Although the northern Chinese have adapted themselves readily to the changeable climate, southern Chinese have shunned it. More by force of habit than for ecological reasons, some of the inhabitants grow paddy and upland rice, but the principal crops are (in tons):

Kaoliang (sorghum)	5,000,000
Soy beans	4,500,000
Millet	3,880,000
Corn	2,720,000
Wheat	1,050,000

There is also considerable livestock and a sizable commercial timber cut. During the thirties foreign trade reached the impressive value of $688,000,000, the greater part of it, naturally, with Japan. Among the exports, agricultural products occupied the most important place, with soy beans and bean cake comprising 42 percent of the total. China has taken some of this food, especially millet, kaoliang, and corn; but the amount was small in comparison with her needs. To what extent the destination of food shipments may shift is difficult to predict; but Manchuria, like Formosa, is a potential source of supply; and unlike Formosa it can still accommodate a few more settlers from the south, although the fertile basin of the Liao River near Mukden is already overpopulated, and the Sungari basin in central Manchuria is also well filled. Northwestern Manchuria is a dry and inhospitable upland consisting of the Great Khingan Mountains, but the area is tapped by the old shortcut of the Trans-Siberian Railway, which reaches Vladivostok via Harbin. One lasting benefit of Russian and Japanese penetration is the network of railroads, but the road system is still primitive.

From this regional review it is evident that the Chinese people are pushing the capacity of the good earth close to the practical limit of productivity, and that rather little help can be anticipated from the large areas outside the Great Wall. Formosa and Manchuria, the only places with anything to sell, can help make up part of the deficit in food; but as a cash customer, China has a desperately low rating. The most urgent need is to raise mass purchasing power, rather than to lower it, as the government has been doing by indiscriminate use of the printing press. The need must be met through some medium other than agriculture; and manufacturing is the most obvious possibility, for the export of raw materials will not suffice under the best of circumstances, and events of the past few years have seriously narrowed the field. The development of nylon has cut the demand for bristles and for silk; the destruction of manufacturing plants in the coastal cities and the collapse of Japan have sharply reduced the market for coal and for metals. Unfortunately the current domestic demand for these raw materials is almost nil. During the war, publicity was given to the migration of industry from the conquered coast to Chunking and other inland cities beyond the reach of the invaders; but in 1943, following the transplanting of equipment from 650 manufacturing establishments, together with 12,000 skilled workers, David Nelson Rowe [5] concluded that, at most, "the total number of factories employing thirty or more workers and using mechanical power is about 2,000. Taking the most optimistic estimate possible, they can not employ more than 500,000 men. This is probably double the actual total of industrial workers in Free China."

Even with complete rehabilitation of the industrial establishments around Shanghai, Tientsin, Hankow, Nanking, and other coastal manufacturing centers, industrial employment in China will be less than it is in such a tiny but highly industrialized country as Belgium. With Japan no longer in a position to control or to compete with Chinese industrial output, the possibilities are tremendous, but the handicaps are appalling. There is little native capital, and foreign capital, having been badly burned, will hesitate to put its money too close to the fires of internal political instability and the external heat of Communism. More especially is this true in the Mukden-Liaotung Peninsula of Manchuria, where industrialization made greatest pre-war progress, but where Soviet influence is strongest and the extent of Soviet control is not yet known. Whatever the obstacles, China's salvation lies down the path of industrialization, along which can be found a promising array of raw materials.

Topping the list in importance is coal. No one knows exactly how large China's coal reserve is. Estimates range from a modest 250,000,000,000 tons to 1,000,000,000,000 tons. The difference, despite its size, is academic, for even the smaller figure is adequate for industry on any scale for a long time to come. Most of the provinces of China Proper possess some of this reserve, and coal has actually been mined in excess of 100,000 tons a year in eight provinces. Commercial operations have been tied initially to transportation facilities, but once started, they have encouraged the develop-

[5] David Nelson Rowe, *China's Military Potential and the Enforcement of Peace,* Yale Institute of International Studies, 1943 (privately distributed).

ment of additional means of transport. In consequence of this tendency, coal extraction is greatest in Manchuria, especially in the Fushun district, notwithstanding the facts that much of the coal will not coke and the reserves are less than those in Shansi and Shensi provinces, in the interior of northern China Proper. Likewise, in Hopei Province, use has been made of the railroads serving Tientsin to develop the second largest mine in China, with an output in excess of 5,000,000 tons a year. The total output has never exceeded 27,000,000 tons, so far as official records are concerned, but unofficial estimates indicate that an additional unrecorded tonnage of 10,000,000 is mined for domestic use, making a total in normal times nearer 40,000,000. Approximately one-third comes from Manchuria, another third from the adjacent provinces of Hopei, Honan, and Shantung, and the balance comes from scattered mines, most of which have local means of transportation available. China thus ranks sixth or seventh among the coal-producing nations, but there is no reason why she cannot ultimately rank among the first five. The most important resource—fuel for industry—she has in abundance.

Oil appears to have by-passed China, for in all the mainland territory actually or nominally under Chinese rule, only 500,000 barrels are being produced (1944). With the reacquisition of Formosa, the output will be quadrupled, but this is still a meager supply in a country of magnificent distances, where truck and car transportation are now more important than rail transportation, and where the building and extension of roads must be a major feature in any program of development and progress. Exploration has not been complete, even in those districts from which petroleum has been obtained, but there are few indications of its presence, and nothing comparable to the signs that drew oil men to the Middle East. There may be more oil in China's future, but there is nothing to suggest a large enough domestic reserve for self-sufficiency.

The water-power potential is far more encouraging, even though little modern use has been made of it. It is estimated that more than 20,000,000 continuous horsepower are available, nearly all of it in the mountains of southwestern China and in the Tibet foreland. Half of the reserve is attributed to the Yangtze River system, in which the magnificent gorges and such tributaries as the Min, which falls 9,000 feet in 150 miles, offer unrivaled sites for dams and reservoirs. A goodly part of the potential lies within economic transmission distance of large populations which can be trained in its industrial use. Silt will be a problem wherever reservoirs are built, hence installations will be costly in both construction and upkeep. It is a major misfortune in the geography of hydroelectric energy that only a few large interior cities will directly benefit— the partially industrialized cities of the coast are too distant and must continue to utilize steam plants powered by coal.

It is extremely helpful to a nation with industrial ambitions to have the triad of energy resources present within its own political boundaries; but of the three only coal is critical, and only oil can be imported at reasonable cost. Relative scarcity of petroleum in the Western Pacific is a handicap, but all that part of the world except the Netherlands Indies shares it. China is no worse off than Britain, or Germany, or France and, like them, must put petroleum and petroleum products high on her international shopping list, even in a minimum program of development. The presence or absence of less easily transportable items, like iron and cotton, is of far greater moment.

The reserves of iron ore in China have been carefully but not exhaustively studied, and there is a tendency to give them a low rating. There are no individual deposits of tremendous size, and a lot of ore is low grade. Of the latter, containing 30 to 35 percent metallic iron, there are approximately 300,000,000 tons proven, and 475,000,000 tons probable, or a total of 775,000,000. More than 700,000,000 are conveniently localized in Liaoning Province in southern Manchuria and have served as the principal source of raw material for the Anshan ironworks near Mukden. Better grade sedimentary or oölitic ores are found in smaller quantities—a proven tonnage of 30,000,000 and an additional probable tonnage of 65,-000,000, or a total of 95–100,000,000 tons, ranging from 50 to 55 percent metallic iron. High grade magnetite deposits with 55 percent or better metallic iron will yield an assured output of 75,000,000 and probably 10,000,000 tons more. Ores of all grades thus total nearly one billion tons, or about enough to last the United States ten years at peak production.

A great deal of breath and ink have been wasted debating the adequacy of these reserves. Until modern furnaces were erected and operated by the Japanese near Mukden, less than 2,000,000 tons of ore were extracted annually, and even subsequently the annual rate has not greatly exceeded 4,000,000, with a peak pig-iron production of 2,400,000 in 1942. If full capacity of all furnaces were restored and the maximum pre-war mine extraction resumed, there is a twenty-year supply in sight. The Lake Superior district is operating with a proven reserve of only twelve to fourteen years. True, it may be one thing to sustain an old, established industry on a depleted reserve and quite another to rebuild and expand one on too slender a margin, yet the risk is worth taking, provided China can regain full economic control of Manchuria. The siliceous ores of Liaoning have always required beneficiation, and magnetite from Hupei has been used for this purpose. Similar ore from the Philippines can be brought in at no greater expense and can be utilized for the dual purpose of beneficiating and stretching the Chinese reserves. The need for heavy industry as a solid base in any program of development is too urgent for any other course to be pursued.

A quick check of other minerals reveals that, as in the case of iron, China is neither well nor badly off. Among the ferrous alloys are manganese and tungsten—in fact, China is the world's most important source of tungsten—but there is a glaring deficiency in the base metals, copper, lead, and zinc. The best outputs which have been recorded are 2,345 tons of copper, to which may be added 4,000 tons from Formosa; 11,000 tons of lead; 39,500 tons of zinc. Although there may be reserves, especially of copper, in the remote Tien Shan Mountains of Sinkiang and possibly in the Altai Mountains of distant Mongolia, there is no prospect of bringing such deposits—if they exist—into a budding industrial economy. Whatever Sinkiang and Outer Mongolia may produce will find its way into the U.S.S.R. for a long time to come, if only for the practical reason of geographic propinquity. China does produce substantial quantities of antimony, tin ore and metallic tin, and gold, together with inconsequential amounts of silver, mercury, and arsenic. Some of the nonmetallic minerals deserve mention: cement and structural materials lead, as they do in most countries, and the output of salt is noteworthy. Only the United States and Germany produce more, and Chinese production normally runs very close to Germany's. Talc and soapstone also feature, and smaller yields of clay, kaolin, alum, soda, phosphate rock, and sulphur suggest that the figures might be larger and the list longer if domestic demand warranted more extensive mining and quarrying operations.

Like India, China has limited space to devote to commercial crops, but she raises almost as much cotton as India, and a great deal of tobacco. Soy beans, potatoes, and grains are used largely for their food values, and negligible amounts of these products find their way into the chemical and plastic industries. Likewise, there is so much pressure on the land for human food that animals have a minor place in China's economy. But there are many swine, which yield bristles, as well as meat; and in the sparse pastures of Mongolia, Sinkiang, and Tibet there is a large enough yield of wool to make China a minor exporting nation. With cotton, wool, silk, and a little leather, China has the raw materials for a larger and more modern light industry than now exists. Unlimited manpower and electrical power should enable her to succeed Japan as the Far East's principal producer of textiles. As in the case of the metals, there are finite limits to the supply of raw materials, but China has a long way to go before she can build up her light industries to the point of absorbing the visible supply.

China's plight is as serious and as urgent as India's, but the prospects of a reasonably rapid resolution of her problems are considerably more dismal. India has a firmly rooted heavy industry which was not wrecked in the war, and which is in condition to serve as a cornerstone in an Indian industrial economy. Japanese destruction and Russian "reparations" have left China little with which to start. Although India can easily create political and economic instability that will match China's, internal conditions in the latter country are hardly conducive to investment, particularly in the face of inflation, low consumer purchasing power, and utterly inadequate transportation. Yet China has coal and iron, hydroelectric potential and cotton, and limitless manpower. Surely this is enough of a backlog for a start, though the start must be from scratch. The alternative is national deterioration through malnutrition, and through attrition within and without from the predatory forces of Communism.

Japan

War propaganda has left Americans with a confused picture of Japan and its people. On December 15, 1941, Japan was hopefully viewed as a nation without resources and as a "pushover" from a military standpoint. One year later disillusioned Americans were being told that she was a powerful enemy, and that the Pacific phase of the war might last ten years. The truth was not exactly in between these two extremes, for Japan's exhaustion was a *fait accompli* before the atomic bomb descended on Hiroshima. Her strength lay in manpower and in a carefully accumulated and shrewdly marshaled military machine, built up in preference to durable civilian goods. Replacement was more than the industrial facilities of the country could handle, even with supplementary stocks of coal from China, iron ore from the Philippines, and rubber from Malasia. However, the Japanese had a full appreciation of the problems and the strength of overpopulation, and had gone farther toward the solution of these problems than any other overcrowded country in Asia, not even excepting Java. So long as the outlets chosen were industry and commerce, the Japanese did well. When military conquest became the outlet, Japanese economy tottered; and, in defeat, it is so badly shattered that the fragments may never again fit together into an integrated whole.

Approximately 75,000,000 people are living in a country whose land area is no greater than that of Montana. Although warm ocean currents bring ample rainfall and equable temperatures except in the northern island of Hokkaido, the islands are so mountainous that only 15.5 percent of the available land is arable. On the 16,-000,000 arable acres nearly two-thirds of the population is trying, rather successfully, to raise all the food Japan needs, and it is a commentary on the industry of the rural population that they meet 95 percent of the nation's food requirements. The rice crop is 20 percent as large as China's, with sweet potatoes, barley, and wheat comprising the most important accessories from the standpoint of acreage and yield. Although it cannot be farmed, the rugged land is not idle, for it grazes 2,000,000 cattle and a smaller number of sheep, and it yields a large timber cut, charcoal and fagots, and bamboo. The urban population (nearly 26,000,000 in 1938, and more during the war years) has been engaged chiefly in commerce and industry. In the latter category there were four industries, each with a billion-dollar volume—metals, machines and machine tools, textiles, and chemicals—which used the services of 2,500,000 industrial workers in an estimated total of 3,215,000 (1938). Other activities which kept large numbers of the population employed were fisheries, shipping, and mining—and military service.

Japan's export trade reveals the extent to which industrialization had progressed. In 1939, her exports exceeded $900,000,000 in value, 21 percent of them consisting of textiles and 14.2 percent of raw silk. Machinery and processed foods also featured, but silk was the only important product shipped in quantity without processing. On the other side of the ledger, the import list reveals her glaring deficiencies in the raw materials which kept her factories busy: raw cotton came from the United States, India, and China; wool from Australia; coal from Manchuria and China Proper; wood and wood pulp from the United States; iron from the Philippines. The busy merchant marine, which exceeded 5,600,000 tons gross weight when Japan attacked Pearl Harbor, scurried over the earth to get bulk raw materials for the insatiable industrial machine, and to deliver its products (plus one precious raw material, silk), four-fifths of which were consigned to the United States. Japan's principal contribution was labor and equipment, but lest it be thought that she had nothing else to offer, a quick survey of her domestic raw materials is in order.

Basic in manufacturing is power, and Japan has it. Oil, it is true, has been a problem, for the meager yields of the wells in Sakhalin, even when combined with production in Formosa, were utterly inadequate. But there is coal—at least eight billion tons of it within the confines of the Japanese islands—and the Japanese proceeded to extract it in increasing amounts, until an annual output of 40,000,000 tons was attained. Though scarcely 10 percent as much as is mined in the United States, this represents a much higher ratio of production to reserves, and it was obtained economically solely because of the low value placed upon mine labor. Many thin and broken seams of mediocre fuel were utilized, even in the most productive coal province in Northern Kyushu. But there was not enough coking coal to meet the requirements of heavy industry, even though these requirements were reduced to a minimum by using a high proportion of scrap in preference to iron ore. Hence the deficiency was met by imports from China. Ultimately the demand for steel so far exceeded the capacity of native mills and raw materials that the Japanese built or acquired furnaces on the mainland within economic reach of better coal and iron.

For light industry there is electricity. Small islands breed small streams, and mountains breed steep ones. So, with typical Japanese frugality, virtually every stream with a perennial flow has been harnessed to furnish twenty-seven billion kilowatt hours for the textile mills, machine tool and chemical plants, railroads, and homes. The power plants usually have a generator capacity in excess of low-water stage, so power is rationed during dry spells; but none is wasted, and its full use has saved many a ton of coal for more vital needs.

The country falls far short of being a warehouse of metals, and actual production makes Japan appear richer than she is. The most optimistic estimate of iron ore reserves places the total at 80,000,000 tons—not even a year's supply in the United States. Yet the small deposits are mined, and heavy demands were made upon modest reserves in Korea. Ore was also obtained from the Philippines, and one of the objectives in occupying Manchuria was control of the iron ores of Liaoning. Low grade though they are, they were the major source of virgin metal for the Anshan steel mills near Mukden. Even greater dependence was placed upon scrap, and if no other credit can be given the Japanese, it must be said that they worked assiduously to clean up the world's junk piles in the years before the war. Of the ferrous alloys, 80,000 tons of manganese have been produced in a single year, but Japan had to scour the world market for all the others.

For ordinary peacetime needs, Japan has sufficient copper. Normal production is under 100,000 tons, and even though the mines were pushed beyond the efficient productive limit during the war, the deposits were not exhausted, nor were they depleted to the point of economic inutility. Some silver is produced incidental to copper recovery, and there is also lead (35,000 tons in 1944), zinc (60,000 tons), gold (900,000 ounces, plus 900,000 from Korea and 145,000 from Formosa), tin, pyrite, and sulphur. Moderate quantities of bauxite were obtained from the mandated islands, but they furnished only part of the raw material for Japan's aluminum industry, which, during the war, developed a capacity of 136,000 tons, counting small reduction units in Formosa and Korea.

This quick survey suggests that the Japanese studied the economy of the United Kingdom and found it good. With an insular location and a diet in which fish play a large part, a merchant marine was indispensable. Plenty of people and a reasonably adequate supply of energy made manufacturing an inevitable development. Soon the industrial establishment outran the supply of raw materials, and it was operated on the basis of imports and of foreign sales of manufactured products, which kept 3,000,000 factory workers busy at substandard wages. At this point the Nipponese began to regard ownership of the sources of raw materials as more desirable than peaceful purchase. It was not long before the Rising Sun shone, not only on southern Sakhalin, Korea, Formosa, and the mandated islands, but also on Manchuria and some of China. Now, stripped of her territorial loot, she finds herself able to produce 95 percent of her food requirements and to carry on a shrunken textile industry utilizing silk and rayon. Whether she will be permitted to resume the importation of cotton, wool, coal, and metals in excess of domestic needs is doubtful. Here is the only pool of skilled labor in the Far East, and with the necessary raw materials it could rebuild Japan to a position of preeminence within a quarter century. Much as the Far East needs industrial workers, it does not need a rejuvenated Japan.

Asia in Perspective

The cards are not all dealt in the Western Pacific and Far East. Japan is out of the game, and whether she gets back in depends entirely on the victors in the recent war. While she played, she played a hard and desperate game, sparing neither her people nor her skimpy resources; and in the final hand she played for table stakes and lost. But it is clear the industrial economy which Japan so laboriously built was a sound venture in a region that contains nearly 30 percent of the earth's people, and it must and will be rebuilt. If the rules of the game are followed, it would be rebuilt in China in the belt of coastal country extending from Hankow and Shanghai on the south to Mukden and Dairen on the north. Here there is coal and an unsurpassed labor potential. Here, too, there is iron, and cotton, and silk.

True, there are glaring deficiencies in raw materials, but they are even more glaring in any other locale. The Philippine Islands lack coal, and even Japan had to go to the mainland for iron and coking coal. Siberia east of Lake Baikal and the Lena River is a thinly settled, frigid frontier, with unsatisfactory coal resources anywhere near the coast. The Bureya Basin, which lies 550 miles from Vladivostok in an airline and over 750 via the Trans-Siberian Railway, is reputed to contain a reserve three times greater than the entire reserve of Japan, and nine billion more tons are supposedly present in scattered deposits in the Maritime Province of the Far Eastern region. Iron ore has been found in the Amur valley about 300 miles south of the Bureya coal beds, and blast furnace operations were started shortly before the war—partly to give the Far East a small steel industry to meet basic needs, but also to discourage the Japanese from pushing across the weakly defended Manchurian boundary. But there is not the happy conjunction of raw materials, labor, cheap transportation, and market that is found in northeastern China and southern Manchuria. Hence with Japan out and the Philippines inadequately supplied with coal though well supplied with high-grade iron ore, the choice of a successor to Japan must fall on Russia or China. Probably, though not necessarily, it will fall upon the nation which has economic control of Manchuria. Soviet reluctance to withdraw from Dairen suggests that the significance of economic domination in Manchuria is fully appreciated. It is the misfortune of the Chinese to be utterly unprepared for the lead which lies within their reach.

Major industrial roles are possible in only two other parts of Asia—India and Western Siberia. The proximity of coal and iron in northeastern India and their convenience to Indian markets and to the port of Calcutta are assets which mark this district as the center of heavy and auxiliary industries for all of southern Asia.

ASIA'S MINERAL RAW MATERIALS

A substantial fraction of Asia's mineral wealth stays in Asia. Steel industries in western Siberia, India, China, Japan, and the Amur Valley have absorbed most of it, and some oil moves from Burma, the East Indies and the Persian Gulf to Asiatic consuming centers. Other metals and a great deal of the oil leave the continent. West of Singapore the movement is predominantly westward to Europe; whereas east of the Straits raw materials tend to make their way to the United States.

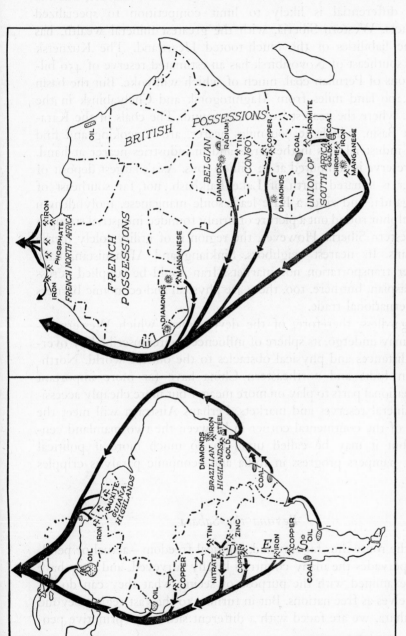

MINERAL PRODUCTION IN AFRICA AND IN SOUTH AMERICA

In Africa only the Union is utilizing locally mined raw materials; and, even here, possibly shipments greatly exceed domestically used minerals in value. Except in French North Africa, movement is via the Atlantic to Western Europe and, to a lesser degree, to the United States. In South America metallic wealth is neutralized by the lack of good industrial fuel and by geographic and political disunity. Although the United States dominates the mining industry, a substantial fraction of the production normally makes its way to European industrial centers.

Australia, with its excellent coking coal and iron ore may possibly compete in either the southern or eastern Asiatic market, but the price differential is likely to limit competition to specialized products. Western Siberia, with the greatest mineral wealth, has all the liabilities of the much-tooted Heartland. The Kuznetsk Basin southeast of Novosibirsk has an estimated reserve of 450 billion tons of Permian coal, much of which will coke. But the basin lies 1,200 land miles from Magnitogorsk and Chelyabinsk in the Urals, where the nearest iron ore is found. The coals of the Karaganda Basin, which lie 600 miles nearer, are noncoking and find other industrial uses in the Urals and in industries nearer at hand. The reserve is computed at 52 billion tons. Asia's largest deposit of copper is situated north of Lake Balkhash, not far southeast of Karaganda; and zinc, a little lead, some manganese, molybdenum and sulphur round out a picture of firmly founded industrial strength in Western Siberia. However, the region is of utility solely to the Russians. Its nearest neighbors, Sinkiang and Afghanistan, have neither transportation nor markets. Iran could be supplied across the Caspian, but here, too, there are physical and economic barriers to international trade.

Regardless, therefore, of the development which Western Siberia may undergo, its sphere of influence is circumscribed by overland distances and physical obstacles to the outside world. Northeastern India and northeastern China have far more important international parts to play on more meager but more cheaply accessible mineral reserves and markets. Perhaps Australia will meet the needs of the continental corner in between the two mainland centers, but it may be called upon to do much more if political unrest hampers progress in India and economic paralysis cripples China.

Beyond the Sahara

In the route thus far traveled, political freedom—actual or pending—pervades the many countries briefly surveyed, and they have been examined with the purpose of seeing what they can do for themselves as free nations. But in turning to the vast region beyond the Sahara, we are faced with a different situation—primitive peoples who lack Western political ideals and the numerical and military strength to achieve independence. The two countries, Liberia and Ethiopia, which have attained freedom, have done so more through the indulgence of the Great Powers than through any acts and policies of their own. Of the others, only one shows separatist inclinations, but in the Union of South Africa it is not the 6,600,000 Bantus who are seeking independence but a vocal segment of the 2,000,000 native-born whites, who believe South Africa will be better off if it seeks its own fortunes outside the orbit of the British Empire. The Royal visit of 1947 softened them, but their feeling may be intensified as they watch Britain flounder in economic chaos. The rest of Central and South Africa is dependent, and it may profitably be examined in the light of what it can contribute to world economy. There is little else it can do but live, and living does not come hard in most of the continent.

The Sahara insulates the Mediterranean countries from the Equatorial region, and the transition takes place within French West and French Equatorial Africa, with corresponding though not entirely similar changes in the east within the Anglo-Egyptian Sudan. In east-west zones that start roughly at latitude 17° north and closely approximate lines of latitude, the rainfall steps up equatorward, from something approaching zero as a limit in the Sahara, to 100 ($\pm$) inches in the Gulf of Guinea and in the countries occupying the southwestern corner of the bulge—Sierra Leone, Liberia, and the Ivory Coast. The fact is, little of equatorial Africa receives precipitation in excess of 80 inches, but rainfall is heaviest along the Guinea coast and heavy enough in the central Congo basin. Except for an upland of moderate elevation in French Guinea, the Guinea coastal region is low and steamy, and the same must be said of the central Congo basin. Elsewhere there is seasonal relief from the heavy rains and high humidity.

Save for the French colonies and the Belgian Congo, the political pattern is the common one of toe holds along the coast. In some instances the foot has followed the toes, notably in the cases of Nigeria and Angola; but whether the administration of Gambia (4,068 square miles) or of Portuguese Guinea (13,944 square miles) should be written up as a profit or off as a loss, is problematical. The political distribution of holdings may be noted:

	Area (sq. mi.)	Population
French		
West Africa	1,800,000	15,000,000
Equatorial Africa	960,000	3,500,000
Cameroun	163,000	2,500,000
Togo	22,000	780,000
Total French	2,945,000	21,780,000
British		
Gambia	4,000	205,000
Sierra Leone	28,000	2,000,000
Gold Coast	92,000	4,000,000
Nigeria	372,000	23,000,000
Total British	496,000	29,205,000
Portuguese		
São Thome	372	60,000
Guinea	14,000	425,000
Angola	488,000	3,300,000
Total Portuguese	502,372	3,785,000
Spanish		
West Africa	118,832	140,000
Belgian		
Congo	902,000	10,250,000
Liberia	43,000	2,500,000
Grand Total	5,007,204	67,660,000

This is not the place for a detailed analysis of the economic life
of the individual West African colonies, but a few generalizations
are pertinent. In combination they have a foreign trade amounting
to $525,000,000, of which $227,000,000 represent the value of imports, in contrast with $297,000,000 in exports. The last two figures
demonstrate that this section of Africa renders greater service to
the mother countries as a source of mineral, agricultural, and forest raw materials than as a market for goods. Only in the French
possessions, in Gambia, and in Togo is there a balance of trade unfavorable to the colony. Nigeria and the Gold Coast are really
valuable properties, with the Belgian Congo and French West Africa

functioning almost as effectively. These four account for 85 percent of the extracolonial trade.

The production of agricultural, animal, and forest foods is the principal occupation of the 67,660,000 people in this part of the world, and they raise enough and to spare. The cacao industry, for example, has all but moved out of its native haunts in Ecuador and into the Gold Coast and Nigeria. Palm nuts, kernels, and oil feature in every export list. But, from this point on, the lists diverge, and no one product acquires outstanding importance except the million tons of sugar produced in Angola. Cotton achieves local importance, especially in the Congo; a little coffee is raised for domestic consumption and export; and rubber accounts for 80 percent of the value of Liberia's outbound trade. Rice is distinctly a minor crop—secondary even to millet, although there are many sites where climate and topography are ideal for its growth. It is not, of course, the mainstay of the native diet, nor is there any too much manpower to raise a crop for export. But the crowded rice paddies of Asia linger in the mind as the diametrical opposite of the West African landscape. In the former the rain forest has been overwhelmed by sheer numbers of people; in the latter the forest overwhelms, or at best restricts, the comparatively thin population. Its unchallenged hold on the Guinea coast and parts of the Congo suggests new worlds to conquer as a means of combating malnutrition and starvation in the overcrowded lands of Asia.

Through the years since West Africa was explored and partitioned, its mineral wealth has become of increasing importance, and now it outranks in value all other products that leave the several colonies. Diamonds and gold are usually associated with South Africa, but in terms of carats—though not in terms of the higher monetary value for the quality gem stones obtained in the Union —the Belgian Congo, Gold Coast, Sierra Leone, and Angola individually have outproduced South Africa in many years, and the Congo has done so consistently, as the accompanying tabulation (in metric carats) illustrates:

	1940	1945 [a]
Belgian Congo	9,600,000	10,386,000
Gold Coast	825,000	1,000,000

	1940	1945 [a]
Sierra Leone	750,000	800,000
Angola	784,270	789,000
Union of South Africa	543,474	1,141,240
South-West Africa	46,578	156,000
Brazil	325,000	275,000

[a] Estimated.

With an output of 12,050,000 carats in 1940, and 13,037,000 in 1945, together with 1,728,287 fine ounces of gold and 3,600,000 ounces of silver (1940), West Africa has been a lure that climate, fever, and boredom have been unable to counteract. Of course, the Union, with its 14,000,000 ounces in 1940, has no peer in gold production, but in regions of comparable area only the Union, U.S.S.R., Canada and the United States excel West Africa.

Although West African diamonds find their way into industry more often than into drawing rooms and opera boxes, and gold is more freely received than given in the world's commerce, West Africa has two more vital mineral contributions to industry—namely, copper and manganese. The copper is regarded with mixed emotions in the copper business. When the first deliveries were made from the hugest deposits ever found, situated along the drainage divide between the Congo and the Zambezi on both sides of the Belgian Congo–Northern Rhodesian boundary, the complacent copper industry was rocked to its foundations. Not long before the event, American copper men, with control of production in the United States and in the Andes, announced there was nothing to warrant a reduction in the price, then fifteen cents a pound. And there wasn't, until central African producers announced delivery costs of approximately four and a half cents. For a time in the thirties a four-cent tariff salvaged marginal and submarginal producers in the United States, but starting in 1937 wars and rumors of war made the copper business profitable for everyone. During the war, mine and smelter production in the Congo topped 165,000 tons; in Northern Rhodesia, 266,000 tons. Although these figures are far below the all-time record set in the United States in 1943, when mine production reached 989,568 tons, and smelters turned out 1,111,458 tons (1942), the fact remains that the largest reserves are deep in the heart of Africa, and that they can disgorge any quantity of the metal the world is in a mood to buy.

The manganese is found in the Gold Coast, and for many years it has been an important source of supply for the steel industry of the United States. Normal output runs 400,000 to 450,000 tons a year, but wartime production attained a peak of 691,000. Other industrial metals are present, among them tin. Congo production was regulated by the International Tin Committee and has never exceeded 17,500 tons, but even this amount was of prime importance during the war. Sierra Leone has some iron ore; French Equatorial Africa, some zinc; Nigeria, a little tungsten; but there is only one other mineral product which has acquired outstanding importance—namely, radium. Radium production was long a carefully guarded monopoly of the Belgium Congo, until it was challenged by the Canadians following the discovery of uraninite on the Arctic Circle at Great Bear Lake. The deposits have acquired even greater interest now that atomic energy can be released more or less at will, but information about reserves and about their atomic value is even more strictly "classified" now than it was while monopoly was the only reason for secrecy.

For the immediate future, West Africa will be little more than an ore bin and a place where special agricultural commodities that make no extraordinary demands on manpower may be grown. Coal is scarce; oil, nonexistent; but there is water power in superabundance. The population has severe limitations both as to quantity and quality for industrial work and commercial farming; and the region has been maligned as no place for a white man. Yet the jungle could be made to yield more rubber than Malaya or Sumatra; if cleared, the jungle and its rainy margins contain as much rice land as India or China; the savanna lands beyond the rain belt can yield far more sugar cane than Angola's million tons and much more cotton than the cramped Nile valley. Unfortunately, agricultural potentialities cannot be measured in terms of acreage, soil, and climate. They require an agricultural population, and in West Africa the greatest deficiency is manpower.

South and East Africa differ from West Africa climatically. The elevation of the plateau, combined with its latitude, produce a subtropical to temperate climate in South West Africa and the Union, and elevation tempers the weather in the tropical latitudes in the block mountains and rift valley country to the north as far as Ethiopia. Mozambique is an exception, and the coastal areas in all of East Africa have as disagreeable a climate as nature could devise. Deserts characterize both ends of the 3,500,000 square-mile belt, with precipitation dropping below ten inches in South West Africa and in what was Italian Somaliland and Eritrea. Between, in Nyasaland, the rainfall increases to a little more than forty inches, and in a narrow strip along the coast of Mozambique near the mouth of Zambezi River it exceeds sixty inches. The east slopes of Madagascar are even more heavily watered but, in general, rain forest conditions are not present. The belt contains the highest mountains on the African continent, but very little of the region ascends above 10,000 feet; and only in Orange Free State and Basutoland, Ruanda-Urundi, eastern Tanganyika and western Kenya, and Ethiopia are there large, unbroken surfaces lying above the 5,000-foot contour.

In comparison with West Africa, this is white man's country, but relatively few white men seem to be aware of the fact. The Union of South Africa contains most of the white people in the regional population, which approximates 45,000,000. South and East Africa is a British preserve, for they control two-thirds of its 3,500,000 square miles. Other nations have been left with relatively small marginal or offshore holdings, like Mozambique (Portuguese), not to mention the all but worthless Italian colonies of Eritrea and Somaliland. Yet the British colonies and protectorates contain something less than half the population. This fact reflects the inaccessibility and remoteness of the Cape-to-Cairo route. Of all the British territorial holdings, only the Union and Southern Rhodesia have been seriously colonized, though the mineral wealth of Northern Rhodesia has prompted some progress there. In the rest of the interior little goes on that is of interest to the outside world. Kenya and Ugandi produce and export some cotton and coffee; Nyasaland contributes small quantities of tea and tobacco; Tanganyika exports sisal. Most of them produce a little gold. But they consist, as they did before the British took over, of self-contained native tribes which want little from the outside and have just as little to give in return. Only Kenya and Tanganyika, with outlets on the coast, have built up any volume of overseas trade, and in combination with Uganda, whose exports clear through Kenya, the value is under $75,000,000.

It cannot be said that the French and Portuguese colonies have done much that is different or better. Madagascar has a more diversified agricultural industry which is more than ample for domestic needs, and she ships surpluses of coffee, vanilla, corn, sugar, and meat to France. She also supplies the United States with a substantial fraction of the graphite it uses, and she mines mica, some gold, and a little phosphate rock. The entire mining industry, however, grosses less than $2,000,000 and, with small native textile, beverage, and food-producing industries, it offers but a minor diversion in an agricultural economy. Mozambique provides the usual example of indifferent Portuguese colonial administration, and the fact that it has anything at all can be attributed to British capital and British interest in its well-being. The interest stems from the fact that Beira is the terminus of the railroad from Rhodesia, and it provides Rhodesians with their nearest port facilities. The capital, Lourenço Marques, is likewise a convenient port for the Transvaal. The country is further distinguished by having a cement mill, brick plants, and rather good motor roads from the coast to Rhodesia and the Transvaal. A few thousand tons of coal are mined, and 10,000 ounces of gold are recovered; but known deposits of graphite, mica, and other minerals are only partially explored, and most are completely undeveloped. Agriculture of the usual tropical types is the main interest of the natives, and sugar and tea are grown in sufficient amounts to feature in the colony's export trade. Somaliland and Eritrea have small populations, important strategic positions, and no economic value. Indeed, the governing nations would make substantial financial gains if they could—and would— give these holdings away.

In marked contrast with its coastal neighbors, the semi-independent state of Ethiopia supports nearly 10,000,000 people. With twenty-seven people per square mile, it has a higher population density than any of the other countries in this region. The land and its climate must take the credit for this situation, because virtually nothing had been done to modernize the country and its way of life from the time Coptic Christianity first became established until the Italians blasted their way into Addis Ababa. Exception might be made of the railway which the French constructed from their Somaliland port of Djibouti to the Ethiopian capital, but travelers on this line may still be willing to subscribe to the above statement. During the few years of Italian occupation more than 4,000 miles of highway, 1,200 miles of telegraph lines, and 2,200 miles of telephone lines gave the country a thin veneer of modernization and the Italians an urgently needed sense of security. It also gave the Ethiopians ideas, for since the return of Haile Selassie, a modern system of education has been introduced, and, so far as the government's resources permit, other improvements have been instituted. The Italians made a strenuous effort to find mineral deposits; geological parties reported mercury, copper, iron, mica, potash, and sulphur, in addition to the rock salt, which was already being exploited to meet domestic demand, and gold and platinum, which were being worked by primitive techniques. The inventory was a sore disappointment to the Italians and an eloquent commentary on the future of the country: a good land for the Ethiopians, to follow their century-old pastoral and agricultural pursuits, but a land that will offer other nations little in the way of worldly goods. Ethiopia is farther ahead by hundreds of millions of dollars of equipment, construction, and engineering works in consequence of Italian conquest, but the tempo of improvement, though accelerated, must now drop back to fit the Ethiopian purse.

South Africa stands out alone as a region with a present and a future. Its past does not strictly follow the customary colonial pattern. It was settled by the Boers, who merely wanted a place to live. However, they had the luck—good and bad—to find diamonds in 1867 and gold in 1884, and soon thereafter the normal course of empire took over the right of way. The British, who forced the Boers out of the Cape in 1836, altered the status of the Boer Republic of the Transvaal to that of a dependent state under British rule in 1902; and while the Boers continued to concentrate on their flocks, herds, and crops, the British proceeded to exploit the fabulous mineral wealth. The rush of the forty-niners is better known to Americans, but neither in California nor in the Klondike has anything been found to match the Witwatersrand. Producing twelve to fourteen million ounces or 37.5 percent of the world's gold output, the Union has kept itself, and to some extent the mother country, solvent for half a century. How long this rate of production can be maintained is uncertain. The mines are deep—as deep as 8,000 feet. The gold continues, and it is primarily a question of engineering skill as to how far man can follow it. Air-conditioning aided greatly when mine temperatures of 110° to 115° were encountered in the lowest levels. But costs are mounting a little faster than technical ingenuity can cut them, and it is probably a matter of a few years—fewer than the fifty of past mining operations—when, one at a time, the individual properties will close as costs become equal to returns.

The perfection of South African diamonds gives them a greater value in the gem market than the stones from West Africa, where, as already noted, production far exceeds the output of South Africa. But rather significantly, it is not the gold, the diamonds, the platinum (60–70,000 ounces of metal and concentrates), and silver (1,400,000 ounces) which bear so heavily on the Union's future, as it is the 23,000,000 tons of coal, the 800,000 tons of iron ore, the 550,000 tons of pig iron, the generous reserves of manganese and chromium, the small but helpful deposits of copper, lead, tin, asbestos, and gypsum. Neighboring South West Africa, Southern and Northern Rhodesia extend the list somewhat and substantially increase the reserves of copper and chromium. These are the ingredients of domestic industry, and it is significant that industrial trends, which were manifest before the war, were accelerated and became more clearly defined during it. Oil is lacking, but it is estimated that there is a coal reserve of 205,000,000,000 tons; and there is a generous water-power potential in several parts of the country and Rhodesia. Iron ore is, unfortunately, rather scarce, though satisfactory for the limited South African market; hence the supply raises no immediate problem. Manganese and alloys for quality steel are present in quantities far in excess of needs; and it may prove feasible to evolve a shuttle system in which vessels calling for bulk cargoes of manganese or chrome or copper can deliver iron ore. How large a market can be economically served is another question, but it may be noted in passing that South American countries are acutely short of coal, and that the run across the South Atlantic is considerably shorter than the voyage from the United States or Britain.

Thanks to the energy of the Boers, South Africa has also staked out an important pastoral and agricultural economy. Although not a large grain producer, it raises sufficient for domestic demand and enough corn for export. Cotton and tobacco are minor crops. Its forte is animal husbandry, and sheep, goats, and cattle are an integral part of the landscape. The Union ranks fifth in wool production, fourth in wool exports. Nearly all the wool clipped leaves the country, but there is plenty of power, if not much labor, for a textile industry. Cattle are also abundant, but not to the extent of ranking the Union with her Southern Hemispheric contemporaries —Argentina, Australia, and New Zealand. Although the subhumid to semiarid climate of the veldt, subject as it is to devastating droughts, favors pastoral activity, a larger percentage of the better-watered land, especially in the highlands in the east, could be converted to farmland and used to grow more grain and larger commercial crops, if and when the demand develops. The country has its limitations, but in no field of endeavor have these limitations been reached.

Latin America

There is an odd similarity between the parts of South America and Africa, even though compass directions are completely turned around. Brazil, with its Amazon Basin and Guiana outliers, resembles the Congo Basin and Guinea Coast. The climates of the two are identical, but one opens to the west, while the other opens east into the moisture-laden trade winds. One has the six- to ten-thousand-foot uplands of Ruanda-Urundi and Kenya-Tanganyika at its back; the other, the ten- to fifteen-thousand-foot Andes. And both have identical functions—they drain off the heaviest regional precipitation which falls on the land surface of the earth. Somewhat analogous to the block-mountain and rift-valley belt of East Africa is the Andes Mountain system—rather narrow at its extremities, but broadening into the Bolivian plateau or highland, which maintains greater elevations over large areas than any of the uplands in Africa. Even the north coasts of the two continents have strong points of resemblance, for the coast range of Venezuela, with its frontage on the Caribbean, is not unlike the higher Atlas Mountains in relation to the Mediterranean. Pushing the analogy still farther, the slender mountain chain of the Isthmus of Panama broadens northwestward into the uplands of Honduras and Guatemala, and, still farther north, into the Mexican Highlands, providing a less obvious parallel with the highlands of the Near East.

Historically analogies could be found between Toltecs and Chaldeans, Aztecs and Assyrians or Babylonians, Incas and Ethiopians, Congo tribes and Amazon tribes; but the historical differences are much more important than rather far-fetched similarities. Latin America was conquered, exploited, and colonized by Spaniards and Portuguese with the same hard-headed objectives and ruthless methods as the empire builders of England, Holland, France, and Belgium. If anything, they were more honest in their ruthlessness and more discriminating in drawing sharp lines between the serious business of subjugating and robbing the natives, the pious duty of saving their souls, and the solace of living with their women, with or without benefit of clergy. Certainly they never built up any myth about South America being no place for white men, nor was there any righteous twaddle about the white man's burden. The result is that white men penetrated every part of South America. Some came no more willingly than "pioneers" have gone into the fastnesses of Siberia, and many were none too fussy about keeping their white blood pure and undefiled. The result is an endless series of contrasts between South America and Africa, even though Darkest Africa is no darker than the jungles of the Amazon, Uganda no more remote than Bolivia. Belem, though exceeded in size by Singapore, is the largest white equatorial city; Lima, though not as large as Madras, is the largest white city within 13° of the equator. Rio de Janeiro is the greatest metropolis lying within the tropics.

The resources of South America are no greater than those of Africa, so the contrasts cannot be attributed to superior natural wealth. On the other hand, South America is far from poor, and, though its resources are not evenly distributed, they are placed in nearly every major region into which the continent may be divided. After the Pope partitioned the New World between Spain and Portugal, the Spaniards focused their attention on the lands within and around the Caribbean and the Gulf of Mexico, and upon the West Coast of South America. Columbus was enchanted with the Antilles, and the seat of Spanish colonial administration was established at Santo Domingo City in Hispañola. His successors were less impressed with the beauties of the islands than they were with the mineral wealth of the mainland, and there followed the conquests of Mexico and Peru. Gold and silver were their chief concern, but, after they cleaned up the golden ornaments which native populations had gathered through the years, they were compelled to mine virgin metal and had to settle for silver. Even today, following thorough exploration and modern methods of exploitation, the whole of Latin America produces less than 2,500,000 ounces of gold, and comparatively few of the deposits were known to the Conquistadores. In regard to silver, however, Mexico mines 75–80,000,000 ounces a year; Peru, 15–20,000,000; Bolivia, 6–10,-000,000; Argentina, Chile, Ecuador, and Colombia, smaller amounts, making a total for all Latin America of 120,000,000 ounces, or very nearly half of the world's supply.

As in every other part of the earth, the lure of the precious and semiprecious metals was lost when the individual prospect of getting rich quickly gave way to corporate enterprise in the more costly exploitation of lode deposits. While Spain ruled, there was little concern about the raw materials of industry; but after independence was won, and after the United States freed itself of its colonial economy, the contributions which the Spanish American countries might make to the growing industrial giant to the north attracted much attention. Interesting as the history of mineral exploration is, only the result can be given space here. The igneous rocks of the Andes seem to have favored copper, though lead, zinc, and tin are also found in appreciable quantities. As a measure of mine capacity, peak production in one of the seven years from 1939 through 1945 is given in the following table, and United States peak production is given for comparative purposes.

	Copper	Lead	Zinc	Tin
Argentina	..	23,800	53,452	1,000
Chile	509,000	4,250	34,438	..
Bolivia	7,274	7,314	5,571	42,500
Peru	44,000	43,171	43,890	123
Ecuador	4,418	..	..	..
Cuba	10,500	..	..	..
Mexico	61,680	273,529	177,000	426
Total	636,872	352,064	314,351	44,049
United States	989,586	498,000	854,844	56

As is usually the case, the search for one thing led to the discovery of others, though in many instances development has been long delayed. When Spain was defeated in 1898, United States interests bought out the Spanish American Iron Company and its rights, and the mining of iron was begun in Cuba, while equally good, if smaller deposits in the Dominican Republican and Puerto Rico have remained unused. Never large, Cuban production has dropped to less than 200,000 tons; and meanwhile, to assure its tidewater furnaces near Baltimore of a satisfactory and dependable supply of ore, the Bethlehem Steel Company opened its own mines in Coquimbo, Chile. As much as 1,750,000 tons have been shipped from this property in a single year; but, uneasy about European sources of supply, the company has also developed deposits south of the Orinoco in Venezuela. The project was suspended during the war, but it has been taken up again, and there is every reasonable expectation that Venezuela will become as important a shipper as Chile. Iron ore is known in other countries, but only in Mexico is there any production—a modest 187,000 tons. Brazil's enormous reserves will be discussed on a subsequent page.

The needs of the steel industry do not stop with iron ore, for alloying metals are equally important, even though the quantities required are small. Several of the Spanish American countries mine small tonnages of manganese, but Cuba is the only one that produces it in quantity. Intensive mining and technological advances which make it possible to utilize lower-grade ores enabled the Cubans to produce more than 300,000 tons in 1943. Modest though this output is, it acquires especial importance as the nearest source of supply for a metal which the United States has always imported. Cuba also did the miraculous job of bringing chromite output up from a norm of 50–60,000 tons to 354,000 tons in 1943, temporarily assuming first place among world producers. Production has not remained—and cannot remain—at this high level, but its significance during the war cannot be overestimated. The output in other Spanish American countries was negligible. Until the war forced the United States into an intensive program of development of low-grade vanadium ores, Peru quite consistently supplied 30 percent of the world's requirements. Between them, Bolivia and Argentina brought tungsten output up to 10,000 tons in 1943, thereby performing the impossible feat of passing the world's most important producer—China.

No account of resources in Spanish America is complete without mention of nitrates and oil, which have both brought transitory prosperity to their possessors. Chile has the distinction of owning the only natural nitrate deposits in the world, and until 1913, all the agricultural, industrial and war-minded nations contributed generously to the Chilean treasury, not to mention the producers, for the privilege of enriching their soils, or of making explosives for industry and war. In 1913, Chile lost her first customer when Germany perfected the synthetic process of "fixing" nitrogen from the air. The monopoly was completely broken when other nations followed suit. Chile went "broke," too, but she still retained a nitrate business. Over 1,600,000 tons are being wrested from the Atacama desert every year, and more than half the output is sold in the United States. So long as it is competitive, and so long as American farmers prefer Chile saltpeter to synthetic nitrates, the business will continue, but with smaller profits and revenues.

Like the nitrates in Chile, oil has done strange things in Spanish America. It started in 1896, when oil was struck in Peru. Successful drilling brought in oil in Mexico in 1901; Argentina, in 1907; Trinidad, in 1909; Venezuela, in 1917; Colombia, in 1922. By 1920, when Mexican output reached 163,000,000 barrels, it looked as if she might, some day, reach and pass the 443,000,000 yield which the United States had attained after 64 years of exploration. The peak was reached, however, in 1921, with 193,400,000 barrels. The decline which followed was sickening to oil men and to Mexican finances, and it was not helped by expropriation. The industry has leveled off at 40–45,000,000 barrels—a respectable figure but a lost hope. Now it is Venezuela's turn. Like Tampico in Mexico, the unattractive and hot little town of Maracaibo was suddenly transformed into a bustling and confused boom town as the surrounding basin was found figuratively to be floating on oil. Poked in a dry and inaccessible corner of Venezuela, it now finds itself controlling the immediate economic destiny of the Republic, and its optimism and cash are transported quickly to the capital, Caracas, by daily plane service. The production curve still swings upward, and there are large proven reserves, but it is a rather fluid foundation for the economy of a nation. In 1945, more than 20 percent of all oil produced came from Spanish America.

	Barrels
Venezuela	323,415,000
Mexico	43,500,000
Argentina	22,885,000
Colombia	22,825,000
Trinidad	21,000,000
Peru	13,748,000
Ecuador	2,664,000
Bolivia	314,000
Total	450,351,000

With so much mineral wealth it is a major misfortune that more of it cannot be put to domestic use, but these countries lack the most vital resource of all—coal. Chile is energetically developing the small deposits near Concepción, but they yield a scant 3,000,000 tons a year. Mexico gets nearly 1,000,000 tons; Colombia, 500,000; Peru, 200,000; Venezuela, 11,000; and Argentina is burning up 60–75,000 tons of solid bitumen in lieu of coal. Less than 5,000,000 tons, none of it too good, for 80,000,000 people! There is water power, but chiefly on the east slope of the Andes. Everywhere, except in Bolivia and Argentina, the population lives on the west slope, and most of the power is hopelessly inaccessible. Even so, it has been only during the past decade that any use has been made of the accessible power.

The oil produced exceeds requirements in Venezuela, Colombia, Peru, Trinidad, and Mexico; and upon the oil of the Maracaibo basin the Dutch have pinned the economy of Curaçao and Aruba, where modern refineries take care of much of the basin's production. Neither hydroelectric energy nor oil is utilized to the extent it could be, despite the persistent reports that Spanish America is rapidly industrializing. Upon analysis, it is not solely the lack of coal upon which the absence of industry can be blamed. From the Río Grande to Cape Horn, the Spanish American people are not industry-minded, nor are the native populations of those countries in which the Indians are essentially unabsorbed. In the course of years a mining tradition has evolved in Mexico, Peru, Bolivia, and Chile; and good mine labor can be found among the Indians and mestizos of these countries. The people are easily trained, and there is no reason why semiskilled and skilled labor cannot be trained for industry as the demand develops. At the moment, however, capital is not abundant, and where it exists, rather little of it finds its way into industrial investments. Furthermore, with only a few praiseworthy exceptions, business standards are low. Nearly all machine-made goods turned out in domestic factories are cheaply and badly made, and a substantial percentage of imported goods is no better. With oil and hydroelectric power, with European imports still scarce and most United States products priced out of the Latin American market, these countries have an incentive to industrialize. The progress made thus far is small in comparison with the possibilities.

From latitude 32° north to 56° south, agriculture and stock raising are still the principal occupations, and over so many degrees of latitude the diversity is so great as to defy description in anything short of a volume. Practically all of the countries raise their own food, but there is some interchange of commodities between the tropical and temperate lands. A few countries have developed specialties, outstanding examples being the sugar crops of Cuba and Puerto Rico, the coffee crop of Colombia, the bananas of the Central American republics, and the cacao of Ecuador. Thanks to climate and topography, Argentina and Uruguay have gone into large-scale food production and have become important sources of supply for the wheat-deficient and meat-deficient nations of Europe and the tropics. Both countries started with pastoral economies, and in Uruguay there is still more interest in stock than in crops, although the country has a higher proportion of arable land than Argentina. The latter has turned to the raising of wheat (over 200,000,000 bushels) and corn (350,000,000 bushels), ranking second to Canada in wheat exports and first by a wide margin in corn shipments. As grain production has increased, attention has turned to the raising of hogs, and exports of pork and pork products rose steadily in the period before the war. For many years cattle were raised solely for meat; and Uruguay, Paraguay, and Argentina staked out first, second, and third places respectively as the countries having the largest number of animals per thousand population, with a ratio of seven and one-half to one in favor of the cattle in Uruguay. Beef still features as a major export item, but now dairy products, and particularly butter, are assuming an important place, especially in Argentina. Sheep are raised both for the wool clip and for mutton; exports of these two products are significant items in Argentina's and Uruguay's overseas trade. Under Perón, Argentinians are being exhorted to industrialize, and already there are food, textile, drug, chemical, leather-working and metal-working establishments, but the dream of establishing a heavy industry is futile. If Argentina had coal, she could make steel, if she had iron. Even the trade agreement with metal-rich Chile will not relieve the raw-materials situation much, for Chile's coal is mediocre. The coal in Santa Catharina and Rio Grande do Sul, Brazil, is not too good or abundant, and it is too urgently needed in Brazil. Argentina's 22,000,000 barrels of oil and rather limited and very remote hydroelectric energy are barely sufficient for her light industrial potentialities; hence she will do better to expand her growing food, textile, and leather industries, and to promote trade in these products. Her future—and it is a promising one—lies in this direction.

Some of the Andean countries, like Ecuador, Peru, Bolivia, and northern Chile are severely cramped by topography or by climate, or by both. Even the ancient Incas recognized the limitations which beset these countries, and they placed first emphasis upon the raising of essential food in administering the affairs of the local Indians. Crops can be grown only in the flat reaches of mountain valleys, which are distributed at every elevation; hence there is an interesting vertical arrangement of agricultural products, ranging from the cacao of the well-watered coastal section of Ecuador to potatoes in the high valleys of Bolivia above timberline but far enough below snowline for tuber growth. In the dry sections of Peru and Chile, every drop of water must be used effectively; and although modern irrigation projects may increase the arable acreage, there is little available land on which larger populations can be accommodated. With restrictions of this magnitude, subsistence agriculture and mining are almost all the future holds.

On the other side of continent, Brazil and the Guianas offer a sharp contrast to the tropical countries of the West Coast. Their combined area of 3,554,000 square miles falls less than 500,000 square miles short of equaling the area of Spanish America, but the popula-

tion is only a little more than half the population of the other countries—about 46,000,000 as compared with 80,000,000. Climate and
topography make the appearance of compactness little more than
an illusion. As if the jungles of the Amazon were not in themselves
a sufficient barrier between northern Brazil and the Guianas on the
one hand, and central Brazil on the other, the Guiana highlands
also interpose a dead space which vies with Arctic Canada, the
Chorski Mountains of Siberia, and Greenland for the distinction of
being the least settled section of the habitable globe. Interior Brazil
south of the Amazon is little better, for the whole of Matto Grosso
(552,000 square miles) has a population density of 0.8. Travelers
on the direct air route from Belem to Rio de Janeiro should have
a vivid concept of the empty spaces in Brazil. Statistically, 83 percent of the population lives in the eastern third of the country.

This lopsided distribution partly reflects the inadequacy of the
jungle as a place to live, although there is a slender thread of settlement along the Amazon, reaching a local peak at Manaos, about 800
airline miles up river. To an even greater extent the higher density
of population in the east registers the deep-rooted reluctance of the
people to move far from the coast. The pioneering spirit is definitely
not an element in Brazilian temperament.

Twenty percent of the population is, in fact, urban, yet the
number of industrial workers is small. Recent statistics are not
available, but the number is less than one million. Yet, in spite of
gregarious tendencies, the country is outstandingly agricultural,
raising all its own food except grain, which it imports from Argentina. It is rated as having more cattle than Argentina, and four
times as many swine; but, with five times Argentina's population,
its meat exports are much smaller (100,000 tons). Most of its large
sugar crop (1,000,000 tons) is consumed at home. For its export
trade it concentrates on a short list of tropical or subtropical
products, chief of which is coffee, comprising 40 to 45 percent of
all exports. In the industrial field, an increasing acreage has been
devoted to cotton, and before the outbreak of the war Brazil's shipments to Germany, Japan, Britain, and France totaled over 1,200,000
bales, or 20 percent of her outgoing trade. Only 450,000 bales were
kept for the domestic textile industry. All the agriculture is carried on in the highlands of eastern Brazil, and the greater part of

it within 250 to 300 miles of the Atlantic Coast. The density of the population increases and its composition changes southward from the state of São Paulo, and there are corresponding changes in the type of agriculture. Cultivation does not cease westward at any natural line or climatic boundary but tapers off within short distances of rail termini. There is still plenty of arable land—not too much in the temperate zone states of Rio Grande do Sul, Santa Catharina, and Paraná, but ample in remote Goyas and Matto Grosso.

Packed away in the state of Minas Gerais there is sufficient mineral wealth to make any nation happy. Most interesting and tantalizing is what is still reputed to be the largest body of iron ore in the world, in spite of detractors and competitive claims emanating from the Ural region of the U.S.S.R. Unfortunately, the deposits lie too far from the coast for export, and the nearest coal that will coke is situated far to the south in Santa Catharina. Little was done with the iron or the coal prior to the war, except for careful study of the deposits and the problems associated with their use. During the war the acute need for steel in South America prompted the United States government, with engineering direction by the United States Steel Company, to proceed with the erection of furnaces and a steel mill at Volta Redonda—a point roughly midway between the iron mines and the coast, to which coal would be hauled by train following boat shipments from Santa Catharina, and from which iron ore in excess of furnace requirements would be hauled to the coast for shipment to the United States. It was hoped that this type of shuttle system would spread costs to the point of making the operation and the export of ore economically feasible. The project was given a lot of Good-Neighbor fanfare, but as of 1947 it had not come off. Little steel has been made at Volta Redonda, and ore shipments reached an unspectacular peak of 420,000 tons in 1941, before the project was fairly started. They have been less ever since. Whether Brazil has a budding steel industry, and if so, whether it can keep it under competitive conditions, and whether the United States has an auxiliary supply of Lake Superior type and grade of ore, are questions to which conclusive answers have not yet been given.

Meanwhile mining of other metals for export continues the even tenor of its way, with considerable war acceleration and postwar deceleration. These include an appreciable quantity of manganese (437,000 tons in 1941, the peak year), titanium (5,000 tons), and smaller tonnages of chromite and nickel. In the light-metal category Brazil has supplied 76,000 tons of bauxite (1943) and the world's largest tonnage of the elusive metal beryllium, not to mention 220 tons of the rare mineral tantalite and many thousands of pounds of quartz with piezo-electric properties for use in military radio equipment. Of more strictly domestic concern is the steady output of 30,000 fine ounces of gold by the old Saint John d'el Rey mine; and, as an important contribution to the vigorous construction and engineering program which has been in progress for more than a decade, 800,000 tons of cement and many other structural materials are turned out annually.

Virtually nothing is done with, or in, the inaccessible Guiana highlands, notwithstanding the fact that, in the Guianas to the north, the Dutch and the British have each mined as much as a million tons of bauxite in a single year; Venezuelans, British, Dutch, and French extract as much as 240,000 ounces of gold—eight times the output of Brazil's lone gold mine; and the Bethlehem Steel Company is exploiting large reserves of iron ore. Nature may capriciously have placed all the mineral wealth on the north side of the highlands, but it is worth a careful look to make sure. Recently American oil geologists have examined oil prospects in the Amazon basin and have pronounced them good, even though there is only a picayune production of 80,000 barrels to support the pronouncement. The examination was part of an ambitious project to develop the basin, but to an outsider the western half of the broad highland extending into Matto Grosso, and taking in parts of Minas Gerais and Goyas, looks more attractive, despite the dreary, trackless, unpeopled wastes that stretch out endlessly beneath him as he crosses the country by plane. Even the Guiana highlands, which reach 8,000 feet in Mount Roraima, look better at least as an initial project. Brazil has much to do. What is needed is a rebirth of the pioneering spirit which first brought the Portuguese to Brazil.

The Current Outlook

A sober survey of resources in the rest of the world is not conducive to pessimism. There are, to be sure, places which have reached the limits of development permitted by the land and its resources. It is difficult, for example, to imagine a much larger population wresting a living from the Arabian desert, or from the Japanese archipelago, or from Puerto Rico or Barbados. There are other countries which can grow only if they change their way of life: China must industrialize, and India must expand her industrial economy, if their peoples are to be salvaged from a future of starvation and substandard living which should have no place in the modern world. But elsewhere there is room for growth—there are unrealized possibilities which suggest that the world has far to go before it reaches the Malthusian limit.

The survey discloses one important fact: there are no undeveloped industrial centers of the magnitude of the Northeastern-North Central States in the United States, of the Midlands, the Ruhr and the Donets Basin of Europe, and the Kuznetsk Basin of Western Siberia. There are, however, secondary centers, whose strategic location marks them for important roles in such industrial programs as may evolve. Northeastern India and northeastern China and southern Manchuria have good industrial prospects, and the Amur valley of eastern Siberia may also support a pioneering population by means of industrial production. In the Southern Hemisphere only eastern Australia and South Africa have adequate coal resources for industry. Brazil hangs in the economic balance, and the success of her nascent steel industry will reflect dire regional needs rather than superior or even adequate resources.

Nature's uneven hand in dispensing mineral raw materials is no more remarkable than the uneven use man has made of them. China, with reserves that would solve most deficiency problems in a country like France, has done almost nothing with her own native wealth, notwithstanding the most urgent need for domestic development. Yet Japan did much with very little—in fact, too much. Even where nature has been more even-handed, the human response has not been the same. The areal distribution of climates has a system and order which the distribution of mineral deposits lacks, yet countries, side by side, enjoying the same climate, similar soils, comparable agricultural potentialities, have never utilized these advantages or succumbed to the disadvantages in the same way or to the same degree. Puerto Rico, as an instance, has acquired a population density of 600 persons per square mile, while the Dominican Republic has only ninety. More phenomenally, the Haitian Republic, sharing the same island and the drier and more rugged third of it, has 300 people for every square mile, though the Haitians have not evolved a comparable standard of living. If these facts point a moral, it is a simple one in principle, but a complex one in its ramifications: the land, the climate, and mineral wealth are not determinants of human progress. They are merely determinants of theoretical limits beyond which the native inhabitants cannot go. Science and technology are pushing the limits back, but they are finite. To what extent a given nation approaches the finite limit of utilization depends upon human factors. At this point in history the geologist and the geographer can provide the assurance that there is still enough to meet the needs of the human race. He must add that the most serious problems are not with nature's gifts but with man's mentality.

SELECTED READINGS

Anstey, Vera. The Economic Development of India. London, 1929.

Bain, H. Foster. Ores and Industry in the Far East. New York, 1933.

Bain, H. Foster and T. T. Read. Ores and Industry in South America. New York, 1934.

Carus, C. D., and C. L. McNichols. Japan: Its Resources and Industries. New York, 1944.

Colby, Charles C., ed. Geographic Aspects of International Relations. Chicago, 1938.

Cressey, George B. Asia's Lands and Peoples. New York, 1944.

—— China's Economic Foundations. McGraw-Hill, New York, 1934.

—— The Basis of Soviet Strength. New York, 1945.

Fawcett, C. B. A Political Geography of the British Empire. Boston, 1933.

Frankel, S. Herbert. Capital Investment in Africa: Its Course and Effects. New York, 1938.

Furness, J. W., L. M. Jones, and F. H. Blumenthal. Mineral Raw Materials. U.S. Dept. of Commerce, Trade Promotion Series, No. 76. Washington, 1929.

Haas, William H., ed. The American Empire. Chicago, 1940.

Holmes, Olive. Latin America: Land of a Golden Legend. Foreign Policy Association, Headline Series No. 65. 1947.

Hotchkiss, William O. Minerals of Might. Lancaster, Pa., 1945.

James, Preston E. Latin America. New York, 1942.

Leith, Charles K. World Minerals and World Politics. New York, 1931.

Leith, C. K., Furness, J. W., and Cleona Lewis. World Minerals and World Peace. Washington, 1943.

Light, Richard U. Focus on Africa. New York, 1941.

Lovering, T. S. Minerals in World Affairs. New York, 1943.

Lyde, Lionel W. The Continent of Asia. London, 1933.

Postel, A. Williams. The Mineral Resources of Africa. University of Pennsylvania, African Handbooks No. 2, Philadelphia, 1943.

Roush, G. A. Strategic Mineral Supplies. New York, 1939.

Simonds, F. H., and Brooks Emeny. The Great Powers in World Politics. New York, 1939.

Smith, Guy-Harold, Dorothy Good, and Shannon McCune. Japan: a Geographical View. New York, 1943.

Spykman, N. J. The Geography of the Peace. Ed. by H. R. Nicholl. New York, 1944.

U. S. Bureau of Mines, Foreign Minerals Division. Mineral Raw Materials. New York, 1937.

Wells, Carveth. Introducing Africa. New York, 1944.

Whittlesey, Derwent. The Earth and the State. New York, 1939.

Stephen W. Reed

WORLD POPULATION TRENDS

WORLD TRENDS in population—its distribution, growth, and changing characteristics—figure more prominently in the minds of men today than ever before. Whether based on fear and suspicion, jealousy and greed, or visions of a peaceful and united world order, the interest in numbers pervades discussion of local, national, and world affairs and leads to courses of action on all levels of organized human activity. The bank teller and his wife consult their budget and debate the advisibility of having another child; municipal taxation experts point to the "shrinking sickness" among the major urban centers of population and advise city officials to seek new sources of revenue; Englishmen read the calm announcement that the birth record of the 1930s, if unchanged, will reduce their country to a nation of 14 million in a century's time—and their reactions belie their phlegmatic character; articulate nationalist spokesmen for the teeming millions of destitute peasants in the Far East press demands for throwing open to immigration the "empty spaces" and trusteeship territories held by European powers; our own Federal Security Agency estimates the increasing costs of social security in a population growing progressively older; other government officials discuss the probability that by 1970 the military manpower of the Soviet Union will outnumber that of the seven largest nations of Europe combined; and a panel of experts of the Food and Agriculture Organization of the United Nations undertake the audacious task of determining the level of global food requirements which would insure an adequately nourished world.

These are but random illustrations of modern man's concern over the population quotient in all phases of human affairs. The relationships between manpower and military strength, literacy and industrialization, malnutrition and death rates, and city life and infertility are discussed and analyzed in homes, offices, and parliaments. And, owing to the growing conviction among men that they hold destiny in their own hand, the belief persists that through understanding of the economic, political, and social implications of global demography the future may be shaped to meet their desires.

In addition to practical inspiration from the common facts of daily life, a broad interest in population has been stimulated by theoretical writings on the subject for at least a century and a half. Two themes, exactly opposite, have inspired much of the literature in this field. The first, presented most forcefully and cogently by Malthus one hundred and fifty years ago, dramatized an apocalyptic vision of natural increase in which all peoples would inevitably breed beyond their food supplies until wars and starvation called a temporary halt to overpopulation. The second, less pessimistic only because it was limited to the Western peoples alone, is much more recent, dating from the last few decades, when declining fertility in Europe raised the fear that the more advanced countries were heading for biological extinction. Having confounded the Malthusian logic, it was argued, Western man would be powerless to stem a precipitous decline. But now, as the rates of decline in the West appear to be slowing down, there is a general feeling that the low point may have been reached. The Malthusian thesis, on the other hand, appears to be the sturdier of the two; for among the majority of the world's peoples, where mortality is still the crucial factor governing natural increase, the pressure of population on food resources shows little sign of diminishing.

The present chapter seeks to draw together some of the more important known facts about world population as a basis for understanding the probable course of future growth. Our statistical knowledge of modern societies is directly proportional to their wealth, literacy, and internal order and political power. Consequently, much less can be said in precise and quantitative terms about population in the backward and less developed areas which are the main subject of this book. Nevertheless, if the latter are to follow the general paths of cultural and demographic development taken in the West, the experience of the most advanced peoples may foreshadow the future of all the world. We have no reason to believe that population trends have any inherent dependency on racial or geographical factors; but there is ample evidence that cul-

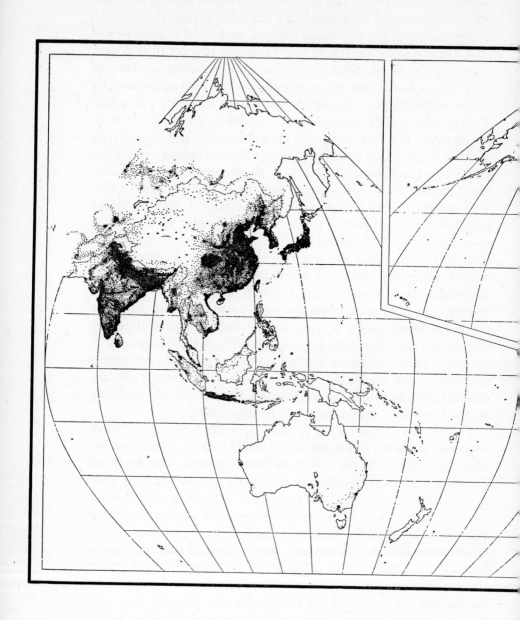

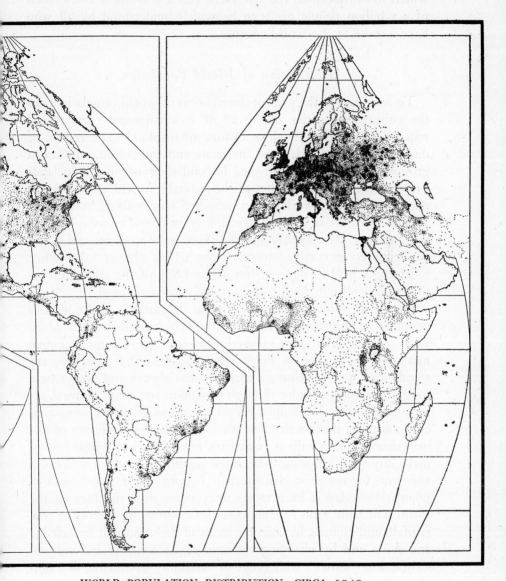

WORLD POPULATION DISTRIBUTION, CIRCA 1940

Based on a Map of World Population Distribution, from the Division of Geography and Car-
tography, U.S. Department of State, 1943

tural conditions influence vital trends. The demographic cycle which has bequeathed the twentieth century world a round total of 2.2 billion people needs to be widely understood by all who have a stake in the world's future.

Distribution of World Population

To account for the present distribution of world population and the cultural diversities among all of its component parts would require no less than a whole history of mankind. The record of dispersions and wanderings, invasions and migrations, of human groups as they pushed outward beyond their original homelands and into the farthest reaches of the habitable world inevitably remains a mere patchwork of reconstructed fragments of history. Ingenious assembly of archaeological, linguistic, and cultural bits and pieces has produced many specific short-term accounts and a few sweeping reconstructive interpretations, which whet curiosity while satisfying scholarly interest, but a vast body of the concrete data that would fill out the record is entirely beyond recall.

But it is possible to assess the present distribution of the world's peoples in descriptive and factual terms which, though qualified by wide and persistent gaps in our knowledge, provide rough bench marks for evaluation of the current situation and contemporary trends. If it may be assumed that man has always and everywhere sought to improve his lot through migration and other cultural adjustments, we may look upon his present distribution as a temporary end point in a process that has gone on since the beginning of human times. That a million years have not sufficed to enable him to make any final adjustment between his numbers and resources is apparent, for the most characteristic feature of the distribution of population today is its extreme unevenness over the face of the earth. The basic ratio of men to land—modified by factors of soil, rainfall, and climate, by existing levels of technological knowledge and skills, and by changing conceptions of comfort and cultural satisfaction—shows a wide range of variability irreducible to a simple formula. The greatest densities of population may be found both in areas surrounding rich mineral resources which have the highest levels of living the world has ever known and in the strictly agrarian

river valleys of India and China where levels of living periodically
fall below the line of subsistence. Areas of low density are also di-
visible into regions like the Canadian prairies and the inland grazing
areas of South America which have Western levels of living, and, on
the other hand, the undeveloped, disease-ridden, and thinly-settled
homelands of the surviving primitive peoples in central Africa and
New Guinea.

Regional concentration of population appears to have marked
the whole history of the race. Areas of more abundant food supply
have always attracted, and permitted, the relative massing of peoples—along the game trails of Magdalenian hunters, in fertile alluvial
plains of the subtropical river valleys, and in commercial and political capitals of ancient states. Modern manifestations of this con-
fluence of peoples do not differ in kind but only in degree from
those of earlier times; technology, trade, and natural increase have
enlarged the areas of continuous habitation and permitted enormous
multiplication of numbers. But the inequalities in regional density
still obtain and give rise to problems of the widest international
significance.

There are today four major regions of heavy concentration, in
which population density far exceeds the average world figure of
40 persons per square mile. All four lie within the Northern Hemisphere, and all but one are located within the Old World, where
the continuing density of population is a reflection of the antiquity
of settlement. The exception, of course, is Eastern North America,
a region which has gained entry into this list only within the last
century, and which still ranks considerably below the Old World
regions in both total numbers and density. As Fawcett defines them,
these major regions contain about three-fourths of the world's present population on a little more than one-eighth of its total land
area: (1) *The Far East* (most of China, Tongking, Manchuria,
Korea, and the Japanese Empire south of 40 degrees north latitude); (2) *India and Ceylon* (bounded by the Indian Desert, Tibet,
and the eastern border of Bengal); (3) *Europe* (bounded by the
60th parallel and the valley of the upper Volga, the Urals, Caspian
Sea and Persian Desert, and the Arabian and Sahara deserts); and
(4) *Eastern North America* (lower Canada and eastern United
States).

TABLE I

The Four Major Human Regions

CONTINUOUS HABITABLE REGION	AREA	POPULATION	
	(in million sq. miles)	*Millions*	*Density*
The Far East	1.7	500(?)	292
India and Ceylon	1.0	400	400
"Europe"	2.8	520	186
Eastern North America	1.9	100	52

Source: C. B. Fawcett, "The Numbers and Distribution of Mankind," *Scientific Monthly*, LXIV (May, 1947), 392.

Judged by density alone, several smaller areas and islands deserve to rank with these major regions. Java, Malaya, Puerto Rico, Formosa, and West Africa are examples of areas of extreme density and rapid growth. Set apart from the major regions by oceans or sparsely settled terrain, their density is owing largely to culture contact with the West and its technology, medicine, and commerce. Favorable local conditions and cultural factors which foster growth permit these relatively isolated countries to have a heavy natural increase, but none of them is directly contiguous with the four major regions listed above.

Although population concentration is not limited to the major regions, their present density is unrivaled elsewhere in the world. These regions have profited proportionately through both outward expansion and the filling in of urban centers and interstitial areas. During the present century the westward movement in North America, to the high plains, the Canadian prairies, and the Far West, has continued; in India the expansion has been into the newly irrigated lands of the Indus Valley; Chinese peasants have poured into Manchuria in one of the greatest modern migrations; Russia has directed colonists and industrial workers into her frontiers in western Siberia; and Europe, with a minimum of marginal land remaining for exploitation, has seen a drift of people from agrarian peasant economies in the east and south toward the industrial centers of the north and west. These movements have been vast in scope, but they constitute little more than internal reshufflings; they have not seriously altered the relative importance of the major regions to one another or to the world at large. The only major historic change in the global balance of world population has been the emergence of Eastern North America during the nineteenth and twentieth centuries as the fourth region of great density.

The likelihood of the eventual addition of other major regions to those already enumerated is problematical. Central and South America and South Africa remain the only continental areas which might witness an eventual concentration of population comparable to that in the Old World and North America. Notestein has constructed a series of hypothetical population estimates by continents for the year 2000 A.D., which, while admittedly based on many unknowns, gives a reasonable forecast of future growth. These are not presented as mere predictions, but rather as an indication of trends to be expected if the conditions assumed actually come about. Table 2 compares the 1940 counted population with Notestein's results.

TABLE 2

World Population by Continental Regions, 1940 and 2000
(In Millions)

REGION	ACTUAL AND ESTIMATED	HYPOTHETICAL
	1940	*2000*
North America	143	176
Europe [a]	541	700
Central and South America	130	283
Oceania	11	21
Africa	158	250
Asia	1,190	1,900
Total	2,173	3,330

[a] Includes Asiatic Russia in the figures for the year 2000.

It seems a fair assumption, therefore, that other regions may in time be added to the present four. South America, for instance, has more than doubled its population in the present century, making it the fastest growing continent of all since 1900. Central America and the eastern coast of South America contain the heaviest concentrations at present, yet even here the average densities are not comparable with those of eastern North America, let alone those of Europe or the Far East. However, if present high rates of natural increase can be maintained, if new resources can be opened to industrial exploitation, and if a balanced economy can be achieved, South America may shortly become at least a minor demographic region of world importance. For no other region in the Southern Hemisphere—including South Africa, or British Oceania—does this prospect appear so probable.

Patterns of World Population Growth

World population can grow in only one way—by a constant and continuing surplus of births over deaths. Whatever physical, biological, and cultural factors may be operating within individual societies to determine momentarily what happens to numbers, the gross totals of those born relative to those who die give a crude retrospective picture of the trend of population growth. By totaling trends for every nation and region having adequate statistical evidence, and by estimating and making necessary analogies for the remaining countries and regions still lacking such data, a picture emerges of the tides of population replacement and growth for the world as a whole.

Striking the balance of world fertility and mortality after a century and more of development in census enumeration and international statistics is not a precise undertaking even today. It is impossible to state within desirable limits of exactitude precisely how many people are or have been alive at a given moment. Accurate statements based on reliable censuses can now be made for probably about 70 percent of the world's people; for the remainder, however, we must depend on a wide variety of estimates which contain an unknown margin of error. The countries of Europe and their satellites overseas have led the way in periodic assessment of their populations; but for many important parts of the world—notably China (which may contain one-fifth of the world population), colonial Africa south of the Sahara, and approximately 30 percent of the Central and South American Republics—current data are far less complete and less trustworthy. Countries which have led the way in the development of science, industrialization, urbanization, mass education, and public health, and which possess relatively autonomous political control—a dynamic process of cultural change which may be labeled "modernization"—have been the ones, for the most part, whose human stocktaking has come farthest; the "backward" countries, exploited colonies, and neglected dependencies have lagged well behind. Indeed, one may say that the development of statistical knowledge and techniques is an integral part of the total process of modernization; and wherever modernization has been inhibited by disorder and revolution, illiteracy, ill-health, and absence of autonomous control, one rarely finds adequate quantitative data wherewith to study population movements.

If in the mid-twentieth century there is still doubt about the exact number of all the world's peoples, how much less accurate must be the reconstructions of world population figures for earlier periods in human history. Yet the study of trends requires a broad over-all grasp of developmental sequences if we hope to understand what directions we are following. "Knowing" the population of a country or region is usually taken to mean that the official figures for the reported population will vary by less than 10 percent around the hypothetical "actual" population. This is a significant criterion of our improving knowledge, but it betrays the tentative character of all conclusions about historic trends in the rate of world growth.

Despite the very great difficulties involved in estimating world population in periods prior to the modern era, several bold attempts have been made to reduce partial enumerations, fragmentary evidence, and broad conjecture for the last three centuries to a series of probable totals which serve to chart the likeliest course of past population growth. Working backward from the known to the unknown, modern students of the problem, such as Beloch, Willcox, Kuczynski, and Carr-Saunders, have established regression lines which take them back to the estimates of reputable seventeenth century scholars like Gregory King and Riccioli. Out of these extensive analyses has emerged an admittedly unverifiable but generally accepted picture of world population growth from 1650 down to the present.

Table 3 epitomizes current conclusions about the world's population history during the last three centuries. The 1650 figure must remain merely a shrewdly reasoned estimate; it is impossible to defend or contradict it on a basis of census facts. What is more important is the indication that, granting the reasonableness and acceptability of the earliest figure, world population as a whole has

TABLE 3

World Population by Continents, 1650–1940
(Millions and Percentage Distribution)

Continent	1650	1750	1800	1850	1900	1940
Europe	100 (18.3)	140 (19.2)	187 (20.7)	266 (22.7)	401 (24.9)	541 (24.8)
North America	1 (0.2)	1.3 (0.1)	5.7 (0.7)	26 (2.3)	81 (5.1)	143 (6.7)
Central and South America	12 (2.2)	11.1 (1.5)	18.9 (2.1)	33 (2.8)	63 (3.9)	130.2 (6.1)
Oceania	2 (0.4)	2 (0.3)	2 (0.2)	2 (0.2)	6 (0.4)	10.8 (0.5)
Africa	100 (18.3)	95 (13.1)	90 (9.9)	95 (8.1)	120 (7.4)	158 (7.2)
Asia	330 (60.6)	479 (65.8)	602 (66.4)	749 (63.9)	937 (58.3)	1,190 (54.7)
World Total	545 (100.0)	728 (100.0)	906 (100.0)	1,171 (100.0)	1,608 (100.0)	2,173 (100.0)
Annual percent growth over previous period		0.29	0.44	0.51	0.63	0.75

Source: Figures for 1650 through 1900 are from A. M. Carr-Saunders, *World Population* (Oxford, 1936), p. 42. Those for 1940 are based primarily upon *Statistical Year-Book of the League of Nations 1942–44* (Geneva, 1945). The annual percentage growth figures are found in K. Davis's article in *World Population in Transition* (Philadelphia, 1945), p. 3.

Note that 139 million Russians are included in the European figure for 1940, and 36 million in the Asiatic; North American figures throughout are exclusive of Mexico; Central and South America are estimated to contain 40 million and 90 million respectively in the 1940 total.

quadrupled in the last three hundred years, and that more than half of this enormous increase has occurred since the beginning of the nineteenth century. As Thompson says, the growth of world population since 1800 has "probably exceeded by 200 or 300 million the growth of population in all ages prior to that time." As a biological species man in the last four generations has accomplished greater growth in numbers than he had in the preceding forty thousand generations. Another significant factor in the growth trend of the last three centuries appears in the bottom lines of Table 3: world population has grown at a steadily accelerating rate in the period for which estimates have been prepared. The percentage of annual increase for the entire 290 years would average only 0.44 percent, yet it is apparent that increase has been constantly above that figure since the beginning of the nineteenth century, and that the rate for the twentieth century thus far shows world population to be growing as never before.

Unless one appreciates the potentialities of these fractions, an increase of less than 1 percent annually may appear so slight as to border on insignificance. Yet fractional rates of growth, when manifested by large population aggregates and maintained over a few generations, give rise to staggering increases. As Knibbs has shown, an annual growth rate of only 1 percent will double a population in approximately 70 years, or slightly more than two generations. If this had been a stable rate in human history, and if man were descended from an original Biblical couple, only 70 generations, or less than 2100 years, would have sufficed to reach the present numbers among all the peoples of the earth. Countless mathematical examples of this sort have been given in support of crude Malthusian arguments over the dangers of such increase. What such illustrations · actually prove is merely that there are terrific potentialities for growth inherent in small annual rates which appear trifling in themselves. Specific cases add credence to the morbid and ever-popular notion that we are heading toward what Kingsley Davis has aptly called "a beehive world" of ten to twenty billion people whose entire activity would be simply a scrabble for emplacement and sustenance. More than half of the world's peoples are growing at current rates above 1 percent; indeed, some of the already congested and backward countries (Formosa, Siam, the Philippines, most

Caribbean and Central American countries, Brazil, and Venezuela) are growing at rates above 2 percent. Figures indicate that Formosa's annual growth rate of 2.5 percent, which obtained during the 1930s, would double her present population of 5.8 million in about 29 years. This is even greater than the rate of growth of the population of the United States during its phenomenal period of expansion in the latter half of the nineteenth century.

The point of these hypothetical exercises and concrete cases lies primarily in the fact that during the last three hundred years, and particularly within the last century, the world has witnessed a burst of population growth that can be likened to an explosion. Mankind grew in numbers during this period as it had never grown before. It is certain that such rates of increase as appear in Table 3 for the world at large, and the much higher rates that are now reported for particular countries and regions, could never before have occurred on a world-wide basis for any length of time. If such growth had been realized in earlier epochs or historic periods, the world would long ago have reached or surpassed its present total of inhabitants, a situation that is unthinkable in terms of the known relationship between food supply, technology, and numbers. Indeed, if the whole span of man's existence on earth is taken as a million years, the smoothed average of the annual rate of increase for world population down to the present would be not more than five-hundredths of 1 percent. This would mean adding but two persons annually for every 100,000 inhabitants. But from the seventeenth to the twentieth century 500 persons were being added each year to the same basic number; and since 1900 the growth ratio has increased to nearly 800 per 100,000.

This is the accepted view of the manner in which world population has grown. There are no grounds for believing that the world had ever before contained more people than it did at the beginning of the modern period of population growth. Archaeology and prehistory, anthropology and ecology, all combine to show that upward spurts in world population have followed upon every major cultural advance in the field of technology and food production. As Sauer has said,

The history of human populations is a succession of higher and higher levels, each rise to a new level being brought about by discovery of more food, either through occupation of new territory or through increase in food-producing skill. The act of expansion into new habitats also stimulated food experimentation with new sources, and frontiers of settlement were therefore likely to generate a sustained growth potential and expansiveness.[1]

It was only after what Childe has called the Neolithic Revolution that our species began to multiply at all rapidly. Some millennia later a second revolution, the Urban, transformed tiny villages of self-sufficing farmers into populous cities, nourished by secondary industries and foreign trade, and regularly organized as states. Under these new conditions the size and density of cities increased, as Childe and other archaeologists have shown from the crowding of urban cemeteries. But the greatest period of population growth was touched off by man's discovery and exploitation of mineral power sources and the organization of new technological abilities through application of the scientific method in the fields of agriculture, manufacture, transportation, and medicine. This Industrial Revolution, still far from having run its course, has been based on applied science; and its most significant cultural effect in terms of population has been to reduce the death rate wherever modernization has occurred. "It is no exaggeration to say," concludes Thompson, "that science lies at the basis of modern population growth."

Granting that world population grew exceedingly slowly during the countless millennia prior to modern times, it is safe to infer that such growth did not follow a smooth and steady progression in every society and in all ages. Upward spurts resulted from the acquisition of new forms of tools, sources of energy, and social organization. Also to be reckoned with are the known fluctuations resulting from wars, plagues, famines, and epidemics in particular societies and regions. In short, where the forces of mortality have been the ultimate controlling factor in population growth, as they were everywhere in the world prior to the eighteenth century and as they are today for more than half of the world's present peoples, violent fluctuations around a very small annual norm of growth have been the rule rather than the exception. Fragmentary evidence culled from selected countries of Europe prior to the eighteenth

---

[1] C. Sauer, "Early Relation of Man to Plants," *Geographical Review,* XXXV (January, 1947), 18.

century indicate that European population as a whole was probably close to stability for several hundred years before the beginning of the modern growth cycle. Plague was a periodic killer on a vast scale, but there is no basis for assuming that the European population in 1700 was any larger than it was in 1600 or much larger than it was in 1300.

The gross figures on population growth since the seventeenth century, impressive as they are for world and continental totals, contain further regional and ethnic differences which are important for closer understanding of the vital processes involved. The figures for continents (Table 3) show that all people have participated in this growth, but in unequal amounts; less apparent is the fact that Europeans and their migratory descendants overseas have experienced the greatest relative increase. From rather less than one-fifth of the world total, they have burgeoned to approximately one-third. In round numbers this means that in three centuries they have increased more than sevenfold while the rest of the world has multiplied only threefold. European leadership in the modern cycle of population growth is owing to Western vigor in making piecemeal adjustments to changing social conditions. There was no master plan which rationally pursued the goal of growth as an end in itself. But through discovery and occupation of new lands and the stimulating effects of permanent contact with the Old World, forces making for increase were unleashed in those territories. Moreover, European conquistadors and imperialists did not foresee the demographic consequences of their activities among subject peoples. Their teaching and example, and their demands and pressures, resulted in changes which started non-Europeans also along the path of the modern cycle of growth. This is apparent in the history of events in countries such as Java, India, and Egypt, where periods of accelerating rates of increase began much later than in Europe itself. The stimulus to growth afforded by the spread of Western institutions is well established, and it remains today the greatest single factor underlying the continuing increase in population. It cannot be regarded as the original or final cause of growth, however, for in Japan and China fairly rapid increase in population had already occurred in the seventeenth century, coincidentally with, if not before, the beginnings of European increase. Therefore, despite the fact that effective contact between East and West was negligible at that period, the population of at least two important Asiatic societies reacted to growth factors which were indigenous. What these were is not clearly known, but political reorganization and civil order must have been important among them.

The Modern Demographic Cycle

In the modern historic period, declines in death rates, followed after an interval by declines in birth rates, have characterized the course of population movement known as the modern demographic cycle. The decline in death rates, wherever found, has been in response to local factors and cultural conditions which invariably determine both the timing and extent of the downward trend. Sufficient cross-cultural similarities exist, however, to permit classification of the principal forces at work. The immediate causes of declining mortality are generally found first in technological improvements affecting the quantity and quality of the food supply, and secondly in political reorganizations which extend and strengthen civil order within expanding peace groups. The multiple forces subsumed under these broad headings, when factored out for independent analysis, give a spurious appearance of rational planning. Philosophical justification of the results of such cultural changes are easily found in retrospect, but the immediate goals of mechanical invention, agricultural improvement, the exploitation of virgin land through discovery, colonization, and settlement, and sanitary improvements in water supplies and municipal hygiene were all limited reactions to economic and social stimuli which, when lumped together, do not add up to a rational policy favoring population growth. The political factors were concerned largely with efforts to preserve internal harmony and civil order, to mitigate the more drastic effects of crop failures, famines, epidemics, and other natural calamities, and to erect a more stable foundation for the individual's right to life through rudimentary programs for social welfare.

The net effect of these changes in countries where they were most marked was to raise a larger proportion of the total population than ever before above a precarious subsistence level where great masses had been easy targets for all of the natural checks to population growth. If we define level of living as the total of observed satisfactions of a society's needs, there can be no doubt that the early nineteenth century in Europe witnessed a significant rise in that level. More food, better sanitation, improvement in preventive medicine, opportunities in new lands, and limited wars gave pragmatic proof of "progress" and the basis for a lusty optimism that permeated all classes.

The epochal discoveries of fire, agriculture, and organized scientific technology mark the most significant human adjustments to life conditions in the history of man, but in some future time we may also look back upon the measures of control over the vital processes of birth and death as being of equal importance. That such control has always been a desirable goal, a felt social need, and an individual aim is not to be doubted; the evidence of all known religious and philosophical systems and medical arts and practices— ancient and modern—bear witness to the pressing burden which forces of nature have perennially levied on the perpetuation of human society. The social and individual pain and insecurity accompanying high mortality, the disruption of work patterns and institutional organizations, and the physical strain on women from excessive child-bearing have always cost more than people have wanted to pay.

The modern growth of population is unprecedented, and came about initially through effective cultural changes which succeeded in forestalling death with greater efficiency than ever before. The decline in mortality, slight though it was at first, permitted more people to live longer and to produce more children. When death rates were universally high, they stood at levels little below birth rates. Consequently, any slight fluctuations in lethal forces making for higher rates—crop failures, famines, and other natural calamities—reacted immediately and detrimentally on natural increase. As Thompson says:

Never before 1800, or a little earlier, had any people enjoyed a relatively long and continuous period of steady death rates, to say nothing of a long decline in death rates leading to a fairly long and steady growth in numbers. A violently fluctuating death rate with a very slow increase of

population over long periods probably characterized most peoples before the advent of the Industrial Revolution.[2]

The immediate and positive response of death rates to relatively
slight improvements in environmental conditions has characterized
the results of modernization wherever it has occurred. The circum-
vention of death and the promotion of longevity have been universal
cultural goals of peoples with the most diverse background of customs and cultural heritage: North European town-dwellers, farmers in British India, and New Guinea natives all readily accept
changes which demonstrably prolong life. Even in primitive societies where killing the aged has been a customary adaptation to
requirements for group survival, this seems more often than not,
according to Simmons, to be a response to environmental and social
necessities rather than to the desires of the victims or a callous disregard for life on the part of the survivors.

But the cultural conditions and technical changes which first
blunted the sharp edge of the sword of death had no coincidental
effects on patterns of fertility. High fertility has been a constant
phenomenon in human society down to modern times. Indeed, it is
an essential and universal adjustment to high death rates, for the
perpetuation of every society demands sufficient births to offset
losses through death. The complex of cultural and demographic
forces which had already proven their worth in maintaining pre-
industrial populations over long periods, under conditions where
death was natural and uncontrolled, are poorly reported. But there
is ample presumptive evidence the world over of deep-rooted institutional determinants—in religion, in economic organization, and
in family structure—which have been sustaining factors ensuring
high and constant fertility. High fertility, until very recent times,
has been the only effective societal answer to the problem of death
at all ages.

The differential and the time lag between the reactions of fertility and mortality to the cultural changes set in motion by the
industrialization of Western society lie near the core of the causes
of modern population growth. Deaths declined in number and frequency, but births continued at their former, prerevolutionary, rates;
and the larger masses of the living gave birth to augmented numbers

[2] W. S. Thompson, *Population and Peace in the Pacific* (Chicago, 1946), p. 23.

of children. There is no evidence to show that improving environmental, biological, or cultural factors led either to desire for or actual production of more children within the family. The explanation is simpler and can be found in the "snowballing" effect of greater numbers of surviving children, who in their turn added more units to the population. In Western countries generally it has been the lag of decades between decline in mortality and later decline in fertility that underlies modern population growth. Population increase therefore has been a function of the differential rate of decline of birth and death rates.

The decline in fertility can now be shown in retrospect to have begun early in the nineteenth century in France, the United States, and Scandinavian countries, at least among the upper classes of those societies. Owing to the youthful age composition of all growing populations, however, this selective decline was masked by gross population totals which showed that such nations still produced more births than deaths. The spread of birth-limitation knowledge and of practices restricting family size was so gradual—and involved matters so essentially private—that records are searched in vain for a detailed understanding of how this vital revolution took place. Even if there were data with which to make a reconstructive history of the diffusion of contraceptive techniques, the motivation which led to the international decline in fertility would still have to be inferred. When the last decades of the nineteenth century and especially the early twentieth century showed beyond doubt that the decline in births was rapidly catching up with the decline in deaths in several European nations, it became apparent that a new balance of vital forces was in process of evolution.

Phases of the Modern Demographic Cycle

Control over the vital processes governing population replacement is an excellent example of the effectiveness of the cultural changes which have marked the evolution of countries and regions in the most recent phases of world history. Every classification does some violence to truth, but, for purposes of comparison and understanding, it is useful to reduce to a few broad categories the many distinct demographic histories of individual societies. The most relevant criterion for such classification now and in the immediate future is the process of population replacement, the net result of the relationship between births and deaths. The phases of the cycle, therefore, may be listed under three major types: the *rationally balanced*, the *transitional*, and the *unstable*.[3] These are to be regarded as stages in a continuum—loosely defined and with certain overlappings when applied on a world-wide scale—that marks the current demographic evolution of peoples in the whole cycle of modern growth of world population. Although there exist borderline cases which might call for the inclusion of a nation in one or the next stage, it is possible to determine roughly from vital rates how far any nation has progressed in the cycle. It is a working hypothesis, not a proven fact, that all peoples everywhere, whatever their numbers, habitat, and cultural heritage, tend to pass from the unstable through the transitional to a stage of rational balance. Western societies have manifested this progression in conjunction with the cultural changes summed up in the term modernization; other non-Western societies such as Japan appear to be halfway through the cycle; and it is assumed that the massed societies of the Orient, under the impact of forces analogous to those which have caused the cycle in the West, eventually will follow the same progression.

The *rationally balanced* societies are those in which, for the most part, modernization has received its maximum expression in the practical application of science to transportation, industry, agriculture, and public health. These are societies wherein the decline of death rates and birth rates has gone on for the longest time. They have succeeded in lowering death rates to a point where fertility rather than mortality determines population replacement. Specifically included among the rationally balanced societies are the nations of northwestern and central Europe (United Kingdom, France, Scandinavia, the Baltic countries, Belgium, the Netherlands, Germany, Switzerland, Austria, Czechoslovakia, Hungary, and perhaps Italy); North America north of Mexico (the United States

[3] The present classification follows those of Thompson and Notestein, whose analyses of current world population are outstanding examples of succinct synthesis. Notestein lists three types of growth: *incipient decline, transitional growth*, and *high growth potential;* Thompson refers to the same three processes by the terms *stationary, expanding*, and *pre-industrial*.

and Canada), and British Oceania (Australia and New Zealand).
While several of these countries have a current excess of births over
deaths owing to the continuing heavy concentration of people in
the reproductive ages and the probably temporary increase of births
due to special wartime conditions, projections of their populations
into the future indicate that their greatest period of growth is nearly
at an end. Some are still increasing slightly, others have already begun to decline. Only by immigration or, more important, by a radical increase in the number of children born per woman, can this
block of countries which led the world in population growth in the
nineteenth century expect to do more than maintain the numbers
which it will reach in the second half of the twentieth century. In
1940 the population of the rationally balanced countries comprised
about 20 percent of the world population.

The *transitional* societies include those countries which, in general, now exhibit earlier phases of modernization, analogous to
those typical of the more developed Western societies prior to the
first World War. Countries of eastern and southern Europe (Poland, Yugoslavia, Bulgaria, Rumania, Greece, Portugal and Spain);
possibly North Africa (Algeria, Tunis, and Morocco); white South
Africa, the USSR, Japan, and eastern South America (Brazil,
Uruguay, and Argentina) are those which have achieved significant control over death rates and a beginning of control over births.
The downward slope of the death rate in Japan between 1921 and
1941, for instance, is reminiscent of that which obtained in England between 1881 and 1901. The birth rate in transitional societies,
although dropping rapidly, still remains relatively high; and the current difference between rates of death and birth, together with the
heavier concentration of people in the child-bearing ages, ensures
that rapid growth will continue in these countries for some decades
to come. It is logical to assume that, as modernization proceeds,
these societies will tend more and more toward the rational balance
characterizing the Western countries. The rapidity of this development depends directly on the speed of modernization. Transitional
societies in 1940 accounted for approximately another 20 percent of
the world's population.

The *unstable* societies are so called because the nearly 60 percent
of the world's peoples which they comprise—most of the world, in

fact—are still the targets of forces of growth or decline which remain outside the present limits of effective rational control. The roster of countries in this classification includes the famine lands, plague spots, and reservoirs of poverty, illiteracy, and disease in the tropics and subtropics: most of Asia (except Asiatic Russia and Japan), Malaysia, the Philippines, and native Oceania; the Near East, Middle East, and Central Africa (including black and colored South Africa); the Caribbean Islands, Central America, and northern and western South America. Much of this vast population is politically and economically dependent on Europe; the greater part of it is incredibly crowded on the available land; most of it fluctuates on the bare edge of subsistence, where slight pressures one way or the other may cause tremendous growth or major catastrophes. The recent history of such countries shows that increased production and the simplest public health provisions can quickly reduce death rates; but it is problematical whether or not such improvements can be maintained over long periods in the face of inertia, lack of capital, and the weak organizations for health which are introduced from outside. Fertility, on the other hand, shows no signs of being affected thus far by these superimposed changes. Hence the potential natural increase among the massive, largely youthful populations in these countries is tremendous. The crucial question, and one to which no final answer can yet be given, is whether or not these peoples can sufficiently alter some of their most deep-rooted customs so as to gain a greater measure of control over their vital processes and thus move toward the goal of rational balance.

Factors in the Modern Demographic Cycle

MORTALITY.—Declining death rates have everywhere been harbingers of population growth in its modern cycle, permitting a greater expansion of numbers over the whole earth than ever before. For as far back as vital records take us in Europe, and possibly also in prestatistical times, the gradual but inexorable downward trend of mortality provides a firm basis for concluding that Western man has signally improved his defenses against the forces of death. He has now reached the point where it lies within his power to control rationally the numbers in his population. The favorable evidence of crude or adjusted death rates, average expectation of life, and infant mortality trends, all combine to impress upon us the increasing security of human life wherever modernization has taken place.

The medieval paraphrase of a Biblical source, "Death is no respecter of persons," indicated a sharper perception of the inevitability of death in periods of crisis than of its normal frequency among large numbers in less turbulent times. The concept of death as a lawless force, striking blindly if not unnaturally, reigned supreme in the minds of men for untold generations until the new idea of statistical probability gained wider acceptance. Surviving primitive peoples with few exceptions do not generally accept the fact of the naturalness of death at any age; more widespread has been the notion that, if only certain rules and practices be scrupulously adhered to, longer life might result. Whether prayer and fasting, ice-cold baths and exposure, or intercourse with fresh young wives have ever turned aside the forces of death or added materially to the life span would be an interesting problem for biological research. But there can be no denying that death is, and within limits always has been, a respecter of persons. Disinterested study of human mortality proves beyond doubt that death everywhere initially seeks out the very young and the aged, males above females, and the malnourished, impoverished, and diseased rather than the well-endowed. In our own times it is clearest of all that death respects those who are born in modern hospitals, who live in surroundings which medical science regards as better than adequate, and whose caloric intake approaches the optimum established by the newer knowledge of nutrition. The currently emerging folk rules, values, and practices based upon the verified relationship between environment and health have proved to contain a positive survival-value which raises such behavior far above any hit or miss customs which man has discovered heretofore.

Mortality has been the most amply reported of the vital processes because of its ramifying social importance as well as its eternal mystery. Whether or not the idea of death from natural causes is a recent innovation in human thought, its incidence inevitably creates sadness, lowers social efficiency, and breaks up functional associations. While death as the ending of a long and useful life may have less sting, as a sudden interruption in middle age or near the height of a person's productive power it seems wasteful, unnecessary, and shocking; and grief, strain, and disruption of human institutions are its normal consequences. Here the significance of the longevity ideal (which is a cultural universal shared by all men) is most apparent, for problems of dependency and orphanhood have painful implications among families deprived of their head, educator, and provider. By comparison, the death of infants and young children breaks fewer social ties and creates only minor disturbances outside the immediate family. Similarly, at the latter end of the life span, the passing of men and women who have moved out of the productive ages breaks fewer bonds, for many relationships of family, work, and place have been allowed to lapse, if they have not already been actively severed in anticipation of death.

An important caveat for the study of mortality is inherent in Thompson's warning that there is no such thing as a "normal" death rate, or a biologically constant lethal force to which all populations are subject. The extent of prevailing mortality in any society depends directly upon the effective organization of cultural factors, such as the commissariat, sanitation and hygiene, and civil order, which lend their support to the preservation of life and a more efficient utilization of human energies. Indirectly, mortality is a function of the age composition of a society; but that, in turn, results from previous forces of birth, death, and, occasionally, migration. "Younger" populations are normally less susceptible to death-dealing agencies than older ones. Thus while we tend to look upon a crude annual figure of approximately 10 deaths per thousand people in the total population of most rationally balanced societies as normal for the present period, crude rates of 15 to 20 per thousand are expected in the less developed, transitional, countries of eastern Europe and Latin America; and rates of more than 30 per thousand are still customary among sizable proportions of the massed populations in the demographically unstable countries of Africa, India, and the Far East. If there ever was a normal world-wide death rate, we are safe in assuming it to have been an extremely high figure by present Western standards, analogous to the widely fluctuating levels found today among the most backward, unenlightened, pre-industrial countries of the world.

With the exception of Sweden, whose vital records go back to the middle of the eighteenth century, trends in the death rate among European nations are scarcely discernible before 1800. From then on, a slowly increasing number of states adopted systems of vital registration and social bookkeeping which constantly sharpened the picture of what was happening to mortality. Where the decline began early, as in western Europe, it was slow and sporadic. Thus, a century ago, the crude death rates in countries of northern and western Europe occupied a narrow range between 20 and 25 per thousand of the total populations. In several of these same countries the birth rate was also declining, but at a somewhat slower pace, thus accounting for the heavy natural increase of the European population in the nineteenth century. In countries outside western Europe, such as Bulgaria, Russia, Japan, India, and Egypt, there is little conclusive evidence that death rates started to decline until much more recently. But once begun, the declines have been more rapid and have contributed greatly to the suddenness of natural increase. In India, for instance, where two consecutive decades (1921–41) for the first time saw a continuing slight decline in mortality, a total of approximately 83 million persons were added to the population.

The importance of knowing the reasons for the historic decline in mortality is manifest in any work concerning the health and welfare of nations. But in studying the possible course of future population growth among the three-quarters of the world's peoples whose trends are still governed by mortality, it is especially important to know how a reduction has been brought about in other countries. From a wide variety of sources—economics, history, medicine, sociology, and political science—a comprehensive picture has emerged of the causal factors underlying the reduction of European mortality during the nineteenth century, and this points the way to what remains to be done in countries which are still seeking the universally desired goal of lower death rates. If we cannot yet directly measure, we can at least appreciate the significance of the adaptive changes which have lowered mortality: (1) the technological revolution in agriculture, industry, and transportation; (2) the successful application of the biological sciences to problems of preventive hygiene and curative medicine; and (3) the growing stability of governments, in both internal and external affairs, coupled with organized concern for the welfare of individuals.

Improvements in agriculture and the opening of new lands as sources of added foods proceeded at a pace that confounded Malthusian theorists because it was greater than the rate of population growth itself. Expanding networks of rapid transportation allowed wider distribution of food supplies and reduced nearly to extinction the chances of local famines and starvation. More efficient food production freed an increasing proportion of agricultural workers from close dependence on the soil; in the eighteenth century every country in the world used about four-fifths of all workers in agriculture to provide basic sustenance, a proportion that has since been drastically reduced in the most advanced countries. Surplus product, furthermore, was turned into capital wherewith to support more effective systems of municipal sanitation and public hygiene. The great public health movement of the eighteenth century—a humanitarian, idealistic, and academic development, as Sigerist describes it —regained its momentum and broadened its scope under the emergency conditions created by industrialization and urbanization. In sum, increased productivity made possible the higher levels of living achieved in European countries after 1800. Powerfully implementing both the greater productivity and the raised levels of living, the greater stability in political organizations and means of social control added materially to the reduction in mortality by making life more secure and supporting the rights of individual citizens.

All of the multiple changes in the culture of western Europe which led to the initial reduction of the death rate were indigenous and interrelated. These changes brought about declines in mortality which were comparatively slow when contrasted with more recent declines in other lands marginal to, or well outside, the European sphere. It is likely that political reorganization and improved civil order have always been effective agencies of reduced mortality for limited periods; death rates probably declined in China under the peaceful reign of the early Manchus, in Japan during the formative years of the Tokugawa shogunate, and in Russia under the tsardom of Peter the Great. However, the more recent declines in non-European countries have come about primarily through wholesale adoption and intergation of European techniques and knowledge.

This borrowing has permitted countries such as the Soviet Union and Japan to reduce their death rates more than twice as rapidly as their Western forerunners. Among other non-European countries and colonial dependencies where the death rate remains very high, the relatively slight, and still quite inadequate, health programs which have been introduced already are having far-reaching demographic effects. But the indigenous basis for reduction has not been established, and the gains that have been registered rest on a more tenuous foundation than do those of Europe.

In the rationally balanced countries and a few others, which now comprise about one-quarter of the world's population, death rates have been brought down to such low figures that little room remains for future improvement. Where death rates continue at a high level, the introduction of modern science and medicine causes them to decline at a rapid pace; but where they are already low, further improvement becomes progressively more difficult to attain. Despite all of the gains in public health and chemotherapy, demographers are forced to the conclusion that the present crude rates in Western countries—around 10 deaths annually per thousand in the population—must inevitably rise. The present rates are deceptive, for when they are adjusted so as to take into account the favorable age distribution that currently characterizes such countries as the United States, the United Kingdom, and the British dominions in Oceania (all of which have virtually identical age structures), it is apparent that the "true" death rates are closer to 15 per thousand. As these populations grow older, therefore, they will approach stability and the age composition will level off; and when this occurs, we must expect the crude death rate to rise to 14 or 15. These are the lowest figures that are statistically possible in a stationary population, unless the actual span of life increases. Since there is no evidence that the ultimate possible span of life, the extreme age to which people can live, can be increased, it is unlikely that this conclusion will be upset. A larger proportion of people may succeed in avoiding premature death, but the age at which the human machine simply breaks down does not seem to respond to anything that medicine and science have yet discovered.

LIFE EXPECTANCY.—The crude death rate is a useful measure because it has been the most widely collected and published of the vital indices. Since it ignores both the sex and age composition of populations, however, it is not an entirely adequate basis for mortality comparison in either time or space. Life tables, which show the average expectation of life for both sexes at birth and in later years, provide sharper definition of the forces of mortality, and furnish what Thompson calls "probably the best single index of economic conditions or levels of living." Based upon observed deaths in a population according to the specified age and sex of those dying within a given period, life tables present with a high degree of mathematical accuracy the statistical likelihood of the average person's chances of survival. The practical importance of these tables for insurance, medicine and public health, and social welfare planning has long been recognized. Among the statistically mature nations of the West such tables have become highly sophisticated tools of demographic description and analysis.

Current life tables for most of western Europe, the British dominions, and the United States have behind them a long history of development. In transitional countries, such as those of eastern Europe, the USSR, Japan proper, and part of Latin America, the basic data for life-table construction, though still deficient in many respects, permit modern techniques of interpretation to deduce therefrom a reasonably accurate picture of prevailing mortality levels by age and sex. In those unstable societies where census figures for age distribution and death registration are available—as in India, the Philippines, Korea, the Union of South Africa (Coloured), and much of Latin America—"intelligent guesses by the technically competent" can be made; but among the populations of China, Southeast Asia, central Africa, and parts of Latin America, where there are neither census figures for age distribution nor death registration statistics, life tables are not yet compilable for any except isolated segments of the population.

The first life table for any country was computed in Sweden for the period 1755–75, and showed a mean expectation of life at birth of 35 years. In 1840 it was 44 years; in 1890, 52 years; and in 1936–40, 65 years. The first figure may well have been the highest in the world at that time, but by present Western standards it would be viewed as catastrophic. It is roughly comparable to, though slightly higher than, the probable current figures for the teeming millions of India and Egypt, and probably of China as well. What was the expectation of life in Europe or elsewhere prior to the modern decline of mortality remains almost pure conjecture, but there is no reason to suppose that the average person born into any society —medieval, ancient, or primitive—ever had more than an even chance of passing his thirtieth birthday. The reduction in mortality which began in the prospering countries of northwest Europe around 1800 caused the expectation of life for the newborn to rise by 1840 only to slightly more than an average 40 years in England and Wales, France, the Netherlands, and Scandinavia; but this figure still remains superior to present ones in the Latin American countries of Mexico, Chile, and Peru (Lima). Further gains in average longevity among the leading European nations were registered during the latter half of the century, and by 1900 the average expectation of life had risen by approximately 10 years to about 50; but in none of the Latin American countries today has this figure been achieved. (The Municipio of São Paulo comes very close to, if it does not equal, that figure.) No country in the world could show a mean expectation of life at birth of 60 years before the twentieth century. Since 1900, however, the gains registered among several of the demographically élite nations have been more rapid than ever before, and have carried the figure upward from 50 years to 65 and better.

The doubling of the average length of life at birth among the technologically advanced peoples of Europe and their descendants overseas stands out as one of the most spectacular events in the modern cycle of world population growth. It clearly indicates the life-saving superiority of rationally balanced cultures over those not yet modernized. In less than two centuries, from one-fifth to one-quarter of the world's people have succeeded so well in cutting down the toll of death that they have created a gulf between their present and former processes of population replacement as great as that which now separates them from native peoples of Africa and the Far East. They have moved quite contrary to the Malayan proverb which says "Let our children die rather than our customs," for they have preserved the lives of the former while profoundly altering the latter.

Although the Western countries have set the pace, their very success in prolonging the average life expectancy has brought them closer to a zone of diminishing returns. Considerable room for improvement yet remains in many of these countries. In the United States, for example, the expectation of life at birth is still eight to ten years lower for the non-white population than for the whites. The colored peoples of the continental United States are, in fact, almost exactly a quarter century behind the whites; during the present century they have been gaining "average years" more rapidly than the latter, but the "racial schism" yet remains. Again, New Zealand, which for decades has been the world leader in reduction of mortality and prolongation of life, provides these benefits in lesser degree to the Maoris than to the white members of the nation. The color bar in South Africa has no clearer proof than vital statistics. And in all countries, it is safe to assume, the underprivileged, diseased, and economically insecure segments of the population, and those who live in regional pockets of poverty, are the ones who depress the averages for expectation of life.

Improvements in all of these instances are socially desirable, practically possible, and admittedly advantageous to the general welfare; but they will not add many years to total life expectancy. Certainly the rationally balanced countries can scarcely expect to boost their average figures above 70 years, unless unforeseen developments in medicine and geriatrics provide effective ways of prolonging the span of life. By contrast, the remaining 75 to 80 percent of the world's population—including all of the transitional and unstable societies—have it within their power to add materially to their average life expectancy if they can and will follow Western models and experience. Already by 1940, the mean expectation of life at birth in the USSR and Japan stood at a level (about 48 years) which the United States achieved in 1900. By wholesale adoption of health measures and allied techniques, such countries have made more rapid progress in reducing death rates than did the nations in which the movement originally began; where improvement started late, it has been particularly rapid. Hence the prospect for all countries which still lag behind the demographic leaders is for further telescoping of a process which required decades in the West.

INFANT MORTALITY.—The decrease in mortality and the increase in life expectancy have not been spread evenly through all age groups. Mortality has declined most rapidly among infants, children, and young adults, the steepest slopes in graphs of such decline being found among those aged 1 to 5; while the slopes are flatter for succeeding age groups. Concomitantly, expectation of life has increased proportionately to a far greater extent among the very young than among those aged 40 or more. Very few years have been gained, on the average, by people who reach the age of 60. The saving of life, in short, has occurred among the youngest age groups where death has always struck hardest. Before the latter half of the eighteenth century probably one-third to one-half of all children succumbed to malnutrition, filth, and contagious diseases prior to reaching adolescence; and infant mortality rates of 300 deaths per thousand live births are presumed to have been common. No country which has reliable statistics today shows such high rates, but current (1936–40) infant mortality rates of 200 to 250 are found in countries such as Chile (234), Burma (206), India (200–250?) and Egypt (203).

At the other extreme are found the more advanced countries, where, by comparison, infant mortality has been reduced to negligible proportions. The 1936–40 figures for New Zealand (32, white; 111, Maori), the Netherlands (37), Australia (39), Norway (39), Sweden (42), Switzerland (45), United States (52), and England and Wales (59) mark the countries which have led the world in improving the life expectancy of their children. They continue to make positive gains, for by 1945 New Zealand, still the leader, had reduced white infant mortality to the record national figure of only 28 infant deaths per thousand live births; and the United States, not far behind, recorded only 36 per thousand in 1946. Many lesser regions—states, cities, and the like—within these and other countries can match the New Zealand figure; but no nation has yet succeeded in bringing home the gains of the most favored few to their entire infant population.

From the few European countries which have kept adequate figures on infant mortality over a period of years, it seems apparent that the death rate during the first year of life partook of the same downward trend during the nineteenth century as did the crude death rate of which it was a part. But, being a more sensitive index than the latter, it showed more violent fluctuations, especially during the middle of the century and among the rapidly changing Western cities where the worst excesses of overcrowding, lack of sanitation, and inadequate child care temporarily obtained. What appeared to be a slow and steady improvement hid a considerable retrogression during the peak period of industrialization of the urban West. By 1900 the worst features of this movement had been relieved by more effective control, but infant death rates even in the best-managed countries were still high (80 to 150); and among the remaining majority of countries they remained at nearly double those figures. Since the opening of the twentieth century, however, remarkable gains in reducing child deaths have been registered, and nearly all countries of the world have lowered infant mortality by one-fourth or one-third. This late but rapid improvement has come about through the virtual eradication of such formerly lethal diseases of childhood as diphtheria, measles, smallpox, and whooping cough, and a tremendous reduction in the prevalence of diarrhea and enteritis. In the field of those organic impairments which normally strike during the first month of life, little progress has been made. Thus, in countries where infant mortality is already lowest, the proportion of deaths attributable to prenatal impairments —congenital malformations and accidents of birth—stands highest. The sanitary and public health measures which thus far have demonstrated immense success in combating infectious diseases seem relatively powerless to cope with problems that arise prenatally. The causes of neonatal (first month) mortality are still obscure and difficult to cope with.

The principal relevancy to population trends of the saving of life in early years lies in the fact that those who may become future parents are the principal beneficiaries. If medicine and improved levels of living prolonged the lives only of those already past childbearing, there would be no possibility of a potential future increase in population. Through reduction of infant mortality during the interwar years, four of the major belligerent nations of Europe— England, France, Germany, and Italy—saved some 3.75 million lives, or about 80 percent of the number of their military dead during the first World War. This alone was not sufficient to reverse the trend of the population decline during the period, but without it they would have been much worse off in terms of replacement possibilities than they are today.

FERTILITY.—At the risk of oversimplification of a very complex process, one may say that the decline in mortality over most of the world in recent times has been a natural and inevitable concomitant of modernization and cultural change, while the later adaptive decline in fertility among a smaller portion of the world's peoples has been a rational, sophisticated adjustment. The same basic forces and conditions—scientific advance, raised levels of living, and improved individual and group security—underlie both sets of phenomena; but the comparison points rather to the fact that mortality has everywhere responded more immediately to those impersonal forces making for a safer and more comfortable environment than has fertility within the family. The enclosure of sewers and the pasteurization of milk, to name two examples, brought great benefits to large segments of the population in urban-industrial cultures of the West without causing more than a very limited number of modifications in daily life; but the concrete steps which must be taken by married couples who decide to limit the size of their family require knowledge, foresight, and perseverance. The saving of life involves a larger number of activities which fulfill many functions over and above those of population replacement; but reduction of births begins at home, in a changing set of cultural values which place a greater premium on economic success, social advancement, and quality of offspring than on mere numbers.

In point of time, then, fertility has been the secondary factor of the two which govern population replacement. And the lag of years or decades between decline in mortality and decline in fertility has been the immediate reason for the massive expansion of world population in modern times. It boosted the rate of natural increase to high levels and greatly broadened the population base in all countries. Only when the decline in fertility catches up with the decline in mortality, as it had appeared to do during the decade of the thirties among the rationally balanced societies, do populations approach completion of that cycle of growth which characterizes current cultural evolution. When mortality was universally high, fertility had to be equally high in order to insure that measure of uneasy natural stability typical of premodern times and of most of the world today; when mortality declined and fertility continued at its usual exuberant pace, the pressures, needs, and changing values of the augmented survivors spurred the well-informed and well-endowed to limit their children and achieve a new balance between the forces of death and birth. Through transmission and inculcation the patterns and techniques of limitation have spread among societies and between the social levels within societies wherever this new balance has become a felt need.

Fertility, or the actual production of children, though it lags behind mortality only in the technologically advanced countries of the modern world, has always and everywhere fallen below fecundity, or the biological capacity to produce. Wittingly or not, all cultures have surrounded procreation with limitations and barriers of custom such as marriage rules, proscriptions on intercourse, and the like, so that we cannot say exactly how many children a normal woman might bear. But the absolute upper limit is a hypothetical figure of little moment. It may have been approached on the frontiers in the New World where early marriage and quick remarriage were the rule and children a real economic asset, but in contemporary Western society it matters little what the limit would be among women who are increasingly able to complete their child-bearing on their own initiative around the age of thirty and after two or three children. As Kuczynski has pointed out, the upper limit of fertility would be reached only if all females in a society throughout their child-bearing period had regular intercourse with procreative males without contraception or abortion. These conditions have never been fulfilled in any society of note. Those who seek to argue on biological grounds that declines in fertility result from dwindling powers of fecundity are no longer taken seriously in the light of the constantly growing body of positive evidence that fertility has been reduced through conscious, willed behavior. Even if human fecundity has somehow been raised (by better nutrition and more healthful surroundings), or lowered (by greater tendency to neuraesthenia and debilitating physiological changes arising in the more hectic tempo of modern life), such changes seem scarcely sufficient to account for the drastic shifts in fertility that have marked the recent demographic history of the Western world.

Except in France, where the modern decline in number of births originated about a century and a half ago, the decline in fertility has been so recent a development—limited on a national basis to the last sixty years more or less—that the materials for its assessment and analysis are fairly limited. Not until the sudden drop in the birth rate among Western powers at about the time of the first World War did the situation become a "social problem." It was then that attention and publicity were first given to the implications of the trend. Consequently, our knowledge of the factors involved is quite limited, making difficult an exact comparative analysis of the movement. Several techniques and tools, such as the crude birth rate, the ratio of children to women of child-bearing age, and the gross and net reproduction ratios, have been invented and perfected to permit rigorous interpretation of good basic data wherever available. But there are few countries which have collected such data over any considerable period of time, so the remaining gaps in our knowledge are wide and deep. Thus while we can speak in broad generalities about international comparisons based on the simpler techniques, the more accurate comparison must wait upon the statistical maturation of the whole world. The most widely available measure of fertility is still the crude birth rate, which shows the number of children born each year per thousand of the total population. Since it overlooks the significant factors of age and sex composition, it may give a somewhat warped immediate picture of natality, and fails to satisfy all of the requirements of modern demography. Nevertheless, despite its lack of precision, it has certain suggestive values for comparative exposition; and since the basic statistics in all but a handful of countries are less than adequate to begin with, it would be futile to use more refined measurements in any widely gauged international comparison.

For convenience, then, the crude birth-rate figures for the modern world may be divided into the following broad classes of fertility: *low* (less than 20 births annually per thousand population), *medium* (20 to 30), *high* (30 to 40), and *very high* (over 40). According to the rates which have been recorded and estimated for countries in the period on the eve of the second World War (1936–40), it appears that only the industrialized countries of northwestern and central Europe, the United States, Canada, Australia, New Zealand, and Uruguay can be classified as nations of low fertility. The Netherlands and Canada (20.4) and Hungary (20.0) show medium fertility according to the arbitrary dividing lines, but all of the remaining countries are ranged between Sweden (14.8) and Finland and Uruguay (19.9). These constitute, of course, the rationally balanced countries, whose net reproduction ratios for the same period were hovering around unity; that is, their populations were virtually stable. Uruguay is a newcomer, if not an exception. Having the densest population of any South American country and a sizable proportion of foreign-born in her population, in that continent she come closest to approximating the demographic position of the others.

The countries of medium fertility include as a solid block those of southern and eastern Europe (with the probable exception of Albania whose unreliable figures indicate a crude birth rate of slightly over 30), the European population of the Union of South Africa (24.9), Japan (28.6), Korea (27.4), and Argentina (24.1). Fertility in these countries is declining rapidly, but it still has not caught or passed the previous decline in mortality, so the prospect is for several decades of rapid natural increase. This is the normal expectation among all transitional countries.

The high birth-rate countries are much less adequately reported, and reasonable estimates must often suffice. These countries constitute approximately 60 per cent of the present world population; and what happens to their birth rates in the future is of the utmost importance not only to their own welfare but also to that of the world at large. Immediately prior to the last war high birth rates (30 to 40) were usual in the Soviet Union; in Burma, Ceylon, and the Philippines of the Far East; and in Jamaica, Puerto Rico, Guatemala, Columbia, Venezuela, and Chile of Latin America. These countries have a high potentiality for growth which is currently being realized; and if the present moderate checks on mortality that have been introduced can be maintained, growth should continue without abating for a generation or more. The very high birth-rate countries, those whose crude rates were above 40 per thousand at

the beginning of the recent war, include the vast problem populations of the Near East and Far East, and Mexico, Salvador, and
Costa Rica in the New World. The 50 million people of Arabia,
Egypt, Iran, Iraq, Palestine, Syria, Trans-Jordan and Turkey appear
to have birth rates that are nearer 50 than 40; China's crude birth
rate is probably between 38 and 47; and India's, for discussion purposes, may be taken as "about 45." Formosa, British Malaya, and
the Netherlands East Indies are also listed in the "over-40" class.
These exceedingly high rates, characteristic of the largely illiterate,
preindustrial, and agrarian societies which we have designated unstable, show no signs of declining. Fertility in these countries has
been untouched as yet by spreading currents of cultural change from
the West. A staggering growth potential remains subservient only
to the unplanned control of custom and the savage forces of high
mortality. Death will continue to be the arbiter of growth under
these conditions for so long as the Oriental peasant peoples fail to
modernize their cultures after the Western model or to make changes
in their family institutions and values.

The current differences in fertility between the countries of
low, medium, and high birth rates result essentially from differences in the timing and strength of the impact of modernization.
Accelerated rates of decline in the birth rate are found among countries where the trends toward industrialization and urbanization
have been more recent. Although the decline in fertility probably
began among restricted groups and social classes in France, Sweden,
and the United States about the beginning of the nineteenth century,
national birth rates of the majority of countries of northwestern
Europe, British Oceania, and America north of the Rio Grande were
not reduced before the 1880s. A century ago the birth rates in England and Wales and the United States, for instance, fluctuated
around 35 per thousand, clearly in what we now regard as the
"high" bracket, where they remained until the last quarter of the
century. By 1900, however, they had declined to "medium" rates
of less than 30; and by 1930 they were well under 20, among the
"lows." The gradual speeding up of the process of decline is equally
apparent in the fact that it took France over seventy years to experience a drop in her birth rate from 30 to 20. The same reduction
was later effected by Sweden and Switzerland in about forty years,

and by England and Denmark in about thirty years. Declines of similar proportion occurred in central European countries still later in considerably less than twenty years.

Knowledge of differential birth rates among nations, and for the same nation at consecutive dates in its history, is essential for summarizing both current situations and past trends. But in the more advanced countries such national rates for the entire population always mask wide variations in fertility among selected groups or classes. Close analysis of these internal differences brings to light a sequential order in the acceptance of new techniques and values which is comparable to the same process of differential modernization among nations. Thus within a low birth-rate country such as the United States, fertility is found to be significantly higher in rural areas than in cities, among "lower" income, occupational, housing, education, and ethnic groups than in "higher" strata. These types of group differences are typical of the majority of rationally balanced and transitional societies today and show that the process of internal change and adjustment has not yet run its course. The apparent decrease in the scope of these variations among some of the most homogeneous and highly industrialized countries in recent decades presages an early approach toward greater cultural and demographic uniformity than has prevailed during the period of most rapid change.

MIGRATION.—The part played by migratory movements in the modern cycle of population growth is too well known to require more than a brief review and current assessment. Migration itself cannot directly alter the world total of population, as do birth and death rates; but its effects on the regional balance of peoples and states, coupled with the stimulus which new environments provide to the cultures of migrants, give it an important place in the demographic trinity. The great wave of European emigration and the lesser ripples from Asia, which represent a massive and largely uncoordinated attempt to redress the discrepancy between population and the world's resources, have created more problems than they have solved. The recent ending of a century of "automatic" adjustment to the legacies of the Age of Discovery shifts the burden from the rural peasant to the mechanically minded laborer, from the pioneer to the planner, and from the individual to the group.

The type of migration that marks the period of mankind's greatest increase has been the relatively free movement of individuals and family groups in process of changing their country of permanent residence and political allegiance. The qualification "relatively free" excludes, of course, the enormous traffic in slaves which antedated the bulk of the movement of Europeans. From the sixteenth through the nineteenth centuries probably 20 million Africans were uprooted from their homes, and a total of 15 million survived the middle passage to reach the New World, although not more than a half million were brought directly to what is now the United States. The qualification also applies to the sizable but unsung groups of bondsmen, indentured workers, and convicts who helped so greatly to set European colonies on their feet. As Carter Goodrich has pointed out, many Americans who claim colonial ancestry might also qualify by further genealogical research as Sons and Daughters of American Indenture.

The overseas movement of Europeans began feebly in the sixteenth century and long continued at a slow pace until it became a swelling tide in the great migration of the latter half of the nineteenth and the early part of the twentieth century. An estimated 65 million persons left Europe down to the decade of the 1930s. A not inconsiderable proportion of this number—probably about 30 percent—eventually returned home; but the movement left the United States with a net gain of approximately 30 million, Canada and Latin America 15 million, and Australia, South Africa, and lesser colonial outposts 6 million persons. Every country in Europe contributed to this host, and all but the smallest sent sizable numbers. Although the nationalities of the stream entering the United States gradually altered (predominantly north and west European to 1895, south and east European thereafter), the rural, agrarian, lower middle classes were at all times most heavily represented not only because they were the largest numerical group in European populations at the time but also because they were the ones who bore the brunt of the population pressures in their homelands.

Economic opportunities in the receiving countries—chances for the immigrants to improve their status—functioned throughout as the principal stimulus to emigration. Other causes, both individual and social, surely were operative at all times, but when there were fewest restrictions on international movement, the flow of peoples was determined to a very high degree by the relative opportunities in the countries of destination. On the other hand, emigration was always greatest from countries which were in the earlier phases of modernization, when the people glimpsed those promises of improvement which their homeland was temporarily unable to fulfill. As long as there remained plenty of good and cheap land in the New World countries, European peasants saw in them the key to their future well-being on the soil; but as the best lands were taken up and as modernization swept across the oceans into the new countries overseas, the more insistent cry came from industrial areas and cities. Hence the latter phase of the great migration not only drew its peoples from those regions of Europe which were delayed in the modernization process, it became a kind of international process of overseas urbanization. In the West, as Forsyth has said, "the main currents of migration in our time flow towards and not away from the thickly-populated areas."

Several factors combined to put an end to the era of the great migration. Restrictive laws in both sending and receiving countries following the first World War and the world-wide economic depression which shrank job opportunities everywhere are the most frequently cited reasons; but other causes, greater in their long-run demographic significance, should not be overlooked. Thus the progress of industrialization and urbanization among countries of northwestern Europe which formerly sent out the majority of the emigrants provided wider opportunities at home through higher production, better levels of living, and greater economic and social security. This permitted a more rapid absorption of redundant rural populations among the original areas of emigration. Secondly, the reduced rates of natural increase provided fewer migrants; and expanding economic systems of the homelands were able to cope more effectively with the natural increase of population. Finally, the waxing strength of economic and political nationalism, expressed in legal rulings as well as in other positive ways, reduced even when it did not totally restrict the desire and opportunities to seek new national homes.

European internal migration, including both the eastward drift of about 4 million Russian settlers in Soviet Asia and the northward and westward drift of a comparable number of laborers from the technologically retarded "pressure areas" of southern and eastern Europe, has become increasingly important in the present century, especially since the first World War. Trade follows the flag, but people gravitate toward industrial opportunity. Mass transfers of population, which exhibit more clearly then anything else the impact of ethnic concepts and nationalistic ideals on population policies, are likewise a postwar phenomenon. Countries such as France, which faces an imminent drop in population, are now taking active steps to forestall decline and bolster their labor supply by encouraging highly selective immigration. If the manifold restrictions under which migration now must operate in such countries can be satisfactorily met, some positive benefits may accrue from these programs both in sending and receiving countries.

Still widely accepted is the belief that the remaining sparsely settled lands, such as the British dominions and Latin America, are in need of mass migration in order to develop further their wide open spaces. Such areas remain empty, however, not for want of attempts to settle and exploit them but rather because at present levels of applied technology they remain marginal and economically unprofitable. The greatly increased costs of pioneer settlement, together with the dubious national benefits to be derived from homesteading and subsistence farming, rule out all large-scale programs of controlled migration on the grounds of their practical impossibility. No one denies that Australia and South American countries are capable of supporting considerably larger populations at no sacrifice of their achieved levels of living. But immigration will be permitted only on their terms, and the terms have become more rigid rather than liberal in their mixture of prejudice and practical considerations. In his revealing survey of the situation in Latin America, Kingsley Davis says, "Latin America cannot attract the kind of immigrants it wants and does not want the kind it can attract; and also . . . owing to its high rates of natural increase it does not need mass migration anyway."

In the Asiatic countries, despite their vastly greater population base, migration within the continent and overseas has always been small in comparison with the nineteenth century dispersion from Europe. In absolute numbers the estimated emigration of Chinese,

Indians, and Japanese to the frontiers and continental peninsulas, the islands of the Pacific, and even to the Americas and South Africa runs into the tens of millions; but this represents a far smaller proportion of their total populations, probably less than 2 percent, than was the case in Europe. In addition to the ethnic antagonisms and legislative restrictions which have dammed up or closely controlled the flow of Asiatics during the last century, the Asiatic countries except Japan have not had the economic surplus wherewith to support large movements overseas, nor are there any extensive and uninhabited countries to which they could go. Someone has pointed out, on the basis of a Japanese example, that even if a newly risen continent became available the combined merchant marines of the world would be unable to cope with the task of transporting to it an annual increase of over 6 million Asiatics. Extension of the cultivated land areas of monsoon Asia outside of Japan, not emigration, is the historic means by which pressures of overpopulation in Asia have been met. The less densely populated lands—Burma, Siam, Indo-China, the Netherlands Indies (as a whole), and the Philippines—still have a sufficiency of potentially arable land to feed their expected natural increase for three or four decades.

Migration certainly was one of the important factors enabling Europe to avoid overpopulation during the last century, but it does not follow that equal effects can be expected from it in Asia. The magnitude of the movement, its duration, and the accompanying processes of industrialization and declining fertility are aspects of the European case that have little chance for successful application in the Far East. Nevertheless, spokesmen for the emergent nationalism of Asiatic countries now seek to make political capital out of migration. Europe's history, the Atlantic Charter, and the philosophy of the United Nations are invoked in the hope that racism and exclusion laws may be abolished to permit free movement overseas in order to help relieve the intolerable pressures of overpopulation in Asia. Dr. Radhakamal Mukerjee, one of India's foremost economists and long a keen student of population, presents in his most recent book a clear demand that the empty spaces of the British Commonwealth be opened for planned colonization and settlement by Indian and Chinese agriculturalists. Although he briefly notes that the control of fertility and not emigration should be regarded as

the major cure for overpopulation, problems of industrialization and
birth control are not elaborated upon at any length. "Freedom from
want in Southeast Asia," he states, "rests largely on freedom of emigration." This is a sincere, idealistic, and altogether important book;
but its overemphasis on migration as a solution to population pressures confuses the issue. The migration potential of Asia is indeed
enormous, as it must be in every land where large rural populations
share rising aspirations and incredibly low levels of living. Whether
or not there will be some easing of international controls to permit
selective emigration to occur in this area defies prediction, but it
would seem most unlikely that concepts of democracy and global
interdependence alone will bridge the gap in levels of living, mechanical equipment, power resources, technical skills, and capital
accumulation that separates the massed humanity of the river plains
in monsoon Asia from the congested, urban-industrial concentrations of Occidental culture.

Emigration could serve only as a feeble palliative to overpopulation in the Asiatic areas; it could not solve the problem. Where mortality remains the only working check on population growth, the
temporary reduction of numbers would lower the death rate. But
chronic high fertility would then assert itself anew, take up all of
the slack left by the outward movement, and start another cycle of
growth. Furthermore, if those who moved out took with them the
same cultural patterns of high fertility, they would simply perpetuate their problems in another land. World population would grow,
but no progress toward a rational balance between numbers and
resources would have been achieved. The world as a whole would
stand to gain little or nothing from the closest possible settlement
of all arable land, unless that is desired as an end in itself. Notestein
suggests that heavy and continued migration from small areas of
high density to large areas of less density, as from Puerto Rico to
the United States if that could be arranged, might facilitate the vital
revolution in that island as it once did in the anomalous case of Ireland in the past. But in a subcontinent, such as India or China, this
adaptation is utterly out of the question; there are neither the financial means, the agencies, nor the destinations in the world today
for handling such masses. Large nations can neither solve nor greatly
alleviate their demographic difficulties by shifting people around

the earth. Individually and collectively they must seek a more efficient utilization of natural and human resources and use every means in their power to aid the overpopulated nations to achieve a rational balance of their vital processes. If a "Marshall Plan" for Europe alone faces such inertia, indifference, and obstruction, where are the statesmen who will dare to extend the concept into those massive problem areas of the rest of the world whose needs are infinitely greater?

URBANIZATION.—One of the most distinctive aspects of the modern cycle of population growth has been the concentration of peoples in great cities and the proliferation of urban places all over the globe. This burgeoning on every continent of masses of people freed from direct participation in the quest for subsistence could never have occured prior to the time when industrialization, occupational specialization, and increased production cut the bonds which tied men to the soil. Cities have existed in every civilization since the days of the first Urban Revolution, but they were almost invariably small by present standards and few in number. Never before the nineteenth century had cities contained so large a proportion of the world's peoples. When the labor of four persons on the land was essential for their own support and the maintenance of one other non-agricultural worker, as was universally the case down to the middle of the eighteenth century at least, nations could support few cities. Furthermore, the mechanical limitations on the collection, exchange, and distribution of food and water by human and animal power alone were added barriers to urban growth. Cities of a million inhabitants certainly represent the extreme upper limit of size in pre-industrial times. If facilities for water transport were available, such numbers might occasionally have been achieved. But London alone among the present capitals, as the great and all-powerful center of trade and commerce of its times, reached that figure before the middle of the nineteenth century.

In his *Population Problems*, Thompson admirably sums up the catalytic effects of the introduction of steam power as one of the principal bases for the second Urban Revolution, that of the last century and a half. He shows not only how the use of steam favored production but also how, thriving on congestion and a tightly knit economic system, it determined the ecology of city life for generations down to the present. The growth of large cities, therefore, defies separation from the modernizing processes. The production of agricultural surpluses permitted the transfer of a significant number of peasants and farmers to non-agricultural pursuits; provision of cheap and speedy means of transport and communication extended ecological boundaries; improving municipal sanitation and public health services reduced the threat of epidemic diseases to acceptable proportions; and political organizations, utilities, and facilities arose in response to the need for cohesion and control.

The spread of the urban-industrial pattern over the whole world in the last 150 years has been an uneven process. Composed of such a variety of interdependent variables, and so abbreviated in time, the trend resembles an almost simultaneous explosion of a number of mine fields rather than a slow spread. In 1801 only 21 "great cities" of more than 100,000 inhabitants had been enumerated. All were in Europe. (New York was the first American city to reach the mark, in 1804, and remained the sole great city in the United States for 35 years thereafter.) In the third decade of the present century, according to Jefferson's calculations, the number of great cities in Europe had risen to 182, and the world total stood at 537. Urbanization has proceeded rapidly wherever there has been technological advance, but the pace has been faster in the newly won lands of Europe overseas than in the more settled mother countries themselves. The development of cities has everywhere outstripped that of the rural districts which such cities draw upon, dominate, and serve. Supercities of more than one million inhabitants have also shown a remarkable proliferation. From 1800 when there were none down to the opening of the second World War, a total of 37 have been enumerated, distributed among the continents as follows: Europe (including Moscow and Leningrad of USSR) 15; Asia 10; North America 5; Central and South America 4; Oceania (Sydney and Melbourne) 2; and Africa (Cairo) 1.

The process of urbanization in transitional countries such as the Soviet Union and Japan has pursued a course broadly similar to that of Europe and the New World but accelerated in tempo. Census data for both countries are limited to the interwar years but cover such considerable transformations in spatial distribution and occupational calling that the effects on city growth of their forced-draught industrialization are clearly apparent. The rapid growth of cities in all parts of the USSR, both in the settled regions of the west and in the undeveloped east, has been, in Lorimer's words, "the major factor contributing to the redistribution of the Soviet peoples." There were 33 cities of over 100,000 population in 1926; by 1939 they had increased to 82. In the same brief period of a dozen years the number residing in places classed as "urban" grew from 26.3 to 55.9 million, a percentage increase from 17.9 to 32.8. This should be compared with the roughly similar course of urbanization in the United States that took place during the 30 years from 1860 to 1890. The 1940 urban population of the United States was 56.5 percent.

In Japan, as Irene Taeuber has shown, the total population classified as "rural" remained fairly constant from 1920 to 1935 and declined slightly thereafter to 1940. During these two decades the total population increased 31 percent, however, to a total of 73 million. Having no readily acceptable outlet for the bulk of the natural increase in international migration, movement within the Empire, or to frontiers of settlement, the "excess" peoples streamed into cities and the big industrial regions. "The trend toward urbanization is most clearly reflected in the growth of cities of 100,000 and over," says Dr. Taeuber. "These cities contained 12.1 percent of the total population in 1920, . . . and 29.1 percent in 1940." This relative growth of Japanese cities is slightly greater than that of United States cities between 1900 and 1940. According to our last census, 28.8 percent of the population was living in cities of 100,000 and over.

The failure of urbanization to make gains among the backward populations of monsoon Asia, Africa, and the Near East is fully understandable in view of factors and processes already mentioned. The great cities and supercities of the Orient, such as Shanghai and Nanking, are treaty ports and capitals of world commerce which have arisen primarily as adjuncts to the trading interests and colonialism of European powers; or else they are the Western-type industrial centers of Tokyo and Osaka. The native populations, with the single exception of the Japanese, have lacked sufficient capital, organizing skill, and nonagricultural job opportunities, to support more than a minimum of urban places without the aid of the industrial nations. The predominantly rural character of the populations in India, China, the Netherlands Indies, and the Philippines can best be seen in figures showing that approximately 80 to 90 percent of their peoples still live a rural existence in which two-thirds to three-quarters are totally dependent on agriculture for their livelihood. The diminution of manpower requirements on the land in these countries has been so slight in the last half-century that natural increase consistently wipes out the gains made.

More significant than redistribution of peoples between the rural-agricultural and the urban-industrial callings have been the revolutionary adjustments which alter the entire way of life, customs, and values of those who change their residence from country to city. Demographically speaking, the most distinctive feature of urbanization in modern times has been the decline in fertility which has accompanied the spatial-cultural shift to urban residence and urban patterns of behavior. Jaffe's general survey shows that differences in the birth rate between cities and their surrounding rural regions have characterized European cultures for upwards of two centuries, and probably originated even earlier. Indices of fertility are unmistakably lower today in metropolitan areas and great cities than in smaller cities and towns and in the rural countryside. This inverse relationship between urbanization and reproductivity underlines the conclusion that fertility is a culturally dependent variable; for, with the exception of the most backward countries such as India, Egypt, and probably China, the *rationale* of lower birth rates of cities stems from the modern institutional organization of industrial society wherein men place a higher premium on the small-family system than on the traditional rural patterns of unrestricted family size. The noteworthy result of this change in values and behavior lies in the fact that in most of the rationally balanced societies today, especially in urban places of more than 100,000 population, people generally fail to replace themselves by a considerable margin. Cities grow by immigration, not by natural increase. Thus American women in great cities had 28 percent fewer children under 5 years of age during 1935–39 than were needed to maintain a stationary population. Even the smallest American cities were failing to reproduce themselves as a class. The experience of Sweden, Germany, and Austria supports the conclusion that the more urbanized a country becomes, the greater is its dependence on the rural population to make up the birth deficit of its metropolitan areas. If such a development continues, the time must come when the country as a whole will fail to reproduce itself. In Great Britain, on the other hand, where the proportions of rural, agriculturally dependent peoples are smaller and in closer physical and cultural proximity to great cities, the rural-urban differentials in reproductivity are less striking. In two great conurbations, Glasgow and Liverpool, urban populations may have been reproducing themselves; but city size, according to Enid Charles, seems less determinative of fertility in this case than the very low levels of living. Improvements in social and working conditions, therefore, will certainly bring these cities into line with the rest of metropolitan Great Britain.

Our knowledge of the actual levels of replacement in cities among the transitional class of societies of eastern Europe, the USSR, and Japan remains sketchy and incomplete, but the fact of their inferiority to rural rates is well established. As is the case with the rationally balanced societies, the evidence clearly indicates that more effective means of family limitation through contraceptive practices are the immediate cause of declining fertility among urban peoples. The underlying values of pecuniary success and ambition, emphasis on the individual in contrast to the familial group, and the desire for freedom from the biological strain and social responsibilities of large families have their genesis in urban-industrial cultures. Wider systems of communication, universal education, and greater physical mobility permit this group of associated needs to seep down vertically through social classes and out horizontally into rural areas, making the distinction between rural and urban ways of life in advanced countries increasingly difficult to draw. The diffusion of knowledge concerning reliable methods of contraception is still in process even in the most advanced countries; consequently we foresee a continuing downward trend in rural birth rates. Therefore any increase in the average number of children in the rationally-planned urban families will have to be considerable if it is to offset the anticipated decline among the dwindling rural populations of industrializing nations.

The situation in India, China, and Egypt with respect to rural-urban differentials differs radically from that in more advanced nations. In these countries, and presumably in the rest of those classed as unstable, the lower ratios of children to women in cities are the result of indirect, institutional, and nondeliberate custom. The planned control of family size may characterize liberal, literate, and emancipated couples who reside in the largest cities, but such elements are so few in number that their example is lost in the crowd. Since these differentials are of long standing, there is little likelihood that, as they stand today, they foreshadow a mass movement toward conscious fertility control.

The Future Growth of Population

The question of how many people the world can ultimately hold has been a recurrent subject for debate ever since man has appreciated the fact of modern population increase. A number of recent attempts to determine the earth's carrying capacity are indicative of the tendency to project the future in terms of biology and the food supply and to ignore or overlook the impact of cultural factors on population growth. The figures for the upper limits presented by representative human geographers runs from East's 5.2 billion to Penck's 13.5 billion or more. Even though such estimates allow for wider application of and improvements in current agricultural technology, the fact remains that the ability to support life at nothing better than subsistence levels is assumed to be the ultimate determinant of population growth. Kuczynski, however, after pointing out that at rates of increase prevailing from 1900 to 1930 the world as a whole would reach a maximum population of about 10 billion in 250 years, adds that under such conditions all human activities and efforts would necessarily have to be directed merely toward keeping the maximum number of people alive. This he regards as an extremely unlikely development.

Hypothetical upper limits such as these are, of course, purely conjectural. They are noteworthy not only because they begin with a figure two and a half times that of our present population but also because of the great latitude between the low and high estimates. It may reassure some, or surprise others, to learn that the world has not yet reached its densest possible habitation. But, actually, all such extremely long-range prediction is chimerical. If the history of the rationally balanced nations provides any single demographic lesson, it goes to support the hypothesis that population growth is a dependent variable, subject to changes in technology, social organization, and the patterns of man's desires. The increasing control over nature provided by science has been concentrated neither solely nor primarily in food-getting activities. On the contrary, the resources of advanced civilizations are used first and foremost in raising levels of living—in gaining more security against disease and death, in better education, and in improved housing. In these factors themselves lie the stimuli which have led on the one hand to the limitation of population growth and on the other to the maintenance and constant improvement of achieved levels of living. In breaking free from his chains of bare subsistence, man finds himself with power to limit growth rationally. It is inconceivable that he will voluntarily surrender this power.

The futility of making long-range population predictions for the whole globe stems from the infinite number of assumptions that must be made regarding human institutions and behavior. The major theoretical problem appears to be whether or not the shift can be made by the unstable societies toward a rational balance of their vital forces. The principal practical problem, and one with far-reaching economic and political implications, is how much net growth these societies will experience before they reach a rational balance of their own. Not more than 20 percent, in the aggregate, of the present world population depends directly on machine industry for its livelihood; but with the tools, skills, and political power which this scattered minority has at its command, it exerts a disproportionate effect on the habits, values, and institutions of the remaining 80 percent. It is quite conceivable that these proportions can be reversed if modernization proceeds as far among the backward nations as it has among the most advanced. But massive forces of poverty and social inertia, ethnic hostility and disease, stand in the path of rapid cultural change in countries which appear to need it most. Kingsley Davis is correct, nevertheless, when he says "today throughout most of the world cultural development is going ahead faster than population growth." The examples of the "Westernization" of Japan and the Soviet Union are simply the most striking of many that might be cited in evidence.

By contrast with the long-range view, short-range prediction of both food requirements and potentialities contains comparatively fewer unknowns and permits reasonable deductions to be drawn concerning the needs and means for some decades ahead. Despite the dislocations and maladjustments of these postwar years, expert opinion agrees that the world can produce all of the food that its 2.2 billion inhabitants require, and in the same process improve world-wide levels of living. Present methods of agricultural production, utilization of power resources, and systems of transportation provide a basis for the conclusion that there is no inevitable reason for famine or malnutrition anywhere. The problem is not one of insufficiency of land or inadequacy of techniques but one of achieving more effective national and international organization by which to make more efficient use of what the world can produce. Of the less than 20 percent of the earth's surface considered suitable for cultivation and as pasturage for livestock, only about 7 to 10 percent (3 billion acres) are now being used to support all of the world's peoples. Deficient diets among a proportion of the population may be found even in the most advanced countries in normal times; and the subsistence populations in most of the world are chronically underfed. Nevertheless, the world's hunger derives not from an absolute increase of pressure of numbers on resources but from distributional inequalities among the more advanced nations and the continuing patterns of overpopulation in the backward countries. In brief, the cycle of world population growth which has been experienced thus far gives greater hope than ever before that an ultimate "bee-hive world" can be avoided.

It must not be assumed that starvation will be abolished "automatically" by means of the cultural changes that are sweeping across the world. Education, capital investment, managerial ability, and last but not least the desire of men to wipe it out must be strengthened globally before such an end is achieved. The recent World Food Survey conducted by the Food and Agriculture Organization of the United Nations has presented a detailed picture of what the world will require for an adequate (not subsistence) diet up to the year 1960. This calls for a production increase of 12 to 80 percent over the pre-War II output of sugar, cereals, fats and oils, meat, and pulses and nuts; 100 and 163 percent respectively for milk and fruits and vegetables. From a world-wide study of soils and current agricultural methods of production, Salter concludes that these increases are attainable in two ways: (1) by more intensive and efficient cultivation of land already in use; and (2) by expansion of the cultivated area in regions of undeveloped soil resources. With respect to crop yield, rice production in the Far East, for instance, shows wide variation in yield per acre; in many important areas such as Java, India, and the Philippines yields are half or less of those in Japan. More surprising, perhaps, is the fact that the yield per acre on United States farms is little better than the world average, and 50 percent less than in western Europe. The extension of cultivated areas, as well as increased yield, will depend upon the economic limits of costs and returns.

Salter believes that a total of a billion additional acres of red soils of tropical and subtropical lands are arable. These are found mainly in South America and Africa and, in lesser amounts, in the greater islands of the Western Pacific. He further believes that 300 million acres of podzols, the ash-like soils of the North Temperate zone— in the USSR, Canada, and the United States—can also be added to the world's present crop lands. Despite all that mechanization, irrigation, and improvement in crops and herds can accomplish to add to food stocks, the major factor governing the potentialities of increased food production in both the old and new land is increased fertilization. Eight times as much phosphate and eighteen times as much potash as are being utilized at present would be required for world-wide production to meet the needs of 1960. Since known resources of these fertilizers are good for another 500 years at least, the problems here are primarily economic.

The paradox in the new relationship between man and the soil is the fact that, as Meyerhoff says, "the few, with mechanical aids, can more effectively feed the many than the many can feed themselves." This conclusion finds ample support in the recent history of all of the more advanced countries; and it provides the basis for realistic programs aiming to reduce the problems of overpopulation in the rest of the world. Modern Greece offers a good example, on a small scale, of a backward nation whose surplus population could be reduced through comprehensive modernization. The recent *Report of the FAO Mission for Greece*, a brilliant survey of cultural and demographic factors in a typical transitional society, outlines a plan whereby Greece in the next generation might change from an overpopulated nation of very poorly paid agrarian workers toward one with less than 40 percent of the workers on farms, far higher output and levels of living for both farmers and city dwellers, and a real national income at least two to three times its present size. The program would require a dual attack through both industrialization and improvement and extension of agriculture, and would raise Greek self-sufficiency to a point where her population growth would benefit rather than detract from her chances to achieve a rationally balanced society.

Modern Greece is but one small nation standing midway between West and East, culturally as well as geographically and also in the pattern of vital trends. Greece has always been peculiarly dependent on world prosperity and trade for whatever measure of prosperity she attains. Her problem, therefore, of "too many people, too little product" has wide political implications. Internally she will require a creative nationalism and unity of purpose to develop the opportunities inherent in her situation; externally there must be international stability and cooperative assistance before her most pressing problems can be solved. But if little Greece's difficulties seem so great, it gives one pause to consider the vastly bigger and more complex problems which face the massed peoples of the Far East whose levels of living are so much lower.

Attempts to predict the ultimate number of people whom the world can contain may be labeled as footless, but short-term forecasting of numbers within advanced nations has recently become one of the most highly developed and useful fields of population research. Although limited to the statistically mature countries, meaningful population estimates can be made on the basis of known cultural and demographic constants which illuminate some of the basic conditions of national and international developments. Trade and industry, welfare services, planning commissions, and governments are shaping their courses of action in accordance with the "reasonable guesses" of demographers as to the future trends in size and composition among a growing number of the world's peoples.

Population estimates are inevitably governed not only by the quality of the available statistical evidence but also by the reasonableness of the assumptions concerning nonpredictable factors. Straight arithmetical extrapolation of known trends is no longer sufficient for accurate forecasting, since recognition has been granted to the interdependency between cultural and demographic factors. Assumptions with respect to levels of migration, effects of war, and changes in political boundaries must be carefully weighed. On the other hand, it has been amply demonstrated that certain aspects of the population equation have an orderliness and regularity which narrow the upper and lower limits of the predictions. Thus, estimates starting from the known age-composition of a society at a given period can carry the several age-groups forward in time, subject to the likeliest rates of fertility and mortality. In a thirty-year projection, for instance, this means that one could begin with all of the older or potentially older people, most of the working population, and a large majority of the future parents already included in the base population. Furthermore, wide comparative study has established the fact that where birth and death rates are high, they will decline rapidly under the forceful impact of modernization; where both are already low, their further decline will be more gradual. Thus by applying patterns of known vital trends to populations whose numbers have been recorded by age and sex, the potentials for growth are made apparent. Space does not allow discussion of the complex methods and analyses involved in this work; the reader is advised to consult the writings of Thompson, Notestein, Lorimer, Glass, and others for detailed explanation of the factors and results.

As has already been indicated, population estimates which can lay claim to reasonable accuracy are limited to most of the rationally balanced and some of the transitional countries. Table 4 presents an oversimplified summary of current figures. The 1970 estimates are subject to qualifications in their original presentation which are too numerous to mention; they are given here as possible midpoints within the high-low probability range simply to indicate the assumed trends of growth for the critical years ahead.

Table 4 clearly shows that the tide of population growth is receding in those countries which saw the start of the modern cycle. It will continue at a very reduced rate in the most advanced countries overseas; but it will still be running strongly in the transitional

TABLE 4

*Estimates of Future Population in Selected Regions and
Countries, 1940–1970*

(in Millions)

	c. 1940	c. 1970
Northern and Western Europe	234	225
United States and Canada	143	176
Australia, New Zealand, Union of South Africa	10.5	12.7
Southern and Eastern Europe	165	192
Union of Socialist Soviet Republics	174	251
Japan	73	88

Sources: F. W. Notestein, *et al.*, *The Future Population of Europe and the Soviet
Union* (Geneva, 1944); K. Davis, ed., *World Population in Transition* (Philadelphia,
1945); F. Lorimer, *The Population of the Soviet Union* (Geneva, 1946); W. S.
Thompson, *Population Problems* (3d ed., New York, 1942).

countries a quarter century hence. What total effects the recent
war will have on these figures, it is too early to tell, but they are
bound to be considerable. War losses are not included because they
are not yet known. But according to Lorimer's analysis of the
hypothetical Russian war losses, that country's 1970 population
would be 222 million instead of 251. The increased number of
births in all countries during the war years will also have some
effect; the United States, for instance, will not reach its anticipated
peak of growth (165 million) until about 1990; and the actual declines which had been predicted for the populations of some north
European countries as early as the next decade may be staved off for
a few years.

For most of the rest of the world, one is tempted to repeat the
despairing comment "sufficient unto the day is the population
thereof," and leave estimates to seers and prophets. In the absence
of statistics comparable in quality to those of the West, the task
of predicting the future populations of Central and South America,
Africa, and Asia must rest largely upon analogy, limited studies,
and broad generalization. With respect to Central and South America, Notestein offers an "illustration" (rather than a "prediction")
of the growth potentials in that region if the decline from high to
low mortality and fertility follows European experience. He makes
a best estimate of the current birth rates in the entire area and

projects the decline in those rates midway between the rates of decline in Germany and Italy; death rates are assumed to be comparable to those generally found in association with such birth rates. From these assumptions a total population of 283 million is deduced, exclusive of immigration, for the year 2000. With respect to Africa and Asia, Notestein assumes annual rates of growth of 1 percent from 1940 to 1970 (slightly higher than, but consistent with, past rates of increase), and 0.5 percent from 1970 to 2000 (slightly lower than any rate since 1850). These assumptions give Africa a population of 212 million in 1970, and 250 million in 2000; Asia would have 1,900 million by the terminal date. Only the course of actual events can prove whether these figures are too high or too low.

The future population of Asia has also been analyzed by Thompson, whose long-standing interest in the demography and cultures of the Far East lends great weight to his conclusions. His approach, while more detailed than that of Notestein, rests also on the trends of natural increase and reaches results that are quite comparable. Using those census data of Southeast Asia since 1900 which he regards as reliable, Thompson assumes for illustrative purposes that the differences between birth and death rates—that is, rates of natural increase—obtaining during the last two or three decades can be maintained until 1990. The results of his calculations on the basis of this assumption are summarized in Table 5.

TABLE 5

Estimates of Future Population in Selected Asiatic Countries, 1940–1990 (in Millions)

	c. 1940	1990
China (including Manchuria)	400	609
India	389	701
Netherlands East Indies	72	170
French Indo-China	24	43
Burma	17	30
Siam	16	60
Philippines	16	46
British Malaya	6	19
Totals	940	1,678

Source: K. Davis, ed., *World Population in Transition*, p. 77.

The question arises whether or not the increases shown in Table 5 are possible. The answer lies solely in the future death rates: if they can be maintained at their present levels (improved upon in China), such growth appears to be inevitable; any considerable improvement would only raise the estimates even higher. Thompson foresees no decline in birth rates in these countries before the end of the century at the earliest. But, as a practical matter, it is difficult to see how death rates in China and India can be maintained at their recent relatively favorable levels in the face of such growth as is indicated by these calculations. By comparison, the other countries listed in the table still have considerable room for expansion; but India and China are already so densely settled, produce so little surplus over and above subsistence, and are so far from achieving effective political stability, that their death rates must seem inevitably to be scheduled to rise, as greater and greater pressure is brought to bear on food resources. Whether this comes about gradually and through general deterioration of conditions, or through a series of sudden catastrophes which will wipe out millions, cannot be foretold. But it is obvious that only with internal order, world peace, and a huge long-range program of subsidized modernization would conditions permitting such growth be assured. If one accepts the concept that "poverty anywhere is a threat to prosperity everywhere," one must also consider the consequences of an improved prosperity in the Far East: aside from unknown political implications of such differential growth, the only demographic threat connected with vast increases anywhere is the danger that countries such as China and India may double their populations without having achieved any real measure of control over their fertility.

The dawn of public understanding of the modern demographic cycle and its implications has brought with it a series of attempts to control patterns of future growth. In place of the traditional, opportunistic approach to problems of population through migration, wars of conquest, and moral suasion, one now meets the concept that growth can be controlled to the end of improving human welfare. Based on the antithetical patterns of increase which divide the modern world into rationally balanced and unstable societies, the programs strive on the one hand to raise fertility to a point which will insure population replacement, and on the other to reduce fertility so that a more equable balance can be achieved between numbers and resources. In the history of Western countries, private and individual interest in family limitation has invariably preceded rather than followed general public agitation over birth limitation and the small-family system. But, as Lorimer says, "The emergence of public population policies is part of the general trend toward organized control of human affairs." Not until the first World War was much done in the field of pronatalist (birth-increasing) policies among Western countries. But shortly thereafter, those in which the drop in birth rates had become most acute embarked upon a series of public policies aimed at forestalling the decline. Democratic no less than totalitarian nations—France and Italy, Belgium and Germany, Sweden, Japan, and Russia—introduced a series of legislative measures designed to raise the quantity and quality of their national populations. Although differing in many details and in extent, programs in these countries advanced from the assumption that the economic burdens of child-rearing are the principal deterrent to larger families. Various "honors" and badges have accompanied marriage bonuses, family allowances, and state-supported maternity services; but with the possible exception of Russia's latest (1944) plan—"one of the most decisive pro-natalist programs ever inaugurated in any country," according to Lorimer—the experience has generally supported the conclusion advanced by Glass that the state "can't buy babies at bargain prices."

Sociologists are far from denying the fundamental importance of economic factors as major determinants of family limitation, especially under the conditions of pecuniary individualism which characterize so much of modern life. And they do agree that other elements in the family institution play a significant part in determining the number of births. Extremists who prophesy the early demise of the Western family grossly exaggerate certain aspects of its changing functions. But it is obvious that population begins at home, and unless values in support of a medium-sized family are more widely accepted, and unless conditions are implemented to make such values attainable, the ultimate decline of the rationally balanced nations is an arithmetical certainty. Will awareness of this eventuality bring about a change in behavior in favor of large families? Will the individual couple come to feel a societal responsibility for its share in the task of replacing the population? These questions defy conclusive answers, but as long as nationalism and group solidarity continue to hold sway as they do today, it seems quite unrealistic to think that people will stand idly by while their society travels a road to biological extinction. The very policies mentioned above indicate the first socially organized attempts to cope with a new situation. That they appear thus far to be unsuccessful underlines our incomplete knowledge of the factors involved; it does not prove that race suicide is a foregone conclusion. By contrast with the millennia of time in which man has been subject to natural increase beyond the realm of simple control, the brief period during which he has escaped from the bonds of his inherited drives is short indeed.

If there are many doubts concerning the adequacy and acceptability of programs designed to raise fertility in the more advanced nations, there is wide agreement on measures which will reduce it in the unstable societies. But policies in the overpopulated countries exist almost entirely on paper; and they are essentially Western in provenience. Lorimer is authority for the statement: "No independent nation or imperial power has ever seriously undertaken measures to check excessively rapid population increase." The time has come, however, when all nations must cooperate in assisting the overpopulated, unstable countries to bring natural increase under control.

So far as is now known, there is no easy way in which this can be done. Bringing birth control to the Chinese, the Indians, the Javanese, and the Egyptians would seem to be the solution. It is the solution; but to attempt to introduce it first would be to reverse the whole course of development through which the rest of the world has gone in its movement toward fertility control. Birth-control techniques are simply means to the end of family limitation; the desire to limit families depends upon revolutionary changes in cultures as they shift from an agrarian to an industrial way of life. The entire set of life conditions which contribute to high mortality and high fertility must alter before the vital revolution can occur. For most of the world this means that ways must be found to increase agricultural production, to introduce and develop industries, to promote trade, to broaden horizons through wider education and promotion of literacy, to improve health facilities and individual welfare services, and to insure political stability and leadership. In short, through balanced modernization of all the world's backward areas a solution to problems of overpopulation may be reached.

SUGGESTED READINGS

NOTE: In avoiding the burden of extensive footnotes, the author is aware that he neglects to indicate the full measure of his dependence on many sources. He therefore takes this opportunity to express his gratitude to the authors of the publications listed below, and to Dr. S. W. Boggs, Special Advisor on Geography, Department of State, for the original from which the map for this paper was taken. Asterisks indicate sources of primary importance for this chapter.

*Carr-Saunders, A. M. World Population. Oxford, 1936.

Chandrasekhar, S. India's Population, Fact and Policy. New York, 1946.

Charles, E. "Post-War Demographic Problems in Britain." *American Sociological Review*, XI (October, 1946), 578–590.

Chen, Ta. Population in Modern China. Chicago, 1946.

Childe, V. G. Man Makes Himself. New York, 1939.

Condliffe, J. B. The Economic Pattern of World Population. Washington, 1943.

Davie, M. R. World Immigration. New York, 1936.

*Davis, K. "Future Migration into Latin America," *Milbank Memorial Fund Quarterly*, XXV (January, 1947), 44–62.

*——— "Human Fertility in India," *American Journal of Sociology*, LII (November, 1946), 243–254.

*——— ed., World Population in Transition (*The Annals of the American Academy of Political and Social Science*, Vol. 227. Philadelphia, 1945.

*Dublin, L. I. "The Trend of the Birth Rate Yesterday, Today, and Tomorrow." *Bulletin of the New York Academy of Medicine*, XIX (August, 1943), 563–572.

Fawcett, C. B. "The Numbers and Distribution of Mankind." *Scientific Monthly*, LXIV (May, 1947), 389–396.

Food and Agriculture Organization of the United Nations, *Report of the FAO Mission for Greece*. Washington, 1947.

*Forsyth, W. D. The Myth of Open Spaces. Melbourne, Australia, 1942.

Glass, D. V. Population Policies and Movements in Europe. Oxford, 1940.

Jaffe, A. J. "Urbanization and Fertility." *American Journal of Sociology*, XLVIII (July, 1942), 48–60.

Jefferson, M. "Distribution of the World's City Folk." *Geographical Review*, XXXI (July, 1931), 446–465.

Kirk, D. "European Migrations: Prewar Trends and Future Prospects." *Milbank Memorial Fund Quarterly*, XXV (April, 1947), 128–152.

*Knibbs, G. H. The Shadow of the World's Future. London, 1928.

Kuczynski, R. R. Colonial Population. London, 1937.

*—— "The International Decline in Fertility." In L. Hogben, ed., *Political Arithmetic*. London, 1938.

*—— "Population: History and Statistics." *Encyclopaedia of the Social Sciences*, XII (New York, 1934), 240–248.

*—— Populations Movements. London, 1936.

Lambert, J., and L. A. Costa Pinto. Problèmes démographiques contemporains. Rio de Janeiro, 1944.

*Lorimer, F. The Population of the Soviet Union. Geneva, 1946.

Meyerhoff, H. A. "The Present State of World Resources." In R. Linton, ed., *The Science of Man in the World Crisis* (New York, 1945), pp. 222–257.

*Milbank Memorial Fund. Demographic Studies of Selected Areas of Rapid Growth. New York, 1944.

Moore, W. E. Economic Demography of Eastern and Southern Europe. Geneva, 1945.

—— "Economic Limits of International Resettlement." *American Sociological Review*, X (April, 1945), 274–281.

Mukerjee, R. Races, Lands, and Food. New York, 1946.

Mumford, L. The Culture of Cities. New York, 1938.

Notestein, F. W., *et al.* The Future Population of Europe and the Soviet Union. Geneva, 1944.

*—— "Population—the Long View." In T. W. Schultz, ed., *Food for the World* (Chicago, 1945), pp. 36–57.

*—— "Problems of Policy in Relation to Areas of Heavy Population Pressure." *Milbank Memorial Fund Quarterly*, XXII (October, 1944), 424–444.

Pearl, R., and S. A. Gould. "World Population Growth." *Human Biology*, VIII (September, 1936), 399–419.

*Population Association of America and Office of Population Research, Princeton University. *Population Index*, Vols. III *et seq.*, 1937–.

Salter, R. M. "World Soil and Fertilizer Resources in Relation to Food Needs." *Science* CV (May 23, 1947), 533–538.

Sauer, C. "Early Relation of Man to Plants." *Geographical Review*, XXXV (January, 1947), 1–25.

Sax, K. "Population Problems." In R. Linton, ed., *The Science of Man in the World Crisis* (New York, 1945), pp. 258–281.

Sigerist, H. E. Medicine and Human Welfare. New Haven, Conn., 1941.

Simmons, L. W. The Role of the Aged in Primitive Society. New Haven, Conn., 1945.

Taeuber, I. B. "Migration and the Population Potential of Monsoon Asia." *Milbank Memorial Fund Quarterly*, XXV (January, 1947), 21–43.

Thompson, W. S. "The Impact of Science on Population Growth." *The Annals of the American Academy of Political and Social Science*, CCXLIX (January, 1947), 111–118.

*—— Plenty of People. Lancaster, Pa., 1944.

*—— Population and Peace in the Pacific. Chicago, 1946.

*—— Population Problems. 3d ed. New York, 1942.

John Gillin

MESTIZO AMERICA

ANYONE traveling southward from the United States is bound to be confused, as confused in fact as Latin Americans traveling northward. Yet acquaintanceship and a serious attempt to comprehend each other's situations usually produce a basis for common understandings. The confusion is due to the fact that vast differences in natural environment confront the traveler, while details in customs and institutions tend to divert him from his usual patterns of behavior. The mutual understandings arise from the fact that the two regions actually practice varieties of Western civilization that are divergent, to be sure, but with certain similarities. The United States citizen traveling through Mestizo America will see a variety of scenery which both in grandeur and in squalor outranges anything he can find in his own country. He will find (provided his travels are sufficiently extensive) jungles, semiarid wastes, high intermountain plateaus, volcanoes in profusion, spectacular mountain scenery, and deserts which he probably never dreamed were a part of America. He will come in contact with individuals of the highest "culture," educated in Europe and acquainted with the more esoteric refinements of European civilization; he will see Indians costumed in anything from loin cloths to quaint versions of seventeenth century lackey uniforms; and he will meet and rub elbows with a great majority of persons garbed in European dress and familiar with many of the patterns which he himself practices. Yet, despite the similarities in custom and dress, there is a difference.

It is a truism that Latin America is a "land of contrasts." The fact that it is so has been much exploited by the writers of travel folders and popular books. Perhaps less considered is the fact that this area is united in a certain pattern of custom, which we may call the general Latin American culture, and that through the patterns of this culture it is almost ready to assume a full, independent role in

world affairs. Much of the confusion of the visitor rests in the fact
that he makes no attempt to understand this organization of life
as a culture in its own right, and that he fails to see the mixed-bloods
or mestizos who predominate in the area as members of a society
ready to play its part on an equal and reciprocal basis with other
societies in the modern world. We shall endeavor to show that the
Latin American culture is, or is on the point of being, a vigorous
expression of the aspirations and way of life of the mestizo race. Our
area cannot be subsumed under facile rubrics, but the potential
virility of its role, the great potentialities of its human and natural
resources all have something in common. If the modern world
chooses to ignore Mestizo America and its future, the passage of
time will bring regret for such a decision. The mestizo countries
are entering the modern world, and let no one ignore this fact.

Latin America is generally considered to comprise all of the politically independent territory of the Western Hemisphere outside of
Canada and the United States, that was originally colonized either
by Spaniards or Portuguese. It is a land of the past for some, for
others a land of the future, and for most a land of mystery. Part
of the mystery and difficulty of comprehension lies in the fact that
the territory we call Latin America is homogeneous neither in natural nor in cultural characteristics. It is difficult to understand in
simple and general terms. Of the twenty nations of Latin America,
three are practically 100 percent white in racial type—Argentina,
Costa Rica, and Uruguay. One is practically 100 percent Negroid
—Haiti. In eighteen countries Spanish is the official language and
in one, Haiti, French or a patois. But the largest country, Brazil,
which has about half the land area of the whole and over one-third
the population, is Portuguese in language and in tradition. Of a
total population of 134,021,440,[1] no less than 19,608,792, or 14.6
percent, are Negroes or mulattoes, and an additional 17,393,983, or
12.9 percent, are regarded as Indians. Perhaps at least 27 percent of
the population, then, is either Negroid or Indian, and of the remaining 73 percent the overwhelming majority outside the one
Negro and three white countries is undoubtedly mestizo, that is, a

[1] Statistical material on this area is neither plentiful nor of high reliability. Figures
given in this chapter are the latest and most reliable known to the author. Even when
derived from official sources they are often admittedly estimates.

mixture of Indian and European elements in varying proportions.

We are concerned here with that portion of Latin America which has sometimes been called Mestizo America. It consists of the 13 mainland republics which are predominantly mestizo, either racially or culturally. They are Mexico, Guatemala, Honduras, El Salvador, Nicaragua, Panama, Colombia, Venezuela, Ecuador, Peru, Bolivia, Chile, and Paraguay. All of them still have extant Indian minorities; all of them were originally colonized by Spain, and Spanish is the official language. The total area of Mestizo America comprises some 3,118,000 square miles, slightly larger than the United States. The population, however, is less than half that of the United States, with about 63,000,000 at latest reports (estimated). The average density is about 20.4 per square mile. The area comprises about 40.4 percent of the total reported for all of Latin America, and the population is roughly 47 percent of the total.

Our concern is the role this group of countries [2] plays, and may play, in the world of today and tomorrow. This is one of the areas of the modern world where change appears to be imminent. After centuries of relative isolation, both physical and cultural, the mestizo peoples of the New World seem now to be on the point of moving into the mainstream of world affairs. Our purpose here is to attempt to assess the possibilities which lie before them and the course which they may take. No attempt will be made to set forth an encyclopedic collection of all the "facts" about these people or their environment. Rather, the point of view will be that of cultural anthropology, with emphasis upon the mode of life, the patterns of custom, their adaptation to the natural environment and to the cultural conditions of the cosmopolitan world.

Here is a new race developing, the mestizo race. And here a new culture is coming into being, a mixed culture derived not only from the Spain of colonial times, but also from the aboriginal cultures of America, from modern North America and western Europe and other cultures of the contemporary world. Not only do the mode

---

[2] This is an arbitrary choice dictated by the following considerations: all of the 13 republics form a continuous territorial block, broken only by Costa Rica, and all still contain important groups of Indians and significant aboriginal cultural influences. If these considerations were not applied, Cuba and the Dominican Republic would also rate as mestizo countries from a cultural standpoint. The Dominican Republic is predominantly mulatto in physical type.

MESTIZO AMERICA

of life and point of view appear to be emerging as a fusion of diverse elements from other cultures, but, in the process of synthesis, local adaptations and innovations have been added, so that the total configuration of the culture seems to be developing aspects of uniqueness. The result seems to be a new way of life practiced and manifested by physical types—the mestizos—which, from the point of view of physical anthropology were quite unknown prior to about A.D. 1500.

The People

The racial stocks of the 13 republics under review are white, Indian, Negro, and mestizo. The accepted distinction between Indians and whites in most parts of the area is one of cultural symbols rather than biologically inherited characteristics. The official figures for several countries make no distinction between mestizos and whites, and those that do so are shown by outside checks to be quite unreliable. The number of racially pure whites without any admixture of Indian or other "colored" factors is certainly very small throughout the area, and, except in a few restricted circles and families, the possession of pure white ancestry is not regarded as socially significant. For practical purposes, therefore, we may consider the racial composition as divided between Indians, mestizos, and Negroids. A glance at the accompanying table will indicate that about 69 percent of the population is classed as mestizo, 4.3 percent as Negroid, and 26.7 percent as Indian. Significant numbers of Negroids are found only on the Caribbean coasts, with proportionately smaller numbers on the Pacific coasts of Ecuador, Colombia, and Peru. In no country do they constitute an important numerical element of the population and, since they occupy low coastal areas not heavily populated by mestizos or Indians, their relations with other groups have not traditionally created a "racial" problem. Except in Venezuela and Panama no strong prejudice against intermarriage or interbreeding with Negroids exists, and it appears that the Negroid element will eventually be absorbed into the mestizo population.

The Indians, on the other hand, form a much more significant element of the population in several of the countries. We may be sure

Racial Distributions in Mestizo America [a]

Country	Negroid	Percent	Indian	Percent	Mestizo-White	Percent
Bolivia	0	0	1,807,820	54	1,725,280	46
Chile	0	0	261,870	5	4,537,432	95
Colombia	1,956,960	30	952,320	10	6,613,720	40
Ecuador	155,275	5	1,273,255	41	1,677,011	54
Guatemala	0	0	1,567,885	55	1,882,867	45
Honduras	25,394	2.2	103,887	9	1,025,107	88.8
Mexico	0	0	6,134,457	29	15,018,864	71
Nicaragua	69,000	5	0	0	1,311,000	95
Panama	119,000	19	55,844	9	456,789	72
Paraguay	0	0	0	0	1,040,420	100
Peru	72,716	1	3,334,936	46	3,864,002	53
El Salvador	0	0	372,936	20	1,490,374	80
Venezuela	355,617	9	395,138	10	3,200,626	19
Total	2,753,962	4.3	16,262,148	26.7	43,843,492	69

[a] Percentages are given in round numbers, and when the incidence of a racial group is less than 1 percent it is tabulated as 0.

Sources: Pan American Yearbook, New York, 1946; *World Almanac*, 1946; *Handbook of Latin American Population Data*, Washington, 1945; George Soule, David Efron, and Norman T. Ness, *Latin America in the Future World*, New York, 1945; latest census publications, statistical abstracts, and statistical bulletins of the various countries, where available.

that the percentages reported are in few cases exaggerated as it has been the policy of many Latin American governments to "play down" the Indian element in official reports and estimates. However, in two countries, Bolivia and Guatemala, the Indians are admitted to be in the majority, and in three other countries, Ecuador, Peru, and Mexico, they are counted as a sizable element, ranging from 29 percent of the population in Mexico to 46 percent in Peru. These five republics are ordinarily regarded as the "Indian countries" of Mestizo America and they occupy the two areas where the highest cultural development and largest populations existed previous to the Spanish conquest. In the other areas, mestizoization has proceeded further. Yet the racial complexion varies from one area to another. Thus Paraguay is rated as practically 100 percent mestizo, but it is generally conceded that the white element in the prevailing intermixture is comparatively small. In Chile,[3] on the

[3] Some Chileans like to assert that their country is predominantly "white." It is probably true that a higher proportion of families of pure European stock live in

other hand, the proportion of "white" genes in the mestizo population is probably much higher, while the average mestizo of Colombia would probably stand midway between these two extremes.

In the so-called "Indian countries" something resembling what
sociologists call a "race problem" exists, although it is distinct in
many respects from the "race problem" of the United States. Where
the Indian element is numerically important it is socially distinguished from the mestizo element, assigned to a lower social status,
denied full participation in social and political activities, and in
other ways accorded the short end of an unequal distribution of
privileges as defined by the dominant culture. Since many of the
mestizos are, however, scarcely distinguishable from Indians in
point of inherited physical traits, how is this social differentiation
maintained? It is in the definition of the relations between the two
groups that the "race" situation in Mestizo America differs from
that in North America.

Although the two groups are often spoken of as races (*razas*) by
the mestizos, the basic cultural definition does not attach much significance to physical features of racial differentiation. Throughout
the area, Indians are distinguished from "whites" or mestizos by
cultural symbols. The most generally recognized of such symbols
for Indians are: living in and identification with a tribally or com-
munally organized native group; habitually going barefoot or wearing sandals; speaking a native language only or in addition to Spanish;
and wearing a "native" costume. (Many of the "native" costumes
of the present day are modifications of Spanish colonial European patterns, but are distinctive from the modern patterns affected by the mestizos.) Cultural difference is also recognized verbally by referring to the Indians as *gentiles* (heathen) if they have
not been converted, as *salvajes* (savages), as *indigenes* or *aborigenes*,
as *gente inilustrado* (unenlightened people), and so on. Since an
Indian's status is defined in cultural and social terms rather than
in terms of physical feature, it follows that in theory at least an
Indian, by discarding his status symbols and acquiring those of

Chile than in most other countries of Mestizo America and that the mestizos themselves tend toward the "white" end of the continuum of mixtures. However, it is
believed that the majority of the population contains some mixture of genes. For
this and cultural reasons Chile has been included in Mestizo America in the present
treatment.

mestizo status, may change his position in society. This is often difficult to accomplish in one generation, but children of Indian parents
do not experience too much difficulty in making the switch, particularly if allowed the opportunity to acquire a Spanish-language education. Furthermore, interbreeding between mestizos and Indians is
fairly common even though formal intermarriage is less frequently
practiced; the offspring are often reared as mestizos. Thus there
appears to be a continual drainage from the Indian into the mestizo
division and all indications are that, barring a change in present attitudes, the Indians will eventually disappear everywhere as a recognized element, just as in several countries they have already been
effectively merged into the general mestizo population.

As an illustration of the "race" attitudes of the current culture,
there are in the highland region of Peru a few groups of relatively
pure Spanish people who, during the troubles preceding Independence, took the revolutionary side, and were forced to flee into
the mountains. For a generation or more they were denied schools
and contact with Europeanized colonial centers and were forced
to take over the costume and many of the customs of their Indian
neighbors, although they did not interbreed with them. At present
they wear Indian costume and speak Quechua and, although they
are much "whiter" in physical appearance than the *cholos* (mestizos) of the highlands, the latter call them "white Indians" and accord them much the same social treatment as that meted out to other
Indians.

In many parts of Mestizo America, Indians and mestizos are not
physically segregated in the rigid manner prevalent among the
Negroes and whites in the United States. The recognized Indians
in many towns live as neighbors of their mestizo fellows and are
not restricted in residence. Tabus on eating in the same restaurants,
trading at the same shops and markets, or riding in the same public
conveyances side by side are in many cases nonexistent. Jim Crow
is not a mestizo pattern. Almost everywhere, however, Indians
practice cultural patterns distinctive from those of the mestizos.
In fact, the Indians usually follow a culture of their own with a
different system of values and integration. Such cultures existing
parallel with that of the mestizos have been called "Republican Native" or "Recent Indian" cultures. Although they have absorbed

some European elements, such as iron tools, the outer forms of Roman Catholicism, factory-made cloth, and the like, their content and orientation are predominantly indigenous. Thus in contrast to the United States, where Negroes and orientals are physical segregated but practice essentially the same culture as the whites, in Metizo America the physical segregation of Indians and mestizos is much less marked, but what might be called cultural segregation is important. This has three effects worth mentioning.

First, since the Indians pursue goals many of which are distinct from those of mestizo culture, competition between the two groups is somewhat mitigated, and frustration in the Indian group has less basis for development than among the "inferior" groups in North America. Second, because of the frequent interaction and contacts between Indians and mestizos and because of the fair frequency of "passing" into the upper status, the mestizo culture absorbs a good many Indian elements, at least in modified form. The mestizo culture is, therefore, far from being a pure European mode of life and thought.

In the third place, the fact that Indians follow their own cultural patterns means, of course, that they do not participate in many aspects of national life. From the economic point of view, for example, they are regarded in most regions as "dead" because their consumption of the products of modern commerce and industry is relatively insignificant; under the colonial economic system their function was the bearing of burdens and the performance of manual labor on the farms and in the mines, and their rewards as a matter of policy were kept barely above the subsistence level. In a modern economy this of course runs counter to the need for an internal market. So long as this situation persists, anyone contemplating the future economic development of our area must bear in mind that, for the present, except for a few types of articles, the consumption potential of a given region or nation is effectively reduced by the proportion of Indians, however high the over-all population figures may be. In most countries the Indians are not sufficiently educated to be literate and they do not participate in political affairs. In short, one may say that in many social and cultural ways the Indians are no more a part of the national society in which they live than are the flocks of sheep or herds of llamas.

Mexico was the first country of the area to launch an effective program of education and welfare work whereby the Indians may be brought into the circle of the national society and culture without destroying their local cultural values and social organization. Peru has recently started a similar program, and Guatemala has announced one. A common aspect of all these programs is the sending of especially trained teachers, prepared in anthropology and linguistics, who instruct the children during the first two or three years through the use of the native language; once the basic concepts and skills of literacy are established the change-over is made to Spanish in order to fit the children to participate with other citizens of the nation.

Although the "Indian problem" cannot be ignored by students of the area, the effective population, so far as participation in the modern world is concerned, is everywhere predominantly mestizo. Through education and interbreeding it is quite possible that the Indians will disappear as a distinct category in a century or so. From the genetic point of view the mestizo population is still in the process of consolidation, showing a wide variety of physical types, associated with the varying amounts of Indian and European elements in the mixtures and depending upon the different physical types of the ancestral elements. Few studies have been made of this mixed population, but no evidence has been adduced to indicate functional inferiority of any kind. Of much more importance than the details of genetic make-up and mixture seems to be the question of cultural development in the region.

The Culture

If we are to understand the actions of Mestizo Americans in the modern world after the event, and also to come closer to reliable predictions of their course in given situations before the event, we must know something of their culture. We use "culture" in the scientific sense, of course, with a small "c," to denote the over-all content and organization of the patterns of customary life, thought, and tradition.

There are three general types or "levels" of practiced culture in Mestizo American territory at the present time, only one of which has important relations with the cultures of the rest of the world. First, there are still a few isolated "primitive" tribes whose mode of life has been somewhat influenced by Europeans but who for practical purposes follow a design for living totally outside the Euro-American orbit of culture. As examples we may mention the Jivaro of eastern Ecuador, the Campa of eastern Peru, some isolated groups in the eastern jungles of Panama and certain nomadic Indian tribelets in northwestern Mexico. Such groups are at present mainly ethnological curiosities, important from the standpoint of historical and general anthropology, but of no significance in understanding modern Latin American life, except for the historical influence they may have had upon a few of its local phases.

The second general type of life has been designated by the term Republican Native or Recent Indian cultures. These modes of life are acculturated or modified versions of aboriginal cultures which have developed during the last 125 years of independence from Spain. They are the patterns which are followed by the bulk of the persons classified as Indians in the preceding section. Several million Indians practice a Republican Native culture of one kind or other in the Andean countries and in Mexico-Guatemala. Such cultures are in closer contact with world civilization than are the "primitives," but, as we have already observed, the systems of attitudes and values do not operate in the same universe of discourse, one might say, as those involved in modern civilization, and the social participation of the practitioners of the Native-type cultures is somewhat restricted. Furthermore, the evidence seems to indicate that the Indian groups of this sort are being steadily drained away into the mestizo population. Therefore, although the Republican Native or Recent Indian cultures, despite the fact that they have extensively influenced the modern forms of life of their areas, are likely, with the passage of time, to pass out of existence as distinct organizations of life and artifact.

The dominant culture of Mestizo America and the one which outsiders must understand if they are to have successful dealings with this part of the world is a variety of Western civilization which we may call the Modern Latin American culture. Sometimes it is described as Mestizo Culture; in Peru it often is called *criollismo* (creole culture); in Middle America, it appears under the name of "ladino culture"; and so on.

Unfortunately this configuration of custom and attitude has not yet been comprehensively studied from the technical point of view. Our information must come from a few intensive community studies plus observations of a more impressionistic nature. Therefore the following brief remarks are to be considered as suggestive rather than definitive.

The difficulty which many North Americans and Europeans have in understanding Latin Americans seems to lie in the fact that the Latin American culture of the mestizos exhibits many patterns of life which on first acquaintance appear to be almost exactly like those with which the outsider is already familiar; for example, literacy and printed matter; Roman Catholicism as the dominant religion; monogamic family of apparently conventional European type; money economy; republican form of government; European styles of costume for both sexes; clock measured time reckoning; European types of tools, house furnishings, and many other artifacts; the use of various European foods, beverages, meats, domestic animals, and crops. These and several other features are all but universal in the Latin American culture and in outward form, at least, they often do not look to be greatly different from analogous traits and institutions of North America and western Europe. The fact that so large a part of the content of the Latin American way of life partakes of the general range of Western civilization justifies us in considering it a variety of this great cosmopolitan pattern of living. It also provides the lines along which Mestizo America may play a role in world affairs and interact on a more or less equal basis with other parts of the world within the orbit of Western Civilization. What, then is "different" or "distinctive" in the Latin American culture which should be understood by those who wish to avoid errors in dealing with its followers?

Despite many similarities, the Latin American culture differs from that of North America, let us say, first in the emphases which are given to the traditional forms, and second, with respect to content. In many situations Mestizo Americans have a view of the world somewhat different from our own. Also their institutions con-

tain many patterns of custom and artifact which are almost unknown in the United States and western Europe. This can seemingly be understood if we take into account that the development
of the Latin American culture was fed from two types of sources
which never figured in the building of United States culture and
which had little or nothing directly to do with the formation of
modern civilization in England. On the one hand, two large and
vigorous native civilizations—the Aztec-Maya in the Mexican-
Guatemalan area and the Inca in the Andean region—flourished at
the time of the conquest. Although the Spanish conquerors succeeded in wrecking these cultures as going concerns (their remnants are to be seen in various local forms of the Republican Native
culture at present), their influence on the mode of life of the conquerors and colonists was more pervading than was the influence
of the somewhat less well-developed Indian cultures on the European settlers of the United States and Canada. Also the native contributions to culture in Mestizo America were quite different in detail from those of the tribes of North America, for the natives of
North America and of Mestizo America were not in regular contact with each other and shared little of their cultures at the time
of the Discovery. Thus it is that Mestizo America has everywhere
taken over a considerable number of elements from the cultures of
the natives, features quite unfamiliar to North Americans other than
anthropologists. These indigenous contributions, of course, vary
from one region to another. Guaraní is the second official language
of Paraguay and has contributed heavily to the colloquial Spanish
of that country; likewise the Spanish of the Andean area is sprinkled
with Quechua words and figures of speech, while both Aztec and
Maya words in considerable numbers have been taken into the
everyday Spanish of Mexico and Guatemala, and, by diffusion, most
of Central America. Native foods and drinks and methods of preparation are almost everywhere a part of the standard cuisine of the
common people; one needs only to think of the ubiquitous *tortilla*,
tamale, and bean paste of Middle America and the almost universal
chicha (maize beer) and potato flour or paste (*chuño*, made from
frozen potatoes) of the Andean region.

Another source of difference between Mestizo modes of life and
our own is that much of the European content of Latin American

culture came from Spain. It is a question as to whether or not Spain even today fully participates in the cultural orbit of western Europe. Certainly it did not do so during most of the colonial period. Many of the great movements which transformed the cultures of such peoples as the English, the French, the Germans, and the Scandinavians into "modern" civilization passed by Spain or were excluded as a matter of official policy. The Reformation, the Enlightenment, the emergence of political democracy, the mercantile movement, the Industrial and mechanical revolutions, the rise of modern science and technology, laissez-faire capitalism—none of these ever obtained a firm foothold in Spain. Furthermore, certain innovations and modernisms which did appear in Spain were forbidden export to the colonies as a matter of political or ecclesiastical policy. Thus for a period of some three hundred years the lands of Mestizo America received a narrow and restricted flow of cultural elements from Europe—and from a rather quaint and anachronistic variety of European culture, at that—feudalistic, monarchical, mystical, non-technological materialistically, medievally scholastic in intellectual pattern, monopolistic and restrictive in economic institutions, and conservative of the medieval forms in all patterns pertaining to the sexes, the family, political forms, and economic activity. Much of the "quaintness" which an outsider of Western Culture sees in Mestizo America today is explained by the persistence of some of these elements and institutions in pure or modified form.

Finally, during the 125 years since they won independence from Spain the lands of our area have been increasingly infiltrated by cultural elements, artifacts, attitudes, and influences from Western Europe and North America. This process has been accelerated for the west coast of South America since the opening of the Panama Canal and everywhere by recent technical improvements in transportation and communication, which have brought the people into more frequent and intensive contact with the non-Spanish world.

From these three sources, then—aboriginal, colonial Spanish, and modern Western Civilization—the cultures of Mestizo America (and most of the rest of Latin America, as well) have been blended and molded. And in the process, as always happens, something new has emerged. The result seems to be a new variety of Western Culture, unique in some respects. As is true to a lesser degree in the culture of the United States, this Latin American culture exhibits regional and local varieties, differing among themselves mainly because of exigencies of environmental adaptation and because of aboriginal historical colorations. It also shows subvarieties associated with the type of social situation in which it is practiced (rural, small community, or urban) and other subvarieties connected with the social category (upper or lower class) in which it is performed. The range of environmental conditions is very great and an almost equally wide range of cultural adjustments is to be found. To the casual visitor or the tourist, it is often difficult to see much similarity between the culture of, say, a mestizo settlement in the jungle and a mestizo town at an altitude of 12,000 feet on the cold intermountain plateau. But residence in the various types of mestizo communities will convince one that a strong common fabric of belief, attitude, and pattern of activity runs through all of them.

We may now attempt to mention briefly certain salient aspects of institutions and customs which one might expect to encounter throughout the mestizo area. Obviously these modes of life deserve a more thorough treatment and a more complete analysis. Perhaps some of the following statements will provoke qualified students to intensify their studies of the pattern of life in this area.

At the start we may say, perhaps, that ideologically the culture is humanistic, rather than puritanical, if such a contrast is permissible. Intellectually it is characterized by logic and dialectics rather than by empiricism and pragmatics. Great value is assigned to the manipulation of words and other symbols, emphasis is laid on form, the symbol is often more appreciated than the thing. Outsiders reared in the empirical and pragmatic phases of Western Civilization are inclined to see these traits as "impracticality," "touchiness," "fine talk that means nothing," and so on. At any rate, the manipulation of symbols (as in argument) is more cultivated than the manipulation of natural forces and objects (as in mechanics). Patterns of mysticism and fantasy are strong among the mestizos, and these patterns show no consistency with those of argumentation, for the worth of the logic or the concepts lies in the manipulation of symbols, not in the empirical investigation of premises. It is partly for this reason, I believe, that ideas from abroad find more ready acceptance on the whole than artifacts and their associated techniques.

Family organization is everywhere, regardless of class, more important socially than in modern North America. Monogamic, indissoluble marriage is the pattern and the fact (because of Catholic influence). But the extended family is more common and of greater functional importance than in the United States. Not only are immediate families larger, due to the theoretical prohibition of birth control (because of Church influence) but the inclusion of several generations and of collateral immediate families through the patriarchal principle is general. These enlarged or extended families often consist of an old man, his sons and their immediate families, their sons and children, and so on, always with the males acting as the heads or agents of the group. They function as units in political, business and social affairs. For this reason it is commonly held that a man without a family of this sort is almost helpless in Mestizo America. Although there are a few exceptions, the self-made man is a much greater rarity in this social system than among ourselves. If one will look into the background of almost any distinguished or prominent personage in Mestizo America, he will find a group of brothers, cousins, or uncles who form a sort of corporation to promote the success of their outstanding representative and, through that, their own advantages. It should not be supposed that only the wealthy and distinguished follow this pattern, for it is quite as important in the lower classes. An orphan is traditionally one of the most pitiable objects in Mestizo America, and parents usually try to make arrangements that their children will, if occasion arise, be adopted by friends or relatives.

Another feature of importance in Mestizo America, which was never so highly developed in the United States and has now almost disappeared there, is the institution of ceremonial kinship. This is managed through the mechanism of godparenthood or ceremonial sponsorship, as it is sometimes called. Everyone in mestizo circles has, in addition to his blood kinsmen, a group of ceremonial kinsmen to which he is bound by patterns of right and obligation; in most parts of the area these are taken very seriously. *Compadres* (males) and *comadres* (females) are persons of roughly the same generation who are bound to each other in a system of this sort, and their relations are of considerably more functional importance than those between godparents and godchildren. The *compadre-comadre* tie is a form of ceremonial friendship, in some cases stronger than that of blood. The basic idea of godparenthood was, of course, introduced along with Roman Catholicism, but many of the native cultures had somewhat similar institutions; this may account for the fact that the institution of ceremonial kinship in many parts of Mestizo America has proliferated and ramified to a greater extent than was ever true of Europe.

Marriage under such conditions is typically defined somewhat differently than in the United States. Romantic love is, of course, in the mestizo pattern, but marriage itself is an important matter, not only for the two partners but also for the extended families involved. Thus all of the blood relatives, and especially the head of the house (that is, of the extended family), take an active interest, and approaches to marriage typically involve formal negotiations between the two families. By careful attention to its marriage alliances an extended family may greatly increase its resources and security as well as those of its individual members.

Thus anyone who wishes to do business on more than a superficial basis with a typical mestizo must know something of his kinship affiliations. What family does he belong to? What are his and its marriage connections? Who are his ceremonial kinsmen? The ceremonial kinsmen do not function as organized groups in the society in the same way that the extended families often do, but the network of an individual's ceremonial friends is sometimes very wide indeed.

Generally speaking, women have less individual freedom than men in social relations. The pattern is that of official male dominance, and the place of married or marriageable women has traditionally been confined to the home and the circle of kinswomen and friends of the family. Friendly contact with men before marriage is supposed to take place only under strict chaperonage; after marriage it is supposed to be confined mainly to the husband, male relatives, and members of the clergy. In the larger centers of late years it has become respectable in some circles for girls to work before marriage, to get a useful education, and to go on "dates" and parties with young men without strict supervision, but in the provinces and in conservative families the old pattern still prevails. Lower class women have always enjoyed more personal freedom and more opportunity for contacts outside the home, especially in work. Partly because it has been a class mark for upper-class women not to work outside the home (and seldom in it), conservatives resist the new tendency for young women to take jobs in offices and to enter the professions.

Possibly because of the practical impossibility of divorce and the social importance of marriage as a means of uniting extended families, the double sex standard enjoys in Mestizo America a semi-approved status not accorded it in the United States. Sex in general is not surrounded by a puritanical aura. Except in very polite society it is often discussed more freely in mixed company than among ourselves. A man, even of high social position, does not lose his standing by visiting a brothel or keeping a mistress, provided he does not neglect his obligations to his family. Here the question of form is important: a man never introduces a prostitute or a mistress to his female relatives (wife included), although they may be well aware of his relationship to her. On the lower social levels many men and women live together on a permanent basis without benefit of clergy; all countries permit legal recognition of the offspring of such unions and most countries make statutory provision for inheritance by common-law spouses, although the effective binding force in such alliances is custom. Such common-law unions usually involve the customary arrangements between the respective extended families.

Consistent with the structure and function of these familial institutions, it is not customary to introduce outsiders into the family circle as guests until they are very well known and highly trusted. The pattern of the man bringing a male friend or acquaintance home to dinner in the bosom of the family, so much practiced in the United States, is almost unknown in Mestizo America. Likewise, married couples not united by kinship bonds of some sort do not entertain each other much in the homes. Married women meet each other in female groups; whereas men carry on their social relations in special institutions, such as bars, clubs, and the like. Social intercourse between men does not imply that their wives are brought into the picture at all.

We shall now proceed to consider other features of the culture, first pausing to discuss certain aspects of the natural environment.

Natural Resources

From the beginning of the Spanish colonization the area which we call Mestizo America was regarded as a source of quick riches either for the mother country or for a small group in local control with affiliations abroad. The earthly welfare of the native or mixed population as a whole was seldom considered, although some attention was paid to "spiritual" well-being. In short, this part of the New World was regarded as a source of wealth to be enjoyed by the few given the privilege of exploiting the area. The approach was that of a true "colonial" economy whose principal reason for existence was to enrich a controlling group abroad and a small upper class locally. The development of a balanced economy and a higher standard of living among the bulk of the population were not considered to be functions of the system. Unfortunately, a large part of the area has continued essentially in the condition of a colonial economy to this day, although time and political changes have wrought some alteration in details. Enlightened local leaders are, however, almost everywhere aware of the need of change in the prevailing system, and change of one sort or another seems to be ultimately inevitable. Before examining these problems we must first sketch briefly the outstanding features of the natural environment.

There has been a recent tendency, even in responsible circles, to look upon Mestizo America as a fabulous land of opportunity, a last frontier where great riches await only the energy and organizing ability of the pioneer. It is true that the opportunities for creating wealth are considerable in many parts of the area, but it is a mistake to consider the present or future economic opportunities of these lands as essentially similar to those of the United States when settlement of our own country began. A soberer view takes into account that most of Mestizo America was occupied and exploited by natives for many hundreds of years before the arrival of the Spaniards in the first part of the sixteenth century, and that persons of European extraction have now controlled the exploitation of its resources for more than four centuries. It is a far older land in terms of large-scale application of human energies to natural resources than is most of the United States.

In a general way we must consider, first, the resources and opportunities offered by nature, and second, the patterns of human activity which have been applied to them. Because our eyes are fixed upon the future as well as the present and past, we shall also venture to assess the interplay of these two factors as a basis for discussing possible or probable changes.

Taken as a whole, neither the landscape nor the climates of this area have much in common with those familiar to the majority of residents of the United States. Northern Mexico is, of course, in large part an extension of the semiarid American Southwest. The central and northern portions of the Central Valley of Chile have a "Mediterranean" climate somewhat like that of central California, and south central Chile has something in common with parts of the woodland area of Eastern United States. Otherwise, because of differences in latitude, altitude, mountain masses, and ocean currents there is little of landscape and climate in Mestizo America closely similar to the United States. Geographers recognize no less than 16 different types of climate in the area and practically every national unit in Mestizo America is a mosaic of natural regions, often differing markedly among themselves.

The land surface of a good portion of the area, one might say, is draped across the backbone of the continent. This is true of all the countries except Paraguay, which has no significant mountains. Technically the figure does not apply to Chile, whose eastern boundary does not cross the continental divide. However, everywhere except in Paraguay great, and often abrupt, differences in altitude exist within each national territory, so that most countries have a series of "vertical" climatic zones. With the exception of Bolivia and Paraguay all have seacoasts. Except for Chile all have comparatively large areas of tropical jungle. Paraguay is the only one which has no intermountain plateaus or highlands.

Mountains are almost everywhere a part of the landscape, except in Paraguay, although they are low in Panama. On the whole they are young mountains, little eroded, which rise relatively rapidly from the low country to altitudes which are unknown in the United

States. In Mexico, Central America, and the west coast of South America there are many active volcanoes, and almost everywhere earthquakes are frequent. In general, then, the terrain tends to be rugged, which renders travel and communication difficult. But one cannot travel far in most parts of the area without passing from one type of natural region to another. Hence the advice to foreign visitors: prepare yourselves for what you least expect.

The variety and diversity of natural conditions in Mestizo America have been reflected in a number of ways in the organization of human life. First, the number of local cultural adaptations has been, and to some extent probably always will of necessity be, larger than in the United States. Second, local peculiarities of custom and point of view have tended to persist because of the difficulties of travel and communication. Third, regional economic specialties and interests are often accentuated. Fourth, political unity has been difficult to attain and to maintain, a fact which has made it easy for individuals or small cliques to stage revolutions and to seize power. The traditional rivalries between the sierra and the coast in both Ecuador and Peru, between Bogotá and Antioquia in Colombia, between Yucatán and Mexico City are cases in point.

When one searches this land for resources useful to man, the picture is again complex. Mestizo America is predominantly agricultural, but it is hardly one of the most naturally favored agricultural areas of the world. It is doubtful that more than about 10 percent of the surface could be usefully cultivated by any known methods. Large areas are nonarable for lack of water, for example, most of northern Mexico and the west coast of South America from latitudes 1° to 25° South. Some of this now unused land may be brought under cultivation by irrigation, but if this is done on an appreciable scale a revision of economic motivations and politico-economic organization will be required. Considerable portions of the North Peruvian coast, now barren, were irrigated in antiquity by the Incas and their predecessors. Other significant portions of the area are covered with tropical forest, a type of land which up to now has proven to be productive, even under plantation management, only by a shifting type of cultivation requiring a high percentage of reserve acreage. Other large areas are mountainsides and hillsides, either without soil or considered too steep to cultivate, and high intermountain basins too cold to grow anything but coarse pasture. There are no broad stretches of open prairies or other level or rolling farmland, such as occur in Argentina, the United States, and Canada. The result is that the farming areas are "patchy." Farming is mainly done in river valleys, suitable intermountain basins and irrigated oases. Under modern culture it has been uneconomical to build or maintain terraced mountainside fields of the kind whereby the Incas enlarged their cultivated acres in Peru. Finally, the land which is arable does not, of course, offer uniformly good soil. Within Mestizo America the volcanic soils are generally the most productive and permanent. Where these do not exist, however, leaching, erosion, lack of fertilizer, and natural infertility cut down the yields.

In spite of the fact that the area cannot be regarded as one of the most naturally favored agricultural regions of the world, experts generally agree that the available land is quite capable of providing a respectable standard of subsistence needs for the present population and even for a considerably larger one, if the economic organization and management of agriculture were properly adjusted to the requirements of the population as a whole. As we shall see later, the land hunger of which we hear so much in Latin America does not seem to be a matter of actual physical lack of adequate soil resources, but rather a condition resulting from defects in the traditional system of land tenure, labor, and management.

Forest products constitute a potential natural resource which is by no means negligible in most countries. The forests of Chile offer some lumber of "temperate zone" types. The tropical jungles of other countries, although not to date amenable to clearing for cultivation on a large or permanent scale, offer natural products useful to man. Stimulus to the exploitation of these items was given by the shortages of World War II. Ecuador, Peru, and Bolivia offer cinchona bark (for quinine) from their Amazonian territories. Almost all tropical forests of the area produce rubber, both of the Hevea and Castilla varieties, mahogany and other woods, and rotenone-producing plants used in insecticides. Chicle (for chewing gum) is produced in large quantities in Yucatán and the Petén. Paraguay produces quebracho (used for tanning). Tagua nuts, ipecac, Brazil nuts, wild animal skins, and various medicinal plants are also collected in parts of the tropical forest areas. Experiments are now under way in some countries looking toward colonization of the tropical forest regions on a basis of subsistence culture combined with the gathering of forest products. One such experiment is at Tingo María, Peru.

All of the countries of this group have coast lines, except Bolivia and Paraguay. According to investigations made during the war the fish supply along most of the coasts is abundant, but on the whole the fishing industry has not been highly developed to date, either for local consumption or for export.

Natural sources of power are not as evenly or conveniently distributed throughout the area as could be desired. The most ubiquitous potential source is water power. It is estimated that about one-eighth of all potential water power on the globe exists within this area. At present only a fraction of one percent is being utilized. Petroleum sufficient to supply all fuel needs for some time to come is already under exploitation in Mexico, Venezuela, Colombia, Ecuador, and Peru, and apparently large untapped reserves of oil lie in the rock structures all along the eastern foot of the Andes. The adequate exploitation of petroleum deposits is, however, a costly business requiring financing either by government or by private capital. Up to the present all countries of the area, with the exception of Mexico and Bolivia, have been content to turn the exploitation of oil resources over to private concerns financed and controlled from abroad. The major part of the petroleum itself has disappeared into foreign markets and the profits into foreign pockets, mostly in the United States and Britain. Coal, the great power resource of the Industrial Revolution, is, generally speaking, of poor quality and inconveniently placed. Before 1940, Latin America as a whole accounted for only one five-hundredth of world coal production. The only countries in the area known to have significant deposits of coal are Chile, Peru, Colombia, Venezuela, and Mexico. It is said that only Mexico and Peru have coking coal, although new discoveries in Colombia may prove to be another source. The largest coal deposits in South America exist in Colombia, with estimated reserves of 18 billion tons. Because of difficulties of transportation and lack of market these have been little exploited. At present the coal production of Mestizo America is insufficient for local needs, and almost all countries import this fuel. Partly because of the scarcity of readily available sources of fuel, most of the mountain and upland forests of Mestizo America have been stripped to provide wood and charcoal, with resulting exposure of the soils to erosion. No deposits of uranium have been reported in the area. In summary, despite the lack of plentiful and convenient supplies of coal, sources of energy in petroleum and hydroelectric power appear to be ample for the development of large-scale industrialization, if intelligently used.

The natural resources which have figured most prominently on the world market are minerals. The young mountains of the area are relatively rich in precious metals and those used in modern industry. Among the outstanding mineral resources mined at present are: copper (Chile, Mexico, Peru, Bolivia), tin (about one-fourth of the world's production comes from Bolivia), vanadium (one of the two richest deposits of the world is in Peru, and a smaller deposit in Mexico), lead, silver (Mexico is a large producer), gold, platinum (Colombia is the second largest producer after Russia), sulfur (in all volcanic regions), borax (Chile, Peru), bismuth (one-fifth of the world's supply is produced in Peru), sodium nitrate (Northern Chile), bird manure or guano for fertilizer (coastal islands of Peru), emeralds (Colombia is the primary world source), iron (Chile, Peru, Mexico, Venezuela, Colombia), tungsten (Bolivia, Peru), antimony (Andean area), and other minerals of lesser importance. Coal and petroleum have already been mentioned.

During the war, as part of its quest for strategic materials, the United States government stimulated prospecting and development throughout most of the area, with the result that new deposits were in some cases discovered and certain minerals not previously exploited may be commercially mined in the future.

Two points must be considered in connection with the mineral riches of Mestizo America. First, as everyone knows, these sources are irreplaceable, so that excessive or wasteful exploitation may bring to a premature end any economy dependent upon them. Second, unless matters are so organized that a considerable share of the products or proceeds of mining are distributed among the local population, mining activity has no significantly beneficial effect upon the area where it takes place.

In summary it appears that ample mineral resources exist physi-

cally in our area both for a thriving local industrial development
and for export, if efficiently managed.

Land and Agricultural Problems

First of all must be considered the question of the management
of the land. The great majority of persons in Mestizo America depend for their livelihood upon working the soil. In general two types
of agriculture prevail with two somewhat opposing objectives: cultivation for subsistence and cultivation for export. There is, of
course, no theoretical reason why these two types of agricultural
activity could not be carried on simultaneously, provided they were
in balance. It is a curious fact, however, that in many parts of the
area, predominantly agricultural, foodstuffs are imported and the
nutritional level is below minimum standards. One of the principal
programs carried on by the United States in cooperation with local
governments during the war was the stimulation of local food production. The immediate purpose was to ease burdens on shipping
needed for the war effort, but it is hoped that the program will have
some permanent effects.

Subsistence agriculture in most regions centers about the aboriginal Indian crops of corn and beans. In higher altitudes such native
crops as potatoes, olluca, and oca also figure prominently, while in
the lowland tropics native manioc is the staple. Among grains introduced since the conquest, wheat is widely used for bread in
some regions, although most of the barley is reported to be converted into beer. Rice is a staple article of diet in many parts of the
area, but in "normal" times considerable quantities have to be imported from the Far East.

Although the Indians had developed a wide variety of maizes, as
well as other domestic food plants, yields per acre in Mestizo America are reported to be low in comparison with the United States.
This seems to be mainly due to the antiquated methods of cultivation and management. A sentimental attachment to the land is a
fundamental cultural pattern among the majority of agricultural
workers in Mexico, Central America, and the Andean area. In many
places a man is lacking in social prestige among his fellows unless
he has at least a small plot which he cultivates. Even workers in the

mines and on the plantations strive to retain small farms, owned or rented by them, which they can cultivate in their own interest. Since the best land has usually been preempted by haciendas, these small cultivators are often relegated to hillside or infertile sites.

Most of the raising of subsistence crops is carried on by methods originated by the Indians, somewhat "improved" by Spanish modifications of the time of the conquest, mainly in the form of iron tools. Although the ox and wooden plow were introduced by the Spanish, they are not universally used, even to the present day, in the production of locally consumed subsistence crops. Even on large haciendas devoted to food production, most of the cultivation is performed by tenant farmers or peones using hand methods, and tools like hoes, foot plows, spades, and digging sticks. Modern farm machinery has been confined mostly to export farming. Recently, certain governments (for example, those of Mexico and Peru), have assisted cooperative groups of small farmers to purchase machinery. The comparative scarcity of animals in hand agriculture eliminates the use of barnyard manure, which in turn reduces the productive efficiency of the soil. The standard method of restoring fertility is to allow the land to lie fallow for several years. It has been estimated that from one-half to one-third of the subsistence land is thus out of production continuously. Commercial fertilizers are used to any extent only on haciendas producing for export. Partly because the small farmer has few animals, rotation with nitrogen-restoring crops like clover and alfalfa, has not been widely practiced. Finally, to mention no further factor, the tendency to force much food farming onto less desirable hillsides, together with the neglect of erosion control, results in low average yields and damage to the soil.

The second type of agriculture, and the one which figures in world trade, is export farming, carried on mainly in large haciendas or plantations, each one devoted primarily to a single crop. In terms of dollar value of exports, the most important products are "dessert crops": coffee, sugar, bananas, cacao. Other important export crops are cotton, sisal, and sheeps' wool. Although production on haciendas is technically more efficient than individual farming, the plantation system has several disadvantages from the point of view of local economy. The tendency to concentrate on a single crop, as has been mentioned, tends to slight the development of a well-rounded subsistence agriculture. Furthermore, it makes the prosperity of the region dependent upon foreign demand and price, which show fluctuations of an often unpredictable character. Witness the gyrations of the coffee market during the 1930s and their unsettling effects upon Brazil, Colombia, and Guatemala. Furthermore the economy of a region dependent upon a single crop is at the mercy of disease, blight, and exhaustion of the soil. Because of the encroachment of banana diseases many once thriving plantation districts of the Caribbean coasts of Central America have had to be abandoned, leaving the former field workers stranded. A blight which attacked the cacao plants during the 1920s and 1930s seriously upset the economy of Ecuador.

The system of land tenure and the organization of the agricultural working force are, of course, inextricably bound up with the exploitation of the soil. The fundamental feature of the Latin American system which has stood in the way of adjustment between the population and its soil resources is the institution of *latifundio*, or land monopolization in the hands of a few individuals, families or corporations. The result has been land hunger in the bulk of the population and many of the symptoms of population pressure in a region where the land physically available for the needs of the people is more than sufficient. Land monopolization has removed large areas from general cultivation to the production of export crops, which may prove profitable to the monopolists but contribute little to the economic well-being of the area. Other large areas have been removed from cultivation for the pasturage of grazing animals, which under a more efficient over-all system would be pastured on hill and slope sites. *Latifundio* also stimulates the buying up of land and withholding it from any economic use in the hope of profiting from a speculative rise in price. Finally, the concentration of ownership of large areas in monopoly enables the owners or managers to dictate the rewards which agricultural laborers may obtain from their work. In most cases these rewards have been confined to a bare subsistence for the worker and his family. His incentive is stifled and he is bound to the land in many regions by a system of peonage or debt slavery.

The historical background of this situation is fairly clear. Spain at the time of the conquest preserved a feudal system of land tenure. It was within the contemporary tradition that basically the same system with some modification in the form of grants of land and Indians (*repartimientos* and *encomiendas*) to faithful servants of the Crown, should be transferred to the New World. It was one of the class marks of a nobleman and a gentleman that he should have landed estates worked by a great mass of agricultural nobodies. A Spaniard of prestige could lend his muscles to the bloody work of war, but not to the menial drudgery of peace. Although many of those who took part in the conquest were menials and even jailbirds in Spain, it is not surprising, when the opportunity presented itself in the colonies, that they should set themselves up as members of a noble class according to the standards of the mother country; which standards, it may be said, have persisted as phases of the mental patterns of the Latin American culture until recent times. Even today when a man accumulates wealth his first thought, in most cases, is to invest it in land. It is for this reason that it has been so difficult to finance industrial enterprises and government securities from local capital in Latin America, a circumstance of which foreign capital has not been slow to take advantage.

In the valley of Mexico and in the highlands of Peru the lot of the newly landed colonial gentlemen was made the easier by the presence of large populations of disciplined Indian agricultural workers who had been accustomed to labor under supervision on state-controlled communal lands (in Peru) or on estates of native nobles (in Mexico). However, neither of these native systems involved the profit motive, debt bondage, or racial-cultural discrimination, all of which with sundry other unpleasant details the new landlords were not laggard in introducing. As Valcárcel and others have shown, group work in the fields under the Incas was carried on in festive mood and for the tangible welfare of the local group and the national society. The Spanish masters shortly reduced the blithe spirits of their peones to sullen compliance in tedium.

Another Spanish institution which promoted land monopoly and land hunger was the Roman Catholic Church. Not only were large original grants made to ecclesiastical organizations for the support of their missionizing activity, but these holdings steadily increased during colonial and also during modern times through contributions made by the faithful. Since land was the principal form of wealth, it was—and in many places still is—customary for the well-to-do to donate land to the Church or to one of its subsidiary organizations in their wills, the proceeds to be used for masses for the repose of the donor's soul. It has been estimated that at one period more than half the total property in each country belonged to ecclesiastical organizations or to the individual clergy. Certainly the Church, its dependencies and officers collectively, became the greatest landowner in Mestizo America. The evidence does not indicate that, except in rare instances, these churchly estates were operated in a manner essentially different from those of private landlords. Since the Church is not an agricultural institution and its representatives were not usually agricultural experts, many large properties were either inefficiently managed or even left undeveloped. The entrance of the Church into a field fundamentally foreign to its function, and its position symbolically, at least, on the side of the land monopolists in the face of abject land hunger among the great masses of the people must be understood by those who profess themselves perplexed by such manifestations of anticlericalism as the "attacks" on the Church emanating from the Mexican revolution of 1911. In most countries of Mestizo America the Church has now been forced to liquidate at least part of its land holdings, a political policy which, it must be said, has been assisted in many areas by a more enlightened attitude on the part of the ecclesiastical authorities themselves.

The Revolution from Spanish political control did not materially mitigate the land problem. In fact, for a century or so after "liberation," monopolization of the land continued to increase. It must be remembered that the Revolution itself and its aftermath were controlled in the event by the creole upper classes, despite the adherence of many participating elements to the principles of the French Revolution, and despite the promises made to Indians and lower class mestizos. Not only did the old colonial estates on the whole remain intact (although in some cases changing hands), but they continued to grow. About 1880 a new element entered the picture, namely, foreign capitalism. The older pattern of family holdings of large estates was changing to one of foreign corporations, financed and controlled in New York, London, Berlin. Almost everywhere in Mestizo America republican governments were induced to make grants of large areas of land to foreign concerns. The usual stipulation was that the foreign corporation would "colonize" the land or "develop" it, and the excuse was that local enterprise and capital were inadequate to the task. In some cases foreign capital did develop areas which had not previously been exploited to any significant extent agriculturally. This was particularly true of the banana plantations of the low, wet tropical regions of Central America. However, the advent of corporation cultivation had profound effects upon the mode of life of the people affected by it. Fruit, sugar, and cotton, for example, when produced for export are most profitably exploited on large integrated estates. Efficient production involves a heavy outlay for irrigation, fertilization, field machinery, and processing or handling plants. Labor is best managed, from the point of view of profits, by concentrating it in barracks or compact settlements under disciplined supervision. Foodstuffs and other necessities are imported rather than produced on the place; consumer commodities are provided through centrally managed company stores or commissaries. Under this system the traditional local organization of the laborers and their families is broken up, the former villages and towns are consolidated into workers' settlements under control of the company, and individual initiative and chances for individual economic progress are practically eliminated. Political freedom is, of course, almost entirely absent.

One of the best examples of this sort of organization is to be found in the large sugar and cotton estates of the north coast of Peru where the arable land, limited by the extent of irrigated river valleys, is divided between a few large owners. One corporation, originally German, is said to control the lives of 50,000 persons living within the boundaries of its estate, which stretches from the Pacific Ocean across the Andes to the Marañon river in the Amazonian drainage. The monopoly of large areas of land naturally places the hacienda in the position of being able to keep wages low, an essential feature of the system. In Mestizo America few of the plantation corporations producing for export, with the exception of fruit plantations which have no serious competition, can compete with the better soils, shorter distance to market, and greater technical efficiency of other areas, except on the basis of low labor cost. These enterprises thus truly become "factories in the field," and the workers undergo a form of rural proletarization almost unknown in the United States and western Europe.

A few figures illustrating the inequitable distribution of land may be mentioned. According to 1937 figures for Chile, 74.5 percent of the properties (ranging up to 123.5 acres apiece) comprised only 4.7 percent of the land. But 1.4 percent of all properties (ranging from 2,470 to 12,350 acres and more) included 68.2 percent of the land. Even more striking is the fact that 0.3 percent of the establishments (all of 12,350 acres or more) represented *more than one half* (52.4 percent) of all farm land. Extreme conditions are illustrated by the municipio of Calle Larga, where 31,500 out of a total of 37,664 hectares (one hectare equals 2.47 acres) were controlled by a single hacienda, and in Los Andes where, out of 70,403 hectares of farm land, 70,000 were controlled by a single *latifundio*. The large pastoral estates (mainly sheep grazing) in southern Chile are mostly controlled by foreign capital. One of them in Tierra del Fuego owns approximately five million acres of land; another, 1.7 million acres; still another, 1.25 million acres.

A recent study of 44,519 representative rural establishments in Venezuela showed that 56 percent of the land was owned by only one percent of the proprietors, while only 6 percent of the land was divided among 66 percent of the owners. In Mexico at the beginning of the modern revolution (1911) one percent of the people owned 70 percent of the arable land. Three related families owned all of the farm land in the state of Hidalgo and one family owned 12 million acres of the state of Chihuahua. Even today, after the government land reforms, 69 percent of Mexico's agricultural lands are reported to be still concentrated in estates of more than 1,200 acres each. One result of this is that Mexico is not self-sufficient in the bare necessities for food and even today imports large quantities of maize and wheat.

These large concentrations of available farm land in a comparatively few hands are characteristic of other countries of Mestizo America with a few local exceptions. For example, large haciendas are not characteristic of El Salvador (the smallest country in our area), parts of eastern Guatemala, the tropical forest regions of the Pacific coast of Colombia, the Antioquia region of Colombia, and the more unsettled portions of the Amazon tropical forest areas of Colombia, Ecuador, Bolivia, and Peru.

The problem created by land monopolization is so acute that most of these countries have at one time or another made gestures toward correcting the situation, and several have taken actual, although up to the present inadequate, steps to cope with it. The most spectacular, and in fact only effective, redistribution of *latifundio* lands has taken place in Mexico, where 45 million acres had been redistributed by 1940, both on an individual and an *ejido* (group) basis. Most of this was done under the Cárdenas regime, accompanied by a coordinated program of credit, technical assistance, public works, and rural education—and also, to be sure, by violent opposition and propaganda from the landed interests, including North American representatives. Elsewhere no really effective measures of redistribution have been taken, but a number of countries have set up mechanisms whereby small farmers may improve their position in a world of big operators. In Peru, for example, even before the free elections of 1945, two types of opportunities were offered to small farmers. On the one hand, communities organized on a communal basis (intended mainly for highland Indians) were recognized as "judicial persons" and were given various forms of aid and protection by agencies of the central government, both in land rights and in marketing of products. On the other hand, groups of small farmers were authorized to form cooperatives, both of the purchasing and selling type, under tutelage and protection of the government. In addition to organizational facilities, instruction in agricultural techniques, loans, seeds, and so on, were offered on a limited basis by the government. Something similar has been attempted in most other countries of the area, although viewed as a whole, on a rather limited basis and in a somewhat timid manner.

The so-called agrarian problem remains unsolved in all countries of Mestizo America, and political programs for agrarian reform are fundamental planks in the platforms of all parties appealing to the masses. None of the apparently alternative solutions appears to be easy. Voluntary dissolution of estates by the present monopolistic owners does not seem likely, in view of their class-cultural concepts of the dignity of landowning, their distrust of other forms of investment, and the profits accruing from the present system, which enables large landowners to maintain a high standard of luxury at home or abroad. Expropriation with compensation by the central government is also universally difficult to carry out, not only because of political opposition from the landed classes, but also because of the practical difficulties either of arriving at agreement on the compensation to the owner or of the central government's being able to finance the payments. A third possibility would lie in taxing the larger land monopolies out of existence, but up to the present this has been in most places politically inexpedient. In most regions the large land holdings do not even pay their share of the cost of government. In many countries land is for practical purposes not taxed, and the expenses of government are provided either by import and export levies, or by direct consumer taxes, excise taxes, and the like. The fourth most discussed solution of the agrarian problem is through political revolution. Distasteful and undesirable as this may be, it is not to be discounted by those interested in Mestizo America. This is still a predominantly agricultural area. The present land hunger may eventually drive the masses to forcible expropriation and redistribution. Such an uprising is bound to carry in its train excesses of various kinds and, if successful, a profound reorganization of economic patterns of all types.

Mining and Industry

The first interest of the Spanish conquerors was in mines and their products, mainly gold and silver, which were prized for their high cash value rather than for their industrial usefulness. The natives of the Andean area and of Mexico-Guatemala had developed mining and metallurgy in gold, silver, and copper, and the conquerors in the first flush of their victory proceded to rob the aborigines of their accumulated stores of precious metals, a project which, ironically, was made the easier by virtue of the fact that the native peoples who had made collections of these metals generally had no idea of evaluating them in cash or as media of exchange. Following this, the Spaniards settled down to the exploitation of mines already opened by the Indians and such others as they could discover. In this process the forced labor of the Indians was driven unmercifully, with the result that large numbers of them died off. The abuse of Indian or mestizo labor has not been entirely stopped to the present day, and constitutes a perennial scandal in certain Bolivian tin mines, for example.

From the sixteenth century to the present, mining has been the principal export industry of Mestizo America. The bulk of the mining enterprises are now owned by foreign capital, technical processes have been improved, and a certain minimum humanitarianism has been generally introduced in treatment of the workers, either by government action or in the interests of maintaining the "stability of the labor force." Low wages are the rule and most of the product as well as the profits are exported abroad.

It is true that some incidental benefits have accrued to the mestizo countries from the present system of mining, but the fact must be faced that most of the activity has been carried on for the benefit of foreign societies and foreign stockholders. Transportation facilities have been built in Mestizo America by the mining companies, but generally speaking, these have not contributed to an integrated network. Railroads, for example, have usually been built from the mines to the port towns, with little or no attention paid to the needs of the intervening territory or to the country as a whole. Wages have been paid to local labor, to be sure, but the income derived therefrom usually proves to be insignificant to the national or areal economy. Wages are low—averaging less than 40 cents a day in many localities. Furthermore, it is not generally appreciated that, despite its importance in world trade, mining gives employment to only a minor portion of the working force. For example, in Peru, one of the principal mining countries, only some 44,000 persons were employed in mining even in the year 1940 when the industry was artificially stimulated by war demands; this group constituted only 1.81 percent of the "economically active population" of the country, according to the official census publication.

In summary, for four hundred years Mestizo America has figured in world economic affairs mainly as a producer of raw materials. Even today it is one of the principal raw material producers for the industries of North America and Europe. Production is controlled in the main by foreign concerns, subsistence agriculture has suffered, wages have been kept low, and the consuming power of the bulk of the population has been severely restricted. All of these features are, of course, characteristic of a colonial and dependent type of economy, lacking in balance and unable, so long as the system persists, to provide the basis for an improved standard of welfare and general cultural development.

If one examines the official employment figures in Mestizo America he will find considerable numbers of persons engaged in "industria" (industry). Most of these so-called industries, however, are small-scale local affairs, such as adobe and brick plants, or are handicraft or household enterprises, such as pottery making by hand or foot-loom weaving. Many of the larger establishments designated as industries are processing plants, such as sugar refineries and packing houses, or producers of "semi-manufactures," such as plants for concentration or reduction of metal ores. In keeping with its function as a raw-materials producer Mestizo America has traditionally relied upon imports of manufactured goods.

About the time of the first World War, however, some impetus was given to the establishment of manufacturing, and the shortages of imports during the second World War have stimulated this development again. Cotton textile factories are perhaps the most ubiquitous industrial enterprises in the modern sense. Much of the cheap cotton cloth used for clothing by the lower classes is woven within the area, and the production of woolen textiles in the Andean countries is also considerable. Other light-manufacturing industries are developing. At the present stage these are mainly devoted to the production of basic consumer necessities—metal tools and containers, agricultural implements, canned foods, building materials and plumbing, rubber products (including tires), pharmaceutical products, bottled beverages, and the like. None of the countries of our group is industrialized in the modern sense, but it is likely that the factory system will continue to spread. At the present time the most industrialized regions are those centered on the cities of Monterrey, León, and Mexico City in Mexico; Antioquia in Colombia; Lima-Callao in Peru; and the four industrial cities of Chile: Santiago, Valparaiso, Concepción, and Valdivia.

The manufacture of heavy producers' goods is little developed. Mexico has a small steel industry mostly installed and operated by foreign capital. Under government auspices Peru is building a steel plant at Chimbote which will have a capacity of about 200,000 tons annually. The successful development of heavy manufactures in Mestizo America would seem to call for intelligent over-all planning involving the entire area, probably the entire Western Hemisphere. A mutually beneficial system of regional specialization in manufacturing and exchange of goods seems to hold much better prospects than attempts at complete national self-sufficiency or autarchy.

In connection with the development of manufacturing another factor in the present situation should be mentioned. This is the rise of a mercantile, commercial, or "business" group in all the countries of our area. The essential interests of this group lie in being able to sell manufactured consumers' goods not abroad, but to the population of the area itself. Spain of the conquest had nothing of this sort to offer, and during the whole epoch of the *coloniaje* trade was operated as a semi-state monopoly. Free enterprise and free competition were foreign to the pattern. Actually they are still somewhat foreign elements in an over-all pattern which often decrees that business be done by personal influence, bribery, class or caste position, political "pull," and the like. But during the present century a small, but expanding, group of traders has been established, at the start mainly by foreign concerns interested in selling their products in Latin America. Gradually a local business class has been built up in all mestizo countries. It is still small, but it has been imbued with the principles of commercial expansion through competition of sales in the internal market. In the course of time this group has expanded its horizon from an exclusive preoccupation with the business of acting as "agents" for foreign firms and has begun to take an interest in local consumer production as well. Obviously the long-term success and prosperity of the group depend upon the expansion of the numbers and purchasing power of local consumers.

In the experience of the writer and others, it seems that the fundamental interests of the burgeoning business class are not yet fully recognized by its members, but in a blind and fumbling sort of way they have tended to oppose the traditional land monopolists, the *latifundistas*, and also the extractors of minerals. The profits of the latter two groups lie in low wages and low standards of consumer wants. The profits of the business class are to be obtained in greater mass income and ever-expanding desires for manufactured goods and commercially supplied services. Thus one may discern in Mestizo America not only a developing conflict between the agricultural masses and the land monopolists, but also a clash of interests between the growing group of businessmen engaged in trade, on the one hand, and the traditional aristocrats of the hacienda and the exporters of mineral products, on the other hand. The fact that the hacienda owners have traditionally tended to "look down upon persons in trade" does not make for a friendlier *rapprochement*. Whether the business class will be able to force a readjustment from the landed "aristocracy" and the mining interests is at best problematical. In a former era of old-style business competition one would have made an affirmative prediction. But in the present period the interests and obligations of the local businessmen are so apt to be tied up with international industrial cartels and monopolies, whose tentacles in many cases extend to raw material producing enterprises, that the businessmen are themselves subject to confusion or paralysis.

All of these economic developments together with increased contact with the outside world have tended to create a consciousness of their "rights" in the minds of the working people. Two tangible results have been the rise of labor organizations and the development of government programs aimed toward the protection and welfare of the workers. The labor movement is, as might be expected, the strongest in the four countries which have gone farthest along the road to industrialization, namely, Mexico, Colombia, Peru, and Chile. These countries also have taken the farthest strides in governmental action, although most of the countries of our group have enacted labor laws far in advance of those in force in most states in the United States. It is true that in some cases (Peru under Benavides and Prado, for instance), these steps have been taken as a means of offsetting the influence of the labor unions. But the fact remains that programs are actually operating in many countries which conservatives of the United States would regard as "socialistic." Not only is the eight-hour day more or less universal (legally) and the usual restrictions on the labor of women and children, overtime pay, compensation for accidents and unemployment, old-age insurance, and the like written into the law, but the leading industrial countries of Mestizo America have taken such steps as the establishment of free worker's hospitals and medical services, the building of modern houses by the government, the establishment of cheap restaurants, compulsory and paid vacations. Even so "backward" a country as Guatemala has its large section of model workers' homes (Barrio Obrero) in the capital city. "Liberal" political parties or coalitions pledged to promote the welfare of the workers and "common people" are in control (in 1947) in Mexico, Guatemala, Peru, Bolivia, and Chile. Although a Conservative was elected president of Colombia in 1946, the Liberal party controls Congress.

Despite evidences, such as this, of the growing influence of the working people, the standard of living remains very low if judged by the standards of the United States or pre-war western Europe.

Standards of Living

The standard of living offered by a culture or group of cultures anywhere may be judged from two points of view. (1) Does the culture provide for the basic necessities of the people? Among these are the fundamental requirements for general health and nutrition essential to the human species everywhere. (2) Beyond this, the standard of living is in part a matter of cultural definition. In other words, what do the people believe is required for the good life and to what extent do they get such things? For example, whether or not automobiles and radios are regarded as necessary depends upon agreement among the people. In those cultures where such items are regarded as necessary or desirable but are actually obtained by only a small portion of the population we may say that, in this sense, the bulk of the population suffers from a low standard, even though their basic necessities are provided. Let us take a look at these matters in Mestizo America first from the point of view of health and nutrition.

The evidence seems to show that a large portion of the population is undernourished and physically ill, according to any standards. Two general factors are involved: insufficient and unbalanced diet, and inadequate scientific medical care, either of the preventive or remedial type. These facts are reflected in the vital statistics. Whereas in 1940 male life expectancy at birth in the United States [4] was nearly 63 years, in Chile it was 35 years, in Peru less than 32, and in Mexico less than 40. Nor do the supposed blessings of city life in this area seem to produce favorable results on life preservation. In 1940 the death rate in New York City was 9.8 (per 1,000 population); in Santiago, 24.8; in Lima, 20.5; in Quito, 21.2; in Caracas, 19.4; in La Paz, 22.4; in Mexico, 23.8. The four leading causes of death in the area today are tuberculosis, influenza-pneumonia, malaria, and diarrhea-enteritis, all of them ailments caused by pathogenic organisms with which modern medical science is able to cope. All of them except influenza-pneumonia have been eliminated from among the four leading causes of death in the United States. In 1939 the tuberculosis death rate by countries (per 100,000 population), to cite a few examples, was: United States, 47.2; Colombia (1938), 46.6; Mexico, 55.4; El Salvador, 61; Guatemala, 64.3; Ecuador, 70.2; Paraguay, 102.4; Panama, 210; Venezuela, 243; Chile, 276. As Charles Morrow Wilson says, "The only reason more Latin Americans do not die of tuberculosis is that more of them do not yet live in metropolises." For the tuberculosis death rate in many Mestizo American cities is truly appalling. Whereas the rate in New York was 49 (in 1939) and in Detroit 44.7, the rate in Santiago was 430, in Lima 435, and Guayaquil 693. It is well known that tuberculosis is a disease which spreads among undernourished people living in squalid conditions.

In the United States a proportion of 10 hospital beds per 1,000 population has long been regarded by experts as a minimum goal for safety. In 1940, Massachusetts had 13.2 and South Carolina 4.2. The Latin American average, however, is estimated to be less than 2 beds per 1,000 population, despite the fact that the average stay of the patient is about twice as long as in the United States. Millions of rural dwellers have no hospital facilities whatever. For example, in 1940 Peru had about 12,000 beds, an average of 1.4 per 1,000 of the population. But the majority were concentrated in Lima where the ratio was 9 per 1,000 population. In Colombia less than one-third of the cities and towns have any hospitals. In Chile three-fourths of

---

[4] It should be understood that when we make comparisons with conditions in the United States we do not consider the latter to be perfect by any means. Imperfect as they are, however, they are familiar to our readers and thus may serve as a touchstone for assessing the situation in Mestizo America.

all the so-called hospitals lack surgical wards. During the war years, the United States government through agencies of the Office of Inter-American Affairs and its successors provided money and personnel to assist in the correction of this situation. A number of hospitals were built and staffed in outlying areas of the various countries, and through scholarships a considerable number of young Latin Americans were brought to this country for training in medicine and sanitary engineering. It remains to be seen whether the governments of Mestizo America will be financially able or willing to carry on and to extend this program. Because of the prevailing low income of the masses, adequate health standards will have to be provided through public agencies in greater proportion than in the United States.

At present the cultural pattern followed by a large share of the common people of the area does not include modern customs of hygiene, consultation with scientifically trained physicians, or the use of modern clinics and hospitals. Most of the treatment which the average mestizo receives consists either of home remedies, patent medicines, or the attentions of medicine men and other magical practitioners. These antiquated and inadequate customs of treatment and attitude can, of course, be changed if medically sufficient facilities are provided and the rewards for their use made clear.

An appalling wastage of life and energy is reflected in the high infant mortality rates. In the United States in 1940 about 48 out of every 1,000 children born alive died during the first year. The rate reported for Bolivia in 1937 was 267; for Santiago, Chile in 1939, 204; for Bogotá (1939) 190.9; for Quito (1940), 182.

Malaria and diseases caused by intestinal parasites are third and fourth, respectively, among the leading causes of death. It is true that the likelihood of infection is much higher in most of Mestizo America than in the United States because of differences in climatic conditions. Nevertheless, modern medical science has the means at hand for the prevention and cure of these diseases, provided the means can be made available to the population in general. In justice, it must be said that most of the governments of the area are awake to their responsibilities in this matter and are taking practical steps. It should also be mentioned that several North American corporations have had the enlightened self-interest to set the pace in preventative and remedial medicine in the concessions which they occupy. This is particularly true of one of the large fruit companies, which has not only brought pathogenic parasites under control on its plantations to a degree comparable to the sanitary regime of the Panama Canal Zone, but has also established a chain of hospitals and a medical service which provide an example for tropical medicine the world over. This policy has not been dictated by charitable considerations but by the realization that it is more profitable to have a healthy and physically efficient working force than to suffer the losses incident upon the wastage of human resources through unnecessary disease and death. The same attitude could profitably be taken by governments in most of Mestizo America.

Unsanitary customs of housing and waste disposal are, of course, responsible for much disease and death. In most of rural Mestizo America, for example, latrines do not exist—human excrement is disposed of in any handy place, the fence corner, the street, the river. Likewise, the fact that polluted drinking water may be a source of deadly infection is very little known, and water is customarily used from whatever source is physically available, with no attempt to purify or protect it. These customs are, of course, partly a matter of tradition. Somewhat similar customs prevailed during colonial and frontier days in North America. In most regions of the United States (with the exception of parts of the rural South, which shows a number of parallels with Mestizo America), such customs and attitudes have been changed through "education." Education, however, usually involves change of mental patterns, the setting up of new ideals. The establishment of actional patterns, putting such ideals into practice, also requires in modern culture a certain level of economic resources. Even though many mestizos may be brought to realize what *should* be done to protect their lives and their health, many of them under present conditions of poverty will be unable to put the knowledge to effective use.

Let us now turn briefly to a consideration of nutrition, since it is well known that diet has an important bearing on health and physiological efficiency. Although the majority of Mestizo Americans are rural dwellers and although we are accustomed to think of rural people as having at their disposal everything needful in the way of food, whatever else they may lack, the facts in Mestizo

America do not support this complacent point of view. The general
opinion of experts assembled at the Third International Conference
on Nutrition held in Buenos Aires in 1939 was published as follows:

The American continent is undergoing a veritable tragedy owing to the
undernourishment which affects with no exception all of the countries
of Latin America. . . . A very important sector of the American world
does not manage to eat the minimum food required for the conservation
of life and for a normal yield of human labor. . . . Even in the most
favorably placed country of Latin America, one fourth of the workers
do not earn enough money to buy the necessary food.[5]

In a League of Nations study of 593 workers' families in 31 urban
districts of Chile, only 27.3 percent of the individuals had as much
as 3,000 calories per day. Another study of the *inquilino* (Chilean
agricultural laborer) reported that the "diet of the inquilino is characterized by an extreme underconsumption of protective foods."
According to the Venezuelan physician Carlos de León, the customary diet of Venezuelan peones in general consists of black beans,
corn bread, and sugar alcohol. On the basis of a considerable
although inadequate series of studies made by various investigators,
it appears that the diet of the average person of the middle and
lower classes is in many cases inadequate even in energy foods (car-
bohydrates). Proteins in the form of meat or fish are rarely consumed, although in Central America, at least, the fairly high consumption of beans and peas may compensate for meat deficiency.
Generally speaking the use of milk and dairy products is not customary; likewise eggs, fresh vegetables, and other vitamin-carrying
foods figure slightly in the diets of many regions. Maize, beans,
bread (in the majority of cases), rice (on the Pacific coast of South
America), and cassava (in the tropical and subtropical regions) are
the staple articles of diet. The widespread use of fresh peppers may
provide a part of the desirable vitamin quota. Another source of
vitamins may be found in the almost universal use of fermented
beverages—chicha, pulque, and the like. However, it seems to be
clear that excessive use of alcohol in most parts of the area repre-

[5] Ministro de Relaciones Exteriores de la República de Argentina, Instituto Na-
cional de Nutrición, *La Tercera Conferencia Internacional de la Alimentación;
Síntesis de sus deliberaciones; Conclusiones que se desprenden.* Quoted in George
Soule, David Efron, and Norman T. Ness, *Latin America in the Future World,* New
York, 1945.

sents not only a reaction to psychological frustration, but also a means whereby the individual stimulates himself temporarily to energy output not provided by his customary diet. In the Andean region coca-chewing serves the same purpose.

We turn now to certain culturally defined indices of the standard of living. In the group of countries under consideration there are, according to late estimates and counts, 422,871 telephones, or a ratio to the total population of 1:148. This compares with a ratio of 1:5 in the United States. The ratio of radios is estimated at 1:51 compared with 1:2.2 in the United States. The ratio of automobiles is 1:192 compared with 1:4 in the United States. (See accompanying table.) It is true, however, that these three items, although

Telephones, Home Radios, and Automobiles in Mestizo America and the United States

Country	Telephones	Radios	Automobiles
Bolivia	2,680	50,000	3,742
Chile	90,943	150,000	45,989
Colombia	42,233	190,000	35,434
Ecuador	8,000	6,800	1,497
Guatemala	2,327	15,000	4,824
Honduras	1,943	8,000	1,569
Mexico	200,000	500,000	138,857
Nicaragua	1,509	8,000	1,435
Panama	6,640	32,000	16,389
Paraguay	3,841	14,500	1,827
Peru	36,344	82,656	29,481
El Salvador	4,411	14,000	3,411
Venezuela	22,000	150,000	38,420
Total	422,871	1,220,956	324,876
Ratio to Total Population	1:148	1:51	1:192
United States	26,381,000	60,000,000	32,557,954
Ratio to Total Population	1:5	1:2.2	1:4

Sources: *Pan American Yearbook*, New York, 1946; *World Almanac*, 1946; *Handbook of Latin American Population Data*, Washington, 1945; George Soule, David Efron, and Norman T. Ness, *Latin America in the Future World*, New York, 1945; also latest census publications, statistical abstracts, and statistical bulletins of the various countries, where available.

desirable, are not considered necessary to a decent standard of living by perhaps the majority of mestizos at the present time. For the average person in Mestizo America a private automobile is regarded somewhat as a private yacht by the average North American.

But these manufactured items, which happen to be among the few for which any sort of statistics are available, indicate something of the desire for and the ability to acquire manufactured goods among the people of Mestizo America. To round out the comparative picture, let us sketch briefly the living conditions characteristic of urban-dwelling, literate mestizos. Such people are "civilized" and they form the economic and political backbone of the area under consideration.

During and following World War II, North American exporters worked themselves into quite a state of ecstasy over the "potential untapped market of 134 million persons" in Latin America which they were going to enjoy once the war had eliminated sundry competitors, such as the Japanese, Germans and British. A cold hard look at average incomes will reveal, however, that under present conditions the proportion of the total population in circumstances which enable them to buy consumers' goods even on the scale of the North American lower classes is very small. In part this is, of course, due to artificial valuation of the currency. Since all of the countries are exporters of raw materials, a general policy has been followed for years of maintaining the local currencies at a low value in terms of dollars or pounds in order to be able to sell cheaply abroad. Postwar inflation of prices without corresponding increase in wages has also effectively cut purchasing power. But even before the war, the real wages of Mestizo American workers in terms of ability to buy even necessities produced in their own countries were far below those of workers in the United States. For example, a study made on the basis of 1939 showed that while an hour's labor in the United States would on the average buy 8.19 kilograms of beans, in Ecuador it would buy only 0.928 kilos; similarly, against the American workman's 2.607 kilos of coffee for an hour's labor, a worker in Colombia, where coffee is produced, could buy only 0.605 kilos.

A more intimate picture of living standards and purchasing power may be obtained by considering conditions in a single country, Peru,

where the writer was able to make personal observations. The
following, unless otherwise indicated, deals with prices and money
incomes there as of 1944. The pattern of living and expenditures is
essentially the same, according to my observations, in mestizo circles of Guatemala, Ecuador, and Colombia, although the statistical
details may differ, and the present picture seems to check with other
studies made.

Let us visit a few of our mestizo neighbors in Lima, the capital of
Peru. Here is the abode of Don Federíco Alvarez (a fictitious name,
but a real case among many I had an opportunity to learn to know
in 1944). Don Federíco is a skilled workman in a textile factory, a
short, good-looking man of about 35. His features show the mixture
of Spanish and Indian traits that characterize the bulk of the population of many Mestizo American countries. He has finished the six
years of elementary schooling and reads and writes without difficulty. Outside the factory he wears a European-type suit, shirt, neck-
tie, shoes, and a felt hat. His dwelling is part of a *callejón*, a common
plan in which a block of adobe apartments is arranged around a small
yard or alleyway with a dirt or cement surface. We pick our way,
through a crowd of children playing in the yard and some women
hanging up clothes, toward the door of Don Federíco's apartment,
which is one of thirty in this *callejón*. One enters directly into a
front bedroom, then goes through a door into a second bedroom
and beyond it into the kitchen. The dwelling, in other words, consists of three rooms arranged in "railroad" fashion; it is paved with
cement. Don Federíco's household consists of twelve persons, himself, his wife, his father, his mother, his four children, and his two
brothers and two sisters. They all live together in the three-room
apartment. Four are wage-earners (Don Federíco, his mother, father, and sister), and among them they make about $45 a month
(300 soles)—a relatively high income for a family of this social
standing.

The apartment is well kept and fairly clean. As is the case in most
Peruvian houses, it has no rugs on the floor. The walls are white-
washed. There is a cold-water tap in the kitchen, but toilet and
bathing facilities are in the small room in one corner of the communal courtyard of the *callejón*. Here one water closet and a large
cement sink with a cold-water tap serve the needs of the thirty

households of the unit. Let us take an inventory of the furnishings
of Don Federíco's house. Each room has a single hanging electric
bulb; the electric bill is about 75 cents (5 soles) per month. The
front bedroom, which during the day is lighted only by the open
front door, contains two double beds, two factory-made chairs, a
night table, a small wooden table, a shelf of wood against one wall,
fourteen colored lithographs in frames, and a small radio. The beds
have rope springs covered with coarse woven mats instead of mat-
tresses and the bed clothes are factory-made blankets which are
stacked up during the day. The second bedroom, which has a sky-
light, contains two more double beds, a table, a trunk, two picture
frames, two wooden shelves, and a dressing table. Each married
couple occupies one bed; Don Federíco's two brothers sleep in one
of the remaining two beds, and his sisters in the other. The children
sleep on the floor, on mats which are rolled out at night and covered
with blankets. This arrangement seems to be somewhat crowded,
but the Alvarezes figure that they are pretty well situated in comparison with poorer families. They, like everyone else, have been
used to crowding all their lives. In the kitchen we find a stove made
of adobe; it is raised to about the height of a table, and the charcoal
fire is built on its top. The kitchen also has a wooden table, a wooden
bench, and a cement sink with a single water tap. The principal
cooking pots are of earthenware, made by the Indians of the mountains and bought in the market, but there are also some enamelware
pans and plates, some china cups, an incomplete set of cheap metal
knives, forks, and spoons, and quite a few tin cans used as containers.
An old hand-cranked sewing machine is also present in the kitchen,
where the señora makes all of the children's clothes.

This apartment costs the family about $2.25 (15 soles) per month.
The only obvious parts of the equipment not produced in the
country are the radio, the sewing machine, and some of the table-
ware. The Alvarezes, however, are steady customers for one other
North American "export," namely, the movies, which they attend
regularly and on which they spend a total of about one dollar per
month.

We could go on from here to visit a large number of homes of the
ordinary people of Peru and other parts of Mestizo America, some
poorer, others more elegant, than that of Don Federíco. But North

Americans find such visits depressing. We prefer to associate with Peruvians who live as we do. In Lima and other capitals we can get to know a congenial group of English-speaking Peruvians who are charming hosts. They have made frequent trips to Europe or North America and have pleasant houses equipped with modern mechanical gadgets of all kinds. They drive American automobiles, and their tastes and interests are much like our own. Many a North American never gets to know any other type of Peruvian, and he therefore comes away with the feeling that Peru is just a Spanish-speaking United States country club. But our social contacts are apt to give us the wrong idea about potential economic development, cultural interchanges, and standards of living which, whether we like it or not, in fact involve the common people as well as the upper class.

It would be very pleasant if all or the majority of Latin Americans lived on a level like that of our "Americanized" Peruvian friends in Lima. However, it is not true. Actually only about 5,000 Peruvian individuals or family units enjoy an annual income of $3,000 per year or more (20,000 soles). Add to this a possible 5,000 permanently resident foreigners in the same bracket, and you have, with their wives and families, a group of perhaps 60,000 people in a total population of about seven million, capable of providing a steady and active market for the ordinary, manufactured goods common in the United States. Some of these persons, to be sure, are millionaires in anyone's currency, but the total group is less than one percent of the population.

If we look about, we see that in most countries there are two population groups left—the Indians and the mestizos. The Indians, as long as they remain in their present status in Latin America, are relatively uninteresting economically for the very reason that, as long as they remain Indians, their design for living does not include many requirements for manufactured articles. The Indian of today may be the customer of tomorrow, but tomorrow may be some time in arriving.

It is, in short, in the large and increasing mestizo population that the customers for increased consumer goods must be found. Our friend, Don Federíco, belongs to this group, which in Peru accounts for 53 percent of the population, and we have noticed that neither his conscious requirements nor his income in their present state seem very promising. He does have the basic patterns of European civilization, however, and, if he had the chance, he would like to add some more.

In order to get some statistical measure of incomes and budgets, Dr. Leoncio M. Palacios and his students in 1940 made a study of 81 families in Lima. The earning members of these families (which comprised a total of 395 individuals) were employed in 18 occupations ranging from common laborer to white-collar office worker. The average annual income per family was $346.93 (2,261 soles), including the contributions of all the earning members of each household. A fifth of the families, however, took in only $229 or less, while only 3 percent of the families had a yearly income of $612 or more. In American money, the average daily wage for a man was about 61 cents and for a woman, 23 cents. The men's wages ranged from a low of 30 cents per day for farm hands (*chacreros*) to a high of about one dollar for white collar workers (*empleados*). Two-thirds of these families lived in dwellings consisting of only one or two rooms, while only one-fifth were as well off as our friend Federico Alvarez, with three rooms. In this particular group of city dwellers, about 51 percent of the total income was spent for food alone, compared to an average of 34 percent for North American workers. Only 13 percent of the Lima workers' income went into housing and furnishings, compared with 28 percent for North Americans. After necessities were paid for, a little more than 3 percent remained for "recreation and social life" and about 3.5 percent for "superfluous" expenditures. The few other reliable studies from different parts of Mestizo America show essentially the same picture.

Political, Religious and Educational Features

We can do little more than mention the fact that Mestizo American society has been traditionally stratified on a class basis; space is lacking to set forth the detailed differences of custom between the various classes. Ownership of more land than a man can exploit by his own efforts has been traditionally the hallmark of upper-class status. With the rise of a business group, some landless but otherwise wealthy families have achieved this rank, but full acceptance has usually had to be accompanied by marriage into one of the old landed families. Generally speaking a middle class has either been absent or very small in these countries, which may have something to do with the comparative underdevelopment of such so-called bourgeois cultural traits as thrift and cash savings, civic spirit, modern social work, and charitable activities, and the like. Industrialization, greater education, the increasing need for white collar workers in business and government, and the development of the professions may all serve to increase the size of the middle class and to define its position more clearly in mestizo society.

There seems to be a greater tendency for members of the small middle classes to make common cause with the lower classes than is the case in the United States. Most of the leaders of "peoples' movements" of recent times have been men of middle class standing. In part this is because the possibilities for upward mobility in the class structure of Mestizo America are so much less than in the United States. The middle class liberal or radical has few opportunities to "work up" in the class structure and thus to become conservative as he savors the rewards of greater wealth and prestige.

Although a "class struggle" can hardly be said to be in progress in the mestizo countries as yet, the leadership given by middle class intellectuals and organizers to the members of the lower classes has begun to awaken the latter to their stake in society. One of the features of social development which promises to be of most significance for the future is the rising self-consciousness of the dispossessed elements who have traditionally been allowed no voice in affairs and few tangible rewards from society. Ever since Independence these elements have periodically burst forth as violent mobs or "armies" under the leadership of military *caudillos;* in "successful" uprisings they have been paid off with a bit of loot and temporary surcease from drudgery, while the leader and his coterie have appropriated to themselves the emoluments of office, come to terms with the upper class groups, and have forgotten the promises made to their revolutionary followers. Although somewhat inconvenient, this system never upset the class system and the control of essential economic and social privileges by the aristocrats. There are signs in many countries now, however, that a new pattern of organization of the masses is emerging, under the leadership of better trained and more socially conscious men who are seriously bent upon permanent reform of the social system.

All countries of Mestizo America are republics in form. The United States constitution has served as a general model: presidents are supposed to be elected at stated intervals for a fixed term, cabinet members, ministers and heads of executive departments are responsible to the president rather than to the legislature, and the administration cannot be voted out of office by the legislature, as in the parliamentary systems of England and France. Bicameral legislatures supposedly elected by the citizens are the most common, and separation of legislative, executive, and judicial branches is theoretically maintained. However, Roman Law is everywhere the basis of the legal system, rather than the English common law. Also, despite certain similarities in form, none of the countries of the area is organized on an effective federal basis like that of the United States, but, on the contrary, all are more centralized. Thus the central administration usually appoints governors of states or provinces and subordinate officials, and takes all responsibility for roads, telegraphs, and schools, even though some countries permit local elections of municipal or other local officials.

As everyone knows, these systems of government have not been notably successful, either in providing stability and tranquility or in promoting the general welfare of their populations. Perhaps it is because they have proved to be republican in form but not democratic in fact. Colombia [6] is the only one of the mestizo countries which has not had a political revolution in this century and it is the only one which since Independence has never been under the rule of a complete military dictatorship. Elsewhere, intrigue, bribery, assassination, and the bloody *coup d'état* rather than the ballot box have at least occasionally been the means whereby the "Outs" changed places with the "Ins." To mention two of the more spectacular examples, Ecuador has had 14 constitutions and, between 1931 and 1940, thirteen presidents; at this writing Venezuela has just inaugurated its twenty-second constitution. For years in many countries the only presidents who finished out their terms (if, indeed, they ever went through the forms of being elected) were

[6] Since this paragraph was prepared for publication Colombia has had a violent revolutionary outbreak, in April, 1948.

those who made themselves dictators in fact if not in form. Among recent or present examples are the late Ubico of Guatemala, the late Gómez of Venezuela, Morínigo of Paraguay, Carias of Honduras, and Martínez of El Salvador. Since forcible revolution has been so often the technique for political change, it has been axiomatic in most Latin American countries that whoever controls the army controls the government. For practical purposes, indeed, the army has been of considerably more political importance than the congress in most mestizo countries.

We cannot attempt an analysis of the factors producing the political instability which has characterized so much of Mestizo America's career since Independence. Among probable factors we may mention the following. Political democracy was actually not part of the culture either of Spain or of the Spanish colonies, so that no background of practice or of thought existed when the outward forms of republican government were set up. The extreme natural regionalism existing in each country has made either a unification of point of view or a coordination of effort very difficult; the political instability of Ecuador, for example, is directly tied up with the fact that the mountain capital, Quito, and its region have little in common with the chief commercial and seaport city, Guayaquil and its hinterland. Extremes of class and relative lack of education for the great mass of the people have often prevented anything approaching democratic participation by the bulk of the population: the majority of the people did not know how to take part in political affairs, and the upper classes would not permit them to do so, anyway. Finally, the role of foreign interests has often been far from a passive one. It is convenient for foreign exploiters not to have to deal with strong "liberal" governments if Latin America is to be maintained as a semicolonial dependency of European and North American industry; truly democratic regimes might look less favorably on the monopolies and concessions made almost solely for the benefit of foreign concerns. For the same reason, European and American businessmen (and their governments) have often looked with favor on amenable dictators capable of "keeping order" and maintaining low wage and tax rates.

One of the noteworthy signs of the emergence of Mestizo America into the modern world has been the decline of political turbulence in certain countries and the appearance of something resembling modern democracy and political maturity. As we have seen, Colombia has usually conducted its political affairs wthout benefit of military strong men and in a relatively serene and democratic manner. The recent free elections in Peru, which resulted in the change of government from a conservative group to a coalition in which the formerly suppressed "radical" APRA party held considerable power; the comparatively peaceful elections of the last three Mexican presidents; the fairly effective if somewhat uneasy coalition of liberal and leftist parties in Chile, which managed to run the country for several years; the present democratic regime in Guatemala —all of these phenomena are exceptions to the old pattern of seemingly endless violence, chicanery, and weakness.

It should be noted that the two strongest political movements in the mestizo area today are both liberal and are both homemade products. I refer to the Party of the Mexican Revolution (which has been in power for 35 years) and the APRA party of Peru founded by Raul Haya de la Torre. Despite their differences, both of these movements are opposed to doctrinaire Marxian socialism, but are oriented toward the achievement of both political and economic democracy. Both are opposed to the slavish adoption by the mestizos of foreign-made political and economic forms and solutions to problems; on the contrary both have attempted to develop political forms and techniques suited to the peculiarities of the history, natural environment, and culture of Mestizo America. The problems of the Indian, of land hunger, of illiteracy, of industrialization, of geographic sectionalism all enter into the political thinking embodied in these movements. Whatever may happen to these two particular organizations, it seems clear that if the Latin American culture of the mestizo area is to reach full function it must develop forms of political expression and controls integrated with the rest of the culture and the needs of the population which it serves.

For practical purposes all but an insignificant portion of the people are nominally Roman Catholics. The influence of the hierarchy and of the Church as an institution has always been strong since the first days of colonization. As a cultural complex, however, the religion of these countries contains some features unfamiliar to many North Americans. (1) Catholicism exhibits a number of traits and

complexes more characteristic of southern Europe than of its manifestations in northern Europe and the United States in modern
times. For example, monastic orders are more prominent and members of the "regular" clergy more numerous; more attention is given
to cults of saints; frequent public parades, adorations, and fiestas
add a note of color seldom seen in North American communities;
sodalities or semi-religious clubs of laymen (*cofradías, hermandades,
mayordomías*) are common and conspicuous; the Church is not
strictly puritanical regarding the small vices of men, so that religious celebrations usually involve a strong recreational component
with a pleasant emotional overtone; dancing, drinking, music, fireworks, gambling and other "worldly" activities are typical features
of religious feast days. (2) The intellectual and philosophical position of the Church has been typically conservative, although we
must not forget that a monk, Las Casas, led a movement in the early
days of the colonization for more humane treatment of Indians and
that a Mexican priest, Hidalgo, organized the first important revolution for political independence from Spain in 1810. (3) Perhaps
more emphasis is given to symbols than is true of North American
Catholicism. At least there is a tendency for common people to regard "religion" mainly as the practice of the approved forms. (4)
Alongside Catholicism are almost everywhere to be found elements
of supernatural belief and practice, in part derived from indigenous
systems, in part survivals from medieval times. Witchcraft, both
black and white, is extensively practiced and believed in; magical
curing is common; evil eye, "bad airs," and throwing of spells are
still "real" to many an ordinary person in the mestizo area. Local
pagan images and sacred places still receive some veneration and
awe.

It is difficult to assess the future of the Church in this area. Certainly it will persist if its officers succeed in integrating religion with
the ideals and drives of the emerging mestizo culture. At present
skepticism is common among laymen, at least on the verbal level,
and laymen on the whole do not participate actively in ordinary
Church affairs other than fiestas and sodalities. Regular confession,
attendance at mass, and the like, are often left mainly to women. If
women are allowed more freedom outside the home and a wider
range of participation in the culture, their devotion to the old rituals

may slacken. Despite some skepticism on the part of laymen and considerable neglect of observances, few persons are, however, willing to sever connections with the Church entirely or to forego its intervention for the salvation of their souls. Opposition centers mainly about the Church's place in secular affairs.

Education was for long an exclusive function of the Church, but since Independence all countries have assumed governmental responsibility in this field. Every country provides free, compulsory, public primary education, although such provisions have been somewhat ineffective to date, judging by results. Statistics of literacy are notoriously unreliable in this area, but estimates of illiteracy for the different countries vary from 50 to 85 percent. Strenuous efforts are being made in most regions to overcome this blockage to social and cultural development mainly through the provision of more and better equipped schools, more teacher training, and adult education. One of the most interesting of these efforts is the "anti-illiteracy" (*desanalfabetización*) campaign launched in most countries with United States aid; it is a method for teaching adults to read and write in a short time.

A frequent criticism of educational policy in Latin America is that it has been dominated by the scholastic ideals of medieval Spain. French nineteenth century Humanism has also had a strong influence. Although universities were established early, they emphasized literature, rhetoric, and philosophy. Theology, law, and medicine were for long the only recognized university courses and the method of teaching was didactic in the extreme. Until 1940 the medical school of one of the leading universities provided its students no laboratory course in dissection and no clinical courses. A student studied anatomy and medicine from lectures and textbooks, and when he went out to "practice his profession" the phrase was a literal description of what happened. Technological training, engineering, and the social sciences were almost entirely unknown in their "practical" phases.

The greatest recent change in education has been a systematic attempt in most countries to emphasize empirical investigations and training in practical techniques. Agricultural and engineering colleges as well as technical training for students on the precollege level are being established on a fairly large scale. The social studies, however, as empirical sciences are still almost unknown. One of the great difficulties has been the lack of adequate libraries and of Spanish editions of modern works in social science. Most of the translations have been confined to the humanities and to the work of Europeans, mainly French. A Mexican publishing house, the Fondo de Cultura Económica, has recently undertaken the translation, publication and circulation of standard English-language works in the social studies, and its example has been followed on a smaller scale by others.

In summary there is some possibility that the higher intellectual culture of Mestizo America will turn out to be a happy blend of humanism with pragmatic science. Certainly much would be lost if the traditional interest in literature, in the manipulation of concepts, in the discussion of human and cosmic values should be entirely discarded in favor of crass "practicality." Yet training in the techniques of science and technology is necessary if the mestizos are to hold their own in the modern world against encroachment and domination by peoples of more "practical" culture.

We see that the Mestizo American area has much to work with and much to overcome. Natural resources are adequate and in some respects outstanding. Difficulties are presented by diversities of natural environment and obstacles to easy communication. An additional burden is the fact that the people of the area have been saddled with the traditions of colonial Spain, the paralyzing framework of an antiquated class structure, and the restrictive grip of a monopolistic system of exploitation dependent on foreign markets. Likewise retarding has been the presence of large numbers of Indians in some part of the area, Indians who have been culturally and economically excluded in considerable part from regional affairs.

All of these are problems which can be overcome. A culture is a system of customs and institutions for solving problems. We have endeavored to show that a new culture is emerging in Mestizo America which bids fair to provide a balanced type of life in the area and furnish the medium for its integration with other active cultures of the world.

SUGGESTED READINGS

Gillin, John. Moche: a Peruvian Coastal Community. Smithsonian Institution, Institute of Social Anthropology, Publication 3. Washington, 1947. Study of a community of mixed, modern Latin American culture with discussion of development of Creole type of life.

Gunther, John. Inside Latin America. New York, 1941. Journalistic review of Latin America with particular emphasis on political and economic trends.

Handbook of Latin American Studies. Cambridge, Mass., 1935—. A yearly review and abstracting of all serious literature on Latin America.

James, Preston E. Latin America. New York, 1942. A thorough survey of human and physical geography.

Redfield, Robert. The Folk Culture of Yucatan. Chicago, 1941. Study of changing way of life in four communities with general remarks on processes of cultural mixture.

Royal Institute of International Affairs. The Republics of Latin America. London, 1937. A fairly thorough review of political geography and trends.

Soule, George, David Efron, and Norman T. Ness. Latin America in the Future World. New York, Farrar and Rinehart, 1945. Particularly good for discussion of standards of living, health, and nourishment.

Steward, Julian H., ed. Handbook of South American Indians. 7 vols. Smithsonian Institution, Bureau of American Ethnology, Bulletin 143, Washington, 1946—. (Vols. 1 and 2, 1946; Vols. 3 and 4, 1948; Vols. 5, 6, and 7 in press). An authoritative summary of the native cultures, physical types, languages and archaeology by more than 100 experts.

Vaillant, George, and others, eds. The Maya and Their Neighbors. New York, 1940. Contains authoritative summaries of the native cultures of the Maya area and Central America.

Van Toor, Frances. A Treasury of Mexican Folkways. New York, 1947. Analysis of many features of everyday life in Mexico tracing origins to native, Colonial Spanish, and modern sources. Many of these patterns are found in other parts of Mestizo America.

Williams, M. W. The People and Politics of Latin America. Rev. ed. Boston, 1945. A standard review of history and political organization.

Charles Wagley

BRAZIL

BRAZIL is one of the great undeveloped areas of the world. Its 3,-286,170 square miles contain 41,565,083 inhabitants. In area it is larger than the continental United States, which has three and a half times more people, and fifteen times larger than France, which has approximately the same number of people. Most of Brazil's population is concentrated within two hundred miles of the Atlantic sea board. All of the large urban centers are situated in the coastal zone where the average density of population is 25 people per square kilometer. The man-land ratio for the entire country is less than 5 per square kilometer, and in western Brazil and the Amazon Valley less than one person to two square kilometers. The three western states of Mato Grosso, Goiás, and Amazonas, and the District of Acre account for less than 7 percent of the total population of the country and for more than half of the total area. In the hinterland, there are rich unexploited deposits of minerals, vast unclaimed areas of potential agricultural and pastural lands, enormous undeveloped reserves of water power, and, above all, space for millions of immigrants.

Some enthusiasts describe Brazil as the "Land of the Future," destined to leadership in the Western Hemisphere, even in the world. They point out that Brazil is on the threshold of a great westward movement, similar to that of the North American frontier between 1870 and 1914. Others are less optimistic, pointing out that, after four hundred years of settlement, there are still only a few people in Brazil and that these cluster along the coast. They call attention to the lack of petroleum and the scarcity of coal, and to the fact that lush tropical vegetation often covers second-rate soil. And, above all, these rather pessimistic observers consider the tropical and semitropical climate of most of Brazil, with its correlated diseases, to be an almost insurmountable barrier to complete conquest of the country. Nevertheless the future of Brazil cannot be predicted with any assurance, because just what factors will make for the success of any region and its people in this highly dynamic world are probably now unknown. But with its vast area, with its great potential wealth, and with the rich cultural heritage of its people, there is no doubt that the country can make an unique contribution to world culture.

The People of Brazil

Even before Pedro Álvares Cabral, outbound from Portugal to India, accidentally reached the coast of Brazil in 1500, Alexander VI, the Spanish Pope, had divided the undiscovered world between Spain and Portugal. The territory east of an imaginary line placed some 470 leagues west of the Azores and Cape Verde Islands was granted to Portugal by the Treaty of Tordesillas in 1494; thus, the area which is now Brazil went to the Portuguese. But the Tordesilla line does not explain its present frontiers.

During the sixteenth and seventeenth centuries, bands of adventurers known as *bandeiras* (banners) left the settlements near the coast, especially in São Paulo, and penetrated into the interior in search of gold, diamonds, and Indians for slaves. These *bandeiras* were similar to the adventurous bands of Spaniards led by De Soto and Cortés, and their exploits were no less noteworthy. Usually fifty or more Portuguese colonists, a Catholic priest, and a few civilized Indians made up the party. Their expeditions took years, and several *bandeiras* never returned from the hinterland. The band stopped at intervals to establish camps and to raise crops for food. Some of them penetrated beyond the low divide of Central Brazil into the basin of the Rio Madeira. *Bandeirantes* from São Paulo attacked the Indians in the Jesuit Missions on the Upper-Paraná and middle Paraguay and they reached Bolivia and Peru in 1649 and 1662. One of them, led by Antonio Raposo from São Paulo, crossed the Andes to the Pacific Ocean, returning after many years to São Paulo by the way of the Amazon River and the Brazilian coast. These expeditions founded settlements in the interior of Brazil and extended the line of Portuguese control to the west beyond the legal frontiers. Finally the Treaty of Madrid in 1750 obliterated the Tordesillas line and recognized the claims of the *bandeirantes*. The frontiers of the Portuguese colonies were approximately those of modern Brazil.

Although these aggressive Portuguese adventurers gained territory from the Spanish, Portugal held her vast colony precariously during the first two hundred years of the colonial era. Portugal was a small country; in the sixteenth century it had hardly more than one million inhabitants. Since Portugal was the leader of a lucrative trade with India, Brazil attracted few settlers from the mother country during these first years. Small settlements were, however, established along the extensive coast—such as São Vicente (São Paulo), São Salvador (Bahia) and Olinda (Pernambuco). The control of the mother country was weak, and, when, between 1580 and 1640, Portugal was seized by Spain, Brazil became fair game for all Spain's enemies. In 1556, the French built a fort on the small island, later called Villegaignon, in the bay facing the Portuguese settlement of Rio de Janeiro, and in 1612 they founded a colony on the Island of Maranhão where the present city of São Luis stands. The Dutch captured São Salvador in 1624 and for many years actually maintained a colony on the Brazilian coast with headquarters at Recife (Pernambuco).

In 1532, the Portuguese introduced sugar cane to the northeast coast of Brazil. It was a huge success; sugar planting attracted larger numbers of Portuguese colonists. Included among them were people of considerable wealth. Sugar became a new luxury to the world in the seventeenth century, and Brazil soon produced most of the world's supply. The northeast coast, centering around Salvador in Bahia, became extremely wealthy. Brazil became a valuable possession and Portugal took a more active interest in its colony. Not until nearly 1700, however, were the Portuguese able to expel the Dutch and the French and to control the entire coast of Brazil.

From the beginning, Brazil has been faced with the problem of a scarcity of labor supply. The Portuguese colonists were few in number and they had no intention of performing manual labor in the New World. During the first century of colonization, they sought to enslave the native Indian population. In contrast to the aboriginal peoples which the Spaniards encountered in Peru and

Central America, the Indians of Brazil were few in number and did not have a highly developed material culture. It is probable that no more than one million American Indians inhabited the entire area which is now Brazil. While most of the tribes were agriculturalists with large gardens of manioc, corn, yams, beans, squash, and other native American crops, their inefficient system of cultivation would not support a large population. Many Indians were enslaved by the Portuguese, but, as early as 1570, the Jesuits obtained a decree from Lisbon prohibiting the enslavement of Indians, except those who were taken as prisoners of war.

For almost two hundred years the Jesuits defended the Indians in Brazil against the colonists eager for slave labor. The Portuguese colonists, who resented the Jesuits, charged them with using the Indians in their mission villages to work for the enrichment of the Order and, from time to time, the Jesuit *aldeamentos* of missionized Indians were attacked by colonists. Finally, in the middle of the eighteenth century the Jesuits were driven from Brazil and, under Viceroy Pombal, laws were made which abrogated Indian slavery under any pretext. The mission villages were ordered converted into towns and hamlets and the Indians were to be incorporated into colonial life. The missionized Indians and the few Indian slaves either merged into the rural population or took to the woods. Especially in north Brazil, the Indian left a strong mark on the Brazilian population and Brazilian culture. It was apparent very early, however, that the Portuguese colonist must needs look elsewhere for a labor supply for his large-scale sugar plantations and for mines.

Brazil, therefore, looked to Africa. The Portuguese had been active in the African slave traffic since the late fifteenth century, and beginning as early as 1538, there was a steady flow of Negro slaves across the Atlantic into Brazil. They came from diverse regions of West Africa; there were Bantus, Yorubas, Ashanti, Fula, and Hausa tribesmen represented among them. During the seventeenth, eighteenth and nineteenth centuries approximately 3,300,000 Negro slaves are thought to have been imported into Brazil.[1] By 1817, of a calculated total of 3,697,910 people in the Brazilian colonies, 2,-887,500 were Negroes (freedmen and slaves), and 628,000 *mestiços*

[1] Artur Ramos, *Las Poblaciones del Brasil* (Mexico, D.F., 1944), p. 119.

(mixtures of white, Negroes, and Indians). There were only 843,-
ooo whites and the rest were Indians.[2] With this tremendous influx
of slaves, the Negro became for a time numerically the most important element in the Brazilian population.

Wherever the Portuguese went—to Africa, to India, to China, or
to the New World—they mixed with the native peoples. As Gilberto Freyre has shown, Portugal itself in the sixteenth century was
a country of mixed racial backgrounds.[3] Many migrations of different peoples had reached Portugal—the Iberians, the Phoenicians,
the Vandals, the Huns, the Visigoths, and finally the dark Mohammedan Moors. Since Moors, who dominated Portugal, brought a
superior culture with them, the Portuguese did not identify dark
skin with a subjugated "inferior" people. The Moorish woman became the ideal beauty of the Portuguese, and it was natural that this
standard be applied to the native women in other parts of the world.
Rather than being repulsed by dark skin the Portuguese found it
sexually attractive. Thus, in Brazil they at first took Indian wives,
and the first century of colonial Brazil was one of mixture between
the European and the American Indian. The wandering *Paulista
bandeirantes* left numerous children, progeny of European fathers
and Indian mothers, who were called *mamelucos* and, in the north
of the country, mixture was frequent despite the efforts of the missionaries to isolate the Indian.

The Portuguese also took concubines from among their Negro
slaves and from such unions grew the class of mulattoes which has
played an important role in Brazilian cultural tradition. Miscegenation between slaves and their masters was not limited to Brazil; but
it seems to have been more frequent there than elsewhere in the
New World and it was accepted as a matter of course, becoming a
subject of pride to Brazilians.[4] As elsewhere, the mulatto was generally the offspring of a white father and a Negro mother. In Colonial times, mulattoes were frequently given special treatment by
their white fathers. They were given easy jobs; sometimes they
were educated and used as administrators on the plantations. Many
were sent by their wealthy owners, who were sometimes their own

[2] Frank Tannenbaum, *Slave and Citizen: the Negro in the Americas* (New York,
1947), p. 8.
[3] *The Masters and the Slaves* (New York, 1946), especially Chapter III.
[4] Tannenbaum, *op. cit.*, p. 121.

fathers, to Portugal to attend the famous University of Coimbra. Gradually the mulatto became freemen and formed a special class, and presently was represented among the professional classes as lawyer and physician. Except for the Jesuit Schools, which at first barred Negroes and "very dark *muleques*," mulattoes and free Negroes were generally accepted in the few schools which did exist in Brazil. In 1586, the King of Portugal criticized the Jesuits and, in a letter, stated that "schools of science ought to be common to all manner of persons, without any exception whatsoever." [5] Moreover, says Freyre, "not only were blacks and mulattoes in Brazil the companions of white lads in the Big House (Manorial House) school room and in the *colégios;* there were also white boys who learned to read with Negro teachers." [6] In the eighteenth century the priesthood was opened to mulattoes and to Negro freemen, some of whom even became bishops. The sculptor Antonio Francisco Lisbôa, known as the Aleijadinho, the Brazilian writer Machado de Assis, and the poets Castro Alves and Gonçalves Dias are a few examples of mulattoes who have been important figures in Brazilian cultural life. At least one president of Brazil, Nilo Peçanha, was a descendant of a mulatto and, even during slave times, several influential figures in court and Brazilian diplomatic circles were of partial African descent.

During slavery, not only mulattoes but many Negroes gained their liberty. Many Negroes reacted violently to slavery and escaped to the interior where they established free rebel communities called *quilombos*. The most spectacular of these communities of fugitive slaves was the *Quilombo de Palmares*, generally referred to as the Republic of Palmares, in Brazilian history. Palmares was situated in the area which is now the state of Alagôas in northeastern Brazil. It lasted from 1630 to 1697. At one time, the "Republic" is thought to have had a population of about 20,000 people—all escaped slaves or their descendants. It was governed by their own ruler who had under him a series of subchiefs who ruled by African law until 1697, when Palmares was destroyed by an armed force of six thousand. Small *quilombos* are said to have been found along nearly all the rivers of Amazonas. Some were near Indian villages, and mis-

---

[5] Freyre, *op. cit.*, p. 407.
[6] *Ibid.*, p. 409.

cegenation between Negro males and Indian women added to the general mixture of Brazilian population.[7]

Escape, however, was not the only means by which the Negro gained freedom; some purchased it, to others it was granted by benevolent owners. It became the custom for a wealthy slaveowner to liberate a slave or two on the occasion of any big celebration, such as the baptism of a son or the marriage of a daughter. Moreover, the slave had a legal right to purchase his freedom at the original purchase price, and could accumulate the money by doing extra work. The story of King Chico, cited by the Brazilian anthropologist Artur Ramos,[8] while it may be in part legend, illustrates this process of purchasing freedom. According to the story, King Chico lived in the beginning of the eighteenth century. He was an African chieftain, it is said, who was captured with all his tribe and brought to Brazil. Except for one son, his entire family died on the slave ship. He and his people were sent to work in the mines of Vila Rica near Ouro Preto in Minas Gerais. There, by doing extra work, he accumulated enough money to buy the freedom of his son, who, in turn, helped raise the funds to buy his father's freedom. King Chico swore "King in my country, King I shall be outside my own country." Then, through cooperative saving, he and his son bought the freedom of all of their people. King Chico set up and governed his own community. He and his people erected a church in the village and, as the legend goes, bought a mine which they exploited as a group. By 1798 there were 406,000 free Negroes in Brazil,[9] and it has been estimated that there were three times as many free Negroes as slaves when slavery was abolished in 1888. A law was already in effect at this time by which all children of slaves were freeborn; even without abolition it was only a question of time until the institution of slavery would have disappeared.

Slavery was a brutal institution in Brazil, as elsewhere, but even during slavery the Negro and the mulatto gained a place in Brazilian life far superior to that of the dark people in the British and Dutch colonies to the north. In the British colonies, freedom for the Negro

[7] Freyre, op. cit., p. 68.
[8] Las Poblaciones del Brasil, p. 178.
[9] Ibid., p. 119.

did not give him the rights of a citizen. The freed Negro or mulatto in the British West Indies could not give evidence in court; they could not serve even as minor public officials; they could not vote; they could not be tried by jury; and they could not bear arms. In North America and the West Indies the freed slave simply lost the protection of his master; his status was little above that of a slave in civil privileges.[10] By contrast, as we have seen, in Brazil the mulatto and the freed Negro were granted the rights of citizens and took part in public life. The tolerance of the Portuguese owners toward a people of darker skin was strengthened by the position of the Catholic Church, which considered the slave to be the moral equal of the master. Even before the end of slavery, therefore, the Negro was beginning to be incorporated into the life of the nation and, with abolition, the process was greatly accelerated.

The fourth major element in the population was furnished by the comparatively recent European immigration. As early as 1819, a colony of Swiss Roman Catholics numbering about 2,000 were settled in a high mountain valley in the province of Rio de Janeiro and, in 1824, a colony of Germans were settled in São Leopoldo, the southern province of Rio Grande do Sul. Several other colonies of Germans and Italians were founded, but the accession of Pedro II [11] who was actively interested in colonization, was the emigration from Europe actively stimulated. Between 1864 and 1866 some 27,-000 immigrants entered Brazil and the number increased each year. Brazil eagerly welcomed the new arrivals, because of the growing sentiment in favor of abolition of slavery and the law of 1871 which doomed slave labor on the plantations. Between 1884 and 1939 Italy furnished 1,412,263; Portugal, 1,204,394; Spain, 581,718; Japan, 185,799; Germany, 170,815; Russia, 109,502; Austria, 85,790; Tur-

---

[10] Tannenbaum, *op. cit.*, pp. 95–96.

[11] In 1808, the Portuguese royal family moved to Brazil to escape Napoleon. Until 1821, Prince D. João, later João VI of Portugal ruled the Portuguese empire from Brazil; on his return to Portugal, he left his son Pedro to govern Brazil. The Brazilians persuaded Pedro to break away from Portugal in 1822 and set himself up as Pedro I of Brazil. Between 1822 and 1889, Brazil was an independent empire. Pedro I was followed by Pedro II, an enlightened ruler and a friend of Victor Hugo, Longfellow, Agassiz, and other leading men of letters and scientists. Slavery was abolished in 1888. In 1889, Pedro II was asked to leave Brazil and the Brazilian Republic was formed. Since formal history is not included in this article, the above facts are given for the general interest and orientation of the reader.

key, 78,455 and Poland, 78,455.[12] Between 1874 and 1939 a total of 4,390,519 immigrants entered Brazil.[13]

This stream of immigrants was not distributed equally throughout Brazil. The state of São Paulo received more than half of them. In São Paulo today most urban industrial workers are of Italian descent and many salaried workers on the great coffee plantations are Italian, recent Portuguese (as compared to old Luso-Brazilian), Spanish, and even Japanese. The Germans, Slavs, and Poles settled principally in south Brazil. Northern Brazil did not partake of this immigraton to any great extent, except for a group of Japanese, settled on a concession at Parintins on the Amazon River, and also some scattered Syrians, and a few Portuguese.

During the same period (1874–1939), according to T. Lynn Smith, the United States received 29,565,400 immigrants, more than six and a half times as many as Brazil received. Furthermore, in 1930 the Brazilian government adopted a quota system, limiting the number of immigrants annually in proportion to the number of the same nationality already in Brazil. In accordance with the 1930 laws, it also became obligatory that 80 percent of the people of a quota be agricultural workers who must remain in rural occupations for at least four years after entry. The number of immigrants therefore fell sharply, and in 1939 there were only 22,668. In 1939, the Portuguese were exempt from the quota system and in 1940 the citizens of the American Republics were also exempt. Under the present regulations, however, a great increase can hardly be expected.

In spite of this relatively restricted immigration from abroad, the population of Brazil has grown at a phenomenal rate during the last half century. As Dr. T. Lynn Smith shows in his excellent analysis of Brazilian population, there has been an increase of 192 percent since the first relatively trustworthy census in 1890. During the same period, the United States with its flood of European immigrants gained only 52 percent. Whereas the population in Brazil was only 14,333,605 in 1890, it rose to 17,318,556 in 1900, to 30,635,605 in 1929, and to 41,565,083 in 1940.[14] Only 9.5 percent of this increase seems to have resulted from immigration. More than 24,000,000

---

12 T. Lynn Smith, *Brazil, People and Institutions* (Baton Rouge, 1946), p. 273.
13 *Ibid.*, p. 268.
14 *Ibid.*, especially Chapter V.

were added to by an excess of births over deaths, in spite of a staggering infant mortality rate throughout the country and a very high
death rate in general. Although the reported infant mortality rates
are thought to be somewhat exaggerated because births are less apt
to be registered than deaths, such statistics do, in the opinion of
trained public health specialists, give us an indication of true conditions. The lowest infant mortality rate reported for Brazil is
from the southern State of Paraná (125 of each 1,000 die before the
first year); this is five times higher than the average for the United
States. In the northeastern states the rate is as high as 502 in Piauí,
335 in Rio Grande do Norte, 295 in Paraíba, and 292 in Pernambuco. It is safe to say that three or four out of ten children who
are born in Brazil die before they reach the end of the first
year.

The crude death rate (that is, the number of deaths per 1,000
population per year) is probably double that of the United States.
The expectancy of life at birth is estimated at only 44.04 years in
the Federal District and 40.13 years in São Paulo, where the best
medical care and the best public health facilities in Brazil are found,
as compared to the life expectancy in the United States of slightly
over sixty years. Contrary to popular opinion, this high death rate
does not result from the so-called tropical diseases but mainly from
diseases which are well known in temperate climates. Tuberculosis
is Brazil's deadliest disease; the mortality rate is 341.1 (per year per
100,000 population) in the Federal District and as high as 479.5 in
Salvador in the State of Bahia.[15] Syphilis has a high incidence
throughout Brazil (the mortality rate in Rio de Janeiro is 57.1) and
it is almost endemic in North Brazil. Leprosy is widespread throughout the country, with a mortality rate as high as 3.2 even in Rio
de Janeiro. Such intestinal parasites as hookworm weaken the population, cause many deaths, and make the people less resistant to
malaria and other diseases. A survey of two rural areas of the country, the Amazon Valley and the Rio Doce Valley (State of Espírito
Santo), showed that 80 to 90 percent of the population were infested with hookworm and other intestinal parasites.[16] These might

[15] I have heard competent Brazilian public health specialists state that these rates
for tuberculosis were probably much under the true figures, if they were known.
[16] Edmund Wagner, "Engenharia Sanitária do Brasil," in *Problemas de Medicina
Prática e Preventiva no Brasil* (Rio de Janeiro, 1946), p. 188.

be controlled by simple but inexpensive improvements in water supply, sewerage, and community garbage control. Malaria is still prevalent in great areas but it is already under control in many places. The record of the successful battle against yellow fever (1910–1920) indicates that even under tropical conditions disease may be controlled. At present, however, it is not only the "tropical diseases" which are ravaging Brazil but diseases such as tuberculosis and syphilis which are common to most countries of the Western world. The statement of a great Brazilian physician that "Brazil is a vast hospital," although it was made many years ago, still holds true. In view of the health conditions that exist throughout the country, it is even more striking that the greatest source of population growth lies in the fertility of the people. Disease, malnutrition, lack of medical care, and lack of adequate public health facilities waste the human resources of the country.

As we have seen, the present population is a mixture of Caucasian (old Luzo-Brazilian and recent European immigrants), Negro, and American Indian racial stocks. A well-known Brazilian anthropologist, Professor Roquette-Pinto, in 1922 estimated that 51 percent were white, 22 percent mulattoes, 11 percent *caboclos* (Indian-white mixture), 14 percent Negroes; and 2 percent Indians. The 1940 census used nine categories of color; until its findings are tabulated and analysed, this estimate of the racial distribution is probably the best available. When set against earlier calculations made in the late nineteenth century Roquette-Pinto's estimates indicate that a "bleaching" process, as Dr. T. Lynn Smith calls it, is taking place.[17] The 1872 census indicated that 38 percent of the population were white; 38 percent were *pardo* (brown or mulatto); 19 percent were Negro and 4 percent were *caboclo* (Indian or Indian-white mixture). In 1890, 44 percent were classed as white, 9 percent as *caboclos*, 15 percent as black, and 32 percent as *mestiço* (mulatto). This trend toward absorption of the darker groups by the white groups is in accordance with the prediction of Brazilian scientists and scholars that the population will soon be "one hybrid Brazilian race."

This so-called "bleaching" is a simple genetic process, molded by the particular social conditions of the country, where for centuries

[17] T. Lynn Smith, *op. cit.*, p. 187.

white men have had more ready access to women than "mixed" or colored men. During the colonial period the owner of the *latifundio* and his sons were the "sires" and the "stallions" of Negro females (to use the words of Gilberto Freyre), and they begat numerous offspring on their young wives and concubines. Even today, upper-class "white" men have frequent extramarital affairs. Because of their prestige and wealth, they continue to have access to women of the lower and darker classes. Not only do white men have large legitimate families but they spread their genes through extramarital affairs. Because of a higher standard of living a larger proportion of the children of upper-class "white" parents survive than do the children of darker and poorer families; thus the white population makes the larger contribution to each succeeding generation. Furthermore, one should not overlook the contribution of several million white immigrants from Europe since 1864; but even without outside immigration, there is a tendency toward the absorption of the darker elements of the Brazilian population by the lighter.

Miscegenation always takes place between racial groups inhabiting the same region, no matter how great may be the antagonism between them. With the relative lack of racial antagonism in Brazil, mixture between the three racial stocks has occurred with unusual frequency. There are mixtures of every conceivable degree and combination of the three basic elements. Only among the descendants of the old aristocracy do people claim pure Caucasian descent and even these "old families" speak with pride of a distant Indian ancestor. This does not mean that it is common to see the blond descendant of a German immigrant married to a Negro; one usually sees a man and wife of nearly the same color. Most Negroes and people of mixed ancestry are poorer and less well educated than people of European descent. Many *mestiços* (both mulattoes and *caboclos*) live in the interior; most of them are descendants of rural farmers and therefore do not have the educational advantages and the socio-financial position of the urban white. The Negro slaves were freed only a little more than fifty years ago and their descendants are still poor. The children of poor people either do not attend school or leave early to earn a living. Therefore, in Brazil we have a familiar picture of the upper class being made up predominantly of the European whites and the lower class of people of darker skin.

As a well-known Brazilian sociologist wrote, "The darker the skin the lower the class, as a rule." [18] In contrast to the United States, however, this is a socio-economic arrangement of classes, not a caste system. In Brazil, when a person of mixed ancestry does climb the ladder of success, he does not come up against an insurmountable color barrier. In the United States, anyone with a small percentage of Negro blood is considered a Negro, but in Brazil as the popular saying goes: "Anyone who escapes being a negro is a white." In some parts of the country the expression *Branco* (white) describes a social status rather than a skin color. The popular expression which Donald Pierson quotes: "A rich Negro is white man and a poor white man is a Negro," although exaggerated as a popular saying is apt to be, expresses the sociological facts of the prevailing racial attitude.[19]

While race does not present a social problem, inequalities in the distribution of wealth, in education and the availability of educational facilities, and in the standards of living are serious barriers to a social democracy. The census of 1940 indicates that 56.38 percent of the inhabitants over 18 years of age are illiterate; in other words, of almost 21,000,000 people over 18 years of age, only 9,000,000 are able to read and write. It also must be remembered that of those classified as literate, a large number are only "semi-literate"—that is, they can write only their own names and read with much labor and little understanding. The situation for the coming generation is not a bright one; Dr. T. Lynn Smith's reevaluation of a study carried out by the Brazilian Instituto Nacional de Estudos Pedagógicos shows that less than two-fifths of the children of primary school age (7 to 11 years old) were in school.[20]

Furthermore, the mass of the population are undernourished. Objective studies have shown that most of the people simply cannot afford a balanced and sufficient diet. Five hundred laboring-class families studied in Recife in 1934 spent 76.1 percent of their total income for food and 18.9 percent for housing, light, and water. The daily consumption per person amounted to only 1,646 calories and the diet was not only deficient in calories but also in proteins,

[18] C. Delgado de Carvalho, "Lectures on Brazilian Affairs," *The Rice Institute Pamphlet*, XXVII, No. 4 (October, 1940), 239.
[19] Donald Pierson, *Negroes in Brazil* (Chicago, 1942), p. 348.
[20] *Ibid.*, p. 665.

calcium, iron, and vitamins.[21] Since 1934 the situation has become increasingly more difficult, especially after the outbreak of World War II with the attendant breakdown of transportation and the rise of food prices over wages. A recent study made of the national income for 1944 showed that agriculturalists (the farmer, the share cropper, and the agricultural wage earner) made up 71 percent of the total gainfully employed; (an estimated 12.6 million) yet they earned only 30 percent of the total national income. Urban employees (who contributed to social insurance organizations) formed 24 percent of the total number and received 20 percent of the national income. Income taxpayers (a total of only 300,000 people in 1944), who comprised only 2.5 percent of the total gainfully employed, received 30 percent of the national income, while another miscellaneous group of 2.5 percent received the other 20 percent.[22] These figures show a concentration of 50 percent of the national income in the hands of 5 percent of the people.

"Factors for the concentration of incomes are not difficult to enumerate," says Spiegel:

The high degree of concentration of ownership, both in agriculture and in manufacturing industries; high interest and profit rates; lack of educational opportunities, designed to improve income-earning capacities of children in low income groups; the tax system; poor social services; and finally, low wages. Important also is the persistent inflationary trend. This factor in conjunction with the narrowness of the home market and low purchasing power of a large proportion of the population accentuates the process of concentration of incomes.[23]

The financial elite of Brazil, in contrast to the uneducated and underfed majority, live in luxury and comfort. The rich travel, send their children to foreign universities, maintain large houses and a large staff of servants, and dip into the arts—as the rich do everywhere. The rather special propensity for luxury which Gilberto Freyre has described for colonial times in *The Masters and the Slaves* is still apparent among the descendants of aristocrats and the new rich who imitate them.

[21] Josué de Castro, *A Alimentação Brasileira á Luz da Geografia Humana* (Porto Alegre, 1937), pp. 134–139; quoted in T. Lynn Smith, *op. cit.*, pp. 349–350.
[22] Henry W. Spiegel, "Income, Saving, and Investment in Brazil," *Inter-American Economic Affairs*, I, No. 1 (June, 1947), 116.
[23] *Ibid.*, p. 119.

Regions of Brazil

In terms of the culture of its people, modern Brazil is more homogeneous than other great areas of the world of comparable size. India, China and the U.S.S.R. contain people of different cultures and of different languages. Even the neighboring countries of Bolivia and Peru contain people of two distinct cultures (Spanish and Spanish-Indian) who speak several different languages. Throughout Brazil, the people share one basic culture pattern—a culture inherited in the main from Iberia but flavored with African and American Indian elements. One language, Portuguese, is spoken over the entire area, except for an insignificant number of forest Indians and a few unassimilated Europeans in the far south. Yet, because the area is so vast and contains such widely different ecological conditions, and because communications have been so poor in the past (and still are) between one part of the country and another, the historical development of the different regions has varied. In some regions industrialization has taken place with great velocity; in others, it has not yet begun. One region is characterized by a large percentage of Negroes in the population and another has a population predominantly European white. The different regions present specialized versions of the national culture.

Brazilians are quite aware of such regional differences. The common man has, for example, stereotyped ideas as to the personality of the *Gaucho* (the man from the southern state of Rio Grande do Sul), of the *Cearense* (from the arid northeast), of the *Carioca* (of Rio de Janeiro), and of the *Paulista* (of São Paulo). There are regional literary movements, like the "Regional Traditionalists" of the northeast coast. Gilberto Freyre, after a trip from his native Pernambuco (northeast coast) to the extreme south of Brazil was moved to observe, in regard to dancing:

The truth is that the country is divided into regions which are distinguished one from the other, not only by the type of popular dance preferred by the people of each region, but also by their manner of dancing some of the traditional or native and even North American and European dances. Because international dances suffer regional stylizations or deformations, they are just as interesting for the psychological and socio-

REGIONS OF BRAZIL

logical study of regions as the native or traditional dances, peculiar to each region.[24]

And, in a lecture in which he defends Regionalism as a "social philosophy" against "excessive national" or "exaggerated international or cosmopolitan" tendencies, he observes, "but Brazil is not simply one natural and cultural region; inside the almost continental immensity of that part of America, nature and culture have their own subdivisions." [25]

Although most students are impressed with the great cultural diversity of the country, social scientists have only recently begun to study its regional differences by means of intensive and objective research methods. There is still, therefore, some variation in the manner of approach. Each student tends to divide the country on the basis of the criteria of his own discipline—that is, a geographer, an economist, a historian, or an agronomist each sees the map of Brazil differently. At present, however, it seems to me that Brazil contains six major regions or "cultural areas"—the Amazon Valley, the Northeast Coast, the arid Northeast, the Industrial Middle States, and finally the Wild West. Each of these is characterized not by one criterion but by a combination of several, including climate, surface features, racial composition of the population, historical past, and distinctive culture patterns.

THE BRAZILIAN AMAZON.—The Amazon Valley is shaped like a great fan with its handle at the mouth of the Amazon River. It reaches far beyond the frontiers of Brazil into Bolivia, Peru, Ecuador, Colombia, and Venezuela, but the major portion of the Valley falls within Brazil. Even outside Brazil, where the language is Spanish rather than Portuguese, the strength of the common ecology has been stronger than differences of nationality and language.

The Brazilian part of the Valley takes in the states of Pará, Amazonas, the northern portion of Maranhão, as well as the Federal Territories of Acre, Amapá, Guaporé and Rio Branco. The dense tropical forest that covers most of the Amazon Valley is by no means a vast swampy jungle. About 90 percent of the area is above flood level. South of the town of Santarem, about halfway from Belem to Manaus on the main stream of the Amazon, a range of hills rises

---

[24] *Região e Tradição* (Rio de Janeiro, 1941), p. 252.
[25] *Brazil, an Interpretation* (New York, 1945), p. 73.

over a thousand feet above the river. To the north near the Colombia and Venezuela frontiers, there are great stretches of grassy plains. A large area of savanna is found north of the main stream and north of the towns of Óbidos and Monte Alegre. Throughout the entire Valley, patches of grassy savannas break the monotony of the forest, and near the Amazon River itself lie wide flood plains covered with coarse grass. Marajó Island, at the mouth of the Amazon, is as large as Switzerland and consists mostly of periodically inundated plains.

The climate of the Valley is surprisingly temperate; Santarem, which is only a short distance south of the equator, has an average yearly temperature of 78.1° F. and at Manaus the yearly average is 81° F. The highest temperature recorded at Manaus from 1911 to 1935 was 99° F. and the lowest was 62° F. The climate is uncomfortable because of the monotony and humidity. The difference between the average daytime temperature during the warmest and the coldest months at Manaus is less than 4° F. The humidity runs as high as 80 percent or even 90 percent. Throughout the Valley rainfall is abundant; in Belem, the average is over 100 inches per year, in Manaus around 80 inches, and in the upper portions of the Valley as much as 100 inches. The rainy season, "winter" as it is called, occurs from January through June; yet actually these are only the months of the heaviest precipitation, for it rains considerably during each month of the year, even during the "summer" dry season. Everywhere along the Amazon it is cool enough at night to sleep with a light cover.

The great system formed by the Amazon and its tributaries has provided man with an easy mode of transportation. The Amazon itself is navigable for ocean-going steamers for more than 2,300 miles from its mouth, and river boats ply all of its major tributaries. Three short railroads total about 238 miles; in 1935 there were only approximately 1,600 miles of automobile roads in the whole Valley. Transportation is always by natural waterways and the people therefore live near the water. There are two cities, Manaus with about 110,000 people and Belem with approximately 290,000 people; there are less than 2,000,000 people in the entire Brazilian Amazon. At least half of these people live along the main rivers or along the small streams (*igarapés*), very much isolated from one another.

They build their houses of planks or straw thatch on high spots along the bank or on stilts out of reach of the seasonal rise of the river. Each house has its own landing wharf. As one travels along the river, one sees only these wharves every half mile or mile apart; on each one there is a stack of cordwood which the family has cut to be sold to the slow wood-burning steamers. Alongside the wharf there is always a canoe, the family's only means of transportation, for the house may be miles from *terra firme* (land above flood level). The house itself is usually hidden in the deep green vegetation and cannot be seen from the river.

Now and again there is a larger wharf and, behind it, a long low wooden building. This is a *barracão*, a trading post, to which the people from the scattered houses in the vicinity paddle every *quinzena* (15 days) to buy or to receive on credit canned goods, salt, a piece of cloth, kerosene, and other necessities of life and where they sell the rubber, Brazil nuts, palm nuts, and pelts which they collect. The customers of a trading post form a rural neighborhood. They are tied to a particular trading post by credit advances. They are generally acquainted and occasionally visit each other. The trader often invites his *freguezia* (customers) to the *barracão* for a dance or for a public celebration such as September 7, Brazil's Independence Day. In general, only the trader has social and economic relations outside this neighborhood. From time to time, he visits a not too distant village or takes the boat to Manaus or Belem. Sometimes the trader maintains a house in the village so his children may attend school. He is generally either in debt to a commercial house in the village or to one of the large firms in Belem and Manaus who send out river boats periodically to pick up forest products and to renew his stock of merchandise. Brazil nuts, palm oils, hardwoods, rosewood oil (used as a base for perfume) timbó vine (used for insecticide), animal pelts, alligator skins, and rubber are some of the products extracted from the forest and exported to the world from Belem, after passing through the hands of several middle men in this credit pyramid. At the bottom, the forest collector remains forever in almost debt slavery. Whenever they are able to pay up, they escape and move on, looking for a more favorable situation, only to fall in debt to another trader. The so-called "migratory in-

stinct" of the Amazon man which many writers have mentioned is
nothing but his constant struggle to escape.

The large trading firms profit from both selling and buying.
When the world demands one of these products of the Amazon
forest, they prosper. The story of rubber is an Amazon classic. Until 1913 or 1914, the world depended almost entirely upon the
Amazon Valley for its supply of rubber. During the second half
of the nineteenth century, a boom took place in many respects not
unlike the Alaska or California gold rush. In the stampede to buy
forest rubber, enormous tracts of land were sold in Belem and
Manaus, sight unseen and with faulty titles. The buyer would later
visit his lands to find out whether he really was the wealthy owner
of a tract of forest producing "black gold" or whether he had
thrown away his money on worthless jungle. If his land had rubber
trees, then labor was his great problem. Raids were made on Indian
villages, and the subsequent enslavement and brutal treatment of
Indians became an international scandal. Great numbers of men
were recruited in the drought areas of northeastern Brazil (see p.
236) and shipped in virtual debt slavery to the *seringais* (rubber
fields) of the Amazon. The owner furnished supplies and assigned
to each man several *estradas* (roads; that is, paths leading to some
150 to 200 rubber trees) each separated from the next by about 100
yards. From the trading post situated strategically at the mouth of
a tributary, the owner kept watch with rifles so that his gatherers
could not escape down river and so that intruders could not go
up river to buy rubber from them.

Money was plentiful in the cities. To the famous Opera House in
Manaus opera companies came directly from Europe to give performances. Both Belem and Manaus boasted cafés, bars, bawdy
houses, and fine chalets. In Santarem, old people who remember
these good days told of men who sent dress shirts to Portugal to
be laundered and of women who ordered their frocks from Europe.
Since the Amazon region was nearer to Europe than to the cities
of south Brazil in transportation time, the wealthy frequently traveled to Portugal and to France and sent their children to school in
Europe. Everyone looked to the future with the firm belief that the
Amazon was blessed by God with a permanent world monopoly

on rubber. As late as 1909, after the oriental plantations had been planted and were already beginning to produce, and just four years before the great rubber crash, Amazon writers were saying:

We need not therefore concern ourselves about Indian rubber plantations which have sprung up in Asia. The special climatic conditions of the Amazon Valley, the new system of treating our product now being applied with such success to our crops of hevea, the vast expanse of our India Rubber districts some of which have not yet been exploited and finally the manifold needs of modern industry, enable us to pay little heed to what others are doing in the same line of business. Indeed, were it not our duty to keep our eyes on the scientific discoveries relating to India Rubber, we could well afford to disregard foreign plantations altogether.[26]

Then, in 1912 the bubble broke. In 1876, an Englishman named Henry Wickham had smuggled some rubber seeds out of Brazil to England. They were planted at Kew Gardens and from these plants the English started rubber plantations in Ceylon and Malaya. As late as 1910, the Far East produced only 9 percent of the world's rubber, but by 1913 they more than equaled the Amazon production and, by 1924, 93 percent of all the rubber used in the world came from these Eastern plantations. Labor is cheaper in the Far East, and the trees, under plantation conditions, were developed to produce more rubber per tree than in the Amazon. By 1940, Brazil exported only 12,000 tons of rubber as compared with over 40,000 tons in 1912. Although, again during World War II, the Allies depended upon the Amazon Valley for a supply of natural rubber, when the Japanese overran the whole area of rubber plantations in the Far East, the development of synthetic rubber and the return of the Far Eastern plantations to the world market again doomed the wild rubber industry in the Amazon.

Following the rubber crash, Brazil nuts, another natural product collected from the forest, became the principal export crop of the Amazon and by 1940, 20,000 tons of nuts and 7,000 tons of shelled meats were exported.

The pattern of life of the rural inhabitant of the Amazon has remained little changed by the ups and downs of prices on the world

[26] *Album do Estado do Pará; mandado organizar por ᶜDr. Augusto Montenegro, Governador do Estado, 1901-8* (Paris, 1910), p. 182.

markets. He continues to eke out an existence from collecting forest products and from some subsistence agriculture. His way of life draws heavily on elements inherited from the Indian. Although only ten to twenty thousand tribal Indians are left in the Brazilian Amazon today, the Indian has contributed more to the life of the Amazon than to any other region of Brazil. The *caboclos* are mainly of American Indian racial stock although most of them have also a few European and Negro ancestors. Only a few Negro slaves were imported into the Valley and the number of European colonists was small. In 1852, it was estimated that "whites" made up only 8.5 percent of the Amazon population, Negro slaves only 2.3 percent, and *mestiços* (probably Negro-white mixtures) 4.9 percent. The rest were Indians or Indian mixtures.[27] Until the late nineteenth century the majority spoke *língua geral*, a modified form of the native *Tupí-Guaraní* language, which the missionaries adopted and taught as a lingua franca; others spoke Portuguese. Even today in the Amazon, so many names of places, of animals, of birds, and so many popular expressions come from the *língua geral* that the educated Brazilian from the south needs a glossary when he reads about the Amazon.

Like most rural Brazilians, the *caboclo* has inherited from the aborigines a wasteful method of agriculture by which forest land must be cleared and burned off each year or two to provide garden sites. His principle food crops—corn, beans, peppers, *cará* (*Dioscoria*, Sp.), peanuts, and manioc—are native to America. In the Amazon, manioc is the staff of life, the basis of all meals: *farinha d'agua*, a flour prepared from the poisonous variety of the plant, is the bread of the region, the necessary complement of any meal. Manioc is prepared as a food drink called *chibé* and as various types of cake called *beijú*. The distinctive dishes of the region, *tacacá* and duck with *tucupí* sauce, use manioc as an important ingredient.

Amazon folklore is strongly flavored by Indian survivals, and the *caboclos* retain many customs and beliefs of Indian origin. Although they consider themselves good Roman Catholics, many believe in American Indian supernaturals and call on medicine men to cure them by traditional methods. They tell of *Mãe d'Água* (Mother of

[27] Virgílio Corrêa Filho, "Devassamento e ocupação da Amazônia Brasileira," *Revista Brasileira de Geografia*, IV, No. 2 (1942), 283.

the Water) and *Zurupari*, a dangerous demon of the forest. They
believe that the *bôto* (a fresh-water dolphin) appears at night as a
handsome young man dressed in white to seduce young virgins, and
illegitimate children are sometimes called *filho do bôto* (child of the
dolphin). They have *pagés*, medicine men, who cure the sick by
sucking or massaging out of the body of the patient an object which
is thought to have been there by magic. As they work their cures,
the medicine men inhale great gulps of tobacco smoke and blow
clouds of it over the patient. They sing and dance, keeping time
with the rattle of a gourd, and as they sing they call their familiar
spirits to aid them in the cure. Sometimes they fall into a trance.
Almost identical practices have been described by early chroniclers
of the sixteenth century for the coastal Indians of Brazil and have
been witnessed by the writer among several Indian tribes. Amazon
caboclo medicine men differ from their Indian counterparts mainly
in the supernaturals they supplicate; he calls not only *Mãe d'Água*,
or *Zurupari*, but also Catholic saints. There is a greater residue of
Indian culture patterns and culture traits in the Amazon Valley than
elsewhere in Brazil.

THE ARID NORTHEAST.—To the south and east of the Amazon
basin, lies the arid *sertão* of Brazil. In terms of climate and surface
features it contrasts strongly with the humid tropical forests of
the Amazon. The *sertão*, inland from the so-called "bulge" of the
Brazilian coastline, covers the states of Ceará, Rio Grande do Norte,
the southeastern portions of Piauí, a southeastern portion of Maranhão, and the western portions of the states of Paraíba, Pernambuco,
and Bahia. It is a region of scrub forest, cactus, and bushes. Low
mesas and a few mountain ranges break the great expanses of flat
broken terrain. With the exception of a few major rivers, the streams
of the *sertão* are dry channels filled only now and again during the
year when the head waters are swollen with rains. Then, the stream
beds fill swiftly and with little warning; in a few hours they are
dry again. The rainfall is very irregular. Periods of drought may
last from one to three years and, even in "good years," the rain
comes in violent showers which turn the countryside temporarily
green. After a few days of hot sun and dry winds, the earth re-
sumes its dusty brown. In a normal year it may rain four or five
times a month in the rainy season (October to April). The annual

rainfall averages between 20 and 40 inches over the entire region; in the interior of the state of Ceará, which is typical of the area, the rainfall averaged 33.6 inches per year from 1910 to 1924. As in most arid regions of the world, there are a few oases, such as the area of the Serra de Baturité and the Chapada de Araripe in Ceará, where irrigation yields magnificent results; most of the *sertão*, however, is not suitable to agriculture.

Carnaúba palms, found in abundance near the coast in Ceará, produce an industrial wax that was fourth among Brazil's exports in 1941. An oil used in paint and varnish manufacture is extracted from the *oiticica* tree, which grows in the interior of Ceará and Rio Grande do Norte. Limited areas of the dry *sertão* produces a high-grade cotton with a long and strong fiber. Some *mamona* for castor oil is planted; recently there has been a movement to plant mulberry trees and to introduce a silk industry in the region. The typical economic activity, however, is grazing. The first Portuguese colonizers who were given land grants here, found the terrain unsuitable for agriculture, so they turned their large estates into cattle ranches. Grazing was not particularly profitable, and they were unable to purchase Negro slaves. By 1860, when the total population of the Province of Ceará was estimated at about 500,000, there were only 34,400 slaves and these few were concentrated along the coast and in the few agricultural oases.[28] The aboriginal people were unusually hostile, but the settlers were able to make use of many missionized Indians to care for their herds and to defend their vast estates. As usual, the Portuguese mixed with the Indians so that the typical *sertanejo*, as the rural inhabitant of this region is called, shows in his high cheekbones, straight black hair, and Mongolian eye form, strong indications of his American Indian ancestry. Only in limited areas, such as the margins of the Rio São Francisco in western Bahia, are Negroid elements noticeable in the population.

A few towns grew up near the headquarters of large colonial estates and in the few well-watered areas but, in general, the population of this arid region lives scattered over the countryside. Although most of the great estates have now been divided, the majority of the people still live on land belonging to cattle ranchers. They serve as *vaqueiros* for the ranch owner, caring for a portion of his

[28] Djacir Menezes, *O Outro Nordeste* (Rio de Janeiro, 1937), p. 58.

herds in return for one fourth of the increase. During "good years" when there is sufficient rainfall, they clear garden sites from the scrub forest and plant maize, beans, manioc, and vegetables. They must fence in their miserable gardens with brush and poles to keep out the cattle. Frequently the gardens are lost for lack of rainfall. Often they are planted in the beds of dry streams to take advantage of the moisture and low water level, but this, too, is precarious; a sudden rainfall in the headwaters of the stream may wipe out the garden in an hour. The people depend, therefore, on meat for the major part of their diet. Because they are not landowners and because of the uncertainty of their habitat, the *sertanejo* is known to be seminomadic, moving about from year to year in search of greater security.

The hostile environment, intensified by droughts, has driven large numbers to seek a living elsewhere. Terrible droughts have occurred from time to time for centuries. They are recorded as far back as 1710–1711 and afterwards for each eight- to fifteen-year-period until the present. The drought of 1790–1794 almost depopulated the province of Ceará and the most famous drought, that of 1877–1879, sent so many refugees to the coast in search of water and food that the city of Fortaleza grew from a small town to a large city of over a hundred thousand inhabitants. At this time, the entire *sertão* was burned black by the sun, cattle died, and there were no crops. One writer estimated that almost 500,000 people died from starvation or from diseases connected with starvation during this one drought period.[29] Refugees, attracted by the beginnings of the rubber boom, began the movement from this region into the Amazon Valley. Since then there have been droughts in 1888–1889, 1898, 1900, 1915, 1931–1932, and in 1942–1943, of greater or lesser intensity. Each time the *sertão* has been turned into a virtual desert and a large number of people have died of starvation or migrated. The exact number of the migrants from the *sertão* to the Amazon is not known but the movement has amounted to at least a half million people since the great drought of 1877–1879; during the last serious drought, in 1942–1943, almost 50,000 refugees went to the Amazon, attracted by the minor rubber boom of World War

[29] Herbert H. Smith, *Brazil; the Amazons and the Coast* (New York, 1879), pp. 416–21; quoted in T. Lynn Smith, *op. cit.*, p. 313.

II. Since 1912, however, the greatest exodus has been to the south, especially to São Paulo, where coffee plantations were in need of labor. At least a half million people entered São Paulo from the northeast; from 1935 to 1940 inclusively, some 326,109 people came there from other states of Brazil and most of these from the *sertão*, and in the years before this period the movement was almost as great.[30]

Despite the ravages of drought and emigration the arid *sertão* is still one of the most densely populated areas of Brazil. The State of Ceará has an average of 14.1 people per square kilometer, Rio Grande do Norte, 14.8; and Paraíba 25.6; whereas the State of Paraná in the south, which has great areas of excellent agricultural land, has only 6.3 people per square kilometer.

Since 1880 the central government has been building dams to conserve water and facilitate irrigation; and since 1910 a federal government bureau (*Inspetoria Federal de Obras contra as Secas*) has been constructing dams, irrigation systems, and roads to forestall the droughts which hit the region about twice in each generation. The government must spend large sums to make these areas habitable, while other regions with more favorable natural conditions are practically uninhabited. However, the *sertanejo* loves his semiarid homeland, maintaining that in the years between droughts it has a fine, healthy climate and that the land in rich. As soon as the news that "it is raining in Ceará" reaches *sertanejos* who have left their lands with a curse, they come trooping back again.

The semidesert has produced a people noted for their fierce courage and enormous endurance, although they are humble and quiet in manner. The *sertão* is a region of fanatic religious movements and famous outlaw bands. The dramatic story of the village of Canudos, which in 1896 led by the religious fanatic, Antonio Conselheiro (Anthony the Counselor) battled federal troops until they were destroyed in a bloody massacre, has been told in a Brazilian masterpiece by Euclides de Cunha, *Os Sertões*.[31] Padre Cícero of the village of Joazeiro was another religious leader of the *sertão*. Although he was excommunicated by Rome, he was almost an absolute spiritual and political leader in the *sertão* and in 1914 led a

---

[30] T. Lynn Smith, *op. cit.*, p. 329.
[31] *Rebellion in the Backlands*, tr. Samuel Putnam (Chicago, 1944).

counter revolt against state troops. Large numbers of pilgrims still flock to the tomb of Padre Cícero each year.

The *sertão* has also produced numerous outlaw bands, one of the most famous of which was led by Virgilio Ferreira da Silva, known as Lampeão. From about 1920 until he was killed in 1938, Lampeão and his *cangaceiros* were the terror of the whole region. Tales of their robberies and murders were tempered with stories of aid that they had given to the poor and needy, in the Robin Hood tradition. Ranchers and villagers gave them help in fear of reprisals, and politicians frequently used the bandit band against their opponents. Like other bandit leaders of this region, Lampeão became a legend and the people of the *sertão* would not believe that he was dead until his head, cut from his body, was exhibited publicly in many towns and finally placed on exhibition in the Museum of the Medical School in Salvador, Bahia. Blind troubadours sing folk verses nowadays in the market places of the *sertão* recounting the adventures of Lampeão, and of other famous *cangaceiros* of the region. Like religious fanaticism, banditry seems to result from the constant frustration of this harsh and uncertain environment.

THE NORTHEASTERN SUGAR COAST.—In the seventeenth and eighteenth centuries, the northeastern coast of Brazil east of the *sertão*, in what is now the States of Pernambuco, Alagôas, Sergipe, and Bahia, was the richest part of the country. The rainfall along this coastal strip is dependable. The land was originally covered by thick tropical forest springing from a fertile red soil. Sugar cane was planted inland from the settlements at Recife and Salvador in the late sixteenth century. Great profits attracted many wealthy families from Portugal and sugar soon made them even richer in the new world. The *Casa Grande* of these colonial plantations, with its own chapel (really its "individual church"), dispensaries, various kitchens, several drawing rooms, numerous bedrooms, and a great veranda was the center of a luxurious and aristocratic patriarchal family life. The *Casa Grande* often had its own Padre, a son of the family who had entered the church. It had its own school for the numerous descendants of the patriarch. Behind these fine structures were found numerous mud and thatched huts (the *Senzala*), the quarters of the Negro slaves, who were the field hands, the nurses, the playmates, the cooks, the intimate chamber maids, and even the concubines of the aristocrats of the *Casa Grande*. Large cities rose along the sugar coast; Salvador and Recife became the most important centers of Brazilian colonial life. The elaborate gold-leafed wooden carving in the church of São Francisco, the polychrome tiles in the church of Saúde and Rosário, and in fact, almost any of more than three hundred churches in the city of Salvador are witnesses to a colonial life of tremendous wealth.[32]

In the early eighteenth century, however, the sugar industry began to decline. Gold and diamonds were discovered to the south in Minas Gerais and attracted many plantation owners with their slaves from the northeastern coast. By the middle of last century the large-scale sugar production of the West Indies was competing with Brazilian sugar on the world market, and the abolition of slavery in Brazil in 1888 took away the labor on which the great plantations depended. Most of the old-style plantations have now disappeared. Commercial companies have installed modern equipment for large-scale production, and modern plantations depend on wage laborers and on seasonal migratory workers who come in from the dry *sertão* to harvest the sugar cane. The northeast coast still produces sugar, but other regions such as Minas Gerais, State of Rio de Janeiro, and São Paulo now compete strongly for the domestic market, and Cuba produces more sugar than all of Brazil. The northeastern coast is no longer the richest center of Brazilian economic, political, and cultural life as it was in the seventeenth and eighteenth centuries. Many of the descendants of the old sugar aristocrats, therefore, look back with a nostalgia to the past and try to conserve, as much as is possible in modern times, some of the traditions of the old aristocracy. People are proud of their family names and pay considerable attention to family genealogies.

The northeast coast, however, is not decadent. Recife in the State of Pernambuco is a busy city of almost 500,000 people. It has cotton textile mills and factories which produce soap, cigarettes, leather goods, fiber for sacking and many other articles; and, surrounding Recife, there are many large sugar *usinas*, the successor

[32] Gilberto Freyre's *Casa Grande e Senzala* (tr. as *The Masters and the Slaves*, New York, 1946) is an exciting socio-historical study of this feudal colonial society based on sugar-raising.

of the colonial *engenhos*. Recife is the busiest commercial center in north Brazil; it has an excellent port, warehouses, and wharves. Salvador in Bahia, although it preserves many colonial monuments, has a population of about 350,000. It is a modern port and the center of a fertile agricultural area, the *Recôncavo*, which produces sugar, tobacco, coffee, castor oil, fruits, and cacao. Southern Bahia, inland from Ilhéus, is the center of cacao production, producing most of the 250 million pounds which is the average yearly crop of Brazil.

Since a very large proportion of the millions of slaves imported into Brazil came to the sugar plantations, there is a strong Negroid element in the modern population of this region. The census of 1890, which was the last one to classify the population by race, showed the population of the State of Bahia to be 20 percent Negro, and 46 percent mulatto. Such figures were averages for the entire state and it must be remembered that in the western portions of the northeastern states, of Bahia and Pernambuco, Amerind-European mixtures predominate. The percentages of Negro and mulatto elements along the coast must have been even higher than these figures indicate. Even today, after more than fifty years of "bleaching" through race mixture, the high percentage of the Negro element along the northeastern coast is obvious even to the most casual observer.

As one might expect, African cultural elements are more numerous in this area than in any other regional culture of Brazil. Africa has contributed heavily to the culture not only of the mass of Negro and mixed population but also of the "white" aristocracy itself. As Gilberto Freyre has described, African influences passed from the slaves to their masters through the close relationships of Negroes and whites. A large number of words of African origin are used and, according to Gilberto Freyre, the soft melodious accent of Portuguese, as it is spoken in this region, is a result of Negro influence.[33] Nursery rhymes, stories for children, and the mythical bogeyman characters used by adults to frighten children are a fusion of Iberian and African forms. African culture contributed much to the diet of the region. Cooking is still done mostly by Negro women, as in slave times. Such dishes as *Vatapá* made with *dendê* oil, peanuts, rice flour, fish, shrimps, and various spices;

[33] Gilberto Freyre, *op. cit.*, pp. 342–43.

Carurú, made with fish, okra, peppers, oil, and various herb spices; and *Acarajé*, beans fried in *dendê* oil, were adopted from African slaves. Even traditional Portuguese or aboriginal foods were modified by the Negro mode of preparing and spicing them.[34] Especially in this region of Brazil, popular music and dance owes much to African culture. The modern *samba* is a development of the *batuque* which was danced by Negro slaves at night in the colonial plantations. The rhythms of the music with great use of drums contrasts violently with the slow sad music of the Brazilian *modinhas* sung to the accompaniment of the *violão* (guitar) in regions of less pronounced African influence.[35]

Perhaps the most spectacular heritage of Africa, however, are the fetish cults called *macumba* in Rio de Janeiro, *candomblé* in Bahia, and *Xangô* in Pernambuco, and corresponding to the *Vodum* of Haiti. Such cults are found along the whole coast of Brazil from São Luis in Maranhão to Rio de Janeiro, wherever there is a concentration of Negroes; but they are strongest in the northeastern coastal region. A North American sociologist, Donald Pierson, estimated that in 1935–1937 there were between seventy and one hundred cults in Salvador, Bahia.[36] The cults are highly organized and follow a complex ritual. They are based on religious beliefs predominantly of West African origin, but now mixed with Portuguese Catholic traditions. The divinities of the cults, the *Orixás*, are recognizable African gods, such as *Xangô*, the deity of lightning, *Ogun*, the god of war and iron, *Yansan*, the divinity of wind, and *Oxossi*, the god of the hunt; each is represented by a small fetish idol. Since the cult members are at the same time Catholics, each of these deities can be identified with a Catholic saint: for example, Santa Barbara as *Yansan* and Santo Antônio, as *Ogun; Exú*, the god of evil is, of course, the Devil. The sacerdotes of the cult are both male and female. Several cults in Bahia are led by female *Mães de Santos* (Mothers of the Saints) but generally it seems that the principle sacerdote is usually a *Pai de Santo* (Father of the Saint). These priests are generally old people with knowledge of the rituals and the ability to recognize the manifestations of the different deities.

[34] *Ibid.*, pp. 459–66.
[35] Artur Ramos, *O Negro Brasileiro* (São Paulo, 1940), pp. 234–36.
[36] Donald Pierson, *Brancos e Pretos na Bahia* (São Paulo, 1945), p. 341.

Each cult has a *terreiro*, a temple, which is generally a large, vaulted straw hut, surrounded by smaller structures in which the altars of the deities are kept and in which the leaders and a few followers live. On specific days dedicated to the particular gods, the members gather to carry out appropriate ceremonies. Each cult has its special ceremonial season, as well as ceremonial days, depending on the deities in the pantheon to which it is dedicated. For example, the ceremonial season of a cult dedicated to *Ogun* is from mid-September to the first week in December. Each Sunday during this period is dedicated to a special deity. During the other months of the year, except during Lent, other ceremonies are celebrated from time to time.[37] During such ceremonies the priest or priestess is aided by male assistants called *ogans* and by female devotees called *Filhas de Santo* (Daughters of the Saint). The entire ceremony is directed by the priest (or priestess) who makes sacrifices of a chicken, rooster, sheep, or a goat, whatever is the appropriate animal for the deity and the occasion, and who arranges the altar of each deity. Especially in Bahia, these sacerdotes of the *Candomblé* cults are highly respected individuals, not only by the members of their own cult but also by the entire Negro population of the region. According to Donald Pierson even a few upper-class whites visit the *Pai de Santo* to ask advice regarding business and political matters and to ask help in curing and preventing illness.[38] The members of the cult treat with a grave formality their leader, who often acts as the judge in disputes between the members. From time to time, the cults have been declared illegal in various cities of the northeast coast but they are such an integral part of the culture that they have survived in spite of occasional legal persecution. Although they are still illegal in several cities, in Salvador, at least, they are now recognized by the civil authorities, who in 1944 issued formal documents of permission for their ceremonial meetings. Numerous culture patterns derived from Africa, as well as from the aristocratic plantation system distinguish the northeast coast from that of the arid northeast, from the Amazon Valley, and from the extreme south of the country.

THE EXTREME SOUTH.—The three southern states, namely, Paraná,

[37] *Ibid.*, p. 342.
[38] *Ibid.*, p. 377.

Santa Catarina, and Rio Grande do Sul, differ from other regions in climate, in physical character of the terrain, in natural products, in racial composition, and in their combined Gaucho-European culture patterns. The climate is temperate, and, although it seldom freezes except in the mountains, there are annual frosts; snow is not uncommon from June through August. The seasons are definitely marked in contrast to those of north Brazil, where they are hardly more than periods of more or less rainfall. The frosts make tropical crops somewhat hazardous. Only in northern Paraná may coffee be planted, although along the coastal lowlands as far south as Rio Grande do Sul, sugar may be raised. Rice, beans, onions, potatoes, tobacco, alfalfa, some wheat, barley, and rye are planted in the south.

The northern limit of frosts in northern Paraná mark the southern limit of the tropical forests. In the states of Paraná and Santa Catarina there are great stretches of Araucaria pine forest, which furnish Brazil with an important source of lumber. The tropical forests of northern Brazil do not offer possibilities for large-scale commercial lumbering; a small area of tropical forest contains a great variety of trees, many of which are unsuitable for lumber. The valuable hardwoods are scattered, one here and another hundreds of yards away. In contrast, the pines of southern Brazil form a homogeneous forest, ideal for the commercial lumber industry. In 1944, the *Instituto Nacional do Pinho*, a Federal Agency of the Brazilian Government which sets the quotas for cutting and replanting of pine, authorized the production of almost two million cubic meters of lumber from these southern forests.

Wherever there are pines, the maté plant (*Ilex paraguayensis*) is also found. Each year groups of collectors move into the forest, where from June to October they strip the leaves from the wild trees and dry them over slow-burning fires. As maté tea is widely used only in Uruguay, Argentina, Paraguay and Brazil, the industry has never been one of great importance. In 1929, however, Brazil shipped almost 85,000 tons to Argentina; recently Argentina has almost doubled its own production by planting maté and in 1940 purchased only 50,000 tons from Brazil. Lumber and maté are distinctive extractive products of the south of Brazil.

The most characteristic landscape of southern Brazil is the great

rolling prairie, the pampa, which opens up to the south of the pine
forest areas and extends from Santa Catarina and Rio Grande do
Sul into Uruguay and Argentina. Grazing is the economic activity
associated with the pampa. The first settlers here were given huge
grants of land, and by accumulation of various individual grants
they were able to build up *estâncias* as large as 20 to 30 square
leagues (320 to 480 square miles). The region was soon noted for
grazing and even today two-thirds of the total area of the State of
Rio Grande do Sul are devoted to pastures. The majority of the
herds are said to be descendants of the scrub cattle of colonial days,
but such well known varieties as Hereford, Polled-Angus Short-
horn, and a black and white variety of *Hollandesa* are now being
bred. In 1938, the livestock population of Rio Grande do Sul was
said to have been 26,613,905 head, including cattle, sheep, hogs,
horses, asses, mules and goats.[39] Although there are several modern
packing plants which export meat to north Brazil and to Europe,
most of the cattle slaughtered in Rio Grande do Sul are still prepared for market by the traditional method known as *charque* (dried
and salted beef). The meat is cut in huge slabs, dipped in brine,
salted, and placed on poles to dry in the sun. Meat prepared in this
manner may be kept indefinitely without refrigeration and may be
shipped to distant parts of the country. *Charque*, or *carne seca* as it
is sometimes called, cooked with black beans or in a stew, is a traditional Brazilian food and the extreme south produces almost 70
percent of all the *charque* consumed in Brazil.

The basic element of the population of these southern states is
European. Few Negro slaves were imported into the region and
these were concentrated in the urban settlements near the coast.
There was some mingling with the aboriginal Indians but compared
to the strength of the Amerind element in the population of the
Amazon and Northeastern *sertão* the Indian element in the southern
population is negligible. The first settlers were Paulista *Bandeirantes;*
later, the Portuguese government introduced settlers from the
Azores. The latter were primarily soldiers and they were settled
in the south to hold the land for Portugal against the encroaching

---

[39] Lúcio de Castro Soares, "Livestock Campos in Rio Grande do Sul," in *People
and Scenes of Brazil* (Rio de Janeiro, Instituto Brasileiro de Geografia e Estatística,
1945), pp. 133–36.

Spaniards from the south. Some Spaniards succeeded in entering, and they, too, merged into the population. As early as 1824, however, colonists from other European countries began to arrive. They were attracted to this region rather than to north Brazil by its similarity in climate and in physical features with the countries from which they came. At first the Brazilian government was eager to have them there in order to guard against a possible northward expansion of the Spaniards. Between 1829 and 1859 more than 20,000 Germans were settled on small farms in the region with the financial aid of the Brazilian government. Beginning in 1850, Herman Blumenau, a German physician, helped a group of settlers to come from Pomerania to an area (now called Blumenau) surrounding the city in Santa Catarina, and by 1870 more than six thousand had followed. In the succeeding years other immigrants of German, Austrian, Swiss, Italian, and Slavic (Russian, Polish, Ukranian) origin came to south Brazil and formed colonies. From 1889 to 1896 about fifty thousand Poles settled in the State of Paraná. Nowadays it is estimated that over 500,000 people in the State of Rio Grande do Sul, 300,000 in Santa Catarina, and about 126,000 in Paraná are of German descent. There are approximately 200,000 Poles and their descendants living today in south Brazil. These three most southern states have drawn more heavily upon Europe for their population than any other region of Brazil.

The extreme south is characterized by two local culture patterns —that of the old Luso-Brazilian people and that of the comparatively recent European immigrant. The contrast is great between the two ways of life. The recent Europeans settled for the most part in colonies and lived aloof from the Luso-Brazilians. The Germans, especially, resisted assimilation tenaciously; they were proud of their German heritage, their German way of life, and their German schools. It was estimated before World War II that, of almost 100,-000 people inhabiting the area around the city of Blumenau, some 75 percent spoke German; around another town of the same zone of Santa Catarina, "Only 5% are Portuguese, 7% German and 88% Italian." [40]

Until recently the German colonists maintained their own schools and teaching was in German, but in 1938 the Brazilian government

[40] Preston E. James, *Latin America* (New York and Boston, 1942), p. 531.

prohibited instruction in any language except Portuguese. The houses of these European colonists are strikingly similar to those of their home countries. In Santa Catarina they are built with walls of brick, steep roofs covered with a flat tile, and large windows, contrasting sharply with the adobe or crude plank houses covered with palm thatch of the rural Brazilians. Recent European immigrants practice better farming techniques and seem to be more permanently settled on the land than the rural Brazilians. They have established one-family farms on which they produce butter, cheese, garden vegetables, eggs, and meat for their own consumption. In Santa Catarina, the Germans grow maize to feed hogs and milk cattle and, in both Rio Grande do Sul and Santa Catarina, Italians have vineyards and produce wine. In Paraná, the Polish colonists have introduced a four-wheeled wagon, similar to the Prairie Schooner of our West, and throughout the region covered bridges attest the presence of these recent European colonists. As Preston James says in his description of Rio Grande do Sul:

The influence of these people (European colonists) in the whole life of the zone can scarcely be measured, for it goes far beyond mere numbers of citizens of European descent. Although the dominant theme of the region is Brazilian, it is a new kind of Brazil set off from the rest of the country by the presence of a considerable number of people who know how to engage in the hard physical work of pioneering in the forests and who are content with relatively modest profits of an economy which is not speculative.[41]

In contrast, the Luso-Brazilians have continued for the most part to be pastoralists. In the zones where they have taken up agriculture, as in the area of rice production inland from Porto Alegre in Rio Grande do Sul, the system of cultivation is typically Brazilian; that is, large estates, tenant labor, and one commercial crop. Most of the great pampa is still devoted to grazing. Although many of the immense estates have now been divided, great *estâncias* with fine homes for the owner surrounded by the miserable huts of the peons are not uncommon. The gaucho, as the cowboy of the pampa is called, is in a sense the symbol of the Luso-Brazilian culture. He is noted for his horsemanship; formerly, he captured cattle with the *bola*, a rope tipped with metal balls which was flung whirling to wrap

[41] *Ibid.*, p. 526.

around the legs of the animal. Meat is the basis of the gaucho diet. It is a gaucho habit to drink maté from a gourd vessel with a silver tube (*bombillha*) many times during the day. Gaucho music is melodic and sung to the accompaniment of a guitar or an accordion.

In his classical study of the arid northeast, Euclides da Cunha contrasted the gaucho, and what might be called the gaucho spirit, with the *vaqueiro* of the northeast.[42] "The southern gaucho, upon meeting the *vaqueiro* at this moment," he wrote, "would look him over commiserately. The northern cowboy is his very antithesis. In the matter of bearing, gesture, mode of speech, character, and habits there is no comparing the two." Whereas the *vaqueiro* is downcast, defeated by his environment, the gaucho "awakes to life amid a glowing, animating wealth of Nature; and he goes through life adventurous, jovial, eloquent of speech, valiant, and swaggering." His clothes "are a holiday garb compared to the *Vaqueiro's* rustic garments"—wide breeches cut something like plus-fours, a colored poncho, a bright scarf, a broad sombrero, and short boots. Traditionally he carries "a gleaming pistol and dagger in the girdle about his waist—so accoutered, he is a conquering hero, merry and bold. His horse, inseparable companion of his romantic life, is a near-luxurious object, with its complicated and spectacular trappings."

Although literally the expression "gaucho" refers to a cowboy, the name is used for all native inhabitants of the State of Rio Grande do Sul. A Brazilian writer remarks, "Generally, we see him [the gaucho] with his typical customs, habits, and psychology in the midst of the plains or in the frontier region, but he also lives in the city, where he shares in urban life without losing the influence of earlier life and his love of the plains." [43] Gaucho music, dances, folklore, sports on horseback, characteristic dress, trappings for his horse, and food habits are part of a cultural complex extending beyond the borders of Brazil into Uruguay and Argentina. Constant contact with these Spanish-American neighbors has given a distinct Spanish flavor to gaucho culture. Throughout Brazil, the gaucho (the native of Rio Grande do Sul) is thought of as an aggressive

[42] *Os Sertões*, pp. 91–92.
[43] Lindalvo Bezerra dos Santos, "People and Scenes of Brazil," *Rev. Bras. de Geografia* (Rio de Janiero, 1945), pp. 137–38.

warrior, somewhat crude, boisterous in manners, and belligerent in contrast to the more courtly, graceful, and retiring Brazilian of the older and more aristocratic regions to the north.

The south of Brazil is an energetic and dynamic region. Porto Alegre, its chief city, is now the third largest industrial city of Brazil, the center of a leather industry, textile factories, breweries, wineries, and meat packing plants. It is an aggressive modern city of more than 350,000 people with a State University, fine residential districts, and a busy commerce. In Rio Grande do Sul and Santa Catarina are found the only coal reserves in Brazil and these deposits are being called upon to furnish fuel for the nascent steel industry of the country (see p. 253). It is generally conceded that the people of south Brazil have a higher standard of living, a more varied diet, and better health than those in the north. Through an increase of births over deaths, the population is growing (360 percent during the last 50 years) more rapidly than in Brazil as a whole (see p. 220).[44] The western forest areas of Paraná and Santa Catarina are considered one of the best areas for colonization in Brazil. The soil, a rich dark red variety called *terra-roxa*, is noted for its fertility and the temperate climate gives relative freedom from tropical insects. Furthermore, the area is not so remote from such markets as São Paulo and Rio de Janeiro as to make the cost of transportation facilities excessive. Large numbers of Europeans will certainly be attracted as soon as they are allowed to enter, and these southern states will in the future play an increasingly important part in the agricultural and industrial life of Brazil.

THE INDUSTRIAL MIDDLE STATES.—The states of São Paulo, Minas Gerais, Rio de Janeiro and a part of Espírito Santo form a region which may be called industrial Brazil. However, it is in no way as highly industrialized as, say, the New England States or the Pittsburg-Detroit area of the United States. Brazil is predominantly an agrarian nation, and agriculture is still the occupation of the majority of the inhabitants of even the more industrialized states. Yet, within this region are found the two greatest cosmopolitan cities of the country, Rio de Janeiro and São Paulo. There are more motor roads and more railways than in any other part of Brazil; commercial farming has been more extensively developed; and most of

[44] See T. Lynn Smith, *Brazil: People and Institutions*, pp. 143-44.

Brazil's industries, in fact all of its heavy industry, are gathered here. Intimate cultural and commercial relations with Europe and the United States are maintained through the excellent ports of Rio de Janeiro and Santos and by numerous international airlines. Most of Brazil's modern universities, research laboratories, trade schools, and cultural institutions are found in this region. In short, except for such cities as Porto Alegre, Salvador, and Recife, it is in this part of Brazil that modern Western technology has been introduced most successfully and from it that modern Western culture is diffused to the rest of the country.

It is a homogeneous region in which rapid industrialization and a modern system of communication has in recent years smothered old local differences. The rising middle class, the industrial worker, and even the farmer, tend to have their opinions formed by newspapers, magazines, and radio, as do the mass of the population in Europe or in the United States. This is, of course, true also for the inhabitants of such other modern cities as Porto Alegre, Recife, and Fortaleza, but they are "islands," so to speak, of industrial culture in regions where older folk cultures have survived to a greater extent than they have in the middle states.

The memory of three distinct local traditions is still cherished. Each of the three states, São Paulo, Minas Gerais, and Rio de Janeiro, has had a different historical development. The State of Rio de Janeiro was developed by sugar planters early in the colonial epoch, Minas Gerais owes its importance to the discovery of rich mineral deposits in the seventeenth century, and São Paulo, the home of the adventurous *bandeirantes*, became economically important in the nineteenth century with the development of coffee planting. Each state developed a local culture pattern and the residue of these differences still is found in the folk culture of isolated sections. These local differences for the most part, however, have by now given away to a standardized Brazilian version of modern machine-age culture. Each of the two great urban cities, São Paulo and Rio de Janeiro, is the center of its own zone of commercial, political, and cultural influence.

The city of São Paulo, with over 1,300,000 inhabitants, is the political and industrial capital of the richest state of Brazil. The fertile southern portion of Minas Gerais falls into the São Paulo focus. The area around the city is said to be one of the most rapidly growing industrial regions of the world. The nucleus is an old settlement, having been established as a Portuguese mission in 1554, but the city is new. In 1883, the old colonial town of irregular, narrow, and mostly unpaved streets had a population of only 35,000 people. By 1890, it had grown to 64,000 and in the next ten years, rose to 240,000, by 1920 reaching 579,000, a little over one-third of the present number.[45] Now it is the most important economic center of Brazil.

This phenomenal development was due mainly to coffee. As the northeast coast of Brazil grew rich when sugar was a commodity for which the world was eager and as the Amazon prospered from its monopoly on rubber, the State of São Paulo went through a cycle of great fluorescence based on one commercial crop. By 1900, Brazil produced three-fourths of all the coffee grown in the world and the State of São Paulo was the greatest producer in Brazil. The city of São Paulo was the financial center, and Santos was the port of this prosperous coffee commerce. The Paulista coffee planters, like the northeastern sugar planters before them, became very wealthy. They built luxurious houses on their coffee *fazendas* and *palacetes* in the city of São Paulo; and many of them owned houses in Paris. The coffee industry attracted immigrants from abroad and from other states of Brazil. Land values boomed. Transportation facilities were rapidly built. As early as 1906, however, there were signs of a decline in the coffee market and the Brazilian government was forced to begin a series of control measures designed to limit the production of coffee so as to bolster the price on the world market. Finally between 1929 and 1931 the price of coffee dropped from 24.8 cents a pound to 7.6 cents per pound and another Brazilian one-crop boom was over.

Unlike the Amazon after the rubber boom and the northeast coast after the sugar boom, São Paulo did not decline economically after the coffee price fell. Wealth gained from coffee, however, stimulated both state and city in other directions. Even before the 1929 crash, planters with foresight began to root up their coffee trees and to sow cotton or sugar cane. By 1935, more than a million and half acres of land were devoted to cotton and the state is now an important exporter of this product. Oranges were planted on a com-

[45] Benjamin H. Hunnicut, *Brazil Looks Forward* (Rio de Janeiro, 1945), p. 401.

mercial scale for export to Europe. With the decline of coffee prices many owners of large estates divided and sold their property; great areas of the state are now occupied by small holders who engage in "mixed farming." [46] The city became a manufacturing center from capital derived from coffee. It has textile mills in which cotton, wool, and jute are woven. It has rayon and silk industries. It produces machinery, clothing, foods, beverages, chemical products, cement, glass, paper, rubber goods, and hundreds of other articles. North American automobile manufactures maintain assembly plants in São Paulo. During the last war, when trade with Europe and the United States was difficult, São Paulo began to manufacture many articles which had always been imported, and became the greatest financial and commercial center of Brazil. Santos, its port, has grown to a city of almost 200,000 people and it is now the busiest port in Brazil.

São Paulo reflects this rapid commercial and industrial development. New streets and avenues are constantly being opened. New factories, office buildings, and apartment houses spring up almost overnight. On the outskirts, new residential districts have developed which would do credit to any city in the world. More than 10,000 building permits are issued annually, and optimistic real estate operators have laid out miles of future residential districts. In the heart of the city there are Italian, Japanese, and Syrian districts of unassimilated immigrants and miles of slums. The city has built a huge stadium to seat eighty thousand people at soccer games and other public spectacles. São Paulo laboratories and scientific institutes are perhaps the most advanced in South America, and the University of São Paulo is developing its faculties with scientists and scholars from abroad. The people are busy, they walk quickly, and they brag of the *bandeirante* spirit of energy and enterprise. They complain that the other regions of Brazil are slow and behind times and a drag on their rapid progress. They sometimes call themselves, with some insight, "the Yankees of South America."

The city of Rio de Janeiro with nearly 2,000,000 inhabitants is the cosmopolitan capital of Brazil with the glitter of diplomats and of foreign travelers from all over the world. The breath-taking beauty of the city and its natural setting are justly famous. Lying at the foot of rich green mountains which rise abruptly from the

[46] Preston James, *Latin America*, pp. 498–99.

bay, it is bordered with long stretches of sparkling white beaches that form graceful curves along the foot of the mountains. It is a city for outdoor living. People climb the mountains to visit the great statue of Christ overlooking the city from towering Corcovado, the Hunchback; they crowd the wide beaches so near their homes. Guanabara Bay, the famous harbor of the city, is dotted each weekend with small sailing boats. The *Cariocas*, as the natives of Rio de Janeiro are called, are famed for the *sambas* which they sing and dance each year at the annual Carnival, as well as for their sharp and subtle sense of humor. Each day in the small cafés along the Avenida Rio Branco, on the Largo da Carioca, or along the small narrow streets, where people sit to sip a penny cup of coffee, new stories and new jokes pass around the tables. Much of the talk in these cafés is about *futebol*, as they call soccer, or about politics. The *Carioca* does not discuss, he argues intensely, the standing of his team, a recent trade of a professional player from one team to another, or the stand of his political party in the coming election, but his views are tempered with a rare sense of the ridiculous. So many stories made the rounds of the cafés during the regime of Getulio Vargas that the President is said to have made a large collection of those about himself. The *Carioca* pokes fun at the more energetic *Paulista*. "The *Paulista* is too busy making money to sit down long enough to drink his coffee," he says. Rio de Janeiro is a beautiful city and its people are warm, gay, and lighthearted in contrast to the bustling *Paulista*.

Rio de Janeiro is not merely an international resort or a tourist mecca; it is a busy modern city. It is the administrative capital of Brazil and the seat of numerous federal agencies and ministries. It is an important commercial, financial, and industrial center. Most companies with business interests in Brazil have offices in Rio de Janeiro. There are textile mills, cigar and cigarette factories, breweries, pharmaceutical laboratories, and many other industries in or near the city. Its port is a close rival to Santos in the flow of goods and in port facilities. Rio de Janeiro is also the hub of a rich mineral and agricultural area made up of the State of Rio de Janeiro, northern Minas Gerais, and a portion of Espírito Santo.[47] All means of transportation focus on the city. In the Paraíba Valley to the west, oranges are produced for domestic and foreign markets and con-

[47] Preston James, *op. cit.*, p. 432.

siderable rice and coffee is grown. Sugar is produced in the vicinity of Campos, and a large portion of Minas Gerais and the neighboring portion of the State of Rio de Janeiro produce beef and dairy products. The northern part of Minas Gerais and western Espírito Santo are an important mining area.

The discovery of gold in Minas Gerais in the seventeenth century and of diamonds in the eighteenth attracted an influx of settlers from the northeast sugar plantations, from São Paulo, as well as from Portugal itself. Rio de Janeiro, as the exporting center, was assured of becoming an important city. The colonial town of Ouro Preto, one of the early mining centers of Minas Gerais, became a political stronghold, for a time outranking Rio de Janeiro. During the gold boom, Minas Gerais was the richest part of Brazil. The fine churches and palatial homes built here in colonial times are only rivaled in the cities of the northeast coast.

The cycle of gold in Minas Gerais lasted about one century (1700–1800), although a few mines still produce enough gold to make it profitable to work them. The famous mine of Morro Velho, for example, which has followed a vein of gold 8,000 feet into the earth is nowadays operated by a British-Canadian company. There are also rich deposits of nickel, manganese, chrome, bauxite, quartz crystal, mica, diamonds and numerous semiprecious stones. Near the headwaters of the Rio Doce at Itabira is found one of the world's richest supplies of iron ore (see p. 264). About 250 miles from the source of iron ore and near limestone deposits of Minas Gerais, at Volta Redonda in the State of Rio de Janeiro, the Brazilian government has built a plant for large-scale production of iron and steel products, with the help of a loan from the Export-Import Bank of the United States. Volta Redonda is being built to produce more than 300,000 tons of steel annually. A new city has been built around this huge plant with a hospital, a health center, schools, shops, a hotel, and houses for workers, technicians, and executives. Volta Redonda, on the railway to São Paulo, about 100 miles from Rio de Janeiro, falls within the Rio de Janeiro area of attraction. Since Rio de Janeiro is the capital of Brazil much of the wealth of the country is concentrated there and heavy industry now being stimulated by the Federal government naturally tends to be placed in the Rio de Janeiro zone.

THE FAR WEST.—The states of Goiás and Mato Grosso, located

in the heart of South America and west of the industrial middle states, are today much like the "Far West" of the United States in the eighteenth century. The area is sparsely inhabited. In 1940, there were only 1,267,134 people in a total area of 2,138,181 square miles, and most of them are clustered along the few railways which penetrate the region. Great expanses are almost completely un-inhabited. The territory between the Xingú and Tapajoz Rivers in Mato Grosso is only partially explored; the few tribes of In-dians here are among the few remaining untouched savages in the world. Several tribes, such as the Chavantes and the Kayapó, still make war on Brazilians.

The lawlessness and violence usually associated with frontier society are found in this "Far West." There are well-known bad men, with several killings to their credit, living unmolested by law. In many small towns of Goiás and Mato Grosso citizens go about armed. Self-styled leaders, such as a ranchowner or a trader, often take it upon themselves to gather together a posse to capture a murderer and he is seldom turned over alive to the legal authori-ties. The "law of escape" is the excuse for executions. Here and there, throughout the region, diamonds or gold are found in the river beds and, when a new *garimpo* (placer mine) is discovered, a boom town springs up, attracting prospectors, placer miners, merchants, prostitutes, gamblers, and others. Registro, on the Ara-guaya river, was just such a rowdy boom town several years ago. Within a year after diamonds were discovered near by in the river bed, the population doubled. Prices were inflated. Murders were frequent. The small town boasted bars, bawdy houses, and gambling establishments. Stories spread of fabulous finds bringing wealth over night to a poor miner and of poor miners being cheated by un-scrupulous traders. When news came that richer finds were being made downriver at Marabá, large numbers of miners and their hangers-on moved there. As frequently happened in the United States, in the Brazilian "Far West" the prospector and the explorer are often followed by the cattle rancher and finally by the farmer.

Much of the "Far West" is good farming and grazing land. Southern Goiás is a high plateau which rises to over 4,000 feet above sea level and gradually drops off to the north in the direction of the Amazon. The Goiás plateau is connected to São Paulo by a

railway, but the service over the single-track line is so slow and
the backlog of freight so great that it takes weeks to make the
run between São Paulo and Anápolis in Goiás. There are a few
motor roads beyond the end of the railway and trucks furnish the
only means of rapid transportation into the hinterland. Gasoline
is so scarce that a considerable amount of freight still moves by
oxcart and by pack animal. Despite lack of transportation, the Goiás
plateau frontier is experiencing a rapid expansion. It is a region
of *campo cerrado*, plains with scrub vegetation, and is still basically
a zone of grazing. Recently, the native herds have been improved
by crossing with the Indian Brahman cattle, or *Zebú* as they are
called in Brazil, which seems to give them an added resistance to
heat and to disease. The plateau is rich in minerals; there are known
deposits of quartz crystal, mica, nickel, chrome, gold, and diamonds.
The oxidized nickel-colbat ore deposits on the Tocantins River at
São José are among the richest known in the world, but, like other
mineral reserves of this area, their exploitation depends upon the
development of transportation. A motor road has been built from
Anápolis to the Tocantins River and it is hoped that soon the ore
will be transported overland to the industrial regions on the coast.

The soil of a large portion of this plateau is *terra roxa*, the same
fertile red soil found in the State of São Paulo and in the northern Paraná. The lack of transportation again hinders large-scale
agriculture, but in recent years plantation owners from São Paulo
have been clearing and planting in Goiás with assurance that in
the near future motor roads and railways will connect this area with
the markets on the coast. The Brazilian government has established
a large agricultural colony of small farmers just north of Anápolis
and in 1946, despite transportation difficulties, the colony is said
to have marketed 160,000 bags of rice (about 140 pounds per bag)
in São Paulo. The wealth of the zone is attracting numerous settlers
each year; according to recent reports (*Time*, April 7, 1947),
almost 50,000 people pass annually through Anápolis into the
frontier beyond. In the future, the Goiás plateau will certainly be
one of the richest agricultural and grazing areas of the country and,
if Brazilian industry ever breaks away from the coast, it would
seem to be a potential zone for industrial growth.

West of the Goiás plateau lie the great plains of Mato Grosso.

There are great extensions of pure grass plains (*campo limpo*) and patches of plains covered with scattered thickets (*campo cerrado*). Near the Rio Paraguay the plains degenerate into swamp lands; to the north, in the State of Mato Grosso, they fade into the thick jungle of the Amazon Valley. The open grass plains country of southern Mato Grosso has received a large number of settlers during the last fifteen to twenty years. A railroad has recently been completed to connect São Paulo with Corumbá on the Bolivian frontier, via the Mato Grosso plains, and within a short time, it will extend to La Paz in Bolivia, stretching from the Atlantic coast to the Pacific. When this is done, Brazil will have access to the Bolivian oil fields around Santa Cruz.

The growth of the city of Campo Grande, now the most important trade center of the region, indicates the rapid development of southern Mato Grosso. In 1910, Campo Grande had only 1,500 inhabitants and now more than 50,000 people live in the county (*município*) of Campo Grande.[48] The new settlers are agriculturalists, but the basic economic activity of the zone is grazing. Ranches are still enormous and cattle run semi-wild over unfenced plains. Together with northern Paraná and the Goiás plateau, the zone offers excellent possibilities for agricultural settlement.

In recent years, the federal government of Brazil has sponsored a policy called "The March to the West" to attract Brazilians from the coast to the undeveloped interior. In 1943, a special agency called the Foundation of Central Brazil was established for penetration and eventual settlement of this vast "Far West." An expedition is now working its way year by year across the almost unexplored area between the Araguaya and the Xingú Rivers. From a base on the upper Araguaya they are pushing a line of communications to Santarem on the Amazon River. The country is to be surveyed from the point of view of permanent settlement and the Foundation has plans for transportation and communications with the coast. Agricultural colonies, new industries, and even cities are in the Foundation's blueprints. With modern technological equipment, this wilderness should not prove too difficult to bring under control and, if Brazil opens up immigration to Europeans, the manpower would seem to be forthcoming. But until these plans are

[48] Benjamin Hunnicut, *Brazil Looks Forward*, p. 43.

actually realized, only the coastal areas of Brazil are being effectively occupied, leaving the great interior, like a hollow core, empty and deserted.

The National Culture

Despite the marked differences which these various regions reveal, a framework of basic cultural uniformity characterizes Brazil as a nation and as a distinct culture area of the Western Hemisphere. According to Gilberto Freyre, this "healthy minimum of cultural basic uniformity" is composed predominantly but not exclusively of Portuguese, therefore of European, culture patterns and values.[49] Since Brazil as a political unit was a creation of the Portuguese, the traditional patterns of government, administration, business enterprise, law, and education are derived from Portugal. Although Amerind influences are strong in one region of the country, African influences in another, and recent northern European influences in still another, it was the Portuguese who were the governors, and in a broad sense, the teachers of Brazil during the period of formation of the national culture. Portuguese settlers formed an important component of the Brazilian people from the Amazon Valley in the north to the pampa in the extreme south. The Portuguese, in a sense, are the common denominator of all Brazil.

Because so many of its basic patterns and values derive from Portugal with its Iberian Latin culture, Brazilian national culture shares many common features with all so-called Latin cultures and especially with those of the Western Hemisphere. Recently Dr. John Gillin of the University of North Carolina, in a most stimulating article entitled "Modern Latin American Culture," [50] described some of the patterns and values which characterize the regions of America which were former Spanish colonies and which distinguish them from Anglo-American culture. The Latin American cultures, according to Dr. Gillin, are Roman Catholic in religion and their Catholicism is Iberian in its emphasis on the cult of the Saints, public *fiestas*, monastic orders and religious brotherhoods (*confradias, hermandades,* and so on).

Philosophically, Latin American cultures are humanistic rather

[49] *Brazil, an Interpretation,* p. 75.
[50] *Social Forces,* XXV, No. 3 (March, 1947), 243-48.

than puritanical; they emphasize logic and dialectics over empiricism and pragmatics. The power of argumentation is more important that the actual objective manipulation of natural forces
and objects. In education, theory is stressed above the details of a
process, and one learns by memory and repetition of the same
subject year after year rather than by experimental learning by
doing. In Latin America, the family is an exceptionally strong and
solid unit. Officially the male is dominant and women are passive
and retiring. There is an exaggerated double standard of sexual
morality. A young man is expected to have premarital sex experience; but should it be known that a girl has made one false step,
her chances of marriage are slight. There is a wide extension of
kinship terms—second and third cousins are considered "cousins"
—and through numerous godparents (the *compadre* system) social
solidarity is achieved with non-relatives. Latin Americans place
kinship and institutionalized relations over personal friendship and
business ties. The patterns of law and legal procedure follow Roman law as developed by the Code Napoleon rather than the Anglo-
American patterns which we know. Latin American society is
characterized by great socio-economic class differences derived
from slavery or colonial feudalism. Towns are built on a "plaza
plan" in contrast to our "mainstreet plan." Houses are generally
placed flush on the street with no front yard. These and many
other culture complexes and elements are common to all Latin
American cultures and, in a general sense, to Brazil.

Yet, the national culture of Brazil is clearly distinguishable from
that of other Latin American countries such as the Spanish-Indian
culture of Bolivia, Peru, and Ecuador, and as the European-Spanish
culture of Argentina, Uruguay, and Chile. It differs from these not
only in specific formal customs and culture patterns but especially
in the singularly Brazilian interpretation and orientation which
common features are given. The result is a different way of life and
a different way of looking at the world. It is a fundamental mistake,
if one is to deal with the countries to the south of us on anything
more than a superficial level, to include Brazil with the Spanish-
American countries as one large group. Attempts to describe Latin
America as a unit have all used the expression "except for Brazil"
with remarkable frequency.

The features which set Brazilian culture off from other local Latin American cultures result from the differences in ecology, differences in the aboriginal cultures encountered in the area, differences in the Portuguese variety of Iberian culture, and differences in the internal historical development of Brazil. The aboriginal culture of the Indians of the territory which is now Brazil was especially well adapted to the semi-tropical and tropical environment. They were few in number in comparison to the great masses of American Indians on the west coast of South America. Unlike the Indians of the West Indies, they were not decimated in the first thirty to forty years of conquest, and the Portuguese newcomers learned a great deal from them. They had an influence on Brazilian national culture out of keeping with their small numbers. Although Portuguese and Spanish civilizations of the fifteenth and sixteenth centuries were similar in their main outlines, the Iberian peninsula is known to have been an area of great cultural and racial diversity. The Portuguese were even less conscious of racial differences than the Spaniards; and they were less warlike and had less orthodox religious fervor than the Spanish conquistadors.

Brazil was the only Portuguese colony in America and, as such, it was isolated from the other American colonies not only by language but by the strict mercantile policy of Portugal, which went to extreme lengths to debar foreign influence. Even today there is perhaps more cultural communication between Rio de Janeiro and New York and Paris than between Rio de Janeiro and Buenos Aires. While the Spanish-American countries of the Western Hemisphere continued to look to Spain as the center of cultural influence (and still do to some extent), Brazil turned to France and to all Europe, as soon as it was politically free of the mother country. Portugal in the nineteenth century was a minor European nation and was looked down on, rather than up to, by Brazil. The fact that the country received more Negro slaves than any other in the Western Hemisphere accounts for the African influences which have given a special tone to Brazilian culture. Brazil was first an Empire before it was a Republic and the class structure of the native Empire with its native nobility, the so-called "Barons of the Empire," was unique in America.

Moreover, Brazil is the only country in the Western Hemisphere in which the people speak Portuguese. Brazilian Portuguese is as different from that spoken in Portugal as American English is from that of England. Many local expressions have been developed and many terms, foreign to the mother tongue, have been borrowed from native languages (Amerind and African), as well as from other European languages. Although there are minor differences in dialect and variations in expression from one region to another, Brazilian Portuguese has such a different intonation and vocabulary from the language of Portugal that there is never any doubt which language is being spoken. With this primary difference in language go a multitude of subtle cultural differences reflected by language, such as modes of address, concepts of beauty, and expression of attitudes. A traditional expression of endearment *"minha nêga"* (literally, "my Negress") used sometimes by a white man to his white wife reflects the peculiar Brazilian attitude toward people of a darker hue and memories of warm personal relations with Negroes.[51] The modern Brazilian manner of expressing superabundance *é mato* (literally "it is forest") can only be understood in terms of the lush forest growths in Brazil. The richness of popular expression is a direct reflection of the varied and rich culture of its people. A visitor from Lisbon has about the same chance of understanding what is said in a Rio de Janeiro café as an Oxford don in an American fraternity house.

Although race mixture is a common phenomena in most Latin American countries, it does not occur to the same extent as in Brazil. Although Brazilians are not without a certain race prejudice, as is shown by the claim of some Brazilian whites that they feel a revulsion from the *catinga do preto* (smell of the Negro), in general, one finds here less emphasis on color as a symbol of superiority or inferiority than elsewhere. Even the caste system of colonial times, with its numerous slaves and its plantation aristocracy, seems to have been tempered by this lack of racial antagonism. During the Empire, men of slave ancestry and low birth rose to high positions in the aristocracy and the monarchial system (see p. 217). There were mulatto Barons and Viscounts, and the Crown Princess herself is said to have made a point of dancing with André Rebouças, a noted engineer and a dark mulatto, when she noticed that a lady

[51] Gilberto Freyre, *The Masters and the Slaves*, p. 418.

had refused to dance with him, presumably because of his color.[52]

In all Latin American cultures there is an emphasis on family ties, but in Brazil it might almost be said that there is a cult of the family. Although present-day conditions—smaller houses, apartments, and industrial life—have brought profound changes in the Brazilian family, it is still a relatively large and decidedly an intimate group. Social life of many Brazilians is carried on predominantly with relatives. There are birthday parties, baptisms, weddings, and family gatherings. The group of relatives is remarkably large; kinship terms are applied to individuals for whom kinship would have been forgotten in other countries. A father's first or second cousin may be called "Uncle" and his children may be "cousins." The spouse of a distant "cousin" is often called "cousin." Beyond any possible kinship connection, solidarity is assured by the godparent relationship (*padrinho, madrinha* and *afilhado*) set up at baptism, at confirmation, and at marriage. It is common in Brazil at marriage for each participant to invite one man and woman to act as godparents at the religious ceremony and a different pair for each in the civil ceremony. The couple then garners eight new godparents at marriage. In Brazil "cousins" and godparents are used to facilitate official and commercial relations; small favors and special consideration may be asked of a *parente* (relative) or of a *padrinho* (godfather). This extraordinary extension of the terms of relationship and the use of ceremonial relationships to extend family ties is considered *muito Brasileiro* (very Brazilian) by Brazilians themselves.

Foods and food habits also differ from those of the surrounding Latin American cultures. Although each of the various regions of Brazil is famous for special dishes, such as the Afro-Brazilian dishes of Bahia and *churrasco* (a barbecue) of Rio Grande do Sul, over most of the entire country *farinha* (manioc flour), black beans, rice, dried beef (*charque*), and coffee are the basis of most meals. *Goiabada* (Guava paste) and *Marmelada* (quince paste) with a piece of cheese are desserts known in every part of Brazil. Except in the maté-drinking area of south Brazil, nothing is more typically Brazilian than the small cups of black coffee, the *cafézinho*, served several times a day in homes and offices. Spain and Spanish-American

[52] Donald Pierson, *Negroes in Brazil*, p. 170.

countries are famous for their late dinners. In Brazil, breakfast is coffee and milk with a piece of bread or manioc cake (*beijú*), lunch is traditionally at 10:30 to 11 A.M. and dinner at about 5 P.M., followed by a light supper before retiring. In Brazilian cities, these traditional hours for meals have been modified by the necessities of modern commercial and industrial life, to conform to the meal hours of Paris or New York.

Numerous other culture patterns differentiate Brazil from the rest of Latin America. The carnival period before Lent, although celebrated in most Catholic countries, is the most important festival of the year to Brazilians, overshadowing both patriotic and religious holidays. The zeal with which the Brazilian people lose themselves in dancing and music for four days and the manner of celebrating carnival is not found elsewhere. The music they sing and the style of dancing is uniquely Brazilian. The music and the dance which is known abroad by the generic term of *samba* (in Brazil there are local terms and local varieties) is quite distinct from the Argentine tango, Cuban rumba, Mexican folk music, and North American jazz. Other festivals, such as *São João* (on June 24), are celebrated in a specifically Brazilian manner. On the great plantations *São João* was the equivalent of Christmas on the southern plantations of the United States during colonial times.[53] There were great dances in the *Casa Grande* and in the slave quarters the Negroes danced their *sambas* around large bonfires. There were special foods, songs, and music for the occasion. Even nowadays Brazilians celebrate the Eve of Saint John by building large bonfires, roasting sweet potatoes, sending up paper balloons, and setting off fireworks.

Brazilian folklore with its complex of *bichos*—such as *quibungo*, of African origin, a horrid creature half human and half animal which swallows children through a hole in its back, *Sacipererê*, a little Negro with one leg who pursues travelers, and *pé de garrafa*, the man with a sharpened leg who lures men into the forest—is a fusion of Amerind, African, and Iberian folklore elements. It is now a truly Brazilian folklore, no longer similar to any of the ingredients. Although Brazilian domestic architecture resembles in a general way that of other Latin American countries, for the "patio" is substituted a backyard-like *quintal* and the internal arrangement of

[53] Gilberto Freyre, *Brazil, an Interpretation*, p. 57.

the Brazilian house with its small room for visitors and its emphasis on the dining room, which serves the family for intimate living, is somewhat different from the typical Spanish-American dwelling. In northern Brazil, people traditionally sleep in hammocks, and even in the south the Brazilian type of hammock is a common fixture in any house. These, and many other cultural traits, too numerous to describe here, are distinctive aspects of Brazilian culture.

Finally, there seem to be a series of distinctively Brazilian personality traits, if we may accept the impressions of travelers and of students. All people who know Brazil and Brazilians agree that different behavior may be expected from Brazilians than from other Latins. Brazilians are said to be more overt and more voluble than the comparatively taciturn Argentinian; they are less proud and less worried about losing face than the Spanish-American. To the Argentinians, Brazilians are not dignified, so they call them "monkeys." Even Brazilians agree that there is something profoundly Brazilian about the personality of José Carioca, the sly, friendly, and talkative parrot created by Walter Disney. Yet, many writers, both native and foreign, mention a certain sadness, a softness, and a melancholy about the Brazilian. "In a radiant land lives a sad people" is the opening line of Paulo Prado's famous interpretative work on Brazil.[54] This is another side of the Brazilian personality. This same author describes the excess of sensuality and the great love for luxury of the Brazilians, and Gilberto Freyre mentions a "gentlemen-complex," that is, an inclination for white-collar work and the professions and a distaste for physical labor, as a personality trait of Brazilians inherited from colonial feudalism.[55] With these traits goes a desire "to get rich quick" and a love of gambling. The economic history of the country is made up of a series of speculative booms, and almost all Brazilians gamble in some form—either in the *jogo do bicho* (a sort of numbers racket), in the federal or state lotteries, or until recently in the luxurious casinos.

Brazilians give a uniquely Brazilian twist to institutions and concepts which they share with the Western world. The Brazilian monarchial system, Brazilian democracy, and Brazilian dictatorship were unlike similar forms of government as they existed in Eu-

---

[54] *O Retrato do Brasil* (5th ed., São Paulo, 1944), p. 11.
[55] *Brazil, an Interpretation*, pp. 62–63.

rope. Even the recent dictatorship, despite its aping of European patterns, never became a harsh system with strict control over the people. Jokes about the dictator, complaints and discussions of the lack of freedom of expression, and rumors of growing opposition were discussed openly in cafés and salons. When the dictator was finally overthrown, it was a typical bloodless Brazilian revolution. As one student of colonial art remarked: "In Brazil, even Christ hangs comfortably on the cross."

Brazilian Potentialities

From a study of Brazil itself, it is probably impossible to predict with any assurance what will be its future development. Many important factors depend not only on Brazil but upon the world at large. What natural resources will be crucial for success in the future, as coal and petroleum have been in the recent past? What raw materials will the world need which are now found in Brazil or which may be grown in its vast undeveloped agricultural areas? Will new tropical products be discovered that will make it worth-while for Brazilians, and for the rest of the world, to inhabit and develop the great Amazon basin? The answers to such questions depend upon the development of Western technology and in-dustry and upon the future direction of international politics. Will the world allow Brazil to make use of its vast resources? Some of the factors that seem to make for the success of a region and for a ris-ing standard of living for its people are known. In terms of these, Brazil has certain advantages and disadvantages; it is potentially a rich and powerful nation, if some of its many problems are solved.

What are these natural advantages? What are some of Brazil's major problems? How can these be solved? On the credit side of the ledger is the wealth of mineral resources which as yet has hardly been touched. Although the richest zone of minerals is in the moun-tain ranges of the State of Minas Gerais, already mentioned (see p. 253), other valuable reserves of minerals are found throughout the country. Brazil has approximately 13 billion tons of iron deposits which is calculated as somewhat more than one-fifth of the total known deposits of iron ore in the world. Brazilian reserves of man-ganese are estimated at 30 million tons. Nickel-cobalt deposits are found in the States of Goiás and Minas Gerais. Chrome is found in Bahia and in Minas Gerais. There are three main fields of tungsten ore, one in the extreme south (in the State of Rio Grande do Sul) and two in the northeast in the States of Paraíba and Rio Grande do Norte. The resources of bauxite, from which aluminum is made, are said to be over 50 million tons; there are eighty known bauxite producing locations in Brazil. Magnesite, the source of metal magnesium used for airplane manufacture as well as for medical supplies, is found in Bahia and Ceará. Zirconium, important in the manufacture of flare signals and blasting caps, is found in several localities and Brazil is the greatest producer of this mineral in the world. Other minerals of which Brazil has important reserves are mica, quartz crystal, various semi-precious stones (such as tour-malines, amethysts and aquamarines), diamonds, tantalite, colum-bite, and cassiterite (tin ore). Without some of these Brazilian metals, the United States would not have been able to carry on its great industrial campaign during World War II. Yet, Brazil might easily produce several times the amount of ore produced at present. Some of the most precious of these minerals must be moved from the mining areas on mule back, and mining techniques are extremely primitive. The lack of adequate supplies of coal for smelting is also a real drawback to making use of these minerals in local industry.[56]

The lack of sufficient supplies of coal and petroleum, which modern industry depends upon primarily for its energy supply, is a serious handicap to Brazil. There are several known deposits in the south, but this coal contains so much ash that it is not highly effective for fuel. Brazil produces only 800,000 tons of domestic coal each year and imports nearly 1,500,000 tons. With the recent development of industry, especially of steel, the need for coal is greater than ever before. The steel mills at Volta Redonda (see p. 253) alone will call for about 500,000 tons of coal per year and at least some of it will have to be high-grade coal from abroad. Even with effective expansion of production in the mines of southern Brazil and corresponding improvements of transportation, it is doubtful if Brazil

[56] See Morris L. Cooke, *Brazil on the March* (New York, 1947), especially Chapter V. I have drawn on this report of a North American Mission on Brazil's industrial problems and potentials for this section of the article. Statistics have also been derived from the annual bulletin of the Brazilian Ministry of Foreign Affairs.

can supply anything near its own needs in coal; the reserves are limited and the needs of industry grow with leaps and bounds. The lack of petroleum is even more acute. Until recently, no crude petroleum was produced. At present, the limited production of the oil field near Salvador in the State of Bahia (about 970,000 gallons up to September during 1942) is insignificant compared with the necessities of the nation (694,092 tons imported in 1940). The Paraná basin in southern and western Brazil as well as a great part of the Amazon Basin are considered potential oil-producing areas, but oil has not yet been discovered in either region in commercial amounts. For the time being, therefore, Brazil must expand its facilities for the transportation and storage of petroleum and coal which it must import.

Because of these shortages, it is all the more important that Brazil develop its great potentialities of water power. South of the arid coast of Natal, most of the rivers flow into the sea off the Atlantic highlands, which rise sharply from the coast. These rivers are potential producers of electric power. The industrial regions of Rio de Janeiro and São Paulo, therefore, have considerable electric power near at hand. Back from the coast, other river systems such as the Paraná with several falls including the great Iguassú falls, offer vast sources of potential hydroelectric power. On the São Francisco River which empties into the Atlantic in northeastern Brazil are the little known waterfalls of Paulo Affonso, only about 250 miles from the growing industrial center of Recife. The electric-power potential of these falls alone are estimated at almost one million horsepower. With the help of a North American technical mission, the Brazilian federal government has developed a plan for the São Francisco Basin similar to the project carried out by the Tennessee Valley Authority. Not only would the control of the São Francisco River furnish unlimited hydroelectric power, but the region, now alternately arid and flooded, is a potential agricultural and industrial area. The potentiality of Brazil's hydraulic energy has been estimated at about 19,000,000 horsepower by the Ministry of Foreign Affairs, which would place it among the nations of the world with the greatest potentialities. By developing this latent source of energy, railroads and factories might be powered by electricity, and the lack of coal and petroleum might be overcome to some extent.

Brazil is potentially one of the world's greatest food producers. Although great areas of the country are semi-arid desert, swamps, or dense tropical forest, it has been estimated that about 80 percent of the total area is potentially productive under present methods of agriculture and stock raising. Great unexploited areas of the southern states of Paraná and Santa Catarina, southern Mato Grosso, and of Goiás offer first-class possibilities for agriculture and pasture. Even the relatively highly populated states of São Paulo and Rio de Janeiro still offer potential areas for agriculture. Yet, Brazil does not produce enough food to feed its population. There are short-ages of basic foodstuffs in Rio de Janeiro and São Paulo as well as in the other urban centers along the coast. Even in rural areas, malnutrition from faulty diet and low agricultural production is the main cause for the lack of energy and the low resistance to disease. A recent study of nutrition problems written by a well-known Brazilian scientist, Dr. Josué de Castro, is aptly entitled *A Geografia da Fome* (The Geography of Hunger).[57]

The causes of this state of affairs indicate the remedy. The widespread and wasteful system of "fire agriculture" inherited from the Indians, the division of the land into great *latifundios* which produce one commercial crop for export, and the lack of transportation facilities to send food from where it is produced to where it is most needed—these seem to be the principle causes for Brazil's low agricultural production. According to a Brazilian senator speaking recently before the constitutional assembly in Rio de Janeiro, only about one-fourth of the national territory is held as rural property—the rest is unexploited and uninhabited lands. According to this same senator, the area under cultivation consisted in 1940 of only 6.5 percent of the area held as rural property, or only 1.5 percent of the entire area of Brazil. More than half of the total area under cultivation was given over to coffee and cotton, that is to say, to export and not food crops.[58] The history of Brazilian agriculture is one of a series of one-crop monopolies and of large estates

[57] Rio de Janeiro, Empresa Gráfica "O Cruzeiro," 1946.
[58] *Ibid.*, p. 295.

producing for export. The relatively few small farmers using anti-quated methods and tools are hardly able to produce enough food for their own consumption. According to the census of 1920, only 15 percent of all Brazilian farms used plows. Because there is usually no way for the small farmer, or the commercial farmer for that matter, to get his products to market, there is little incentive to produce a surplus.

Transportation is one of Brazil's most urgent problems. According to the Ministry of Foreign Affairs, there were only 242,995 motor vehicles in the entire country in 1940, as compared with the millions in the United States. Brazil had only 160,460 miles of motor roads and 21,283 miles of railroads, as compared to more than 3 million miles of motor roads and 227,283 miles of railroads in the United States. Furthermore, the railways of Brazil do not all have the same gauge, thus freight must be unloaded and loaded again if it is routed over different railroads. Motor roads and the extension and electrification of railways are primary necessities for the development of Brazil. With adequate transportation facilities, the agricultural products now available might be distributed where they are most needed, and farmers would be stimulated to produce food for the distant markets of urban Brazil. Modern methods of agriculture are already taught in several agricultural schools and these need only be introduced to the farmer. There is already a tendency for large estates to be divided into small farms, and the federal government has established several extensive agricultural colonies of small farmers in various parts of the country. Land for agricultural expansion in this vast country is plentiful; Brazil, despite its present agricultural plight, might well become an important breadbasket for the world.

Unless the people of a region have the technological equipment to make use of them, natural resources contribute little or nothing to the standard of living. The world has seen many nations with technical know-how outstrip others with much richer storehouses of natural resources. Brazil has numerous excellent technicians and there are several up to date engineering schools, medical schools, agricultural colleges, and scientific institutions, yet the lack of technicians and trained personnel is one of the main obstacles to developing the country. There are only 12 engineering schools, and these have only 5,000 graduates. There are schools of agronomy which have high standards, as the *Escola Nacional de Agronomia* near Rio de Janeiro and the one at Piracicaba in São Paulo, and there are a few trade schools and agricultural schools on a practical level. The number of graduates, however, is insignificant. For lack of teachers and accommodations the entering classes at most of the schools—even at the National School of Chemistry in Rio de Janeiro—are limited to thirty or forty students. Usually there are three or four applicants for each opening in a recognized technical school. Industry suffers from the lack of primary education of the common workman and from the lack of skilled hands. To overcome its percentage of illiteracy, Brazil must develop educational facilities on several different levels. First, until the masses are literate, it will remain difficult to communicate modern concepts of hygiene, of agriculture, of industry, and of social life in general. Primary education therefore should be made available first. More trade schools must be established to provide skilled labor and more schools of agronomy are needed to train not only agricultural specialists but also practical farmers. Higher education must be opened up on a broader basis. Brazilian educators are already aware of the seriousness and urgency of the educational problem; future progress depends heavily upon whether or not these educators are heard by the people and their government.

One eternal problem, namely *a falta de braços* (the lack of people), would not seem difficult to solve. The amazing fertility of the Brazilians has already been described (see p. 220). With the introduction of modern methods of hygiene, the shocking infant mortality and the exorbitant death rates would be reduced and the population would increase. In addition, millions of Europeans are anxious to emigrate. The Brazilian government has recently announced plans for European immigration but no definite action has been taken. Such regions as Santa Catarina and Paraná, Goiás, and southern Mato Grosso might well receive hundreds of thousands of immigrants. With such rich areas available for settlement, there is no reason for Brazil to take up the difficult problem of settling Europeans in the Amazon tropics. Future studies might discover reasons and methods for settling millions of people in the Amazon basin, but until that time, these southern areas have greater priority.

One of the most difficult aspects of the problem has been to find immigrants willing to live in rural areas. Most uprooted Europeans prefer to settle in the coastal cities, but the need is for agricultural-ists and skilled workmen rather than urban tradesmen. With mil-lions of people in Europe ready to emigrate, an aggressive policy on the part of the Brazilian government would bring people to a land where they are needed.

These are some of the principal strengths and weaknesses of Brazil. Brazil is rich in area and in natural resources. With its people of diverse origins and of great racial tolerance, it has the basis for a great social democracy. Modern Western civilization with its em-phasis on technology is available to Brazil for the borrowing. Po-tentially, Brazil has most of the necessary equipment to make it one of the great nations of the world. For the time being, however, "Brazil is rich and Brazilians are poor," as the people sigh with their rather fatalistic sense of humor. Brazil must learn to make use of its natural wealth by developing its human potentialities unless it is to remain forever a land of false promises.

SUGGESTED READINGS

Cooke, Morris L. Brazil on the March. New York, 1947.

Cunha, Euclides da. Rebellion in the Backlands. Tr. from Os Sertões, with Introduction and Notes by Samuel Putnam. Chicago, 1944.

Freyre, Gilberto. Brazil, an Interpretation. New York, 1945.

—— The Masters and the Slaves (Casa Grande e Senzala). New York, 1946. A study in the development of Brazilian civilization, tr., from the Portuguese of the fourth and definitive Brazilian edition, by Samuel Putnam.

Hunnicut, Benjamin H. Brazil Looks Forward. Rio de Janeiro, 1945.

James, Preston E. Latin America. New York and Boston, 1942. See pp. 385–571.

Kelsey, Vera. Seven Keys to Brazil. New York and London, 1940.

Nash, Roy. The Conquest of Brazil. New York, 1926.

Pierson, Donald. Negroes in Brazil. Chicago, 1942.

Smith, T. Lynn. Brazil: People and Institutions. Baton Rouge, La., 1946.

Zweig, Stefan. Brazil, Land of the Future. Tr. Andrew St. James. New York, 1941.

H. J. Simons

RACE RELATIONS AND POLICIES IN SOUTHERN AND EASTERN AFRICA

The Clash of Cultures

FRINGED by long stretches of surf-beaten, sand-barred coast or dense forest belts, possessing few safe natural harbors and navigable rivers, and with access into the interior obstructed by the escarpments of high plateaus, Africa, the most tropical of the continents, was until recently also the most isolated and, to the Western world, the least adequately known of the great land masses. In spite of the centuries-old commerce in slaves—Africa's first and for a long time only important export commodity—Western contacts with the continent south of the Sahara were confined to coastal regions until well into the third quarter of the nineteenth century. The only exception was in the extreme south, where white colonization began in 1652. In 1835 there were fewer than 100,000 whites in Africa, and of these 66,000 were found at the Cape. Even today, more than half of Africa's four million whites are concentrated in southern Africa, with a further million settled along the Mediterranean seaboard. Elsewhere Europeans live for the most part in small, isolated settlements, hardly more than outposts of Western civilization. Yet in little more than half a century the indigenous peoples of Africa, variously estimated to number between 150 and 200 millions, have been integrated into the world society. Ruled more or less directly by European powers, they buy and sell on the world markets, participate in world wars, and rapidly acquire the culture traits of the Western civilization that is the decisive factor in determining the speed and direction of their social development.

This process of change occurs throughout the continent, but it is most rapid and comprehensive in the territories of white settlement. Here, where the African is in close and continuous contact with Western civilization, or to be more accurate, with its African variant, the greatest demand is being made upon his power of assimilation and adaptability. In proportion to the range and intensity of the European economy, the traditional subsistence techniques of the tribal peasant and pastoralist are being superseded by modern productive methods and relations. In satisfying his related demands for land and labor, the European destroys the territorial homogeneity of tribal society, makes wage-earning in labor centers an imperative for an increasing number of its members, and weakens the basis of its kinship, political, and religious institutions. The balance that formerly existed between the human society and its environment is disturbed, some of the consequences being soil erosion, widespread and increasing malnutrition, and much preventable disease. While, however, large-scale white settlement makes change unavoidable for the African, it also creates a framework for new social forms, through the provision of educational facilities, health services, religious institutions, a modern legal and administrative system, and a diversified market economy. The emerging society embraces both the indigenous African and the immigrants or their descendants from Europe and Asia, but while the ethnic groups have common institutions and forms of behavior, they retain distinctive cultural features and are graded into a rigid hierarchy of social classes on the basis of racial characters.

The size of the white settlements and their future prospects are therefore of paramount importance to the indigenous population, and form a convenient point of departure for a survey of the multiracial societies that are in course of development. What factors have determined the present distribution of white people in Africa? There are South African Calvinists who maintain that the presence of Europeans, at least in South Africa, is the result of divine providence, but a simpler explanation can be found. "In the remote past it was the search for precious metals which established contact with Asia and Europe." [1] Before the end of the fifteenth century the Portuguese had developed a lively trade on the West Coast, where the Dutch, French, and British built forts and established factories during the succeeding two hundred years. Not, however, until explorers and traders penetrated into regions habitable by Europeans

[1] Lord Hailey, *An African Survey* (London, 1938), p. 1484.

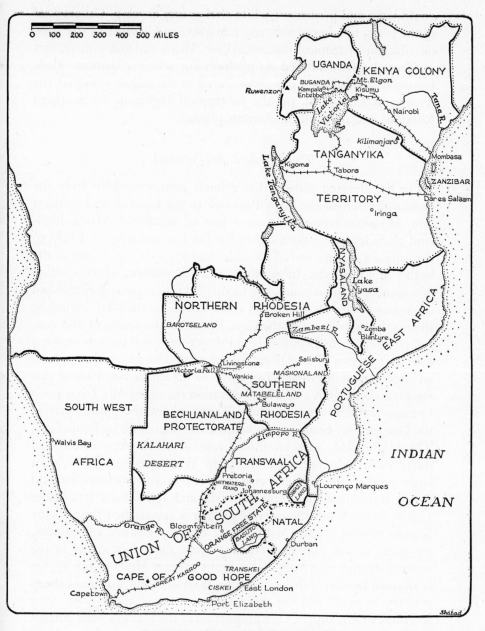

EASTERN AND SOUTHERN AFRICA

could large and permanent white settlements develop. On account of the great, unrelieved heat and humidity, and because of the wide distribution of tropical diseases, West Africa and the greater part of Central Africa offered no prospect for white settlement. More favorable conditions were encountered in the region south of the Tropic of Capricorn, on the subtropical highlands of Southern Rhodesia and on the East African plateau.

The Physical Background

On the elevated plateau that extends uninterruptedly from the central districts of the Cape Province to the equator, and projects into Abyssinia across the eastern part of equatorial Africa, highland altitudes compensate somewhat for low latitudes, and help to determine a climate more favorable than any other in Africa to the health of Europeans. In South Africa the plateau, which consists of a number of great plains, can be divided into regions distinguished according to surface characters and climate. To the west, including the northwestern area of the Cape Province, the southern and central portions of the Bechuanaland Protectorate, and the southeastern districts of South West Africa, lies the vast Kalahari semidesert, with an annual rainfall ranging from five to fifteen inches. East of this region is the High Veld, whose elevation rises gradually from 4,000 to 6,000 feet across the Southern Transvaal and the greater part of the Orange Free State, to culminate in the highlands of Basutoland. The High Veld, with an annual rainfall of between fifteen and thirty inches, a temperate climate, the Witwatersrand goldfields, a large part of the Union's coal reserves, and important diamond mines, is the most important industrial and agricultural area in the Union. In the south the interior plateau is approached through the Karroo regions, an area with a wide range in altitude and rainfall, the mean annual rainfall increasing from about five inches in the western districts to fifteen inches in the eastern. Much of this region is affected by severe droughts and soil erosion, but it is good sheep country in parts, and the sheep farms, many of them 5,000 acres and more in size, supply a big proportion of the Union's wool output.

The diversified tract of country between the plateau and the oceans includes a narrow coastal fringe lying below 1,500 feet.

In the northwest the coastal belt is continuous with the Namib desert of South West Africa; towards the south and east the rainfall increases steadily. A comparatively narrow strip of about 50,000 square miles, running parallel to the coast and extending as far east as the George-Knysna districts receives most of its rain in winter; further east, as far as Port Elizabeth, in an area of about 16,000 square miles, precipitation is fairly evenly distributed throughout the year. This winter rainfall area in the Western Cape, the first to be colonized by Europeans, and the population center of the Colored people, contains the Union's most important wheat and deciduous fruit-growing districts. In the southeastern region, extending from the foot of the escarpment across the Eastern Cape Province, Natal and Swaziland to the narrow coastal belt, an abundance of grass and a rainfall of between twenty and thirty inches provide favorable conditions for mixed farming, with animal husbandry as the main occupation. The part of this tract that falls within the Cape Province has been mainly "reserved" for the original African inhabitants who, though settled more densely than any other section of the Union's rural population, remain preeminently pastoralists and primitive agriculturists. The eastern coastal belt, though only ten miles wide near East London, broadens to a width of over a hundred miles in Zululand and Mozambique. The Natal coast, the only considerable area of low plain in the Union and its most productive region, has a tropical climate favorable to the cultivation of sugar cane, Natal's most important economic crop, and the reason for the introduction of Indians. Today Natal's Indian population almost equals the European. Resembling the northern Natal coastal region, but with a higher elevation, is the Transvaal Low Veld, lying between the eastern Transvaal frontier, along which runs the Kruger National Park game sanctuary, and the Great Escarpment on the west. Subtropical in climate, having a low and uncertain rainfall, and falling within the malarial zone, the Low Veld, though constituting one of the few regions available in South Africa for new settlers, offers uncertain prospects to the white farmer.[2]

[2] See W. Fitzgerald, *Africa* (London, 1934); A. B. du Toit and J. S. Coetzee, *Wêreld-Aardrykskunde* (Nasionale Pers, 1941); J. H. Wellington, "Pioneer Settlement in the Union of South Africa," in *Pioneer Settlement*, ed. W. L. G. Joerg (American Geographical Society, Special publication No. 14, New York, 1932).

Few parts of the Union are climatically unsuitable for whites; the limiting factor is low rainfall rather than high temperatures. Further north, within the tropics, conditions are less favorable and, although high altitudes modify temperatures, the future of the white population can hardly be regarded as assured. In Southern Rhodesia the plateau, a continuation of the High Veld, running southwest to northeast through the middle of the country, stands 4,000 to 5,000 feet above sea level, but only 24 percent of the colony lies above 4,000 feet, considered by some authorities as the lower limit of the area suitable for Europeans. About 89 percent of the white inhabitants are settled on the plateau above or near this limit, and within this region fall the main agricultural and mining centers. Maize and tobacco, the principal commercial crops, are produced largely in the wetter east; cattle-raising predominates in the drier southwestern tract. Gold is worked at many places along the railway, which largely follows the highland belt; chrome iron ore, asbestos, and coal are other minerals produced in large quantities. Flanking the High Veld to the east and west, and situated for the most part in comparatively low-lying country, are the native reserves. Comprising about 30 percent of the colony's land area, they are widely scattered in fairly small patches, badly served by communications, and in large parts inferior to the white settled areas in respect of both soil and rainfall. The low-lying valley regions of the Zambezi in the north, and of the Limpopo and Sabi rivers in the south, are tropical and unhealthy for settlers, but the southern Low Veld is generally free from tsetse fly and suitable for ranching. The western districts, which include a large tsetse-fly area running parallel to the Northern Rhodesia border and a semiarid tract adjoining Bechuanaland Protectorate, fall within the Middle Veld region, but are outside the zone of white settlement.[3]

Central and East Africa

The two Central African Protectorates of Northern Rhodesia and Nyasaland have pronounced physical affinities with the East African plateau. Historically and ethnologically, however, as well

[3] J. H. Wellington, "Possibilities of Settlement in Africa," in *Limits of Land Settlement*, ed. I. Bowman (New York, 1937).

as in respect of their economic and social life, the inhabitants are more closely related to those of Southern Rhodesia. Amalgamation between the three territories has been strongly advocated in recent years, especially by the Europeans of Northern Rhodesia, who see in the proposal a means of limiting the influence of the British Colonial Office on native policy and of strengthening their own relative position.[4] In spite of an abundance of land available to immigrants in the northeastern districts, Northern Rhodesia has proved less attractive to settlers than the countries further south. The bulk of the white community is contained in the central area within 20 to 30 miles of the railway line, which keeps generally to the higher country. On this railway belt between Livingstone and Broken Hill are settled the majority of white farmers, engaged mainly in cattle-raising and the growing of maize and wheat. Beyond Broken Hill, and extending to the Belgian Congo, is the mining industry, the decisive factor in the territory's economy. Apart from government officials, the greater number of Europeans are engaged in mining activities, mainly in the copper belt. The northeastern lobe, connected with the rest of the Protectorate by a narrow neck of land running between Portuguese and Belgian territory, contains two small and isolated European settlements, one at Abercorn on the Tanganyika border, the other in the Fort Jameson district near Nyasaland. Two vast land concessions, having a combined area of six and a quarter million acres, and situated in the Tanganyika district and the Eastern Province, are held respectively by the British South Africa Company and the North Charterland Exploration Company, but few farms in the concession areas have been taken up by Europeans, while many of the inhabited parts of the reserves demarcated in these regions are overpopulated.[5] Another and more important group of reserves is situated in the central area along the railway belt. The western part of the Protectorate, including Barotseland, whose status is protected by treaty, is almost entirely occupied by Africans. Unlike the Bantu-speaking peoples of southern Africa, the majority of Africans in the Protectorate do not go in for cattle-raising, an industry limited by the widespread occur-

[4] See *Rhodesia-Nyasaland Royal Commission Report*, Cmd. 5949, 1939 (London, H.M.S.O.).

[5] *Report of the Commission Appointed to Enquire into the Financial and Economic Position of Northern Rhodesia*, Colonial No. 145, 1938 (London, H.M.S.O.).

rence of tsetse fly to the western part of Barotseland, the central railway zone, and the Fort Jameson and Abercorn districts.

Many of the adverse conditions, such as poor communications, high transport charges, and a small local market, that have restricted the growth of the European agricultural community in Northern Rhodesia are even more conspicuous in Nyasaland. In the latter territory, which, with an area only one-seventh that of Northern Rhodesia, has a larger African population, the density of settlement forms another limiting factor. The Protectorate consists of a strip of land, about 520 miles long and from 50 to 100 miles wide, and traversed from north to south by a deep depression, forming part of the Great Rift Valley. The greater part of the trough is occupied by Lake Nyasa, south of which lies the Shire Highland plateau surmounted by the Mlanje Mountains. On these highlands are concentrated the bulk of the white settlers, who produce most of the important export crops, tobacco and tea. A large proportion of the European population consists of officials and missionaries; commerce is mainly in the hands of Indians, of whom there are almost as many as there are Europeans. In both the Central African Protectorates the supply of African labor exceeds the local demand, and a considerable migration of workers takes place to neighboring territories where wage rates are generally higher.

The outstanding feature of the East African plateau is the Great Rift Valley, a system of trough-like valleys extending, mainly from north to south, along a course indicated by a series of lakes between Lakes Rudolf and Nyasa. On the west the plateau is bordered by a branch of the Rift Valley running through Lake Tanganyika to the Albert Nile; in the north there is a vast region, including nearly half the area of Kenya, of arid and sparsely populated scrubland; on the eastern flank the plateau falls in terraces to a hot and moist coastal plain. Occupying a broad depression between the eastern and western rift valleys is Africa's largest lake, Victoria, whose borderland areas, divided between the three East African territories, contain two-thirds of East Africa's twelve million inhabitants. The plateau itself, exclusive of the Rift Valley and certain of the lake basins, has an average altitude of about 4,000 feet. Most of the areas of white settlement are situated at elevations above 5,000 feet. The mandated territory of Tanganyika, East Africa's largest and most populous territory, has two important regions of high altitude on which Europeans are settled, one in the northeast around Moshi and Arusha on the slopes of Kilimanjaro and Meru, the other in the southwest in the Iringa Province. Owing to the high altitudes at which it can be grown, coffee is, with mixed farming, the principal basis of European settlement in the Northern Province, and both coffee and tea estates have been developed in the southwestern areas. Sisal, the territory's most successful plantation crop, is grown in the northeastern district of Tanga Province along the railway line, under climatic conditions that are unfavorable to white settlement. While European settlers and Indian middlemen have made an important contribution to its development, Tanganyika is regarded as a "Native African territory . . . in the sense that the economy of the territory is predominantly native African." [6] The highland areas, which carry a dense African population, have been virtually closed to further alienation, and "experience obtained from the beginning of European colonization in 1893 points clearly to very definite limitations of white settlement and enterprise." [7] One of the main limitations is the comparatively small area of habitable country and the high density of its African population. Two-thirds of the territory are uninhabited or sparsely inhabited; the one-fifth that is well watered and free of tsetse fly holds five-sixths of the people. [8]

Apart from the highlands, the main centers of African settlement in Uganda are in the Lake Province, where coffee is extensively cultivated by peasants, notably in the Bukoba and Mwanza districts. Cotton, another peasant crop of importance, is grown in settlements established in Uzinga district. The Victoria basin, in which these districts lie, is the most fertile part of Uganda, and contains the main concentration of non-African plantations in the Protectorate. Small groups of European planters, engaged in the cultivation of coffee, tea, and rubber, are also found on the Toro highlands surrounding Mount Ruwenzori. Sugar cane and sisal are grown on a number of estates owned by Indians. The Protectorate, however, provides "the

[6] Sir Sidney Armitage-Smith, *Report on a Financial Mission to Tanganyika*, Cmd. 4182, 1932, p. 16.

[7] Charlotte Leubuscher, *Tanganyika Territory* (London, 1944), p. 42.

[8] C. Gillman, "A Population Map of Tanganyika Territory," in *Report by His Majesty's Government in the United Kingdom on the Administration of Tanganyika Territory for the Year 1935*. Col. No. 113, 1936.

leading example in East Africa of the systematic encouragement of
the agricultural capacities of an unusually intelligent native people in a relatively densely populated country." [9] Cotton, the chief
export product, is grown almost exclusively by peasants on thousands of small plots scattered throughout the country, and particularly in Buganda Province and in the region east of the Nile. Uganda,
with one and three-quarter million acres under cotton and a crop
of 418,000 bales, as compared with 129,000 acres and 28,000 bales
in 1916,[10] vies with the Sudan for second place among the cotton-
producing countries of the British Empire. The country's great dependence on this one product has disadvantages, and in some respects, such as a fall in the general level of soil fertility, the increase
in the area under cotton has been too rapid. On the other hand, the
industry has been largely instrumental in enabling the Protectorate
to pay its way out of revenue and so to escape the burden of a large
external debt.[11]

In Kenya, resembling in this respect southern Africa rather than
Uganda or Tanganyika, the economy has been developed upon the
basis of European enterprise, and the African's direct contribution
to exports in the form of hides, wattle bark, and raw cotton is relatively small. Fully half the African population is concentrated in
two regions: the three Kavirondo reserves in the borderlands of Lake
Victoria, and the three Kikuyu reserves adjoining the main centers
of the European coffee industry on the Highlands. These two regions, with a combined area of about 9,200 square miles, constitute
only one twenty-fifth part of the Colony's land surface area. In
contrast nearly two-thirds of the Colony, including the Uganda
Extension, Turkana, and the Northern Frontier District, is inhabited at an average density of less than one to the square mile.
The Highlands, the most extensive tract exceeding 5,000 feet above
sea level in tropical Africa (excepting the great plateau of Abyssinia)
includes an area of about 16,700 square miles known as the European
Highlands. The land alienated to Europeans, amounting to some
11,000 square miles, is situated along the Uganda railway and its
branches, on the Trans-Nzoia and Uasin Gishu plateaus, and in

[9] Sir Alan Pim, *The Financial and Economic History of the African Tropical Territories* (Oxford, 1940), p. 127.
[10] J. D. Tothill, ed., *Agriculture in Uganda* (London, 1940), p. 190.
[11] S. H. Frankel, *Capital Investment in Africa* (London, 1938), pp. 270 ff.

Area and Population

Territory	AREA		POPULATION				
	Sq. Miles	Year	African	European	Asiatic	Colored	Total
South Africa	472,550	1941 [a]	7,250,700	2,188,200	238,400	844,400	10,521,700
South West Africa	317,725	1936	287,731 [b]	30,677	14		318,422
Basutoland	11,716	1936	559,273	1,434	341	1,263	562,311
Bechuanaland	275,000 [a]	1936	260,064	1,899	66	3,727	265,756
Swaziland	6,704	1936	143,709	2,740	4	701	147,154
Southern Rhodesia	150,354	1941	1,378,000	68,954	2,547	3,974	1,453,475
Northern Rhodesia	290,320	1938 [a]	1,381,050	13,155	596	578	1,395,379
Nyasaland	47,949	1945	2,044,707	1,948	2,940		2,049,595
Tanganyika	360,000 [a]	1944 [a]	5,437,069	16,112	46,558 [e]		5,499,739
Uganda	93,981	1944 [a]	3,926,528	2,553	27,573		3,956,654
Kenya	224,960	1944 [a]	3,825,533	32,054	61,127	21,755 [d]	3,940,469

Source: Official Year Book of the Union of South Africa, No. 22, 1941 (Government Printer, Pretoria); The South and East African Year Book and Guide, 1947 (London); An Economic Survey of the Colonial Empire (1937).
[a] Estimate. [b] Includes Colored. [c] Includes Arabs. [d] Arabs, Somalis, and others. [e] Includes Arabs, Somalis, and others.

the districts of Laikipia and North Nyeri between Mount Kenya and the Rift Valley. Coffee, grown almost exclusively by Europeans, and mainly within a radius of fifty miles to the north and northwest of Nairobi, is the most important economic crop; maize, tea, sisal, and sugar cane are the other chief crops grown in bulk on European-owned farms.[12]

The Hazards of Farming

Throughout southern and eastern Africa, farming remains the biggest single industry, giving employment to the bulk of the population, contributing a large, and in the East African territories a major, portion of the national income, and exercising a far-reaching influence on social and economic policies. As the preceding survey will have shown, two distinct though interrelated fields of agricultural activity exist in these regions. There is the relatively advanced type of farming practiced by the white farmer, usually the owner of what in Europe would constitute a fair-sized estate, employing the technique and equipment made available by modern science and industry, receiving a considerable amount of assistance and protection from the state, and expecting a standard of living comparable to that of the well-to-do farmer in Europe or America. On the other hand is found the more primitive, predominantly subsistence economy of the African peasant, tilling his small fields with the aid of the hoe or, at best, an ox-drawn plough, rearing cattle and small stock for social as much as for economic ends, and often dependent for a large proportion of his family income on wages earned in European employment. Between farmer and peasant exists a many-sided relationship, involving both conflict and cooperation. They compete with each other for land, markets, transport facilities, and government assistance; they look to each other for labor and wages respectively; and both are affected, in different degrees, by price fluctuations, drought, soil deterioration, diseases, pests, and the other factors that make farming in Africa a hard and uncertain occupation. In all the territories under discussion, with the exception of Uganda, Tanganyika, and Nyasaland, the European is the dom-

[12] Cf. Report of the Kenya Land Commission, September, 1933. Cmd. 4556, 1934; An Economic Survey of the Colonial Empire (1937), Colonial No. 179, 1940.

inant figure in agriculture, but he maintains this position largely by means of an elaborate and costly system of protective tariffs, subsidies, and market control schemes.

There is, indeed, reason for pessimism concerning the future of the white farmer, not only in territories where his hold, only recently established, may be regarded as precarious because of lack of experience and a generally undeveloped economy, but also in South Africa, whose experience of 300 years of permanent white settlement under relatively favorable conditions provides a test that may well be considered to be decisive. His advantages are considerable. He is shielded by law from African competition for land in no less than 85 percent of South Africa's total area; he has available a large supply of labor at comparatively low wage rates; exceptionally low railway rates are charged on most farm products; over a million pounds are spent annually on agricultural research, precautionary measures, administration and the dissemination of information; and in the ten years between 1931–32 and 1940–42 the state spent about £25,000,000 in direct assistance and subsidies to farmers.[13] In addition, prices to consumers have been raised by a considerable amount, estimated at not less than £6,000,000 for the year 1939–40, by various protective and price-raising measures applied to the principal foods, including sugar, wheat, maize, butter and cheese, and leaf tobacco.[14] Apart from export products for which there is no appreciable local demand, and perishable products marketed internally and not subject to import competition, nearly every important agricultural commodity is sold at internal prices above the corresponding world prices. These efforts to place agriculture on a sound footing have undoubtedly contributed towards the significant expansion that the industry has made during the past 20 years. In the period 1924 to 1938, the production of the crops most benefited by price-assistance policies more than doubled, as did also the exports of fresh and dried fruit for which relatively favorable prices were paid in overseas markets. At the same time, the number of

[13] *Third Interim Report of the Industrial and Agricultural Requirements Commission,* U.G. No. 40 (1941), par. 88. From 1910 to March, 1936, the state had spent over £71,000,000 from loan funds, of which nearly £20,000,000 had to be written off, and over £41,000,000 from revenue for agriculture. (S. H. Frankel, *op. cit.,* p. 119.)

[14] *Third Interim Report of the Industrial and Agricultural Requirements Commission,* par. 89.

African males employed on European-owned farms increased by approximately one-sixth.

Of the total working population, about two-thirds are engaged in farming activities, or one-third if the 1,750,000 Africans who farm in the reserves and about 750,000 non-Europeans who undertake casual work on farms are left out of account. Yet farming has contributed only one-eighth of the national income since 1934. The outlook becomes even more depressing in the light of comparisons with leading agricultural countries. South Africa, in spite of a policy aimed at self-sufficiency and a large farming community, does not produce nearly enough of most important foodstuffs to provide her own population with an adequate diet. New Zealand, in contrast, can satisfy her food needs with the output of only 6.4 percent of the gainfully occupied population; Australia does the same with 9.7 percent, and the United States with 23.4 percent.[15] With low productivity go a heavy burden of debt and a small average income. Farms are overcapitalized; the estimated total mortgage indebtedness of £100,000,000 amounts to one-third of the total value of European-owned farm land, calculated at an average value of £3 per morgen.[16] The available data show that most farmers are on a low standard of living as compared with Europeans in urban areas.[17]

In attempts to explain agriculture's poor showing, stress has been placed on the unfavorable physical conditions. Two-thirds of the country is so dry as to be suitable only for extensive ranching. "An optimistic estimate of the Union's future arable land would be in the neighborhood of 15 per cent of the surface. Naturally, this must include a great deal of indifferent land at present marginal to cultivation." [18] Less than 6 percent of the surface is actually under the plough. Some authorities are optimistic about the prospects for animal husbandry, but it too has been severely handicapped by poor grazing, erratic and high seasonal rainfall, prolonged droughts, and a high incidence of diseases and pests. Land has been irrigated, but

[15] *Ibid.*, par. 9.
[16] Department of Agriculture and Forestry, *Reconstruction of Agriculture: Report of the Reconstruction Committee of the Department of Agriculture and Forestry*, 1943–44, par. 9.
[17] See *Census of Europeans, 6th May, 1941: Report on Structure and Income of Families*, U.G. No. 28–45.
[18] H. D. Leppan, *The Agricultural Development of Arid and Semi-Arid Regions with Special Reference to South Africa*, (Central News Agency, 1928), p. 81.

seldom with economic results, and it is considered unlikely that more than a million morgen, forming less than one percent of the total area of farm land, will ever be brought under irrigation.[19] To these drawbacks must be added the disadvantage of high transport charges, the absence of inland waterways, and a sparse and scattered population.

The pessimistic note sounded in recent official publications is perhaps due less to a perception of the country's natural limitations than to an unexpressed disbelief in the possibility of eradicating from the existing economic structure and the organization of the industry those defects that are the principal cause of its malaise. Dictated by political rather than economic considerations, the desire to achieve self-sufficiency and to conciliate the farming section of the electorate has led to the adoption of protective and price-assistance measures without safeguards to insure sound land usage. In consequence, soil erosion is considered by some authorities to be a greater menace in South Africa than in any other country. Farmers look to a paternal government to shield them from the consequences, not only of nature's harshness, but also of their own ignorance and inefficiency, and to insure an abundant supply of non-European workers, who, illiterate, badly housed, fed, and paid, have neither the incentive nor the opportunity to acquire and apply the skill that agriculture so badly needs.

It is unlikely that efforts to find a remedy will lead to the abandonment of protection and artificial stimuli, or to a return to the free market and unrestricted competition. If state assistance continues, and that it will do so is almost certain, it will have to be accompanied by an increasing amount of state supervision. The Department of Agriculture's Reconstruction Committee, while asserting that private ownership of land must be retained, recommends a policy that, put into effect, would virtually eliminate private enterprise in agriculture. The Social and Economic Planning Council, skeptical of the efficacy of state-owned enterprises and state controls, believes that farming could be improved by other means, but that even an increase of 100 percent in output and net farm incomes would leave the industry in a relatively unfavorable position, in which one-third of the working population, exclusive of the re-

[19] Department of Agriculture and Forestry, *Reconstruction of Agriculture*, par. 26.

serves, would produce only one-fifth of the national income. The Council's proposed solution is to facilitate the rural exodus.

Hand in hand with the strengthening of the farming industry must, therefore, go a policy of economic development elsewhere in order to effect a shift from the less to the more remunerative occupations, possibly for the whole increment in the rural European population and for a large proportion of the non-European increment as well.[20]

The Settler in the Tropics

In the Rhodesias and the East African territories, European farming, though still exploiting the advantages of cheap, abundant land, unexhausted soil and low wage costs that characterize pioneer settlements in Africa, also have to contend with the corresponding disadvantages of poor transport facilities, small local markets, and inadequate knowledge of agricultural requirements and possibilities. If in these territories state aid to farmers is on a smaller scale than in South Africa, the reason is not that they need less, but that governments have less to give. The tendency of official policy, especially marked in the period subsequent to the slump of the early thirties, is in the direction taken in South Africa. In Southern Rhodesia marketing boards now control the maize, dairy, pig, and tobacco industries; internal maize prices are stabilized at a figure higher than the export parity; bounties are paid on the export of high-quality beef and dairy products; government financial assistance is given for livestock improvement and irrigation works; and substantial reductions have been made in debts to the government for land sales, fencing, wells, and interest charges.[21]

Developments in Northern Rhodesia have not justified the early optimistic view taken of agricultural possibilities. In the twenty years following the impetus given to settlement after the 1914–18 war, the number of white farmers remained almost stationary, at a figure ranging between 300 and 400. Attempts to grow an export crop on a large scale as an alternative to maize have failed. In 1936 maize producers were subjected to a control board, formed to di-

[20] *Report No. 4: The Future of Farming in South Africa*, U.G. No. 10 (1945), pars. 115–16.
[21] *Rhodesia-Nyasaland Royal Commission Report*, pars. 119–21, 239–41. Cf. Lord Hailey, *An African Survey*, pp. 818–20, 1382–84.

vide the local market in a fixed ratio between European and African producers, and to export the surplus, while representative farmers' organizations expressed opposition to the introduction of more pro-- ducers. Shortly afterwards, however, the nemesis of "scarcity" economics overtook the Protectorate. In a prolonged period of poor harvests, maize production dropped from 437,000 bags in 1937 to 152,000 in 1941, the European share of the output declining from 242,000 to 120,000 bags. At the same time consumption rose by 42 percent, and the deficiency had to be met by imports.[22]

In Kenya, the storm center of British colonial policy in Africa, the prospects of farmers have been the subject of strongly conflicting viewpoints, and the extent of government aid is in dispute.[23] According to one authoritative opinion, expressed at the time of the 1930 slump, white settlement had proved to be a "hothouse plant," nourished by "adventitious aids, or even artificial aids," and while there was a possible future for large-scale plantations, the attempt to develop farming colonization as an economic business on its own was unlikely to succeed.[24] The Joint Committee on Closer Union [25] considered that the progress made by the settlers did not compare unfavorably with that of other countries at a similar stage of development, but that "the figures and arguments submitted merit serious consideration before any policy of further intensive white settlement is adopted." The nature of the assistance provided to farmers since the 1931 slump has been summarized by Sir Alan Pim:

Apart from the protective import duties and the special railway rates, more especially the preferential rating of local as compared with imported articles, a good deal of direct assistance has been given to farmers, particularly to cereal farmers. The help given includes rebates, mainly of railway rates and port charges, to the extent of £81,000 in 1929–30, and £47,000 in 1932, together with agricultural advances totalling £101,000, and loans of £116,000 to the cereal industries.[26]

With the upward swing of the demand for agricultural products during and after the war, faith in the future of European farming

[22] *Address to Legislative Council by H.E. the Governor,* 6th December, 1941.
[23] See the discussion in *Race and Politics in Kenya* by Elspeth Huxley and Margery Perham (London, 1944).
[24] Sir Edward Humphrey Legget, in evidence before the Joint Committee on Closer Union in East Africa, vol. ii, *Minutes of Evidence,* pars. 3499 ff.
[25] Vol. I, *Report,* 1931, par. 60.
[26] *Financial and Economic History of the African Tropical Territories,* p. 119.

has been revived, and renewed attempts are being made to strengthen the white population by encouraging the immigration of settlers into the Rhodesias and Kenya.[27] Future prospects will depend largely on the possibility of reducing the disparity between the prices of agricultural and manufactured commodities, and of avoiding a recurrence of the pre-war depressions. On the basis of the available evidence, the cautious observer will be inclined to share Lord Hailey's doubts "as to whether European agriculture will do more, even in good times, than make possible a very modest living as a return for hard work and the incurring of grave risks of loss of invested capital, and whether in bad times it must not prove a recurrent charge upon the revenues of the governments." [28]

The African Peasant

In any attempt to assess the economic position of the peasant population, considerable importance must be attached to the legal and administrative measures that prevent Africans and Europeans from competing on the open market for rights in land. Complete territorial segregation does not exist, for the farmer is an employer of non-European labor even when, as is usual in Southern Africa, he and the members of his family are also land workers. Except in Uganda and Tanganyika, where the system does not operate, segregation takes the form of demarcating "white" and "black" areas, and of prohibiting the members of one racial group from buying or leasing land in the area reserved for the other group. Though developed in response to a variety of conditions in the different territories, the reserve policy had a general origin in the struggle between the intruding whites and the indigenous people for land, which was also a struggle for the best land. In Natal, the first colony to adopt a considered and extensive reserve system, the aim was both to protect the colonists against the numerically dominant African population and to insure an adequate supply of labor for the farms. The Transkeian Territories and Zululand, on the other hand, owe their existence as "native" areas to the well-organized Nguni

[27] See *Colony and Protectorate of Kenya: Land Utilization and Settlement; a Statement of Government Policy*. Sessional Paper No. 8 of 1945 (Nairobi), par. 49.
[28] *An African Survey*, p. 1395.

society that deflected the main stream of colonists into the interior. The tribes of the Transvaal were not able to offer an effective resistance, and here reserves were marked out, mainly under British pressure, to safeguard them against the total loss of their hereditary lands. In Rhodesia and Kenya, again, the administrations made white settlement an objective of policy at an early stage of occupation, and demarcated reserves in order to determine the areas available for colonists. More recently, in all the territories, as the African adapted himself to the new conditions and showed an ability and desire to buy up land formerly occupied by his ancestors, the reserve system acquired an additional function of protecting the European against competition.

The reserves also protect the African inhabitants against European competition, and for this reason many critics of the policy hesitate to challenge the principle of segregation, but concentrate on attempting to increase the share allocated to the African. These are not, however, separate issues. With political and economic power concentrated in the European, race differentiation involves race discrimination. In proportion to population, the African share is much smaller than the area set aside for Europeans. Fertility of land and access to water supplies are more important than area size, but even if it is assumed that the reserves are no worse off in these respects than the European zone, the Africans in the area allocated to them can never hope to be anything more than small peasants, while the Europeans are intended to have the status of well-to-do farmers.

If not related to the carrying capacity of the land, comparisons of population density are misleading. Even crude density figures, however, tell a tale when they reveal contrasts as startling as those that exist between the areas of white settlement and the reserves. No amount of qualification could eliminate the significance of the difference between the 1,978 acres that constitute the average size of a European farm in South Africa and the 60 acres available for residential, arable, and grazing purposes to an average family of five in the reserves, or between the average densities per square mile of 14.09 in all rural areas, including the reserves, and 57.2 in the reserves alone.[29] More generous provision has been made for African

[29] Social and Economic Planning Council, *Report No. 9: The Native Reserves and*

settlement in other territories, yet areas of high population density occur in all. Densities recorded range from an average of 38 per square mile in Swaziland reserves [30] to 119, 156, and 240 in specific areas in Northern Rhodesia; [31] in the Kikuyu reserves in Kenya the average is 480, and in parts of Kavirondo density rises to as much as 1,000.[32]

Most of the reserves are overcrowded; of those that are not, few can absorb immigrants as well as provide for the natural increase. In any event, since occupation rights are limited by the traditional system of land tenure to members of a particular tribe or local group, congestion cannot be relieved by a redistribution of population. "The one reiterated cry of the Natives in the reserves is that the land is overcrowded, and that more land is required," remarked the chairman of the South African Natives Land Commission of 1915.[33] That may be said with equal force of African opinion today, not only in South Africa but in all territories where reserves have been demarcated. The dominant European attitude, on the other hand, is that the peasants suffer, not from scarcity of land, but from indolence and primitive methods of cultivation. This is the viewpoint, typical of that held by numerous official investigators in the different territories, which was adopted by the 1915 Commission: "While, therefore, it is quite true that most of the reserves are overcrowded, this is only so because of the uneconomic manner in which the land is occupied and cultivated, and the overstocking which goes on without let or hindrance, save for losses through drought and disease." [34]

Their Place in the Economy of the Union of South Africa, U.G. No. 32 (1946), p. 9; Social and Economic Planning Council, *Report No. 4*, par. 23; *Official Year Book of the Union of South Africa, 1941*, p. 992.

[30] *Report on the Financial and Economic Situation of Swaziland*, Cmd. 4114, 1932, pp. 6–7, 29; *Swaziland: Annual Report of the Department of Native Land Settlement*, 1946, p. 9.

[31] *Report of the Commission Appointed to Enquire into the Financial and Economic Position of Northern Rhodesia*, p. 65. Cf. *Labour Conditions in Northern Rhodesia*, Colonial No. 150, 1938, p. 39.

[32] *General Aspects of the Agrarian Situation in Kenya*, p. 8; *Report of the Commission Appointed to Enquire into and Report on Financial Position and System of Taxation of Kenya*, Colonial No. 116, 1936, par. 7. Cf. *The Kikuyu Lands* by N. Humphrey and others (Nairobi, *Government Printer*, 1945).

[33] *Natives Land Commission. Minute addressed to the Honourable the Minister of Native Affairs by the Honourable Sir W. H. Beaumont*. U.G. 25-'16, par. 38.

[34] *Ibid.*, pars. 42–43.

Africa's Depressed Areas

Like some other aspects of African culture, peasant agriculture exemplifies the principle described by Professor Toynbee as "the intractability of institutions." [35] In this instance, however, the "source of disharmony" proceeds from external forces, and not from an internal process of change such as led up to the agrarian revolution in Europe. When land was abundant, the peasant's methods, including the burning of bush, shifting cultivation, and periodic migrations to new pastures, achieved a satisfactory balance with the natural environment, and at least avoided serious injury to the soil. In the limited areas of the reserves, the traditional safeguards against soil exhaustion can no longer be employed, and the new techniques required are imperfectly understood and infrequently applied. Many of the defects of the medieval agriculture of Europe, and also those noted by Arthur Young in his survey of the common fields at the beginning of the last century, are conspicuous in the reserves. The arable fields are small, scattered and variable in size; enclosures are rare; crops are not rotated and the single-crop maize, or millet or some other cereal predominates; fertilizers or manures are seldom used, and fields are usually cultivated for several seasons until they show signs of exhaustion. The majority of peasants still use the hoe; where the plough has been introduced, it has made cultivation easier but, by encouraging concentration on cereals, its effects have been detrimental to the land and to the people's nutrition. Some authorities, notably agricultural officers in Northern Rhodesia, now incline to the belief that the traditional methods of farming should not be condemned outright, but there is no serious difference of opinion about the broad lines of criticism.

If one weakness has been stressed above all others, it is that of overstocking. Much has been written about the African's practice of raising cattle for social rather than for economic ends; the importance of quantity rather than quality for purposes of ritual killings and dowry payments and as a medium of exchange; the significance of cattle as a source of meat and milk; and the social prestige attached to the possession of large herds. Because of the poor quality of the stock, the failure to improve the animals by breeding,

[35] *A Study of History* (London, 1939), IV, 133.

and the absence of fodder reserves, milk yields are low and the
value of animal products is extremely small.[36] No convincing evidence has been advanced to show that the majority of peasant families possess more cattle than they need for nutrition and draught
purposes. The practice of pasturing animals on common grazing
lands, and on arable after harvest, permits of a wide range in the
distribution of stock. In South African reserves a few individuals may
be found in a district who own 50 head of cattle or more, but many
families own none.[37] Agricultural officers in Southern Rhodesia
estimate that the peasant family needs at least eight head of cattle,
but actually owns an average of only 4.6 head. In the Transkeian
Territories the average is 5.77 head per family. The common belief among Europeans that Africans are unwilling to sell cattle and
hoard stock as a miser hoards money may also be found to stand
in need of correction. Sir Philip Mitchell reminds us that "the problem is at least as much a problem of economics and transportation
as tribal customs and beliefs." [38] And, he adds,

my experience in both the Tanganyika Territory and Uganda has taught
me that the alleged reluctance to sell stock is a great deal less strong than
is generally represented, while the appetite for meat is almost unlimited.
But without Government help in organising and controlling the sale
and distribution of stock, in developing small butchers' shops throughout the villages and appropriate forms of dairying, the problem is beyond
the capacity of the African native stock owner and small trader to tackle.

Overcrowding, overstocking, and unsuitable methods of husbandry have resulted in widespread erosion and diminishing fertility of the soil in South African reserves. Conditions have deteriorated during the past twenty-five years to the point where,
with few exceptions, insufficient food is produced for the minimum
requirements of the inhabitants, and supplies must be augmented
by purchases from local stores. On the assumption that other foods
are eaten, and leaving only a small margin for wastage or consump-

---

[36] Cf. *Report of the Witwatersrand Mine Natives' Wages Commission*, 1943, U.G.
No. 21 (1944), pars. 144, 152; S. L. Kark, "Cattle and Milk in a Native reserve," *Race
Relations*, XI, no. 2 (1944), 30.

[37] *Report of the Witwatersrand Mine Natives' Wages Commission*, p. 10; Social
and Economic Planning Council, Report No. 9, p. 23 Cf *Report of Native Production
and Trade Commission* (1944), par. 66.

[38] *General Aspects of the Agrarian Situation in Kenya*, p. 7.

tion by animals, minimum grain needs are estimated to be 2.75 bags per head of population per year. In the Transkeian Territories, one of the best reserves, the peasants produced an annual average of only 1.5 bags of grain between 1939 and 1942, and imported on an average another 288,000 bags a year. Inclusive of imports, the average amount available per head was only 1.6 bags. In the Ciskeian reserves, less than half of the total grain requirements are produced by the inhabitants during a good season. Vegetables are even scarcer than grain; like fruit, they are regarded as a rarity. Even more serious is the deficiency in milk supplies. As for other territories, although information is less complete, the available evidence suggests that the peasants are hardly better off than in South Africa, even where more land is available. Many parts of East Africa, for instance, are no longer self-supporting in food, and depend largely on imports from European farms.

For an appreciation of the problem presented by the reserves, the low standards of production should be expressed in standards of health, housing, education, and social amenities. Statistics of livestock and farm products are more detailed and accurate than those of the human population. In no territory are vital statistics of rural Africans collected, and general descriptions are bound to give a misleading impression of conditions in particular areas. Such regional studies as have been made, however, substantiate the conclusion that the reserves for the most part are depressed areas, rural slums as they have been called, in which people stagnate in poverty, illiteracy, and ill-health. The income of £40 to £50 a year on which an average reserve family of five or six persons has to live in South Africa is probably higher than the incomes of peasants elsewhere. At least 30 to 40 percent of the reserve inhabitants of South Africa have never attended school; in South Africa and in Southern Rhodesia more than half the number of African children of school-going age are each year receiving no education at all. Educational standards are lower in most of the other territories.

The great bulk of the people live in primitive huts, usually built of wattle with clay, windowless, with no outlet for smoke, and infested with vermin and rodents. Huts are frequently overcrowded, partly because of shortage of building materials or labor, partly because of hut tax. Earthen floors are the rule; beds, tables, chairs and other articles of furniture are found in only a small minority of huts. Sanitary facilities are seldom found; supplies of pure or even polluted water are inadequate for domestic purposes. In the absence of reliable morbidity statistics, the incidence of various diseases cannot be indicated, but the South African material shows that enteric, typhus, plague, leprosy, tuberculosis, malaria, bilharzia, and syphilis are widespread. As reported by the Social and Economic Planning Council:

One of the most striking features about health conditions in the Native Reserves is the prevalence of debilitating conditions which prevent the inhabitants from producing their maximum efficiency. It is also noteworthy that practically all diseases prevalent in the Native Reserves are associated with overcrowding, poor standards of personal hygiene, inadequate sanitation, impure water supplies, malnutrition and general poverty. To this extent, they are all preventable.[39]

The rate of increase of African population is believed to be high, but the reason is that birth rates are high; available figures for the death rate are rarely less and often more than twice as high as that for Europeans. Medical and nursing services are inadequate. South African statistics show an average of one medical practitioner to 21,500 persons in the reserves.[40]

Migrant Workers

Even this low standard of subsistence cannot be maintained on the basis of peasant agriculture alone. In South African reserves few households depend solely on agriculture and, as opportunities for gainful employment of other kinds are negligible in the reserves, a large and constant migration of men takes place to mines, farms, and towns. The extent of the migration is an index of the peasant's failure to make a living out of the soil. Expressed as a proportion of the male population between 18 and 54 years of age, the number of men absent from the various South African reserves at the time of the 1936 census ranged from 41.6 to 69.0 percent. In other territories at least a third of the adult male population is away at work; in some districts the proportion is as much as 60 or 70 percent.[41]

[39] Report No. 9, par. 133.   [40] See below, pp. 317 f.
[41] See Social and Economic Planning Council, Report No. 9, p. 45; Report of Native Production and Trade Commission (1944), p. 11; Rhodesia-Nyasaland Royal Commission Report, Chap. XV; Lord Hailey, op. cit., p. 702.

For this migration a variety of reasons can be suggested, though not enough is known about the causal factors to allow their relative weight to be assessed. There is evidently a pull towards the employment centers as well as a push from the reserves. Town and mine offer adventure, new experiences, escape from the monotony of the village, opportunities for education and advancement, and other advantages to outweigh the discomforts and risks of the journey into the unknown.[42] On the other hand, it can be shown that manpower is the chief, often the only export from tribal areas, and that only by working for a wage can the peasant pay for imported goods, education, church dues, and taxation, and the other items in the family budget for which cash is needed. According to the calculations of the Witwatersrand Mine Natives' Wages Commission, for instance, the average Transkeian peasant, working nine out of twelve months on the mines, earned £20 1s. 7d. in wages if employed as a surface worker, as compared with £4 5s. received for the sale of produce from his holding in the reserve, and £13 10s. 2d. representing the monetary value of his own produce consumed by the family. The total income of £37 16s. 9d., it was estimated, fell short of the family's minimum requirements by an amount of £17 0s. 7d.[43]

Could not the peasant, however, produce more on his land and grow export crops to supply his cash needs? That possibility is ruled out for the landless families in the reserves, estimated to number 20,000 in the Transkei and 13,000 in the Ciskei, as well as for the large number of families in all territories whose allotments are too small to provide the cultivator with a living. Migration takes place on a large scale also from other regions, where tribal lands have not been restricted;[44] in these parts, at least, the peasant may be thought to have a freedom of choice between industrial employment and production on his own account. Only a series of intensive regional studies could establish whether such an option does exist, or whether, in spite of the relatively abundant land at his disposal, the peasant is forced by such conditions as inadequate transport,

[42] I. Schapera, *Migrant Labour and Tribal Life; a Study of Conditions in the Bechuanaland Protectorate* (London, 1947).

[43] *Report*, p. 19.

[44] Cf. "Migrant Labour in Africa and Its Effects on Tribal Life," by Margaret Reed, *International Labour Review*, XLV, No. 6, 605-31.

poor communications, uncertain markets and prices, to obtain cash by working for a wage. There can be no doubt, however, as to the necessity of a radical change in the system of land tenure and husbandry if the African is to be placed in a position to devote his full time to agricultural pursuits.

One obstacle to such a development is the self-perpetuating character of migrant labor. At one time General Smuts strongly recommended the general adoption of this method of securing labor for European enterprise on the ground that the movement to and from employment centers achieved the dual purpose of spreading civilization among tribesmen and preserving their own cultural values.[45] In reality, by bridging the gap between town and country migration accelerates the process of culture change in the reserves. If the worker on his return home does undergo a refresher course in tribal culture, he is also an agent for the spread of Western concepts and values. The absence of a large proportion of men in the prime of life is incompatible not only with tribal integrity, but also with a harmonious and well-balanced society of any description. The migratory system is not the only, but it is believed to be the main cause of what the South African Social and Economic Planning Council describes as "the breakdown of Native family life. Widespread prostitution and marital instability, adult crime and juvenile delinquency, venereal disease and sexual perversions, are amongst its effects." [46]

In addition, the loss of manpower reacts adversely upon agriculture in the reserves; the migrant worker has little incentive to improve his methods of husbandry; and the women, upon whom the burden of cultivation falls, are unable to give the land the needed attention and skill. Instead of agriculture being the mainstay of the reserve population, it becomes a subsidiary occupation whose yield steadily diminishes while the period spent by the men in employment centers tends to increase. The process has gone so far in South Africa that the migrant worker serves an average period of 13.6 months at a time on the Witwatersrand mines, in contrast to the period of three to six months worked in 1903, while 16.8 percent of the men return to the mines after spending "what may be re-

[45] *Africa and Some World Problems* (Oxford, 1930), p. 60.
[46] *Report No. 9*, par. 185.

garded as no more than a holiday visit" of from one to three months in the reserves.[47] On the basis of facts such as these, the Social and Economic Planning Council concludes not only that the migratory system is "morally, socially and economically wrong," but that "as far as the Reserves in particular are concerned no real progress towards their rehabilitation is possible as long as the migratory system of peasant labour is encouraged." [48]

This admission, coming from an authoritative body charged with the drawing up of blueprints for South Africa's future development, marks a significant change in the attitude that has determined policy for the past century. The reserves have been regarded, not as the homes of an independent, self-contained peasantry, but as the reservoirs of labor for the European entrepreneur, from which he could draw workers at need, without incurring the responsibility and cost of housing and feeding the women and children, the unemployed and unemployable, or of providing the education and social services needed by an urban community. What, then, is the explanation of the change of heart? Humanitarian considerations play a part; more important, however, is the realization that the progressive deterioration of the reserves and its effect on health standards threaten to deplete the potential labor force, and that the growing gap between the production and minimum family requirements of the peasantry can be bridged only by raising wage levels or improving methods of husbandry.

Program for Rehabilitation

All past efforts to arrest the impoverishment of the reserves have been too limited to free the peasant from dependence on the labor market. Additional land has been set aside for African occupation,[49] though not enough to put an end to overpopulation. Other measures adopted include anti-soil erosion works, the reduction, under compulsion if necessary, of livestock, the improvement of the quality of stock, the provision of water supplies, the eradication of tsetse

[47] *Report of the Witwatersrand Mine Natives' Wages Commission*, pars. 208–9.
[48] *Report No. 9*, par. 185.
[49] Under the South African Land and Trust Act, 1936, the Swaziland Native Land Settlement Proclamation, 1946, the Southern Rhodesia Land Apportionment Act, 1941 (superseding the Land Apportionment Act, 1930), the Kenya Crown Lands (Amendment) Ordinance, 1942 and Native Lands Trust Ordinance, 1938.

fly and the education of the people in sound methods of agriculture. More radical methods are now being applied or discussed in most territories. Emphasis is placed on the resettlement of reserves, and the prescription of land usage in accordance with plans drawn up by experts and carried out under close supervision. The South African scheme provides for regional planning committees, the establishment of rural villages, the zoning of residential, arable, and grazing areas, the limitation of population and stock, fencing, soil conservation, irrigation and water supplies. In Kenya it is proposed to transfer families from overpopulated areas to regions made habitable by the elimination of fly and the improvement of water supplies.

Drastic as these schemes are in some respects, they do not go far enough in the most vital aspect of the agrarian problem. Nowhere is it proposed to reform the archaic system of land tenure that, enforced by the administrations, limits the peasant to the cultivation of a ten-acre plot or less. Production can be appreciably increased, but not to the point of giving the peasant family a satisfactory standard of living, on holdings of the size now being cultivated in most regions. As the results of the Glen Grey quitrent tenure in the Ciskeian and Transkeian districts have shown, individual ownership and the registration of title carry with them no advantage as long as the peasant is bound by the rule of "one man one plot." Indeed, individual titles have proved to be a serious obstacle in the way of carrying out desirable reforms. It is not the Western type of ownership that is significant, but the associated right to buy and sell land or to mortgage it. Land scarcity and production for the market will, if not obstructed, promote individualism and the sale of land in violation of tribal custom, and bring about differentiation in the amount of land held and a landless peasantry.[50] No administration has, however, been prepared to give legal sanction to these changes, and regulate them as part of a carefully devised policy. Rather than face up to the problem of providing for the families without land, or of meeting the cost of surveys and registration of titles, the governments have nominally enforced the traditional system of landholding and tolerated such modifications as resulted from the new conditions. As to the unsatisfactory con-

[50] A. Philips, *Report on Native Tribunals* (Nairobi, 1945), Chap. XXV.

sequences of this neglect there can be no doubt, but it is not a simple matter to find an alternative.

One possible form of development is that adopted by Southern Rhodesia in the "native purchase areas," where, eventually, about 24,000 farms averaging 250 acres each will be available for sale or lease to the energetic and ambitious person who wishes to rise above peasant rank. In view of the white farmer's record, however, it is difficult to be enthusiastic about the prospects for the African who attempts to work a medium-sized farm with a negligible amount of capital and inadequate knowledge. There is also the possibility of introducing a cash crop, such as cotton or coffee, which, as has been shown in Uganda and Tanganyika, can be produced on small holdings, but it is by no means certain that the peasants in these territories have a higher standard of living than Africans in South Africa whose wages supplement their income from the land. If the future lies with large-estate farming, highly mechanized and scientifically organized, one alternative to peasant agriculture may be the plantation, such as is now being developed in East Africa for the production of groundnuts. There are a number of objections to the adoption of this course, however, including the danger of overconcentration on a single product, the political influences wielded by highly capitalized private enterprise, and the transformation of the peasant into a wage earner. If policy should be aimed at keeping the African on the land and enabling him to employ modern technique and equipment on an adequate scale, the most hopeful, if not the only, course is to encourage collective or cooperative farming by whole communities.

It cannot be assumed that administrations generally will be able to carry out a policy of allowing the reserve population to achieve a satisfactory standard of living on the land. Opposition to effective rehabilitation must be anticipated from several quarters. European taxpayers are unlikely to accept responsibility for the initial capital expenditure, which, together with a portion of the maintenance charges in the early stages, will have to be met out of public funds. In the Eastern and Central African territories money has been provided from the Colonial Development Fund, but on an insufficient scale to finance a rehabilitation program of the required dimensions. Internal resources are small in most territories; the diversion of a substantial part to the reserves would involve considerable adjustments in the economy, as well as political repercussions. The peasant would not be able to pay for the improvements, at the outset or later on, unless he were in a position to give all his time to growing foodstuffs and export crops on his own land. It cannot be supposed that employers as a class would assent to a scheme resulting in a big reduction in the labor force. Nor would white farmers readily accept the African as a competitor in a small market. The policy followed in Kenya of prohibiting Africans from growing arabica coffee,[51] and the restriction placed in Southern Rhodesia upon the African's share of the maize quota, indicate what the reaction would be to such competition. To these difficulties must be added the inevitable opposition of the peasants themselves, who, as experience with stock-limitation and soil-conservation measures shows, combine with conservatism a deeply rooted distrust of European intentions. Suspicion may be dispelled, inertia and the deadening weight of tribalism overcome, if rehabilitation is made part of a wider program of social reconstruction, bringing education, opportunity for self-development, and a greater share in government to the people. All these changes, however, involving as they do a radical revision of the African's status, would encounter strong resistance from the dominant European minority.

Mines and Industries

In South Africa and the Rhodesias, agriculture's poor showing has been offset by the exploitation of mineral resources. Mining, not farming, is the mainspring of the economic system in these countries. The industry has been the chief factor responsible for the growth of large European communities, big towns, and secondary industries; and it has determined the direction of railway lines. In 1875, at the beginning of the mining era, South Africa's white population was 332,000; in 1904 it had risen to 1,117,000, of which number 528,000 are estimated to have consisted of immigrants.[52] Owing to the great output of gold and diamonds, the value of mineral pro-

[51] The ban was lifted in 1939, but Africans may grow coffee only in limited areas in Meru and Kisii.

[52] C. G. W. Schumann, *Die Ekonomiese Posiesie van die Afrikaner* (Nasionale Pers, 1940), p. 68.

duction, to which these two minerals contributed 92.4 percent in 1940, is higher per capita than in any country elsewhere in the world. Gold mining alone accounts for almost one-fifth of the national income, two-fifths of the public revenue, and three-quarters of the exports. The industry therefore enables South Africa to pay for motor cars, machinery, electrical equipment, raw materials, clothing and other imported goods, and provides for a large part of state expenditure, including assistance to farmers.

Diamonds and gold, for all their importance in the economy, have not been used to a large extent as raw materials for local industry. With the encouragement of the state, diamond cutting has developed since about 1928; gold is refined on the Witwatersrand; and some mining stores, notably explosives and acids, are manufactured in South Africa. The indirect stimulus given to manufactures is considerable; the gold-mining industry provides an important market for local products, as well as the foreign exchange needed for imported machinery and materials. Producing, however, wholly for export, and anxious to keep working costs as low as possible, the gold-mining groups have not been concerned to promote the development of industry. The mining of base minerals, many of which occur in large quantities,[53] has been discouraged by heavy transport costs and low market prices. The resources needed for a big metallurgical industry exist, but capital has not been available for its development. The local iron and steel industry, for instance, like the railways, was developed with state capital. In all, mineral raw materials do not play an important part in South African industries; the value of diamonds, copper, iron ore, chrome ore and other minerals used locally was probably not more than £1,500,000 in 1934–35 and £3,000,000 in 1938–39.[54]

In certain directions industrial expansion has been extremely rapid in the past thirty years; more so, indeed, than the rate of expansion in Canada, Australia, New Zealand, and the United States. South Africa is today the most industrialized country in Africa, and possesses the most diversified and best-balanced economy. Manufacturing received a big impetus during the first, as also during the second,

[53] For a survey of mineral resources see A. J. Bruwer, *South Africa: Fundamentals of Reconstruction* (Johannesburgh, 1945), pp. 116–120.

[54] Board of Trade and Industries, *Report No. 282: Investigations into Manufacturing Industries in the Union of South Africa*, 1945, par. 89.

World War, and the protectionist policy initiated by the National-ist-Labour Government in 1925 provided an additional stimulus. In the period 1917–18 to 1939–40, the gross value of manufactures increased by 260 percent, that of mining by 166 percent, and of agriculture by 32 percent. The number of persons employed in manufacturing industries increased from 123,842 in 1916–17 to 413,492 in 1941–42, and their contribution to the national income rose from 9.6 percent to 19.0 percent.[55] These advances, however, are less impressive when viewed in relation to the low level from which they began. Both in the degree of industrialization and in standards of productivity South Africa has lagged behind the other dominions.

South African industries are for the most part of the "lighter" kind, not highly mechanized, operating on a small scale, and largely dependent on protection afforded by tariffs and dumping duties. Only a small proportion, constituting in 1939 just over 4 percent, of manufactured products are exported, while the local market is re-stricted by the small population and the low purchasing power of the great majority of non-Europeans. In addition to the disadvan-tages experienced by all countries in the early stages of industrializa-tion, there are others associated with the concentration of capital and energy on the dominant gold-mining industry, with the rigid caste-like stratification which denies non-Europeans adequate edu-cation and confines them to less skilled and badly paid types of work, and with the social pattern that places a premium not on efficiency and productive capacity, but on the possession of a white skin.

In the other territories, absorption into the world economy is a process of less than two generations, and the nature of the mineral resources has not yet been fully explored. Mining has played a dominant role in the development of Southern and Northern Rho-desia, providing about 80 percent and 95 percent respectively of domestic exports. Asbestos is produced on a big scale in Swaziland, the improvement of the diamond market has stimulated production in South West Africa, where output reached the value of £1,114,395 in 1945, and significant quantities of gold are produced in Kenya and Tanganyika. In Uganda and Nyasaland mining is of minor im-portance. Appreciable quantities of base minerals and metals are be-

[55] *Ibid.*, pars. 35, 37, and 55.

lieved to exist in most of the territories, and some of them may prove
to be as richly endowed in this respect as South Africa. The tendency has been, however, to concentrate on the exploitation of gold
and diamond deposits, and the further stage of developing base minerals has been reached only in the Rhodesias. By far the largest
expansion outside South Africa has taken place on the Northern
Rhodesia copper belt, where minerals to the value of more than
£12 million were produced in 1944, copper alone accounting for
over £10 million.

Although the war gave an impetus to the growth of manufacturing industries, they are still in their infancy in most of the territories. Only Southern Rhodesia can be said to have entered upon a
period of industrialization, though development is bound to be slow
in view of the small requirements and purchasing power of the local
market. In addition to processed food products, clothing, furniture,
cement and tobacco are the principal commodities manufactured,
and a small export trade in them is carried on with neighboring territories. As in South Africa these industries receive a considerable
measure of protection. Manufacturing development is at a much
lower level of development in Northern Rhodesia and Nyasaland,
and is confined largely to the processing of agricultural products
for export. Of the East African territories, Kenya has made the
biggest advance in industry, but even here the value of goods manufactured from local products or imported materials has never been
considerable.

With the exception of South Africa, and to a small extent Southern Rhodesia, the various territories under discussion are primarily
producers of minerals and agricultural products, and importers of
manufactures, mainly consumers' goods. Yet, without industrialization, a balanced economy cannot be achieved, and opportunities for
the employment of people in overpopulated areas must continue to
be inadequate. The unremunerative character of agriculture and
the low wages paid to farm workers do not offer much prospect
for a substantial improvement in the African's standards of living,
while the relatively small value of agricultural exports limits the
amount of possible imports. Existing resources appear adequate for
industrialization; power is available either in the form of coal or
water, and mineral deposits are extensive. None of the schemes pro-

posed by the British or colonial governments, however, makes provision for industrial development; the emphasis is placed on the increased output of agricultural products and raw materials used in
Britain's manufactures. To take one example, although cotton has
been grown on a big scale in Uganda since 1922, and contributes
about four-fifths of domestic exports, its cultivation has not stimulated local industry other than ginning, while Uganda's imports of
cotton piece goods, blankets and threads constitute one-fifth of
total imports.[56] The complex of relationships inherent in the colonial system is not conducive to the growth of manufacturing industry in the colonies; industrialization, as South African experience
has shown, is a consequence not merely of mining development,
but to a greater degree of the achievement by a colony of a real
measure of self-government.

The Labor Market

In all the territories under discussion, Africans, either as migrants
or permanent workers, supply the bulk of unskilled and semi-skilled
labor for farms, mines, factories, domestic service, and other occupations. Their absorption in the Western economy represents a
process similar in its cultural significance to the experiences of the
English yeoman and agricultural laborer during the great agrarian
and industrial changes of the eighteenth century. The African, however, has to take a leap into an alien world, ruled by men of a different race, speaking a different language and observing unfamiliar
customs, and subjecting him to the restraints and inhibitions of
the color bar. These peculiar features of the economic revolution in
Africa have given it a direction and pattern unlike that of the corresponding upheaval in other continents.

The transition to wage earner has followed much the same course
in the various territories. After the initial period of conquest or subjugation, workers were available from tribes whose lands had been
sold or given to Europeans and for whom reserves had not been
set aside. Additional labor was provided by men who spontaneously
left the tribal areas to work for cattle or money wages and to buy
European merchandise. These sources were rarely adequate, and

[56] *An Economic Survey of the Colonial Empire* (1937), pp. 23-25.

pressure was put on men in the reserves: directly, as under the thinly disguised slavery of the "apprenticeship" system practiced in the Transvaal Republic, or through administrative officers, as in Southern Rhodesia and Kenya, or in the form of compulsory labor on roads and railways as in East Africa; indirectly, by imposing head and hut tax or by sending recruiting agents into the reserves. Direct pressure is no longer needed in South Africa; even the mine recruiter is an agent for forwarding rather than for procuring labor. In some of the other territories, workers may still be legally impressed for porterage or public works, and during the war, industrial conscription, forcing Africans under the threat of penal sanctions to work for private employers, was introduced in Southern and Northern Rhodesia, Kenya, and Tanganyika. As the discussion of the reserves has shown, however, there are now few areas outside Tanganyika and Uganda where the majority of families are not dependent on wages. In some territories there is not enough work for all who want it, or the wages offered locally are not attractive to potential workers, with the result that men migrate to employment centers in other territories.[57]

While breaking down the isolation of tribal communities and making wage earning a necessity, employers have also attempted to direct the flow of labor into certain channels and to restrict the mobility of the workers. The controls take the form of pass laws, developed historically in South Africa as a measure of protection for the white pioneer during the period of armed conflict, and subsequently used to prevent Africans working under contract on the mines from changing their employment or returning to the reserves before the expiration of the contract period. The pass laws today form one of the keystones of "segregation": tying the labor tenant to the farm under the Native Service Contract Act of 1932 and the Proclamation 150 of 1934; restraining the growth of the permanent urban African population under the Natives (Urban Areas) Act; giving the mines a virtual monopoly of recruiting and a means of enforcing contracts under the Native Labour Regulation Act; protecting agriculture and mining from the competition of secondary industry and commerce for African labor; and preventing this competition from raising African wages. An even more compre-

[57] See International Labour Office, *International Labour Review*, XLVI (1942), 45.

hensive type of control operates in Southern Rhodesia [58] and Kenya,[59] in the form of compulsory registration of all males over the age of 14 or 16 years respectively, and the issue of certificates containing particulars of employment which must be produced on demand. The enforcement of these regulations involves a "harassing and constant interference with the freedom of movement of Natives [which] gives rise to a burning sense of grievance and injustice," [60] and leads to a large number of convictions, amounting in South Africa to 125,000 a year. It is significant that pass laws, though recommended at times, have not been introduced in territories such as Uganda, Tanganyika, and Nyasaland, where the European employer exercises less influence on official policy.

The statutes restricting the movement of Africans, and the masters and servants laws which operate in most territories,[61] contain penal clauses that make desertion from employment or breaches of discipline a criminal offense. These substitutes for the more usual sanctions available to employers in other countries, such as dismissal from employment or a claim for damages, have their origin in the African's inexperience of the master-servant relationship, his ability, owing to the unspecialized nature of his work, to change his employment with relative ease, his retention of a base in the reserves on which he can fall back if unemployed and where his family can subsist in partial independence of his wages, the reluctance of employers to provide incentives by improving conditions of work, and above all his inferior status. With the growth of a body of permanent workers in the towns and the increasing dependence of the migrant workers on wage earning, prosecutions under the labor laws have steadily diminished in urban and mining centers in South Africa. The penal sanctions retain, however, their significance in relation to the farm workers, who, undergoing a transition from the position of share-cropper and labor tenant to that of a full-time wage earner, and deprived of many of the privileges, such as the right to pasture stock, that made agricultural employ-

[58] Under the Southern Rhodesia Native Passes Act and Native Registration Act.
[59] Under the Kenya Native Registration Ordinance, the Resident Labourers Ordinance, and the Employment of Servants Ordinance.
[60] Union of South Africa, *Report of the Inter-Departmental Committee on the Social, Health and Economic Conditions of the Urban Natives* (1942), par. 305.
[61] Lord Hailey, *An African Survey*, pp. 659 ff.

ment attractive, now tend to migrate from the farms to the towns. The farming community, suffering a chronic shortage of labor and unwilling or unable to compete for it with other classes of employers, has strongly opposed attempts to abolish or relax the pass laws and masters and servants laws.[62]

Most of the conditions giving rise to legislation of this kind would disappear if the forces set in motion by the European economy were allowed free play. The tendency is for Africans to settle permanently with their families in the towns. The number of Africans living in South African urban areas increased by 287.1 percent from 439,707 males and 147,293 females in 1921 to 1,094,322 males and 590,890 females in 1946. The increase in the number of females in particular is a reflection of the growth of an urbanized African community. If the process of urbanization were allowed to take its course, a stable body of workers would develop which, being wholly dependent on wages, would respond to the incentives and conform to the standards of discipline found in mature industrialized societies. There is the additional advantage, recognized in the Belgian Congo, of an increase in efficiency and productivity on the part of the permanent wage earner. In South Africa the need to achieve stability in the labor market has been emphasized by numerous government commissions,[63] but the government has preferred to tighten controls and extend them to new areas such as the Western Cape.

The migrant's possession of a source of income in the reserves has the paradoxical effect of strengthening his bargaining power while at the same time depressing wages below minimum standards of living. That is to say, in so far as wage rates are adjusted to the relation between the supply and demand of labor, he uses his favorable position to obtain a wage income sufficient to bridge the gap between subsistence needs and the income derived from production in the reserves. This level of wages is clearly inadequate for the land-

[62] For the farmers' attitude see the *Report of the Native Farm Labour Committee* (1937-39), Chap. III.
[63] E.g., the Interdepartmental Commission on the Labour Resources of the Union, 1930; the Native Economic Commission, 1930-32; the Industrial and Agricultural Requirements Commission; the Social and Economic Planning Council; and the Board of Trade and Industries. Cf. Southern Rhodesia, *Report of Native Production and Trade Commission* (1944), par. 240.

less migrant and the urbanized worker. In reality, however, employers show a marked resistance to any tendency for wages to increase in response to labor scarcity. Farmers call upon the government to make labor available, and the mine owners, in addition to recruiting foreign workers, have virtually eliminated competition among themselves for African labor by agreeing that no mine shall pay more than a specified maximum average. As a result, no substantial increase has taken place in mine wage rates for Africans since 1914. The employers' resistance to granting wage increases is all the more effective because of the poor state of trade-union organization among Africans, the prohibition of strikes by Africans,[64] and their exclusion from the industrial legislation providing for the settlement of disputes and the negotiation of agreements between employers and employees. A Wage Board has fixed wage rates for numerous categories of employees in commerce and industry, but not at subsistence levels where unskilled workers are concerned.

Unskilled wage rates in South Africa are said to be on a level with those paid in Italy, and skilled wages with those of Canada.[65] The comparison illustrates the peculiarity of the wage structure found in Southern, Central, and Eastern Africa, where skilled work is reserved largely for Europeans. In Western Europe, and the British dominions, the spread between the highest and lowest rates of wages is in the neighborhood of 30 percent, and seldom exceeds 50 percent; in South Africa it is several hundred percent; In many South African industries the skilled worker earns as much in a day as the unskilled man earns in a week. European workers, including semiskilled and unskilled, in private industry, received an average annual wage of £279 in 1941–42, as compared with the average of £60 paid to Africans. This unusual disparity evolved in the early period of mining and industrial development, when artisans were scarce, and had to be attracted from abroad, while unskilled labor was abundant and cheap. In course of time, traditional practices became part of the color-caste organization; differences between skilled and unskilled occupations came to be regarded primarily as differences between European and non-European. Only in the Cape, where

---

[64] Under Masters and Servants Laws, the Native Labour Regulation Act, and a War Emergency Proclamation which is still in force.

[65] Board of Trade and Industries, *Report No. 282* (1945), par. 120.

Colored people supplied the bulk of craft labor for many generations, are non-European artisans found in large numbers.[66]

The industrial color bar, as the barriers are commonly known, consists of a number of conventional and legal devices to give the European a monopoly of skilled work. Few statutes contain specific discriminations against the employment of non-Europeans, the most notable instances being the Mines and Works Act Amendment Act of 1926, in terms of which regulations have been issued preventing Africans and Indians from qualifying for certain specified occupations, and the Industrial Conciliation Act, which excludes Africans from the machinery established for the settlement of industrial disputes. An Apprenticeship Act, operating in conjunction with the virtual absence of facilities for the industrial training of non-Europeans, has the effect of confining apprenticeship to Europeans with few exceptions. More important than these laws, however, are the monopolistic policies pursued by the craft unions and the preference of employers for European skilled workers, particularly when the same wage rates are fixed for all workers doing the same class of work.

Competition between racial groups occurs not only as a result of the upward movement of non-Europeans, but also through the impoverishment of a section of the European community. That position has existed for several generations in South Africa, where the rural white population has also undergone a process of rapid cultural transformation, resulting in the migration of a large proportion to the towns. The "poor white" shares the political and legal privileges of the superior caste, but for want of ability or opportunity is unable to rise above the economic level of the lower-class non-European. In an attempt to repair this damage to the caste barriers and to reduce the area of interracial contact on an equal plane, the state has adopted a "civilized labor policy" of employing Europeans as unskilled workers at subsidized wage rates. Further industrialization, it is realized, must result in a reduction in the proportion of work held to be skilled, now probably higher in South Africa than in any other industrialized country, and a considerable increase in the number of semiskilled workers of all races. Desirable though this

[66] See H. J. Simons, "The Coloured Worker and Trade Unionism," *Race Relations*, Vol. IX, No. 1 (1942).

development appears on economic grounds, the ultimate political and social consequences of the equalizing effects are not acceptable to the European ruling class. The possibility of introducing horizontal segregation in industry where vertical segregation has broken down is now being explored. The establishment, partly with state funds, of a textile factory at Kingwilliamstown near the reserves, in which only Africans will be employed as operatives, is an illustration of the proposals now being made for a policy of decentralization in industry, involving the transfer to the reserves of some of the industries making cheap standardized wares for native consumption, as well as for a policy of residential separation and racial parting in factories.[67]

Regarded from a different viewpoint, the industrial color bar is an indication of the extent to which the non-European has been drawn into industry and of the challenge he offers to racial discrimination in this sphere. If discriminatory practices in other types of occupation have received less prominence it is not because they are absent, but because the possibility of competition appears extremely remote. Moreover, while many employers and other influential sections oppose the industrial color bar on economic grounds, the same groups usually assume without question that the exclusion of non-Europeans from professional, managerial, and other white-collar occupations is both inevitable and desirable. There are non-European artisans, principally Colored and Indians, who work for an employer or on their own account but, in a number of occupations traditionally associated with middle-class status, European monopoly is very nearly complete. In the public services and on the state-owned railways, non-Europeans with few exceptions are employed only as laborers, messengers, or in other subordinate posts. Banks, insurance companies, business concerns, and other privately owned concerns under European control similarly exclude non-Europeans from all but the most menial occupations. According to the industrial census of 1941–42, 30,549 Europeans were occupied as working proprietors, managers, accountants, and salaried personnel, as compared with 118 Africans, 177 Colored and 560 Asiatics. Similarly, in the professions non-Europeans are represented by an insignificant number. The 1936 census returns show that there

[67] Board of Trade and Industries, *Report No. 282* (1945), par. 135.

were 9,090 Europeans, 14 Africans, 23 Colored and 25 Asiatics following the profession of advocate, attorney, architect, civil engineer, dentist, chemist, or medical practitioner. The non-European is not shut out from these occupations by a legal barrier; he is excluded because there are no educational facilities, or because he cannot avail himself of those that exist; because professional men refuse to article non-Europeans; or because the government employs only Europeans in the important posts of the public service.[68] The rule that no non-European is to be placed in authority over Europeans gives the European a virtual monopoly over all positions of structural importance in the social hierarchy.

In spite of this rigid caste system, the more mature and diversified South African economy opens up wider opportunities for the majority of Africans than exist in territories further north. Money wages of skilled workers are higher, both the standard of living and the wage rates of unskilled workers are lower, than in South Africa. There is consequently a migration of white South Africans to the Rhodesias, and a migration of Rhodesian Africans to South Africa. The South African pattern of race relations is being reproduced in broad outline in the Rhodesias, where a strict color bar prevails on the railways and copper mines.[69] In the other British colonies, however, the number of white artisans is small, and there has been no state intervention to restrict the entry of Africans into skilled occupations. On the contrary, since Indians provide the bulk of skilled labor, and as European hostility towards the Indian is pronounced, the tendency has been to train Africans to oust the Indian artisan. To continue the comparison with southern African conditions, the East African territories may be said to afford greater scope for a limited number of Africans in the railway service, government departments, the building trade and industry. During the war South Africa looked to the European population to provide the additional skilled labor force required; in East Africa it was the African who received training as tradesmen, telegraphists, radio operators, dental mechanics, and clerks. In administration and commerce, on the other hand, the new generation of Kenya-born Europeans is pushing

[68] For information about the position of the non-European in the public service see the *Fifth Report of Public Service Enquiry Commission,* U.G. 53 (1946).

[69] R. Welensky, "Africans and Trade Unions in Northern Rhodesia," *African Affairs,* XLV, No. 181 (1946), 185–91.

downward in competition with Indians for clerical and other subordinate positions. It is possible, then, that a stratification of the European community will lead to the adoption of the South African measures for safeguarding a specifically European standard of living against the African and Asiatic.

Aids to Welfare

An analogy, useful for some purposes, can be drawn between the changes now taking place in Africa and the effects of what is commonly known as the industrial revolution in Europe and America. It is not a parallel, however, that should be taken very far. In the Western world, the new economic organization was a culmination of earlier developments, the result of a process of internal growth that reacted upon the whole of the society. There was much dislocation, but also a large amount of correlation between changes in economic, political, legal, family, and other institutions. In Africa, the impetus to change comes from agencies external to the indigenous society; the degree of correlation is small, and the extent of disequilibrium great. The European, concerned primarily with the exploitation of natural resources, concentrated his capital and energy on railways, farms, and mines, and neglected the building of schools, houses, and hospitals for the African. Considerable effort was expended in converting the peasant into a wage earner; very much less attention was given to his physiological and cultural needs. Money could be found for economic ventures, but not for social services. Provision has been made for the African's needs, not concurrently with economic development, but subsequent to it, even though the effects of the time-lag have been harmful both to the African and to economic progress.

If the services required could be arranged on a schedule of priorities, first preference would have to be given to education. The school is an indispensable medium for preparing the African to cope with changes that take place through other means. In providing educational facilities, the authorities have had to start from scratch; to find a literary form for the spoken word, as well as books, teachers, and buildings. The first schools in all the territories were opened by missionaries, as part of their proselytizing work, and they still play a dominant part in non-European education. The state's contribution has consisted of grants-in-aid and the establishment of a central administration, operating through an inspectorate, to bring about uniformity and the observance of minimum standards. Progress might have been slower if the governments had assumed sole responsibility; they would, for instance, have been more sensitive to European opinion hostile to non-European education. The present arrangement, however, is unsatisfactory, in that it militates against coordination and systematic planning. The case for state provision and control of schools has been greatly strengthened by the large growth in the government's share of expenditure on education.

No more than a superficial examination of school statistics is needed to show that an enormous leeway has to be made up. The position in South Africa, where both European and non-European education has reached a higher level than that attained in other territories, may be taken as a guide. Education is compulsory only for European children, who are required to attend school until reaching the age of 15 or 16, according to provincial regulations, or until passing standard six or eight. Although both primary and secondary education is free for Europeans in all except one of the provinces, only 55 percent of children aged 16 are in school, and not more than 11 out of every 100 pupils matriculate. With regard to university education, however, better provision has been made for the European community than that existing in most other countries. In 1940 the number of university students, including non-Europeans, was 11,220; in 1945 it rose to 14,220; and in 1946, as a result of the influx of ex-servicemen, the peak figure of 20,186 was reached.

A considerable expansion of facilities for the education of non-Europeans has taken place in the past two decades. The number of pupils has trebled and the expenditure quadrupled. In 1925 there were 188,400 African and 64,000 Colored and Asiatic children at school; in 1945 the numbers had increased to 588,000 and 200,000 respectively. State expenditure on African primary and secondary education increased from £605,500 in 1934–35 to £2,540,000 in 1946–47; while in the latter period another £860,000 was spent on meals for African schoolchildren. In spite of this progress, only about 7 percent of the African population attend school, as compared with 16 percent of the Colored and Asiatic and 20 percent of the European population. Of African children in the age group 7 to 14 years, 30 percent are at school in any one year, compared with 77 percent of Colored and Asiatic [70] and 100 percent of European children.

In all non-European schools a large turnover is experienced, and the majority of pupils leave school with a very elementary education. Accommodation and equipment, generally adequate in state schools, are inadequate and unsuitable in most of the mission schools. There is much discrimination to the disadvantage of non-Europeans in respect of school fees, boarding grants, transport facilities and grants, the provision of books, hostel accommodation, the salaries and training of teachers, and the ratio of pupils to teachers. [71] These qualitative differences are reflected in the comparative costs. State expenditure on European primary and secondary education in 1940–41 amounted to £22 11s. 3d. per pupil, as compared with £7 spent on every Colored and £3 0s. 4d. on every African pupil in the Cape Province, where provision for non-Europeans is most generous.

Higher education for Africans, including a course for training medical aids, is provided at the South African Native College at Fort Hare, with a current enrollment of over 300 post-matriculation students, some of whom come from outside South Africa. Non-Europeans are also admitted to courses at the Universities of Cape Town, Witwatersrand, and Natal, but are excluded from other university institutions. In addition, of some 3,000 students registered in the University of South Africa more than a fourth are non-Europeans. It is with respect to technical training that non-European education has made least progress. Four schools have been established for training Africans as agricultural demonstrators, non-Europeans are admitted to a small number of courses at the Cape Town and Witwatersrand technical colleges, and several missionary institutions train Africans in carpentry, building, printing, bookbinding, and other trades. But the numbers involved are insignificant. The Social and Economic Council's comment on the inadequacy of facili-

[70] Exclusive of Asiatics in Natal, of whose school-age population 45 percent are estimated to be at school.

[71] For a comprehensive survey of facilities available for each racial group see *Report of the Provincial Financial Resources Committee*, U.G. No. 9, 1944.

ties for non-European vocational education is that unless they are improved "the Union is doomed to a losing competitive struggle against the mentally developed labour of the Western countries, against the awakening Eastern races and even against other parts of Africa." [72]

In other territories, African education is at a lower level in most respects than in South Africa. The number of pupils enrolled may be proportionately greater, but the type of education provided is usually inferior. The proportion of children of school age attending school at any one time ranges from 18 percent in Kenya to 50 percent in Nyasaland. The great majority of pupils are found at "kraal" or "bush" schools, many of which because of their low standards do not qualify for government grants.[73] A large proportion of pupils receives no more than a kindergarten education, attendance is irregular, the rate of elimination extremely high, and very few pupils reach the secondary standards. Provision for higher education is made at Makerere College in Uganda, which gives courses in medicine, agriculture, veterinary science, and teaching, for matriculated students from East African territories. Technical training in such trades as carpentry, building, and leatherwork is given at missionary institutions in all the territories, and at government schools in most of them.

The discrepancy between the state's expenditure on European and on African education, which was seen to be a marked feature of the South African system, is even more noticeable in some of the other territories. In 1938 the government spent £6 per head of population on non-African education and 1s. 5d. on African education in Southern Rhodesia; in Northern Rhodesia the corresponding figures were £5 and 8d., and in Nyasaland 11s. and 3d.[74] In Kenya, government expenditure per European pupil in 1944 was estimated to be £23 6.61s., per Indian £5 5.30s., and per African 12.25s.[75]

[72] Report No. 2, U.G. No. 14 (1944), par. 119.

[73] Cf. Southern Rhodesia Report of Native Production and Trade Commission, pp. 16–19; Rhodesia-Nyasaland Royal Commission Report, pp. 85–89; C. Leubuscher, pp. 82–89; R. Hinden, Plan for Africa (Allen & Unwin, 1941), p. 107; H. B. Thomas and R. Scott, Uganda (London, 1935), Chap. XXI; S. and K. Aaronovitch, Crisis in Kenya (London, 1947), Chap. VIII.

[74] Rhodesia-Nyasaland Royal Commission Report, par. 203.

[75] Kenya, Legislative Council Debates, 2d ser. XXI, Part II, 1945, pp. 6–7.

Health and Housing

To the African's handicap of illiteracy is added a considerable burden of ill-health. He does not eat enough to sustain health, and what he eats is often deficient in protein, vitamins, and calcium. In South Africa, malnutrition is widespread also among other sections of the population, and the National Nutrition Council, established in 1940, has stated that "malnutrition is an urgent national problem of enormous extent. . . . If this insidious deterioration process is much prolonged, the cumulative result on the future labour supply of the country and on the quality of its people must be calamitous." [76] Food deficiencies have been found in all the other territories.[77] The undernourished African has to grapple with a host of tropical diseases, many of them associated with poor diet, poor housing, impure water, and the absence of effective sanitation. Malaria occurs in most parts, with the exception of highland regions and the more southerly districts of South Africa. Sleeping sickness is found throughout Central Africa and in East Africa. Hookworm, bilharzia, and other worm diseases are serious in most territories; in East Africa over 90 percent of the population are believed to be infected with one or more kinds of worm. Outbreaks of formidable epidemic diseases, such as smallpox, typhus, and typhoid fever, are frequent in most territories. Cases of leprosy occur throughout Africa, though compulsory isolation has reduced its prevalence. The worst scourges are syphilis and tuberculosis, both of which seem to have been unknown in southern Africa before the European era. In South Africa it is estimated that 20,000 persons die annually of tuberculosis, the incidence of which is many times higher among non-Europeans than among European.[78] It has been calculated that the incidence of yaws and syphilis in East Africa is 60 percent.[79] Although vital statistics of rural Africans are not recorded in any territory, there is abundant descriptive and some statistical evidence of the high incidence of diseases. The frequently quoted saying that "disease knows no color bar" is misleading. In South Africa the non-

---

[76] First Report, 1944, U.G. No. 13-1944, p. 18.
[77] See E. B. Worthington, Science in Africa (London, 1938), Chap. XVII.
[78] The Senate of South Africa, Debates, 22d April, 1947, col. 1297.
[79] Lord Hailey, An African Survey, p. 1144.

European death rate is believed to be about twice that of the European. In 1942 the percentage infant mortality rate among Europeans was 4.75, among Asiatics 8.84, and among Colored people 17.67. The comparable African figure is unknown, but estimates of infant mortality in urban areas have ranged from 20 to 40 percent, and an investigation carried out in the reserves of the Cape Province in 1937 showed that mortality during the first year of life was approximately 25 percent, during the second year 33 percent, and by the time the age of 18 was reached, 50 percent.

African diseases are bred out of the African's poverty. Unless standards of nutrition, housing, sanitation and education are raised far above existing levels, very little headway can be made against disease. Nor can any territory provide curative treatment and preventive services on anything like the scale required as long as national incomes remain at the present figures. Health services are grossly inadequate in relation to African needs; they would be inadequate even for a society with the health standards found in Europe. As recently as 1938, in a relatively advanced territory like Southern Rhodesia, no medical facilities were available for about a quarter of the African population. In Northern Rhodesia, at the same period, more than half the territory was without medical attention. In the three East African territories, government hospitals before the war provided a total of about 5,300 beds for an aggregate African population of not less than 11 millions, an average of about one bed for every 2,000 persons, as compared with one bed available for every 150 Europeans. In South Africa a ratio of one bed for every 200 of population has been accepted as representing adequate provision for general hospitalization. Accommodation for Europeans in provincial and mission hospitals and in nursing homes in 1941 conformed to the desired standard; for non-Europeans, however, the beds provided amounted to an average of one for every 920 persons.[80] In respect not only of hospitals, but also of the number of medical practitioners and nurses, clinics and health centers, South Africa is far better equipped than other African territories. Yet the National Health Services Commission of 1944 found that personal

[80] *Report of the Provincial Financial Resources Committee*, par. 378. For Africans the Committee accepted a standard of one bed per 600 units on grounds of expediency.

preventive services were rudimentary and incomplete; personal curative services were totally inadequate for the great mass of the people; health services generally were distributed mainly among the wealthier sections who needed them least, and were but poorly supplied to the underprivileged groups. The Commission proposed a national health scheme, financed by means of a special tax, and based on a network of 400 health centers. Up to the present, the main developments have been the authorization of a hospital tax on Africans alone, the establishment of 16 health centers, seven of them in the reserves, an increase in the number of whole-time district surgeons bringing it up to 40, provision for free hospitalization in some provinces, greater facilities for training non-European medical practitioners and nurses, and the elaboration of a centralized health administration.

For all their inadequacy, health services represent a great advance over the tribal methods of treating diseases. This cannot be said of the housing accommodation provided for the majority of Africans. In this respect, standards have tended to deteriorate under European rule. Attention has already been drawn to the primitive nature of the traditional hut in the reserves. The tribal village, however, in addition to possessing an aesthetic charm, did avoid overcrowding, while the frequent rebuilding of huts reduced the extent of vermin infestation. These qualities have largely been lost in areas where villages are permanently sited because of the scarcity of land and of building material. It is a growing practice, though one contrary to tribal rules, for adults to share the same huts with young persons, and in the permanently established village, the absence of sanitary controls becomes a menace to health.

These conditions are duplicated on most European farms, where the laborer usually builds his hut out of local materials such as grass and sods, wattle and daub. On larger estates of the plantation type, employing migrant workers for the cultivation of sugar, sisal, tea, wattle, maize, and other commercial crops, the tendency is to build barracks which, apart from affording protection against rain and wind, have little claim to merit. Similar compounds, or one-roomed huts, are also found on most mining properties; with some exceptions, such as the Wankie collieries in Southern Rhodesia and certain of the Northern Rhodesia copper mines, it is not the practice in British territories to provide accommodation for the families of African miners. The mine compound, unattractive though it is and devoid of any of the qualities of a home, does at least conform to a prescribed standard of hygiene. It is within the urban areas and on their boundaries that the slum conditions characteristic of industrialized countries have been reproduced with most harmful effects. Where towns are small and Africans are able to live on their own lands in the vicinity, as is the rule in Uganda and Nyasaland, the number housed within the urban area can be kept at an insignificant figure by limiting the right of residence to domestic servants and other persons employed in the town. In most of the large urban areas in all the territories, however, the problem of housing the African has been found insoluble.

South Africa once again provides an extreme example. In 1919 it was estimated that 10,000 new houses were needed in urban areas for Europeans and a similar number for Africans. The latest figures show a shortage of 60,000 houses for Europeans and another 120,000 for non-Europeans. Restricted by the segregation policy to a limited residential area, prohibited in most towns from buying or leasing land, unable to obtain credit facilities from building societies and other financial organizations, the African, and to a lesser extent the Colored and Indian, are dependent upon the accommodation provided by local authorities or the central government in "locations" and townships. In some of the new locations the standard of housing is satisfactory; in the majority, families regardless of size have to live in two- or three-roomed houses which "are from the first day of their occupation overcrowded and therefore slums as defined in the Second Schedule to the Slums Act, and escape condemnation as such only by reason of the specific withdrawal of Native locations from the purview of the latter Act." [81] For the non-European communities, water supplies, sanitation and other environmental services are usually inadequate.[82] Poor though these standards are, they compare favorably with the conditions under which large proportions of the non-European urban community live in privately owned houses. Forced out of the towns by high rents, scarcity of housing,

[81] *Report of the Inter-Departmental Committee on the Social, Health and Economic Conditions of Urban Natives*, par. 94.

[82] Social and Economic Planning Council, *Report No. 8: Local Government Functions and Finances* (1945), U.G. No. 40 (1945), p. 13.

and the segregation laws, many have settled in the peri-urban areas
in shacks of wood and corrugated iron, clay and wattle, or poles
and sacking, with primitive sanitary arrangements and poor cooking facilities. As many as 100,000 Africans live under these conditions around Johannesburg alone.

Similar slums have developed in and about the urban areas of other
territories. Accommodation is often of the barrack type, a survival
of earlier days; or Africans are allowed to build their own huts on
stands rented from the local authority. Both types of housing have
proved unsatisfactory.[83] It is now recognized that primitive housing
is as undesirable in the towns of Africa as in the cities of Europe, and
that the African cannot out of his own resources provide himself
with a satisfactory standard of accommodation. Attempts to design
a type of house that would be adequate and at the same time within
the African's means have not been successful. Either the standard
must be lowered, which is what occurs in the slums and in a great
deal of the housing provided by employers and public authorities,
or rents have to be subsidized out of rates and taxes. The second
alternative has been followed to a varying extent in most territories, but not on a sufficiently large scale to eliminate the slums
that have been inherited from an earlier period, nor even to keep
pace with the growth in the urban population. Although a sub-
economic housing scheme was instituted in South Africa as far
back as 1930, the rate of construction of subsidized houses averaged
only about 2,000 a year up to 1944. Since then greater efforts have
been made, including the creation of a new central housing authority with extremely wide powers, and the adoption of a national
building program of £40 million. In spite of these provisions, only
7,000 houses for Africans were built in 1946 under schemes financed
by the government.

One of South Africa's peculiar difficulties is that Africans, de-
barred from skilled trades by the color bar, have to live in houses
whose costs are largely determined by the living standards of European employers and artisans. In the face of strong opposition from
the building trade unions, the government has embarked upon the

[83] See *Labour Conditions in Northern Rhodesia* (1938), Colonial No. 150, and
Labour Conditions in East Africa (1946), Colonial No. 193, both by Major G. St. J.
Orde Browne.

training of African artisans, who are to be employed at reduced wage rates on building houses for Africans alone, but by the middle of 1947 only 50 had been trained, while 81 more were undergoing training.[84] The experience of other territories where Africans are employed as artisans in the building trades indicates, however, that the great obstacle to adequate housing is not the relatively high wage of the European worker but the poverty of the African.

Labor and Social Welfare

The presence of a large body of European wage earners, well organized and politically influential, has the effect of narrowing the non-European's field of employment to the less skilled types of work. The disability is offset to some extent by his participation in the benefits of an advanced standard of labor welfare legislation which the African unaided cannot hope to secure. In South Africa industrial workers of all races derive advantage from the collective-bargaining and wage-fixing machinery, and from statutory provisions insuring a 46-hour working week, limitations on overtime, increased pay for overtime, annual holidays with pay, confinement allowances, protective clothing, cloak-room facilities, and safety appliances. Compensation is payable to non-Europeans as well as to Europeans, though on a lower scale to Africans, for injuries sustained or diseases contracted during the course of employment. Unemployment insurance, recently established on a national basis, covers employed persons regardless of race.[85] The two fields of employment, agriculture and private domestic service, in which the number of European wage earners is negligible, are significantly excluded from the whole range of labor welfare, while separate and on the whole inferior provision is made for the African miner, who is also excluded from unemployment insurance and from the provisions for paid holidays and increased overtime pay. In the British colonies, with industrial development at a low level and a predominating African labor force in agriculture, mining, domestic and government service, labor legislation has scarcely advanced beyond the primitive stage of regulating and enforcing master and servant con-

[84] *House of Assembly Debates*, 2d June, 1947, col. 6348.
[85] Unemployment Insurance Act, 1946.

tracts, and of implementing international conventions relating to forced labor, recruiting, written contracts, and penal sanctions. A large number of statutes concerning labor conditions was enacted during the war, but the main improvements consisted of relieving the severity of the penal sanctions applied to contracts and of greater protection for juveniles.

In yet another respect—the alleviation of poverty—South Africa is far in advance of the rest of the continent. One of the striking developments of the past decade in South Africa has been the expansion of welfare services, including the creation of a separate government department of Social Welfare, and their extension to non-Europeans. Although the cruder forms of "poor relief," such as the discretionary issue of pauper rations or the subsidization of rents, are still common, the main trend has been towards the introduction of regular payments to beneficiaries falling within prescribed categories. A survey made in 1940 showed that "ethnic discrimination in the social services of the Union is the rule rather than the exception, and does not excite much comment. There are fewer social services available to the non-European than to the European, and many of those that are available are far from adequate. This is true both of private and State services." [86] Old-age pensions were payable to Europeans and Colored persons only, invalidity pensions to Europeans only, state-aided butter and milk schemes were not available to Africans, nor to Asiatics in Natal, and institutionalized treatment of most kinds was provided on a far more generous scale for Europeans. An estimate of the annual expenditure in 1941–43 on all forms of social insurance and assistance measures showed that out of a total of £9¾ million, Europeans received more than £8¼ million, the Colored £¾ million, and the Africans only £598,000.[87] It is doubtful whether there has been any significant change in these proportions, but effective recognition has been given to non-European claims. Since 1944, blind-persons, old-age, and invalidity pensions have been extended to members of all the non-European groups, the basic rates per annum being fixed at £42 for Europeans, £21 to £24 for Colored and Asiatics, and £12, £9, and

[86] E. Batson, "The Social Services: Discrimination and Counteraction," *Race Relations*, Vol. VII, No. 2 (1940).
[87] *Report of the Social Security Committee*, U.G., No. 14 (1944), par. 79.

£6 for Africans in cities, towns, and urban areas respectively. In 1947 the payment of family allowances was instituted, but to the exclusion of Africans. Race discrimination is still observed, with regard to both the scope and the quality of the services provided, and the standard of relief is meager; yet these recent developments are important because they entail an admission that non-European poverty is the concern of the whole community.

Implicit in that admission is a challenge to the basis of South African society. There are limits to the extent to which mass poverty on the African scale can be mitigated by a transfer of income from the privileged section. Professor Batson points out: "Long before relief-transfers in the Union could become big enough to have an appreciable direct effect upon the burden of poverty, their effect upon the contributing groups would transform the present capitalist industrial system from top to bottom." [88] The distribution of wealth is correlated with the distribution of political power, and few European taxpayers would be disposed to accept a lower standard of living in order to raise non-European standards. Considerable opposition has been shown, and is still manifested, to the extension of unemployment benefits, old-age and invalidity pensions, and school feeding schemes, to non-Europeans. There is also the economic difficulty: with an average national income of only £35 per head per annum, South Africa cannot afford social-service expenditure commensurate with Western standards. It has been calculated that, to achieve the necessary minimum of social security by 1955, the national income level would have to be raised by at least 50 percent over the 1938 level.[89] It is conceivable that this target can be attained, but only if the structure of society is modified so as to allow the non-European to develop and apply his abilities.

In Transition

Measured in terms of standards of living, the development of new forms of social organization, educational statistics, and the output of literature, the non-European has progressed farther in South

[88] "The Social Services and the Poverty of the Unskilled Worker," *Race Relations*, Vol. VI, No. 4 (1939).

[89] Social and Economic Planning Council, *Report No. 2*, U.G. No. 14 (1944), par. 29.

Africa than in any other territory south of the Sahara. It should not be supposed that there has been complete assimilation into a common culture. Most Africans retain traits that distinguish them from other groups. They still recognize tribal and clan affiliations, marry according to their customary laws, practice the traditional ancestor cult, and retain a strong belief in the efficacy of magic. Governmental organization has been adapted to these peculiar features. A separate Native Affairs Department is concerned exclusively with Africans, a special system of courts, staffed with Europeans, applies native law in civil cases between Africans, and tribal chiefs occupy a minor position in the administration of the reserves. In spite of the cultural heterogeneity, however, the anthropologist is already able to discern a "common South African civilization, shared in by both Black and White, and presenting certain peculiarities based directly upon the fact of their juxtaposition." [90] Cultural differences no longer coincide, as they did in the early period of white settlement, with racial differences. The Colored people, and a large number of Indians, have been fully absorbed into the Western type of community. All South Africans are bound together in a single political and economic structure. Europeans and non-Europeans frequently do the same kind of work and have a similar economic status, receive the same kind of education, belong to the same churches, or trade unions, marry under the same laws, live in the same type of house, speak the same languages and play the same kind of games. They may even, as the war has shown, share a common set of loyalties, and cooperate in defense of an ideal acceptable to individuals of different races.

It is civic status, and not culture, that is correlated with racial origins. Europeans and non-Europeans at similar cultural levels have different rights and obligations. The differentiation is not absolute. All persons, for instance, are liable to pay income tax if they fall within the tax-paying categories. There is, however, differentiation between Europeans and non-Europeans, and between different groups of non-Europeans, in most of the consequences normally attached to citizenship status.[91] Only Europeans are liable to con-

[90] I. Schapera, *The Bantu-Speaking Tribes of South Africa* (London, 1937), p. 386.
[91] See H. J. Simons, "Disabilities of the Native in the Union of South Africa; *Race Relations*, Vol. VI, No. 2 (1939).

scription during war and to compulsory military training during peace. Only Europeans may be elected to Parliament and, except in the Cape Province, to Provincial Councils and Municipal Councils. Only Europeans possess adult franchise rights.[92] There are no non-European judges, magistrates, or commissioned officers in the police and armed forces. Differentiation involves restrictions upon the right of Africans, and in some provinces of Indians and Colored, to acquire land and to trade. It involves a ban upon the movement of Indians from one province into another, and legal restrictions in the form of pass laws upon the movement of Africans. It involves racial separation, always to the disadvantage of non-Europeans, in residential areas, schools, churches, political and trade union organizations, trains and buses, hotels, cinemas, restaurants, recreation grounds, and public conveniences.

An authoritative body of European opinion, as we have seen, believes that the process of assimilation should be accelerated by encouraging the townward movement from the reserves, facilitating the absorption of non-Europeans in industry, and extending the scope of education, health and social services. The admission that the cultural barriers between white and black have broken down beyond repair, and that the future of all racial groups in South Africa must be worked out within the framework of a single society, is implicit in the comment of General Smuts: "Isolation has gone and segregation has fallen on evil days." [93] Within this society, however, the majority of Europeans are determined to maintain their dominant position by enforcing the pattern of race discrimination. It is this relationship between white and black that General Smuts had in mind when, during the debate on the Asiatic Land Tenure and Indian Representation Bill, he claimed that "South Africa has decided once and for all, that our complex society will be dealt with on separate lines. We have done it in the case of the natives, and we are going to do it in the case of the Indians." [94] What is envisaged in this passage is a single society rigidly divided into hierarchical

[92] In July, 1943, registered parliamentary voters consisted of 1,282,605 Europeans, 46,535 Colored and Indians, and, on a separate roll, about 12,000 Africans.

[93] *The Basis of Trusteeship in African Native Policy* (S.A. Institute of Race Relations, 1942), p. 10.

[94] *House of Assembly Debates*, 25th March, 1946, col. 4173.

orders, membership of which is determined at birth by racial characters.

Assimilation has not led to a relaxation of color bars. Indeed, a history of race relations in South Africa would show a tendency for discrimination to increase as the cultural gap between white and black diminishes. Africans in the Cape were removed from the common voters' roll in 1936, eighty years after they had been given the franchise on the same terms as Europeans. The 1946 legislation prohibiting Natal Indians from acquiring land or residing outside specified areas was a reaction to their successful adaptation to South African conditions and their ability to hold their own in competition with the European. Illicit sex relations between Africans and Europeans were made illegal in 1926; today there is persistent agitation among a section of the white population for the illegalization of marriages between Europeans and non-Europeans.[95] These trends, however, reflect an awareness of the threat that assimilation presents to the color bar. Already a considerable number of non-Europeans are superior in education and economic standards to the lower-class Europeans. The dilemma was pointed out by Mr. Hofmeyr, Deputy Prime Minister, when he accused the Leader of the Nationalist Opposition, Dr. Malan, of

seeking to do simultaneously two things which are ultimately incompatible. He wants to maintain a vertical colour line, on the basis of a policy of separation as he advocates it. He wants also to be just and generous to the extent, for instance, of giving non-Europeans a full university of their own as adequate as any of our European universities. But that university of his is going to set up trends which must inevitably destroy that vertical colour bar he is setting up.[96]

The reality of this conflict has been grasped by a school of "totalitarian" segregationists, who now demand nothing less than the repatriation of all Indians—four-fifths of whom are South Africans by birth—and the expulsion of all Africans to an unknown destination.

To the non-European, race discrimination is part of the order of things, as inevitable as class differences appear to people in other

[95] See *Report of the Commission on Mixed Marriages in South Africa*, U.G. No. 30, 1939.
[96] *House of Assembly Debates*, 23d January, 1947, col. 11095.

countries. Aspects of segregation—the pass and tax laws, the inadequacy of land in the reserves, police raids—are sources of constant dissatisfaction, but European domination is too pervasive and intensive, too traditional, for serious challenge. Moreover, the rapid expansion of the economy has provided the ambitious and able non-European with opportunity to improve his personal position in spite of the color bar. The greatest resistance to segregation has come from groups, such as the Natal Indians or the Colored teachers, who are being forced back to a lower level or who feel that further progress cannot be made under existing conditions. The non-European, however, is ceasing to be a passive factor in politics. Political education has been spread through agencies like the African National Congress, the Natal Indian Congress, trade unions, teachers' leagues, and the press. A section has become politically conscious. The demand for an extension of the franchise is being raised. Significant developments have occurred since the end of the war: the "passive resistance" of Natal Indians to segregation and their rejection of the scheme for a communal franchise with representation by Europeans; the decision of the African members of the Native Representative Council (the body created to advise on African opinion) to suspend its sittings until the Government has proposed a policy acceptable to them; and the current campaign to boycott the election of the seven Europeans who represent Africans in Parliament. These events may be due to a postwar reaction, or they may be a prelude to a popular struggle of non-Europeans against race discrimination.

In the British dependencies, the imperial government has exercised a restraining influence on the growth of a caste system. Southern Rhodesia, with a status corresponding to that of a dominion, though nominally a colony, must be excepted from this generalization. The Crown retains the right to disallow legislation imposing special disabilities on Africans, but the colony's pattern of race relations follows closely that of South Africa. The official policy of "parallel development" for white and black is in reality one of segregation and discrimination. As in the South, the prospect of non-European progress tends to intensify discrimination, a recent instance being the Prime Minister's proposal, withdrawn after debate, to exclude Africans from the franchise and to substitute representation by two Europeans in Parliament. The reason given for the motion was that, although there are fewer than 200 Africans on the roll and no more than 2,000 possess the high qualifications prescribed for the vote, African voters may be in the majority in the next two or three decades.

In the Central and East African territories, the future of the Europeans themselves is uncertain. It is doubtful whether they will ever command a numerical strength that would enable them to enforce a caste system. The well-established Indian communities in these dependencies have in the past offered effective resistance to complete white domination, and the participation by Africans in local government according to indirect-rule principles may be expected to hasten their political development. For some time to come, however, the decisive factor must remain the nature of imperial colonial policy. Up to the present, the British Government, while acceding to a wide range of discrimination, has set its face against the transfer of full political authority to the settlers. In Kenya, for instance, though the doctrine of the "paramountcy of native interests," adopted in 1923 by the British Government as the basis of policy, was subsequently replaced by the concepts of "trusteeship" and the "dual mandate," [97] the demand of the colonists for representative government has been resisted on the ground that it could not be representative of any but the white minority. The declared policy of the imperial authorities is to extend the franchise in different parts of Africa, and to put Africans on legislative councils. Africans, as well as Indians, are members of the legislative councils of Kenya, Uganda, and Tanganyika, and a similar step has been proposed for Northern Rhodesia and Nyasaland. Even though the European representatives may continue to form a majority for many years to come, the direct participation by the non-European in central government constitutes a sharp break with the South African tradition that has prevailed in the past.

These tendencies, as well as events abroad, are bound to have repercussions in South Africa. The criticisms at the United Nations of South African native and Indian policy made a deep impression on wide sections of the South African population, both European and

[97] For an examination of their significance, see M. R. Dilley, *British Policy in Kenya Colony* (New York, 1937), Part III.

non-European. Non-Europeans are beginning to show an interest in the advance towards self-government in West African dependencies. South Africa, which formerly took the lead in social and economic progress, may find that more rapid advance is being made by the African in other parts of the continent. The changes that are taking place in the East, the emergence of new independent non-white states, the challenge to universal white domination, and the passing of the old imperialism, may be expected to weaken the structure of South Africa's caste society. In South Africa itself, as we have seen, strong economic and social forces are tending to undermine the bases of race discrimination. It is conceivable, however, that the dominant European minority will devote its energies to a strengthening of caste barriers, even at the cost of retarding its social progress and surrendering its own liberties.

SUGGESTED READINGS

SOUTHERN AFRICA

Burger, John, pseud. The Black Man's Burden. London, 1943. A popular, critical review of policies and conditions.

de Kiewiet, C. W. A History of South Africa. Oxford, England, 1941. Special attention is given to events after the discovery of diamonds and gold.

Hoernlé, R. F. A. South African Native Policy and the Liberal Spirit. University of Cape Town, 1939. An analysis of race domination and the concept of trusteeship.

Hunter, Monica. Reaction to Conquest. London, 1936. Culture change in a South African tribe.

Macmillan, W. M. Complex South Africa. London, 1930. An analysis of African and European poverty in rural areas.

MacCrone, I. D. Race Attitudes in South Africa. Witwatersrand University, 1937. The historical background to race discrimination and a psychological analysis of attitudes.

Marais, J. S. The Cape Coloured People. London, 1939. The emergence and assimilation of the Coloured people.

Schapera, I., ed. Western Civilization and the Natives of South Africa. London, 1934. The most comprehensive study, in the form of essays, of present-day conditions and culture contact.

Schapera, I., ed. The Bantu-speaking Tribes of South Africa. London, 1937. A review by various writers of the traditional culture and of the changes under Western influence.

van der Horst, Sheila T. Native Labour in South Africa. London, 1942. Traces the absorption of Africans into the European economy and analyzes the effects of industrial colour bars.

CENTRAL AND EASTERN AFRICA

Davis, J. Merle, ed. Modern Industry and the African. London, 1933. The effect of mining development on African life in Northern Rhodesia.

Dilley, Marjorie R. British Policy in Kenya Colony. New York, 1937. A critical review.

Hinden, Rita. Plan for Africa. London, 1941. A comparison of policies and conditions in Northern Rhodesia and the Gold Coast.

Huxley, Elspeth and Margery Perham. Race and Politics in Kenya. London, 1944. A lively discussion, in the form of letters, of race relations in Kenya.

Leubuscher, Charlotte. Tanganyika Territory. London, 1944. A survey of economic and social policy under Mandate.

Leys, Norman. The Colour Bar in East Africa. London, 1941. A study of policies in British East and Central Africa.

Thomas, H. B. and R. Scott. Uganda. London, 1935. Standard work on Uganda.

Wilson, Godfrey. An Essay on the Economics of Detribalization in Northern Rhodesia. Northern Rhodesia, Rhodes-Livingstone Institute, 1941. (Parts I and II). Economic and social policies on the copper belt, and the effect on rural Africans.

William R. Bascom

WEST AND CENTRAL AFRICA

Resources

WEST AND CENTRAL AFRICA cover an area two-thirds larger than the United States, with about half its population. West Africa lies along the Guinea Coast and extends north into the Sudan region below the Sahara; it includes Liberia, Portuguese Guinea, French West Africa (with Dakar and its dependencies, and the seven colonies of Senegal, French Guinea, Ivory Coast, Dahomey, Niger Colony, French Sudan, and Mauritania), the Trust Territory of French Togoland, and the four colonies of British West Africa (Nigeria, Gold Coast, Sierra Leone, and Gambia). Central Africa, as defined here, includes the Trust Territory of French Cameroons, the four colonies of French Equatorial Africa (Gabon, Middle Congo, Ubangi-Shari, and Chad), Spanish Guinea (including Rio Muni and the island of Fernando Po), the Belgian Congo, the Belgian Trust Territory of Ruanda-Urundi, and Angola (including Cabinda). The British Trust Territories of Togo and Cameroons are treated as parts of the Gold Coast and Nigeria for administrative purposes and will be so considered here.[1]

The whole area is deficient in developed sources of power. No petroleum deposits have been developed and coal is mined only in Nigeria and the Belgian Congo. Electrification has proceeded farthest in the Belgian Congo. Of the cities which have electricity, most rely on fuel-operated generators, but a number of the mining companies have their own hydroelectric power. Three great river systems, the Congo, the Niger, and the Senegal, and smaller rivers

[1] The section on West and Central Africa was mimeographed by the author and forwarded to the governments concerned for comment. Limitations of space make it impossible to include all those received, but the more important corrections and criticisms have been incorporated into the text and footnotes. This should not be taken to mean that the governments have endorsed the interpretations and opinions expressed here.

such as the Volta and Gambia provide vast but virtually untapped sources of hydroelectric power. Any real industrialization must depend on the development of these resources. Meanwhile, petroleum products and coal are imported.

Minerals are the most valuable natural resources. In 1941 the Belgian Congo, Gold Coast, Sierra Leone, and Angola were the world's four largest diamond producers, and West and Central Africa accounted for 94 percent of world production. The third largest gem diamond on record was discovered in Sierra Leone in 1945, but most of the diamonds are industrial. Gold, which is very important to the local economies, is exported from most of the territories.

The Belgian Congo, which produces two-thirds of the world supply of cobalt, is also the fifth largest producer of copper. Nigeria and the Belgian Congo expanded tin production during the war, both exceeding Siam, the third largest pre-war producer. The Gold Coast is the fourth largest producer of manganese and its large bauxite deposits were also developed during the war. In 1941 Sierra Leone exported over a million tons of iron ore from what are said to be some of the richest deposits in the British Empire.

The Belgian Congo was the world's largest pre-war producer of uranium ore, and Nigeria provided 98 percent of the world supply of columbite. The Congo's exports of tantalite, another rare mineral of strategic importance, were small compared to Brazil's, but the Congo ranked second in world production. Senegal was the third largest producer of ilmenite. Mineral exports of minor importance include silver, platinum, palladium, and the ores of chromium, zinc, lead, cadmium and tungsten.

West and Central Africa produce two-thirds of the world's cocoa. The Gold Coast, the largest producer, and Nigeria, which ranks third, together account for about half of world production. The area is also a major producer of vegetable oils and oil seeds, which are processed into soap, shampoos, margarine, salad oils, candles, glycerine, and other products, in Europe and America. More than 1,500,000 tons of palm kernels, palm oil, and peanuts (known in West Africa as *arachides* or ground nuts) were exported in 1938, with lesser amounts of peanut oil, peanut cake, sesame seed (called sem-sem or benniseed), shea nuts, shea butter, cottonseed, cotton-

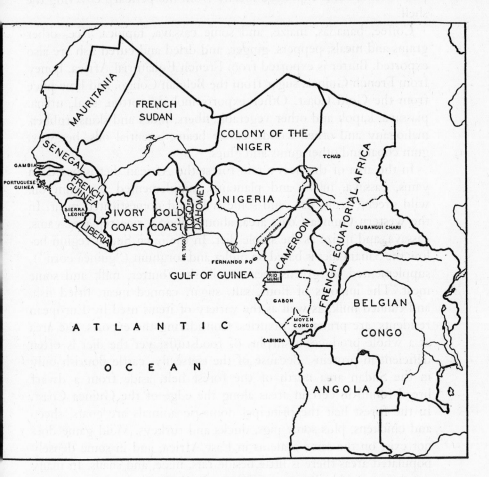

WEST AND CENTRAL AFRICA

seed oil and copra. Palm kernels are obtained by cracking the hard black shell of the nut of the oil palm; palm oil, which differs from palm kernel oil, is expressed locally from the pericarp covering the shell.

Coffee, bananas, maize, and some cassava, tapioca, rice, other grains and meals, peppers, ginger, and dried and cured fish are also exported. Butter is exported from French Equatorial Africa, honey from French Guinea, sugar from the Belgian Congo, and lime juice from the Gold Coast. Other exports include cotton, sisal, urena, piassava, kapok and other vegetable fibers, hides and skins, rubber, mahogany and other woods, castor beans, essential oils, beeswax, gum copal and other gums and glues.

In the area of the heavy rain forest the African diet is based on yams, cassava, maize and plantains, supplemented by palm oil, wild greens, fruit, fish, and some wild and domesticated meat. In the western Guinea Coast area, about Liberia, rice replaces yams, cassava, and maize as the staple diet. In the open Sudan region below the Sahara diet is based on millet and sorghum ("guinea corn"), supplemented by greens, peanut oil, shea butter, milk and some meat. The imports of flour, salt, sugar, canned meat, dried fish, and canned milk, as well as the variety of items used by European residents, are primarily luxuries. Considering the exports, the area as a whole produces a surplus of foodstuffs; yet the diet is often deficient in proteins. Because of the tsetse fly, cattle flourish only in the Sudan area north of the forest belt, aside from a dwarf breed found in certain areas along the edge of the Guinea Coast. In the forest belt the principal domestic animals are goats, sheep and chickens, plus some pigs, ducks and turkeys. Wild game does not exist on the same scale as in East Africa, and in some densely populated areas there is little beside rats, mice, and snails. In many groups vegetable oils are more important than meat or fish as sources of protein. Vitamins are obtained from green vegetables and fruits, vegetable oils, and in the forest belt from palmwine, the fermented sap of the oil palm and other palm trees. Large quantities of starchy foods constitute the bulk of the African diet. In parts of the area, governments are investigating and trying to improve African nutrition through such measures as the immunization of cattle against tsetse flies, mixed farming with plows and fertilizers, the introduction of new crops, and the development of fisheries.

The fertility of African soil has often been overrated. A large part of the area is desert or subdesert, and Mauritania supports only about one person per square mile. Even in the tropical forest belt, heavy rainfall and moderately fertile soil mean heavy labor in clearing and weeding the land. Some areas where a dense population prevents adequate fallowing, such as central southern Nigeria, are faced with severe erosion problems. Large-scale mechanized agriculture cannot be adapted to the forested area unless the land is cleared, nor to the flatter subdesert area until irrigation projects, such as that begun by the French in the Niger valley near Timbuktu before the war, have actually been completed. Aside from the export crops, increased production of foodstuffs can be absorbed by the local population.

Between the desert regions of northern French West Africa and the tropical rain forests running along the Guinea Coast and up the Congo basin, there are marked differences in climate. In the former area temperatures are high during the day and low at night, while rainfall and humidity are extremely low. In some parts of Niger Colony and French Sudan the annual rainfall drops below 10 inches a year, while the average recorded at Debundscha, below the Cameroon Mountain in the bend of the coast (near Buea) is 358 inches, making it one of the wettest spots in the world. Throughout most of the tropical rain forest the rainfall is between 50 and 150 inches a year, while temperatures generally range between 65° and 95°.

Along the Guinea Coast the forest areas have two main seasons. The rainy season (June–September) has heavy rainfall, high humidity and median temperatures, with the moisture laden winds blowing north from the Atlantic. The dry season (November–February) has little rain, low humidity, and wider variations in temperature, with a dry wind ("Harmattan") blowing south from the Sahara. In some parts there is a second short dry season in August. The farther north from the Guinea Coast one goes, the more the effects of the Harmattan and the subdesert climate are felt. South of the equator the seasons are reversed. Climate varies with altitude as well as latitude, and Jos in Nigeria and Dschang in French Cameroons, both of which are about 4,000 feet above sea level, have been established as European resorts for local vacations because they are cooler and more healthful.

Area and Population of West and Central Africa

Colony	Capital	Area	Estimated Population	Source
Nigeria (including Cameroons)	Lagos	371,393	23,000,000	1931 Census
Gold Coast (including Togo)	Accra	91,843	4,000,000	1931 Census
Sierra Leone	Freetown	27,925	2,000,000	1931 Census
Gambia	Bathurst	4,132	200,000	1946 Census
BRITISH WEST AFRICA		495,293	29,200,000	25,106,303
Dakar and Dependencies	Dakar	60	200,000	1941 Census
Senegal	St. Louis	77,730	1,800,000	1941 Census
French Guinea	Conakry	96,886	2,200,000	1941 Census
Ivory Coast	Abidjan	184,174	4,000,000	1941 Census
Dahomey	Porto Novo	43,232	1,500,000	1941 Census
Niger Colony	Niamey	499,410	2,000,000	1941 Census
French Sudan	Bamako	590,966	4,000,000	1941 Census
Mauritania		323,310	300,000	1941 Census
FRENCH WEST AFRICA	Dakar	1,815,768	16,000,000	15,582,535
French Togoland	Lome	21,893	700,000	1938 Census
French Cameroons	Yaounde	166,489	2,800,000	1938 Census

Also appearing in the right column (Source), population figures: 19,928,171; 3,160,386; 1,768,480; 249,266; 165,188; 1,723,068; 2,117,705; 4,047,041; 1,424,220; 1,944,190; 3,794,270; 366,853; 780,497; 2,516,623.

Region	Capital		Year			
Gabon	Libreville	93,218	1943	Census	409,852	400,000
Middle Congo	Brazzaville	166,069	1943	Census	740,772	600,000
Ubangi-Shari	Bangui	238,767	1943	Census	947,316	1,000,000
Chad	Ft. Lamy	461,202	1943	Census	1,432,869	2,000,000
FRENCH EQUATORIAL AFRICA	Brazzaville	959,256			3,530,809	4,000,000
TOTAL FRENCH AFRICA		2,963,406			22,410,464	23,500,000
Belgian Congo	Leopoldville	904,757	1946	Registration	10,702,859	10,500,000
Ruanda-Urundi	Usumbura	20,535	1946	Estimate	3,596,997	3,500,000
TOTAL BELGIAN AFRICA		925,292			14,299,856	14,000,000
Angola (including Cabinda)	Luanda	481,351	1940	Census	3,738,010	3,700,000
Portuguese Guinea	Bissau	13,944	1940	Census	351,089	300,000
TOTAL PORTUGUESE AFRICA		495,295			4,089,099	4,000,000
Spanish Guinea	Santa Isabel	10,828	1942	Census	170,582	200,000
Liberia	Monrovia	43,000	1947	Estimate	1,500,000	1,500,000
TOTAL WEST AFRICA		2,389,898			43,320,424	47,700,000
TOTAL CENTRAL AFRICA		2,543,216			24,255,880	24,700,000
TOTAL WEST AND CENTRAL AFRICA		4,933,114			67,576,304	72,400,000

Population

West and Central Africa have roughly half of the total popula-
tion of Africa. Nigeria's population is the largest in Africa and the
third largest in the entire British Empire, following only India and
Great Britain herself. While the largest number of people are under
British control, 60 percent of the area, equal in size to the United
States, is French. West Africa has almost twice the population of
Central Africa, although its area is slightly smaller. Even greater
variations in the density of population occur within these regions.
Belgian Africa has three and a half times as many people as French
Equatorial Africa, though its area is slightly smaller. The Portu-
guese colonies, with the same population as French Equatorial
Africa, are only half as large. British West Africa, with the same
area as the Portuguese colonies, has over seven times as many people,
and more than twice the population of Belgian Africa. French West
Africa has twice the area of French Equatorial Africa, and four times
as many inhabitants.

Population statistics from this part of the world must be accepted
with caution, even when they are not admittedly estimates. Com-
menting on the 1901 census in Gambia's *Official Handbook*, Archer
states, "To this total must be added at least 40 per cent. for the Protectorate, as the people are very shy and object strongly to their
numbers being taken." [2] Since that time census-taking techniques
have been refined, but even in the larger colonies government staffs
are still far from adequate for accurate enumerations, and it is economically advantageous for Africans not to give the true number
of taxable adults.

Even with the crude statistics available, three broad trends are
apparent. There is a tendency toward greater urbanization; the population is increasing; and there have been migrations to territories
offering higher wages and higher standards of living. Migrations
across political boundaries, such as those from French colonies and
Liberia to British West Africa, do not appear to be as large as
some have claimed. A considerable part of the apparent decrease

[2] *The Gambia Colony and Protectorate: an Official Handbook* (London [1904?]),
p. 137.

in population in the Belgian Congo and French Equatorial Africa, attributed to the Concession System and the introduction of diseases by Europeans, may be accounted for by excessive estimates in the early days.

Birth rates, death rates, and life expectancies are no more reliable. Vital statistics are seldom available outside a few urban areas where medical facilities are the best, and they are not representative of the area as a whole. In the Gold Coast the registration of births and deaths is required in 37 urban districts, representing only 11 percent of the total estimated population. In other colonies the figure is even lower.

Malaria, black water fever, yellow fever, relapsing fever, dysentery, diarrhoea, tuberculosis, typhoid, typhus, cholera, small pox, gonorrhoea, syphilis, yaws, leprosy, sleeping sickness, bilharzia, elephantiasis, guinea worm and other filariasis, and various other fevers, intestinal parasites and skin diseases have combined to give West Africa its unenviable reputation for an unhealthy climate. Out of 140 men on the first British ships to visit Nigeria (1554), only 40 survived to return to England. The high European mortality rate, which earned for West Africa the title of the "White Man's Grave," was largely due to malaria. It was not reduced until after Baikie's trip up the Niger and Benue rivers in 1854, when quinine was used for the first time in Nigeria and not a single member of the crew was lost.

Health hazards for whites have been greatly reduced, and further advances may be expected from the wartime discoveries in tropical medicine, particularly in the field of antimalarials. They have not, however, been eliminated. It is still customary for British officials to spend eighteen weeks of leave in England for health reasons, after every eighteen months of service on the Coast. In that they are not part of "White Man's Africa," the diverse climatic, economic, cultural and political regions of West and Central Africa can justifiably be considered as a unit apart from the rest of the continent. Real colonization has been discouraged by disease and only about 162,000 Europeans (0.2 per cent of the population) are found in the entire area. There are about 44,100 (1.2 percent) in Angola, 35,800 (0.3 percent) in the Belgian Congo, 47,400 (0.3 percent) in French West Africa, and 35,000 (0.08 percent) in the remaining territories. The latest Nigerian census (1931) showed 4,952 "Europeans," including Americans, and 490 "Asiatics," most of them from Lebanon and Syria. It has been cynically suggested that the real reason for the difference in land policies in British West and East Africa has been the mosquito.

In the far larger problem of extending the benefits of medical science to some seventy million Africans, only the beginnings have been made. Smallpox is fairly well controlled through vaccination, and in parts of the French Cameroons the average death rate from sleeping sickness in terms of the total population has been reduced from 29 percent in 1919 to 0.5 percent in 1944. Government medical programs in the British, French, and particularly the Belgian colonies represent marked advances, but on their present scale even the combined efforts of government and missions are inadequate to solve the problem. Purified water systems are found only in the largest cities, while sewage disposal and other sanitation facilities are minimal. Even in Ibadan, Lagos, Accra, and Freetown—the four most important cities in British West Africa—night soil must be removed by hand, except for a few homes which have private septic tanks. In 1938, Nigeria had only 3,966 hospital beds for twenty-three million Africans.

Economics

Wealth and income in Africa present a wide spread, with Europeans at the top and Africans at the bottom. Within these two classes there is considerable variation and occasional overlapping. Some African lawyers and doctors in British West Africa earn more than the majority of government officials. Before the war, the Emir of Kano, the highest-paid Nigerian chief, received $42,500 (£6,000 salary and £2,500 establishment allowance) annually, while British administrative officers began at £400 and the Governor of Nigeria was paid £8,250 plus allowances. Such figures, however, obscure the wide gap separating the wages of Africans and Europeans. It is only in exceptional cases that an African earns as much as the lowest-paid European employee. The 1946 report of the official Enquiry into the Cost of Living in Nigeria states "it is doubtful if the normal family income in Nigeria even now exceeds £15" or $60 a year.

Below the African chiefs, traders, and professional men comes the mass of African population, including the wage laborers, the producers of export crops, and the subsistence farmers.

In 1936 French West Africans could avoid "prestation labor" by paying 7¢ (1.14 francs) a day. From 1936 through 1938 the average daily wage of Gold Coast miners was 35¢ (1s. 5d.) a day and in 1938 farm laborers in the Nigerian cocoa belt earned between 12¢ and 34¢ (6d. and 1s. 4½d.). Incomes were undoubtedly lower, and probably wages as well, in the areas which supplied the cocoa and mining laborers. Minimum daily wages in Lagos, where wages are high, were fixed at 18¢ (9d.) in 1934 and raised to 25¢ (1s.) in 1937, and 51¢ (2s.7d) in 1947. During the war wages increased through most of West and Central Africa, but not enough to offset the rising prices of local products and imports, and in Nigeria not without a major strike against the government. In 1945 Firestone employees in Liberia received a minimum of 18¢ a day, plus subsidized food prices. In 1947 the minimum daily wage in Leopoldville was fixed at 32¢ (14 francs). In 1941 the minimum wage in French Equatorial Africa was 2.8¢ (1.25 francs) a day, while laborers in mines and forests generally received 7–9¢ (3–4 francs).

Even the producers of the valuable cocoa crop seldom prosper. In the Gold Coast village of Akokoaso, the average annual cash income was $107.12 (£21 8s. 6d.) per family, of which half was required for debt charges. Of the 201 families surveyed in 1932–35, 125 families were in debt, and 60 percent had earned incomes of less than $50. Another study estimates that 30 percent of Gold Coast farmers had pawned their lands in return for long-term loans, while 75 percent pledged their crops in advance.

Below the wage laborers and producers of cocoa and other export commodities are the subsistence farmers, who provide their own food, clothing, and housing. These often earn money through the sale of local foodstuffs or handicraft or through part-time employment, but in many cases their cash income is negligible. In the community of Ozuitem in southern Nigeria, annual cash incomes of 16 individuals in 1938–39 ranged from $16.83 to $92.53 (£3 8s. 9d. to £19 1s. 9d.) for men, and from $5.73 to $28.54 (£1 3s. 5d. to £5 16s. 6d.) for women. In Akokoaso, where one family had an annual income of £200, 19 families were listed as having no income whatsoever. In the village of Umor in southern Nigeria the average annual cash income in 1941, based on a survey of 81 cases, is estimated at $5 (25s.) per household, of which 19s. is accounted for by palm kernels and palm oil.

"Non-Europeans" are found at both ends of the economic scale. Some of the wealthiest individuals are Syrians and Lebanese, whose fortunes have grown out of small trading operations, started with little capital and gradually expanded by shrewd trading and an interest in small sales, frugality and a willingness to begin on a low standard of living, and an ability to adapt themselves to African patterns of trading.

Of the export industries in African hands, the most important are cocoa, palm kernels, palm oil, peanuts, and other vegetable oils and oil seeds. Here Africans are the entrepreneurs as well as the laborers. The production of European-owned plantations is relatively small and Europeans serve mainly as middlemen, purchasing the crops and exporting them. The Gold Coast cocoa industry was developed by Africans, when Tetteh Quarshie smuggled the first seeds from Fernando Po in 1879. For the initiative and enterprise which made the Gold Coast a larger producer than the entire Western Hemisphere of a crop native to Mexico, Africans have never received due credit. Between 1905 and 1936 exports were increased from less than 5,000 tons to their peak of 311,151 long tons. African farmers received little assistance from government until cocoa had already become a principal export crop and the foundation of the Gold Coast's economy. The future of West African cocoa, however, may now depend on whether science can discover effective controls for the diseases which threaten the bearing trees.

Mining and to a lesser degree logging are essentially European industries. One of the three largest mahogany producers on the Gold Coast and numerous smaller producers are Africans but elsewhere, like the African mining companies, they are of negligible importance. The production of rubber, sisal, coffee, and commercial bananas is dominated by European-owned plantations.

African-owned industry has had little capital and has had to operate with little equipment. Mechanization, where it is found, is generally confined to the large European companies. The degree of mechanization presents startling contrasts. Some gold mines on the Gold Coast are highly mechanized, while elsewhere Africans may be seen panning gold in calabashes. Against the modern equipment of Lever Brothers' plantation in the Belgian Congo are the primitive extraction techniques which account for the bulk of palm oil and palm kernel production. Some American logging equipment, including Caterpillar tractors, Hyster cruiser arches and Athey wagons, is operating in West Africa, but gangs of laborers still haul logs by hand to the roadside. In contrast to the installations of Firestone in Liberia and of rubber companies in French Equatorial Africa and Cameroons are the simple hand rollers and home-made drying sheds in which so much wild rubber was processed during the war. Similar contrasts are presented even between the European tin-mining companies of northern Nigeria. Within sight of the steam shovels, drag lines, and hydraulic equipment of some companies, others employ thousands of African laborers equipped only with picks and head pans. Electromagnetic separators are used in Nigeria to treat tin ore and columbite, and some manganese is sintered before export from the Gold Coast, but smelters are found only in the Congo.

In local industries also the Belgian Congo leads, followed by French West Africa and Angola, with British West Africa, French Equatorial Africa and French Cameroons ranking above Portuguese Guinea, Spanish Guinea and Liberia. Industrialization on any appreciable scale, however, has scarcely begun. European trading companies have been accused of stifling the development of local manufactures in order to preserve their export and import trade. Regardless of the cause, West and Cenral Africa are exporters of basic raw materials and with few exceptions they are dependent on imports for manufactured goods.

A textile factory in the Congo manufactured nearly 21,000,000 meters of piece goods for local consumption in 1946, but cloth is still one of the major import items of the area as a whole. Factory-made soap is produced in Belgian Congo, Angola, Nigeria, and French West Africa. Angola and Nigeria manufacture cigarettes and Leopoldville and Accra have breweries. Soda water, soft drinks, and dairy and bakery goods are produced in a number of centers; ice cream and macaroni in a few towns. Brick and tile factories are found in Belgian, French, and Portuguese territories, and cement is produced in the Congo. There are saw mills in many parts of

the forest area, but the bulk of wood exported is in log form. Printing presses and machine shops are found in a number of places, and
in Leopoldville river boats for the Congo are built. The Congo has
tanneries and a factory which produced 403,392 pairs of shoes in
1946. It also produces sulfuric acid, acetylene, and compressed
oxygen and has distilleries and refineries. Sugar, flour, cured fish,
furniture, and souvenirs are also manufactured in some places on a
scale beyond that of crafts. With few exceptions this includes the
manufacturers for local consumption, aside from the work of tailors,
carpenters, tinsmiths, blacksmiths, mechanics, and the pottery,
basketry, weaving, leather work, brass casting, ivory carving, and
woodcarving. This is the area which produces the arts and crafts
for which Africa is famous.

Foreign trade is the life blood of the European populations and
the colonial governments, most of whose revenues are derived from
customs. The economy as a whole, however, still has its real foundations in the production of subsistence goods for local consumption.
The bulk of the food and housing, and a considerable portion of
the tools, household utensils, and even clothing are produced by
Africans, using techniques which have changed little since the time
of the first Portuguese explorers.

Complex African economies, characterized by guild specialization, intra- and intertribal trade, large markets, middlemen, and the
use of cowrie shells, metal, and other forms of true money are
found throughout most of the area. Most families engage in the
basic economic activity of farming, while some produce more
than they need for sale in the market. Special wholesale markets
are found where agricultural products are purchased for resale in
the regular markets, which are attended by thousands of individuals.
Trade in farm produce is generally in the hands of women, but
Hausa and Yoruba men can be found trading in most of the large
cities between the Senegal and the Congo. African traders, whose
skill has earned the respect of many Europeans, operate for profits
or commissions or both. Hunters, fishermen, traders, and specialists
in the many arts and crafts are organized into guilds which are entered through an apprentice system. Several forms of taxation, interest, and "pawning" (a form of loan whereby the creditor receives the production of an individual or a piece of land instead of

interest) are known. In Dahomey direct taxes based on individual wealth were levied in the days of the native kingdom after an annual census of population, livestock, and granaries.

The remarkable growth of exports and imports has involved increasing numbers of Africans in the complexities of the world market—as consumers, as producers of export commodities, as government employees, and as wage laborers for European and African employers. Prices of local produce in African markets are now affected by booms and depressions and by population changes such as the influx of European troops and civilians during the war. Internal trade has expanded with the development of transportation and with the peace which followed the end of the slave wars. The growth of the class of wage laborers and the increasing amounts of African lands devoted to export commodities have created a greater demand for local foodstuffs in the markets. Economic recessions and retrenchments or isolation through war can cause inconvenience, dissatisfaction, and serious decreases in standards of living, but where Africans have not lost their rights to the land the basic essentials can still be provided by subsistence farming. When opportunities for wage labor disappear, or when world prices for export commodities fall to a point which virtually prohibits production, they can still support themselves on their own farms.

Before the arrival of Europeans all land was the property of African individuals, tribes, or kinship groups. Unoccupied or "vacant" lands, and the oil palms and other planted trees on them, were also owned. In some tribes land could not be permanently alienated according to native law. In the early days of colonial expansion little regard was shown for African property concepts.

Under the Concession System, developed in the Belgian Congo under King Leopold II, all land outside towns and villages except that actually under cultivation was claimed as the property of the concessionaires, and the use of products of the forest by Africans was defined as theft. When the Belgian government assumed control of the Congo in 1908, freeholds totaled more than 96,525 square miles, not including immense uncalculated areas where only monopolies of forest products had been granted. Early thirty-year grants of monopolistic rights over forest products covered two-thirds of French Equatorial Africa, while those still in effect in

1899 amounted to 335,907 square miles. In 1899 two companies held concessions covering 45,174 square miles in the German Cameroons, where the Concession System also spread for a period.

Morel, one of the critics of these injustices and of the brutalities which accompanied the Concession System, concluded that the fundamental issue, which would decide whether "African peoples develop along the lines of freedom, or along lines of serfdom," was the preservation of African land tenure. It is not difficult to see that where their lands have been alienated Africans may be faced with the necessity of accepting wage labor or starving. The economic freedom of choice between working for wages, producing for export, and simple subsistence farming is lost, bargaining power is reduced. Large-scale alienations of land, through outright purchase or through long-term leases, can be as effective a means of recruiting low-paid native labor as excessive taxation or production quotas.

Reviewing conditions in French Equatorial Africa in the 1920s, Buell concluded that, "When private companies monopolize millions of hectares of land from which all competition is excluded, and when the government obliges the natives living in these areas to pay taxes, it is inevitable that the natives will be forced to work for these companies, and that they will be underpaid, whether for their labor or for their produce." Under these circumstances, permitting or even guaranteeing Africans the legal right to farm, hunt, and gather within alienated lands does not provide adequate protection of their economic status. The amount of land actually alienated to non-Africans, therefore, is a better test of native land policy than the legal conditions under which leases and sales are permitted.

In spite of government attempts to liquidate the Concession System, European-held lands still seem to be largest in the Belgian Congo and French Equatorial Africa, followed by the Cameroons Trust Territories, British West Africa, Liberia, Portuguese Guinea and French West Africa. The native reserve-system of South Africa is lacking except in Angola. According to the 1933 report of Angola's Survey Department, lands held by Africans amounted to only 597 square miles on tribal reserves and 23 square miles owned by individuals, out of a total of 481,351. Possibly most of the remaining land is state property in legal theory, as in French West Africa. Concessions in Portuguese Guinea totaled 524 square miles in 1947. No information is available for European-held concessions in Angola, or for alienations in Spanish Guinea.

By 1936 the total freehold concessions in the Belgian Congo had been reduced to about 20,370 square miles, excluding Katanga, Kivu, and all leases. The Katanga Company has one-third representation and receives one-third of the revenues from an area of some 173,745 square miles administered by the Comité Spécial du Katanga, in which the Belgian Government holds two-thirds interest. The government also holds controlling interest in the 30,888 square mile concession for colonization administered by the Comité National du Kivu. Huileries du Congo Belge, the Lever Brothers' subsidiary, was to receive freehold rights of up to 1,353 square miles in 1945.[3]

The original temporary concessions in French Equatorial Africa were reduced to 214,286 square miles in 1912, by granting freehold rights to 9,844 square miles. Further reductions took place up to 1929 when the remaining original concessions expired, but no recent figures are available. In the Cameroons the Germans had reduced their original concessions to 5,792 square miles by 1913, but additional concessions were granted. More recent figures are not available, but 219 of the 362 individual concessions held under the Germans were sold at auction by the French after they took over their share of the former mandate. Under French administration 350 square miles had been granted in concessions by 1935, of which 84 square miles were held in "definitive" rather than "provisional" right.

Liberia granted the Firestone Company the right to lease up to 1,565 square miles for 99 years for rubber production and including mining and logging rights. Firestone pays the government 6¢ an acre for land actually developed. Only 313 square miles have actu-

---

[3] The Belgian Congo government states in letter No. 20520/2.455/INF. of Dec. 29, 1947, that at the present time traditional African concepts of land tenure are respected. The sale of African lands is permitted subject to certain regulations, and only "vacant" lands may be appropriated or alienated by the state. "Vacant" lands are understood to mean only those which are not used for habitations, sedentary or shifting agriculture, or the gathering of wild foods and forest products. No lands are declared "vacant" until after an investigation; to allow for expansion each village is customarily given an area three times the size of the area it inhabits and cultivates. The letter further comments that Morel's criticisms of the Concession System have been refuted.

ally been leased and 125 placed under cultivation, but the plantation represents the largest rubber operation in the world. In encouraging the Firestone negotiations, the United States Government in effect sanctioned the plantation system and supported the outside capitalist as against the independent African farmer. In 1945 Liberia entered an agreement with Landsell K. Christie of New York for the development of the Bomi deposits of 5–8,000,000 tons of high grade iron ore. In 1947 the formation of the Liberia Company to develop Liberia's other mineral resources, cocoa, vegetable oils and other agricultural products was announced. The company is to be owned and operated by Stettinius Associates, and the Liberian Government will receive 25 percent of the initial $1,000,000 issue of stock. It is proposed to base cocoa and coffee production on small African farms rather than large, foreign-owned plantations. Considerable amounts of African lands have been acquired by Americo-Liberians. Land may be purchased by them at fifty cents an acre or acquired in a variety of other ways. Total alienations of this kind are not known, but government officials own large quantities of land.

Both British and French West Africa refused to adopt the Concession System. Although according to law all "vacant" lands in French West Africa are owned by the state, only 582 square miles were reported as alienated at the beginning of 1935. These alienations, of which 222 square miles were on a "definitive" basis, are mainly for banana, rice, and coffee plantations in the Ivory Coast and French Guinea.

Alienations to Europeans in British West Africa can only take the form of leases, except in those areas which are technically classed as "colonies" rather than "protectorates." In Nigeria, Sierra Leone, and the Gambia, the "colonies" have a total area of only 1,706 square miles (1931 population 435,812). Minor freehold rights were acquired by Europeans in the protectorates before the governing legislation was passed. This legislation, incidentally, was protested by both British and Africans as interference in traditional African land rights. Freeholds outside the Colony in Nigeria are limited to lands vested in the Niger Company when their Royal Charter was revoked, and are described as small in area. Total alienations by leases and freehold in Nigeria, Sierra Leone, and Gambia are not known, but they probably exceed those in French West Africa, which has no large-scale mining industry.

In the Gold Coast, where both the Colony and Ashanti with a total of 48,316 square miles (1931 population, 2,149,440) are classed as "colonies," leases totaled 11,861 square miles in 1947, mostly for mining or timber. There were no freehold concessions. In British Togoland and Cameroons, alienations to non-Africans are now restricted to leases, but there are about 540 square miles of private estates in the Cameroons originally granted by the Germans. Alienations have been small in both French and British Togo.

The British West African governments have successfully opposed the introduction of the plantation system. In the 1920s Lever Brothers tried to obtain freehold rights for palm-oil plantations in southern Nigeria. In spite of the personal intervention of Lord Leverhulme, Governor Sir Hugh Clifford was able to maintain the government policy. Later the United Africa Company leased less than 20 square miles for rubber and palm oil, the total area in European plantations in Nigeria today, and in the end the Nigerian palm oil trade suffered through the competition of Lever Brothers' plantations developed in the Belgian Congo and other areas. This instance deserves special mention because it is sometimes maintained that British colonial policy always follows the line of British business and capital.

Although the British have been successful in maintaining Africans in the role of producers and entrepreneurs rather than wage laborers on foreign-owned plantations, the economic plight of the cocoa farmer shows that they have not found a touchstone which protects the African's economic status. Costly litigation, exorbitant rates of interest, and discounts on crops pledged in advance ate into an income kept low by competition with other low-wage areas, numerous brokers and commissioned middlemen, speculation, and a marketing situation in which 98 percent of the Gold Coast crop was purchased by 13 European trading companies, with one company taking about half.

The major commercial interests in West Africa are the United Africa Company and other subsidiaries of Lever Brothers, who produce Lux, Lifebuoy, Rinso, Swan, Spry and Pepsodent, and who dominate the world market in vegetable oils and oil seeds. As a recent series of articles [4] has shown, the U.A.C., Africa's largest enterprise and the world's largest trading company, accounts for more business than any of the other companies of the diverse and farflung economic activities of Lever Brothers and Unilever, "the largest corporation outside the U.S. and one of the half-dozen largest in the world." The U.A.C., referred to by Africans as "The Octopus," and its affiliates have the major share of West African trade in cocoa, fats and oils, imported goods, and other commodities except minerals. An agreement by 12 of the 13 Gold Coast companies to pay uniform prices for cocoa led to the cocoa "hold-up" or "strike" of 1937–38. For over five months African producers in the Gold Coast, and some in Nigeria, boycotted European imports and refused to sell over 200,000 tons of cocoa, until the buying agreement was suspended through the intervention of a special investigating Commission.

Government attempts to develop cocoa marketing cooperatives in British West Africa since 1931 have not met with much success. Less than 3 percent of the Gold Coast cocoa crop and less than 4 percent of the Nigerian crop were marketed by cooperatives in 1936–37. With the objective of stabilizing prices and eliminating some of the unnecessary speculation and middlemen, Nigeria and the Gold Coast each established a Marketing Board in 1947 to purchase cocoa through trading companies operating at fixed commissions to sell cocoa through a Company which it established in London. Profits in good years are to be used to subsidize prices in bad years and to finance research on cocoa diseases.

British policy in West Africa has differed from that of the Belgians, French, Portuguese, and the British in other parts of Africa on the question of European plantations, but there is no apparent difference with respect to mining. Mining concessions, which also require African wage laborers, account for most of the land alienations in British West Africa. Furthermore, when the Royal Niger Company's charter was revoked in 1899, the government assumed its debt of £250,000 and made a cash settlement out of Nigerian revenues of £565,000. It also agreed to pay the company 50 percent of all government royalties on minerals within specified areas of northeastern Nigeria. These payments, amounting to £159,579 in

[4] "The World of Unilever," by Gilbert Burck, *Fortune*, December, January, February, 1947–48.

the fiscal year 1943–44, are made to one of the U.A.C. subsidiaries and are due annually until January 1, 1999.

In her book, *Plan for Africa*, Rita Hinden has given a remarkable analysis of what happens to the mineral wealth of the Gold Coast. Between 1929–30 and 1936–37 the Ashanti Goldfields Corporation paid its investors dividends amounting to 938 percent, an average of 117 percent a year for the eight-year period. In these eight years, investors received dividends totaling 938 percent. Of the average of all mineral exports from 1936 through 1938 (£5,377,000), only £2,000,000 was spent in the Gold Coast. In the year 1937–38, the gold and diamond mines showed a net profit of £2,561,556 and actually distributed £1,957,167 to stockholders. Profits are of course by no means as spectacular in other West African industries, but a considerable share is enjoyed by overseas stockholders. The amounts received by European employees and government officials in Africa are not high by American standards, while the share of the African laborers and producers is minimal.

Where mineral resources are irrecoverably lost to the Africans with every ton of ore that is exported, the distinction between freehold rights and long-term leases becomes academic. There is a real question as to what minerals will be left when the present leases have expired. Meanwhile African landowners receive small payments according to their leases with the mining companies. One of the richest gold mines in the world, operated by the Ashanti Goldfields Corporation, which had a net profit of £1,042,064 in 1937–38, is leased for £400 a year. The British West African governments have maintained that they have no authority to interfere in these leases, which they regard as private contracts between Africans and the mining companies. Perhaps the most revealing portion of *Plan for Africa* is the quotation from the records of the Gold Coast Legislative Council of a debate in 1939 on the question of increasing payments of mining companies to African land owners. Unfortunately, only two brief portions can be quoted here.

One is the Governor's comment on a previous reference to the profits of the Ashanti Goldfields Corporation:

We must remember that when we read in the papers that a mine has made so much profit, this profit has to be divided among numerous shareholders who in many cases only receive three or four per cent. on their capital. In fact, as a general rule, they would probably have done far better if they had invested their money in a first-class Government security.

The other is by a European unofficial member:

If the people of this country are so convinced about the huge profits made by the mines and want to have a share in this prosperity, there is nothing to stop them doing so. They, like others, have a perfect right and full liberty to invest money in the mines and thus take a share in the industry . . . and if any shares are wanted, I have a few myself which I would be quite willing to sell.

Social Organization

The fundamental kinship unit in African society is the clan,[5] which is found in most tribes in this area. Members of the same clan or sub-clan generally share distinctive food taboos and are prohibited from intermarrying. Traditions about clan origins are extremely sacred, and are rarely revealed to outsiders. Clans are not usually localized, but cut across large sections of the tribe. Lineages, portions of a clan living together in a given locality, are the most important social and economic unit. They may own land as a group and assist each other financially in time of need. The loyalties which bind an individual to his clan members, and particularly to the members of his lineage, are the strongest bonds in African society.

Polygyny obtains throughout the area, with the possible exception of the Pygmies of the Congo forests. Outside the Islamic groups, the number of wives is limited only by an individual's wealth and the number of available women. In some groups, where economic surpluses and populations are small, polygynous families are exceptional. Marriages are generally effected by the transfer of property from the family of the groom to the family of the bride, but it by no means follows that wives are purchased as property of the husband. Wives may sue for divorce, own their own property and in some tribes even sue the husband for property which he has broken or destroyed. The Dahomeans recognize thirteen different forms of marriage depending on how, when, and by whom the bride wealth is paid. Polygyny and the payment of bride wealth have

---

[5] The clan includes all individuals who trace descent unilaterally, whether it be patrilineally or matrilineally, from a common ancestor.

survived the attacks of missionaries. In some colonies, such as Nigeria, two forms of marriage are legally recognized: "native marriage" which is subject only to tribal tradition and may be polygynous, and "Christian marriage" which is performed by the Church and, at least according to law, is monogamous. Ironically, missionaries have been partially responsible for weakening the family. Attacks on "native marriage," bride wealth, and polygyny have led to an increase in adultery and premarital relations.

Rank and status are very important, with the upper classes enjoying social and economic privileges as well as respect and deference. Chieftainship is generally hereditary within the clan, but all male members of the royal clan may be potential rulers. Aside from clan membership, ascribed status is based on age and sex. While women enjoy a high legal status and may influence the decisions of their sons, brothers, or husbands, the important political offices are filled by men. Within the clan, status depends primarily on age, and within polygynous families on seniority. Since the end of the slave wars, wealth and property have become increasingly important as the basis of achieved status in place of military record.

The colonial system has superimposed a new set of social classes on the already highly stratified societies, with the ranking administrative representatives at the top. In British West Africa the social distinctions within the European group are most clearly to be seen in the large government centers such as Accra and Lagos, but are generally disregarded where there are few Europeans. Following the members of the political or administrative service, in approximate order, are those in the legal and medical branches, police, education, forestry, agriculture, customs, posts and telegraph, veterinary, public works and railway. Within these branches position and income are important. The head of the Forestry Department may be a frequent dinner guest with a high-ranking political officer, but a member of the Railway Department seldom dines with a political officer of his own status. Position and income determine rank within the parallel and overlapping series which includes banking, shipping, trading, and mining employees and which starts at a somewhat lower level. The rank of missionaries generally falls below both these groups. Within these various occupations, social status depends also on nationality, with English and Scotch followed by British from the Dominions, Americans, French, and Greeks. Below these are the Syrians, Lebanese, Indians, and other "non-Europeans," who in turn rank above all but the most important chiefs and wealthy Africans. Although these distinctions are a British innovation, they have been grasped by the class-conscious Africans, as they are observed in different towns, and can be followed when the occasion calls for it.[6]

A new group, generally referred to as the "detribalized" Africans, has also emerged under contact. This includes a well-to-do, highly acculturated upper-class of business and professional men, often educated in European universities. Clerks, office messengers, and domestic servants who have had some primary education constitute the middle class. Below them are unskilled and semiskilled laborers who may have had no formal education, but who also have left their homes for jobs in the larger cities. The "detribalized" Africans have often given up their tribal religious and social practices, either of necessity because of isolation from their kinsmen, or of choice because of their aspiration to European ways. They have usually learned a European language or the dominant African language of the city in which they work. Aside from the upper class, it is doubtful how many of this group are "detribalized" in the sense that they will not return to their former way of life when their economic circumstances permit.

"Detribalized" Africans are only a small part of the population, which is composed of a large number of ethnic groups of varying size, each with its own language and culture. In the Belgian Congo close to 150 distinct groupings have been listed. In Central Africa the more important groups include the Kongo and Yombe near the mouth of the Congo river, the Mbundu and Chokwe of Angola, the Luba of southeastern Belgian Congo, the Kuba, Songe and Rega of central Congo, the Ruanda and Rundi of the Belgian Trust Territory, the Azande and Mangbetu of northeastern Belgian Congo, the Mongo and Ngala of northwestern Congo, the neighboring Mandja and Baja of central French Equatorial Africa, and the Wute and Pangwe or Fan of the French Cameroons.

Among the more important West African groups are the Hausa

---

[6] The Government of the Gold Coast comments that this paragraph "is regarded as bearing no relation to post-war conditions in the Gold Coast."

(6,000,000) of northern Nigeria and Niger Colony, the Fulani or Fulbe or Peuhl (4,000,000) scattered through French and British West Africa, the Ibo (4,000,000) of southern Nigeria, the Yoruba (3,500,000) of southwestern Nigeria and eastern Dahomey, the Akan (3,000,000) including the Ashanti and Fanti of southern Gold Coast and the Baule and other Agni groups of southeastern Ivory Coast, the Mandingo (3,000,000) including the Kpelle, Vei, Mende, Susu, Koranko, Khasonke, Malinke, Sarakole and Banmana of Liberia, Sierra Leone, and western French West Africa, the Mossi (1,500,000) of northeastern Ivory Coast, the Ewe (1,000,000) of Togoland and Gold Coast, the Kanuri (900,000) of northern Nigeria, and other smaller but important groups such as the Dahomeans, Edo, Ekoi, Ga, Ibibio, Ijaw, Jukun, Kissi, Kru, Nupe, Senufo, Songhai, Temne, Tiv, Tuareg, Tukulor, and Wolof.

West and Central Africa are characterized by the great African kingdoms and empires which existed long before the arrival of European explorers. Along the Guinea Coast were Benin (Edo), Ife (Yoruba), Oyo (Yoruba), Dahomey, and Ashanti (Akan). North of these in the Sudan were Bornu (Kanuri), Mossi, and the vast Fulani empire stretching from the mouth of the Senegal to Lake Chad. The Fulani empire was preceded by the Hausa empires and in still earlier times by Ghana (Sarakole), Songhai, Melle (Malinke), and others known today from Mohammedan records. Comparable kingdoms, such as Kongo, Loango, Luba, and Kuba were also found in Central Africa.

The political history has been one of the rise and fall of dynasties, and of borders and allegiances changed by civil wars and battles between rival states. States of varying size have expanded, contracted and then spread again. They have included in some cases peoples of different languages and cultures, and in others only parts of ethnic groups. In contrast to the Fulani empire, which ruled Hausa, Nupe, Yoruba, and other subjects in Nigeria and numerous groups in French West Africa, are the Ibo, sharing a common language and culture in spite of local differences, but divided into some 500 autonomous political units.

Below the king or "paramount chief" were a large number of court officials and assistants with both political and ritual functions. Some of these usually constituted the king's advisors and in many cases they were made responsible for outlying provinces. In Ashanti and Benin the Queen Mother enjoyed special status. The cities and districts within the kingdom were ruled by chiefs who exercised considerable autonomy in administering their areas. They were in turn assisted by their own councils, ward chiefs, precinct chiefs, and lineage heads, and by chiefs of the smaller towns within their administrative areas. Civil and criminal cases were tried before formally constituted courts of law presided over by the chiefs.

The person of the king was in a sense sacred and surrounded by many taboos. Both subjects and subordinate chiefs showed deference in his presence. From the time of the first Portuguese until the establishment of colonial rule, traders and representatives of European governments treated African kings as heads of foreign powers. As late as 1875 the British Consul, representing the Foreign Office rather than the Colonial Office, wrote in advance to King Ja Ja requesting permission to enter the "creeks" of the Niger Delta at Bonny, where Ja Ja ruled over a small area about Opobo. The chiefs along the Guinea Coast collected export duties ("comey") and harbor fees from the trading ships, and even after King Ja Ja had been exiled to St. Vincent in 1888 he was paid "comey" by the British Government. The trade monopolies exercised by some kings over imported goods, as well as local produce, were the reason for many of the early "punitive expeditions" in southern Nigeria, including that which resulted in Ja Ja's deportation.

Political Administration

The development of the British policy of "indirect rule" is inseparably associated with Nigeria. Although Lord Lugard is regarded as its founder, the beginnings of indirect rule date back to the period of the Royal Niger Company, when Sir George Goldie wrote "If the welfare of the native races is to be considered, if dangerous revolts are to be obviated, the general policy of ruling on African principles through African rulers must be followed for the present." The difference between direct and indirect rule is partly one of intent, since all colonial powers have at times, as a matter of expediency, found it necessary to condone African customs and to delegate some of their responsibilities to African chiefs.

In "indirect rule" the utilization of African administrative organizations and the authority of traditional chiefs is a matter of deliberate policy. These are sanctioned by the central Nigerian government as "Native Administrations" and "Native Courts" and their realms of competence are defined by legal statute. The powers of the Native Courts are limited by the maximum sentences they may impose, which vary from three months imprisonment to the death penalty, the latter subject to confirmation by the Governor. Over 90 percent of all criminal and civil cases in Nigeria are tried in Native Courts according to tribal law. Within the realm of powers delegated to Native Administrations, the District Officer has in theory only the status of advisor, though in practice it is sometimes difficult to distinguish between advice and instructions and though District Officers may review cases and even, in some areas, serve as a court of first appeal.

Increasing amounts of the direct taxes on Africans are turned over to the Native Administrations to spend on local roads, hospitals, schools, water works, electric plants, and the salaries of chiefs and Native Administration employees. Kano Emirate receives 70 percent of direct taxes levied within its area and had a revenue of over £206,720 in 1936–37, while in southern Nigeria 23 native treasuries received 50 percent of the taxes and had estimated revenues of less than £1,000. In 1946–47 the expenditures of the 225 native treasuries amounted to £3,315,219, equal to almost 30 percent of the expenditures of the central Nigerian government.

Indirect rule in Nigeria was developed through experience with the large and well-organized Hausa states of the north. It was later extended to the Yoruba kingdoms and other large states in the Western Provinces. In the Eastern Provinces its application has been delayed by the large number of small political units. It has also been extended to other British colonies in Africa and has had an increasing influence on Belgian and French colonial policy. Indirect rule has meant that Africans could continue to be governed by their traditional chiefs according to their own laws and customs. It has also meant the rule of large African populations with a minimum of European officials, and a reduction of administrative expenses to a point where Britain did not have to contribute to the colonial budgets.

In the formulation of indirect rule both Goldie and Lugard were influenced by British experience with the independent feudatory states of India. Today Nigeria is beginning to face some of the same consequences as India. The educated Africans are beginning to demand representation in Native Administrations and the democratic election of chiefs. Some chiefs, such as the King of Benin, have been attacked personally in the newspapers for failing to represent their people, for having sold out to the British, or for failing to support the nationalist movement. Many government officials in Nigeria are aware of the inherent inconsistency between the policy of indirect rule and Britain's proclaimed objective of preparing Nigeria for democratic self-government. Education is accepted as a fundamental method of obtaining this objective, but indirect rule provides no place among the traditional rulers for Africans who have been educated. Democratic representation is also accepted as essential, but in theory indirect rule does not allow for a change in the rules of succession or a modification of the powers of the chiefs. In fact, the authority of hereditary chiefs has usually been strengthened by government support. In practice, various compromises, not yet crystallized into formal policy, have been worked out. An increasing number of chiefs have attended school, and chiefs' councils have been given more authority and enlarged to include educated representatives.

On the other hand the status of chiefs has been basically altered by delegating to the Governor the power to remove them from office, rather than relying on the traditional procedures for deposing undesirable and unpopular rulers, and less directly by the Governor's authority to approve or reject candidates selected for vacant chieftainships. These powers have tended to reduce the chief to the status of a civil servant and to shift his loyalty from his people to the administration, which can in effect hire and fire him. As a result, there is actually less difference between the indirect rule of the British and the direct rule of the French than appears on the surface.

Under direct rule in the French colonies, African chiefs function as agents of the French government; authority in local affairs is vested in the *Commandant du Cercle*. Chiefs were frequently appointed for their knowledge of French, their loyalty, or other considerations of usefulness, regardless of the traditional ruling groups and patterns of succession. Although there is a clear distinction between "la justice française" and "la justice indigène," the "native" tribunals are not the traditional African courts and do not represent the chief and his council, and African customary law has been adapted to French legal concepts.

In 1921 the Governor-General of French West Africa declared "we have committed the fault of breaking up completely the native social structure in place of improving it for the purpose of serving our administration," and expressed the hope that use would be made of whatever might have remained. Three years later the reconstruction of African political organization was advocated by the Governor-General of French Equatorial Africa, and under Félix Eboué this became official policy in 1942.

I intend to see to it [he wrote in a circular to administrators] that the practise of placing in power upstarts, menials or native "tirailleurs" whose services must be rewarded, is discontinued once and for all (there are a thousand other ways to reward them). Moreover, it is my desire that the legitimate chiefs be sought out, wherever our ignorance has allowed them to go into hiding, and that they be reinstated in all their external dignity. . . . [The perfect administrator will discover and educate the traditional chiefs;] he will then have in them reliable assistants who will relieve him of most of the actual work, and his role will merely be to suggest, to advise, and to control.

At a conference in Brazzaville in 1944, Eboué's principles were accepted by the Governors and Governor-Generals of the French territories.

Where British administrative theory aimed at cultural self-determination, that of the French originally aimed at the assimilation of Africans to the status of French citizens. All Africans born within the four "communes" of Saint Louis, Dakar, Goree, and Rufisque were automatically made citizens under the 1848 constitution of Senegal. Outside these areas, in the "protectorate" of French West Africa, citizenship could be acquired by satisfying certain requirements as to age, criminal record, education, monogamy, a "civilized" way of life, military service, service under a

French employer, and devotion to French interests. African "citizens" were granted franchise and had a different legal status than African "subjects." They were likewise held for only 18 instead of 36 months of military service.

These attractions apparently did not outweigh the disadvantages of French inheritance and marriage laws, and were not sufficient to cause any significant number of voluntary nationalizations. Out of a total of 97,246 African citizens in French West Africa in 1944, only 4,741 resided outside of Dakar and Senegal. Many of the citizens in the communes are uneducated and have not adopted French culture, since they acquired their status automatically and are governed by a special code of laws which permits polygyny and other African laws and customs. This special status has been very attractive, particularly because of the terms of military service. As a result various methods of acquiring it were devised, including having mothers visit Dakar for their confinements so that their children might be born within the commune.

In recent times the emphasis has shifted from "assimilation" to "association," a policy aiming at the creation of a native elite in an intermediate position between the unacculturated Africans and the fully assimilated African citizens. The functions of the elite are conceived as assisting in administration and as diffusing French culture to other Africans after filtering out elements unsuited to African conditions. The 1944 Brazzaville conference reaffirmed this position when it called for expanded secondary and professional education, which is "essential for the development of native élites who ought to be called upon to hold an ever greater number of positions in business, industry and the administration."

Portugal and Spain have both been influenced by French colonial policy. Direct rule is practiced but neither assimilation nor association have proceeded as far as in French West Africa. In the Belgian Congo direct rule was practiced in the early days but in 1906 and 1910 the foundations for indirect rule were laid. In 1933 the chiefs' functions as traditional authority and as emissary of the Belgian Government were clearly distinguished. In the latter capacity he can act only on superior orders, except to protect public health, order, and security. In the former he is subject only to advice and to a possible veto by the administration. At the same time provision was made for native treasuries with incomes derived from additions, not exceeding 20 percent, to the poll tax collected by the central government.

In Liberia, although it is an independent state, Africans have a status which differs little from that of colonials. They are administered through Americo-Liberian District Commissioners according to a policy of indirect rule which is marred by reports of individual extortion. The Americo-Liberians represent about 1 percent of the population. Liberia is not a reliable basis for judging the future of other African territories, since it does not represent an example of African self-government. Politically it resembles the Union of South Africa in that a large African majority is ruled by an immigrant minority which controls an autonomous government. The fact that the majority and minority in Liberia are racially related has not prevented revolts by the Africans in 1832, 1852, 1856, 1875, 1893, 1910, 1911, 1916 and 1932, in several of which the government received "moral support" from American naval vessels.

Aside from the Native Administrations, municipal bodies have been established in the larger French and British cities like Dakar, St. Louis, Freetown, Accra, and Lagos, with authority to collect property taxes and administer various city affairs such as sanitation, markets, electricity, and water works. All these bodies have African majorities, but it is only in the four communes of Senegal that all members are elected. Property and other qualifications restrict the number of eligible voters in British territories. In the Belgian Congo no Africans are represented in the municipal council of Leopoldville.

The British colonies have been ruled under the system of Crown Colony government. The Governor is assisted by a Legislative Council and an advisory Executive Council. As a colony progresses on the path toward dominion status, its inhabitants are given increasing representation in these councils, their powers are increased, and suffrage is extended. The Executive Councils in British West Africa consisted entirely of government officials until 1942 when African members were appointed by the Governor in Nigeria, Gold Coast, and, later, Sierra Leone. In Gambia the legislative council consists of six government officials, one appointed European, two appointed Africans, and one elected African.

In 1946 the Gold Coast received a new constitution through which for the first time the government officials are a minority in the Legislative Council. The council consists of six senior government officials, nine Africans elected by the Joint Provincial Council of the Colony, four Africans elected by the Confederacy Council of Ashanti, five Africans elected by qualified voters in Accra, Sekondi-Takoradi, Cape Coast, and Kumasi, and six members, who may be either African or European, appointed by the Governor. The Governor presides and retains veto power, as he does in all other British West African colonies, but he does not vote. The authority of the Legislative Council was also extended to include Ashanti. The Northern Territories of the Gold Coast are now the only part of British West Africa still subject to legislation enacted by decree of the Governor and the Executive Council.

In 1947 Nigeria also received a new constitution, giving Africans a majority in the Legislative Council for the first time. The new council consists of 16 senior British government officials and 28 "unofficial" members. Eighteen of the "unofficial" members are Africans selected by newly created Regional councils; two are African chiefs appointed by the Governor from the council for the Western Provinces; one is an African appointed by the Governor to represent the "Colony"; three are appointed by the Governor to represent interests not adequately represented, including European business; and four are Africans elected by ballot, three from Lagos and one from Calabar.

Eastern, Western and Northern Houses of Assembly have been created, and a separate House of Chiefs with 23 members for the Northern Provinces. Three chiefs are appointed by the Governor to the Western House of Assembly, but no chiefs are represented in the Eastern Provinces. The other African members in the three Houses of Assembly are selected by the Native Administrations from among their own members (other than chiefs), with minorities appointed by the Governor to represent areas without Native Administrations, African commercial interests and other groups not otherwise represented. In each of the four Regional councils Africans hold a majority over government officials and the latter have no vote in the selection of representatives to the Legislative Council. Besides serving as an electoral college, the Regional councils have certain powers over financial and other affairs within their areas.

Under this "Richards Constitution" government officials have relinquished their majority in the Legislative Council; the representation of the protectorate with 96 percent of the population has been increased from 15 to 48 percent; and the authority of the Legislative Council has been extended to all of Nigeria. The unique feature of the constitution is its attempt to bridge the gap which in the past has separated Native Administrations from the central government. The new constitution, however, has been denounced by the nationalists as inadequate for the present state of advancement of Nigeria.

A new constitution for Sierra Leone was proposed in 1947, providing a Legislative Council of seven government officials, nine Africans elected by the Protectorate Assembly, one African chosen by the elected members of the Protectorate Assembly from among its appointed members, one African appointed by the Governor, two Europeans appointed by the Governor to represent business interests, and three Africans elected by ballot in the Colony. This would also provide an unofficial African majority.

French Senegal, with its four communes and large number of African French citizens, represents a unique situation. A Colonial Council of 62 members includes 26 who are elected by the French citizens, 18 elected by African chiefs, and 18 elected by French "subjects" who are ex-soldiers. Before the recent addition of the third group, when three-fourths of the Colonial Council were generally Africans, the Africans were divided on issues, with the chiefs upholding government policy when it was attacked by the citizens. Increases in taxation of African subjects and prestation labor were supported by the chiefs, who drew government salaries and whose tax rebates were increased by such measures.

Although the powers of the Council are specifically defined and restricted, it was able to force the government to withdraw increases on registration taxes within the communes in 1922–23 in spite of a government majority made possible by the support of the chiefs. The Council has authority to give advice, to "deliberate" on five specified subjects including taxation, to legislate on thirteen specified subjects of lesser importance, and to approve the budget of the Senegal. Its deliberations must be approved by the Governor-General in Council of Government, who may request it to "deliberate" again if he does not approve its decisions. Its legislation may be vetoed by the Governor-General within two months; otherwise it automatically comes into effect. Its greatest power comes through approval of the budget, though even here "obligatory" expenses including expenses of the administrative departments, services of debts, contributions to the governments of France and French West Africa, and secret funds are excluded. The Council can withhold about half the expenditures of government and block the imposition of new taxes. As a result its African and other unofficial members enjoy more authority than those of any other consultative assembly in Africa.

Councils of Administration serve as advisory bodies to the governors of other French West African colonies, to the governors of French Cameroons and Togo, and to the Governor-General of French Equatorial Africa. A comparable Council of Government advises the Governor-General of French West Africa. All these councils, except in Niger Colony, Mauritania, Togo and Cameroons had elected African representatives and unofficial majorities, but little authority. The 1944 Brazzaville conference recommended that these councils be abolished and replaced by Regional Councils based on African political structures and by representative Assemblies, composed of Europeans and Africans elected by universal suffrage wherever practicable. The new councils and Assemblies were to be consultative except for "deliberative powers" concerning budgets and new public works.

French colonial structure has also been markedly altered at the top as a result of new trends in French thinking. Formerly all legislation for the Empire was enacted in France and only adapted to the local situation by the colonial governments. The Empire is now conceived as a "French Union" in which each territory will enjoy representation in the central Constituent Assembly and considerable local autonomy. "The evolution of the colonies must never tend to a separation from the mother country, but . . . must be aimed at strengthening the bonds which tie all colonial territories together with Metropolitan France, with a view to forming one single national entity, one and indivisible." After the 1945 elections the Constituent Assembly included ten representatives of French West Africa, six from French Equatorial Africa, 48 from other territories of "France Overseas" not including Indo-China, and 522 for France. Previously Senegal alone had been represented by one member of the Chamber of Deputies. M. Blaise Diagne, who in 1914 became the first African Senegal Deputy, served in this post for many years.

In the Belgian Congo the Governor-General has only emergency legislative powers. His edicts cease to be valid after six months if not confirmed by legislation from Belgium. One African was made a member of the Governor-General's advisory Council of Government in 1947. The Governor of Angola has an advisory council of five government officials and five unofficial members of Portuguese nationality, chosen by recognized economic organizations. In Portuguese Guinea the advisory council includes elected representatives of commerce and the municipalities. The Portuguese councils must be consulted on all legislative measures, but may not make proposals increasing expenditures or reducing revenues.

Liberia has a President, Cabinet, Senate, House of Representatives and a Constitution patterned after that of the United States. Africans have the legal right to participate in politics, but property and literacy qualifications have kept political control in the hands of the Americo-Liberians. The Africans received political representation for the first time in 1945 when provision was made for the election of one member of the House by each of the three provinces of the interior. The vast majority of Africans are still ineligible to vote for Senators or the President.

Religion

Native religions in West and Central Africa are marked by a complexity which parallels that of the political and economic organization. Along the Guinea Coast a large number of deities are recognized, each with its own mythology, priests, worshipers, and rituals. Cycles of elaborate rituals, usually centering around an annual "festival" which may last a week or more, are performed. Parts of the festivals are secret and may be witnessed only by the head priest and his principal assistants; at other times public ceremonies may be held which attract thousands of participants and observers.

Although West African religion is polytheistic, most individuals worship only one god. Every individual recognizes the existence of numerous deities and may propitiate them at the advice of a diviner, but he has a special relationship to the one he "serves." This may be the one worshiped by his father or mother before him; it may be selected for him on the advice of a diviner; or it may be chosen in other ways. Propitiation of other members of a pantheon, such as the wives, brothers and sisters, and associates of a major deity, may also be required. A few individuals worship more than one god, and even before contact with missionaries, a few did not "serve" any. In many tribes a formal initiation is an essential part of joining a cult group. This may be even more elaborate than the annual ceremonies, involving the seclusion of groups of initiates for many months while they are instructed in myths, songs, dances, and other aspects of ritual and theology.

Worship of a deity may be assumed because of a desire for some blessing which the god can bestow or to avoid evil consequences if the god is neglected. All deities have the power to grant children, the most common request and the main goal in the lives of most Africans. They can relieve sickness and insure good luck, good harvests, and wealth. They are also able to intervene in human lives by punishing those who do not fulfill their obligations. African religion, however, is not dominated by fear. The gods are conceptualized as capricious, but not difficult to please; they are amenable to persuasion and praise, but are powerful enough to be very dangerous if they become angry.

Some deities are known only within a circumscribed area, while others are worshiped over wide areas by thousands of followers. Gods of lightning, of iron, of the earth and the waters are found under different names in many tribes. Myths tell that the gods lived on earth in the remote past and that from them many human families have descended. As a consequence the student often finds it difficult to draw the distinction between deities and ancestors, but it is usually clear in the minds of Africans. Ancestor worship and the deities are dominant factors in the integration of the social structure, the former in particular providing the sanctions of the clan system.

Humans are believed to have not merely one, but four or five separate souls. *Rites de passage* are important here as in other parts of the world, particularly those surrounding death, when the individual

joins the ancestors. Elaborate puberty ceremonies are more characteristic of Central than of West Africa, but throughout the area
recognition is given to the transition from childhood to the status of
adult. Marriage is primarily a civil affair emphasizing the contractual
features that are binding on the respective families, but divination
and the propitiation of family ancestors and deities are necessary to
insure its success. Food taboos and rituals are also associated with
pregnancy, childbirth, and naming. Any abnormal circumstance at
parturition is interpreted as an omen. Throughout the area, twin
births are grouped in a special category, but whereas some groups
believe them endowed with special powers that bring good fortune,
others consider them as dangerous and evil. The treatment of twins
and of their parents varies according to the beliefs that prevail.

Magic is another very important aspect of religion. Charms and
medicines are classified into categories based on their purpose and
the method in which they are applied, and they seem almost countless in number. Over 3,000 dealing only with physical well-being
have been published in Yoruba. Others are employed to win cases
in court, to bring success in business or love affairs, to protect travelers or the inhabitants of a house, to make thieves noiseless, invisible
or impossible to capture, to transport someone immediately to another town, to bring wealth or general good luck, or to prevent un-
specified disasters. Their efficacy is derived from the power given
to each particular charm by a deity or deified ancestor. They are
dispensed by specialists who are paid for their services and who
guard their secrets so carefully that in some cases their knowledge
dies with them. Belief in witches, ghosts and other spirits, and in
dreams and omens also exists.

A position of special importance is occupied by the diviner, who
in a sense integrates the various forms of worship. Diviners may
be consulted by any individual, regardless of cult affiliation, to determine which deity requires a sacrifice, what the sacrifice should
be, or what charm should be compounded to avert disaster. Several
methods of divination are practiced within a single tribe. Some permit the diviner to adapt his predictions to his knowledge of his
client's personal situation; others are based on thousands of fixed
verses committed to memory through years of study.

A large part of the population, particularly in West Africa, adopted Islam before European contact. As a result, African rules of inheritance, polygyny, and political structure, as well as religion, have been modified, while Islam has been modified in turn. Charms made of verses copied from the Koran are used by non-Islamic groups. In contact with both Moslems and Christians, Africans have shown a readiness to try new rituals and charms, but a reluctance to abandon their own. The persistence of African religious beliefs even in the New World shows the tenacity of African religion and its great importance in African life.

Catholic missionary activity began with the arrival of the first Portuguese. In the fifteenth century the King of Benin agreed to bring his people into the Church in return for a Portuguese wife, but the effects were not lasting except for the adoption of the cross as a symbol and crucifixion as a form of human sacrifice. In other areas also, nominal "converts" soon resumed their traditional forms of worship when the Portuguese withdrew. Protestant missionary activity, which has also met with resistance, is little more than 100 years old. The first Protestant missionaries in Nigeria, a mulatto from Sierra Leone with two Fanti assistants, arrived in 1842. Except in the British colonies and Liberia, the predominant Christian religion is Catholicism.

Neither Catholics nor Protestants have made any appreciable headway in Islamic areas. Elsewhere the figures in Christian tradition have been fitted in with African gods, and in spite of missionary attacks on African "superstitions," most church members still participate in rituals for the gods or ancestors, make sacrifices, and consult diviners and medicine men. At a Yoruba ceremony which was attended by several hundred people, about ten individuals were pointed out who were *not* church members. In spite of the steady growth of the number of nominal Christians, they are far less numerous than either the Moslems or the so-called "pagans" who worship the traditional African gods.

Where traditional African sanctions of behavior have been destroyed or weakened under missionary influence, usually nothing has taken their place. There are among the educated young people some who now look upon all African beliefs as superstitions, but who have not substituted for the African codes of morality and ethics those of Christianity. They have become individualistic and materialistic, apparently recognizing no controls of behavior except legal punishment. Crime and adultery have increased in the areas of greatest European contact; clan responsibilities and loyalties have been weakened; and the authority of the elders has been undermined.

While these developments present important problems for the future, they should not be exaggerated, since the traditional sanctions are still effective for most Africans. They cannot be attributed exclusively to missionary criticisms of African customs, though they are not as pronounced in tribes which have adopted Islam. They are as apparent among Nigerians educated in government schools as those in mission schools. Teaching a simple fact, for example that malaria is caused by the bite of a mosquito, may raise doubts not only about native explanations and remedies for one disease, but also about the validity of the entire system of traditional beliefs.

Missionaries were the pioneers in the field of African education. The first government school in Nigeria was not established until 1909, and except in the French colonies, government has relied upon missions to administer nearly all the schools. Missions can claim most of the credit for British West Africa's reputation for good education. There seems, however, to be a growing resistance toward missionary activity in British colonies.

Educated Nigerians, many of whom were themselves trained in mission schools, have reacted to missionaries in various ways. Some take a strong negative position as regards missionary attempts to change their forms of marriage, dress, and amusement. They regard the opinion, freely expressed by many of those connected with missions, that African religions are superstitions which should be destroyed, as unjustified and ethnocentric. They find it difficult to reconcile this point of view with the teaching they receive that all gods are one, and with the use of names of African gods to refer to the Christian deity in Bible translations, hymns, and sermons by the missionaries themselves. Nor can they reconcile the belief in a single Christian God with the rivalries between the various Christian sects. Finally, missionaries are suspected of being effective if unintentional agents of imperialism. Some of these views are shared by the uneducated Africans, but most literate Africans believe that the good admittedly done by some missionaries is not sufficient to justify the presence of foreign missions in Africa.

Education

Except for the Islamic tribes of the Sudan where African language was written in an Arabic script, the peoples of this part of Africa were without writing little more than 100 years ago. In the Islamic tribes reading and writing were, and still are, taught in classes headed by mallams, but the instruction is usually confined to texts from the Koran. Arabic script has also been used to record local history, and Timbuktu's early fame as a center of learning spread even to Europe. As in the groups without writing, however, the greater part of the individual's learning was acquired verbally or by example. This is largely true even today when schools have been established and many of the languages have been reduced to writing.

African patterns of child care and child training, children's games imitating the activities of adults, evening sessions of riddles and story telling, the use of proverbs to emphasize the reason for criticism or praise or to make a point stick in a child's mind, and the first fumbling attempts of children to assist their parents in cooking, farming, trading, weaving, and many other occupations are all important parts of the process by which adult knowledge, beliefs, and attitudes are acquired. Much of this learning is informal and incidental to other activities. In the apprenticeship system, the West African cult initiations and Central African puberty initiations, the situation is directed specifically toward instruction and training.

It is inaccurate to speak of Africans as uneducated simply because they have not had formalized schooling of a European type. However this usage is difficult to avoid in the English language and where it has been adopted here to avoid clumsy paraphrases, it should be understood that "uneducated" Africans may possess tremendous stores of knowledge, and frequently far more real wisdom and human understanding than the young "educated" African who has spent four or five years in school. "Education," as one of the main issues and major problems in Africa today, does not concern the traditional African patterns of learning, which still operate effectively with the possible exception of the young men who leave home at an early age to enter school or work for Europeans and may grow up ignorant of large portions of their cultural heritage.

The schools are still largely in the hands of missionaries. In 1945, 99 percent of the students in the Congo and Ruanda-Urundi and over 90 percent in British West Africa, excluding the Koranic schools, were in mission schools. In Liberia the proportion was 61 percent in 1944, in French Equatorial Africa 52 percent in 1945 and in French Togo 50 percent in 1937. Only 19 percent of the students in accredited schools in French Cameroons in 1944 were under missionaries, but this figure does not include 75 percent of the school population who were in non-accredited schools. Both Belgium and Britain have followed the policy of subsidizing mission schools which conform to certain academic standards, giving rise to the distinction between "assisted" and "unassisted" unofficial schools. Mission schools are also subsidized in French colonies, but here official schools and government educational programs have been more important. In British West Africa both government and missions charge school fees which may be as low as 5¢ or as high as $25 a year; still there are far more applicants than can be admitted. The fees for university studies at Achimota college ($250 to $360 a year including board and lodging) are beyond the means of all except wealthy families, and over half the students rely on the assistance of scholarships. French schools are free, but admissions are limited.

Spanish Guinea has provided education for the greatest proportion of the population. The Belgians have also made education available to large numbers of Africans, but with very small government expenditures per African student and relatively low academic standards. In British and French territories the standards have generally been higher and more has been spent per student, but fewer Africans are enrolled in school; recently both British and French have made plans for a broader educational program which more closely resembles the Belgian pattern. The percentage of Africans in school is no higher in Liberia than in British and French colonies, while academic standards are lower. The Portuguese colonies have spent the most per school child, but their educational system is intended primarily for Europeans and accommodates only a small percentage of the African population.

In 1934 Angola spent $43.90 for each child enrolled in school, but the total number of students was only 6,254 and only 14 of the 472 students in secondary school were African. Gambia spent $28.40 per student in 1946, Portuguese Guinea $26.60 in 1946, French Equatorial Africa $18.50 in 1945, followed by Sierra Leone (1946), Gold Coast (1945–46), French West Africa (1946), French Togoland (1935), Spanish Guinea (1945), Nigeria (1943–44), and Liberia (1944) with between $5 and $15. Government expenditures per school child were $3.46 in the Belgian Congo (1946), including 3,273 European students; for African children in unofficial schools, 99 percent of the total enrollment, they were $1.70 in the same year. They were $1.05 in French Cameroons (1938) and 57¢ in Ruanda-Urundi (1944).

Government expenditures on education, per capita, were 87.2¢ in the Gold Coast in 1945–46 (as against 33.8¢ two years earlier), 59.3¢ in Spanish Guinea (1945), 39.6¢ in Gambia (1946), 29.3¢ in Belgian Congo (1946), 27.2¢ in Portuguese Guinea (1946), 19.6¢ in Sierra Leone (1946), and 14.1¢ in French Equatorial Africa (1945), using population estimates above. In all other territories they were less than 10¢ a year, while they were only 4.0¢ in Liberia (1944) and 3.7¢ in Ruanda-Urundi (1944). Between 3.5 and 6.5 percent of the budget is devoted to education except in the Gold Coast (8.0 percent in 1945–46) and French West Africa (1.7 percent in 1946).

Of the total population 8.7 percent was enrolled in school in Spanish Guinea (1945), 8.5 percent in Belgian Congo (1946), 6.4 percent in Ruanda-Urundi (1944), 5.0 percent in French Cameroons (1944) including 3.8 percent in non-accredited schools, 4.7 percent in Gold Coast (1945–46), 2.9 percent in Nigeria (1946), followed by Sierra Leone (1946), Gambia (1946), French Togoland (1935), Portuguese Guinea (1946), French Equatorial Africa (1945), Liberia (1944), and French West Africa (1944–45), with between 0.5 and 1.5 percent. In Angola only 0.17 percent of the population, including European students, were enrolled in 1934.

These figures are not strictly comparable because of the marked expansion of education in recent years and because the cost of school buildings is not listed as an educational expense in British West African colonies, and other educational costs are concealed. Although they do not include the expenditures of missions, Native Administrations, and other non-government organizations, they represent total enrollments in so far as these are available. The combined facilities of mission, government, and all other schools accommodate less than 2,500,000 Africans, or about 3 percent of the population.

Even in British West Africa, teachers' salaries are too low to attract the best-qualified Africans. Of the 27,860 teachers employed in Nigeria in 1946, 24,000 were untrained. In southern Nigeria 61.9 percent and in Sierra Leone 51.2 percent of the 1935 enrollments were in classes below Standard I, the first primary grade. Only 1.1 percent and 5.4 percent respectively were in grades above Standard VI. The total number of enrollments in secondary schools in all British West Africa in 1942 was 11,670.[7] In 1947, 705 British West Africans were studying in British universities, 244 on scholarships, and a few others in American universities. About 250 were taking university courses in British West Africa. Achimota (Prince of Wales College) near Accra is the outstanding college in all West and Central Africa. It represents an original investment of £600,000 and receives an annual government appropriation of £64,000, but less than 100 students have been enrolled at the university level. Although these numbers are very small, British West Africa has produced more lawyers, doctors, journalists and clergymen than any other part of the area.

The subjects taught in primary school vary somewhat from one area to another, but generally they include reading, writing, arithmetic, history, and geography, hygiene and nature study, domestic science and arts and crafts, gardening, and, in mission schools, religious instruction. Manual training, agriculture, nursing and elementary medical training, engineering, teacher training, and other practical subjects are taught in special post-primary schools or in some cases by the technical departments of government.

The curricula in the secondary schools and colleges of British West Africa have not always been adapted to the local geographical and cultural background. Greek and Latin, English history, and British flora and fauna were studied, and until recently it was necessary to import biological specimens from England. British education officers recognized the desirability of fitting education to the special needs of the local situation, but all efforts in this direction

[7] Cf. Angola, p. 371.

were opposed by Africans until the British universities modified their matriculation examinations to provide for the increasing number of students, in all parts of the Empire, who were taking them. These examinations are necessary for continued study abroad and important because of the better opportunities for jobs and the increased prestige enjoyed by those who possess certificates from British universities. European employers as well as Africans give inferior ratings to certificates from local schools.

Objection to adapting courses to the African situation persisted, however, even after the change in the matriculation examinations, because of the fear that this may be used as a device to maintain Africans in a subordinate position. With reference to the Union of South Africa the Report of the Interdepartmental Committee on Native Education, 1935–36, states frankly: "The education of the white child prepares him for life in a dominant society, and the education of the black child for a subordinate society." Though it is understandable that Africans should wish to be satisfied that no dangers lurk behind courses of instruction especially designed for them, it is difficult to see how learning Latin or studying plants and animals which they will never see after they leave the classroom is of any real benefit to most African students. There is a difference between teaching subjects which are useful in an African environment and withholding essential knowledge or teaching obedience, loyalty and submission to European employers and officials. Nevertheless, as long as educational programs are drawn up by alien rulers, however honorable their intentions, any deviations from European patterns will be suspect.

The implementation of a modern educational policy in Africa, Morris Siegel has said, would "tear down the very foundations of the colonial system or else precipitate grave disorders in the territories." [8] Where there is no intention of granting Africans eventual independence or of permitting them to rise to equal status with Europeans, there are good reasons for restricting educational facilities, even when colonial budgets might permit improvement and expansion. The amount and the kind of education provided depends on both the wealth of the colonies and their natural resources and foreign trade, and on the objectives of those providing the educa-

[8] The *Journal of Negro Education*, XV, No. 3 (1946), 562.

tion. Religious objectives may be as important as colonial policy. In mission schools religion is emphasized, occasionally to such a degree that they provide little more of general education than many Koranic schools.[9]

The Belgian system of education has been characterized by J. S. Harris as one in which

The government, in conjunction with the mission schools, has now achieved an educational program which feeds clerks and skilled labor to its administration and private enterprises. With the exception of the specialized training for agricultural and medical assistants, there is no higher education comparable even to our high schools. . . . Only if it is intended that the Congo peoples remain permanently under European tutelage, can a disjointed system of education which denies them any effective training beyond the rudimentary and limited vocational levels be justified.[10]

In the Congo and French colonies many British West Africans are still employed as clerks.

The importance of practical studies is emphasized by the Belgians and the French. The two essential aspects of education in French Equatorial Africa have been outlined by Jean de la Roche as

First, it would be a terrible mistake, fraught with disastrous consequences, to equip African youth with intellectual training only. . . . Second, though nothing now prevents an African youth from choosing an intellectual career, the majority of students are needed to carry out practical work of vital and immediate importance.[11]

[9] The British Colonial Office comments, "The conclusions of this paragraph do not agree with British Colonial policy and the facts given are not supported by current events," and refers to developments discussed below under Present Outlook.

[10] *Journal of Negro Education*, XV, No. 3 (1946), 425–26.

[11] *Ibid.*, p. 402. The Belgian *Ministère des Colonies* comments: "Such considerations as those of Harris and de la Roche are incorrect since, in maintaining a diversified professional education under a state of continuous improvement such as the development of the country requires (medical and agricultural assistants, clerks, technicians, etc.), the Belgian Congo is now attempting to create a non-specialized secondary education of the European type (*collèges*) directly preparatory to studies of a university character (*collèges universitaires*). All utilitarian ends are disregarded in this education, which looks essentially toward the formation of a native lay elite (doctors, veterinarians, agronomists, engineers, etc.) to supplement the religious elite (Catholic fathers and Protestant pastors) which already exists. The Belgian Colonial Government, being constantly aware of the dangers for the Blacks themselves in a program that would result in the unconsidered formation of an intellectual group which would have no way of making an honorable living, considers that it is essential to act cautiously in this matter, and limit the number of colleges

The French policies of assimilation and association have been reflected in education. The French language is not just a subject of study; it is the medium of instruction. It is required so that Africans will "think like Frenchmen" and is an important instrument for the dissemination of French culture. This has not been altered under the new French policy. The 1944 Brazzaville conference agreed that "Instruction must be in the French language. Use in teaching of local dialects is absolutely forbidden, in private as well as in public schools."

The British objective has not been to replace African by European culture, but to permit a new culture to develop which combines the most desirable features of both traditions. In line with this policy of cultural self-determination, of which indirect rule is an aspect, students are first taught to read and write in their own language (the "vernacular"). Later they learn English, and at upper primary and secondary levels English is used as the medium of instruction. The Portuguese and Spanish follow the French pattern, while in Liberia, government schools teach in English, and mission schools use African languages initially. In the Congo the problem is complicated by the large number of African languages and by the fact that both French and Flemish are official languages, but African languages are used at lower levels and French is used as the language of instruction in higher primary education.

The choice between these two alternatives is partly a matter of policy and partly based on practical considerations, such as the availability of teachers and textbooks, the number of individuals speaking the same language, the number of African languages and the extent to which they have been reduced to writing. Teaching in African languages shows a greater respect for African institutions and for the African's right to determine his own future, and it is also the most effective method of giving elementary education. It may, however, perpetuate a situation where Africans cannot communicate with each other because of linguistic differences and be a sharp barrier to further education. Only a very limited number of books have been translated into African languages and at pres-

---

in terms of existing needs for them. It is recognized, however, that the continuous evolution of a country necessarily involves the development of instruction in the humanities and university education."

ent European languages are the only practical media of advanced education.

The French have taught in their own language and have attempted to spread French culture as rapidly as possible, but they have, unlike the British, maintained two separate educational systems. In Native Schools the curriculum, course outlines, and textbooks are intended to provide training for life in Africa, and neither French history nor Latin are taught. Separate schools for Europeans and a few Africans follow a curriculum which is identical with that of schools in France.

Racial Relations

Discussions of the attitudes and relationships between European rulers and their African subjects can seldom be accepted as objective because they are so frequently used as ammunition for attacks on or the defense of colonial policy. Since these relationships are not easy to describe and are in themselves indelicate subjects, documented analyses are less common than flat statements which cannot be evaluated. In spite of obvious dangers of error, the issues involved are important enough to warrant an attempt to answer the questions: How do Africans feel and behave toward their European rulers, and how are they treated in turn? These points will be taken up in reverse order.

The range of variation, depending on the nationalities, the tribes, and the individuals and situations concerned, is not wide. Nowhere is there as complete racial segregation as in the Union of South Africa or the southeastern United States, on the one hand, or social equality on the other. Discrimination does not consistently follow racial lines, but in all the territories, Africans clearly occupy a subordinate position. This is true even in Liberia, where the concepts of color bar and racial discrimination are not pertinent. Some of the acculturated Africans have been accepted in Liberian society and have intermarried with the Americo-Liberians, but there is a wide social, political and economic gap between the Americo-Liberians and the Africans of the interior.

It is also true in the Portuguese colonies, despite the social acceptance of mistresses and mulattoes. It appears that, while there is less discrimination in terms of color, there is also less consideration for African cultural traditions and economic freedom, and more physical brutality reminiscent of the Concession System than in any other part of Africa. In Portuguese colonies, as in French, Spanish, British, and Belgian, Africans may be kept standing in line while a European is waited on ahead of turn, or while an official chats with a casual European visitor who has just dropped by. The African's chances for social, political, and economic advancement are not improved because mulattoes have the same legal status as Europeans, one which is different from his own. The number of mulattoes in the Portuguese colonies, 28,000 to 44,000 whites in Angola and 2,200 to 1,400 whites in Portuguese Guinea, is no indication of social equality, which does not exist between a man's African mistress and his legal European wife.[12] The Spanish pattern in general resembles that of the Portuguese.

The French colonies, particularly in French West Africa, are regarded as having little racial discrimination. Africans may attend the French schools and sit side by side with French children. Buell

[12] The Government of Portuguese Guinea comments: "The picture of the European whose black mulatto mistress and sons are denied social equality is obviously erroneous. There is no racial discrimination in Portuguese Guinea, as the number of mulattoes shows. Anyone who is qualified may rise to any position, regardless of color.

"One of the brightest names in the history of Portuguese Guinea is Honório Barreto, an African from Guinea, who was its governor in the first half of the 19th Century. We do not believe that at that time any other colonial power sent Africans to Europe to study or let them become governors of the colony from which they came. Other Africans from Guinea serve today as Chief of Customs, Inspector of Commerce, and as *Administradores*, a position equivalent to the French 'Commandant de Cercle'. Colored people hold very important positions in Portugal and Cabo Verde, which lies off the West Coast of Africa. In Cabo Verde Portugal has accomplished the task of total assimilation in a way that has not been equalled in any other part of Africa. Most of the mulattoes in Portuguese Guinea come from Cabo Verde, many of whose inhabitants emigrate because of overpopulation.

"In Portuguese Guinea neither mulattoes nor Africans have less rights than Europeans. All 'natives' (*indigenas*) may become 'citizens' (*cidadaos*) and enjoy absolutely equal rights with European 'citizens', if they desire and if they have developed sufficiently to live under Portuguese laws. 'Natives' are protected by special legislation which respects tribal law and tribal organization. Authority is respected and there are no nationalist movements because there is neither economic exploitation nor racial discrimination. Perhaps the reason for this mutual respect is the fact that Portugal has spent five of her eight centuries of history as a colonial power in Africa. While in many parts of Africa the natives call the Portuguese 'whites' (*branco*) and other Europeans by their respective nationalities, in Guinea any African who has adopted European ways of life is called 'white'. This shows that to Africans, to be white is not a matter of color, but of civilization."

describes French women selling fish to Africans in the markets and waiting on Africans and Europeans alike in the stores and restaurants in Dakar. Geoffrey Gorer, on the other hand reports, "The colour bar is extraordinarily strong in Dakar. Negroes are practically never seen in the cafés, restaurants and hotels." In 1944 there were 4,108 mulattoes to 47,425 Europeans in French West Africa. Sons of African women, born in or out of wedlock, have been sent to France for an education by their French fathers, but Eboué's statement of policy devotes considerable attention to those in French Equatorial Africa whose fathers have not recognized or provided for them.

It is sometimes believed that there is no discrimination against Africans in French colonies because Negroes have risen to high positions. Many of these have been West Indians and the outstanding example, Félix Eboué, former Governor-General of French Equatorial Africa, was born in French Guiana. Racial ties have not prevented such individuals from identifying themselves with the French rather than the Africans, as can be seen from Eboué's words, "We shall not ensure their happiness by applying to them the principles of the French Revolution, which is our Revolution." As Siegel points out, the policy outlined by Eboué calls for teaching the Africans "their political, economic, and social place, which in every case is one of inferiority and subordination to their French masters."

The position of African "subjects" who constitute 99 percent of the population cannot be judged by that of New World Negroes or African "citizens." The latter also enjoy a superior position and a few like M. Diagne, the former Senegal Deputy, have risen very high. Nevertheless the unequal salaries of Africans and Europeans was one issue on which at least as early as 1928 chiefs and citizens in the Colonial Council of Senegal united against the government. Buell states, "While the French have accepted the 'equality' doctrine, they have shown no intention of turning over the country to the blacks, nor even of associating the blacks upon a basis of equality with the Europeans in the administration of the country. There are fewer natives in the administration in French than in British colonies." The 1944 Brazzaville conference calls for an increased Africanization of the administrative service, and defines the future relationship between Africans and Europeans as follows:

Our entire colonial policy will be based upon the respect and the progress of the native society, and we shall have to accept fully and absolutely the demands and consequences implied by this principle. The natives may not be treated as devoid of human dignity, they can be subjected neither to eviction nor to exploitation. However, the colonies are destined, by their very nature, to be inhabited jointly by both Europeans and natives. Although our policy must be subordinated to the full development of the local races, we must also give European activity the place to which it is entitled.

The prerequisite for the progress of the African continent is the development of the native populations. The activity of the Europeans and other non-Africans in the colonial territories of Africa must conform to this condition.

On the other hand, this progress of the African continent, as it is being contemplated, cannot be achieved in the near future without the collaboration of non-African persons and enterprises to a much greater extent and in greater proportion than at the present time. Consequently, all necessary talent, ability and services will be duly enlisted and utilized. . . .

All the various trades must gradually be taken over by the natives. The Governors-General and the Governors of the territories shall establish, within a brief period, an inventory of the enterprises which will be opened progressively to the natives.

Comparison between British and French relations with Africans are often misleading. There have been no special schools for the few British children; at the last census there were only forty European and American children under fifteen in all Nigeria. British women do not customarily work in stores or restaurants in West Africa, and a part of the difference is due to British reserve, which may express itself in lack of warmth in social and personal relations even among themselves. Africans, furthermore, are seldom seen in French cafés in Abidjan and Douala. Although there is more fraternizing and less racial discrimination in French than British colonies, the differences are less marked than they appear on the surface.

Regardless of policies in other parts of British Africa, as far as West Africa is concerned statements that Africans cannot travel first class, or that British never eat, dance, or sleep with Africans are untrue. Railways, buses, theaters, restaurants, hotels, and other public places are open to Africans and Europeans alike. The refusal of the Greek manager to accommodate an African in the Bristol Hotel in Lagos in 1947 brought protests in the African press, questions in the House of Commons, and official condemnation from the Governor of Nigeria. Except during the war, when automobiles were scarce, however, the British seldom found it necessary to travel with Africans on the buses in Lagos and Accra. For economic reasons Africans rarely patronize European hotels or restaurants, and seldom pay for the higher-priced seats occupied by Europeans at the motion picture theaters. Africans likewise travel third class much more often than second or first. The explanation for this may be found in the differential between first class and third class railway fares which, for example, in Nigeria are 4d. as against ¼ d. a mile, respectively.

There is little more racial discrimination in British West Africa than in the northern United States, but its existence cannot be denied. There are separate latrines, hospitals, and cemeteries for Europeans and separate residential areas in the larger cities. The official Nigerian policy, as announced by the Governor in 1947, is to end discrimination in European hospitals and residential reservations. There are separate hospital beds for Europeans in French West Africa and separate European hospitals in the Belgian Congo. Separate residential areas, usually justified on the grounds of health, are also found in French and Belgian colonies and presumably in the Portuguese and Spanish colonies as well.

Cafés, restaurants, and bars are important in the social life of the French and Belgian colonies, but British entertaining is done at home or at the club, the traditional center of social life in the British colonies. The "European" club is a private institution, open only to members and their guests. The exclusion of Africans is not comparable to the refusal of American restaurants to serve Negroes, but discrimination is evident in the absence of African members and African guests. In Lagos and Accra, clubs with mixed membership have been established to promote better understanding between Europeans and Africans.

Public dances which were popular with the British troops during the war are also attended by Africans and Europeans in Glover Memorial Hall in Lagos and King George V Hall in Accra. These functions are not frequently attended by the British at the top of the social scale, but the entire European community, from the Chief Commissioner of Ashanti down, traditionally attends an annual mixed dance in Kumasi. Africans are also invited to some of the garden parties and other large functions given by British officials and they may be included in small dinner parties on lower social levels.

In the larger centers where clubs and the presence of European women make an active social life possible, there is generally less contact than in towns where a few Europeans are isolated. Social contacts also more frequently involve individuals at the lower end of the European and at the top of the African social scales than they do those at the opposite ends. The differences between individuals within the British community are even more pronounced than those between the British as a group and the French. At the higher levels of British society social contacts with Africans are often motivated by political considerations, rather than a desire for closer social relations or personal friendship. British and African social activities are effectively separated, but by economic factors and individual preference and personal invitation, and not by barring Africans from public places.

Although the number of mulattoes in British West Africa is very small, African mistresses are common. Some British have even expressed the belief that all single men, as well as some married men whose wives are with them on the coast, keep mistresses or visit African prostitutes. The main difference between the British and the French or Portuguese in this respect is that the British do not carry on such relationships openly. The few individuals who make no attempt to conceal them become the subject of gossip and are ostracized by most of the British community, while a man whose mistress may pass as the wife of his cook is still accepted socially as long as this arrangement is not a matter of common gossip. British men consider it bad taste for these relationships to be called to the attention of British women, who in turn pretend to ignore them. Their French neighbors regard such niceties with a certain amusement. In some cases children and mistresses are provided for by inheritance, but since these unions have no legal status even when they are permanent, such arrangements are on an individual basis. As might be expected from the pattern of secrecy imposed on

British-African matings, legal marriages between British men and African women are unusual, but they do occur.

Another distinction between the British and French colonizing methods is that while the former have without intent created a larger and better educated "élite" than the latter have been able to do through deliberate policy, they have not treated them as an élite. British policy has been directed toward the welfare of the mass of illiterate Africans, and has looked upon the educated Africans with suspicion. As a result they have alienated the most articulate group, while the French have maintained its loyalty. Since the only African opinions which reach America and Europe are those of this minority group, judgments of colonial policies have frequently been distorted.

When the African trade unions criticized the Nigerian Government for paying higher allowances to some 93 Nigerians who held "superior posts" in 1945, the government replied that it was a "cardinal point of Government policy to appoint Africans to superior posts and to apply to such posts conditions of service similar to those attached to European posts. Had the Government not extended the payment of local allowances to such posts, . . . Government would have laid itself open, with some justification, to a charge of discrimination." But most Africans in government service still received much less than Europeans. The *Nigeria Civil Servant*, as quoted in *The Eastern Nigerian Guardian*, shows 1,631 European employees in 1944 earning £1,077,390 as against 14,866 Africans earning £988,640.

Aside from the educated élite which will be discussed later, the attitudes of the mass of the African population do not appear to be significantly affected by the varying treatment they receive from Europeans. There is no real evidence that the social aloofness of the British in West Africa results in less loyalty or devotion, on the one hand, or that it brings greater respect, as some British officials believe, on the other. From the large number of laborers from Portuguese colonies who remain in the Union of South Africa, where racial discrimination is undisguised, it can be seen that social equality may not be regarded by Africans as the most important consideration.

The mass of Africans look up to Europeans as superiors, fear

them for the military strength of their governments, and respect
them for the technological accomplishments of Western civilization.
There have been few native revolts, and violence against Europeans is rare except in the port towns in connection with robbery.
There are few parts of the African bush where it is not safe for a
European to walk at night alone and unarmed, and there was no
violence even when feelings ran high during the Gold Coast cocoa
strike.

Where colonial rule is not oppressive, the lives of millions of
Africans in small villages are almost unaffected by the acts of government. They pay taxes, buy and use European goods, and may
produce goods for export; some member of the family may even
work for Europeans in the mines or on the coast. But they may see
a European only once a week or perhaps once a year, and they have
little knowledge of what policies are being formulated in the capital.
To them the price of peanuts or palm kernels and the cost of kerosene or cloth have more importance than questions which lie within
the government's power to decide. They may be aware of the differences between their own country and its neighbors—in taxation,
military recruitment, wages and prices, land and labor policies, racial
discrimination, and educational opportunities—but they would find
grievances against whatever foreign power ruled them. Few Africans have the loyalty and devotion to their rulers that is sometimes
attributed to them.

A Yoruba man in one of the important towns in the cocoa belt
of Nigeria once explained that his people were thankful to the
British for three things. First, they had suppressed the slave trade
and ended slave raiding, so that it was safe to travel about the country and go to the farm at night without fear of being captured. Second, they had built a reservoir and installed water works in the town,
so that people were no longer killed in fights for water during the
dry season. Third, they had imported tin for roofs, so that it was
no longer necessary to thatch the houses every year.

Personal acquaintance has taught some Africans that all Europeans have their foibles and are not above making mistakes. Personal idiosyncrasies often become crystallized in nicknames which
spread rapidly and may be remembered after the proper name has

been forgotten. Since they are usually uncomplimentary, they are seldom known to their owners. Africans may also swear at passing Europeans in English or French after they are out of earshot or when it is certain that they will not stop, or they may insult them in an African language which the European does not understand. Still more subtle and still safer is the practice of addressing Europeans with conventional African greetings or with complimentary terms, such as "My master," while modifying the intonation so as to change the greeting to a taunt or sneer. Traditional songs of defiance and derision are composed about some Europeans, who may remain unaware of the meaning behind a melody which is being hummed. Verbal reprisals of this sort are frequently directed against Europeans in general and may be employed when the European is a complete stranger and too ignorant of African ways to take offense. In the areas where there has been the greatest contact, they are more commonly and more openly used.

Personal contact with Europeans has also taught Africans that in many ways Europeans are more naïve and easier to deceive than members of their own group. Regardless of social rank or official position, Europeans are to a large extent at the mercy of their African servants and assistants. That there is so little actual dishonesty under the circumstances is surprising. The European who regards Africans as simple and childlike puts himself at a further disadvantage. This belief is sometimes deliberately cultivated by African employees, who find in it an easy excuse for the evasion of responsibility.

The Yoruba of Nigeria, who disapprove of all marriages outside their own group, make no exception for sexual relations with Europeans. In one interesting case a Yoruba who had married a British woman while attending an English university was disowned by his father when he returned to Africa. The Yoruba, who employ all the forms of verbal reprisal mentioned, have the reputation in British circles of being impudent. Yoruba clerks sometimes refuse to carry out orders of their European employers and in other ways behave in a manner which the British regard as insolent. A second form of behavior is to be found among the Hausa of northern Nigeria, who are outwardly ingratiating and obsequious, although they are said to maintain their personal pride and self-respect. Most British prefer the Hausa to the Yoruba, Ibo, and other southern Nigerians.

The Ibo come closest to a pattern of true racial equality in their dealings with Europeans. Even in their traditional role of "house boys" they treat their European masters much as a European servant would, receiving instructions politely and without insolence, but not hesitating to raise reasonable objections. On the other hand they may ignore the commands of a European to whom they have no obligations. Regardless of tribal differences, the British like and feel more at ease with the older Africans who have had the least contact with Europeans, and they are least adept in dealing with the younger acculturated group. A member of the Nigerian Secretariat has been quoted as asking, "Why is it that we do so well with primitive people, but once they become educated, we seem to fail?"

Acculturation

Acculturation has been taking place in West Africa for centuries. Islam is believed to have reached the Sudan area of northern French and British West Africa more than 1,000 years ago. The earlier civilizations of Byzantium, Rome, and Egypt may also have influenced West Africa, and may themselves have been influenced by African cultures in return. Disregarding these contacts across the Sahara, the possible circumnavigation of Africa by the Phoenicians in the seventh century B.C., the Carthaginian contacts in the sixth century B.C., and the unsubstantiated French claims to have reached the Gold Coast in the fourteenth century A.D., West Africa has been in contact with Europe for 500 years.

In 1441 Portuguese explorers marketed African gold and slaves in Lisbon; by 1472 they had reached southern Nigeria. After Columbus discovered America, the early interest in gold, ivory, pepper, and precious woods faded in the face of the slave trade, which reached its peak in the early nineteenth century. The slave trade left its unmistakable mark on both Africa and the New World, where the influence of the Yoruba, the Dahomeans, the Ashanti, and some Congo tribes is still evident, particularly in Brazil, Cuba, and Dutch Guiana. In the latter nineteenth century, trade in palm

oil and other products gradually replaced the traffic in human beings, and the present political boundaries were established. In 1870
90–95 percent of West and Central Africa was still independent of
European control, but by the end of the century all except Liberia
had been acquired by foreign powers.

From the first arrival of Portuguese explorers, Africans have been
eager to obtain goods of European manufacture. Trade, in which
Africans were already experienced, developed rapidly. Exports, including slaves, grew steadily to pay for the increasing imports for
African use, which today range from cheap perfumes and beads to
automobiles, refrigerators, and tractors. The desire for cloth, kerosene, lamps, hardware, bicycles, and other European merchandise
is still the real incentive which causes Africans voluntarily to abandon subsistence farming for production of export commodities or
to enter employment as wage laborers. When imports were cut off
from French West Africa during the war, there was unrest and
production dropped. In British West Africa cloth, soap, and cigarettes were easier to obtain than in England, because the British
cut their own rations so that the production of strategic African
materials might not suffer.

Two American crops, cocoa and peanuts, have become major export commodities. Two others, maize and cassava, have become of
fundamental importance in the subsistence economy. In addition,
Hevea rubber, tobacco, American varieties of cotton, pineapples,
avocados, papaya and other plants of American origin have become
a part of the African economic picture. Even in the field of material culture, Africans have been selective in their acceptance of
European culture. Although textiles constitute one of the most
important imports, West Africans refuse to buy if cloth is shoddy
or colors fade or run, and they are discriminating in their choice of
color and design. The *Gold Coast Handbook* advises traders that
"an acquaintance with local tastes is desirable before shipping.
Tastes for patterns and colours vary in different districts, and goods
which obtain a ready sale in some districts are useless in others."

European contact has also affected African religion, social organization and political institutions, as has been indicated, but in
these fields European culture has had less appeal to Africans. Even
outside the areas of Islamic influence, true conversions to Protes-

tantism and Catholicism have not occurred on a large scale. Mis-sionary attempts to modify marriage have resulted either in apathy or antagonism. Western medicine has been accepted for the treatment of smallpox and a few other diseases, but as one educated African explained, it still has much to learn about tropical diseases. For these, for childbirth, for good luck, for success in business, court trials and the other important aspects of life in which Western medicine does not pretend competence, Africans still use the charms and medicines of their forefathers and make sacrifices recommended by their diviners. Pidgin languages, including Afro-English and Afro-French dialects, have become important as *lingua franca* along the coast. These are basically European in vocabulary and African in grammar and idiom.

The British policy of cultural self-determination has permitted Africans to continue to live by their own laws and customs, where these were not "repugnant to morality or justice," to be governed by their traditional chiefs and administrative organizations, and to use and learn to read and write their own languages in school. The British attitude, rather than being one of respect for African traditions as it has often been described, is more accurately one of tolerance of African customs and a respect for the African's right to choose his own way of life. Their hope is that the best and most useful features of African and European culture will be incorporated into a new pattern suited to the environment and to the people under present-day conditions, and they have felt that this can best be accomplished through slow and gradual change.

Cultural self-determination, however, like indirect rule or the use of African languages in school or the adaptation of education to African needs, can be used to keep Africans from achieving equality and independence, although none of these are sinister per se. Speedy acculturation and rapid education are opposed both by those who are sincerely apprehensive of the undesirable consequences of too rapid change, and by those who wish no change at all in the African's present position. Policies based on a respect for the African's right to develop his own institutions, suited to his own needs, may be identical with those whose basis is a belief that Africans are racially inferior and incapable of assimilating European culture.

At the other extreme, the French policy of assimilation, which grew out of the French Revolution's goals of *egalité* and *fraternité*, was a direct denial of the African's liberty of self-expression. It was a form of "cultural imperialism" which tried to impose French culture upon Africans without any reference to their desires. Delafosse, the French anthropologist, wrote in 1921 that the French tyrannically suppressed native institutions in the name of racial equality, acting as inhumanely as those who preached racial superiority. "We believe," he eloquently protested,

that human societies, even though established outside of our microcosm, and upon a basis of which we are ignorant, and which we are often incapable of understanding, should not be suppressed from the earth by the mere will of another society, whatever it may be, and that we have no right either to reduce them to slavery or to impose upon them laws and customs which they reject, which have not been fitted to them, and the forced adoption of which may lead them to death.[13]

For both assimilation and association the retention of African customs was a measure of failure. But the policy of association, in actual practice, permitted most Africans to live their lives almost undisturbed; in theory their customs were expected to change more slowly. Eboué's policy, based on French experience in Morocco, resembles cultural self-determination. Nevertheless the emphasis in French policy remains chiefly on political institutions, and their hope is still for the rapid spread of Christianity and French culture, and the "modernization" of African customs. The 1944 Brazzaville conference did not mention "respect" for African religion and showed little for bride wealth and African marriage. "As regards more particularly large-scale polygamy, it is indispensable that this scourge of Continental Africa be resisted by the Administrators with all the means at their disposal."

Belgian policy, although following the British in some respects, has a character all its own. Africans are regarded as one of the Congo's most valuable economic resources.[14] Maximum African

[13] R. L. Buell, *The Native Problem in Africa*, II, 89.
[14] The Government of the Belgian Congo comments: "This statement, made by superficial observers, and the conclusions drawn from it are *absolutely incorrect*. It does not correspond at all with Belgian colonial policy or its philosophy, which considers the native above everything, not as an economic value but as a human value. It is in the idea of respect for the human individual that one must find the basis of Belgian policy in the Congo. If this policy of considering the native first

manpower, operating at full efficiency, is needed to exploit the mineral wealth and other natural resources. Accordingly the Belgians have provided the best medical facilities and made certain that African wage laborers receive good food, good housing, and enough education to fit them for European employment. In several of the major economic enterprises government and business are partners in the exploitation of the natural resources. Cultural self-determination, as it is practiced in the Congo, seems to have been adopted as a practical means of avoiding native discontent with little emphasis on eventual African self-government.

Regardless of these variations in policy, there is no area where African customs have either remained unchanged or completely disappeared. In some areas the early eagerness to adopt European ways is already passing. In some cities the Yoruba, who have constituted the white-collar workers of Nigeria for almost a century, passed through this phase some time ago. The white suits, white shoes, and white sun helmets in which the young educated men of Ife posed for a photograph about 1920 had been put away by 1937, together with their morning suits and dinner jackets, and some of the same men had returned to African wrap-around cloths. Most Nigerian office workers wear shirts and shorts or trousers at work, putting on their African clothes when they return home. In 1947 the nationalists proposed that they wear African clothes to work, as the Hausa generally do. A similar movement in the thirties renounced all European merchandise which could not be produced locally by Nigerians; it did not gain wide or lasting sup-

as an economic asset is that of certain large companies, it is not that of the government.

"In other words, the policy of the Government of the Belgian Congo is very largely progressive, and aims at improving the situation of the natives both as wage laborers and as farmers. Social legislation is well developed. The wage laborer is protected in the case of industrial accidents and occupational diseases; a family allowance is given him; and salaries and working conditions are supervised by Government. Since labor unions have been organized, there have been practically no more strikes. Native associations have been recognized and are functioning satisfactorily. For the farmers the government is creating a financially well-endowed Welfare Fund to develop agriculture, hygiene and education among rural populations by improving individual and group health, native techniques and the utilization of labor, the marketing of products, the purchase of goods necessary to the community, and to raise the moral and intellectual level of the masses. These projects are in addition to the already considerable accomplishments of the government services. Natives participate directly in many administrative bodies, notably in the *Conseils de Province* and the *Conseil de Gouvernement.*"

port, but its slogan, "Back to the Land," is still to be seen in Lagos.

As forms of protest, nativistic religious movements have been of minor importance compared with political parties, labor unions, and newspapers. In 1914 a Nigerian who called himself the Second Elijah led a nativistic revival in the Niger Delta which gained thousands of followers. He denounced witchcraft, immorality, and European gin, and later the Europeans who produced the gin as well. He proclaimed that Europeans were not really children of God; had they been so they would not have had to build a bridge across the Niger, for the waters would have parted for them as they did for the children of Israel. Elijah was finally arrested for sedition, but out of his movement grew the Delta Church. Other leaders of this type have arisen in Nigeria, including one whose amplifiers and loudspeakers were confiscated during the war, but their influence has been localized.

A similar movement was led by the prophet Simon Kibangu, a carpenter in the Congo who had become a Baptist. In 1921 he was directed in a dream to work miracles, raising the dead and commanding the lame and the blind to be healed in the name of Christ. As his fame spread, European medicines were thrown away and African charms and shrines were destroyed. Drumming and dancing were tabooed, and polygynous chiefs accepted monogamy when they were baptized. His followers said they were tired of giving contributions to Europeans and set up their own schools and churches, creating a demand for hymn books and Bibles which for some time could not be satisfied. Wage laborers left their work to be cured by the prophet. Native farms were deserted to such an extent that the government feared it might have to close the main railways, because its African employees could not obtain food. When Wednesday was declared the day of rest, European commerce was further disorganized.

Minor prophets, said to have been influenced by "radicals" from British West Africa, gave Kibangu's movement an anti-European appearance. Among the prophecies ascribed to them were that all white men would be killed by fire from heaven on October 21, 1921, and that the Congo Africans would be delivered from white oppression by American Negroes or, alternatively, by the second coming of Christ. Arrested in June, Kibangu escaped through another "miracle." Martial law was imposed on the Africans, and the European community lived in fear until Kibangu and his followers were rearrested in September. A military court imposed the death penalty on Kibangu, but this was commuted to life imprisonment. Nine other prophets received life sentences, others shorter terms, and about a hundred followers were exiled to other parts of the Congo.

In the Ivory Coast in 1914 William Wade Harris, a Liberian, proclaimed one Savior, one God, and the Bible as His book. He preached the dignity of labor and obedience to authority, denouncing alcohol, robbery, and African religious beliefs. Through his influence African shrines were destroyed and African priests converted or driven away. His followers, according to the French, spread rumors that the whites were about to leave and that taxes would be reduced. They were arrested and Harris was sent back to Liberia. Eleven years later the Wesleyans assumed responsibility for the churches of the "Harris Christians," who still numbered over 30,000. All of these movements were essentially puritanical and Christian in character, denouncing traditional African beliefs and customs. Their alleged anti-white aspects were secondary and it is difficult to interpret them as reactions against European culture or protests against colonial rule.

Strikes, boycotts, and political action have become increasingly important. Under difficult conditions, Nigerian civil servants carried out a successful strike against the government in wartime, holding up exports of strategic materials and costing the British administration prestige. Eventually they won a 50 percent increase in wages, although the 1945 strike was renounced by the former union leaders and declared illegal by the government. From the point of view of acculturation, an interesting note is that masses of union members took an African oath, on the Earth, not to abandon the strike. Traditional oaths were also taken during the 1937–38 cocoa "hold-up" not to sell cocoa or buy European merchandise.

In contrast to the government's policy during this strike, has been the appointment of government officials in British West Africa and elsewhere throughout the Empire, whose duty it is to organize and promote the growth of African trade unions. The labor movement in British West Africa antedates these appointments and one of the tasks of the officials has been that of welding numerous small unions into large and more effective organizations. Union memberships number about 80,000 in Nigeria and 15,000 in the Gold Coast. The number of industrial wage earners is still less than 5 percent of the population, but it is increasing, and the development of a sound labor movement at this time will benefit Africans in the future.

Nationalistic Movements

The 1945 strike and others in Nigeria, the Gold Coast and Sierra Leone, and the cocoa "hold-ups" and boycotts in 1937–38 and 1930–31 show the increasing effectiveness of European economic techniques in African hands. Strikes were also reported from the Congo during the war, but leadership in economic and political action clearly rests with the nationalist movement in British West Africa. This fact cannot be explained in terms of greater oppression or exploitation by the British; if anything the reverse is the case. Nor are there grounds for regarding Portuguese or Spanish rule as more enlightened because nationalist movements have failed to develop in their colonies. It appears, rather, to be the result of better education, plus the rejection of the educated "élite." In part the development of a nationalist movement, in this case at least, is a measure of good—not bad—colonial policy.

Within British West Africa, political leadership passed from Sierra Leone to the Gold Coast in the thirties, and to Nigeria in the forties. Developments in Nigeria today, which are closely followed by the unions and political leaders in Accra and Freetown, set the pattern for the whole of British West Africa, and perhaps in time will determine the course of events in all West and Central Africa. Reorganizations, changing names, and differences in points of view of Nigerian trade unions, political parties, and political leaders, important as these are on the local scene, only conceal a fundamental unity in objectives.

Influencing all these groups is Nnamdi Azikiwe, or "Zik" as he is called, the most important leader in West and Central Africa. Zik, an Ibo who was born in 1904, studied in the United States for ten years, principally at Lincoln University in Pennsylvania. In 1934 he returned to West Africa to serve as editor of the *African*

Morning Post at Accra until he was convicted of sedition, a charge from which he was later cleared in a higher court. Zik then went to Lagos where he established Zik's Press and began to publish the *West African Pilot* in 1937. Later he added *The Eastern Nigeria Guardian* of Port Harcourt, *The Southern Nigeria Defender* of Warri, *The Nigerian Spokesman* of Onitsha, and *The Daily Comet* of Lagos.

Early in the 1945 strike the government suspended publication of the *Pilot* and *Comet*, but Zik's editorials and coverage of the strike were distributed to the Lagos subscribers by the *Guardian*, published by the Port Harcourt staff but printed on the *Pilot's* press. The initial reaction that Zik was finally to receive a lesson which many British felt he deserved gave way to a feeling that his resourcefulness had placed the government in an embarrassing position. British prestige was further decreased among Africans, and Zik's personal position further enhanced, by his unsubstantiated claims of a government plot against his life during the strike.

These incidents and the assertions in West Africa, America, and England that freedom of the press had ended in Nigeria, obscure the previous British record, whose tolerance can only be appreciated by examining the outspoken criticisms of British rule which have appeared almost daily in the African papers in Nigeria, Gold Coast, and Sierra Leone. Only one example, from the *Ashanti Pioneer* in Kumasi, may be cited here,

NIGERIA EXPLODES!

The Lagos Section of the Nigerian Governmental and Municipal technical workers are on strike! So read the B.B.C. news on Monday, July 2, 1945.

Yes, the Nigerians are learning to talk in the language that the Occident understands—economically speaking. And the strike is hinged on wages!

To hell with wages in the Colonial World! No one who has been observing, intelligently, the Nigerian renaissance in Trade Unionism and National Consciousness could have predicted otherwise.

And the storm has broken. And it has been sweeping on since June 22, 1945! And it continues. And the strikers are relentless! And the Government procrastinates. And the explosion reverberates!

It will be echoed, far and wide—in the hearts of exploited millions in the Colonial World, in India, in South Africa and elsewhere, where impudent, self-styled Master Races perpetrate Man's economic Inhumanity to Man!

Only in British territories have such comments been permitted in African newspapers.

Zik's papers have consistently carried on an outspoken campaign against social and economic discrimination against Africans, attacking the marked differences in salaries and allowances received by Africans and Europeans in government service, even when both hold the same degree from British universities, and charging that there is a ceiling beyond which Africans cannot hope to advance. In reply, the government has pointed to Africans serving as judges in the Supreme Court, as administrative officers, and in other "superior posts" in government. The British have blamed Zik for injecting the race question into West African politics, as a result of his education in the United States; others point out that Zik cannot be held responsible for developments in Freetown, where anti-white sentiments are most openly expressed, prior to his return to Africa.

"Zikism," as the program of Zik and his followers has been called, has among its main points increased education, Africanization of the civil services, the industrialization of Africa, and self-government. It criticizes the slow rate of progress toward these goals and the deviations from official policy, rather than the professed objectives of the British in Nigeria. Russia, England, America, and the French Revolution are studied by the educated group, but the greatest influence on the nationalist movement is the example of India, after which the proposed return to native dress and the recent boycott of the Legislative Council have been patterned. Recently Zik's papers have featured crimes of violence, prostitution, and economic inequalities in England in answer to the argument that Nigerians require the civilized guidance of Britain until they are ready for self-rule.

Officially "Zikism" emphasizes the inherent values in African culture and urges their retention in the pattern of the future. Actually many of the educated Africans look down upon the "old-fashioned" and "superstitious" ways of their fathers, though as a group they regard this condescending attitude as wrong. They are also aware of the difficulties of walking the tightrope between European and African ways of life. The desirable features of European cultures have been clearly defined, but references to African traditions which should be retained are vague. African art, music, and folklore, which have prestige in European eyes, are usually mentioned.

Even in these fields "African art" may not be the traditional African forms, but any art done by Africans. More interest is shown in paintings and in carvings which combine African and European techniques and conceptions than in African masks and figurines. The musical emphasis is on current popular songs, or jazz and other European music, rather than the excellent music of the religious ceremonies, and the same is true of the dance. Aside from proverbs and the traditional histories of various cities, educated writers have largely neglected traditional folklore. Essays and poems are composed, while the vast wealth of African folk tales, for example, remains unrecorded. In passing it is worth mentioning that British policy, which also stands for the encouragement of African art, has stimulated the use of European techniques, has assisted in finding markets for adapted forms and useful objects such as book ends, purses, and cushions, but attempts only to preserve masks and other traditional objects with the exception of Benin bronze casting. The point is not whether these new forms are good or bad, but that the creation of those which made African art famous is largely neglected.

Through his newspapers Zik has spread his political philosophy and won the following of the educated youth of Nigeria. His ideas have also spread by word of mouth among the illiterate Africans, but the true extent of his influence is unknown. Zik and the other political leaders claim to speak for Nigeria as a whole, while the government maintains that they represent few outside the educated minority. Since no impartial attempt has been made to determine the extent of their support, these claims cannot be evaluated. They were not settled by the decisive victory during the 1947 Legislative Council election under the new constitution, because this was limited to Lagos and Calabar and also because of the small number of qualified voters. Zik and two other candidates of a party affiliated with the "N.C.N.C." (National Council of Nigeria and the Cameroons) were elected as representatives for Lagos, a city of more than 200,000. Zik received the support of two-thirds of all *registered* voters, but this was only 3,573 votes out of a possible 5,379. Having run for election to prevent "persons who have little backing" from representing Lagos, the three representatives temporarily boycotted the Legislative Council in protest against the new constitution.

Relationships between government and the nationalists have deteriorated steadily since the 1945 strike. The unions have been warned that in the event of another general strike, less leniency will be shown to workers as regards reinstatement and back pay. A new labor law, described as anti-strike legislation, has been enacted, but several localized strikes have occurred, including miners in the tin fields and a strike of United Africa Company employees in Burutu during which government police fired into a crowd of workers in June, 1947, wounding two or three Africans. Talk of a strike of teachers for higher pay was in the air for almost a year.

The Nigerian Government has been criticized in the British press for having permitted itself to be falsely accused in Zik's papers, and the Governor and the British press have charged the Nigerian press with "irresponsibility." Several members of Zik's staff have been sued for libel, and the editor of the *Comet* was sentenced to eighteen months imprisonment in 1947 for having suggested that police should aim to miss when ordered to fire on Africans. The Nigerian Government has continued a wartime propaganda publication to offset Zik's influence, and it has been reported that the London *Daily Mirror* plans to establish a paper in Nigeria for the same purpose.

While the government has adopted a sterner position, the nationalists have shown increased determination. In his speech following his election as President of the N.C.N.C., Zik said:

Under my leadership, you should expect plain speaking and the natural consequences. Today I might be with you, but that is no guarantee that I would not be in jail or in exile or in the death-house tomorrow. You must be prepared to suffer heavy blows from the enemy; you must be prepared to make sacrifices in order to guarantee for Nigeria a nobler heritage. In other words, we are entering a new era in the political history of Nigeria, as from today, under my leadership, and you must be prepared for the worst.

Following their boycott of the Legislative Council in 1947, the N.C.N.C. delegates visited England to present their case to the British people and to the Colonial Office. Failing to receive any satisfaction, Zik announced, "We are the last delegation that will come to Britain from Nigeria and the Cameroons. We will beg no more." Zik's influence is reported to have declined after his unsuccessful mission, and the possibility of a split in his following between the Ibo extremists and the Yoruba moderates has been mentioned.

One of the serious problems the nationalists are facing arises out of the mutual suspicion and distrust with which the ethnic groups of Nigeria now tend to regard each other. Of the three which account for over half the population, the Ibo and the Yoruba have been active in the nationalist movement, but the Hausa and many of the smaller groups have remained largely outside. Tribal loyalties and cultural and linguistic ties have been the foundation of the support of political leaders in the past, but may prevent colony-wide solidarity. The nationalists are aware of the importance of unity, but unless it can actually be achieved Nigeria, like India, may be split on cultural and religious lines.

Even more important as a basis for a strong nationalist movement is unity between the small "educated" minority and the "uneducated" Africans, with whom the government contends the N.C.N.C. has lost touch. This will require more than lip service toward the value of African culture. If there is not to be disillusionment with the educated group as leaders, it must mean an end of condescension toward traditional African customs, and a sincere and studied attempt to represent the interests of the majority, and not only the wage laborers and the élite. What happens in other colonies will depend in large part on the success or failure of the nationalist movement in Nigeria.

The most recent outbreak in Accra may possibly provide a common rallying point sufficient to shift the nationalist leadership back again to the Gold Coast. On February 28, 1948, rioting in Accra ended with 22 or more Africans dead and over 200 wounded, after which two sloops were summoned from South Africa, troops were flown in from Nigeria and others were held in readiness at Gibraltar, and a censorship "more rigid than in wartime" was imposed. The outbreak apparently resulted from two independent causes, according to the report in *The Economist*. A.W.A.M. (Association of West African Merchants) stores had previously been boycotted by Africans in protest against high prices and alleged favoritism to European customers. After negotiations in which a 33⅓ percent reduction in prices is said to have been promised by A.W.A.M., the boycott was called off. On February 28, Africans crowded the stores to buy goods they had gone without, but they discovered that many items were not available, that few prices had been reduced, and that some prices had been raised. Crowds gathered in the street to voice their protests.

On the same day a procession of African ex-servicemen was marching to the Governor's residence to present a petition. Having served in Burma and on other fronts, they had had little success in finding civilian employment in jobs for which they had been trained during military service. They had obtained previous permission to present their petition and were proceeding, unarmed and in an orderly fashion, when their right to enter the Governor's grounds was questioned by a police guard. In the excitement of the argument that followed, an ex-serviceman was shot and killed. Aghast, his companions raised their voices in bitter protest, and in fright the police dispersed them with tear-gas. Still complaining, the veterans returned to the center of Accra, where their indignation mingled with that of the frustrated shoppers. Wholesale looting and violent rioting followed, which was quelled by the police only after bloodshed. A general strike was ordered by the Trade Union Congress of the Gold Coast, following the arrest of six of the veterans and official charges of Communist incitement of the outbreak.

Present Outlook

Both the British and the French are committed to increased educational programs stressing mass education, and expansion is already apparent. The French plans aiming at universal compulsory education were drawn up in conferences at Dakar and Brazzaville in 1944. They involve a proposed expansion of educational facilities over forty years in French Equatorial Africa until government expenditures in 1984 are 14 times what they were in 1944, and over thirty years in French Cameroons until government expenditures are 72 times their amount in 1938, and in French West Africa until there are 25 times as many classes and pupils as there were in 1945.

British plans have been outlined in two White Papers prepared by Parliamentary Commissions. One calls for the development of university education in West Africa. The other calls for mass education and the rapid elimination of illiteracy (at present about 95 percent); it envisages the education of both children and adults and is not conceived as replacing present educational programs. University and mass education are related because of the present shortage of trained teachers. A University of Nigeria is to be established an Ibadan, and a Gold Coast University College is to be provided by extending the functions of Achimota.

The success of these constructive programs will depend on the measures taken to overcome the shortage of teachers and to solve the problem of finance. The French conferences recognized the necessity of subsidies from France. In its Colonial Development and Welfare Acts of 1940 and 1945, Britain has already reversed its long-standing policy that each colony must be self-sufficient. Under its ten year development plan (1946–56), Nigeria's educational budget will be increased by about 50 percent.[15] This should make real improvement possible, but it is insufficient to achieve the objectives of the White Papers or to solve Nigeria's educational problem.

The British plan for mass education has a great deal to recommend it, although it may be misinterpreted because the average standard of education would be lowered by teaching greater numbers for shorter periods. Reading and writing, taught in African languages, can be achieved without undermining African cultures, destroying traditional sanctions of behavior, or necessarily resulting in sudden acculturation. Until mass literacy is achieved, Africans will not be able to follow the affairs of their country adequately, express and

---

[15] Of the £55,000,000 contemplated for Nigeria's development plan, £15,000,000 will come from Nigerian revenues, £17,000,000 from loans, and £23,000,000 from British revenues under the Colonial Development and Welfare Act. Of this total, £7,673,000 or $30,692,000 a year is allocated to education, plus whatever additional amounts Britain may spend on higher education in Nigeria.

enforce public opinion, or protect themselves from exploitation by local or foreign interests. The African press is capable of reaching almost the entire population in their own languages, if they could read and if circulation were large enough. Mass literacy is essential to political progress. It will make it possible for the true majority to make its will known and for government to present its case to the population, but it may serve to increase the power of Zik and any other political leaders who may follow him. Both the British and French proposals can only mean decreasing the period of political dependency, and in view of their past policies, there seems adequate justification for waiting to see what is actually done instead of accepting the paper plans at face value.

None of the governments, however, should, for long, maintain their educational programs at the present low level, and medical and other social services must also increase. A recent trend in education which will continue to grow is the development of African-financed and African-directed schools in British West Africa. The nationalists are already calling for a Nigerian university, apart from that projected by government.

As the nationalist movement spreads, so will racial consciousness. Indeed, it is apparent that race is already becoming a real point of tension. Africans are pressing, with increasing success, for the Africanization of colonial civil service, and both the British and French are committed to more rapid progress in this direction. It seems too late for the British to win the allegiance of the African élite, as the French have, by turning over to them the role of ruling other Africans.

The United Africa Company is also employing increasing numbers of Africans in managerial positions, and during the war it announced its intentions of shifting from retail to wholesale trade, in which African merchants would have a more important position. Among educated Africans there is a growing recognition of the fact that capital is necessary to the development of African-owned enterprise, and that the share of the African laborer is small compared to that of the investor. African-owned business enterprises are due to increase.

A significant recent development was the inauguration in 1946 of a series of international conferences to discuss problems common to the territories of West and Central Africa. A French-Anglo-Belgian conference on public health was attended by observers from Liberia and Portuguese Guinea, and Anglo-French representatives have discussed veterinary and communication problems. Between 1947 and 1950 eight French-Anglo-Belgian conferences to which other powers will be invited have been planned to discuss agriculture and soil conservation, rural economy, forestry, nutrition, labor, education, health and problems of rinderpest, trypanosomiasis and the tsetse fly. These attempts to solve common economic and social problems jointly and to coordinate some aspects of the technical services may pave the way for effective approaches to problems which do not end at colonial boundaries and which cannot be solved by any one colony alone. They may also, conceivably, lead to a gradual disappearance of some of the differences which now distinguish the administrations of the European colonial powers.

Continued economic development of Africa is inescapable. There is much talk about plans for the industrialization of Africa, but how far these plans will be prosecuted is still to be seen. The United Africa Company and the Mengel Mahogany Company of the United States have planned plywood factories for Nigeria and the Gold Coast. The U.A.C. obtained approval for a factory to process cocoa butter, such as the American Rockwood Company wished to erect in Lagos during the war. However, the "groundnut scheme" for Nigeria proposes that peanut oil be extracted in England although the peanut-cake by-product is needed in West Africa for cattle feed. Many government plans for economic development are being held up by the lack of personnel, by the world shortages of materials, and by the lack of dollar credits.

How far these African countries will go in the direction of becoming exporters of manufactured goods instead of raw materials, and of becoming self-sufficient even in commodities whose raw materials are produced locally, only time will tell. We must also wait to see whether or not even a socialist government in England will be able to direct the benefits of the mineral wealth of Africa to Africans instead of European stock holders. In 1946 the British government stated officially that colonial mining enterprises should be for the benefit of the community at large, and set forth arguments for vesting all mineral resources in the colonial governments. In Nigeria recent legislation along this line has been strenuously opposed by the nationalists. There are no signs of socialization of mineral resources in the Belgian Congo or other colonies.

The next year or so should see a postwar boom such as that which occurred in West Africa in 1920, if not prevented by war, an international economic collapse, or government controls of dollar exchange. This time, exports and imports may be much larger because of recently developed and recently discovered resources. The integration of Africa into the world economy will continue, with an increasing effect of overseas economic conditions on African life, and a growing recognition in Europe and America of the extent of their dependence on the natural resources of West and Central Africa.

SUGGESTED READINGS

Azikiwe, N. Liberia in World Politics. London, 1934.

Buell, R. L. The Native Problem in Africa. 2 vols. New York, 1928.

——— Liberia: a Century of Survival, 1847–1947. University of Pennsylvania, African Handbooks, No. 7. Philadelphia, 1947.

Burns, A. C. History of Nigeria. London, 1929; 3d ed., 1942.

Forde, D., and R. Scott. The Native Economies of Nigeria. Ed. M. Perham. London, 1946.

Hailey, M. An African Survey. London, 1938.

Harris, S J. See Thompson, C. H.

Herskovits, M. J. Dahomey. 2 vols. New York, 1938.

Hinden, Rita. Plan for Africa. London, 1941.

Lugard, F. D. The Dual Mandate in British Tropical Africa. Edinburgh, 1922.

Meek, C. K. Law and Authority in a Nigerian Tribe. London, 1937.

Michiele, A., and N. Laude. Notre colonie. 13th ed. Brussels, 1946.

Morel, E. D. The Black Man's Burden. New York, 1920.

Nadel, S. F. A Black Byzantium. London, 1942.

Orizu, A. A. N. Without Bitterness. New York, 1944.

Paulme, D. Organization sociale des Dogon. Etudes de sociologie et d'ethnologie juridiques, Vol. 32. Paris, 1940.

Perham, M. Native Administration in Nigeria. London, 1937.

Rattray, R. S. Ashanti Law and Constitution. London, 1929.

—— Religion and Art in Ashanti. London, 1927.

Roche, Jean de la. See Thompson, C. H.

Siegel, Morris. See Thompson, C. H.

Spieth, J. Die Ewe-Stämme. Berlin, 1906.

Talbot, P. A. The Peoples of Southern Nigeria. 4 vols. London, 1926.

Tauxier, L. La Religion Bambara. Paris, 1927.

Thompson, C. H., ed. "The Problem of Education in Dependent Terri-
tories." The Journal of Negro Education, XV, No. 3 (1946), 263–578.
See particularly the contributions of Jackson Davis, Jean de la Roche,
J. S. Harris, and Morris Siegel.

Torday, E., and T. A. Joyce. Notes ethnographiques sur les peuples com-
munément appelés Bakuba. Annales du Musée du Congo Belge, Eth-
nographie, Anthropologie, Series 3, Vol. II, No. 1. Brussels, 1911.

Verhulpen, E. Baluba et Balubaïsés du Katanga. Antwerp, 1936.

van Wing, J. Études Bakongo, histoire et sociologie. Bibliothèque Congo,
No. 3. Brussels, 1921.

Westermann, D. Die Kpelle. Göttingen, 1921.

Carleton S. Coon

NORTH AFRICA

NORTH AFRICA, from the Atlantic to the Egyptian border, is a strip of country of varying width separating the Sahara desert from the Mediterranean Sea. It includes six political provinces, all ruled by European powers: the Tangier, Spanish, and French zones of Morocco; Algeria, Tunisia, and Tripoli. One is run by an international committee. One is a Spanish protectorate; two, French Morocco and Tunisia, are French protectorates; one, Algeria, comprises three departments of France itself, while the sixth, Tripoli, Italian before World War II, is now without formal assignment. In none of these countries is native self-government more than a fiction.

From the western anchor of the Atlas mountains—the sea cliffs between Safi and Agadir in Southern Morocco—a series of three chains, of unequal height and broken in places by low passes, curves from south-by-west to north-by-east to their eastern end in Tunisia along a stretch of over a thousand miles.

To the south lies the Sahara, a world of drought and scanty vegetation, supporting a small sedentary population in oases and mountains, and a few camel nomads in between; to the north lie the fertile farm lands of Morocco and the Algerian coast. In Morocco still another mountain range, the Riffian extension of the Spanish Sierra Nevada, is raised between the plain and the sea, into which it disappears in the Galiya peninsula, at Cabo Tres Forcas. The intermontane valleys and coastal plains furnish an easy route from east to west; but refuge areas branch off to north and south.

At both ends of Northwest Africa, where the fingers of the Atlas Mountains dip into the sea, fertile valleys lie between them. At the west is the Sous, comparable to parts of Southern California, and to the East the two main valleys of Tunisia, that of the Oued Medjera, from Ghardimou past Suk el Arba'a and Medjez el Bab to Tunis itself, and that of the Oued Zeroud, past Kairwan to Sousse.

Morocco and Tunisia, with their north-south coastal plains and intermontane valleys, are richer far than Algeria in soil and agricultural possibilities. Aside from the valley of the Chelif and the coast around Algiers itself and Bône, Algeria contains little bottom land. In the high plateaus, in the Setif region, is a semiarid region suitable for cereal cultivation on the Dakota style.

Agriculture is the chief natural resource of Northwest Africa. The products and techniques employed are dependent on the climate of this region, which is Mediterranean. There are three distinct seasons: winter, with cool, rainy weather, from October through February in the wetter areas; the spring, beautiful, green and clear, from March through May; and summer, hot and dry, for the rest of the year. The vegetation of the treeless plains varies from a lush, flowery meadow cover in early spring to that of a bare desert in August and September. In the mountains and certain other areas forests still remain, with pines and cedars at the higher levels and evergreen oak and cork trees on the lower slopes.

Farther east, beyond the lands made fertile by the uplift of the Atlas Mountains, the Sahara approaches, and in some places reaches, the sea. From the Gulf of Gabes to the Nile Delta there is no arable land save a narrow strip near Tripoli itself, the hills and coast of Cyrenaica, and a number of oases, including Jalo, Jaghabub, Siwa, and Kufra. This eastern half of the North African coast is almost without rainfall except in favored spots, and can hold but a twentieth of the population of Morocco, Algeria, and Tunisia.

In all of North Africa, natural power resources are slight. No coal deposits have been discovered. Oil is obtained in small quantities from a half-dozen wells near Suk el Arba'a in French Morocco. A small well near Kifane, in the Riffian tribe of Gzennaya, just north of Taza, has proved commercially useless. Water power has not been exploited. Although the mountains give speed to the streams, the flow is seasonal in most cases. Most of the power utilized in modern North Africa is generated by coal and by petroleum products introduced from outside.

In metallic deposits, North Africa is richer. The Spaniards mine iron at Azgangen in the eastern Rif, near Melillia, shipping the ore out in British vessels which bring in coal from Wales. The French mine phosphates in the foothills of the Middle Atlas and ship them

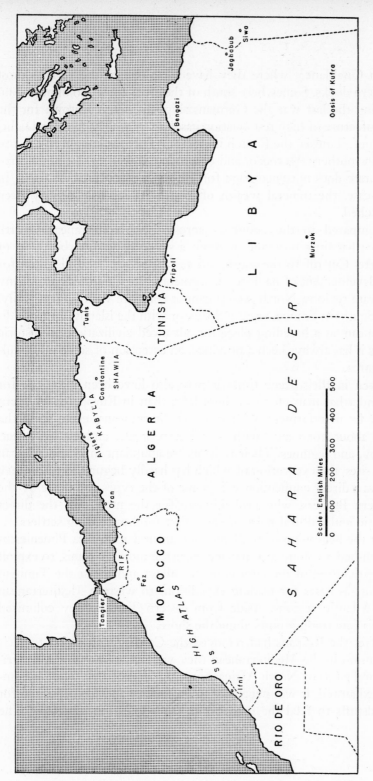

NORTH AFRICA

from Casablanca, where they have built one of the world's most modern docks. Somewhere south of the Atlas they also mine cobalt. During the last war the Germans shipped this to Europe, for the manufacture of high test aviation gasoline, as well as steel. In Djebel Hallouf, Tunisia, the French mine lead. There is known to be copper in southern Morocco, and other ores as well. The French Protectorate does not encourage foreign exploration in these parts. In any case, the mineral wealth of North Africa has scarcely been scratched.

Compared to the countries across the Mediterranean, North Africa has little coastline, without a single natural harbor west of Bizerte. Cut off to the south and east by desert, and ill-suited for the development of navigation, segmented by mountains into many separate regions, North Africa cannot compare with Greece, Italy, and Southern France, or even Spain, or with the islands of the Mediterranean, as a breeding place for advanced civilization. In historic times it has always been a province, an extension of some other culture area.

Even in Pleistocene times it may also have been marginal, for Neanderthal man clung on here later than in Europe or Western Asia. Its first farmers, who were Mediterraneans, came from the East about 3000 B.C., with pigs, sheep, goats, and cattle, wheat, barley, and legumes. Their remains are abundant in North African cave sites, the excavation of which has hardly begun. Their pottery shows a direct connection with some of the types still made by the modern Berbers, who probably are for the most part the direct genetic and cultural heirs of these first food-producing settlers.

At the beginning of the Iron Age, around 1000 B.C., Phoenicians established a colony at Carthage, near the modern Tunis, to exploit the agriculture of the intermontane valleys and to use the Tunisian harbors as bases for trading expeditions to western Mediterranean and Atlantic regions. Aside from eastern Tunisia they colonized only a few trading posts along the shore.

After the Romans had removed the Carthaginian threat to their expansion in the Mediterranean they were able to exploit the agricultural lands of North Africa. Since the soils of Italy were becoming exhausted, it was necessary to go farther and farther afield for breadstuffs to feed the growing urban population of Rome. The

Berbers in Tunisia and most of Algeria submitted to Roman rule, and many became Christian. The Romans had less success in Morocco, where they conquered but a small area, from Tangier down to Rabat, and inland to Meknes.

During the Roman period, colonists from Italy and Greece settled in Tunisia, and there were Greek colonies of earlier date in Cyrenaica. Jews entered North Africa in Roman, and probably also Phoenician, times, and established quarters in the towns and villages in the mountains. The Vandals held Tunisia for a few decades, but succumbed to the climate and to the Romans. Around the fourth century A.D. someone introduced the camel into North Africa. Veiled riders began to harass the Romans, who with only horses and asses could not compete in mobility with these enemies equipped with superior transportation.

From this blow the Romans never recovered. When, in the second half of the seventh century A.D., the first wave of Moslem Arabs burst into North Africa, they rode straight through the valleys and along the coasts to the Atlantic, finding little resistance. The natives of North Africa have always looked to each new conqueror as a deliverer from the oppression of their immediate predecessors, only to learn time and again that there is little to choose between masters.

The early Arabs in North Africa were city men from Hedjaz and Yemen, learned and pious. They founded cities, built mosques and universities, and left the country districts in the hands of the Berbers. The considerable European population of eastern Tunisia was probably converted to Islam and absorbed into the general amalgam. Only the Jews resisted Islam, and kept their ethnic identity as well as their religion. Their speech, however, they lost, shifting to Berber and to Arabic where and as most convenient.

The second Arab wave struck in the 12th century. It consisted of a vast tribal migration. Two confederacies, the Beni Hillal and the Beni Soleim, were ejected from the Syrian desert for repeatedly plundering pilgrim caravans. These rough and ready illiterate Bedawin (Bedouins) linguistically different from their predecessors who spoke the Arabic of the Koran, pitched their tents on both sides of the mountains, and drove out some Berbers while they intermarried with others. They and the products of their mixings became the forebears of the present-day rural Arab population of North Africa. With many other elements they also contributed to that of the cities.

In 1492, when Ferdinand and Isabella threw the Moors and Jews out of Spain, most of the former and many of the latter sought refuge in North Africa. Some of the Moors were absorbed, but others formed tight little colonies, like that of Sheshawen in Spanish Morocco, which still remain intact. Their Arabic contains a number of Spanish roots. The Jews went to the cities where they also formed special colonies. Their speech was and is old Andalusian Spanish. It serves to distinguish them from the longer-seated Arabic and Berber-speaking North African Israelites.

During the 16th century the Ottoman Turks conquered North Africa up to the Moroccan borders, and remained more or less in power for three centuries. First, however, they had to drive the Spaniards out of the seaports. After this expulsion and a parallel ejection of the Portuguese along the Atlantic coast of Morocco—the Spaniards never gave up their footholds from Ceuta to Melilla—the great age of piracy began.

At this very time the slave trade was bringing new genetic elements into North Africa. Some of the slaves were Negroes, marched or ridden overland from the Sudan, while others were white; English, French, Spanish, and other captives taken by sea raiders. The Spaniards sold American Indians to the Moors, and carried some Berbers in turn to the Americas. At least one shipload of Wampanoags (New England Indians defeated in King Phillip's War) were sold on the block at Salé. From the newly established Turkish lands to the east, no doubt Slavs and Circassians and Greeks also reached the North African slave marts.

In 1786, the Sultan of Morocco, first of all old-world monarchs, recognized the independence of the United States and presented our government with a palace at Tangier, which still serves as our legation. Neither he nor his successors had any control over the sea raiders a thousand miles to the east, whom our navy defeated under Stephen Decatur in 1815.

In 1830, the French government, seizing on an insult to the French consul by the Dey (the Turkish sultan's regent), landed troops at Sidi Ferrouch outside Algiers, the same spot where the American and British soldiers came ashore on November 8, 1942. In 1881 they conquered Tunisia. In 1912 they began their conquest of Morocco, while the Spaniards began to spread out from their crown possessions of Ceuta, Peñon de Velez de Gomera, Isla de Alhucemas, Melilla, and Islas Zafarinas. By 1934, after many checks and reverses, the French and Spaniards had subdued all the natives of North Africa to the Tripolitan border. In Tripoli itself, the Italians, who drove the Turks out of their last foothold on the southern Mediterranean in 1912, had beaten the natives well into submission by the beginning of World War II.

In 1939 North Africa was a continuous belt of European possessions, whatever the official form of government in each section, with a million and a half Christians in residence, the successors to the Romans. The tide had turned again—Phoenician-Roman; Arab-European. The next move from the East, Arab nationalism, is still in the stage of ominous preparation. The French fear it as the Romans did those first veiled riders on camels.

Through all these swings back and forth, these invasions and infiltrations from the East and then the North and then the East again, the Berbers, whose ancestors had lived there for two millenia before the Phoenicians arrived, have changed very slowly and very little. Today they inhabit most of the mountain slopes and upland valleys of North Africa, the inaccessible hideouts off the main invasion routes, the lands in which the annual rainfall is most constant, the means of livelihood most secure. But they are not all farmers.

Peoples of Berber language and tradition practice all of the means of livelihood known in North Africa. The majority of them keep very busy with the annual cycle of ploughing and sowing, reaping and threshing, irrigating gardens and pressing olives, tending their miniature terrace plots with a combination of skill and care that has excited the admiration of many observers. These are the Berbers who occupy the mountain lands. Divided into many tribes and confederacies widely separated from each other, they include the western Riffians, the Shluh, the Kabyles, and the Shawia. A few still remain in lowlands suitable for dry farming, despite the competition of Arabs and Europeans. Among them may be reckoned the eastern Riffians and the Zenatan peoples who live between the Moulouia River in Eastern Morocco and the Algerian border.

Up in the Middle Atlas mountains, which in reality do not seem like mountains at all, but a high region of forests and meadows like some of our national parks in the Southwest, live Berbers who are part-time farmers, part-time pastoralists. The French call these people *transhumants*. During the winter they used to live in castles of pounded earth, down in the lower reaches of the valleys, at the foot of the mountains, and there they would plant their crops and pasture their sheep. In the spring, after the harvest, when the snows had melted on the high meadows, they would migrate to summer pasture, living in black tents like Bedawin, and leaving only a few old people and servants to guard their winter residences. In the Algerian *Hauts Plateaux*, others follow the same means of livelihood.

Out in the Sahara the famous Tuareg, who entered the desert from the Middle Atlas after the introduction of the camel, are full pastoral nomads, or were until the French subdued them, and brought the bones of their ancestress Tin Hinan to the Algiers Museum.[1] They lived off the flocks of dependent tribes, off the grain grown by serfs in oases and mountain fields, and off the tribute and plunder exacted from trans-Saharan caravans. Their stock in trade was a combination of speed, endurance, bravery, and skill in political organization.

Even the whole catalogue of livelihoods from mountain tillage to desert pillage does not cover the occupations of Berbers. With the rise of the caravan trade in post-camel times, colonies of them settled the principal oases of the Sahara, such as Ghardaia and Tidikelt, and became able merchants. In Tunisia the Berbers of the Isle of Jerba have devised a clever means of supporting the excess population. Many of the young men migrate to the towns and cities, where they set up small shops, work long hours, earn considerable fortunes, and then go home to marry and live lives of relative ease. In Algeria, the Berber Mzabites from Ghardaia oasis do this same thing, as do the Soussis in Morocco. The Sous contains whole cities of Berbers, not only Tardouant but also Tiznit and Agadir; here the visitor will see streets of goldsmiths and jewelers, spice merchants and weavers, with none of the usual oriental crafts and trades omitted. Although some of the workers are Jews, Soussis work at each trade. Up in the tribe of Taghzuth, in Spanish Morocco, where the valleys are steep and cold and forests of cedar and pine stand

[1] Via the United States. They were retrieved by Professor Reygasse of the University of Algiers.

near, whole villages of Berbers live by tanning and leatherworking, and in the old days many made gun barrels. By trading these products they obtained foodstuffs from the peoples on the flatter lands below.

If by chance you drop in to the house of a friend in Casablanca, French, British, or American, for a drink before dinner, it will probably be served you by a clean, tidy little man with excellent manners. He will be a Soussi, and his wife will be cooking dinner in the kitchen. In Algiers, the waiter in the Aletti who wears a shiny black tailcoat is not a Frenchman, not a Spaniard or Italian, as you may think. He is a Kabyle from the mountains. He speaks French, Arabic, and Berber, all almost equally well. He may have a brother working in France.

I have been waited on by a Kabyle in a Boston hotel, seen Berber acrobats in a Boston theater, and watched a Berber holy man lick a hot iron in the side show of Ringling Brothers at the Boston Garden. One need not go to North Africa to see Berbers. They get around, they go amazing distances, often without revealing that they are Berbers at all, and very often they come home. When they do, one would think that the life of the Berber villages would be profoundly changed. It is not. The Berber leaves his foreign ways with his shoes at the doorstep. He has been doing this ever since the days of the Phoenicians and the Romans.

In North Africa there are probably ten million Berbers, compared to six million Arabs, and one and a half million Europeans. Berbers constitute over half of the entire population, they work at every kind of trade and profession which they are allowed to learn, and could be completely self-sufficient, even if every Arab, Jew, and European in North Africa were to disappear tomorrow. This could be said of no other element of the population. We hear much about the Arabs and the European Colonists, but of the most numerous people in North Africa, our popular press is singularly silent.

Although the Berbers could conceivably live alone, their ability to do so is the result of the many skills learned from the different invaders who brought with them metal-working and irrigation, and such things as olive trees and the machinery used in oil making. Conservative as they are, the Berbers will take over new techniques of which the usefulness is evident; some of them run their own bus

lines in Morocco. The point is that they learned centuries ago how
to grow the most food possible on tiny mountain terraces, and the
size of their fields and the slope of land prevent the introduction of
modern machinery. However, almost every Berber house in some
districts contains a Singer sewing machine, which requires no special
source of power and no materials which the Berbers cannot obtain.

The lack of change in the work activities that give them their
food tends to make the Berbers conservative. Despite twelve centuries of Islam, most of them have kept their old traditional law,
because it fits their way of living better than Koranic law. In addition to the lunar calendric feasts of Islam they continue to celebrate
their old solar festivals, because the critical points of time in the
agricultural cycle go by the sun and seasons. Only when they adopt
some new means of livelihood can the Berbers be expected to show
change in other departments of living.

It is hard to speak of Arabs in contrast to Berbers, because the two
peoples have lived in juxtaposition so long that no fine line can be
drawn between them. Some tribes of pastoralists are Berbers whose
ancestors adopted Arabic speech only a few generations ago, others
speak Berber, but claim descent from holy men from Arabia. Many
of these claims are probably true. In the mountains entire villages
and tribes of farmers and craftsmen have lost their Berber speech
for various reasons. On the whole, however, the majority of Arabs
may be divided into two classes, country people and city people.
The country Arabs are either pastoralists, transhumants, or sedentary farmers, depending on the amount and seasonal distribution
of rainfall in the special regions which they inhabit. To shift from
one to another or to a third of these ways of living is easier than
might be expected.

Both country Arabs and country Berbers participated, before the
establishment of European rule, in a single economic system. They
lived in villages or camps, and raised the food, wool, and most of
the other raw materials that they needed. They traded much of their
hardware from blacksmiths resident in their settlements. The blacksmiths also butchered large animals for division among families.
These blacksmiths were Negroes, or men of Negroid appearance.
They were supposed to have come from the Sahara or southern

oases. The blacksmiths had kin of their own, to whom they eventually returned, and did not intermarry with local people.

Sometimes itinerant well-diggers would appear, looking for business, and musicians, too. Peddlers brought manufactured goods, some of it made in Europe. But most of the buying and selling took place in markets. All over North Africa, markets were held six days a week, Friday excepted. The biggest were usually the Thursday markets, where people prepared for their Sabbath needs. These markets were so spaced geographically that any person, no matter where he lived, except in the most remote mountains and desert stretches, could walk or ride to one, do his business, and get home again, on any of the six profane days of the week. In these markets he could dispose of his surplus crops and livestock and sell the produce of his spare-time handiwork, such as wooden spoons. His wife could sell eggs, and pots if she knew how to make them. She might also sell the woolen cloth that she wove. In the transhumant and pastoral tribes, wool was an important cash crop, both raw and in the form of blankets and rugs. The women wove these in upright looms, which did double service as tent poles.

At these markets the farmer and his wife would buy pieces of meat, and vegetable foods, such as melons and oranges, pottery vessels, and metal objects made by local smiths. All of these objects, being locally produced, were merely passed around within the geographical range of the market, and permitted a certain amount of local specialization. Professional merchants would take away surplus staples, as grain and raisins, to sell in larger markets in the city. Or the farmer might make the trip himself if he lived near enough and were not afraid of being robbed or arrested.

Besides circulating local products, the markets served to distribute imports from special spots within the country, or from foreign parts. All salt came from a very few places, such as Azni in the Grand Atlas, where there are salt springs and professional salt makers; all millstones likewise, came from the Caves of Hercules near Tangier, where hereditary stoneworkers still peck them out of special coarse, hard stone. The quarries have been worked continuously since Neolithic times.

From the world outside North Africa came hard cones of white sugar, specially made in France; cheap white cotton cloth, from England and America; fancy brocades to be used for men's turbans and women's clothing, mostly from France; gunflints from Brandon; candles, spools of thread, matches, needles, files, gun locks, old bottles, blocks of graph paper, which for some reason were preferred for writing, and many other products from many quarters. In country markets one even sees agate and moonstone finger rings made by the Jews in distant Yemen and peddled as amulets to ensure the birth of male children. Spices of every variety are there, brought from India, and frankincense from the Hadhramaut. If a visitor made a complete list of every product in a North African market with its country of origin, most of the known world of commerce would be represented. Some of the commodities were produced only outside North Africa. These were essentially raw metals, candles, matches, gunflints, the essential mechanism for firearms (stocks and barrels could be made locally), and spices. Some came from the Eastern world, some from the North. Both sources, so important in North African history, were represented.

The vendors in these markets included local people selling their own wares, along with special traders handling the imported materials. They went from market to market, and had no other business. From time to time they would go to a city to unload their purchases and replenish their stock. Over in the meat section the butchers were congregated. These were either the local blacksmiths or their relatives. They would kill animals brought to them and sell the meat on commission. One would also see weighers, special officials who carried legal weights and balance scales, and were ready to prevent or settle disputes. Professional criers stalked about, shouting announcements, which emanated from a small house or tent alongside the market area. In this house or tent sat the important men of the neighboring villages, and it was they who governed the market. Very often they held meetings there, decided political questions, tried cases, and collected fines. Now and then one would see an eccentric holy man wandering between the stalls, either a private psychopath or a member of an itinerant order. The marketgoers paid him little attention.

The political organization of country people was the product of extensions of kinship and geographical propinquity. Although the question of *who marries whom* was answered in different ways in different places, the family which resulted was always an economic unit. The mountain Berbers in particular lived in large extended families consisting of sets of brothers and their male descendants, wives, descendants' wives, and unmarried daughters, all along the line. Such a family rarely extended beyond three generations, and when the brothers of the senior generation died it broke into its component elements. Beyond the family was the kin, often coterminous with the village. The larger villages usually consisted of several such kins, and in Algeria, in the larger Kabyle villages, these were divided into ṣofs (Fr. çof) or moieties, paired for all political and ritual activities.

The village itself was based on a number of related kins living together on the same unit of land. To these were usually added refugees from other regions who had married and settled down there, the local fqih or head of the mosque, who was also schoolteacher, and a blacksmith-butcher and his family. Only those who belonged to old, long-settled kins had any political power. According to the old Berber system the boys meeting in the schoolhouse—a part of the village mosque—had the right to hear petty cases and impose fines.

The heads of kins and specially chosen delegates represented the village in deliberations which concerned several villages, which were held at a larger, intervillage mosque. The only permanent political unit, however, was the village, for villages could fight each other, while within the village there was no warfare. As occasions warranted, units of larger and larger size would come together, meet in council, and elect leaders. Through the repetition of such meetings, lines of habitual organization were formed into intervillage groups, named tribes, and confederations.

This was the Berber system. The Arab system was parallel in its progression of groupings from village to "fifth," or cluster of villages, to "fourth," a grouping of these, to "tribe," and to confederation. The difference lay in the system of government. Instead of a council the Arabs had a *moqaddem* for the village, a *sheikh* for the fifth, a *khalîfa* for the fourth, and a *qaid* for the tribe. These men exercised arbitrary power. Sheikhs usually had deputies, a secretary and a number of *mokhazniya*, or policemen. The *qaid* of a large tribe usually had a small court of his own, often with slaves and even a jester.

The distinction between the Berber and Arab systems was far from absolute. Before the European intervention many of the Berber tribes had lost their free councils, through the rise of dominant families of chieftains like that of the Glawi. Among both Berbers and Arabs there was a distinction between first-class citizens and the members of degraded occupational castes, the blacksmiths, butchers, market weighers and criers, well-diggers, and professional musicians. Both Arabs and Berbers owned slaves, although among the mountain Berber farmers these were so rare as to be inconsequential in the social structure. To the pastoral Berbers, particularly the Tuareg, and to the oasis people, owning slaves was economically very important, for most of the heavy manual labor was done by them. The hotter the climate the easier it is for white people to delegate muscular effort to black people, if they are available.

To the most casual visitor, it is apparent that the city dwellers are a breed apart from the country folk, and particularly different from the mountaineers. The country men spend their lives in the sun and open air, and habitually walk several miles each day. They are bronzed and erect, with strong fingers and bulging leg muscles. Their faces are lined, and their eyelids narrowed from the glare of the sun. When away from home they glance about them keenly, always on the alert for trouble.

The city men are, as even the casual observer can see, a different breed. Living always in the shade, they are startlingly pale; whereas the country men are brown as Kanakas, the townfolk are white as Londoners. Some are lean, others are fat; all seem to have delicate hands. In Fez, Tunis, and other cities, many are blond, and albinos are common. The reasons for these striking differences in appearance are not hard to find. City people seldom expose the skin to the rays of the sun. The men sit all day in tiny booths, selling their wares or working at their benches, and at night close the shutters and walk fifty feet or so to their houses. Except for that entailed in their work processes, they get almost no exercise. Their food is full of starch; their favorite drink, of which they may take ten cups or more a day, is a sirupy sweet mint tea. This way of living has been going on for generations. Before they came to North

Africa, the ancestors of many of them lived in a similar way in Arabia and Iraq. If we include ancient, pre-Islamic civilizations, some of these men may have genes derived from two hundred generations of shopkeepers, living and dying in narrow, sun-shaded streets. That they look and behave differently from the farmers and herdsmen is scarcely surprising.

The organization of North African cities is similar to that of European cities in the Middle Ages. The city is surrounded by a wall, with several gates. Inside, houses are built right up to the wall, so that some use it to save masonry. Other walls divide the city into wards, with huge metal-faced wooden doors that are kept closed at night. Since the city has been added on to at various times, some wards are old, others new; if it is a royal city, a number of palaces, set in gardens, will be appended by special walls to the city itself.

There is always a large open space, inside one of the principal gates, which serves as market place. Here country people come every day to expose their wares, and merchants from the inner part of the city will sit also. Snake charmers chant and bite the heads off adders, storytellers, magicians, and dancing boys hold the attention of crowds. Some of the performers are comedians, imitating the actions of different animals; others merely sit and play flutes dreamily. Outside the walls a special market is held, usually on Thursdays, although the day may differ from city to city. Here livestock and rugs, just in from the mountains, change hands. Here one comes to buy a good riding mule, keeping out of the way of other prospective customers who are trotting up and down, putting the animals through their paces.

In every quarter of the city stands at least one mosque, with its high, square minaret, golden ball, and hook for the white flag which indicates the time for prayer. One of these mosques, founded by some sainted leader of old, may be holier than the others. To it people make pilgrimages from distant places, and certain spots on its wall may be worn and rubbed smooth by the lips of the devout. In larger cities, there will even be one or more Medáris (sing., *Madrasa*) or Moslem universities.

Within the city people are engaged in family life, manufacturing, trade, ritual life, education, and politics. Many of the poorer families live in single rooms, in little houses set at the bottoms of gloomy alleys. A richer family will have several rooms, often a complete house, including two or three floors. Still richer families will own or rent complete houses built around courtyards. While all these houses look alike from the outside, with their blank walls of mud-dried brick or plaster, inside the owners lavish what wealth they can command on luxurious fittings. In the richer houses, the central court is a garden, set with bitter orange trees and bushes of jasmine to sweeten the air. In the middle bubbles a fountain, whose waters cool the air and irrigate the ornamental trees and shrubbery.

The main room of such a house faces the garden and is farthest removed from the street. It is a long, narrow, high-ceilinged room, perfectly designed for maximum coolness in hot weather. It has no windows, but half of the front wall is open. A curtain may be drawn for shelter from strong breezes, and in rainy weather, the wall can be closed entirely by a huge ornamental door. This is the main room, where the master entertains his guests. It is fitted with wall benches capped with cushions and backed with mats and ornamental cloths. Inside, at one end, stands a huge brass bed. Nowadays the room is lighted with one bare electric light bulb hanging on a long cord. Next the door is often a radio.

A poor family will consist of a man, his wife, and several children, with perhaps one or more aged grandparents or other dependents. If they live in a single room, they will hang a piece of cloth across from wall to wall, to screen off the end for the seclusion of the womenfolk when outsiders enter. In a wealthier family, there are special rooms for the women, and guests are not admitted there. If the women enter the main room while guests are present, they do so quietly and unobtrusively, to wait on table. Rarely, in a rich and poorly regulated household, an observant guest placed in a strategic position can see the younger womenfolk peering around a doorway and tittering.

The rich household includes the man, his wife or wives, his sons and unmarried daughters, and sometimes his son's wives as well. It will also include one or more servant girls, who are the daughters of poor people, or countryfolk; these girls wait on table. Then there may be a cook, and a gardener-handy man, who answers the door and runs errands. In the old days, these latter were often slaves.

A city Arab does not ask another about the health of his wife.

When men visit each other, the womenfolk retire to their quarters. When a woman has to go out in the street, she wears a veil. If she is the wife of a wealthy man, a servant will go with her. Well-to-do women lead very sheltered lives in the city. They spend most of their time in the house, with occasional journeys to the walled-in part of the roof. They may direct their servants to prepare the food, and even do a little cooking themselves. Usually they do not go marketing, but send someone else. The rest of their time may be spent embroidering, sewing clothing for their children and themselves, and visiting or receiving the visits of other women. At these times they comb and wash their hair, pretty their eyes with kohl, henna their hands, and otherwise beautify themselves. In this pursuit they spend as much time as women anywhere else.

All outsiders who have had the chance to know them will agree that nearly all the city women in North Africa, and indeed in other Moslem countries as well, are eager to learn about the world outside the family walls, and crave to be told stories. Most of them are illiterate. They are not allowed to go to the movies, in towns where movies exist; when the husband comes home tired from a day of haggling or arguing in the market or shop, he is greeted with a broadside of questions, "What happened today? Whom did you see? What is going on?" and shortly after, if not at the start, "Come on, now, tell me a story."

Very often when a European tries to get a definite commitment from an Arab, the latter puts him off, saying, "I will tell you tomorrow." The European will go off muttering, "These natives can never make up their minds," or "He is a shifty fellow, plotting some deceit," when as a matter of fact the poor Arab simply wants to go home and ask his wife.

The life of a city Arab goes through a cycle of crises, each with its attendant ceremony. When a boy is born, his father gives a big feast to which he invites kinfolk on both sides of the family, and prominent friends. Often the father will ask in a poor man to serve as butt of the other people's jokes. These the poor man can endure, since he fills his belly with fine food and basks under the eyes of the rich and great. He can also hear and watch the hired musicians and dancing girls whom the boy's father has brought in.

Later on the father will have his son circumcised, and a gay procession of mules and men will march to the mosque, to the blaring of shawms. There will be another feast. When the time comes for the boy to be married, the father will choose a wife and make all of the arrangements, and when finally the shy couple is united and left alone to make each other's acquaintance, another feast is laid on. After the boy is married, if he falls ill, some friend or relative may hire a dozen men to take turns praying continuously for him for a week. When he dies his kinfolk carry him to the cemetery inside a light coffin supported by poles on the shoulders of men, like a litter of the dead. For many nights his womenfolk tear their hair and weep.

To belong to an old family is just as important in a North African city as it is in Boston or Philadelphia. Each city has its roster of great names, the Fasis and Kanounis and the like, and its *nouveaux-riches*. Some of the great families are entitled to the spiritual rank of *sherif*, or descendant of Mohammed, and others try to fabricate genealogies to place themselves in this class. Social prestige through giving bigger and better feasts is the ambition of some families. Others, who already have the coveted prestige, maintain it simply and quietly by almsgiving and pious deeds.

Most of the city people make their living by industry or commerce. Some are potters, tanners, weavers, dyers, silkworkers, blacksmiths, coppersmiths, silversmiths, goldsmiths, jewelers, swordssmiths, carpenters, turners, cabinetmakers, inlayers, and leatherworkers. The last named may further specialize in bags, slippers, and belts. Every art and craft found in medieval Europe, or in India or China, has its special men at work in little shops along the North African city streets. Each craft has its own section, with its own elected guild head, who represents this craft in the council of guilds, and to the government. The members of each guild know each other. They recognize some masters as more skilled than others. They set bottom prices, and although each man is entitled to what he can get if some stranger or simpleton is willing to pay the first price asked, anyone who allows himself to be beaten down below a certain level will get in trouble.

This is even more important in the case of the merchants, who do nothing but buy and sell. While a craftsman will sell at retail as well as wholesale, he prefers to spend his time working at his trade, particularly since the act of selling in North Africa is such a lengthy process. The merchant is not happy if he and his customer reach an agreement at once. He must argue, praise his goods, crinkle the leather, roll in and out the bolt of cloth, put on a tragic expression when a lower price is bid, and even weep when the transaction is finally completed. Although much of this elaborate acting is put on nowadays for the benefit of Europeans, North Africans still get excited, haggle, and show emotion among themselves. Commerce is not the dry business of an American chain store. As with everything else these people do, it is an art.

When one leaves the street with its small booths and shops, and steps inside the doorway of a larger establishment, one finds a difference. The building is a large one of three or four stories. A central court yields sunlight. There is one main salesroom, but offices have been partitioned off. The walls are lined with shelves, and these are filled with bolt upon bolt of cloth. This is the establishment of a great commercial family. The head of the house sits in his private office, while the clerks are his sons and nephews. A brother may be away for several months, in Manchester or Lille, buying up stock. This is where the retail cloth merchants get their material, much of which they take out to country markets.

In another large establishment you may see row upon row of Singer sewing machines. The man who is fortunate enough to hold the agency is protected by the American government, for his company cannot afford to have him and their property subject to the arbitrary seizures of some pasha. The cloth merchant likewise is protected by the British. In still another establishment, a father and his sons may be weaving silk on a Jacquard loom, by some secret process kept within the family for generations. They too enjoy immunity: the old father once rescued an American consul from bandits, and was suitably rewarded.

Goods that are sold in large quantities, as bolts of imported cloth, or that are unique, as the sewing machines and the bridal belts of silk that come from the Jacquard loom, pass hands without haggling. This is also true of perishables to be consumed at once. In a market restaurant a cup of tea is worth just so much, no more or less, and the same is true of a doughnut or a piece of pastry. But if you sit sipping tea in a small restaurant or merchant's shop—for they often give their customers tea—you will see and hear the auctioneers. A young man walks rapidly down the street, shouting "Fifteen! Fifteen!" He is carrying a freshly dyed hide, brilliantly green. Some one says, "Sixteen!" The young man darts over to the shop whence the word emerged, sees who spoke it, and then moves on, crying "Sixteen! Sixteen!" He strides through all the streets in which people might be interested in buying dyed leather, until he has received a bid that no one else will raise. Then he sells to the highest bidder. This system of auctioning is one of the reasons why all the men practicing one craft or selling one kind of goods congregate in the same streets.

Here and there, in an open space near the meat market, a dozen poorly clad men squat. Each holds a basket. They are porters, and will carry your purchases anywhere for you, for a fee. If you pass there often, you will form the habit of hiring the same porter, and hence each one builds up his special group of clients. If you are friendly, the porter will tell you all the latest gossip, because he moves about widely and knows what goes on before it happens. Another prime gossip is the barber, who shaves his customers' heads in their own houses. In the old days special eunuchs went about from harem to harem, shaving the intimate parts of women, and they were gossips too.

Some elderly women who have the entrée to rich households now and then serve as go-betweens, escorting young wives out during the daytime, when husbands are away, to secret rendezvous with lovers. Amorous young men need not, however, risk this dangerous expedient. Every Arab city has its brothels. Prostitutes also congregate in the *fnádiq*, or caravanserais. A *fonduq* is a combination boarding stable and hotel, where travelers lead their mules and camels, once they have reached the city. Cities vary in morality. Some are famous for homosexuality, and boys are trained for this, dressed and painted like girls and taught to dance in special ways. In others, the practice is frowned upon. In some of the country districts, within the last twenty-five years, perverts have been burned alive.

North Africa is an essential part of the Moslem world; it is called *el Maghreb*, the west, in contrast with Syria and beyond, which are called *esh Sharq*, or *el Mashriq*, the east. In it live some of the holiest of *shorfa* (pl. of *sherif*), and two spots have long been especially holy, Kairwan in Tunisia, and Fez in Morocco. The holy ones of Kairwan, the Fatimid Caliphs, transferred their seat centuries ago to Egypt. The holy ones of Fez, dynasty after dynasty, have remained in Morocco. The present Sultan of Morocco is Mulai Mohammed. His dynasty originated in Tafilelt, an oasis on the southern side of the Atlas, where it forms the head of the western Saharan caravan route. He is the holiest man in North Africa. Most Moslems of Morocco, Algeria, and Tunisia consider him their spiritual leader. While he has not yet claimed the title Commander of the Faithful, which goes with the assumption of the caliphate, many believe that he will, and more feel that he should.

The orthodox practice of Islam is carried out in North Africa by the majority of the people. Orthodox Islam has no hierarchy comparable to those of Rome and Lhasa. It is essentially a lay religion. Each community builds and supports its own mosque or mosques, chooses its own leaders. Each mosque has its staff, consisting of a *fqih* or *moqaddem*, who leads the prayer and may also act as muezzin in calling to prayer, and a number of assistants, including students. In a city like Fez the leaders of the mosque are organized into a group, called *'Uleyma*, or learned ones. It is they in concert who make decisions, as for example when to begin the fast of Ramadhan.

In the Maghreb as elsewhere in orthodox Moslem countries, the duties of the faithful are five. These, the "Pillars of Islam," are: Profession of Faith, Prayer, Almsgiving, the Fast of Ramadhan, and the Pilgrimage. The Profession of Faith is the recitation of a compound declarative sentence, *La ilah ila Illah, wa Muhammed rasul Ullah*, meaning, "There is no god but The God (Allah), and Mohammed is the messenger of The God." The recital of this rhythmic and alliterative sentence admits an individual into the faith of Islam and the company of Moslems.

A faithful Moslem prays five times a day, at dawn, noon, midafternoon, sunset, and nightfall. He must do this in a state of ritual purity, facing Mecca, which in North Africa means the east. Since purity is destroyed by sleep and by contact with impure things such as dogs, corpses, urine, excrement, sexual fluids, and the like, it is considered safest to wash each time before praying. The devout person washes his hands to the elbows, his feet, and his ankles. If necessary he can wash with sand in place of water, but in North Africa this occurs but rarely. In praying he prostrates himself, twice at dawn, four times at noon, four at midafternoon, three at sunset, and four times again at nightfall. The usual prayer begins with the *takbir*, or recital of the words *Allah Kebir* (God is Great); then follows the *fatiha*, the opening verse of the Koran; then comes the *shahada* or profession of faith, as above; then the words *salat 'ala 'n Nebi* (a prayer for the Prophet), then *salam*, followed by *Amîn* (Amen). To extend the prayer, other verses of the Koran may be interpolated.

Moslems may pray separately, or in groups led by an *imam*. Out in the country, or along the highroad, each man watches the sun and decides when it is time to pray. In the cities the voice of the muezzin is heard high above the noise of the streets, and the devout enter mosques. Friday prayer is held at the mosque at noon, and men only are present. An ideal congregation includes forty or more men. The *imam*, or *fqih*, preaches from a pulpit. In his sermon he is obliged to mention the head of the state, usually the sultan. After the sermon he leads the congregation in prayer.

On Friday, shops are open. It is not a day of rest. After returning from the mosque and eating lunch, many worshipers go back to work. But in the morning, in preparation for the visit to the mosque, poor people who have no adequate bathing facilities at home like to go to the public *hammam*, or steam bath, after which they dress in clean clothing. Moslems in North Africa like to be clean; if they are dirty, it is because they have no soap, no money to go to the *hammam*, and no change of clothing.

Every year each Moslem is supposed to donate a part of his income to charity. It may be paid in kind, and the amount varies from ten to twenty percent, depending on the kind of goods given. Some men resolve this obligation by handing out coins to the poor at their doors every Friday morning. Others redeem slaves. Still others make donations to local mosques, where the supplies and cash are used for repairs, for the maintenance of the ritual specialists, for food to be given travelers, students, and refugees living at the mosque, and for weapons to be used in the holy war, should one arise.

Once every year the holy month of Ramadhan arrives. Since its date is set on the lunar calendar, it creeps around the seasons. It may come in the rainy winter with its short days—for most of North Africa lies in the same latitude band as the territory from Savannah to Richmond, or Tiajuana to San Francisco—or in the baking-hot summer with days three or four hours longer. The length of the day is important in Ramadhan since the whole point of the ceremony is that all good Moslems are expected to fast from sunrise to sundown, from the moment that it is possible to distinguish a black thread from a white one to the moment when it is no longer possible to do so. Persons who are ill, pregnant, engaged in holy warfare, or off on necessary journeys, may break the fast, but should make it up later. Fasting involves a complete prohibition against eating, drinking, smoking, the use of perfumes, sexual relations, and even bleeding. Rich people do little work on Ramadhan. Many of the shops are closed. In others the proprietors may be seen quietly dozing, or reading aloud from the Koran. The poor people, however, and the farmers, must keep on working just the same, even in the heat of the harvest. Needless to say tempers grow short through the day, and many quarrels occur an hour or two before sundown. They also grow short through the month, and quarrels reach their climaxes near its end. It would be interesting to know how faithfully Ramadhan is kept among the Moslems of Kazan, Russia, at 55° north latitude, in summer.

Ramadhan does not always begin on the same night. In each major city, the *'uleyma*, or learned ones, representing the principal mosques and universities, sit up in a tower, looking for the silvery curve of the new moon. When they spy it they pass the word to those below, and a man touches a match to a cannon. The fast is on. Drums boom from the quarters of the city, and from the villages around. In the old days the fast might begin on different days at different places, dependent on the cloudiness of the sky, for if the *'uleyma* fail to see the moon at their first vigil, they repeat each night until they succeed. Nowadays, however, it is the *'uleyma* of Fez who decide for all Morocco, and telephone their discovery all around. I do not know for certain how it is worked today farther east.

In September, 1942, when everyone was tense with the expectation of invasion and bloodshed, the Spanish thought to help the Moslems of Tangier in announcing Ramadhan. They placed cannon secretly, after dusk, at various stations in the city, and when the new moon appeared, let go with a salvo which shook the ancient walls and terrified many of the European inhabitants, particularly those who did not know about Ramadhan. The local *'uleyma*, with their ears to the phone, were chagrined, and so were the ordinary Moslems next day, for the *'uleyma* in Fez had not seen the moon, clearly visible in Tangier, at all.

Every night during Ramadhan the cannon is fired to tell the faithful when to break their fasts. Many of them sit on their roofs or balconies, faced with a bowl of water, a bowl of a special thick, nourishing soup called *hareira*, and often a package of cigarettes. They eat, drink water, tea, and coffee, and make merry until late in the night, snatch a few hours of sleep, and are awakened by another cannon shot which gives them time to eat a filling breakfast before dawn, after which, unless they have to work early, they return to sleep. On the last night of Ramadhan they begin the celebration of a great annual feast called variously *el 'aid es sghir* and *bairam*.

Ramadhan is a very critical time in Moslem-Christian relations. Not only does it produce physiological effects on the Moslems which make them more sensitive than usual to criticism and injustice, but it serves as a symbol to reinforce their mutual relations in a common front against outsiders. To a hungry, thirsty man driving a recalcitrant donkey into the market about four o'clock on a hot afternoon, the sight of a sleek Christian sitting in a sidewalk cafe sucking at a long, ice-cooled drink is infuriating. If ever there is to be trouble, that is the time.

The fifth duty of a Moslem is the pilgrimage to the holy places of the Hejaz: Mecca, where the Messenger of God was born, and Medina, where he died, as well as other spots where noteworthy events occurred. There are special rules, such as wearing the *ihram* or seamless garment, the circumambulation of the *Ka'aba*, and special walks from place to place. Although the pilgrimage has its own month, *Dhu 'l Hijja*, the pilgrims must also keep all of the rules of abstinence which pertain to Ramadhan.

In the days of Mohammed when the Moslems were limited to

the neighborhood of Mecca and Medina, the pilgrimage was not unduly difficult, and it was expected of every one except poor people and slaves at least once in a lifetime. With the expansion of Islam to such distant points as Morocco and China, it became increasingly difficult, for the means of travel did not improve from the time of Mohammed until the nineteenth century. Hence in North Africa as in other outlying Moslem areas it was the unusual person who went on the pilgrimage, and when he reached home, often after several years, he was greatly honored. It would be difficult to estimate the proportion of pilgrims (*hojaj;* pl. of *hajji*) among the adult male pre-French population of North Africa, but it is safe to say that it was many times as great as it is at present.

It requires no divination to explain this. The pilgrimage is the one act, more than any other or all others, which ties the Moslem world together. Most Moslems are illiterate. Their countries are widely separated, and only through Christian agency have they the use of such modern facilities as steamships, postal services, telegraphs, and telephones. If every year men from Morocco, Algeria, the Sudan, Java, India, the Hadhramaut, Yemen, Albania, Bokhara, Bengal, and Western China can get together in a place saturated with holiness, can perform together humble rites, in identical clothing (for all have discarded their varying costumes for the *ihram*), and under the leadership of the same holy masters, the unification which results is hard to overestimate. While waiting at Jidda, on the road, and afterward at the seaport for transport home, these men can compare notes. What are the French doing in Morocco? *Ulad el harram* (Children of that which is forbidden), they are doing thus and so. And in Djibouti? The same thing, may God curse them. And in Syria? Even further abominations.

Whatever else they do, the European powers, particularly those with a bad record among Moslems, must keep their subjects or protégées away from the pilgrimage. And so they do. Not only do the Moslems from different quarters get together, but they do so in a free Moslem country, which has its own powerful and pious Moslem king. The idea of self-government is contagious. Nevertheless to keep all Moslems away from the pilgrimage causes more disturbance than letting a few go. Each year the French send a delegation of Moslem leaders to Mecca, with all transportation paid for by the government. In 1946 the delegation from Algeria, alone, filled one commercial passenger plane and one small steamship. Naturally the pilgrims were all men whose political views were acceptable to the French authorities, or those whom the authorities wished to placate.

So much for orthodox Islam. Strongest in the cities and in some of the more thickly populated rural areas, it is weak in a number of outlying parts. In the Middle Atlas the transhumant Berbers have, until recently at least, been poor Moslems, because they lacked instruction; very few *foqaha* had ventured into the forests and upland meadows to teach them. Among some of the tribes, some men were not circumcised until late in life and then only when they intended to make their first trip to the city. Many of them did not know how to pray. Some of the Tunisian and Algerian Berbers, the Jerbans and Mzabites, preserve the Kharejite heresy, picked up from missionaries from Syria and Egypt nearly a thousand years ago and now out of fashion nearly everywhere else.

More important than the ignorance of the Middle Atlas Berbers or the heresy of the Jerbans and Mzabites is the strength of the religious brotherhoods in North Africa. These form a constant target for nationalist agitators. A religious brotherhood is a group of men who owe allegiance to some individual holy man who lives in or near the tomb of a holy saint of old and serves as his successor. In most cases the head of the order is the lineal descendant of its founder.

For example, Sidi Mohammed ben Aissa, who was born in 1465 A.D., was a great theologian, teacher, and healer. He drew about him a band of disciples to whom he imparted some of his holiness and supernatural powers. He died in Meknes. There his followers erected a tomb over him. The order of Aissawa, or followers of ben Aissa, arose after his death. His followers spread it far and wide, and separate chapters were established in most of the important cities and many of the tribal districts of North Africa. Once a year the Aissawa used to hold celebrations in each city on the occasion of the Saint's birthday, but the big festival of course was in Meknes. On this day the regular work of the town ended, and for several days following. The whole town prepared itself for the invasion of devotees, just as an American city makes ready for an American Legion convention.

The Aissawa are divided into a number of subgroups, including the *Diyab,* or jackals, who wear jackal-fur costumes and smear their faces with blood. They leap about comically, stealing food from the shops to feed the other delegates. Since they do a big business at this feast anyway, and since retaliation would follow resistance, the shopkeepers count on a certain amount of theft and accept it in good fun. The ordinary devotees, who wear long hair, dance through the streets chanting monotonously Alláh Alláh Alláh, and working themselves into a frenzy. Many become glassy-eyed and froth at the mouth. The feast also includes a *fantasia* or cavalry show, with mounted riflemen dashing at full gallop down an enclosed field, rising in the stirrups and firing a salvo with flintlocks, then wheeling and riding back to reload.

The Aissawa are one of the most numerous of brotherhoods, and put on probably the single most spectacular show. I saw the last big one in Meknes in 1926. After that the French forbade it. Another spectacular sect is that of the *Ḥamadsha,* who chop at their own heads with axes and bludgeon them with clubs, all to a monotonous chant and repetitive bagpipe music. Each brotherhood attracts a different stratum of people. The Aissawa are ironworkers, porters, and other low-class cityfolk, plus members of the wilder tribes in the mountains. Some are Arabs, some are Berbers, and many are Negroes. The Derkawa, however, whose shrine is in Tangier, are upper middle-class craftsmen and merchants, carpenters, turners, coppersmiths, and wealthy farmers, sheikhs, and councilmen in the country. They shave the upper lip and wear full chin beards. They are probably the most respectable and powerful order in the Maghreb. In contrast to them one may observe the humble brotherhood of Gnawa, also centered in Tangier. The Gnawa are racially full Negroes, very black and broad-nosed. They are said to come from Rio de Oro. They wear rags and comic headdresses, belts covered with cowrie shells, and leather sandals. In their hands they carry pairs of iron clappers. Wandering through the streets of the towns, singly or in pairs, the Gnawa sing to attract a crowd. Once a few people have paused to see them, the Gnawa break into a fast jazzy dance, clicking out the time on their clappers, and singing a little song. They collect the few coins given them, bow and bless the audience, and move on.

To the east, in the oases of Kufra, is centered another brotherhood, as holy and as powerful as that of the Derkawa. It is the Senussi sect. It differs from the brotherhoods farther west in one respect. While the western sects are associations, staffed by a cross section of the population, and one man may be an Aissawi and his neighbor a Rehali, and up the street live a couple of Derkawa, there is an all-or-none character to Senussi doctrine. Like the Wahabi movement in Arabia, and the old Mahdi sect in Somaliland and the Sudan, it is puritanical and it is militant. Everyone in the street, in the town, and in the oasis, must belong. It brooks no rivals. Needless to say, this sect has been in trouble with European authority.

In North Africa as in all Moslem areas education is the concern of religious specialists. In the country the local *fqih*, who serves as village prayer-leader and clerk, teaches the young boys, and sometimes girls as well. He holds school in a special room of the mosque, and in it makes his pupils learn how to read and write, and to recite the Koran. Each student recites out loud, and all at once. The master stands over them with a stick, listening to each in turn, and correcting their exercises, which they write on slabs of wood. The parents pay tuition by feeding the schoolmaster in rotation. Not all can afford this, and not all boys learn to read and write. Most of those who do, learn just enough to make out the meaning of legal papers, particularly deeds of land purchase, and to keep rudimentary accounts. Those who cannot read at all ask the *fqih* to do it for them.

In every village some boys want to go on with education. They repair to the larger mosque which serves a number of villages for Friday meetings, and where the councilors meet, if it is Berber country. Here the *fqih* may be a specialist in some branch of learning, such as religious history or Koranic law. The students who gather here to study under him may be fed by their families, or by the people who live near by. They also make a little money on the side by sewing and embroidering men's clothing, and writing amulets for the fearful, revengeful, and lovelorn.

When these students, known as *tolba*, have at length learned all the teacher has to tell them, a group will move on to another mosque where a specialist in another subject can continue their instruction. Some of the *tolba* end up in a city, studying at one of the universities along with the sons of rich merchants and dignitaries. From each teacher, as he leaves, the student obtains a written certificate, if his work has been satisfactory. Some of the students go back to the villages and become *foqaha* (pl. of *fqih*). Others become big merchants, or government officials. Still others hang around the universities for a long time living on charity, for they like the life and hate to leave it, like the "perpetual graduate student," who is the plague of our universities. Others become members of the religious sects that grow up around the larger mosques and about the personalities of individual teachers, and wander about from place to place, providing musical entertainment, writing amulets, curing the impressionable and living off the credulous, until their welcome is worn out or they make some slip and find it best to move on.

In the old days each city was ruled by a *pasha*, whose Turkish title was used in both Ottoman and Sherifian territory. He held court in his own house, where citizens could come to him with complaints. His duties included maintaining order through a police force, collecting taxes, and judging cases of a secular nature. Since this work was extensive, he deputized much of it to subordinate officials. These included stand-ins or *khulafá* (pl. of *khalifa*), who would take his place while he was away, indisposed, or too busy to see all applicants; a *qadhi* to judge cases involving breaches of the *shariya* or Koranic law; a *moqaddem el ḥauma*, or ward-leader, to keep order in each ward; a *muhtassib* to be responsible for the market place and to supervise the guilds; and an *'amin el-mustafadet*, or local tax-collector. The police force was dressed in special uniforms and armed with sticks.

In the city, resident *quyad* (pl. of *qaid*) cared for the needs of the country people of different districts when they came to town, and kept them out of trouble. Since some of the countrymen did not even speak Arabic, these *quyad* were able to regulate their buying and selling and make considerable profit from commissions.

In each city and immediately outside it, a large amount of real estate belonged to the *ḥabus* of each major mosque, a religious, charitable trust. This property was acquired through legacies of rich men who left portions of their estates to charity. It may be directly compared to our Carnegie libraries. Agricultural lands belonging to a *ḥabus* were worked by poor, landless farmers, who received much larger proportions of the produce than they would if they worked for private owners. The rest of the produce went into the *ḥabus* treasury, along with rents from urban buildings.

Each *ḥabus* was administered by a council of trustees, known as *nawazhir* (sing., *nazhir*) or overseers. It formed a stabilizing influence in the community. As far as one knows, in most cities, we may presume that the *nawazhir* exercised their trust with reasonable honesty. Since the very purpose of the *ḥabus* was to level out differences of wealth and maintain equilibrium through smoothing over the areas of friction between classes of the population, and since the whole concept of the *ḥabus* was religious, it would be much safer to do one's grafting elsewhere.

Relations between the city government and the national government were qualified by existing means of transportation. Before the Europeans arrived the only way to go from one city to another was by horse or mule. It took days to cover distances which we now count in as many hours. Before 1912 merchants who wanted to make the trip from Rabat to Casablanca would congregate with loaded animals about a *fonduq*, or inn, and wait until enough had come together for safety. They would hire soldiers to protect them, and start out. Each night they would camp, setting up round white tents. On the fringes of this caravan one would see Jews, dressed in black from cap to slippers, riding on asses. They were not allowed to take part in the formation of the caravan, but clung to it for protection. Once in a while, when the procession reached a flooded river, and no one knew whether or not it was safe to cross, the others would "jew the ford." They would drive one Jew on his donkey out into the stream, and force him forward, screaming and protesting. If he crossed safely, the others would follow.

With transportation on this level, it is no wonder that each city was virtually autonomous. The pasha was supposed to pay an assessment in taxes to the national government each year, but he was required to show no books. How much he should exact was up to him and his tax collector. It is also no wonder that in order to keep

the various pashas in hand and his country united, the ruler made
an annual round of prolonged visits to his several capitals.

In discussing the national governments of the Moslem states of
North Africa before the Europeans took over, I shall follow that
of Morocco closely, because it is the one that survived the longest,
that still is most nearly preserved, and that has been described in
the most faithful detail. The others, although tributary to the Turks,
were probably very much the same.

Morocco was known officially as the Sherifian Empire; its territory as *maghreb el aqsa*, The Far West. Its ruler was the Sultan.
He was at the same time the spiritual head and temporal leader. He
must be descended from the Prophet and, at the crisis of succession,
must be approved by the *'uleyma* of both Marrakesh and Fez. Although the extent of his temporal domain varied from time to time,
his spiritual leadership was unchallenged. He had, and still has, the
right to summon even the members of dissident tribes to the *jihad*,
the holy war.

With no roads at all in the modern sense, and with overland travel
limited to riding and packing animals, it was impossible for any sultan to keep all of his domain under his political control. It was
divided therefore into *bled el makhzen*, or government country,
and *bled es s'iba*, or difficult, rebellious country. The word for rebelliousness and that for lion are almost identical; what is difficult
country for the government is lion country for the rebels. "Rebel"
is hardly an appropriate word, for, as 'Abd el Krim's deputies
pointed out in the Oujda conference of 1924, some of it had never
been conquered by any sultan. Religious penetration had gone much
farther than political power.

In the Bled el Makhzen, the local *quyad*, or tribal chiefs, visited
the sultan's court from time to time, where they were received with
suitable pomp and ceremony. They turned over their taxes and
received gifts in return. Sometimes they left sons in the court to
be brought up in a refined, urban manner, while doubling as hostages
for the good behavior of their fathers. Now and then the *qaid*
of a distant tribe would fail to appear; no tax money would come
from him that year. It might be worth while to send an expedition
against him or it might be better to let him stay in dissidence. The
most consistently dissident were of course the larger aggregations

of Berbers. The Riffians never paid taxes, and few officials of the makhzen dared enter their tribal territory. Now and then when a palace rebellion was afoot, the plotters would take to the Rif and try to raise an army; this was done by Menebhi around 1900. Since he too was against the government, he was well received and passed safely from tribe to tribe.

The Shluh of the Middle Atlas, who live in large, consolidated villages like the Algerian Kabyles, were ruled by a few great *quyad*, of whom the *Glawi* was and still is the greatest. Besides being the *qaid* of the Glawa tribe, this near-potentate is also the pasha of Marrakesh. As an elector of sultans and absolute chief of a private mountain principality, he has great bargaining power both with the makhzen and with the Europeans, into whose days he has survived.

The sultan of Morocco ruled with oriental simplicity. He had no civil service, but chose his wazirs personally. His treasury was all of one piece—there was no division between public funds and palace appropriations. He appointed pashas and *quyad*, and was himself commander of the army. This was composed of mercenaries, including an imposing black guard in special uniforms, and tribesmen who paid their rent on imperial lands in military service. In a holy war, the fighters would be organized by tribes. During the 1890s the sultan hired a Scot, Kaid MacLean, to reorganize his army and drill his troops. At my last visit, members of the MacLean family were still in Morocco, operating a hotel in Tangier.

Besides these civil and military officials the sultan also appointed religious specialists and leaders, the *qdhá* responsible for the observance of the Koranic law, and the *nawazhir* who controlled the *ḥabus*. During the seventeenth and eighteenth centuries he also maintained a navy, which was concerned with raiding and what the Christians called piracy. Indeed, many of the Moorish pirates were renegade Christians, and this was also true of those who put out of Algiers, Tripoli, and other Mediterranean ports.

The fiscal system of the empire was a compromise between the needs of a large and moderately elaborate state, and the ideas of Mohammed himself on the subject of finance. The prophet considered fixed, secular taxation a source of evil, and held the same views on interest. He preferred voluntary contributions to be used for charitable purposes, somewhat like our Community Chests and

Red Cross. These contributions included the *zika* (classical, *zikat*), or purification, and the *'ashur*, or tithe. The first was a contribution of cash or movable property, such as animals, in addition to the latter, which was a tenth of the harvest or annual income. The individual Moslem was supposed to pay these not to the pasha or his representative, but to the *qadhi* (pl., *qdhá*), who collected them for the sultan. Apart from the *qdhá*, some of the heads of religious brotherhoods and independent *shorfa* (sing., *sherif*) collected tithes and contributions separately, and kept them.

The revenues collected by the pasha and his deputy, the *'amín el mustafadet* or often more simply *'amín*, were of a different nature. These included gate taxes, imposed on all produce brought into the city, market taxes, fines, and property seized for civil and political offenses. They also included customs duties and poll taxes upon Jews and Christians. The sultan's government received taxes paid by conquered tribes—those which had been in the *bled eṣ ṣ'iba* and had been forced into the orbit of the makhzen. The rationalization here was that the sultan had taken their land and they must pay, often with military service, for the right to cultivate it or graze their flocks on it. Needless to say, men such as these were the best warriors the sultan could get for his army. Tribes which rebelled repeatedly, and which made nuisances of themselves by plundering caravans over and over again, were sometimes uprooted bodily and moved to new territories where they would be surrounded by strangers and located near the seats of authority.

The *ḥabus* bore the expense of much of the municipal administration, and particularly did it support the universities and schools. Poor students ate food which the *nawazhir* provided. They also took charge of the upkeep of the schools and of the faculties, but since the property had been willed, many curious and unusual duties fell on these dignitaries. Just as in our civilization a person may leave money for a home for indigent cats, so in North Africa the feeding of birds might become the duty of the *ḥabus* officials.

Such, in a nutshell, was the civilization of North Africa at the time or times when the Moslem empires and kingdoms were in full flower. Aside from the substitution of Islam for Christianity, the absence of a graded religious hierarchy, and the identification of the church with the state, there was no difference of any consequence between these states, with their cities and guilds and religious foundations and universities, pilgrims and saints, and tough tribesmen in the hills, on the one hand, and the nations of Europe in early Renaissance times on the other. In fact, historians agree that much of the impetus which spurred the Renaissance Europeans on to a greater complexity was derived from North Africa and points east, through the Moorish kingdoms in Spain, and the "Saracen" outposts in Sicily and Southern Italy. When well-educated Europeans who should know better say, and I have heard them, *"Les indigènes ne sont pas assez evolués"* to rule themselves, learn the technical skills of modern Europe and America, and do all the things we do, they are talking through their képis and homburgs. The North African civilization is just as highly technical, just as complex, just as advanced in every way, as that of Tzarist Russia which hordes of immigrants brought to Western Europe and America before and after the revolution. It is just as complex as that of the Chinese, whose capacity for self-rule is a modern axiom. From the standpoint of physical anthropology, the North Africans are *more* highly evolved, on the whole, than Northwest Europeans. All we need do to see that these things are so, is to offer them adequate education.

The Turks lost Algeria in 1830, Tunisia in 1881, and Tripoli in 1912. During the two to three centuries of their rule in these North African countries, they made no cultural contribution of moment. One can see their influence in the baggy pants and braid-encrusted waistcoats of the Zouave uniform, in the fezzes and pointed slippers worn even in Morocco, which was outside their political sphere, in some of the dishes on the North African menu, and in the borrowing of certain words, such as *pasha*. Two things North Africa needed, improved transportation and mechanized industry. Neither of these things could the Turks give, and they left the countries in the same state of technological advancement in which they had found them, little better off than in the days of the Roman Empire. They did not even found Turkish colonies. So few were the Turks who came to North Africa that one has a hard time discovering any of them there today.

The conquest of Morocco by France and Spain began in 1910. The dates for the termination of Moslem rule in the other countries are given above. Since these dates the number of Christian Europeans has grown from almost zero to about 1,400,000. The total for Christians and Jews is roughly 1,800,000. Only in Morocco can the figures for Europeans be separated from those for native Jews. Here we are told that the Jews number 180,000. Hence it is only reasonable to set their number at 400,000 for all of the North African countries.

Figures for about 1936 give the following:

	Moslems	Christians and Jews	Total
Morocco	6,700,000	580,000	7,280,000
Algeria	6,250,000	1,000,000	7,250,000
Tunisia	2,400,000	200,000	2,600,000
Tripolitania	850,000	50,000	900,000
Total	16,200,000	1,830,000	18,030,000

These are very rough figures, particularly for the former Italian colonies. More recent figures for Algeria show 9,500,000 Moslems and native-born Jews. Unofficial, reliable estimates indicate that these numbers may be much too low, that there may be as many as twelve million Arabs and Kabyles alone, with nearly two million Europeans and foreign-born Jews. Regarding the European population, thousands of refugees poured into North Africa during the last war, and it is a question how many of them have left. Census-taking in North Africa is no easy job.

In Algeria, between 1922 and 1935,

the birth-rate for Europeans fell from 28 per 1,000 to 21 per 1,000; it rose for natives from 25 per thousand to 35 per 1,000; the death rate for Europeans fell from 19 per 1,000 to 14 per 1,000; for natives from 21 per 1,000 to 18 per 1,000; finally, the rate of increase for Europeans fell from 9 per 1,000 to 7 per 1,000 while it rose for natives from 4 per 1,000 to 17 per 1,000.[2]

If we accept the figures tabulated above, which are surely not an overestimate, we see that in one hundred and twenty years the French have raised the European (and Europeanized Jewish) population to one sixth of the size of the Moslem population. The latter has of course also grown, and is still growing. In Tunisia and Morocco, held for seventy and thirty-five years, respectively, the ratio

[2] Werner Cahnman, *The Review of Politics* (Notre Dame, Ind.), VII (July, 1945), 343-57.

is twelve to one. In Algeria, where the French have been settled the longest, where Europeans are both absolutely and relatively the most numerous, where the chance for assimilation has been the greatest, political unrest is today the strongest, and the situation the most critical, of all North African countries.

The principal object of this series of papers is to evaluate the acculturation of the various non-European peoples of the World to the civilization of their European masters and mentors. In North Africa we can do this most clearly by narrating a few events which have taken place in the last few years.

On January 19, 1944, the representative of a political party handed the Governor-General of Morocco at Rabat a paper signed by fifty-seven of the most respectable and intelligent Moslems of the country, merchants, teachers, government officials, journalists, and the like (see p. 456). Three days later, French troops fired on a crowd at Fez, killing over eighty Moslems.

After one hundred and fifteen years of living together in closest intimacy, during which time Algeria had been made into three departments of Metropolitan France, a number of Moslems paraded in Setif before British and American soldiers, carrying an historic flag. The French fired without warning, and a riot ensued. Then the French army brought up tanks and planes, and destroyed forty-one Moslem villages and camps, from Setif to the sea. The dead were hastened into mass graves, and it is hard to tell how many were killed. One eyewitness account numbers them at thirty-five thousand, but even if this is exaggerated tenfold, there were three thousand five hundred too many.

On April 7, 1947, in the middle of the civilized city of Casablanca, where American soldiers and sailors had been killed by French gunfire four and a half years earlier, a number of Senegalese infantrymen "went berserk" and murdered sixty-five persons in the streets, at random. It is said that of these sixty-five, sixty-one were "Moroccan citizens." What the other four were the newspapers did not say. The reason for this slaughter is given as "a trivial incident."

What is wrong?

I could quote page after page of legally worded political documents, prepared by both sides, but they would not contain the answer. Nor can this state of affairs be blamed, so conveniently, on

German or Communist agitators. Nor on the half-dozen American scholars who have tried their best to learn Arabic and to study the native civilizations of North Africa objectively. Even they are not guilty, handy as their guilt would be. Neither they nor their government want trouble in North Africa. They may, however, help to find the cause, for this lies in a combination of all fields, of which the political is but a small part. In the pages that follow, I shall concern myself with the activities of the French in all of their North African regions, colony and protectorates alike, drawing my examples where I find them. This will be principally from Morocco, which I know best and where they have had their greatest success.

Let us begin with the utilization of the landscape. From the very first the French have been concerned with the deforestation and soil erosion so visible all around. Algerian highways are lined with beautiful rows of eucalyptus and other trees, and the same is true in Morocco and Tunisia. Wherever they have gone the French authorities have sought to preserve and increase the natural forests, by establishing *gardes forestières*, enforcing new legislation against indiscriminate felling, circling, and mutilation of the trees, and against the practice of burning over the brush each year to make pasturage. In Northern Tunisia, in the region of Cap Serrat, this prohibition appears to have caused a decrease in population, since the land is too sandy for tillage and now the grazing is reduced. However, in the preservation and increase of timber resources and in their effect on rainfall distribution, these measures taken by the French government are wise ones, to the advantage of all.

The lands which the Moslems themselves found most productive did not interest the French. These are the terraced mountain slopes, where the rainfall is relatively constant from year to year and the yield per acre high. What the French wanted were the flat lands which the Moslems used mostly for grazing. For example, the fertile Chelif valley was hardly tilled at all. With American agricultural machinery, the French, and particularly Alsatian, farmers who secured the land from its Arab owners ploughed deep furrows, harrowed the soil, shook out the grass roots, and planted vines. This valley is as favored for wine growing as the Napa Valley of California.

In the foothill region of the Middle Atlas, the country of the Beni

Mtir is also suited for viticulture. But the Beni Mtir were pasturing their sheep there and did not want to give up their country. All of a sudden, in 1926, they were required to pay their taxes ahead of time. It was too early for the sheep-clipping. They could not raise the money. Their land, once a fine source of mutton which Moroccans could eat, of hides for their belts, bags, and slippers, and of wool to make their clothing, now turns out quantities of wine each year. Most of this is drunk in Europe by Christians. It is possible nowadays to find a few men of the Beni Mtir working in the vineyards.

Besides the winelands, the French wanted wheatlands. On the windy plain around Setif, as on the rolling hills from Oujda to Oran, it grows high and full-eared. Here the French have acquired most of the land, and are using modern machinery. On the coastal plain of Tunisia and up the valley to Suk el Arba' and Ghardimou, one sees vast estates, with wheatfields interspersed between olive yards and orange groves, and big French houses in the midst of clumps of shade trees. On the Moroccan coastal plain, and inland as far as Fez, the land is mostly planted in grain.

We must remember that North Africa was the breadbasket of the Phoenicians, then of Rome, before the Arabs arrived. During the last century it has also been the breadbasket of France. It has produced great quantities of wheat, wine, and olive oil for export, as well as hides, wool, and mutton. The French have greatly improved agricultural techniques in the lowlands by the introduction of power machinery and by scientific methods of farming. A splendid citrus-fruit industry has grown up in Tunisia and in Morocco, to such an extent that fine oranges, tangerines, and grapefruit are sent to France. In the Sous valley of Morocco, Frenchmen grow tobacco under cloth, and also raise *primeurs*, such as fresh strawberries and artichokes, for the Parisian market. Furthermore many of the succulent soles eaten in Paris were caught off Fedhala.

Some of the holdings along the main highways of North Africa still belong to the Arab tribesmen, who work them mostly by hand. The greater number, however, are French. The French farmers are hard workers and their fields are models of tidiness and organization. The ways in which the French colonists acquired these estates vary greatly. Many paid fair prices for them, legitimately, to previous owners. Others settled on *ḥabus* lands, at the expense of the native charities, and when they improved the property it was theirs. Still others went into "partnership" with Arab owners. Arabs are forbidden by Koranic law to participate in usury, or pay or receive interest at any rate. Unable to invest in stocks and bonds, rich Arabs have long been in the habit of buying shares in agricultural properties and collecting shares of the crops. When the Europeans first came in, they were immune from the arbitrary and often crippling taxation of the native government. Hence many an Arab landowner was more than willing to sell to a European a 50 percent share of his property. In return he was protected by the immunity of his European partner. In the early days of Moroccan colonization, many adventurers took part in this kind of speculation, and they were not all French. A number of Englishmen also had their hands in this business. The Americans apparently did not know about it.

It would be very difficult to obtain accurate figures covering individual incomes and the distribution of wealth in North Africa. The richest men are undoubtedly some of the big *colons* with their huge, almost feudal estates, and some of the Jewish financial families like the Smadjas. Among the Moslems a few great chieftains whom the French have retained in power, like Sidi el Hajj Thami el Glawi, are probably very rich, and so are some of the American protégés, agents of the Singer and other companies. The rest, however, are abysmally poor. In the country they have little to do with money, raising their own food and weaving their own wool, and bartering in the local markets. It is a question whether or not they are poorer than they were before the French came. Whatever the comparative figure for the average Moslem, the variation must have increased, for many of the automatic mechanisms for the maintenance of economic equilibrium have ceased to operate, through the seizure of mosque lands in the country and the changes in status of the *ḥabus*.

During the recent war there was a tendency for native farmers to lose their lands and go to work for the *colons*, because of the scarcity of cloth and other vital commodities. The *colons* had control of the cloth and would distribute it to their workmen, while the others ran the risk of going unclad. I have seen Moslem women nearly naked in Tunisia through lack of cloth. I have given cloth to Moslems at Cap Serrat, and learned that it was taken away from them by French soldiers. The distribution of American goods in North Africa was a farce. In Oujda in 1943 while I was a member of General Clark's staff, I discovered that in Berkane, near by, a store selling American shoes, sent there for impartial distribution, had a sign up reading *Interdit aux Juifs*. There was no need to write *Interdit aux Mussulmanes*.

Returning to the subject of the utilization of the soil—we see that the Europeans pushed the Moslems off the more desirable lands by one means or another, just as we pushed the Indians onto reservations and then found ways of seizing or renting the best strips along the rivers. All moral aspects aside—as nations we are all tarred with the same brush—the North African Moslems are very different from most North American Indians. The Moslems are literate people with an ancient civilization and strong ties to other parts of the world; they are farmers and craftsmen, with all of the techniques our ancestors knew before the days of exploration and discovery of the sixteenth and seventeenth centuries. They are tied together by a common ritual language and a common religion. And they are so numerous that they will always form the majority of the population. If we are to compare them with American Indians, it should be with the Quechua-speakers and Aymaras of Peru and Bolivia, who after four centuries of conquest outnumber the descendants of the Spaniards.

The "natives" whom the French and others found in North Africa were highly skilled people. Some were very clever farmers, and the world contains none better at irrigation and the care of fruit trees. Others were good herdsmen, wise in the ways of animals; still others, in markets and towns, skilled craftsmen, artists at copper and leatherwork and at creating fine glazed pots with delicate designs in blue and green faïence. These men, with their nimble fingers and fine coordination of hand and brain, can learn with ease any technique known to modern man. They could make radio tubes or handle television sets. In 1926 I measured a fifteen-year-old Riffian boy in Ajdir who had been in charge of Abd el Krim's telephone service. By some underground channel they had received the instruments, wires, and batteries from Tangier. This boy, with others, had set up the machinery and had strung wires to Targuist, and thence over the mountains to the south. The line was used when

Abd el Krim's forces were advancing on Fez. The lad had all the enthusiasm for his telephone set that any American boy has for his radio, his model airplane, or his stripped-down "jaloppie."

The French soon saw the beauty in Arabo-Berber arts and crafts. They set up a special department of the government to see that these skills should not disappear. In Morocco the *Bureau des Arts Indigènes* held, before World War II, annual exhibitions and awarded certificates to the best craftsmen. In many a leather shop of Fez, you will see these certificates framed on the wall, a source of honest pride to their owners. Thanks to the French, fine pocketbooks and slippers are still made in Fez, cushions in Marrakesh, and pottery in Safi. In Sheshawen the Spaniards have done the same thing, specializing in textiles.

General Lyautey, like every other human being with any eye for beauty, and with any sense of human dignity and awe, was entranced by the vision of Fez, white and delicate, couched in its well-watered valley, a city which any infidel could see was holy. And holy he left it. Far outside its walls on the western plain he build a new Fez, a European town with government buildings designed in a neo-Moorish style, a happy blend of old shapes and tones with new conveniences. In Rabat, the new town, which was made the seat of government, rose outside the walls of the old. In Marrakesh, in Meknes, in all of the cities of Morocco this rule was followed. Even today you can walk down the vine-shaded streets and hear the hammer of the coppersmith and the cries of the leather auctioneer. You can smell the tart savor of cummin from an open door where someone is cooking.

To General Lyautey, and to other Europeans, it seemed a splendid idea to preserve the *status quo* in North Africa. The natives were Moslems and did not want any interference with their religion. Since their religion permeated all their daily thoughts and actions, these should not be disturbed either. By preserving intact the old cities, a medieval life could be carried on in them forever. Tourists could come from England and America to be shown these marvels by gallant French officers resplendent in natty red képis. They could then carry home trunkloads of pocketbooks and bales of rugs, leaving bales of dollars behind.

There seemed no reason to encourage machine industry in North

Africa. There was plenty of that in France. It would be best to use North Africa as a source of raw materials and a market for French manufactures. The Phoenicians and the Romans had set a precedent. And that is what happened. Even in Algiers there are many warehouses but few factories. In Casablanca one sees the phosphate works, but that is only part of the business of exporting raw materials. In Tlemcen, in 1944, a small woollen mill was about to be opened. Industrialization in a colonial region is an unpleasant business of low wages and poor working conditions, but at least it teaches people modern power-machinery skills, and that is a beginning. In North Africa it has barely begun.

In the garages of the modern cities, the man who gets under your car to drain out the oil is probably a Soussi. He is as skilled a mechanic as he is allowed to become. There is no school in which he can learn the automobile repair business, and he is not encouraged to compete with European labor, of which there is a surfeit. The Soussi can drive a car as well as you or I, but very few Moslem drivers are seen. That too is a privilege reserved as far as possible for Europeans.

The farmer would like to know something about fertilizers, about mechanical pumps for irrigation, and a number of other things that would increase his yield per hour of labor; he would like also to improve his cattle and sheep by the use of a good stud. In a few places, stud and veterinary services have been opened, but they are far from adequate, and the farmers are too poor, and for the most part too ignorant, to use them. What they need is agricultural schools, and these they do not have.

One would think that it would be a pleasure to teach such bright and receptive children as the sons of the city Arabs and mountain Berbers all about the world, but they have little chance for instruction. In the cities are schools for the sons of Arab notables, where they learn impeccable French, and in the mountains some of the Berber children learn good French too. The rest of them pick up French of sorts anyhow, using the second person singular in addressing everyone, the infinitive as the only verb form, and tossing back in innocence or ignorance the filthy expressions directed at them. During the war their vocabularies were increased by English obscenity. Besides the French language, they may learn a certain amount of French history and law. Law is the important subject. The Moslems who are sent to Paris to study seem to concentrate on law. There must be some free choice in this, and the only reason one can devise to explain it is that they are preoccupied with their inferior position under the French and want to better it by legal means.

Law is the shadow of other disciplines; it is the codification of the rules of behavior which people must follow if they wish to avoid a disturbance. It is hardly the subject for people who wish to change their social condition. They would do much better to study physics and chemistry, industrial and sanitary engineering, biology—theoretical and applied—agriculture, and medicine. But instruction in these subjects is unavailable or inadequate. They cannot even study English.

For the French, however, educational opportunities are not lacking. There is an excellent Lycée at Casablanca. At Oujda, General Clark took over a girls' lycée for his Fifth Army Headquarters. The University of Algiers is a renowned institution, and has included on its faculty such great scholars as Gautier the geographer and Reygasse the archaeologist. In Rabat an *Institut des Hauts Études Berbères* prepares officers to serve as colonial administrators. They have there been taught by such illustrious experts in North African civilization as Michaux-Bellaire and Laoust, the Moroccan counterparts of Henri Basset in Algiers. These men have issued the splendid journal *Hespéris*, found in many American libraries. In Rabat is also located an *Institut Scientifique Cherifien,* staffed with competent geologists, biologists, and the like. In the government Museum, Armand Ruhlmann devotes his very competent efforts to prehistoric archaeology. One could hardly say that either science or education were neglected by the French.

But the truth of the matter is that the Moslems have had little share in this development. Whereas, in Egypt, Moslems have taken over archaeology and have published excellent studies in physical anthropology, geography, and other subjects, nothing comparable has happened in North Africa. The Moslems in North Africa are no less competent in intellectual pursuits than their Egyptian brethren. They have had neither the stimulus nor the opportunity.

While one might cite a school here and there, or an educational program of sorts, nothing has been done to compare with the modernization of the Moslem universities in Egypt or the activities of the American University of Beirut. It is in the field of education for Moslems that the French have failed most visibly. Naturally it would have been futile to try to educate, along modern lines, mature men steeped in an ancient culture. But it would have been most easy and most intelligent, in a long-range sense, to have educated young people, born under European rule, in modern subjects, particularly in those fields in which religious conflict is at a minimum.

In the field of transportation the French have done very well indeed. Realizing the need of good roads if one is to conquer and remain in power, they collected hosts of Arabs and Berbers to sit along the roadsides cracking rocks with little hammers, at seven francs a day. When this hand-crushed stone had been spread, on came the macadam. Before World War II the road system and the filling-station equipment in French Morocco compared favorably with that of the United States. Since most of the cars were American, one had to look twice on the road between Rabat and Casablanca to make sure one was not driving down the coast of southern California. One saw almost as many Arabs riding in cars in California as on the Moroccan road.

In the two fields of family structure and religion, the French have been scrupulously careful to interfere as little as possible. They have been so sensitive to the idea that the faith of the Prophet must be carried on without change that they have made laws forbidding a Moslem to drink a glass of wine. While this may prevent trouble among soldiers and demoralized tribesmen, it is a source of embarrassment to a cultivated and sophisticated graduate of a European university. It violates the basic human right that people should be allowed to worship God in their own way. And it makes more trouble than it prevents.

In connection with family structure, the subject of intergroup marriage arises. In North Africa there has been very little intermarriage between Moslems and Christians. The religious barrier serves to reinforce the economic and political cleavage. I know an American Negro in Tangier married to a Sephardic Jewess. I have seen a Frenchman living in an extremely isolated spot in the mountains married to two Berber girls, apparently very happily. In the Grand

Atlas a number of Frenchmen live in a state of marital contentment with Berber wives. In Algeria, Kabyle men are sometimes
found living with French wives in their native villages. In Spanish
Morocco, there has been a considerable intermarriage in the Melilla
region between Riffian men and Spanish women. These examples
are all, except for the first, concerned with Christian-*Berber* unions.
These are rare, but commoner than unions between Christians and
Arabs. I personally know of only two of these which produced offspring, the marriage of the Sherif of Wezzan to an Englishwoman,
and that of a Fezzi cloth-importer to a Scottish lady. In both cases
the children considered themselves members of the Moslem community. By and large, marriages of this kind are very rare. The
Christian, Jewish, and Moslem communities keep themselves geneti-
cally distinct.

Missionaries have tried, from time to time, to break these barriers
down. Most of the missionaries have been British and American;
most of them have had very little success. In Tangier they have
taught a few street urchins to sing hymns and play soccer; in Kabylia
they have made a few converts. Christian ethics suit these people
less than Moslem ethics. Besides, the distaste for everything Christian has already been firmly implanted before the missionaries arrive.

In the political sphere the French have been overactive. The first
peoples whom they conquered were everywhere Arabs. As they
moved inland, they perpetuated the Arab form of government, by
setting up *quyad* and *khala'if* and *shuyukh* and *moqaddemin*, all absolute and each taking orders from the one above him in the hierarchy.
At the top of it the French officials could conveniently fit. This
system gave little latitude for free representation and justice. It
is often argued that the Arabs never had that before and are not
used to it, hence why should they receive it now? This argument
falls of its own weight. Furthermore, the premise is not true. In the
old days if a local leader became too oppressive someone killed him;
today no one would dare to do so.

As they moved farther inland, the French began to meet Berbers.
Since they had already learned Arabic on the way, they continued
to speak this language rather than try to learn an entirely new one,
and the Berbers, many of whom knew some Arabic anyway, quickly
accommodated themselves. Thus the Christians spread the language

of the preceding wave of conquerors where the latter themselves could not. The Berbers had a rather formless, intricate kind of government by councils and representation, which made them hard to deal with. It was much easier to pick some man who had helped the French conquer his own people, and reward him with the office of *qaid*. He would be unpopular, but he could be depended on. And so on down the line. To the Berbers, freedom had died. When the Christians entered, the jinns left the land.

Later on the French government found out the difference between Arabs and Berbers. More than a few French scholars spent years studying Berber dialects and Berber civilization, and wrote splendid and authoritative linguistic, historical, and anthropological monographs, at the cost of physical hardship and at the risk of their lives. These men made fast friends among the Berbers. But few government officials listened to them. Only when it seemed a bright idea to encourage Berber languages, Berber traditional law, and Berber civilization in general, as a means of checking the Arabs, through the policy of divide and rule, did these scholars really come into their own. Hence the Berber *dahir* (decree of the Sultan) of 1934, which infuriated the city Arabs and resulted in the formation of the Istiqlal or National Independence Party.

As time went on the French found out that their benevolent system of preserving the natives as a little historical gem, an anachronism for the delight of tourists, while a modern world of European North Africa grew up around them, was not working. And most of them, apparently, did not know why. Most of them still do not know. As trouble grew, the French redoubled their security methods. The Deuxième Bureau, the Sûreté, and a dozen other agencies with varying degrees of secrecy came over from France or arose on the spot. Millions of francs and more millions of units of human nervous energy were spent, and are still being spent, in spying on everyone and everything. Every native who has spoken to a foreign European must be hustled to the station and questioned, over and over. Every American who stays more than a few days, who is not obviously a tourist; his room must be searched while he is out. People must sidle up to him in the cafe and engage him in conversation. For some reason difficult to understand, the French in North Africa seem to consider the United States their Number One rival. An expression frequently heard in Morocco is *les maîtres futurs du pays*, in reference to Americans.

From all of this it is abundantly clear, at least to me, that the French government, whatever its motives, has made a number of basic mistakes:

(1) in trying to keep North Africa as a source of raw materials on which she could unload processed goods;

(2) in trying to keep the Moslems on a medieval level of technology;

(3) in restricting their liberties and spying on them so thoroughly;

(4) in being afraid of them.

The concept of a division of labor between industrial homelands and agricultural dependent areas was shown to be untenable years ago. That was the principal economic fact behind the American Revolution. That is why England and the Netherlands have had so much trouble in Asia, and France, again, in Syria and Indo-China. It is much more profitable to process the raw materials near the source of origin, and it creates much less disturbance with the local population.

The idea that intelligent human beings, who are skilled craftsmen and on the same level of cultural complexity as the ancestors of the French themselves before the Industrial Revolution less than six generations ago, can be kept on reservations like Plains Indians and trotted out to put on a show for the tourists is unsound and bound to make trouble. (I do not recommend this for the Indians either.) The Moslems can see what is going on around them. They want to drive automobiles, too. They don't want to be told by infidels that they must observe their own religion.

Like anyone else, if they are treated as equals and trusted they will behave accordingly. They do not like being constantly hauled into police stations and questioned. They do not like to have their meetings broken up; they do not like it when they are prevented from printing what they want to say in their newspapers. When they get four years in jail for reading a forbidden pamphlet, this does not endear the French government to them.

What the younger ones want is to be modern, to learn what the Europeans and Americans know, to take over our skills and professional competences. This will not interfere with their religion, any more than it interferes with ours. They are not telling us to go to church.

Along with all of these restrictions and with the resentment automatically engendered by them, comes fear. The French see the handwriting on the wall of colonies in general. They have lost Syria. They are having trouble in Indo-China. The Dutch have lost their hold in Indonesia, the British in India and Egypt. What the French have done is no more foolish than the deeds of the others. It is easy to be wise in retrospect, and about someone else's protectorate.

All Frenchmen are not stupid, selfish, or timorous. In more than one isolated outpost of the *Bureau des Affaires Indigènes,* the visitor will meet, if he is allowed to go there, a charming, cultivated, learned, sincere young man, living in the middle of a tribe of Arabs or Berbers, speaking their language intimately, listening to their troubles every day, respecting them and receiving from them respect, and even parental and filial love. Up on the pass of Tighime in Corsica I once saw a French officer in a conspicuous red cap sit smoking a cigarette while two German pursuit planes, swooping low over the gap, shot at him. When urged to duck, he replied, "I must set an example for my men." His men were Berbers from the Middle Atlas. He spoke to them in Berber. They were willing to die for him. Many of them did. No finer man than he ever lived. His name is Colonel de la Tour. In 1947 he became *Chef des Affaires Indigènes* at Rabat.

Up in the Spanish Zone of Morocco one of the leading officials of the *Asuntos Indigenas* is an equally fine man, who has lived with the Riffians ever since the surrender of Abd el Krim in 1926. When I first saw him, during that year, he was living alone in a tribe that had just given in. Some of them told me that were he any other Spaniard, they would kill him. This young man they could not bring themselves to touch. He talked with them as a brother, in their language, and sat on the ground with them while the wandering musicians played their flutes and guitars. He was a real man. That he had been born a Christian seemed a curious mistake, but it was probably all part of God's plan. And thank God he is still alive, for there is need of such a man.

Needless to say, all Frenchmen and all Spaniards do not measure up to the two I have chosen as examples. Few men of any nation could. The Moslems easily recognize the difference between Christians who are well-bred and those who are ill-mannered, grasping, and bumptious. Among the colonists in North Africa, the latter are in the majority. The situation might have been the same if the members of any other nation had settled there. How many of the G.I.'s stationed in North Africa had any regard for the "Ay-Rabs" or any concept of their lives and problems?

What is wrong here as nearly everywhere is not people but the system. If the Europeans had had enough sense to build schools and train the Moslems in technological disciplines, in agriculture, animal husbandry, transportation, and the use of machine tools of all kinds, then the study of law and other more abstract subjects could have come later. If they had kept the younger generations of Moslems interested and busy and had let them participate in the development of their own countries, if they had protected them by preparing them to protect themselves, they would have succeeded in their mission. And France would have gained mightily in the end.

The political situation in North Africa at the present moment is in such a state of flux that anything one could write would be out of date before it reached the printer. Working from east to west, we know almost nothing about the former Italian colony, except that the French seem to be in command in the oases and the British along the coast. The Russians have publicly stated their demand for this territory, and censorship seems to be in force. In Tunisia, the French are still ruling through the puppet-show of the new Bey, set up to replace the old one deposed during the war for German sympathies. The Neo-Destour (Constitution) party is being run from exile by Habib Bourguiba. In Algeria the principal native party is the PPA, *Partie du Peuple Algerien,* headed by a leader named Messali, a native of Tlemcen, who spent several years in exile in Brazzaville. Five or six members of this party have lately been members of the French Chamber of Deputies. Another party is the *Parti du Manifeste Algerien,* headed by Farhat Abbas. Their aim is the autonomy of Algeria with its own government and its own flag, within a French union. Eleven of this party were elected to the Deputies early in 1946, but when their program was rejected they failed to run again, and in the elections of November, 1946, no member of their party was returned. There is also an Algerian Communist party, under Omar Ouzgane, which has not particularly prospered.

In Morocco several different nationalist movements have now been merged under the Istiqlal Party. The nationalist leaders in exile—Ahmad el Fasi, Ahmad Belafrej, and others—have been released and some of them have gone to Cairo. The party had one free organ, the weekly newspaper called *Le Voix du Maroc*, published in Tangier by Moktar es Sadik Ahardan. It ran for a year and three months, in 1946–47, before it was suppressed. The Sultan has begun to take an active part in the Moroccan independence movement. On April 9, 1947, after many attempts, he succeeded in visiting Tangier, and making a public speech in which he stated the aspirations of Moslems in Morocco, their desire for an united empire, and their interest in the Arab League. This event was closely followed by the resignation of Gov. General Erick Labonne and the appointment of Gen. Juin in his place. Another event of equal if not greater moment was the escape of 'Abd el-Krim to Cairo on May 31, 1947. Since then the venerable Riffian strategist has collected about him a staff of Nationalist leaders, who hail him as chief. No one knows where this will end.

Meanwhile, the French Government has sent Yves Chataigneau, a career diplomat with ambassadorial rank, to Algeria as Governor-General. So solicitous is he of the welfare of the Moslems that many of the *colons* call him Mohammed Chataigneau behind his back. He may be able to help the critical situation out, but he is greatly handicapped by all that has gone before, and by the short-sighted attitude of most of his countrymen. In Morocco the French are making concession after concession, and trying hard to remedy what seems to be an impossible situation.

Despite these last minute attempts at patching the dyke, the flood will come. Relations between Europeans and Moslems are at the breaking point. Through second choice, the Tunisians, Algerians, and Moroccans are looking to the Moslem nations of the East for inspiration and support. If, a few years ago, they had had their choice between the technology of the West and its liberal intellectual outlook, on the one hand, and the more recent modernization of the East on the other, most of them would have chosen the West. But the choice has been denied them, they have been treated as children or as second class adults, *pas assez evolués*, and reaction has set in. Despite the appointments of Chataigneau, Juin, and de la Tour, it may be too late to mend the situation.

For any reader who may by chance wish to know what it was that the Arab handed the Governor-General on January 19, 1944, I append the following. The English translation was made by the Moslems themselves.[3]

Rabat, Jan. 11 1944

Ambassador Gabriel Puaux,
Resident General of the French Republic in Morocco,
Rabat

Sir,

By virtue of the mandate conferred upon us, we have the honour to inform you that we have deposited in H. M. Sidi Mohammed, our beloved sovereign's hands, the resolutions, copies of which are attached herewith.

We beg your Excellency kindly to transmit said copy to the President of the Committee of National Liberation and to bring to his notice the appeal we are making to the liberal and comprehensive spirit which animates all the French of the Resistance Movement, so that the Moroccan question be settled according to the righteous principles dominating all international relations.

We are convinced that our wish will meet with the cordial approval of General de Gaulle and your Excellency.

We also beg you to note that our Movement, aiming at the emancipation of our country within the framework of legality, is in nowise contrary to the legitimate interests of France in Morocco.

We think that the moment has come for France to reward the Moroccans for the blood they have shed, are shedding and will, if necessary, shed in the future, for the triumph of her ideal and her own liberty.

Your Excellency's obedient servants,

For the Supreme Council of the Independence Party,
OMAR ABDELJALIL
M'HAMED ZEGHARI
MOHAMED GHAZI

[3] *Istiqlal Party Documents* (Documentation and Information Office of the Istiqlal Party, Paris, September, 1946), pp. 1-5.

TEXT OF THE PROCLAMATION OF THE ISTIQLAL PARTY

The Istiqlal Party (Independence Party) which includes the members of the ex-National Party and independent personalities;

Considering that Morocco always constituted a free and sovereign State and that it preserved its independence during thirteen centuries, until the time when, under particular circumstances, a regime of Protectorate was imposed upon it;

Considering that the main object of this regime was to endow Morocco with a series of administrative, judicial, cultural, economic, financial and military reforms, without prejudice to the traditional sovereignty of the Moroccan people, under the overlordship of its King;

Considering that the Authorities of the Protectorate have substituted to this regime a regime of direct administration and arbitrary power for the benefit of the French residents, whose plethoric functionaries are to a large extent superfluous; considering that no attempt has been made to conciliate the divergent interests in presence;

Considering that owing to this system, the French Colony has been enabled to seize all public powers and take possession of the natural resources of the country, to the detriment of its inhabitants;

Considering that the regime thus established has tried to split by divers means the unity of the Moroccan people, that it has prevented the effective participation of Moroccans in their country's government and that it has deprived them of all public and individual liberties;

Considering that the present circumstances are unlike those of the Protectorate's beginnings;

Considering that Morocco efficiently participated in the World Wars on the side of the Allies; that its troops have recently accomplished exploits which aroused general admiration, in France as well as in Tunisia, Corsica, Sicily, Italy, and that even greater participation on other battlefields, to hasten France's liberation, is expected from Morocco;

Considering that the Allies, who are shedding their blood for the cause of Liberty, have recognized, in the Atlantic Charter, the right of peoples to freedom, independence and that recently, at the Teheran Conference, they have proclaimed their reprobation of the doctrine, according to which the strong shall dominate the weak;

Considering that the Allies have manifested, on several occasions, their sympathy towards the peoples whose historical patrimony is less substantial than ours and whose degree of civilisation is on a lower level than Morocco's;

Considering finally that Morocco constitutes a homogeneous unit which, under the lofty leadership of its Sovereign, takes cognizance of its rights and obligations in internal as well as international matters and appreciates the benefits of democratic liberties which are in conformity with the principles of our religion and have served as a basis for the constitution of all Moslem countries.

DECIDES:

A.—Concerning general policy:

1.—To ask for the Independence and territorial integrity of Morocco under the leadership and guidance of H. M. Sidi Mohammed Ben Youssuf, whom the Almighty shall exalt;

2.—To solicit H. M. to enter into negociations with the interested Nations tending to the recognition and safeguard of this independence, as well as to the fixation, within the framework of national sovereignty, of the legitimate interests of aliens residing in Morocco;

3.—To ask for the adhesion of Morocco to the Atlantic Charter and for its participation in the Peace Conference.

B.—Concerning internal policy:

To solicit H. M. to take under his high patronage the reforms indispensable to insure the country's prosperity.

The Istiqlal Party leaves to H. M. the task of establishing a democratic regime similar to the form of government adopted in the Moslem countries of the Orient, safeguarding the rights of all elements and all classes of Moroccan society, and defining the obligations of each.

Rabat, Moharrem 14, 1363 (January 11, 1944).

For all sections of the Istiqlal Party and all regions of Morocco
[Signed]

MOHAMMED LYAZIDI, Member of the Executive Committee of the former National Party

HADJ AHMED CHERKAOUI, Member of the Supreme Council of the ex-National Party, Director of the Rabat School

HADJ AHMED BALAFREJ, Member of the Executive Committee of the ex-National Party, B.A., Graduate of the Hautes Études of the Sorbonne, Director of the Institute Mohamed Guessous

MOHAMED GHAZI, Member of the Executive Committee of the ex-National Party, Editor of the Review *Rissalat el Maghreb*

ABDELKRIM BENJELLOUN TOUIMI, B.A., B.L., Judge of the Cherifian Supreme Court

ABDELKEBIR EL-FIHRI EL-FASSI, Judge of the Cherifian Supreme Court

ABDELJLIL EL-KABBAJ, Inspector of the Habous, Honorary President of

458 CARLETON S. COON

the Association of Former Students of the Moulay Youssef College
of Rabat

ABDELLAH ERRAGRAGUI, Secretary of the General Library

MESSOUD CHIGUER, Secretary of the Makhzen Central

EL-MEHDI BEN BARKA, President of the Association of Former Students
of the Moulay Youssef College of Rabat, Member of the Government
Council, Professor of the Imperial College and the Lycée Gouraud

[Here follow the names of forty-seven other men, from every class and
calling, including some of the most distinguished educators, jurists,
and government officials in Morocco.]

SUGGESTED READINGS

Bates, Oric. The Eastern Libyans. London, 1914.

Bernard, Augustin. "Rural Colonization in North Africa (Algeria, Tunis,
and Morocco)." In *Pioneer Settlement*, American Geographical Society of New York, Special Publication No. 14. New York, 1932.

Brunel, René. Essai sur la confrérie religieuse des 'Âissâoûa au Maroc.
Paris, 1926.

Brunn, Daniel. The Cave Dwellers of Southern Tunisia. Tr. L.A.E.B.
London, 1899.

Cahnman, Werner. "France in Algeria." *The Review of Politics*, VII
(1945), 343–357.

Cline, Walter B. "The Arab Tribal Community in a Nationalist State."
The Middle East Journal, I (1947), 18–28.

Coon, C. S. Tribes of the Rif. Peabody Museum of Harvard University,
Harvard African Studies, Vol. IX. Boston, 1931.

Crist, Raymond E. "Land Tenure in Tunisia; Inter- and Intra-National
Implications." *The Scientific Monthly*, LII (1941), 403–415.

Despois, Jean. "Types of Native Life in Tripolitania." *Geographical
Review*, XXXV (1935), 352–367.

Fogg, Walter. "Changes in the Lay-out, Characteristics, and Function of
a Moroccan Tribal Market, Consequent on European Control." *Man*,
XLI (1941), 104–108.

—— "The Importance of Tribal Markets in the Commercial Life of the
Countryside of North-west Morocco." *Africa*, XII (1939), 445–449.

Forbes, Rosita. The Sultan of the Mountains; the Life Story of Raisuli.
New York, 1924.

Gautier, E. F. "Native Life in French North Africa." *Geographical Review*, XIII (1923), 27–39.

—— "Nomad and Sedentary Folks of Northern Africa." *Geographical Review*, XI (1921), 3–15.

—— Le Passé de l'Afrique du Nord; les siècles obscurs du Maghreb a Paru en 1927. Paris, 1937.

—— Sahara, the Great Desert. New York, 1935.

Ghirelli, Angelo. El Norte de Marruecos. Melilla, Morocco, 1926.

Guyot, R., R. Le Tourneau, and L. Paye. "Les Cordonniers de Fès." *Hespéris*, XXI (1935), Parts I–II, 117–240.

Halpern, Manfred. "The Algerian Uprising of 1945." *The Middle East Journal*, II, No. 2 (1948), 191–202.

Hanoteau, A., and A. Letourneaux. La Kabylie et des coutumes kabyles. 3 vols. Paris, 1872–73.

Hilton-Simpson, M. W. "The Influence of Its Geography on the People of the Aures Massif, Algeria." *Geographical Journal*, LIX (1922), 19–34.

Jackson, E. P. The Moroccan Atlas: a Study in Mountain Geography. Michigan Academy of Science, Arts, and Letters, *Papers*, XVIII (1933), 209–237.

Laoust, E. Mots et choses berberes. Paris, 1920.

Liebesny, Herbert J. The Government of French North Africa. African Handbooks, 1. University of Pennsylvania Press, University Museum, 1943.

Lissauer, Abraham. "The Kabyles of North Africa." *Annual Report of the Smithsonian Institution* (for 1911), pp. 523–528.

MacKay, Donald. "The French in Tunisia." *Geographical Review*, XXXV (1945), 368–390.

Meakin, Budgett. The Moors. London and New York, 1902.

—— The Land of the Moors. London and New York, 1901.

—— The Moorish Empire. London and New York, 1899.

—— Morocco-Arabic Vocabulary. London, 1891.

Mellor, F. H. "The French Protectorate in Morocco." *The Geographical Magazine*, II (1935–36), 173–193.

Merry del Val, A. "The Spanish Zones in Morocco." *Geographical Journal*, LV (1920), 239–349, 409–422.

Michaux-Bellaire, A., "Essai sur l'histoire des Confréries marocaines." *Hespéris*, I, (2d Quarter, 1921), 141–159.

Montaigne, Robert. "Evolution in Algeria." *International Affairs*, XXVIII (1947).

"Social Life and Customs in North Africa." In *Butrava*, No. 8 (1944), pp. 10–14. Reprinted from *Official Guide for U. S. Military Personnel, World War II*.

Tharaud, Jérome, and Jean Maraud. *Fez, ou les bourgeois de l'Islam.*
Paris, 1930.

Le Tourneau, R., L. Paye, and R. Guyot. "La Corporation des tanneurs
et l'industrie de la tannerie à Fès." *Hespéris,* XXI (1935), Parts I–II,
117–240.

Westermarck, E. A. Ceremonies and Beliefs Connected with Agriculture,
Certain Dates of the Solar Year, and the Weather in Morocco. Hel-
singfors, 1913.

—— Pagan Survivals in Mohammedan Civilization. London, 1933.

—— Ritual and Belief in Morocco. London, 1926.

Wysner, Glora. The Kabyle People. Hartford, Conn., 1946.

"The Zkara, a Christian Tribe in Morocco." The American Geographical
Society, *Bulletin,* XXXVI (1905), 343–345.

F. L. W. Richardson, Jr.
James Batal

THE NEAR EAST

Introduction, by F. L. W. Richardson, Jr.[1]

THE NEAR EAST, now also called the Middle East, is an area of continual surprises and bewildering contrasts. To the archaeologist, it is magnificent ruins in living squalor; to indigenes, disease and poverty; to imperialists, power and wealth. For Christian pilgrims, it is Jerusalem, the church of the Holy Sepulchre where Moslems keep peace among rival Christians; for world Jewry, it is Palestine reborn. In Egypt the rich international set find winter sun, palace parties, and gambling casinos; while tourists take in camel riding, pyramids, Luxor, and stomach dancers. Those who explore the desert discover isolated fortresses that turn out to be monasteries millennia old, with living monks, or discover an exquisite blue mosque with golden dome rising from a fetid town.

Geographical and Historical Setting

The Near East, a region larger than the United States, commands a central world position astride three continents. It extends north-south from the Black and Caspian Seas to the Indian Ocean and Negro Africa and west-east from North Africa to India and Tibet. There are nineteen countries, colonies, dependencies, or principalities in the Near East.[2] Seventeen of these lie wholly within the area and two partly within it—one, the USSR south of the Caucasus Mountains, and the other the northern part of the Anglo-Egyptian

[1] The writer gratefully acknowledges the assistance of James Batal for his valuable suggestions and editing. Likewise the writer is indebted to several other friends, associates, and teachers for their criticisms, particularly to Professor Philip Hitti of Princeton University, Prof. E. A. Speiser of the University of Pennsylvania, and Laureston Ward of Harvard University.

[2] Not counting the tri-partition of Palestine passed by the United Nations in November, 1947.

Sudan bordering Egypt. (See Table 1.) Of these nineteen political units, all are Arabic-speaking except Afghanistan, Iran, Turkey, and the USSR. With the exception of Lebanon and Russia, all are predominantly Moslem and most proclaim Islam as their state religion.

To a considerable extent the Near East has been isolated by formidable barriers from the peripheral populations of the three continents it straddles. In early times these barriers kept the Near East sufficiently free from invasion to allow its peoples to have created what was at one time perhaps the finest world civilization. Between the Near East and the rest of Africa stretches the world's greatest desert, the Sahara; between the Near East and much of Asia are the world's highest mountains, the Himalayan and Tibetan wall. The world's largest system of inland seas, the Mediterranean, Black, and Caspian, plus the Caucasus Mountains separate the area from Europe. Throughout history three gateways have carried practically all the in-and-out traffic: first, the fording of the narrow, river-like strip of water between Turkey and Europe; second, the grassy desert between Iran and Siberian Russia; and third, the inland seas, notably the Mediterranean. With improvements in transportation technology, increasingly more people have poured through these gateways. The first was forded from earliest times; horsemen pounded through the second more than a millennium before Christ; and by Greek times the Mediterranean was becoming less and less a barrier to peoples and more and more the most used highway in and out of the Near East. In fact in modern times the Mediterranean has been like a giant pipe-line pouring Western peoples, armies, technologies, ideas, and intrigues into the Near East.

Although its area is vast, the territory of the Near East is largely sea and desert. The bulk of the population live crowded on the oases which dot the desert like green islands in a brown sea. (See Map 1.) Like an archipelago these verdant islands stretch from the thickly cultivated European mainland to humid central Africa. Each "island" contains a population ranging from a hundred to millions. (See Map 2.) Cultivated Egypt is reputed to be one of the most densely populated regions in the world.

The effective cell unit of political organization, controlled by one or more cooperating overlords, has always been the cultivated "island"; that is, city, state, or combination of adjoining "islands"

Humid northern and
southern border areas

European
cultivation

Tropical jungle
and forest

Mountains and culti-
vated islands

Oases valleys mountains

Rivers foothills

Equator

Raisz

MAP 1. NEAR EAST ENVIRONMENT

A scattering of green cultivated "islands" in a water and desert sea, lying between heavily
cultivated Europe—Turkey and humid Africa.

—Babylon, Persepolis, Baghdad, Palmyra, Petra, Baalbeck, Damascus.

TABLE I

Area, Population, and Government of Near East Political Divisions

Country	Area (sq. mi.)	Population (estimates)	GOVERNMENT	
			Form	Date Established
ADEN				
1) Colony, and	75	50,000	British Colony	
2) Hinterland	112,000	700,000	British Protectorate (incl. about 25 native rulers)	1839
3) AFGHANISTAN a	250,000	10,000,000	From Absolute Monarchy to Constitutional Monarchy	1922
4) ANGLO-EGYPTIAN SUDAN a	323,000 b	2,000,000 b	British and Egyptian Condominium	1899
5) BAHREIN Is.	225	80,000	British Protectorate (a ruling Sheikh)	1820
6) EGYPT	386,000	18,000,000	Constitutional Monarchy	1922
7) IRAN a	628,000	15,000,000	Constitutional Monarchy	1925
8) IRAQ	143,000 (?)	4,000,000	Constitutional Monarchy	1932
9) KUWEIT	2,000	70,000	British Protectorate (a ruling Sheikh)	1898
10) LEBANON	3,500	1,000,000	Republic	1941
OMAN COAST				
11) Muscat Oman	82,000	} 600,000	British Protectorate (a ruling Sultan)	early 1800s
12) Trucial Oman and Qatar	?		British Protectorate (about 8 independent Sheikhs)	1820
13) PALESTINE	10,000	1,600,000	British Mandate c	1923
14) RUSSIA a	..	..	The 25 million Moslems in the USSR are not included here.	
15) SAUDI ARABIA	700,000	4,500,000	Absolute Monarchy	1926–1932
16) SYRIA	58,500 (?)	3,000,000	Republic	1941
17) TRANSJORDAN	34,700	400,000	Constitutional Monarchy	1946
18) TURKEY a	297,000	19,000,000	Republic	1923
19) YEMEN	100,000	4,000,000	Absolute Monarchy (Theocracy)	Post-World War I
Total	3,130,000	84,000,000		

a The three non-Arabic-speaking countries entirely within the Near East (Afghanistan, Iran, Turkey) and two countries partly within the Near East (USSR, Anglo-Egyptian Sudan) are in various degrees buffer states or transitional in culture to regions contiguous to them; that is, Turkey is transitional to Europe, Afghanistan and Iran are buffer states between India and Russia, and the northern part of the Anglo-Egyptian Sudan is transitional to Negro Africa. The USSR, and the most strongly transitional—Afghanistan and the Anglo-Egyptian Sudan—are excluded for the most part from the present discussion. Turkey is included only in the over-all treatment of the region, but excluded from the second section devoted mainly to a description of the Arab core of component states.

b Including only about one-third.

c Future status under debate in the United Nations, October, 1947.

TABLE 2

Foreign Trade of Near Eastern Countries with Different Regions

(Imports and Exports Combined, in Thousands of British Pounds, as of 1937; for Iran only, as of 1935–36)

PART I: BY PERCENTAGE

	Near East	Rest of Asia	All Europe	Western Europe	USSR	Balkans Poland	Miscellaneous	Not Listed	Total
EGYPT	4	10	76	70	1	5	7	4	
PALESTINE AND TRANSJORDAN	17	3	70	59	..	11	6	5	
SYRIA AND LEBANON	29	10	47	42	1	5	9	5	
IRAQ	10	23	48	47	1	..	13	6	
IRAN	9[a]	12	58	39	19	..	8	13	
TURKEY	3	2	78	70	6	3	15	2	
TOTAL OF ABOVE COUNTRIES	8	9	68	59	5	4	9	5	

PART II: BY TOTAL VALUE

	Near East	Rest of Asia	All Europe	Western Europe	USSR	Balkans Poland	Miscellaneous	Not Listed	Total
EGYPT	2,528	7,200	53,763	49,446	472	3,845	4,905	2,621	71,017
PALESTINE AND TRANSJORDAN	3,098	530	13,104	10,987	..	2,117	1,134	836	18,702
SYRIA AND LEBANON	4,365	1,503	7,043	6,245	70	728	1,315	747	14,973
IRAQ	1,544	3,505	7,235	7,169	66	..	1,877	979	15,140
IRAN	3,336[a]	4,217	20,444	13,700	6,744	..	2,791	4,479	35,267
TURKEY	1,200	730	32,010	28,820	2,150	1,040	6,260	900	41,100
TOTAL OF ABOVE COUNTRIES	16,071	17,685	133,599	116,367	9,502	7,730	18,282	10,562	196,199

[a] Almost entirely oil exports to Egypt.

Sources. EGYPT: Economic Research Institute, *Statistical Handbook of Middle Eastern Countries* (Hamadpis Liphshitz Press, Jerusalem, 1944), p. 71. Exchange rate, 975 £.E. per 1,000 £.B.
PALESTINE AND TRANSJORDAN: *The Statesman's Yearbook*, 1939 (New York, Macmillan Co.). Exchange rate, 1 £.P. per 1 £.B.
SYRIA AND LEBANON: *Report on Economic and Commercial Conditions in Syria and Lebanon, 1936–1938* (Department of Overseas Trade, H.M. Stationary Office, London, 1938, No. 710), pp. 29–31. Exchange rate, 641 £.S. per 1 £.B.
IRAN: *Report on Economic and Commercial Conditions in Iran, 1937* (H.M. Stationary Office, London, 1938), pp. 31–32, 35. Exchange rate, 80.5 rials per 1 £.B.
IRAQ: League of Nations Publications, 1937, II (3) Quarto vol., A. No. 19, *Economic Intelligence Service* (Geneva, 1938), p. 216. Exchange rate, 1 dinar per 1 £.B.
TURKEY: *Ibid*, p. 348. Exchange rate, 6.16 £.T. per 1 £.B.

But despite this ever-present island loyalty, the disposition of seas, deserts, and mountains has grouped these green islands into four areas peripheral to one common center. The four peripheral areas are Arabia, Egypt-Sudan, Turkey, Iran-Afghanistan. Through thousands of years of history, each has had relatively greater independence, homogeneity, and continuity than their common center comprising Palestine, Transjordan, Lebanon, Syria, and Iraq. The latter have been fought over ceaselessly and handed from conqueror to conqueror. As shown on Map 2 this geographical and historical grouping of city states into nations has been likened to four houses around a common courtyard.

The semi-isolation of these four houses (Map 2) has been furthered by the fact that all are contiguous to large, fertile regions —Turkey to Europe; Iran-Afghanistan to India and Russia; Egypt to the Sudan, to Ethiopia, and by sea to Europe; Arabia to Ethiopia. However, only in the case of Turkey and Iran has the actual connection been important.

Excepting Arabia (for which there are few statistics), each of the "houses" is strongly tied to its rich neighbors by trade (see Table 2). Egypt and Turkey are tied to Europe with 76 and 78 percent of the foreign trade, respectively, going to European markets; Iran to Russia, with 19 percent, or three times the percentage of any other country's Russian trade. All these "houses" have very small percentages of foreign trade with other Near East countries: Egypt 4 percent, Turkey 3 percent, and Iran 9 percent. The latter, however, would be far less were it not for petroleum exports to Egypt, a new and special development.

In contrast, the five central courtyard countries trade proportionally more with other Near East states, an indication of their closer internal ties—Palestine and Transjordan combined, 17 percent; Syria and Lebanon, 29; and Iraq, 10. Iraq with its location at the head of the Persian Gulf has by far the largest trade with Asia outside the Near East. Twenty-three percent of Iraq's total foreign trade was with Asiatic countries, especially Japan and India.

Although Europe dominates the trade of all Near Eastern countries, (Transjordan excepted),[3] geographic location distinctly

---

[3] Fifty-six percent of the foreign trade of Transjordan, for example, was with Palestine in 1938 (*The Statistical Handbook of Middle Eastern Countries*, p. 123).

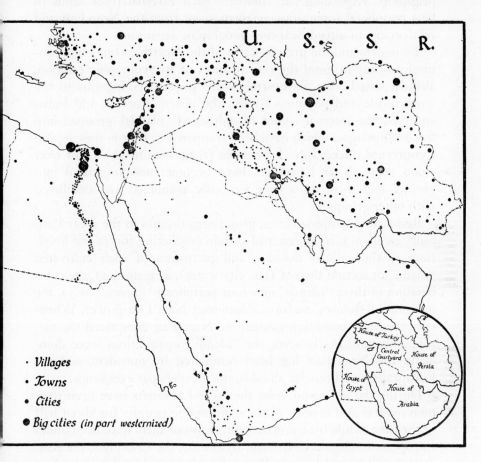

MAP 2. VILLAGE AND URBAN POPULATION

Centers of population concentrated in cultivated "islands." The distribution of these centers can be conveniently laid out into four areas peripheral to one central area—likened to four houses around a central courtyard (see inset).

Based on maps 1:4 million, Asia, G.S.G.S., London.

weights all trade connections. The "houses" favor contiguous external powers, and the courtyard countries their Near Eastern neighbors. Regarding the "houses," their external trade tends to be a peripheral force drawing them away from the Near East and reinforcing an already existing geographic separation.

To understand the present status of the Near East one must continue to bear in mind the two features of its geographic position, already stated: one, its central world position contiguous to the more fertile and populous Russia, the rest of Europe, and India; and two, the internal layout of cultivated "islands" grouped into five archipelagos, which have been compared to four houses around a courtyard. Geography has set the frame within which, for over 5,000 years, Near Easterners have become mass organized into groups, and these units have variously combined and conflicted with other groups.

Reduced to simplest terms, group organization in the Near East, today as always, is the result of certain conflicting forces and loyalties: 1) the cellular isolation and patriotism of each cultivated "island" or section thereof (i.e., city states), as against, 2), the combination of these "islands" into four peripheral "houses," or 3), the merging of "houses" to form a dominant Near East power. Whenever the latter condition existed, the Near East conquered the surrounding world; whenever the "island" organizations were dominant, the Near East has been conquered by outsiders, and the Europeans have been, by all odds, the invaders par excellence.

During the last 2,000 years the Near Easterners have been their own masters and masters of the surrounding peoples for about half of the time, while for the other half, many of them have been under European domination—Roman, Byzantine, the Crusaders, the Russians, English, and French. Power has alternated between the East and the West. The last alternating cycle began a little after 1300 when Near Easterners had expelled Crusaders, driven out and absorbed Mongols, and in turn themselves dominated much of Europe, North Africa, and India. The energy behind this drive finally dwindled, and, beginning about 1800 with Napoleon, Europe began to inundate the Near East with armies, navies, traders, administrators, and modern technologies. Map 1 illustrates four progressive stages in this, the most recent European domination of the Near East.

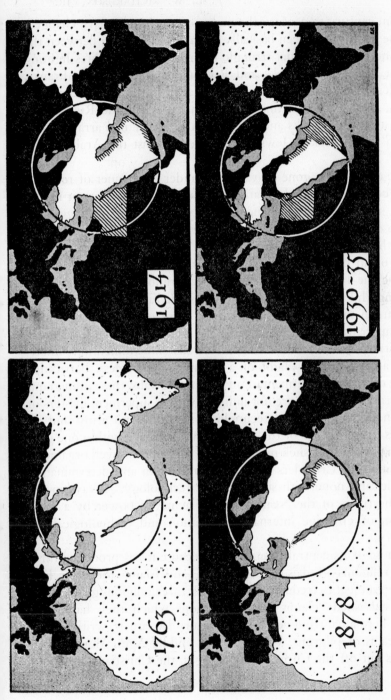

MAP 3. PROGRESSIVE EUROPEAN ENCROACHMENT ON THE NEAR EAST

Circle encloses Near East Area. Black indicates territory controlled by Europeans and Russians; white, territory controlled by Near Easterners, and stippled, territory independent of all these. Before 1800, Near Easterners were masters in the Balkans, North Africa, and Northwest India. Since about 1800 to 1930–35, Europeans have been steadily wresting control from practically all peoples around the Near East and many within it.

From century to century the names of contestants have changed, but the conflict has been basically the age-old struggle between the East and the West. That struggle has resulted in Israelites fighting against Philistines; Greeks and Romans against Persians; Franks, Cruaders and young European powers against Moors, Saracens, and Turks; and, in modern times, Western European and Russian imperialists against Turks and Persians. And finally it is arraying European and American Jews of the West against the Arabs of the East.

The conflict has gone on variously under the banner of religion, trade, conquest, and colonization, but whatever the name, it results in the same basic phenomenon, namely *foreigners divesting local leaders of an authority they are unwilling to relinquish*. Seen in this light, the present-day Zionist program is but the most recent outbreak of an old, old struggle. Ironically enough, it is, in reverse, the one the Hebrews themselves faced originally against the sea-invading Philistines. Inevitably, to Arabs, the Zionists are the crowning burden in the century-and-a-half curse of Western conquest and imperialism, with the added sting that indigenous Jews are a minority easily controlled. Little do Arabs (not to mention Americans) appreciate the plight of a decimated, homeless, barricaded people on another continent. To Palestinian Arabs, these immigrant, Western-trained Jews are not unlike invading supermen, masters of modern organization and technology. Probably no army and no peoples in the history of the whole Near East have within one generation poured in in such numbers, certainly none as well organized, none since the Crusaders with the avowed aim and determination to settle, and none with such superior technology. Always in the past, the rulers of the Near East have achieved power by a combination of effective internal organization and technological superiority.

Because of its central position and because it has produced three world religions—Judaism, Christianity, and Islam—the Near East has been subject to frequent invasion. All three religious groups have long established "claims" or connections which they have constantly kept alive from century to century by rituals, writings, and by streams of pilgrims. When, in the Middle Ages, Moslem conquerors cut off the Christian pilgrimages, the Christian world was understandably inflamed. With famine and pestilence in their midst, and with church, military, and commercial leaders spurring them on, little wonder the masses joined the Crusades. When, through pogroms, millions of European Jews were massacred, burned, and butchered, little wonder that they too turned to the land which they have always prayed to restore to the glories it achieved in the days of Solomon. This yearning became more acute as other countries closed their doors to Jewish refugees.

Thus the Near East suffers from world disturbances brought on not only by itself, but by peoples beyond its borders and over whom it has no possible control. The Mongols, driven by droughts, progressive dessication, political troubles, and the like, started on their world conquest and poured into the Near East. In recent times, the Near East has been a battle ground in two world wars, in the most recent of which the independent states remained neutral, with certain short-lived exceptions. Even the United States was drawn in when American troops manned the Persian railroads and numerous airports throughout the area. And America is leasing in Arabia what is perhaps the world's biggest oil pool. How free is the Near East?

In its role as a center of world peoples, powers, and intrigues, the Near East has alternately been a power welling up and overflowing its boundaries with peoples, religions, and armies, and, as at present, a vortex or power-vacuum into which has poured the peripheral pool of world's peoples, armies, ideas, and technologies. Map 4 is an attempt to symbolize the historical role the Near East has played in the old world alternating between a source of power and power-vacuum.

In this ceaseless internal and external power struggle, it has been mainly the overlords and the city dwellers, a mere handful of the population, who have benefited or suffered. The mass of the people, by and large, have been little affected. The bulk of the present eighty million or more are still impoverished and disease ridden. Incapacitated and debilitated by eye infections; parasitic diseases such as malaria, hookworm, bilharziasis; epidemic diseases such as typhus, typhoid, and cholera; and venereal diseases—it is small wonder that they put their trust in God, in Allah, or even propitiate the Devil, as is the practice of one group. The peasants can put little trust in their earthly overlords, and they themselves have neither knowledge

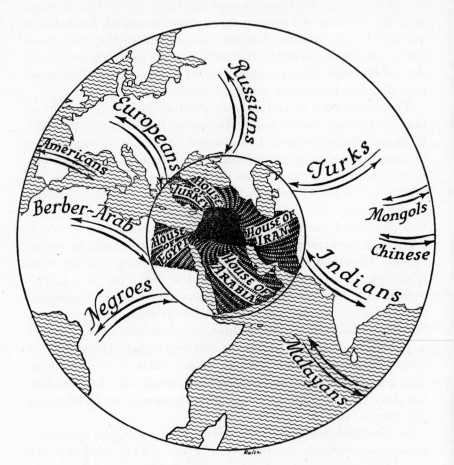

MAP 4. CENTRAL WORLD POSITION OF THE NEAR EAST

Outer circle includes the Old World; inner circle, the Near East; solid black center represents the most frequently contested area within the Near East, here called the Central Courtyard.

Arrows in the outer circle indicate the movement of peoples in and out of the Near East. The Near East has been alternately a vortex into which world populations pour and a fountainhead from which Near Easterners themselves overflow to regions outside.

of germs nor of medicine. To them this condition is merely a part of life, sanctioned, or at least tolerated, from on high. To Western eyes, theirs is a spineless submission, the blind faith of ignorance, but to them religion is strength, security, and salvation. If one offers to take a peasant from one of the backward countries to a doctor to cure his growing blindness, he is very likely to refuse treatment, saying with superior, resigned conviction, "If Allah wants me to be blind, blind I'll be."

Withal, the Near Easterners have attained an adjustment that many of us fail in attaining, an adjustment in which villagers and tribesmen are more contented, if not more cheerful, than many groups in our own Western world, and in which nervous disorders are a rarity.

World Importance

With a land area largely desert, with people impoverished and disease-ridden, why since the beginning of time should anyone want to stay in the Near East who had any other place to live or work? And especially why have Westerners been trooping in in such large numbers for the last 150 years? Can it be oil, as the Near East is one of the world's richest petroleum centers, if not the richest? Britain, Russia, France, and now the United States, all have a stake in its oil development. Great Britain is dependent on Near Eastern oil and, at our present rate of consumption and known reserves, the United States may well be dependent upon it in the near future. However, oil is a recent arrival on the Near East scene. Western European nations long before the oil age had been jockeying for power within the Near East.

In addition to oil, several other products are important in present day commerce—Egyptian cottons, Iraq dates, Persian carpets, Turkish chrome, Palestine citrus fruits. Other items still present are reminders of by-gone days of splendor—frankincense and myrrh from Arabia and pearls from the Persian Gulf. In Roman times the Near East was important as a granary, but grain raising became unimportant here after the steel plough made the rich prairies and grasslands of the world into wheat fields.

A glance at Table 2 reveals that over two-thirds of the imports and exports of eight Near East countries is with Europe. But the total foreign trade is not great; for example, the Netherlands Indies and British Malaya combined, with about the same population and a fourth the territory, carry on almost twice the volume.

A glance at Map 5 demonstrates that the major reason for the importance of the Near East is not to be found within the area itself, but outside. Note its central geographical position, between torrid Africa, raw-material producing Asia and manufacturing Europe, areas which contain more than half the world's population (550,-000,000 Europeans and Russians, and 400,000,000 Indians, to say nothing of the uncounted Chinese and Africans).

With the Suez Canal connecting the Mediterranean and the Red Sea, the Near East is the funnel through which passes practically all commerce between Europe and the East. In 1937, the net tonnage passing through the canal totaled 35,269,000, representing 69 per-cent of the net tonnage of all foreign trade entering American ports.[4] This is the commerce stimulated by Western development of Eastern raw materials—originally spices and luxuries, more recently rubber, ore, tea, cotton, grain—most frequently shipped to western Europe for consumption or manufacture into finished products. Some of the latter, such as cotton cloth and tires, are then reshipped throughout the world, many of them to regions which supplied the original raw material. It is a trade that has enriched every country fortunate enough to control a good share of it—Venetians, Portuguese, Dutch, English, to name only a few.

The surest way to get rich from this trade is to control all the sources of supply and all the routes. Now in the air age, with the advantages of a desert climate, minus rain and fog, the central Near East is the focus of world air routes. To control the water and air routes requires political power. And to achieve and maintain political influence usually requires political intrigue. No power is secure with other powers pressing for advantages. As Speiser has said, "Participation in the doings of the Near East has come to be equivalent of a seat on the world's geostrategic exchange." [5]

There is an additional reason why the Near East is important

---

[4] *Statesman Yearbook,* 1939, for the Suez Canal; for the U.S.A., Statistical Office of the United Nations, *Monthly Bulletin of Statistics,* No. 5 (May, 1947), p. 77.

[5] E. A. Speiser, *The United States and the Near East* (Cambridge, Mass., 1947).

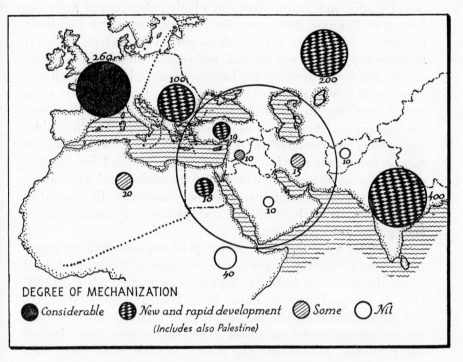

DEGREE OF MECHANIZATION

● Considerable ⊕ New and rapid development ◐ Some ○ Nil
(Includes also Palestine)

MAP 5. POPULATION AND TECHNOLOGICAL DEVELOPMENT

Population: Size of circle is proportionate to population. Numerals refer to population in millions.

Technological Development: Degree of Mechanization refers to industry, transport, agriculture (petroleum production is excluded).

Note that the Near East is a funnel for Old World trade, the focus of sea highways between populous, industrialized Europe and heavily populated India. Also note the small population and technological backwardness of the Near East as compared with Europe, Russia, and India.

today. Its countries are in the forefront of the new independence movement among Asiatic and African peoples. Following World War I, the eventual granting of independence to Iraq, Syria, Lebanon, Palestine and Transjordan was a basic policy agreed to by the League of Nations. But in part because the powers felt that these countries would at first not be able to govern themselves, the novel mandate experiment was devised, whereby an individual country, responsible to and agent of the League, would administer each territory until it was able to stand alone. The Treaty framers went out of their way to affirm the altruistic nature of the agreement: the mandate is a "sacred trust of civilization"; and, again, the mandatory power as trustee of the League of Nations "will derive no benefit from such trusteeship." The system supplied for the first time in history the principle of public trusteeship and national responsibility to an international supervisory authority. By the end of World War II, of the mandated countries only Palestine had not been granted independence.

The world is seeing the intensification of this independence movement right in the heart of the colonial empires of those who helped frame the mandates. India has already been granted her independence. In Burma, Indonesia, Indo-China, Tunisia, Algeria, Morocco, and Madagascar there are restless stirrings for freedom. Most Moslem peoples are in the van of the independence movement, including Pakistanis, Indonesians and North Africans, with their Near Eastern brothers encouraging them on in the Security Council meetings of the United Nations (summer of 1947). The Near East is the hub of Islam (see Map 6), which numbers close to three hundred million followers, almost twice the population of the United States. We sometimes forget that at one time the Moslems were the dominant world power, controlling the whole Near East, ruling or terrorizing at different times practically all of southern Europe from the Black Sea to the Atlantic, as well as much of Asia, and dominating the Netherlands Indies as aggressive traders and missionaries.

It all had humble beginnings. In the Islamic year 1 (our 622 A.D.) a handful of Arabs around Mecca and Medina launched a career of conquest and conversion which has not stopped to this day. Tribal warfare among believers was tabooed, so with unbridled fervor

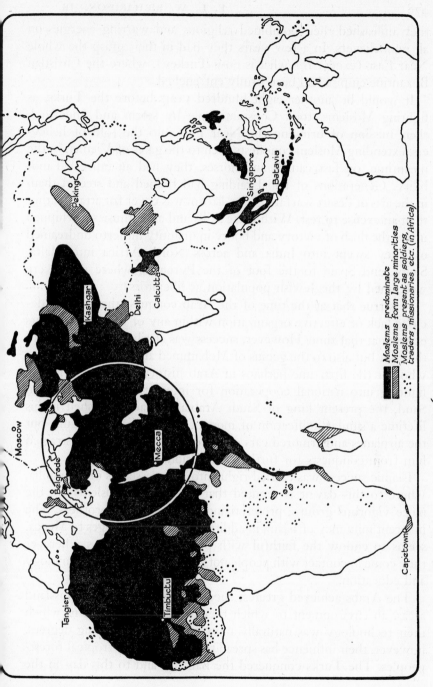

MAP 6. THE MOSLEM WORLD

Based on a map from P. K. Hitti, *The Arabs, a Short History*. By permission of the author and of Macmillan & Co., Ltd., London

Moslems predominate
Moslems form large minorities
Moslems present as soldiers, traders, missionaries, etc. (in Africa).

Moscow
Belgrade
Tangier
Timbuctu
Capetown
Mecca
Kashgar
Delhi
Calcutta
Peking
Singapore
Batavia

they unleashed their combined religious and warring energies on all nonbelievers. In a few years they had in their grasp the whole Near East (except for what is now Turkey), where the Christian Byzantine empire had been firmly entrenched.

It would be another eight hundred years before the Turks, as fighting Moslems, took Constantinople by storm and undertook their mission of bringing the Near East into the fold of believers, extending Moslem conquests even to the gates of Vienna. In the meantime, on fast camels and horses, they had an unrivaled mobility. Generations of tribal raiding had trained and steeled them in the arts of desert warfare. A sandstorm was cover for attack rather than an excuse to rest. With these skills and an enthusiasm whipped up by the flush of victory and booty in quantity hitherto undreamed of, they swept into India and across North Africa into Malta, Sicily, and Spain to the foot of the Pyrenees, where they were welcomed by the Jewish population as liberators.

It is true that at the time of the Arab conquest there was a decided lack of effective organization within any of the other powers existing at that time. However, success was attributable not only to this fact but also to the genius of Mohammed and the early Caliphs, who for the first time perhaps in Arab history turned internecine fighting into national cooperation for international conquest. Ibn Saud, the present king of Saudi Arabia, has accomplished in his lifetime a similar unification of most of the Arabian peninsula, but the airplanes and armored cars of European powers have restrained him from conquests for further unification.

Islamic success was carried right into the grasslands of Mongolia, where to this day it is reputed that Mongolian Moslems are the more vigorous groups, proselytizing their Buddhist brothers. The inherent militancy of Islam, in addition to the pilgrimages to Mecca, seems to endow the faithful with proselytizing success wherever they come in contact with people not superior to them in technology and education.

The Arabs achieved greatest success in the desert and grassland areas, an environment to which they were adjusted and in which their technology was naturally most effective. In varying degrees, however, their influence has spread to temperate and tropical forest-peoples. The Turks conquered the Balkans, and to this day in the evergreen forest of Yugoslavia there are villages with wooden mosques and minarets. In the Netherlands Indies there are 80,000,000 Moslems. The Indonesians, who are mostly Malayans, were converted to Islam by aggressive Arab traders in the thirteenth century. For centuries previously these traders had sailed to Southern India, borne by two favorable winds—the summer monsoon, which carried them with sails full in one tack from South Arabia to Southern India; and the winter monsoon which was equally favorable for the return sail. This was the middle lap in the three-course ocean trade from the Indonesian spice islands to the growing towns of Europe, urban markets gradually becoming free from continual feudal fighting and armed banditry. Here in Europe a lively trade in spices was growing, not only for use as condiments but principally as drugs, a quackery unexposed until the advent of scientific methods. Hindus controlled the first lap from Indonesia to Southern India, Arabs the second to South Arabia and the Near East, where much of the goods was consumed. Sometime around the thirteenth century, Arab traders succeeded in wresting the first from the Hindus, and as usual they set about proselytizing. The Arabs thus controlled the first two courses. The third lap through the Mediterranean was controlled by the Venetians. Later in the fourteenth century, they added a fourth to Western Europe by way of Gibraltar and the Atlantic, and thus themselves controlled two laps. Both Arabs and Venetians grew rich on their trade until the sixteenth century, when all trade routes to the East were short-circuited by the circumnavigation of Africa first by the Portuguese and soon after by the Dutch and the English. Although nowadays most trade between Europe and the East flows through the Suez Canal, virtually all of it is carried in Western bottoms. The Near Easterners have become the victims of a Western ship technology and trade organization that was fathered by them or for which they were at least largely responsible in first developing.

Despite the loss in trade, riches, and political power, the Moslems have managed to maintain the number of converts almost intact to this day, Spain, Sicily, and a few other places excepted. In many parts of central and southern Africa (as indicated on Map 6) Moslem missionaries are steadily winning converts among Negroes. Their success is due to the fact that, unlike Christian missionaries, many do not maintain a superior racial attitude, but on the contrary intermarry among their converts, live intimately with them, and treat them to a greater degree as equals, socially and religiously. All this, despite the fact that south Arabian traders and other Moslems have for centuries profited from a rich slave trade, successfully stamped out by the British barely one generation ago.

In the arts of peace, too, the Arab conquerors made great progress. They introduced the new tool of political toleration. The three groups, Christians, Jews, Sabaeans, were required to pay a small tribute, far less than that of pagans, in return for protection. In architecture, painting, and embroidery, Moslem works are among the masterpieces of the world and were of considerable influence in early Italy. In the fields of science, philosophy, and mathematics, the scholars of the Near East were the best in their day, carrying the torch of learning while Europe of the Middle Ages slumbered. Prior to this period of Near Eastern ascendancy there were other eras in which the Near East led the world in cultural development: the first extended for 3,000 years, perhaps even longer, before Alexander the Great; the second has covered about 1,000 years of the 1,300 since Mohammed. Will there be another time?

Every day most Moslems publicly turn toward Mecca, and many of them, five times a day, recite "There is no God but Allah and Mohammed is His Prophet." For one month a year, they fast during daylight hours. Their faith prescribes the regular giving of alms. Once a lifetime, all who can are supposed to make the pilgrimage to Mecca, upon the completion of which they receive the title of Hajji, which gives to all, whether prince or pauper, an enviable prestige.

From all corners of the old world, the pilgrims come, nearly 200,000 strong before World War II, all kinds—rich and poor, yellow, white, brown, and black. Of all world religions, Islam seems to be the most successful in demolishing barriers of creed, race, and nationality; the line is drawn between believers and nonbelievers. On the common ground of faith these Moslems meet and reinforce a somewhat waning solidarity. World War II brought the annual pilgrimage practically to a standstill. Since then the transportational difficulties are slowly being overcome, and prominent among those who succeed in overcoming them is the contingent of pilgrims from among the 25,000,000 Moslems of the U.S.S.R.

The Near East's solid front against the Zionists, its grievances against the British, and the solidarity of Pakistan Moslems against the Hindus are familiar topics in today's newspapers. But perhaps we are less aware that the 80,000,000 Indonesians, fighting for freedom, are overwhelmingly Moslem. Before World War II, they formed the largest contingent of Meccan pilgrims, 30,000 annually. It is not irrelevant to point out that their island domain, including British Malaya and associated islands, produce about half of the world's tin and almost all the world's rubber. Before World War II the Netherlands Indies produced as much petroleum as Europe. In 1938 it made up 12½ percent of the total production of the entire Eastern Hemisphere. And as though by some trick of history, they together with their Near Eastern and North African brothers, watch over the strategic "tollgates" of the old world's waterways —Gibraltar, the Dardanelles, the Suez Canal, and the Malaccan Straits (Singapore). Who knows, perhaps the Moslems will once again dominate the world?

Technology and Organization

In this conflict between the Eastern and Western world, he wins who either surpasses the other in technology or effective group organization or both. Consider the European conquest of the Near East in World War I. Technologically, Europe commanded battleships, airplanes, mechanized industries, modern armies; the Near East had fleets of sailing dhows, a Turkish navy hardly worthy of the name, practically no mechanized industries, and a poorly equipped Turkish army.

The European powers numbered their populations in tens of millions, many of them effectively organized into business-finance corporations, world-wide trade combines, governments, hospitals, scientific bodies, modern universities, as well as religious and philanthropic bodies. In contrast, during World War I, Near Eastern countries had populations of around 10,000,000, if that many, ineffectively mass organized in (1) a corrupt government or (2) religious bodies, and practically nothing more. The Near East had no world-wide trade combines of any consequence because the West had wrested world trade from her centuries ago. The industrial revolution had not been diffused to her, so she had neither the technical means of production, communications, and transport, nor the scientific knowledge which makes possible the large-scale modern business and scientific organizations. Her businessmen carried on small family businesses; modern hospitals were few and far between; and most doctors were Christians or Jews. Primary education was conducted in the mosques and consisted in memorizing the Koran, writing, and a little arithmetic. University education revolved around theology and metaphysics. Technical education was given by the guilds through the apprentice system. All this while she was the prisoner of her theocratic state—a collusion of political and religious leaders—a technique of mass organization she had fathered thousands of years before.

Now her theocracy was decrepit with old age, corrupt, but picturesque: a theocracy in which heirs-apparent were confined to harems till they ascended the throne, sometimes in a doddering old age and usually with the experience of an adolescent; in which eunuchs had power; and in which chief turban-winders and nightingale-keepers padded the palace payrolls.

Compare this state to that of a few centuries earlier, when Turkish armies met no equal in Europe because they alone were organized with conscripted forces and because their troops displayed an unmatched discipline. Even a thousand years ago when, with religious and political zeal combined, soldiers mounted on fleet Arab horses and camels met no equal in Europe or Asia. Then their theocracy was a uniting force, while their opponents had neither the organization nor technology to impede them.

From circumstantial evidence it seems that in the Near East, the practice of irrigated agriculture first made necessary the mass organization of people. It is similar to a modern factory production line where each successive operation is dependent on the ones preceding. Take for example ancient Mesopotamia, now Iraq. While the yearly floods are running, anyone who fails to keep his section of the banks high and firm and in good repair incurs the danger of ruinous floods bringing starvation not only for himself but for every family, tribe, village, and city along the river. In those early times, up to tens of thousands of people were involved—a large mass organization even today.

In the low-water season, these same people would be busy digging out the canals which the silt-laden waters yearly choked. And throughout most of the year, officials meted out the irrigation water allotted to each canal or user. There was danger also from enemies or saboteurs who could cut the diversion dams or headwaters of the dam or weaken the banks or levees which contained the flood waters. Consequently a military or police system was inevitable. Thus it is possible that irrigated agriculture made necessary the first mass organization of people not only in the Near East but also in India and China. Let us Westerners not grow smug in attributing our present superiority to our supposed "innate" or "God-given" racial superiority. At present the skill of effective mass organization and the torch of technical progress are both in our hands, but they were not a little over 500 years ago, and who knows where they will be even 100 years hence?

Present Disorganization

The Near East is divided and subdivided into 19 separate political units—nations, protectorates, colonies, and so on—and 26 units counting such closely associated territories as the Islamized Horn of Africa and Cyprus.[6] The population of almost all 26 political units is splintered into a number of minority and tribal groups many with an intense internal loyalty. It is a region of nation against nation, and within one nation, group against group, and cross-cutting nations cooperating groups with stronger loyalties to themselves than to their nation. The national rivalries are similar to those in the Balkans and Ireland. The group conflicts are similar but usually more intense than those in the United States between old Americans and immigrants, Catholics and Protestants, Gentiles and Jews, labor and management, Negroes and whites. And the groups with loyalty cross-cutting national ties are similar though less effective than world Communists.

The present 26 individual political units are listed on Table 3.

[6] Omitting the recent subdivisions of Palestine.

TABLE 3

Number of Political Units in the Near East [a]

Geographical Regions	Around 1800	1914	1947
I. NILE VALLEY from Mediterranean to confluence of Blue and White Nile.	1. EGYPT (nominally under Turkey; Sudan conquered 1821)	1. EGYPT (nominally under Turkey, British administered) 2. ANGLO-EGYPTIAN SUDAN, British-Egyptian condominium, former predominant	1. EGYPT (strong British influence) 2. ANGLO-EGYPTIAN SUDAN same as 1914
II. ASIA MINOR Aegean to Caspian, Caucasus and Black Sea to Syrian desert and Mediterranean.	2. TURKEY 3. IRAN (see 3)	3. RUSSIA 4. TURKEY 5. IRAN	3. RUSSIA 4. TURKEY 5. IRAN
III. EASTERN MOUNTAINS AND PLATEAUS Persian mountains to Indian mountain wall, and Persian Gulf to Siberian plain.	IRAN (see 3) 4. AFGHANISTAN	IRAN (see 5) 6. AFGHANISTAN	IRAN (see 5) 6. AFGHANISTAN
IV. CENTRAL COURTYARD Northern Syrian desert and fertile crescent.	TURKEY (see 2)	TURKEY (see 4)	7. IRAQ (former British administered League of Nations Mandate) 8. LEBANON (former French administered League of Nations Mandate) 9. SYRIA (former French administered League of Nations Mandate) 10. PALESTINE (British administered League of Nations Mandate) 11. TRANSJORDAN (former British administered League of Nations Mandate)
(*Arabian Peninsular*) V. CENTER—grassland plateau	5. WAHHABI ARABS	7. NORTHERN TRIBES 8. SAUDI ARABIA	12. SAUDI ARABIA

VI. WESTERN—highland rim

VII. EASTERN—lowland rim

VIII. SOUTHERN—highland rim

Miscellaneous

IX. AFRICAN HORN
East of Ethiopian mountains and plateaus.

X. CYPRUS—Mediterranean island.

XI. SOCOTRA—Island off African Horn.

TURKEY
WAHHABI ARABS (see 5)
WAHHABI ARABS (see 5)

6. LOCAL CHIEFTAINS (including Sultan of Kishin; see 10)
WAHHABI ARABS (see 5)
7. MUSCAT (British Protectorate.)

8. LOCAL CHIEFTAINS, TURKEY (see 2) (nominal), shore near Red Sea opening
MUSCAT (see 7) conquered Indian Ocean ports early 1800s
9. ETHIOPIA

TURKEY (see 2)

10. SULTAN OF KISHIN (see 6)

TURKEY
9. ASIR
SAUDI ARABIA
10. TRUCIAL OMAN (British Protectorate)
11. BAHREIN ISLAND (British Protectorate)
12. KUWAIT (British Protectorate)
13. ADEN (British Colony)
14. ADEN HINTERLAND (British Protectorate)
15. MUSCAT OMAN (British Protectorate)
16. BRITISH SOMALILAND (British Protectorate)
17. FRENCH SOMALILAND (French Protectorate)
18. ITALIAN SOMALILAND [b] (Italian Protectorate)
19. ERITREA (Italian Colony)
20. ETHIOPIA
21. BRITISH ADMINISTRATION (nominally under Turkey)
22. BRITISH PROTECTORATE

---

13. YEMEN
SAUDI ARABIA (see 12)
SAUDI ARABIA (see 12)
14. TRUCIAL OMAN (British Protectorate)
15. BAHREIN ISLAND (British Protectorate)
16. KUWAIT (British Protectorate)
17. ADEN (British Colony)
18. ADEN HINTERLAND (British Protectorate)
19. MUSCAT OMAN (British Protectorate)
20. BRITISH SOMALILAND (British Protectorate)
21. FRENCH SOMALILAND (French Protectorate)
22. ITALIAN SOMALILAND [c]
23. ERITREA [c]
24. ETHIOPIA
25. BRITISH-COLONY
26. BRITISH PROTECTORATE

---

[a] Political units are numbered in arabic numerals. A continuous political unit in more than one geographical region is numbered only once. The eleven geographical regions are numbered in roman numerals.

A political unit is a more or less continuous land area under a single authority, either bounded by a large body of water or other land areas under different authorities. (A protectorate is counted as a separate political unit.) To refrain from excessive complexity; (1) Independent local Chieftains, Sultans, Sheikhs, etc., in one geographical region or within one protectorate are counted as one political unit. (2) All islands are excluded except Cyprus, Socotra, and Bahrein. Each is counted as a separate political unit only if it is governed by a local authority different from that of the nearest mainland.

[b] Northern part.

[c] Status September, 1947, awaiting Italian Peace Treaty.

(Small protectorates and the like with independent chieftains are counted as one unit.) Compare this to the United States with one single authority and an area comparable in size. In fact the Near East is a rival of Europe for the dubious distinction of being the most politically fractured region in the world. Europe has twenty-five countries not counting Iceland and the several small principalities, duchies, etc. like Luxembourg, Andorra, Monaco. Instead of lessening, political fracturing in the Near East has progressively increased in the last 150 years creating more states and potentially more national rivalries. Around 1800, there were ten political units; in 1914, twenty-two; and in 1947, twenty-six.

Table 4 was compiled to bring out the degree of group splintering within political units. This table is restricted to five contiguous political units that are perhaps the most splintered of all Near East territories. These five territories comprise the central courtyard previously referred to—Palestine, Transjordan, Lebanon, Syria, and Iraq. Here live about 10,000,000 people split into almost twenty-five large organizational fragments of 10,000 people or more each. Every conceivable kind of Christian, Jewish, and Moslem sect is found and many pagan, most of them too small to include in Table 4: semi-nomadic Arab Christians with tents for churches; "good" Samaritans now no more than 100–200 strong, representatives of ancient Israel still clinging to their old practices and speaking Aramaic, the language of Jesus; representatives of modern Israel, political terrorists streamlined with the latest innovations in technical equipment, public relations, and legal procedures. But perhaps the most surprising of all are the handsome, hard-drinking, devil-worshiping Yezedis who solemnly rationalize that God will hurt no one so they had better expend their energies on keeping the devil at peace.

Some of these ethnic splinters are survivals of once large groups, important in bygone days. There are the Sabaean or Mandaean moon-worshipers, now silversmiths in most cities and towns of Iraq. From their ranks in the third century A.D. sprang the founder of Manichaeism whose attempt to blend the doctrines of Zoroaster with those of Christ met with such success in the Roman world that the Emperors vigorously suppressed it, but it still continued to trouble the Christian churches well into the Middle Ages. There are the Nestorian Christians who once boasted 230 bishops from Ceylon to Siberia, and from South Arabia to the great grass plains of Mongolia, where in the thirteenth century the Nestorians were prominent in the court of the Khans as doctors and administrators of the far-flung realm. Now they number a few tens of thousands and it was only thanks to the British that they escaped massacre and complete annihilation immediately after World War I.

In spite of all the persecutions and bloodshed by Moslems in the name of Allah, they have shown throughout their history a tolerance far greater than most realize, particularly for Christians, Jews, and Sabaeans. This goes back to the original teachings of Mohammed who held in great respect these believers in revealed religions who preceded him and with whom he came in contact. They were exempted from land and capitation taxes, and in civil and criminal judicial matters, they were subject almost entirely to their own spiritual heads except when Moslems were involved. This toleration is therefore in large part responsible for the rich heterogeneity of ethnic splinters still existing in the Near Eastern world today.

However quaint these groups may appear to us, they have cursed the Near East with unabsorbed and often fanatically independent groups and minorities, usually revolutionists against any who ruled, and frequently quislings for those who aspire to rule. The British used the Arabs against the Turks. The Italians and Japanese courted the Yemenites as a counterweight against the British and the rest of the Arabians. The Russians stirred the Kurds against the Turks, Iraqis, Iranians and indirectly against the British. The French played Maronites against Moslems.

For the Arabs, the Zionists are the most threatening of any minority group. Not only do Zionists comprise one-third of the population of Palestine and have been steadily gaining through immigration, but they are masters of modern technology and mass organization methods. Even taking the 10,000,000 people in the five countries of the central "courtyard," here the immigrant Jews together with indigenous Jews number close to 1,000,000 or one-tenth the whole population, which makes them the largest non-Moslem minority (see Table 4). The Sunni village and urban majority are only 4,000,000 and at that divided between five countries. What makes Zionism more unbearable to the Arabs is that

TABLE 4

Important Moslem and Non-Moslem Groups in the Central "Courtyard," Having 10,000 or More Members (Syria, Lebanon, Palestine Transjordan, and Iraq: Total Population, 10,000,000)

MOSLEMS

Name	Approximate Number	Percent of Total
1) Sunni-Arabs (urban and village dwellers)	4,000,000	40
2) Shiite (urban and village dwellers)	2,500,000	25
3) Sunni-Arabs (nomads)	over 500,000	
4) Sunni-Kurds	over 500,000	
5) Alouites	over 100,000	15
6) Druzes	over 100,000	
7) Turcomans	over 100,000	
8) Iranians	over 100,000	
9) Ismailites	over 10,000	
Total Moslems	8,000,000	80

NON-MOSLEMS

Name	Approximate Number	Percent of Total
JEWS		
1) Western Immigrants	over 600,000	Little less than 10
2) Eastern Indigenous	over 100,000	
CHRISTIANS		
3) Western Administrators Business Men, soldiers	?	
4) Maronites	over 100,000	
5) Greek Orthodox	over 100,000	
6) Chaldeans	over 100,000	
7) Armenians	over 100,000	
8) Greek Catholics	over 10,000	Little more than 10
9) Eastern Orthodox	over 10,000	
10) Syrian Catholics	over 10,000	
11) Jacobites (Syrian Orthodox)	over 10,000	
12) Nestorians	over 10,000	
OTHER		
13) Mandaeans or Sabaeans	over 10,000	
14) Yezidis	over 10,000	
Total Non-Moslems	2,000,000	20

NOTE: As census taking is both rare and inaccurate in the Near East, any compilation of population statistics is open to doubt. This table was constructed by taking the best estimates from sources below.

For each of the 5 territories, the estimates of the size of the constituent groups was usually based on old censuses or estimates. To adjust for the present day, the following rough method was used: In a given territory, group A for example in 1927 comprised 100,000 out of a total of 1,000,000 people or 10% of the population of the territory. Say then, the latest total population census or estimate for the same territory is 1,500,000, so group A has been computed as numbering 150,000 people (i.e., 10% of the total), unless definite information to the contrary is known.

Sources: for Syria and Lebanon: Said B. Himadeh, *Economic Organization of Syria* (Beirut, 1936), Chapter I, on Population. For Palestine and Transjordan: Sir Harry Luke and Edward Keith Roach, *Handbook, Palestine and Transjordan, 1934*; Frank W. Notestein and Ernest Jurkat, "Population Problems of Palestine," *Milbank Memorial Fund Quarterly*, Vol. XXIII, No. 4 (October, 1945), pp. 307–352.

For Iraq: Hans H. Boesch, "El 'Iraq," *Economic Geography*, XV, No. 4 (October, 1939), especially table, p. 342; Iraq Directory.

western Europe is dumping on them its own minority problem, administered by Great Britain and financed largely from the United States. In this day of large powers, 40,000,000 Arabs need a strong friend and protector—an ally to side with them—and they will find one.

Five thousand years as a world center with waves of conquerors and proselytizers parading in and out has left the Near East with the accumulated remnants of shattered groups like a beach littered with shipwrecks. None of these groups has been either sufficiently dominant or persistent to wipe out the differences. Thus far the Arabic language and Islamic religion have been the most effective unifiers. But even within Islam itself, rival sects have long been pitted against each other. Conquerors have been unable to rule large areas for long. Invariably large political units have split up. For example, the Wahhabis unified most of Arabia around 1800 (see Table 3), but it later re-splintered. Internal dissensions had weakened it within, and from without came Great Britain as a protector of autonomous coastal chieftains perpetuating the political fragments of the moment. Ibn Saud, the monarch of expanding Arabia, and descendant of former Wahhabi Arabian conquerors, has himself reconquered and unified much of it. But more than once British gunboats, airplanes, and the like have prevented him from incorporating further territory. Pax Brittanica in Arabia has encouraged and brought about commercial development, elimination of commercial slavery, internal reforms within individual states, but it has done little to lessen political fracturing within the peninsula.

Even America, in the name of education, has inevitably added to the process. In Turkey and Lebanon, colleges that were originally financed largely through private American contributions and still have Americans on the staff are the outstanding modern educational institutions in the whole Near East. Many students, educated in the better ways of life and thought, graduate to find opportunities limited. The technical backwardness, the contrast of enormous wealth and extreme poverty, the group hatreds, nepotism, the apathy of peasant millions and overlords alike for change, all this is disheartening. Some of these "over-educated" products of the West form small listless "middle-class" groups talking progress, but either powerless, unwilling, or thwarted in their attempts to change a millennia entrenched system.

A second reason at the root of the Near Eastern disorganization is the segregation of the population into small scattered settlements, forming four peripheral groups and one central one, likened to four houses around a central courtyard. The Near East is not a continuous cultivated land mass; it is figuratively an archipelago of cultivated "islands" (see Map 1).

Up to a generation ago, practically the only communication between the "cultivated islands" of the Near East was by native sailing ships, so-called dhows, and camel caravans. Just as the towns along the coast are seaports with their sailors, so the towns facing the desert are desert ports, where one meets camel-riding nomads holding themselves in a tall, straight haughty manner. Only after World War I was the 600-mile desert crossing made in an automobile from the Mediterranean to Baghdad in Iraq. By camel caravan it takes six weeks, by automobile close to a day and a night, and by plane a few hours. Before World War II, a paved road was partially completed to supercede an indefinite desert track. Cars still traveled only in convoys as precaution against Bedouin raids and in case of breakdown. With the going from one country to another, with the inevitable customs, delays, and the ceremony and care attached, this desert crossing was much like an ocean voyage.

It is little wonder then that each isolated settlement has developed its own group loyalty, and its own customs and leaders, with varying degrees of national loyalty to the particular "house" in which it is located. Those of the "courtyard" have had little loyalty beyond their own local group.

The third basic reason for group fragmentation is the existence side by side of three groups, each engaged in totally different economies; agriculturists, nomad stockherders, and fishermen, each of course based on an adjustment to a different environment. People in these groups usually have as little to do with one another as possible and frequently raid or steal from each other.

As though these differences were not enough, each group has been, so to speak, blessed with the curse of great commercial possibilities by expending a little extra organized effort. Irrigated agriculture on the alluvial plains of Egypt and Iraq give tremendous returns when properly cultivated and regulated; desert nomads have controlled the overland trade routes; and finally the fishermen and coastal dwellers had much of the world's ocean and sea trade at their doorstep, which they could either carry, control, or pirate, and in times past, they effectively did all three. The opportunity for great wealth divided each of the three groups internally between the rich benefiter and the toilers, and resulted in a kind of feudal system notably among the agricultural group. Among the desert nomads with their system of joint family responsibility, tribal councils, and the like, chief use of the wealth and prestige gained by one tribe was to attack other tribes.

These three basic economies have naturally been supplemented by the artisans, producers of all manner of utensils, tools, clothing, and luxuries. To this day they can be seen plying their trade in the picturesque, smelly bazaars, which are roofed to keep out the hot oriental sun. With the introduction of the modern factory system, the number of these artisans is diminishing. As in most countries, the factories are concentrated in the cities, drawing artisans and peasants from the communal security of their little villages to become puppets of the new get-rich-quick, but uncertain, industrial system. These urban factory wage earners are naturally creating a new group, and only time will tell the extent to which they will further fracture Near East society.

Technical means of communication and travel have been inadequate to bring together frequently the various historical, insular, and occupational groups. People diffused throughout a circular region 2,000 miles in diameter cannot keep in close touch by camel, horse, and dhow. The difficulties merely of keeping currently informed and of policing have been enormous.

Modern large-scale medical, industrial, agricultural, and other organizations can be the rallying points for the merging of people from different groups, but such modern organizations require a technical knowledge and ability that few Near Easterners possess. Efficient hospitals and public sanitation departments are impossible without medical knowledge and training; large factories, water power, and most big irrigation projects are impossible without advanced engineering training; large-scale agricultural extension is impractical without local scientific experimentation. With some exceptions, these have barely made a beginning in the Near East.

Thus, the technological basis for merging people into modern large-scale organizations has largely been lacking in the Near East. Loyalty and cooperation have usually been limited to one's family, customers, small community, religion, and state; and frequently loyalty to the latter has not been great.

The number of groups and minorities within the Near East has complicated the evolution toward large, unified states. Each group is a little nation unto itself. For example, even before partition the Jewish Agency in Palestine operated much like a government with its own chosen leaders, its own "state" religion, its own schools, its own language, its own script. Now in varying degrees, this same system applies to most if not all of the other groups. These five "courtyard" countries are in a sense not sovereign nations; each is a federation of little nations and groups.

Many of these groups, in addition to having their own "state" religion, their own schools, also raise taxes or tithes among their adherents, administer their own system of justice regarding such family relationships as marriage, divorce, inheritances. All are endogamous. The people of many are physically differentiated, and superficially differ one from the other in any number of ways including dress, manner of speech, occupation. In material culture, architecture, and the like, there are frequently also considerable differences.

The European "master race" (and this is true also of Americans, wherever present) are no exception, as they, too, retain their own schools, religions, language, script, endogamy, physical differentiation, dress, and mannerisms. They have their own clubs from which "natives" are with rare exceptions excluded. What is generally not known, is that Westerners have been protected by their own system of courts. These were abolished by the Turks before World War I and by Iran in 1928. In Egypt, where they will terminate in 1949, they are called "Mixed Courts," made up of judges from several different European countries and at present (1947) presided over by an American.

The semi-independent nation-like organization of most of these groups makes them loyal to themselves only. Within a state, such a group can become both a powerful and a dangerous minority in the hands of self-seeking politicians and powers and can either be bribed or terrorized into voting as desired, making public demonstrations and the like. This is one danger of importing Western democracy into an Eastern setting.

By and large during the last few thousands of years, the perpetuation of "independent" groups, and perhaps even the creation of new ones, seems to have been a stronger tendency than the merging of groups. Every ambitious leader who has tried to weld minority groups into larger and larger units has inevitably met with resistance—not only during conquest but after decades and generations of rule. Is it any wonder, then, that to build and maintain a state, a small minority has had to impose its rule over all the other minorities? And is it any wonder that to maintain its control, the ruling clique has frequently buttressed its power with religious sanction—creating, in short, a theocracy?

Thus in the Near East with groups against group, a state of any size tends to be a house divided against itself. In contrast, we Americans today have inherited a degree of unity unusual in any large state, and even so we have only achieved a limited democracy. Near Easterners today are the unfortunate heirs of a millennia-old fragmented social system in an age when independent group survival depends on large political units. Plagued with this inheritance, how effective is democracy for them now? Can such a fragmented social system become unified mainly through democratic action? Or must the seeds of democratic rule be sowed to reap at a later harvest? In the world as it is today, these are vital questions and important for all of us to ponder.

THE COMPONENT STATES, BY JAMES BATAL [1]

The Southern Rim

THE ONE SPOT in the world where centuries count little, even though it contributed toward the early development of civilization, is the rim of states that half-circle the Arabian peninsula. This arc, com-

---

[1] The writer gratefully acknowledges the valuable guidance and advice of F. L. W. Richardson, Jr.

posed of mountains, desert, sandy plain, and tropical jungle, has a coastline nearly as long as the combined 4,883 miles of the Atlantic, Gulf, and Pacific coastlines of the United States. Starting at the mouth of the Persia Gulf this rim includes a number of entities.

Kuwait is a principality, occupying 1,950 square miles of almost unrelieved desert. Ruled by an Arab sheikh, this tiny area would have been swallowed up in 1919 by the stern Wahhabite revivalist, Ibn Saud, had not guns from British planes taught the desert chief to respect the treaty of protection which England had securely forged with this sheikhdom in 1898. The Kuwaitians, nearly 80,000 of them, depend almost entirely upon the sea for their living; Kuwait city itself is a major port for the transshipment of goods brought there from India for the Mesopotamian valley countries.

Trucial Oman comprises a coastal plain on the Persian Gulf. Inland, its sandy sparsely settled wastes seldom rises above 500 feet to merge with the Rub Al-Khali (the Empty Quarter). Along the coast are numerous capes whose inhabitants are principally fishermen; the area nearest to Bahrein Island is famed for its pearls, while Bahrein itself is a veritable pool of oil wealth. This entire coastal strip has been little affected by Westerners, largely because of the extreme tropical climate. Minor tribal chiefs, independent of the Sultan of Oman, administer what government there is, under protective treaties with the British.

Oman, better known as Muscat-Oman from the name of its principal port, is an independent sultanate. Its 82,000 square miles stretch from the Musandam peninsula in the Persian Gulf southerly to the Hadhramaut on the Indian Ocean. Mountain ranges that climb to 9,900 feet, with bare granite and limestone walls that frequently descend rather abruptly to the sea, mark its general topography. This area is inhabited by some half million people, mostly Arabs. Along the Indian Ocean the population contains a strong Negroid element. Between the mountains are lateral valleys whose fertile reaches are cultivated by settled Arabs and some seminomadic Arabs. There they raise sheep, horses, goats, and camels. Oman has an annual rainfall of about 10 inches which, along with a tropical climate, helps to yield abundant vegetation. Tamarisks, oleanders, milk-bush, and acacias grow profusely in the coastal area while in the interior the dates for which Oman is most famous are cultivated along with cereals, vegetables, and fruits. Oman is ruled by a sultan with a British protective treaty helping to subsidize him.

Aden is another British protectorate which occupies the south-central part of the Arabian rim. One must not confuse Aden, the city, with Aden, the province, in whose 112,000 square miles are a number of states in protective treaty relations with the British since 1839. Within Aden province lies the Hadhramaut, at whose easterly end is Dhofur, the one spot in the Mediterranean desert belt where dense tropical vegetation flourishes as a result of the abundant rains brought by the Indian monsoons.

Throughout the entire Arabian rim the only legislative council imitating democratic procedure is found in Aden City, a British crown colony of 48,638 population, mostly Arabs. This council legislates only for Aden city. The sultans and the tribal leaders rule the rest of Aden province according to their ancient tribal customs or Islamic traditions. Some 630,000 persons, some of them wild non-Arabic speaking tribes in the Dhofur region who have ringlet hair and paint their bodies blue, dwell in the Hadhramaut.

A narrow but lofty valley, situated between the encroaching sands of the Rub Al-Khali and the coastal mountain ranges that reach often to 8,000 feet, gives the Hadhramaut a distinctive setting. It is a valley strung with many towns, significant for their intense agricultural activity. This is the region where skyscrapers were first introduced to man—buildings five and six stories high carved into the hillsides.

Yemen is a medieval mountain kingdom at the western tip of the Arabian rim. Its 78-year-old ruler, Zaydi Imam Yahya (reported to have died in January, 1948) came from a long line of holy ancestors reaching back a thousand years to the Prophet Mohammed. The world's remaining vestige of a priestly kingdom has tried deliberately to isolate itself completely from western influences. Only in 1946 did the Imamate finally agree to admit foreign diplomats officially, and since that was an extremely unusual occasion, modern plumbing was installed in the royal guest house for the benefit of the American negotiators.

Yemen broke out of its medieval cloister when it was elected the 57th nation in the United Nations in September 1948. The Yemenite kingdom, with its 100,000 square miles of territory, mostly mountainous and inhabited by some 500,000 people, is the most fertile part of the Arabian Peninsula. It is famous for its highly terraced agriculture and for having first introduced coffee to the world (in the 16th century).

Yemen's inhabitants are sharply divided into readily recognized social classes, a setup that has all the attributes of the ancient Sabaean social system:

1) divine origin (the imam and his relatives);
2) hereditary sheikhs and landowners;
3) free farmers (these comprise the bulk of the population); and
4) the workers (Hojjeri, usually Negroid in appearance; many of whom live primitively in caves in the escarpment area).

All these states have several things in common. The climate is tropical. Except where the Arabian desert cuts into it, the peninsular rim is buttressed by mountain chains that rise to 10,000 feet. Sufficient rain falls on all these mountains. Alluvial soil is washed down the valleys, making possible intensive cultivation of cereal and vegetable crops in some places and the growing of coffee orchards in the Yemen and fruits in Oman. In the plains that branch inland from the mountains toward the desert, nomadic herdsmen raise sheep, goats and camels. They come to the settled communities to exchange the products of their animals for the dates and cereals and a few manufactured articles which they may need for a Bedouin life.

Despite the fact that 6,000,000 people, living along the rim, offer a potential market for manufactured goods, manufacturing industries in any of the sultanates are as rare as snowstorms in the neighboring Rub Al-Khali. The masses live simple and elemental lives as did their ancestors when they traveled in camel caravans along the fringes of the desert, bearing the precious incense and myrrh from the frankincense country of the Hadhramaut. The one agricultural pursuit that made life tenable despite the aridity of the desert— camel raising—is losing its life-sustaining source for the Bedouins, who see the automobile from the West displace the desert burden beast of the East in the transportation field that was exclusively the camel's for centuries. This change has created hardships for the camel herdsmen of Oman.

Another factor that has retarded the technological development of the Arabian rim is this: it is cheaper to employ human labor than to buy machines. A landowner or village sheikh has no need for expensive mechanical equipment for agriculture. These sheikhdoms have had no money to surface roads, and of what use would it be to pave them when nobody (except the Westerner) was in a hurry to get to some other place? With only two miles separating a natural linking of Muttreh and Muscat, it remained for the initiative of the British military to cut a road between the mountains to connect these populous Omanee cities.

The tropical climate has taught the population its centuries'-grounded lesson: there is no need to hurry to get things done; life will go on just the same. With such a basic philosophy, there has been little incentive to establish electrical production plants for the propulsion of an industry that doesn't exist even in the blueprint stage. It is cheaper to harvest wheat or barley by the means of hundreds of hands than to buy a mechanical reaper which would merely throw out of work people whose lot has been inured to elemental wants since recorded history. Only the government people and upper classes in the larger communities rely upon motor vehicles for transportation. The commoners continue to travel on foot or ride a donkey or camel, if they happen to be economically that fortunate.

Wherever modern medical or sanitation agencies are found in this Arabian rim they are of Western origin. The American missionaries not only founded hospitals in Muttreh and Kuwait but also wandered into the mountain valleys among the Bedouins, with mobile clinics to heal the sick and to teach sanitary habits. Disposal of sewage has scarcely been changed from the dawn of the Mohammedan era. In Sana'a, the walled-in capital city in the Yemen highlands, sewage flows down the middle of the main street. Municipal drinking water systems are a rarity everywhere.

Throughout the rim, life revolves about the tribal system. This finds its extremes among the almost aboriginal, non-Arabic-speaking groups that live in the miniature tropical jungles of Dhofur. In Yemen, however, the feudal links of society are most strongly tempered. Sabaean, Minean, Katabanian, and Hadhramaut kingdoms flourished here in ancient times. Accounts of their splendor are found in the Biblical stories of Solomon and the Queen of Sheba. Even the impact of the Ethiopian conquest of Yemen in the sixth century and, before that, of the Jewish kings who ruled that terraced highland domain for a short time, failed to maintain whatever gains they may have made in the social order. A road along the Yemen coastal plain is called "The Road of the Elephants," a reminder of the failure of the elephant expedition which Ethiopians launched to conquer that area in the same year Mohammed was born. The mountains, paralleling the Arabian peninsula, have served as insuperable barriers in preventing outsiders from entering the Arabian desert, thus at the same time helping to maintain the purity and character of the Arab race in its desert core.

Unlike the desert tribes who live in tents made of goats' hair, the wilder tribes in Hadhramaut dwell in caves, while in other areas the serfs and lower classes live in simple stone abodes.

Unity and disunity characterize the states. Each sheikhdom or sultanate is a unit by itself. But taking the whole as a group, each state is separated from another by the very same geographical forces that bring homogeneity within itself—the weather, mountains, desert and sea, each playing its important role as a natural boundary.

What has made Yemen a united medieval kingdom for more than a millennium is, in the first place, its topographical features. Composed of high mountain ranges, reaching sometimes to 10,000 feet and at times rising almost precipitously from an extremely narrow coastal plain (when there is any plain at all), the area has been difficult of access to foreign intruders. Once when an African army, mounted on elephants, tried to conquer the Arabian peninsula, it was this very geographical obstacle that helped to block their progress and enabled a Persian fleet with a flanking attack movement to annihilate the invaders.

The sheer land rise from the Red sea coast rendered it impossible, certainly in ancient and medieval times, to scale the rugged Yemen mountains with wheeled vehicles. Even the patiently trudging donkey finds it difficult to haul commerce over mountain trails. The Rub Al-Khali desert makes it practically foolish for man to penetrate it from that forbidding boundary. Thus have the people in Yemen been left quite alone to develop their lives untouched by the modern gadgets that form so essential a part of the Westerner's everyday living.

With the exception of a small Jewish minority, whose history reveals traces of a direct link with the days of Israel's glory, the entire population is strongly bound through its Iman to the Shiite sect of Islam. The ruler himself claims divine descent from Ali, Mohammed's nephew who married the prophet's only surviving child, Fatima. With the population submitting to an absolute feudal monarch who vests his rule upon the spiritual and temporal interpretation of the Koran, it was thus not difficult to keep Yemenites united under the banner of one strong religion—Islam.

Yemen is believed to have introduced the use of coffee to the world, but it was in the plantation of the new world that coffee became an important economic crop. Even the seaport of Hodeida offered little to attract foreign intervention in Yemen; that came only after the opening of the Suez Canal.

Much the same may be said of Oman, a mountainous land also cut off on the north from the interior of Arabia by the Rub Al-Khali and on the south by a generally narrow coastal plain facing the Indian Ocean. Several different tribes inhabit Oman. One, which gained its freedom from the Sultan of Oman after a seven years' war, is the Anjar, fiercely independent. The other tribes pay nominal respect to the Sultan. Muscat, one of the two principal harbors in the Persian gulf, has given Oman its prominence for the Western world, while its fisheries and pearl industry attract trade especially with India. Tribal customs have helped to keep the inhabitants united within a rather small land area. Islamic faith and tribal laws have contributed to the development of an internal leadership that has kept these people closely knit for centuries. Because of the intensely humid climate, the lack of exploitable raw resources and the absence of convenient means of communications, Oman, like Yemen has been practically unspoiled by the external powers.

Another state whose unity has remained untouched is the Hadhramaut, actually a mountainous valley continuously inhabited by tribes who are cut off from Arabia by the extension of the barren desert on the north and the Arabian sea on the south. This uninviting geographical locale has not only kept Westerners in almost complete ignorance of the country, but in the Dhofar flourishes a miniature tropical jungle rarely penetrated except by its aboriginal looking tribesmen.

Whereas Western nations have forced their way into other newly developed parts of the world and have imposed upon them colonial institutions, the Arabian peninsula rim has remained singularly free from this encroachment. But another kind of penetration has taken place—in almost every case in the form of "friendship" or "protective" treaties by Great Britain with each state. Peculiarly, such treaties have not interfered with the unity of each state; on the contrary they have strengthened the hand of the tribal or religious leader to the extent that for more than a century Britain has been able to carry on trade without friction from competing foreign powers, although in an earlier century, Portugal, and in more recent times, Turkey and Russia, tried to wrest control from Britain. Strangely enough, English money has helped to keep the peoples in the rim united to their own ruler, under their existing status quo.

The "House" of Arabia

The Near East, for centuries playing an insignificant role in the accelerating development of the modern world, has within the narrow limits of the past three decades graduated from a rags-to-riches role. Nowhere in the world, perhaps, is such a transformation better illustrated than in the very core of the Near East itself, previously described as the "house" of Arabia. Here is a nation that within the span of two recent World Wars has developed, from a poverty-stricken desert land, into the world's richest oil pool, although its form of government—a tribal kingdom—has scarcely changed from ancient times. From this seemingly barren core of Arab civilization have poured the waves of nomads that gave to the entire Near East a common culture through its Arabic language and a new religion that not only inundated that entire area but overflowed through the Mediterranean sea to the Atlantic ocean and eastward to the farthermost islands in the southwest Pacific.

Tribal existence, the dominating mode of life for Saudi Arabia from time immemorial, has changed only slightly in recent years, and that change has been caused by two forces: first, the autocratic rule of a former tribal chief, Abdul Aziz Ibn Saud, who for the first time in history succeeded in uniting the desert tribal chiefs; secondly, to the advance of Western technology that at this very moment is alchemizing the desert sands into grains of gold through the medium of petroleum exploitation.

The "House" of Arabia is the core of the world to which more than 300,000,000 faithful adherents of Islam turn, their faces toward Mecca, the birthplace of Mohammed their prophet. This is the core that in the first World War sustained the Arab revolt which led to the eventual establishment of seven internationally recognized Arab states, from a domain formerly under the Turkish Ottomans, although the seeds for political freedom of this area had been sown decades before in the "courtyard" states.

Saudi Arabia, an area of some one million square miles with a population of less than ten million people, is principally desert. Except for the Hijaz on the eastern coast of the Red Sea and the Nejd in the northern part of the desert—provinces where about four inches of rain fall a year—and except also for the Red Sea coastal mountain province of Asir, which gets more than ten inches of rain a year, Saudi Arabia is nearly barren of moisture. It is so barren, in fact, that its southernmost part nearest to Hadhramaut properly bears the name Rub Al-Khali. Most of Saudi Arabia's inhabitants are Bedouins who live in tents in the desert about ten months of the year; many of the rest are seminomads or village settlers who dwell close to the oases that nature has sprinkled in the region closest to the Red Sea and the Persian Gulf areas.

The Bedouins have accustomed themselves to their bleak environment, struggling with nature in their eternal quest for grazing lands for perennially wandering herds of camels and goats. Such a life has made them the most independent of peoples anywhere in the world, since their wants are few and their movable tents may be easily shifted about at will. Their lives are intensely centered about the family, which with its larger growth into tribes has produced a loyalty of kinship that democracy can seldom penetrate. For centuries the desert tribes made their living by raising herds of camels and goats and trading the produce from those animals—milk, melted butter, hides, and meats—for the grain and dates of the settled villagers, or, until recent times, lived by raiding or protecting the caravans whose routes followed the most ancient of the world's crossroads, across the northern or southern edges of the Arabian desert.

But Ibn Saud, who came out of his ancient family's domain in Riyadh in the center of the desert to unite all the Arab chieftains under his enforced leadership, is gradually changing the nomad's mode of life. He is inducing many of the tribesmen to settle in permanent villages near the oases or in rain-fed areas where agricultural production can thrive. He has imposed the Western concept of law and order over that entire area, by reducing the individual power of the tribal chiefs and by transferring that control under a state authority of which he is the head. He has established a well-organized army which, while it still uses camels as the principal means of locomotion, is gradually being mechanized with motor vehicles. He has solidified tribal loyalty by his several intermarriages with the daughters of tribal chiefs, since his Moslem religion permits him to have more than one wife.

Although Saudi Arabia is within the area of the world's earliest recorded civilization and although Europeans penetrated into almost every part of the Near East with their culture and technology, this Arabian kingdom had remained singularly free from Western encroachments until the past dozen years. An American engineer, writing in 1946, reported that Jidda, the most important commercial city and port in Saudi Arabia, contained only about fifty American and European families. The development of the Arabian oil fields by the California-Arabian Oil company has caused the establishment of a community of some 2,000 Americans on the Persian Gulf coast, the largest number of Western settlers in any part of the Peninsula. Otherwise the Western influence in the mode of life of the Saudi Arabians has been practically nil, despite the fact that it was Westerners who introduced the modern standards of commerce, schools, hospitals, water systems, paved highways, automobiles and airplanes—all this in an area where Bedouins still struggle in the ancient manner to avoid drought and starvation.

The impact of Westernization is most noticeable in another respect. For centuries the Sherif of Mecca depended, for his national income, upon the revenues obtained from pilgrims to Mecca in the Red sea province of the Hijaz. Ibn Saud added that province to his desert domain when he defeated King Hussein in 1924, the same Arab leader who had led not only his tribesmen but the entire Arab world in a successful revolt against the Ottoman Turks on the side of the Allies in World War I. Since faithful Mohammedans come annually on this pilgrimage, they contribute a large share to the Meccan ruler's income. In times of war, the Sherif of Mecca would find himself without this substantial source of income. It so happened to Ibn Saud in the second World War, but he scarcely felt the loss, since royalties from the oil companies were now the primary source of income. Moreover, this form of Westernization has added another boon to Arabian economy: the oil companies have given employment to large numbers of nomads and thus have directly helped Ibn Saud in his program to settle the Bedouins on the land.

Natural resources and industrial development are the norms that today determine the importance of a nation in its relation with the rest of the world. The "House" of Arabia is fashioned mostly from desert and steppe with a few oases not too far distant from coastal mountains. Search anywhere in Saudi Arabia and you will not find a navigable river. Its streams are merely wadi beds which in the rainy season are briefly filled with water. While some areas could be transformed into farms through irrigation, there are no streams to tap, save in the narrow belt of Hofuf in the province of Al Hasa, which a thriving underground source has turned into a fertile date paradise for the seminomads who have settled there.

Unlike its neighboring states in the "courtyard," whose economies are decidedly agricultural, Saudi Arabia has done little, comparatively, in producing its own food needs. Its agriculture is largely animal husbandry—camels and goats—and these are raised without modern scientific application. Even the wheat and the dates which, along with milk, form the basic diet of the desert tribesmen are obtained by exchanging for them the products of the herds. As the automobile has altered the lives of most people in all parts of the world, likewise has it done so among the Bedouin tribes, although less perceptibly. Bedouins used to find ready markets for their camels in Cairo, Damascus, and Baghdad, but now the motor truck is steadily displacing the camel as the commerce carrier. Now the jeeps, carrying law-enforcement officers, skim over the desert sands with the result that the tribal marauders no longer carry on their caravan raids or smuggling with the same uninterrupted success as they have done from Biblical times until the present generation. All this change within a single generation!

Whereas the agricultural pattern in Egypt tended to keep the people united, since in the Nile valley agriculture was dependent upon an interlocking irrigation system, the virtual absence of any sedentary agriculture and the dissemination, over vast spaces, of tribes respecting only their own fealties, have kept Saudi Arabia from being united, until Ibn Saud's ascendency to power in the early 1920s. Paved roads also serve to unite peoples: formerly there was no need of highways, since the camel could reach the most inaccessible place. The single railroad, from Damascus to Medina, was largely patronized, not by the penniless desert roamers, but by the city and town dwellers making their annual pilgrimage to Medina and Mecca.

In this entire expanse of 1,000,000 square miles, there are few mineral deposits. Gold and silver mining has been undertaken on a rather small scale by an American syndicate at Mahab Dahab, about 250 miles northeast of the Red Sea port of Jidda. The scarcity of mineral resources is indicated from the fact that for centuries the tribesmen acquired their metal cooking utensils from Damascus; the principal currency was the Marie Therese dollar, minted in Austria. The limited output of the Mahab Dahab mine and the vast production of the oil wells on the Persian gulf side of the desert, both developed within the past fifteen years, are Saudi Arabia's prime resources. It is oil that has raised her stature to important world rank, especially to the extent that the Near East now plays a more vital and significant role in the international affairs of the United States—a role whose importance was hardly conceived before the beginning of World War II.

The Saudi Arabian desert has been a centuries-long testing ground for the survival of the fittest without benefit of medical science. Up to recent years, whatever health institutions existed in that land were founded by missionaries, principally Americans. Not so long ago the royal house of Saud called an American missionary physician to give medical aid to a female member of the family—a revolutionary departure from the Mohammedan taboo that males should not attend the illnesses of Arab women. It is significant that such a change should come from the most orthodox of Islam's sect, the Wahhabites.

Even though the tribesmen are Moslems, they do not generally practice polygamy. Unlike the women in the settled villages, the women of the desert seldom, if ever, wear veils. They live simple and rugged lives; they are not of the harem-kept breed of the rich city sheikhs. The desert population has remained fairly stable in numbers throughout the centuries, with the tribal feuds and wars, drought and hunger causing more casualties than disease.

With a population overwhelmingly nomadic, naturally there was small need of railroads to connect nonexistent commercial or industrial centers. The only skyscrapers are the steel skeletons stuck into the brown sands to disembowel the desert of its oil wealth. These derricks are the only evidence of industry throughout this desert kingdom, and their presence is completely foreign-inspired and namely, American.

Saudi Arabia has no government that can in any degree be compared to a democracy. Outwardly King Ibn Saud reigns as an autocratic monarch. Actually his government is a patriarchal one. There is no legislative assembly, no system of civil or criminal courts. Ibn Saud sits as ruler, lawmaker, and judge. Any of his tribal subjects may come before him with his plea which the ruler hears much in the manner of a tribal chief in ancient centuries. There is no appeal from the king's decision. Since there is no civil or criminal code, justice is administered according to the Koran.

Since oil has forced this kingdom to deal with modern powers, Ibn Saud has expanded his government beyond the desert's rim. He has followed the Western pattern of establishing embassies to deal with foreign powers. But even in this expansion, he has retained the patriarchal character of his government by filling most of the foreign posts with his sons, as well as choosing his own kin to hold the key positions in the important ministries at Riyadh or the bureaus in Jidda and Damman.

Not until the oil wells gushed their liquid gold was wealth a thing generally known in Saudi Arabia. Its primary beneficiaries, even today, are members of the royal family and the exploiting foreign promoters. The masses of the people still exist day-by-day on the barest necessities of life: two meals daily based primarily on a loaf of bread soaked in soured goat's milk, or dates with bread. Meat is rarely eaten, and generally only upon religious feasts or such celebrations as weddings.

The nomad's wealth is limited to the clothes on his body, his tents and their appurtances, the food supply for which he barters once a year and, most important of all, his herd of camels or goats. With such a simple economy the trade fluctuations in the world's markets have had little effect in the past; but today, with the market demand for camels steadily declining and with Ibn Saud's determination to settle as many nomads on the land as possible, a more complex way of life is in the making—somewhat attuned to Western methods.

It is indeed another paradox that the present inhabitants of the land which gave to the Arabs that purity of language found in the Koran should themselves be so universally illiterate. But this is understandable when it is remembered that education can hardly keep pace with a people who are constantly on the move in search of grazing lands. Where education does thrive, it is found among the villagers and townsmen. Schools have been established only in recent decades, those in the Persian Gulf area by the American missionaries at first, and now by the oil companies. Ibn Saud has recognized the importance of education by helping to found schools in the settled areas and also by sending Arab youths abroad to study, especially in America.

The "Courtyard" (Palestine, Lebanon, Syria, Transjordan and Iraq)

Turkey, Iran, Saudi Arabia, and Egypt may be likened to "houses" facing a courtyard that embraces Palestine, Lebanon, Syria, Iraq, and Transjordan. This courtyard has been the battle arena of Ramesis II, Joshua, Nebuchadnezzar, Cyrus, Alexander the Great, Pompey, Constantine, Omar, Richard Coeur de Lion, Haroun Al Rashid, Salim the Grim, Genghis Khan, Mohammed Ali, Allenby, Feisal, Wilson, and Catroux. Egyptian, Hittite, Phoenician, Assyrian, Chaldean, Hebrew, Persian, Greek, Roman,

Arab, Mongol, Turk, British, and French armies have trampled over it. After the end of the first World War the victorious Allies, dealing with the "courtyard" as if it were an exclusive real estate development, carved it into the independent Arab states of Lebanon, Syria, Iraq, and Transjordan, and a Palestine mandated territory.

Here man finds the earliest recorded traces of his civilization. Here the world's most ancient trade routes crossed as the world's mightiest air lines do now. Here were nurtured in turn Judaism, Christianity, and Islam, that flung their outposts to remotest havens in the world. Here occurred the amalgam of a dozen different races, the first reformations in religious creeds, the flowering of social orders that even today are undergoing great flux. In this same "courtyard," man has always struggled for his bare existence, just as foreign nations have persistently competed for his political and economic domination.

This "courtyard" of 237,961 square miles, with a population of more than 10,000,000 [2] people, is slightly larger than the states of

	Sq. Mi.	Population
Iraq	116,000	4,500,000
Lebanon	3,475	1,025,000
Syria	73,587	2,800,000
Transjordan	34,740	400,000
Palestine	10,159	1,800,000
Totals	237,961	10,675,000

Arizona and New Mexico,[3] with much similarity to them in topography. More than half is barren mountain and desert. Mountains rim three sides. In the Lebanon on the west end and in Turkey on the north, the peaks tower close to 10,000 feet, while beyond the Iraqian border the ranges frequently rise to 12,000 or 14,000 feet. The mountains receive ample rains, from 10 to 12 inches on the Mediterranean coastal area to as much as 50 inches in the Elburz chain north of Iraq. The courtyard's southern area is an extension of the Arabian desert, over which poured successive waves of Semitic immigration. The Arab conquest of the courtyard after

[2] The figures, taken from *Information Please Almanac*, for 1947, are as follows:

[3]

	Sq. Mi.
Arizona	113,956
New Mexico	122,634
Total	236,590

Mohammed's death (632) was not a conquest of race or language or culture. It was merely a conquest of religion that combined both a temporal and spiritual regime.

The settlement pattern follows a rather simple scheme: the principal cities and towns are found either in the lower mountain valleys or at the junction of the plains with the mountains, in effect, oases along the anciently established trade routes. Damascus is such a city as the very origin of its name indicates, "Pearl of the Desert." The Syrian desert is punctuated with ancient oases settlements although nomads still roam its wastes as in Biblical times, seeking pasture for their herds.

PALESTINE.—Palestine has been a land eternally torn by strife. A Holy Land for Jews, Christians, and Moslems, it has seldom been allowed to live in peace with the rest of the world. For centuries it was a part of Greater Syria under the Ottoman Empire. In its long history some nineteen commissions of varying political and religious hues have wrestled in vain to find a peaceful solution to the problem of its statehood. Torn by bloodshed in the establishment of the world's three dynamic religions, its peace interrupted by seven waves of crusaders, it has been crossed and crisscrossed with the clashing arms of the East and the West. As this book goes to press, Arabs and Jews are waging open warfare.

Except for its religious shrines, Palestine is not a highly valuable prize. Two fertile plains—one running along the Mediterranean coast and the other cutting inland across the country (Esdralon)—furnish whatever agricultural richness it possesses. In this tiny land of 10,159 square miles (about the size of New Hampshire) is found the world's deepest land gorge, the Dead Sea, 1,300 feet below sea level. This abyss cuts through the plateau that advances eastward into Transjordan. The southern part of Palestine below Beersheba, called the Negev, is actually an extension of the Arabian desert. Rainfall in the winter and spring is sufficient for growing crops, especially along the maritime coast, which is richly cultivated with citrus orchards. Snow occasionally and briefly covers the tops of the Judean hills, bleak-looking ranges denuded by ancient man and kept bare by hungry goats. The lack of forests is significantly indicated by the fact that wood is a major import (13,161,000 cubic feet in 1941). For years the citrus industry provided the country's largest export item, four-fifths of its income coming from that source.

The Jordan river, rising in the Lebanon mountains, is Palestine's principal stream. Aside from its Biblical lore, its present importance is underscored in a power development project similar to the Tennessee Valley Authority. Another river, the Yarmuk, forms the northern boundary line with Transjordan. In all, five streams flow into the chemically laden Dead Sea, at whose northern extremity is an extracting plant.

The one thing that has given homogeneity to the "courtyard" is its Arabic language, but a wedge has been driven into its solidarity in the form of the introduction of Hebrew and English as official Palestinian languages, imprinted on all currency and used in official government business. This wedge was first made possible by the British government's pronouncement of the Balfour Declaration on November 2, 1917, and later through the League of Nations which designated Great Britain as the mandatory power and incorporated the essence of the Balfour Declaration into the terms of the mandate. This deal was arranged despite these facts: that the Arabs had inhabited the land for thirteen centuries and their ancestors before them for many more centuries; that the British, through McMahon in Cairo, had promised King Hussein of the Hijaz, independence for all Arab people; that the Arabs had helped the British to liberate Palestine from Turkish sovereignty.

A Jewish national home was imposed on the Arabs without consulting them, without their permission, and at a time when Palestine was inhabited by some 500,000 Arabs as compared to 55,000 Jews. Today the population consists of 1,200,000 Arabs and 650,000 Jews, with some 400,000 of the latter entering the Holy Land by enforced immigration.

A tragic consequence of the Balfour policy has been that the British mandatory power, instead of preparing the Palestinians to govern themselves (as the Arabs were prepared in Iraq) has neglected that international trust to the complete dissatisfaction of both Jew and Arab and in its stead established in effect a police state. The British formally ended their mandate on May 15, 1948, even while the United Nations was occupied in lengthy debate as to the future political status of the country.

Education had been practically ignored in Palestine under the Turks. Since World War I it has been promoted largely as a private enterprise. Of schools administered by the government 404 were for Arabs, with an additional 177 private Moslem schools and 189 Christian institutions, with a total enrollment of 90,748 pupils. The Jewish schools numbered 442, with an additional 309 private institutions, having a total enrollment of 86,626 pupils. Under their own aegis the Jews have expanded education to include five teacher-training colleges, 26 secondary and six trade-vocational schools, besides a Hebrew University in Jerusalem with a steadily increasing enrollment now exceeding 650.[4]

Three decades of British occupation, however, have resulted in the development of the transportation system, so that in addition to the section of the Hijaz railway, built by the Turks for the Meccan pilgrimage trade, there are 302 miles of broad-gauge railroad tracks and over 2,500 miles of roads, of which 1,451 miles are all seasonal highways.

Except for those established by French, Italian, and German missionary groups, Palestine had almost no medical institutions under the Ottomans. Malaria was prevalent, especially in the neglected marshy plains. Private Jewish enterprise, the government, the Arabs themselves, and the Rockefeller Foundation have all helped to free Palestine from malaria. The Arab schools are playing a major role in improving the health of the inhabitants, largely through compulsory hygienic courses for the schoolchildren.

Several factors conspire to keep Palestine disunited. First, is Western imperialism. Britain needs oil to maintain her empire's lifeline. The United States, too, wants that oil for strategic purposes. The Holy Land serves as the indispensable link between the East and the West with the principal bastion at Haifa, a British-developed port important as the terminus of the oil pipe-line that originates several hundred miles away in Kirkuk, Iraq. Once united under a common Arabic language, culture, and religion during the greatest glory of the Islamic empire, Palestine today is torn between the sharply contrasting cultures and ideologies of the Arabs and the Jews.

[4] All school figures are for 1941-42 and are from *Information Please Almanac* for 1947.

It is a curious anomaly that while the Arabs and the Jews are united in one thing—their opposition to English control—they are disunited within their own ranks. Their internal leadership is frequently rent with sharp dissensions. To complicate the problem, the Bedouins who, clinging to their ancient traditions and manner of living off the desert, become a serious problem to the settled population when drought drives the nomads into their communities.

The divergencies of the many religious groups have also kept the Holy Land stewing with trouble, especially when anniversary feast days roll around. The Caliph Haroun Al Rashid, practicing the democracy of his Mohammedan faith, acknowledged to Charlemagne, Christian rights in that Islamic domain. For two centuries, Crusaders disrupted the unity of Palestine. In more recent decades Christian sects have so failed to keep religious peace that a Moslem family was entrusted with possession of the key to the Church of the Holy Sepulchre in Jerusalem.

The Mosque of Omar, second only to Mecca as a Mohammedan shrine, and the Wailing Wall enshrined in Judaic hearts as the enduring relic of Solomon's temple, frequently arouse religious fervors that break out of the ancient-walled-in city of Jerusalem and spread tension to all parts of Palestine. At the moment this is written, Palestine is once more torn with strife, Arabs killing Jews and Jews killing Arabs as a result of the United Nations' General Assembly's recommendation of November 29, 1947, that the Holy Land be partitioned into Arab and Jewish states.

LEBANON.—A Christian oasis in a Moslem world! That is Lebanon, where dwelt the ancient Phoenicians—civilization's first recorded seafarers; the one Christian citadel in the Near East which refused to succumb to Islamization; a tiny republic (about the size of Rhode Island and Delaware) which had always been an intimate part of Syria until imperialist France sliced it into an independent state in that great power-politics game of "divide and rule." This is the land which under the impact of Westerners—especially French and American missionaries—spearheaded the Arab renaissance: a movement that within the past three decades has flowered into seven independent Arab states and the more recently established Arab League whose over-all political, economic, and social program also envisions freedom for Arabs in North Africa.

Lebanon is named after the mountain range that rises from a narrow plain on the eastern shore of the Mediterranean, its 3,475 square miles including also the wide and level Beka'a plain with, on its east, the lower range of the anti-Lebanon mountains forming the boundary with Syria. The coastal mountains reach a peak of 9,900 feet within sight of Beirut, the capital and principal seaport. Rainfall, principally in the fall and winter months, averages 22 inches a year. The warm Mediterranean winds deposit an ample supply of water for the citrus and banana and olive groves along the coast, for the vineyards on the mountain slopes and the cereals on the plains.

Lebanon's agriculture is much like Palestine's in this respect: its mountain sides appear neatly ribboned with terraced gardens. Once the home of the majestic cedars that were cut down in pre-Biblical times to be hewn into ships, the mountains have since remained deforested, this condition being abetted by goat herds.

Lebanon is especially famous for its superior grapes and apricots that, along with the olives, grow in the valleys of the mountains. While much of her cereals are produced in the Beka'a plain, wheat is nevertheless imported in large quantity from the Hauran in Syria. Unlike her neighboring states, Lebanon depends primarily on rainfall for crop production, since there are no rivers of importance for irrigation. What the Lebanese call rivers—the Litani, Bardoni, Ibrahim, and Kalb—are really nothing more than small mountain brooks. For many years silk cocoons were a principal export. Since Lebanon was under the orbit of French influence—even from the time of the Crusaders—it naturally followed that Lebanon provided much of the raw product for France's silk factories in the early years of silk weaving. One of the factors that gave rise to strong emigration from Lebanon half a century ago was the decline in the silk industry after the Japanese usurped world-wide control of that market.

Whether it was the mountain barriers that halted the spread of Islam or the dominant strain of Ghassinid Arabs whose spirit of rugged independence made them impervious to a new religious dogma, or whether it was the impact of Western ideologies which penetrated through the eastern Mediterranean gateway into the Near East, Lebanon has remained a Christian island in a Moslem sea.

It is significant that the renaissance which the Arabs are witnessing had its start about a century ago in a movement that owes its impetus to Western missionaries, especially the French Catholics and the American Presbyterians and Congregationalists. Even prior to that period, the Westernization of the entire Near East had been launched when France obtained a concordat from the Turkish sultan in 1535—a right to protect the foreign colonies of traders in the Ottoman empire—which privilege was later expanded to become the first of modern capitulations permitting foreign powers to protect Christians in sovereign lands other than their own. These capitulations later made possible the exploitation of the Arab world for the benefit of Europeans.

Compared with Western standards, agricultural production in Lebanon is technologically backward, due primarily to the poverty of the peasants who have not the financial means to buy farm machinery, since even in good times they barely eke out an existence from the eroded and stony land. Strenuous efforts are being made, especially through the government-sponsored exchange of students and through instruction in the grade schools, to train farmers in the use of scientific equipment. Those who can most easily afford to purchase it are the rich landowners, but they are most often absentee owners who care little about the peasant's welfare, being content to take their large proportion of whatever he produces. Since the majority of the deputies in the Lebanese Parliament consists of absentee landowners, there seems little hope for amelioration of this condition.

Although the Crusaders chose Lebanon as the main artery through which to launch their religious and economic conquests in the eleventh and twelfth centuries, the means of communication remained about the same until recent times. Beirut, Tripoli, Sidon, and Tyre have been the principal water gateways from the West into the Near East, and the mountain trails (only slightly improved from the days of the Phoenicians) have continued to provide the main inland routes. Modern highways, as they are known in Lebanon today, together with the development of railroads, telephone, telegraph and airports, were enterprises sponsored largely by the French, in order to facilitate military control of the land.

Compared to her sister Arab states, Lebanon has always been

farther advanced in the maintenance of health standards. Here again the credit belongs in large measure to the missionary zeal of Westerners who found a dominantly Christian Arab population eager and willing to accept Western health standards based upon science—in contrast to the more hesitant Moslem Arabs who clung with unquestioning fealty to ancient superstitions and religious fetishes. It must be likewise admitted that the establishment of clinics, hospitals, and medical training schools, was in the first instance for the protection of the Westerners themselves. There is scarcely a city in the Near East whose physicians do not include graduates of the French or American medical colleges in Beirut. Many other physicians can also boast of education in European or American institutions.

Tribal living is the exception, rather than the rule, in Lebanon. People generally dwell in stone houses, whose red-tiled roofs remind one of southern Italy or Spain. Stone is Lebanon's most abundant building material. Homes are furnished much as they are in any American or European community; in the larger towns they are connected with water mains and sewerage systems, as contrasted with the stark and elemental simplicity of the mud huts of Egypt and the tented villages of the desert areas.

The family, rather than the tribe, is the social unit in Lebanon. While strong family loyalty persists, there is nevertheless a greater freedom of social intercourse among Lebanese Christians than among the orthodox Moslem families either in Lebanon or in the neighboring states. With reference to the status of women, while courtship is not widespread as it is among Westerners, the Christian girl in Lebanon is not compelled to accept a suitor picked out for her by her parents, as her Moslem peasant sisters are, in most other parts of the Near East. Family loyalty is so strongly entrenched that it has prevented industrial and commercial development on a wide scale. In business relations, trust is strictly confined within the family; that is the primary reason why great corporations have never developed successfully in the Near East.

Nature, religion and the foreign powers have conspired together to keep Lebanon disunited. Nature, through her mountains, has provided settlements in the fertile valleys or along the coastal or inland plains, but she has at the same time kept these communities separated by tortuously accessible approaches. However, the expansion of highways, due to the steadily increasing use of automobiles within the past quarter of a century, is proving a stronger link toward unity. By means of the automobile one can now cross the width of Lebanon within two hours or its length within three hours.

Religion has served to impede the unity of Lebanon's population. The Maronites are strongest in numbers among the Christian sects, but equally forceful in many communities are the Greek Orthodox, the Greek Catholics, the Roman Catholics, the Gregorians. On the Moslem ledger, the Sunnites outnumber the Shiites by a small margin. To make the religious mixture still more indigestible there are the Druzes in the central area that overlaps into Transjordan and Syria, and the Alouites in the northern part that is contiguous to the Latakian province in Syria. Throughout the centuries the Ottomans were able to rule Lebanon with comparative ease, simply by playing one religious group against the other, causing bitter feuds that at times degenerated into massacres. Even in most recent years, France employed the "divide and rule" technique by spawning rivalries between the Maronites and the other Christian sects. In earlier years even Great Britain was not immune from such schemes, as she relied upon the Druzes to cause religious turmoil so that the attempted control of Lebanon by any foreign power against Ottoman authority would remain precarious. Russia, too, under the Czars championed the cause of the Greek Orthodox, who looked to the Muscovite prelate for protection.

With such rivalries between powers and divergencies of loyalties between religious groups, it was not any wonder that a native leadership failed to develop that would gain the harmonious support of most Lebanese. Even as recently as 1946–47, American newspaper reporters with a strongly biased support of the Zionist cause sought to destroy Lebanese-Arab unity by reporting that the Christian sects in the mountain republic held views divergently opposed to their Moslem brothers with respect to an independent Palestine.

The lip service which France paid to Lebanon as an independent country finally came to fruition as an exigency of World War II, when Vichy France sought to hold the mandated territory in obeisance to the Axis. After the British had destroyed this false position by conquering the area in the summer of 1941, General Catroux proclaimed France's previous pledges of Lebanese sovereignty; a gesture which DeGaulle's government impudently tried to repudiate in 1945. Two years earlier, the United States, Great Britain, and Russia (among other powers) had recognized Lebanon's independence. The French attitude, together with her bullying use of military force, as she did in her heyday of colonial conquest, finally gave the Lebanese the one rallying point long needed to unite her people to free themselves from foreign exploitation.

Lebanon has a parliament modeled on Western lines: a chamber of deputies chosen at a free election in representative districts; a president, elected by the deputies; a cabinet, whose prime minister is named by the president and whose members in turn are chosen by the former with a fair cross-section of the various religious groups represented in the ministries.

SYRIA.—Syria may be described as occupying the very center of the Near Eastern "courtyard." Its area of 58,456 square miles is about the size of Michigan, but its population of 2,860,411 is slightly more than half of that American state. It is inhabited today by a racial stock that includes a strongly predominant Arab population, besides Kurds, Turks, Armenians, Circassians, and French.

Syria has been mankind's most fought-over battlefield, despite the fact that a rather barren desert preempts over half its area. Of some 12,500,000 acres suitable for cultivation, actually only one-fifth of it is now under cultivation.

Syria's principal cities reflect the roles imposed on them by nature. Damascus, the capital, is situated on the western edge of the Syrian desert, actually an oasis on the Baradi river that spills into the desert. Hama, farther west near the Lebanon mountains but in the Orontes valley, is renowned as a Bedouin trading center. Aleppo, second largest city and also on the Orontes, has been a major trading center on the ancient northern caravan route. The anti-Lebanon and Kurdish mountains form Syria's western and northern boundaries with Lebanon and Turkey, and it is in this area that most of Syria's population dwells. The Jebel Druze in the southwest near Transjordan, containing the great wheat-producing district of the Hauran, rises to 5,800 feet and tapers away in gradually declining steppes to the Euphrates valley. The rest of Syria is largely steppe-land and desert, with Saudi Arabia on the south and Iraq on the east.

With the exception of the mountain and valley areas, Syria receives hardly more than four inches of rain a year. The summers in the hinterland are extremely hot and dry, where the bleakness of the desert is broken by oases settlements, such as the ancient Palmyra (once the seat of Queen Zenobia who in the third century ruled over the "courtyard").

The Euphrates river, whose source is in the Taurus mountains in Turkey, cuts through from the north to flow along Syria's eastern border. Along its northern reaches are more oases-like communities, while the interior is inhabited largely by nomadic groups that have come in persistent waves from among the Anaza tribe in the Arabian peninsula, constantly refreshing the Arab bloodstream for centuries.

The seminomads live rather settled lives about the oases. Besides growing wheat and barley, they and the Bedouins raised 2,492,000 sheep, 1,583,000 goats, 490,000 cattle, 287,000 donkeys, and 54,000 camels in 1943. Stock-raising is the principal industry, but this husbandry is still conducted along primitive lines, although Syria is making strenuous efforts to provide agricultural education to its farm population. It is common, on rural highways leading to major cities, to see flocks of sheep or goats traveling scores of miles on foot to a market center. Livestock transportation via railroads is rarely utilized, since in the first place few railroad lines reach into the interior of the livestock producing areas. It is also cheaper for the rich sheikhs to employ human beings at low wages to lead their flocks to market, even though the trip may consume days.

Syria in ancient times was the granary of the world, and today she retains that position in the Arab world. Where tractors and agricultural machinery have made their appearance, they are generally found on the huge estates of wealthy landlords. The hired fellaheen and the farmers, who own only a few feddans of land, have rarely benefited from the scientific discoveries and the labor-saving machines of modern agriculture. The part of the Euphrates river that flows through Syria offers possibilities of increasing the amount of arable land through irrigation but, until the wealthy landowners are willing to support such a huge government venture, progress of this kind is unlikely to occur.

Railroad facilities are found primarily in the fertile crescent that connects the principal cities along the Orontes and Euphrates valleys with the eastern Mediterranean seaboard at Beirut, Tripoli, and Latakia. Rail lines stretch from the Aleppo gateway northward into the Taurus mountains to Turkey or southward into Baghdad. Railroad development, as in Lebanon, has been of very recent origin and promoted by French interests.

Highways, too, owe their modernization and increase to France, who held the mandate of Syria from the League of Nations from 1920 until world-wide public opinion finally forced her eviction in 1946. Road development was designed especially to knit closer together the strategically situated French garrisons that kept Syria under French domination, at the same time not ignoring their importance to trade outlets. Telephones and telegraphs, water mains and sewerage systems and airports are Western importations that have benefited city dwellers almost exclusively. The villagers and nomads have yet to enjoy these modern conveniences.

In the desert areas the system of tribal living continues much as it did when Damascus was the seat of the Moslem empire in the eighth century. The peasants who raise the bulk of the wheat, barley, millet, tobacco and mulberry crops (the major ones in Syrian export trade) still live under the feudal lordships of the wealthy sheikhs and absentee landlords who, incidentally, compose the vast majority in the Syrian parliament.

The desert, the variety of religions, the several races indigenous to her soil, the invading foreign powers—each has worked in its own way to keep Syria disunited. This Levant "courtyard," like most of the then-known civilized world, was once ruled for 89 years from Damascus, the seat of the Ommayyad dynasty. Through all the centuries the desert has intervened to keep the settled population widely separated, while the nomadic tribes have moved constantly about in search of evanescent blades of grass.

Although the Moslem faith predominates overwhelmingly, it has in itself had its disunifying influences from the time that the Shiites broke away from the Sunnites to establish the Caliphate capital in Baghdad. The Sunnite sect provides nearly 70 percent of the Islamic groups scattered about the country. In the Latakian province dwell the Alouites who make up 11 percent of the religious population. In the most fertile part of Jebel Druze dwell another different Moslem group—the Druzes who contribute 3.1 percent to the religious mixture. The Greek Orthodox add 4.6 percent to the conglomeration, the Armenian Orthodox 3.5 percent and the Syrian Orthodox, Syrian Catholic, Greek and Armenian Catholics, and Israelites contribute 7.9 altogether.

Besides the present Arab stock, Syria contains Kurds in the northern mountain areas. Armenians, Turks, Circassians are spread out along the Orontes valley. The Arameans, remnants of the once dominant language group in the Levant, live along the Euphrates. The failure to assimilate these groups has kept Syria from that essential unity necessary for a strong, free land.

This country has been history's most abused political pawn. With its distinctive religious creeds and clannish racial fealties, a strong leadership loyal only to the greater service of Syria was impossible of development. The foreign powers, knowing this situation, took advantage of it to impose their authority. It is significant, however, that in the nationalist movement, Syria played a vital role in weakening the Sultan's power in 1908 and later participated in the coup against Turkey in 1915. After the province was freed from Turkish control following World War I, France tried to keep Syria in her pawn-like role. Under pressure from the United States and Great Britain, France was obliged to leave Syria in 1946, the first definite break in colonial imperialism. And it was left to the Syrian Faris Bey Al-Khouri (imprisoned by the Turks in 1915 for his part in his country's revolt against the Ottomans and twice imprisoned by the French) to preside as chairman of the Security Council in the United Nations in August, 1947, which heard the plea of the Moslems in Indonesia to be freed from another colonial power—the Netherlands.

Since Syria gained her freedom, she has made primary education compulsory, spending 9 percent of her budget for schools. She has founded a university in Damascus with faculties of law and medicine; she has tried to help solve the medical problems of her widely separated population by providing them with mobile health clinics; she has encouraged expansion of hospitals; she has enacted laws to protect labor, especially in the cities.

But Syria has not the latent resources of an industrial state to finance huge, nationwide, undertakings. Her income is very limited. What industries she does have are mainly of the consumer variety: flour, fruits, soap, oils, brass utensils, tobacco, textiles and shoes. Most of her industries are handicrafts, designed primarily to meet the needs of local consumption. While the supply of potential labor is ample, the presence of the necessary natural resources, such as minerals for the development of industry on a large scale, are missing. For the immediate future, at least, Syria seems destined to retain an agricultural economy.

Syria, long under the domination of Western powers, has aped the democracies by establishing a constitutional form of government, built around a parliament with a chamber of deputies whose members are elected every four years. The deputies elect the president and he chooses the prime minister. A cabinet of ministerial department heads serves as the executive branch of the government. Syria has charted her course along democratic lines.

TRANSJORDAN.—Transjordan, with 34,740 square miles, is about the size of Indiana, yet its population is little more than that of Indianapolis (386,972). Originally a part of the Turkish province of Syria, Transjordan came into being as a result of the "real estate" exchanges by the Allies after World War I. As an inducement to revolt against the Moslem Turks, the British offered Sherif Hussein, the protector of the Islamic shrines of Mecca and Medina, freedom and statehood for the Arabs. The Arabs revolted in 1915–16 and helped to defeat the Turks. The Allies then carved Palestine out of the ancient area known as Syria and gave the mandate to Great Britain. The latter in 1920 chiseled another chunk from Palestine, named it Transjordan, and eventually called Abdullah, second son of Hussein, to be its ruler. This action may have been designed to appease the Meccan sherif. After a quarter of a century of mandatory existence, Transjordan, which is still dependent on British bounty for its economic support, was granted full independence by Great Britain in March, 1946, when Abdullah yielded his emirate to assume the kingly crown with English blessing.

Why so much importance should be attached to this backward spot in the Near East is, of course, best interpreted in its relation to Palestine. Moreover, Transjordan's location acts as a buffer state to any Pan Arabian ambitions that the Wahhabite Ibn Saud may have; he faces Hashimite rulers both in Transjordan and Iraq.

Transjordan's population contains about 120,000 seminomads and 50,000 nomads. Of 454 settled places, only four may be classified as urban centers. About one-fourth of the entire population live in ten communities that have municipal councils. The remaining 300,000 are scattered in settlements along a fertile strip bordering Palestine, or roam the desert with their movable black tents. Control of the area passed from the Bedouin chiefs, who had ruled it for centuries, to the Turks in 1894. Thirty years previously the Turks had taken possession of the area around Amman, the present capital. The Turks ruled this land as part of the province of Syria.

Of Transjordan's total area, 80 percent of it (representing about three times the state of Vermont) is nothing more than steppe and desert. Actually, it is a plateau, whose height varies from 4,500 feet on the gorged edges of the Dead Sea land rift to 1,500 feet where the plateau molds itself into the desert that sweeps eastward from the Syrian and Arabian deserts toward Iraq. One has only to travel 233 miles to cross Transjordan at its greatest length or breadth—a distance comparable to an automobile drive from New York to Washington.

Transjordan's mountains in the Jordan valley area get about 20 inches of rain a year, principally from October to March. The rainfall tapers off to about five inches in the eastern and southern desert zones. A few oases interrupt the barrenness of this arid expanse which in the winter time offers pasturage for the sheep, goats, and camels raised by the nomads and seminomads as their principal means of livelihood. Most of the settled population lives in the 31-mile wide strip west of the Hijaz railway line.

The desert is inhabited a few months each year by about 30,000 Bedouins, who have maintained a precarious routine for many centuries. They lead their herds in search of pasturage and, when it vanishes in the desert, bring their flocks to the settled communities. Their husbandry has not kept even slight pace with the progress made in the Western world.

The seminomads number about 100,000. Preserving their customs and manner of living along Bedouin lines, they continue to live in movable tents even though they have now become cultivators of the soil. They differ from the nomadic tribes in the growth of subtribes within their own organization and with it the increase in petty chieftains.

For centuries Transjordan has been divided by conflicts between nomad and seminomad, the former regarding the settled dweller as his natural prey (especially in times of severe drought) and the latter holding the Bedouin as his perpetual enemy. It was probably from the desire to reconcile these groups and bring them together in mutual understanding that Abdullah was empowered in the first years of the mandate to establish a tribal control board. The scheme was abolished in 1931 when it was decided to place the desert under the more effective administration of an officer of the Arab Legion, run by the British.

Transjordan's ethnic background, traceable directly to the Arabian peninsula from which it has received freshening streams throughout the centuries, is far purer than the strain of Arabs in the adjoining countries, Saudi Arabia excepted. Most of the settled populace, as well as the nomads and seminomads, are descendants of Bedouin tribes. Of an estimated 400,000 population, some 7,000 are Circassians, 30,000 are Christians, and 3,000 represent miscellaneous racial groups, including a small number of Negroid strain who probably came into this region as slaves from the Sudan.

Of the Arabs, 90 percent are Moslems with the vast majority in this group Sunnites. In their profession of Islam, the settled Arabs are far more devout than the Bedouins whose constant wandering with their herds has ingrained in them a rugged individualism.

Of the Christians, the majority adhere to the Greek Orthodox Church. The Catholics include the Roman (Latin) and Greek (Melkite) rites. These congregations trace their origins back to Byzantine-pre-Islamic times. There are also small numbers who have become Protestants through the proselytizing efforts of American and British missionaries. Christians are found not only in the principal towns of Salt and Madaba and in the leading villages but also among such distant tribes as the Habashneh and Magali. In appearance they can hardly be distinguished from the Moslem tribesmen. Polygamy is permitted in their mode of life, but it is rarely practiced among them.

The Christian tribesmen have no priestly hierarchy, since they choose their priests from among themselves, much as the Moslems select their imams to lead them in Friday prayer in the mosque. As for churches, these do not exist according to Western concepts; their churches are in their tents.

It is indeed a paradox that while the Arabs and their ancestors have inhabited the Transjordan region for centuries as an almost 100 percent indigenous population, it was left to an influx of immigrants from the region of the Caucasus mountains to develop the country economically and politically. Some 1,000 Chechen Moslems migrated into Transjordan in 1864 from their mountain homes in Daghestan, rather than submit to the Christian rule of their new conquerors, the Russians.

The Circassians immigrated after the Turkish-Russian war of 1877–78, to escape Christian rule. They settled mostly in Amman, the capitol of Transjordan. The Turkish sultan gave them preferential treatment, exempting them from taxes and conscription for a number of years. Since in their homeland they were familiar with modern methods of agricultural production, and since they brought farm equipment new to the Arabs and at the same time introduced the use of horse-drawn vehicles, they were able to turn undeveloped tracts into sown fields. They introduced new methods in tending livestock. Thus they helped to lift Transjordan's economy from its depths of poverty. The Circassians are Shiite Moslems.

About 1,400 miles of good roads connect Transjordan with the principal towns in Syria, Palestine, and Iraq. Some highways grew from military necessity while others developed in response to the increasing demand for motor travel. Part of the pilgrimage railroad, which the Turks built from Damascus to Medina, passes through Transjordan. Besides building new roads, the British mandatory helped to establish educational and health systems. Illiteracy is very widespread among the nomads.

The government is a constitutional monarchy. Assisting the King is a cabinet of ministers composed of department heads and a Legislative Council of twenty elected members. The country is divided into districts; each unit is governed by an administrative officer aided by a council composed of two officials and four nonofficial members. According to the treaty of mutual aid signed on March 22, 1946, Britain is permitted to train, equip, and provide financial assistance and officers for the Transjordan Frontier Forces, to which, together with the Arab Legion, is entrusted the defense of the country. The treaty gives Britain the right to maintain her own troops in Transjordan as well as several bases for her air force. These are privileges of tremendous importance to a Britain which is being slowly edged out of Egypt and away from the Suez canal, for it gives the London government an opportunity to further secure her world trade lifeline in the new shuffle of international politics.

Transjordan's resources include undeveloped deposits of iron, manganese, and ochre. Salt is already being extracted from its side of the Dead Sea. Wheat, fresh fruits, wool, and livestock are the chief items of export, in exchange for sugar, tea, and hundreds of miscellaneous items. That there are immense possibilities for the development of its untapped resources may be indicated in the persistent demands in some Zionist quarters that any creation of a Jewish state in Palestine should eventually embrace Transjordan as an integral part of it.

IRAQ.—The land between the Euphrates and Tigris rivers, regarded by some Biblical scholars as the original Garden of Eden and known from earliest times as Mesopotamia, today bears the Arabic name of Iraq. Of the six Arab states that were created after the first World War, Iraq was the first to become internationally recognized when its constitutionally organized monarchy was admitted as a member of the League of Nations in 1932.

Situated at the eastern end of the Levantine "courtyard," Iraq (116,000 square miles) is about the size of the combined states of New York, New Jersey, Pennsylvania, Maryland, and Delaware, but its estimated population (4,500,000) is less than that of Maryland. Iraq is primarily an agricultural country; more than 60 percent of its population gain their living directly from the land. Of the entire area more than half is arid desert; slightly more than one-fourth, principally along the river valleys, is potentially cultivable. A vast irrigation network once covered the latter area, but was completely ruined by the Mongols about the time they destroyed Baghdad in 1258. At the present time only 6,500,000 of the 20,-000,000 cultivable acres are irrigated. Although the government is making some progress in restoring the irrigation system, the poverty-stricken fellah receives little benefit, for the reclaimed land is usually gobbled up by the already rich sheikhs and absentee landlords. As in ancient times, Iraq continues to play her role as a world's important granary. When the shipping crisis cut off the Near East from the rest of the world and when famine tormented India again, it was Iraq's immense wheat and barley crops that not only helped to reduce the crisis but also helped to feed the Allied armies on Near Eastern soil, thus freeing Allied ships for other essential war needs. Except for sugar, coffee, and tea, Iraq produces practically all her essential foodstuffs, and this in spite of the fact that primitive methods of cultivation are still practiced as they were a millennium ago when the Abbassid dynasty spread Arab fame over the world. While the government has set up agricultural experimental stations and expanded agricultural education, the native wooden plough with its metal shoe and the sowing of seeds by hand still remain in almost universal use.

Rainfall occurs almost exclusively in the northern part of the country, where the mountains form a sort of retaining wall as they stretch from the Turkish to the Iranian borders. In that area, between Mosul and Kirkuk, are the wheat and barley granaries which depend entirely upon a rainfall that averages 16.71 inches a year. The rains occur mainly between December and March; the precipitation in southern Iraq is too slight to permit a regular agriculture without irrigation. It is extremely significant that mankind's earliest known civilization should have started on land where irrigation was necessary for maintenance of life. It is in this irrigated area that Ur, the most ancient city in the world and the birthplace of the Prophet Abraham, was reputedly built some 6,000 years ago. The significance of irrigation to Iraq's life may be seen in the fact that 6,500,000 acres are cultivated in that manner as compared to 3,000,000 acres in the dry farming zone.

Iraq has three main geographical divisions:

1. The desert that encroaches upon the Euphrates river from the western direction of Syria, Transjordan, and Saudi Arabia—a hard gravelly plateau in its northern part and a great stretch of sand in the south. It is sparsely settled with Bedouin tribes constantly on the search for the precarious pasture for their sheep and camel flocks. An intense heat fans over the southern part of this plain below Baghdad, hugging close to the 120-degree mark in the summer time.

2. The plains, where the vegetation derives from the alluvial soil deposited by the Euphrates and Tigris rivers between the desert and Iran's Zagros mountains. Flying over this twin-rivered valley in a plane, one can see the awesome specter of what centuries of neglect have done to an anciently established canal system. It is in this same valley that the bulk of Iraq's population lives, the cultivation being confined to river banks and their radiating canals.

3. The mountains in the northeast corner, where dwell the Kurds who cultivate the valleys and graze their goats and sheep on the ample verdure of its hills. The area where the mountains merge into the upper plains contains the Semitic plain-dwellers and the non-Semitic mountaineers—the Arabs, Kurds, Turks, Arameans and Yazidis.

Iraq's population statistics are difficult to obtain, because the Arabs have not forgotten that the census was a device used by the Turks to levy taxes or impose military service. The Arabs compose 80 percent of Iraq's population, with the Kurds representing the largest single minority at 17 percent. The Jews, whose origin is traceable to the Babylonian captivity, number close to 100,000. The ancient Assyrians, whose forbears once proudly dominated Asia Minor, have dwindled to a mere 25,000. They were brought into Iraq by the British from Iran and Turkey and speak an Aramaic dialect. Northeast and west of the oil producing region of Mosul live the Yazidis, a secret sect whose main shrine is at Shaikh Adi and whose religious tenets include the propitiating of the Devil.

The religious complexion of Iraq is equally interesting. Almost 95 percent of the people are Moslems with the Shiites composing 54 percent of the population and the Sunnites 41 percent. Karbala, the city where Mohammed's grandson, Hussein, was assassinated, and Najaf are shrines as holy to the Shiites as Mecca and Medina. The Sunnite stronghold is among the Kurds. There are close to 100,-000 Christians, most of whose villages are in the Mosul province. The Christians are also divided into various sects, the ancient Nestorians (Assyrians) and Jacobites being the principal units, while others include the Uniate branches of these two, the Chaldeans, and Syrian Catholics.

The nomadic tribes inhabit the western desert. They live in tents woven of wool, or of goat or camel hair. They live mainly on bread, dates, and the soured milk of their animals, exchanging live-stock and animals skins for wheat, cloth, and elementary utensils.

The settled tribes or seminomads live mainly along the banks of the rivers, raising crops and stocks and remaining near their fields until after the harvest, when they resume their nomadic habits. Their homes are mud huts. In the rice area in the southernmost plains live the marsh Arabs, their dwellings a veritable forest of reed huts built upon piles driven into the water or on islands in the marshes.

Most of Iraq's stable villages are found in the north, where the dwellers may be agriculturists or carry on handicrafts. Some villages serve as distributing centers for the tribes. Often villages are entirely owned by the extremely wealthy absentee landowners or by Aghas (the sheikh or chief of a Kurdish tribe).

The cities are mainly commercial towns for the transshipment of merchandise to neighboring communities and for the home or export trade of the agricultural, animal, or mineral products of Iraq. The few industries are found in the cities where the merchant class includes large numbers of Syrians, Jews, Armenians, and other non-Moslems.

Nearly three-quarters of the population are peasants scattered through villages and small settlements, while another 20 percent are pure nomads. Politically oppressed by the Mongols, Persians, and Turks for nearly seven centuries, it is small wonder that Iraq has been a backward state. What progress it has made has come in the past quarter of a century since the British freed it from the Turks.

The oil pipelines that stretch across the barren desert to Tripoli and Haifa are the major bloodstreams rejuvenating this once fabulously medieval land. The British developed the petroleum fields that now finance the modernization program of Iraq's young government. The number of schools has been increased from 75 for the entire country under Turkish suzerainty to more than 1,000 today; women teachers receive the same salaries as men; and compulsory education has been introduced.

Iraq's first modern hospital was founded in 1872, but from that time until the first World War medical progress was appallingly slow. It was the foreign missionaries who founded medical institutions in the main towns. It was the British army of occupation which laid the foundation for Iraq's present health system—this in a land where a thousand years ago there were no medical schools anywhere in the world more famous than those in Baghdad, where Avicenna and Rhazes taught, and whose influences so profoundly affected the development of medicine and science in the West. Within the past quarter of a century a new medical college has been founded, with schools of pharmacy, nursing, and midwifery. Such endemic scourges as malaria, hookworm, bilharziasis, trachoma, and tuberculosis are being fought. In the past deaths caused by malaria have reached as high as 50,000 with as many as 700,000 people suffering either a new infection or reinfection.

The young government, founded by Feisal, third son of Grand Sherif Hussein of Mecca, divided the country into 14 provinces with a chief health officer directing the campaign to raise the health standards of a people who for centuries have been the victims of disease primarily because of their poverty and ignorance, the parents of disease. The magnitude of Iraq's health problem is indicated in the fact that in 1946 there were only 528 doctors for the entire country. Were it not for the boon that has come from its oilfields, Iraq would probably have continued to be fatalistically at the mercy of endemic diseases.

While nature seems to have decreed that Iraq must remain basically an agricultural country, the government has made efforts to stimulate new industries by offering tax relief. Among 70 manufacturing concerns benefiting from government concessions are 20 brick and tile producing plants, eight cigaret companies, six woolen textile plants, three cotton ginneries, four distilleries, six mills, and three soap factories. Many handicrafts flourish by ancestral methods. While this industrial record is nothing much to boast of, at least it is a start, even if that start owes its principal support and some of its financing to Western interests.

Oil production remains by far the principal industry and wealth producing export item, not even excluding the date groves along the 100-mile stretch of the Shatt Al Arab river from the Persian Gulf to Basra, the area that produces 80 percent of the world's date trade.

Many forces contributed to keeping Iraq disunited until Feisal, whom the French had previously purged as king of Syria, showed his great statesmanship after he had been chosen in a plebescite by the Iraqis in 1921 to head their government.

The thinly spread population, the lack of raw resources, the intervening desert, the limited means of communication restricted to three rivers and a recently established railroad, the diversity of races with their innate loyalties, the clashing ideologies of several religions, especially the feuding supporters of the Shiites and Sunnites, the feudalism of the rich sheikhs and landowning classes, the economical and political exploitation of the native by foreign powers—all these have been obstructions which this new Arab state has had to hurdle.

Iraq cannot be expected to become a great power until these conditions have been ameliorated. The frequent feelers that such Pan-Arabists as General Noury As Sa'id Pash have sent out for the establishment of a greater Syria under a Hashimite dynasty led eventually to the founding of the Arab League and give further evidence of some of the many rifts in the fabric of unity in that area.

The "House" of Egypt

Although it has a constitutional monarchy with an elected chamber of deputies, a senate, a cabinet of ministers headed by a prime minister—all this modeled on the Western conception of democracy—Egypt remains essentially a feudalistic state. To realize this fact, one has only to travel the 900-mile length of the Nile valley and see the abject poverty of its fellaheen (peasants), who number 13,000,000 of the country's 17,000,000 population; this in a land which 5,000 years ago had attained a high civilization while Europe was still a barbaric outpost.

What ironical travesty that the land which claims to be the cultural and spiritual center of the Arab and Moslem world should itself hold a population that is 80 percent illiterate, living in squalor and disease and hunger, while a few thousand live off these exploited masses in greatest comfort.

In Egypt, it is cheaper to hire human beings than to buy an animal to do the work. It is not unusual to see two or three men harnessed with ropes and pulling a boat through a canal when the breeze is not sufficient to stir the vessel's own sails. This is the land where barefoot men often replace donkeys in the shafts of two-wheeled carts, tugging farm produce to the market.

If the Nile River were to dry up suddenly, Egypt would vanish. This 4,000-mile stream, longer than the combined length of the Missouri-Mississippi rivers, rises in Central Africa and flows northward into the Mediterranean, cleaving the monotonous expanse of the desert that covers North Africa from the Atlantic Ocean to the Red Sea. It deposits a rich alluvial soil which from time immemorial has given life to a population whose maximum density per square mile in the overcrowded Nile delta is claimed to be higher than in any country in Europe. Through its course in Egypt the Nile is not fed by a single stream.

Egypt has an area of 386,000 square miles, about seven times the size of the New England states, but its cultivable area (8,600 square miles) is hardly greater than the state of Massachusetts; 99 percent of its population is concentrated along the extremely fertile Nile banks. Large irrigation works have tremendously increased agricultural production in the past century, so that two or three crops a year are possible instead of the lone crop under the Nile flood or "basin" system of Pharaonic times. The great irrigation works of the past half century—the Aswan dam and the several barrages that cross the Nile—have enabled the already wealthy and absentee landowning classes to profit still more from the engineer's skill, whereas up to the beginning of World War II, the fellaheen worked for as little as eight cents a day.

The population falls into at least eight distinct groups: the palace, Al Azhar University, aristocracy, wealthy middle class, lower middle class, leaders among the fellaheen, illiterate fellaheen, and illiterate townsmen. These may be regrouped into (1) the comparatively well educated, totaling a few hundred thousand persons and including the palace crowd, Azhar graduates, the aristocracy, the well-to-do middle class, and those from the lower middle class who happen to be well educated; (2) the barely literate, totaling about 2,500,000, and including the village 'omdahs (mayors), sheikhs, and heads of influential families, who can read and write but have not received an education in the Western sense; (3) the more than 12,000,000 illiterates—the fellaheen and the lower strata in the towns.

In Egyptian elections, all votes are generally controlled votes, the only question being who will do the controlling. The fellaheen and the poor townsmen are completely dominated by the other groups, as they have been from the beginning of time. Since they have faith in their 'omdahs or village notables, they take election advice from them, the village "bosses" having previously accepted orders from other groups more powerful than themselves. The fellah has no conception of democracy. He knows that through the harshness of fate he is absolutely dependent on the 'omdah as protector, and votes as his protector wants him to vote. The agents of the dominant classes make a special tour to the villages at election time and pass the word along to the little "bosses" as to how the masses should vote.

The fellaheen as a group are 95 percent Moslem. In Upper Egypt a few hundred thousand of them are Christian Copts. All are primitive tillers of the soil. Generally they are superstitious, and hence great followers of tradition. They distrust foreigners especially. Their "natural" leaders exert a pressure which they gullibly accept.

The people of the Nile lead a wretched life. Physically and mentally they are little better off than the animals they raise and with whom they often share their one-room mud houses. They have been in this condition for the past four thousand years. Any reasonable person who has the courage to present a bold, practical plan for reform is denounced as a communist and, as in Iraq where the landholders are solidly intrenched, runs the risk of spending a few years in prison. What is worse, out of sheer ignorance the fellaheen are unwilling and probably unable to help themselves.

Gloomy as this analysis may seem, the Egyptian government is making steady and definite progress to alleviate the conditions of the masses. A significant first step was the establishment of the Ministry of Social Affairs a few years ago. Since then many laws have been passed to better the educational, health, and economic standards of the masses.

The Westernization of the Near East can be said to have edged its way first through Egypt's door. It started, in a definite sense, in 1798 with Napoleon's invasion and conquest of the Nile valley, a conquest that was destined to carry him through the Near East until his march was checked at Acre in Palestine, by the British fleet, aiding the Turks. The Napoleonic interlude paved the way for Mohammed Ali, an Albanian general, to take over the suzerainty of Egypt under the Turkish sultan. Mohammed Ali, founder of the present royal dynasty in Egypt, realized that if the Arabs were to have an army strong enough to resist future invaders, it must be trained and equipped like a European army. Therefore he sent his officers to the continent to learn scientific techniques and political organization. Since armies must have medical care for their sick and wounded, he founded the first modern hospital in Egypt, Kasr El Aini. He built roads and canals. He founded schools. He established a system of government, modeled on the Western idea of responsible department heads. The men he sent to Europe returned to Egypt and helped to spread Western culture in a land where Al Azhar university had long propagated an entrenched orthodox Islamic culture.

Missionaries—French, British and American—filtered into the Nile valley, to add the influence of their schools and health institutions to the Westernization process. Where once the center of Arab culture and learning was in the Levant, after the middle of the nineteenth century it began to shift to Egypt, as the Turkish sultans more and more restricted the movements of the Lebanese, Syrians, and Iraqis, who thereupon left their lands for the greater freedom in the Nile valley. This exodus into Egypt as a haven from Ottoman oppression attained its richest flowering at the time that the British had gained control over Egypt and not only permitted the Levant Arabs to settle in the Nile valley but also used them in key government posts, in preference to native Egyptians who were illiterate or untrained in government work.

The British seized control of Egypt in 1882 to protect the empire's lifeline to India. The key to this control was the Suez canal, which DeLesseps started digging in 1859, a task that took 10 years. Its importance as a world artery was convincingly impressed upon his queen by Disraeli, then British prime minister, who engineered the deal for control of the canal stock. It is this artificial waterway that has been the futile goal of the Germans, the Turks, and the Italians in their struggle to dismember the British Empire. It is this artificial waterway that keeps England so solidly anchored in Egypt,

its army now camping on the banks of the Suez as Egyptian statesmen in the summer of 1947 appealed in vain to the United Nations
Security Council to order their eviction.

As a result of Western health standards, an orderly process of
government, and the development of cotton as the principal export crop with its resultant increase in employment of the fellaheen,
the population of Egypt has more than doubled since England intruded in her affairs. Of all the Arab states, industrialization is making its greatest and most rapid strides in the Nile kingdom. Within
the space of five miles one can visit a textile plant at Mehalla
el Kubra in the Nile delta, where thousands of natives operate
Western-made machines, and then drive out in a few minutes time
to a canal where he can still see the Saqqaras, the ancient water
wheels of the pre-Christian era, the water buffaloes or donkeys,
their eyes blindfolded as they faithfully plod in a circle, drawing
water for the fields where the cotton grows. Along the Nile also
grow barley, wheat, rice, dura, berseem, beans, dates. The peasants'
daily diet is based on beans and dates when they are in season. Meat
is a food that the poor can rarely afford to enjoy, generally only on
weddings, religious feasts or national celebrations.

It would appear from a percentage point of view that Egypt is
the most Moslem of the Arab countries we have been studying. The
population, according to the 1937 census, was divided into 14,-
552,695 Mohammedans, 1,303,970 Christians, and 62,953 Jews. In
this, as in other Arab lands, the population is splintered into many
religious and racial groups. Most of the Moslems are Sunnites, among
them Arabs in the Delta area near the Sinai peninsula border whose
ancestors emigrated there from the Arabian peninsula. There are
also Sudanese, whose presence in the Nile seems to have added a
Negroid strain to the native Egyptians, especially in Upper Egypt,
and the still different racial stock of the Senussi in the Libyan border
area; some Berbers have also filtered in.

Among the Christians the largest single group is the Copts, about
one and a half million of them, who have clung tenaciously to their
faith despite Moslem pressure. Occupying mainly the Upper Nile
Valley, they live in tiny villages and the larger towns, and while
playing a decidedly minor role in the politics of Egypt, have nevertheless served as clerks and accountants in government offices. Some

have held cabinet posts. It is among the Copts that Protestant missionaries have gained few converts they have made in nearly a century of effort.

Egypt is one country where the trade and commercial life has been largely dominated by foreign groups. The Syrians and Lebanese (thanks to the establishment of the printing press which the American missionaries brought from Malta to Beirut in the 1820s) founded the first newspapers and magazines that now circulate throughout the Arab world from Cairo. Many of the principal department stores and businesses were founded by them. They have helped to expand Arabic education in this most ancient of valleys, but they have never been really assimilated, since the Moslem masses dislike them because of their Christian background.

The largest non-Arab racial group is the Greeks who are engaged in commerce, principally as owners of small stores, restaurants, hotels, and bars. Few of them command the respect of the Egyptians, although about 10,500 of some 75,000 have become naturalized Egyptian subjects. The next largest group, the Jews, are found almost exclusively in Cairo and Alexandria, where they are in business, ranging from money changing and keeping of small shops to high finance and ownership of impressive department stores.

The foreign culture that has left the deepest impress on Egypt is the French. Many upper-class Moslems send their daughters to French schools rather than to American or British, because of the cloistered life and at the same time the excellent instruction given by the nuns. The names of the months are Arabic forms of the French, daily newspapers and magazines of wide circulation are printed in French, and, until recently some of the government reports were printed in French as well as in Arabic.

The Italians had large colonies and were active in Egypt's business prior to World War II, while the Armenians compose another racial group that is engaged primarily as artisans, although a few hold minor government posts.

Toward all the foreign racial groups, the Moslem feeling is one of toleration; a feeling which occasionally gets out of control when nationalists stir the students to incite the lower classes against the English, particularly, or the Greeks and Jews.

In keeping with his design to modernize the army, Mohammed

Ali sponsored the establishment of spinning, weaving, glass, armaments, and shipbuilding industries, but these died away under his successor. Up to World War I only the sugar industry had prospered.

When the first global conflict closed the Mediterranean and caused Egypt to shift from its principal crop, cotton, the necessity was seen for diversifying employment opportunities. Sugar, weaving, spinning, soap, furniture, iron, leather, and perfume industries were launched but many of these failed after the war, in competition with the higher industrialized plants of the West. Since then, however, sound progress has been made, but still on a rather small scale, in the industrialization of Egypt. This progress followed as the foreign powers eased the severe restrictions which they had imposed in commercial treaties. An idea of the extent of the development of industries, since Bank Misr was established in 1920 with a primary aim of helping in the founding of business, is indicated for the period from 1930 to 1939 with this growth: 145 establishments ginning cotton (136 of them British financed); 162,843 tons of sugar produced locally and fully meeting the country's needs in 1938; production of 159,500,000 square meters of cotton pieces, meeting 40 percent of the nation's needs; also cement plants, alcohol, leather, cigarets.

The largest industrial development in the Near East is the spinning of cotton yarns. Egypt has a total of 69,000 spindles, giving employment to 20,000 workers in that branch alone. While Egypt is decidedly an agricultural country, its expansion in industry may be indicated in the 609,735 persons employed, a figure representing 8.2 percent of all gainfully employed persons. The industrial process has been markedly increased both during and since World War II. As one of the wealthier Arab nations, which profited immensely from the recent war, Egypt bids well to stay in the forefront of the Near Eastern industrialization movement.

The Eastern Flank: Iran

Perhaps because of its majestic mountains, the ancient Parthians called Iran "Land of the Nobles." Certainly if one were to cross its 800-mile width from the western exposure along the Tigris valley to the eastern proximity to the Indus valley, or plumb its 1,400-mile length from the northern Caspian shore to the southern extremity along the Persian Gulf coast of Baluchistan, one could not help but be awed by mountains that emerge from plateaus 4,000 feet high to become snow-capped peaks, towering, like pinnacle-shaped Mt. Demavend, to 18,000 feet. Iran's great land mass of 628,000 square miles would cover the block of American states from Maine through Minnesota and thence southerly along the Missouri-Mississippi rivers through Tennessee to the Atlantic coastline of Virginia.

The population, estimated at 16,000,000, comprises a polyglot of races, but the language which gives this area its distinction from the rest of the Near East is the ancient Persian. Iran has many kinships with its neighboring Arab lands. The script is Arabic, but the spoken words are Persian, although both languages have borrowed slightly from each other. Secondly, the vast majority of the inhabitants are Moslems of the Shiite sect, the Sunnites generally being represented by the non-Persian language groups, including large groups like the Kurds who dwell in the Elburz mountain area; the Armenians, in Azerbaijan: the Arabs, in the Persian Gulf delta; the Lurs, those hardy, Alpine-like tribesmen in the verdant Zagros mountain valleys; the Baluchis, south of the Persian Gulf, from the coastline to the Indian border, with a racial background seemingly blended of Arab, Indian, and Negroid. The northwestern part of Iran is inhabited by the Turkomen, among whom the Mongoloid influence is also apparent. Jews, aggregating some 40,000, center chiefly in the larger cities.

The most nomadic of Iran's peoples are the Arabs and the Baluchis, in the Khuzistan and Baluchi provinces, respectively. They retain medieval traits in political and social organization. The Lurs and the Kurds are the major tribes in provinces so named, and only within the past generation have the Kurds tended to adopt a sedentary life. Other mountain tribes include the Bakhtiaris and the Quashqais. Tribes have played a dominant role in Iranian life since remotest time.

Iran is, in effect, rimmed by mountain ranges that frequently rise above 10,000 feet, from a broad central plateau of bleak steppes and stony desert—150,000 square miles of land whose northern part is called the *Dasht i Kavir,* and southern part, the *Dasht i Lut*—terrain as barren as Rub Al-Khali and equal in area to the states of Washington and Oregon. Except for some minor streams that empty into the Caspian Sea or disperse themselves within the bowl-like interior, rivers in Iran are rare, the only navigable one being the Karun which rushes through the southern plain into the Persian Gulf. In many areas water is relatively scarce; in many villages its distribution is controlled by a specially designated waterman answerable to a group of elders called the "graybeards," who form a sort of city council.

The very nature of Iran's geographical setting has prevented it from achieving the conditions of unity that are the first requisites toward the making of a great, powerful nation. Nature has fashioned a mammoth amphitheater in Iran: its center, the barren desert previously described; a plateau 4,000 feet above sea level; two major mountain chains—the Elburz and the Zagros—whose 10,000 to 14,000-foot barriers encircle the country on three sides. There are really only two easily accessible entrances into Iran, at the mouth of the Persian Gulf in the south and the corridor in the northeast through which swept the conquering Turkish and Mongol hordes.

Iranians do not live in contiguous communities, but as broken links in a human chain. Their villages are found mostly at the junction of the mountain valleys and the plains. Since they dwell vast distances apart, it naturally follows that there has been little incentive toward a strong unity of diverse racial strains.

The Persians have bowed beneath the invading swords of the Assyrians, Babylonians, Egyptians, Greeks, the nomadic tribes of the Arabs, Turks, and Mongols. And like the Arabs, they have at times absorbed their conquerors, adopting the best in the arts and politics and religions of the foreigners and adapting or assimilating these features to their own fabric of life. At the time of Cyrus the Great, the empire was extended through the Near East to the Grecian mainland. Persian influence later was so great that when Alexander the Great vindicated his ancestors, the Hellenic warrior chose two Persian daughters to be his legitimate wives.

The ancient crossroads of world trade have also passed through Iran, since it was nearer than Arabia as the connecting land link to northern India and China.

Iran is definitely an agricultural country. Most of its people live in the numerous communities that abound on the well-watered mountain slopes near their junctions with the fertile plains and valleys. On these slopes the people raise wheat and garden crops in the summer while moving their herds farther up the mountain sides to pasture on the rich verdure made possible by the melting winter snows and the copious spring rains. In some of the higher altitudes as much as 50 inches of rain falls in a year. In the winter the nomads return to live in the valleys and plains.

Iran has made one rather unique contribution to the development of agriculture. It is called a "qanat," an artificial canal which carries water underground, sometimes as far as twenty miles. The "qanats" are bored at the foothills of the mountains and are designed to preserve the water from evaporation and wastage. The "qanats" irrigate crops many miles distant. The summer season is generally dry, especially in Southern Iran, during the four-month period when the Shamal blows—a strong, hot, dust-laden wind.

Whether it be on the mountain sides or in the alluvial plains, agriculture is largely carried on with ancient quaintness—a pair of bullocks or donkeys drawing a plow usually fashioned from an iron shoe that has been attached to rough-hewn wood. At harvest time, bullocks still trample upon the dried stalks to thrash out the wheat after it has been harvested with hand-powered sickles. While ancient methods of farming generally prevail throughout the land, modern techniques have, however, penetrated in recent years; these are found not among the tribes but rather in the market-garden districts close to the cities where Western civilization has made its greatest impress.

Iran produces a wide variety of agricultural products, from rice, tea, tobacco, and fruits on the Caspian shore to dates and the usual tropical fruits in the Persian Gulf area. Cotton cultivation has been introduced but primarily to supply British demands, although a few native factories for the spinning of yarn and weaving of cotton goods for home consumption have sprung up, primarily with government support.

Only in more recent decades has some progress been made toward industrialization, and that at best a feeble and elementary one. While the development of industry began only a quarter of a century ago, it was left to World War II to give it a somewhat vigorous impetus. Carpet making, since the early nineteenth century, has been and continues to be the principal export industry of products made by man, but even its importance is beginning to recede as its artisans forsake the poorly paid and laborious task of hand weaving for the greater monetary rewards that come from the exploitation of petroleum, whose products give Iran its greatest income from exports.

These new industries have helped in extending the urbanization of many of Iran's areas: 23 cotton textile mills with 120,000 spindles, giving rather steady work to 7,300 persons; 9 woolen mills with 25,000 spindles, employing 3,200 persons; 25 hosiery factories, with 1,000 on their payrolls; 7 sugar refining plants, giving jobs to 50,-000; 6 match factories with an output of 100,000,000 boxes; 15 soap and glycerine plants, 7 tanneries, 15 installations producing alcoholic beverages, 6 canning firms, a cement mill, a glass factory, an 8,000-spindle silk mill and two factories that make sacks. Many of these establishments are state-owned or supported, for without government aid Iran has not had the financial means to establish industries on a scale and scope to which the West is accustomed through its free enterprise system.

The country is considered rich in mineral resources but these, too, have been relatively undeveloped. What exploitation of natural resources has taken place has been sponsored by the Iran Department of Mines, but the major and most lucrative development has come through the Anglo-Iranian Oil Co., Ltd., a United Kingdom petroleum concern with an authorized capital of £33,000,000. What a boon petroleum is to Iran's economy is indicated in its £23,000,000 export in the last full normal year before World War II, an amount nearly three times as great as the total of all other exports.

Iran lacks a well-knit transportation system, because of several factors: First, high mountains fringe the country like a bulwark against the outside world. There are no navigable rivers (except the Karun) to connect the great cities for commercial communication. The great central desert intercedes as an impassable barrier.

Highway and railroad development thrive with industry and since Iran, for centuries, has been primarily agricultural, there has been little incentive to development. Admittedly richer in undeveloped resources than any of its Near Eastern neighbors, Iran's growth as an important power has been retarded by the lack of transportation, which in many rural areas still depends on donkeys or the cart or the camel for shipment of goods.

The renaissance in transportation may be traced to Western influences, particularly to the rivalry between Czarist Russia, longing for a warm outlet through the Persian Gulf, and Britain, determined to safeguard her Empire lifeline to India from Russian envy. The first railroad lines were laid by a Russian company in northern Iran and with British support in southern Iran. It was not until just before World War II that Iran finally could boast of a railroad traversing the entire country from the Persian Gulf to the Russian border, construction of which was shared by Swedish, German, and Czechoslovakian companies during two decades. American Lend-Lease during the war years added to its modern equipment.

Highway construction, however, had been underway since the first World War, so that today a fairly good network provides motor vehicular connection to most major cities. In the smaller villages and among the tribes, animal power remains the popular means of transport.

Air travel, introduced by British, German, French and Dutch companies before World War II, brought Iran into closer contact with European centers, although the foundation for aviation within that country had been laid by the Anglo-American Oil company. Today a network of air routes connects Iran's principal cities with the outer world.

Except in the larger cities, what few sanitation systems there are in Iran cannot begin to compare with their Western counterparts, but nevertheless steady progress has been made. There was practically no modern medicine in Iran until 1873, when the Shah founded Iran's first modern college after having sent some of his younger medical subjects to France to study medicine. Today Iran has a medical college. Hospital facilities, rare or nonexistent only a few decades ago, are expanding rapidly in all parts of the country.

Tribesmen have always maintained a secure foothold in Iran's general history and particularly in its political life. What makes Iran different from the Arab countries is this: Most of her tribes are found in the periphery of the central plateau in the mountain ranges. Whereas Arab tribes come from the desert to beyond its fringes and then return to the desert, Iran's nomads roam from the mountains to the plains and back to the mountains, seeking pasturage for their herds of sheep, according to the seasons of the year.

When the British discovered oil in the Bakhtiari hills in the early 1900s, they paid large sums of money and provided arms to the tribal chiefs to keep their tribesmen under control while the British exploited the oil concessions unmolested. Sons of the Bakhtiari chiefs were sent to England to be educated. Among them later were found some of the leaders in the revolution of 1906 that led to the permanent establishment of a constitution, under which form of government Iranians today live.

At various times in Iran's history one or another of the tribes have gained supreme political control, the more recent one of longest tenure being the Kajars whose dynasty ruled for more than thirteen decades up to 1925. The Kurds, noted for the bravery and spirit of independence, were nomadic until comparatively recent times. The Lurs are considered the most restive of Iran's tribes; they live in brown square-shaped tents (like the Arab models) in the fertile Zagros valleys and until recently were wont to sweep down upon the sedentary villages for plunder. The Bakhtiaris are nomads, too, who move their flocks in the summer to graze on the mountain slopes and then return in the winter to the plains. The Baluchis seem to have racial kinships mostly to India but to a great extent, also, to Arabia with additional evidence of African immigration indicated among many through their Negroid features. The Turkomen tribes inhabit the northeast steppes that stretch from Iran into central Asia; they live in round tents, like those of the Mongols in inner Asia, instead of in the square-shaped tents of the Arabs.

Religion has been another disunifying factor. While most Iranians are Shiite Moslems, the non-Persian language groups are generally Sunnite Moslems. Iran, too, is the land where such world religions as Zoroastrianism, Mithraism, and Manicheism germinated. All three have made lasting impressions upon the thinking of the West, the latter two religions being of a missionary character long before Mohammed founded Islam. Today, Zoroastrianism, which is not a crusading religion, still flourishes much as a closed corporation in that it does not seek converts; one must be born within that religious group to be a member. About a century ago (1844) another universal religion was founded in Iran, the Bahai faith. At present it has relatively few adherents. Then there are the Christians, represented mainly in the Armenian and Nestorian groups, plus some 40,000 Jews.

The feudal systems among the different tribes have had such a grip on individual groups that their massing toward a national loyalty has been exceedingly slow, as witness the revolt of the Azerbaijans which was successful for a few months in 1946 until their rebellion was finally liquidated by government forces from Teheran. Tribal uprisings against the established government have occurred frequently in Iran's long history. The loyalty of the groups is primarily to the tribe rather than to a distant central government. The motor vehicle, the plane, the railroads and an expanding network of roads is gradually changing this aspect of Iranian life.

External powers have also kept Iran disunited. Foreign intercession has plagued the land from the time of Alexander the Great. As recently as the winter of 1946 Russia supported the establishment of an autonomous state in Azerbaijan, while the Western powers backed the sovereignty of Teheran. Westerners want a strong stable government in Teheran since that is the surest way for them to continue to exploit the oil resources originally developed by Western capital. The huge oil refinery at Abadan on the Persian Gulf stands as a symbol of a busy Western technology in a hitherto placid East. Iran's modern prominence in the council of nations is attributed mainly to the courtship which the Western powers especially apply to her oil.

Iran existed untarnished by European civilization until about the middle of the nineteenth century when the West intruded with her products of the industrial revolution. Previously Iran had attained important stature only as a buffer state when England gained control of India in the eighteenth century and Russia cast envious eyes in that direction, too. The rivalry between these two European powers heightened in the last century and with the United States

Near Eastern policy today quite frankly concerned about the region's oil reservoir, Iran once again finds herself in a rather gingery position, politically, especially in view of the present "cold war" between Russia and her Balkan satellites on the one side and the Western nations on the other side.

The British, ever alert to expanding trade and business to further their industrial expansion at home, constructed the first telegraph lines in Iran in 1864, thus awakening this slumbering land into rapid communication with the faster moving world. Financial troubles have stalked Iran ever since Western business imposed its markets upon her. The British obtained the concession to establish a national bank in 1872. This was succeeded by an imperial national bank a few years ago. In 1944 the Iranian government requested an American economic commission to help evolve a financial system more atuned to modern times.

When the Shah in 1906 granted the first constitution, the document was inspired and drawn by Iranians who had been indoctrinated with Western ideas, either through education in Europe or America or in mission schools, and by Iranians who had been in contact with Western governments. The constitution today provides for a Majlis (a national consultative assembly) whose membership represents the various classes of Iranian society—the princes, clergy, tribes, landowners, businessmen, farmers, and the religious and racial minorities. The parliament is composed of 162 members, each elected to a two-year term. The real governing power, however, is not the Shah but the Prime Minister, who, as the constitutional executive, chooses his own ministers to form a cabinet that has the responsibility of government. Commissions govern in the larger cities and towns and the "council of graybeards" perform a similar function among the tribal villages.

The French have also left their imprint on Iranian life, especially in the fields of art and learning. Iranians have gone to France to study medicine, engineering, law, and modern art. Iran realized in the 1870s that if she were to compete on a military basis with foreign powers she had to revamp her antiquated army and equip it properly. This she did by sending her officers to train in France. Later she invited Russians, British, Swedes, French, and Americans to aid in the modernization process. Such an influx of foreigners in key government positions served to disunite Iran (although their primary assignment was to unite it), as each foreign group played the game of power politics in the interest of his native country.

A special mission of American educators was also called in to help reorganize the educational system. Iran became a pawn of the foreign powers again during World War II when Britain and Russia occupied it to prevent a Nazi *coup d'état*. Adding emphatic proof to Western intrusion is the Teheran pact which Churchill, Roosevelt, and Stalin signed in 1943, guaranteeing Iran freedom from foreign interference in its internal affairs after the defeat of the Axis. Russian troops were finally withdrawn in the spring of 1946, but only after implied and direct complaints had been made by the United States and Great Britain to the United Nations. While on the surface it appears that Iran is exercising her sovereignty as a free, independent state, actually, she is being subjected to pressure power politics as the Western nations veer toward solidifying a military bloc to meet whatever threat they seem to suspect in the Soviet Union's growing dominance in the East.

SUGGESTED READINGS

Ancient Times

Albright, W. F. From Stone Age to Christianity. Baltimore, Md., 1940.
Breasted, J. H. Ancient Times. Boston, 1935.
Turner, Ralph E. The Great Cultural Traditions: the Foundations of Civilization. 2 vols. New York, 1941.

General Background and Problems

Ireland, P. W., ed. The Near East: Problems and Prospects. Chicago, 1942. Note especially the essays by H. A. R. Gibb on the sociological basis of the Arab movement.
Speiser, E. A. The United States and the Near East. Cambridge, Mass., 1947.

Western Civilization in the Near East

Ben-Horin, Eliahu. The Middle East. New York, 1943.
Hourani, Albert H. Minorities in the Arab World. New York and London, 1947.

Kohn, Hans. Western Civilization in the Near East. New York, 1936.

Parkes, James. The Emergence of the Jewish Problem, 1878–1939. Royal Institute of International Affairs, London, 1946.

Ross, Frank A., C. Luther Fry, and Elbridge Sibley. The Near East and American Philanthropy. New York, 1929.

Agriculture, Industry, Geography

Cressey, G. B. Asia's Lands and Peoples. New York, 1944.

Keen, B. A. The Agricultural Development of the Middle East. London, 1946.

Lowdermilk, Walter Clay. Palestine, Land of Promise. New York, 1944. A soil conservationist looks at one country in the Near East.

Proceedings of the Conference on Middle East Agricultural Development. Middle East Supply Centre, Cairo, 1944.

Weinryb, Bernard. "Industrial Development of the Near East." Journal of Economics, LXI, No. 3 (May, 1947), pp. 471–499.

Worthington, E. B. Middle East Science. London, 1946.

Review of Commercial Conditions for Persia, Iraq, Turkey, Egypt, Palestine. London, c. 1945. Individual booklets for each country, published for the Department of overseas trade.

Islam and the Arabs

Antonius, George. The Arab Awakening. Philadelphia, 1939.

Arnold, Sir Thomas, and Alfred Guillaume, eds. The Legacy of Islam. Rev. ed. London, 1943.

Hitti, Philip K. The Arabs, a Short History for Americans. Princeton, N.J., 1943.

Jeffery, Arthur. "The Political Importance of Islam." Journal of Near Eastern Studies, Vol. I (Chicago, 1942).

Toynbee, Arnold J. "The Islamic World." Survey of International Affairs 1925, Vol. I (London, 1927).

Van Ess, John. Meet the Arab. New York, 1943.

Arabian Desert and Mountain Coastal Rim

Coon, Carleton S. Southern Arabia, A Problem for the Future. Papers of the Peabody Museum of American Archaeology and Ethnology, Harvard University, Studies in the Anthropology of Oceania and Asia, Vol. XX, 1943.

Ingrams, W. H. Arabia and the Isles. London, 1942.
Lawrence, T. E. Revolt in the Desert. New York, 1927.
Musil, Alois. The Manners and Customs of the Rwala Bedouins. American Geographical Society, 1928.
Philby, H. St. J. B. Arabia. New York, 1936.
Twitchell, K. S. Saudi Arabia. Princeton, N.J., 1947.
Wilson, Sir Arnold. The Persian Gulf. London, 1928.

Egypt and the Anglo-Egyptian Sudan

'Ammar, Abbas M. The People of Sharqiya, Their Racial History, Serology, Physical Characters, Demography, and Conditions of Life. Royal Geographical Society of Egypt, Cairo, 1944.
Crouchley, A. E. The Economic Development of Egypt. London, 1938.
Elwood, P. G. The Transit of Egypt. London, 1928.
Gautier, E. F. Sahara, the Great Desert. New York, 1935.
Lane, Edward William. An Account of the Manners and Customs of the Modern Egyptians. 1836.
Ludwig, Emil. The Nile, the Life Story of a River. New York, 1937.
MacMichael, H. A. The Anglo-Egyptian Sudan. London, 1934.

The Levant—the Western "Courtyard": Lebanon, Syria, Palestine, Transjordan

Bell, Gertrude. Syria: the Desert and the Sown. London, 1907.
Himadeh, Sa'id B. Economic Organization of Syria. Beirut, 1934.
Hourani, Albert H. Syria and Lebanon: a Political Essay. London, 1946.
Infield, H. F. Cooperative Living in Palestine. New York, 1944.
Luke, Sir Harry, and Edward Keith Roach, eds. Handbook, Palestine and Transjordan. London, 1934.
Nathan, R. E., O. Gass, D. Creamer. Palestine: Problem and Promise. Washington, D.C., 1946.

POLITICS (Among the many books on Palestine politics from all points of view, the three following are noteworthy)

Hanna, P. L. British Policy in Palestine. American Council on Public Affairs, 1942.
Jeffries, J. M. N. Palestine, the Reality. London, 1939.
Palestine: a Study of Jewish, Arab, and British Policies. 2 vols. Esco Palestine Study Committee. New Haven, Conn., 1947.

Iraq, the Eastern "Courtyard," and Iran (Persia)

Boesch, H. H. "El 'Iraq." *Economic Geography*, Vol. XV, No. 4 (October, 1935), 325–361.

Corry, C. E. The Blood Feud. London, 1937.
A story about the Marsh Arabs.

Haas, W. S. Iran. New York, 1946.

Iraq: an Introduction to the Past and Present. Edited by a Committee of Officials of the Kingdom of Iraq. Baltimore, Md., 1946.

Ireland, P. W. Iraq: a Study in Political Development. New York, 1938.

Lloyd, Seton. Twin Rivers. London, 1947.

Luke, H. C. Mosul and Its Minorities. London, 1925.

Reference Books, Surveys, Journals

Admiralty Handbooks. London. On the individual countries.

Encyclopedia Brittanica. Note especially histories and descriptions of Arabia, Egypt, Iran, and Turkey.

Jewish Agency for Palestine. Statistical Handbook of Middle Eastern Countries. Jerusalem, 1945.

Middle East Journal. Middle East Institute, Washington, D.C. 1947–.

Steinberg, S. H., ed. The Statesman's Yearbook. Ed. S. H. Steinberg.

Toynbee, Arnold J., ed. Survey of International Affairs. Royal Institute of International Affairs. London, 1920–.

Daniel and Alice Thorner

INDIA AND PAKISTAN

I. Introduction

STRANGERS approaching India [1] usually expect to find it hard to understand. But if ever there was a country which appreciated and rewarded a genuine endeavor to understand it, that country is India. The stranger who is willing to exert a little effort may be surprised to find how much sense he can make out of India and how clear are the outlines of many of its problems. The present essay is offered in the belief that India is intelligible and desires to be understood. But this is only a sketch, and the reader who desires to improve upon it will find at the end some suggestions as to how this may be done.

THE IMPERIAL SETTING.—The dominant influence in the shaping of modern India has been its connection with Britain. The British empire in India began right after 1750 and lasted a little less than two hundred years, thus enduring exactly as long as the Mogul regime which preceded it. Unlike the Moguls the British never settled down in India. To them home was always 5,000 miles away in Britain, from which left successive batches of green young officials and to which retired old and sun-baked veterans. In all the history of empire there is no parallel to the rule of so populous a land for so long a period by aliens from a tiny island so far away.

During most of the period of British rule in India, Britain was the leading economic and industrial power in the world. Her great advantage in technique and resources made her the strongest nation of the day. As the possessor of the world's premier navy and merchant marine, Britain had instruments through which she could transmit her power and make it felt. British officials and soldiers in

---

[1] As this essay is concerned primarily with institutions and developments up to August 15, 1947, the term "India" is used to refer to the entire country before its division on that date into the Dominions of India and Pakistan.

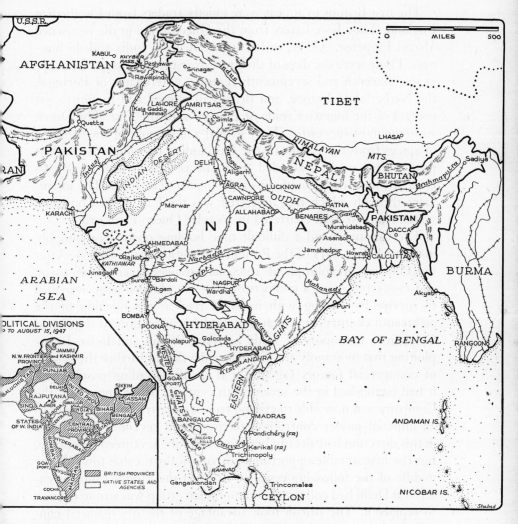

INDIA AND PAKISTAN

far-off India acted with an assurance and masterfulness rooted in the material superiority of their home country over any of its rivals or possessions.

The first Britons to appear were chiefly traders from Elizabethan England. They bore letters from the Queen praying the renowned Mogul Emperor, Akbar, to let them remain in India and do business. Those were the days of the great overseas expansion of Europe in the sixteenth and seventeenth centuries. Merchants of Portugal, the Netherlands, France, and Britain all plotted and fought to get control of the lucrative trade of Europe with the East. The Dutch were the most successful. They gained a monopoly of the handling of spices from the islands of the East Indies, on which the highest profits were made. Excluded from the islands, British merchants made the most they could out of the mainland of India. The instrument through which they operated was the East India Company, which had first been chartered in 1600 by Queen Elizabeth.

The Company had a difficult time holding on to its monopoly in the turbulent days of the Puritan Revolution (1640–1660). After the Glorious Revolution of 1688, its privileges were taken away and conferred upon a competitor. But the old company managed to survive by joining with its chief rival at the opening of the eighteenth century, at which time Parliament reaffirmed the united Company's monopoly of trade between Britain and India. In exploiting that monopoly the Company developed further the system of commercial factors (agents) and fortified trading posts which it had established in the seventeenth century. More important, the Company was now able to expand from a mere trading enterprise to an imperial power controlling territory and revenue. Aspirations in this direction had already been voiced in the seventeenth century, but the first suitable opportunity to satisfy them came towards the middle of the following century. By then the Mogul Empire centered at Delhi had collapsed, and no Indian group was strong enough to replace it. The British took advantage of the disorganized conditions that prevailed to make themselves masters of eastern India, including the rich provinces of Bengal, Bihar, and Orissa. Similarly, they acquired control of the key southern city of Madras and the coastal areas to the south of it. In both of these regions they inflicted military defeats upon their only serious European rival, the French, who also had aspired to territorial control of India.

The conquest of important areas by the East India Company set off a quarter of a century of unlimited plunder. This took three forms: outright extortion from local potentates, arbitrary exactions from Indian merchants, and drastic increase of the customary levies upon the peasantry. Company officials were astounded at the ease and speed with which they were able to amass impressive fortunes. The effect upon the population, however, was expressed in 1789 by Lord Cornwallis, who after his surrender at Yorktown in 1781, had been named Governor-General of India: "I may safely assert, that one-third of the company's territory in Hindostan, is now a jungle inhabited only by wild beasts." [2]

These processes debilitated the economy of India so rapidly that Parliament and the East India Company became deeply alarmed lest India's value as a colony be utterly and irretrievably ruined. In the closing years of the eighteenth century Lord Cornwallis put an end to a number of the worst abuses, reformed the administration of the country, and reorganized the land revenue system of eastern India. Not long thereafter Parliament sharply modified the economic power of the East India Company. Under pressure from manufacturers and merchant houses—whose numbers and strength had been increasing rapidly during the industrial revolution then taking place in Britain—Parliament in 1813 ended the East India Company's monopoly of trade and threw it open to the public. Cheaply manufactured British cotton goods now descended like an avalanche upon the markets of the more accessible Indian cities. This first great wave of British machine-made exports had a devastating effect among the urban artisan classes. It virtually wiped out handicraft centers like Dacca and other cities, which for centuries had been world-famous for their fine muslins and other textiles. By the 1830s the Governor-General was reporting to the East India Company that the bones of the cotton weavers were bleaching the plains of Hindustan.

As India's economy weakened, commerce began to stagnate. The

[2] Lord Cornwallis's Minute of September 18, 1789, reprinted in House of Commons, *Fifth Report on East India Affairs, 1812*, Appendix, p. 473.

great mercantile houses of London, Liverpool, and Glasgow, together with their British correspondents in Calcutta and Bombay, called for the placing of Anglo-Indian trade on a new basis. "India," stated one of their chief spokesmen, the founder of the banking house of Grindlay and Co., "can never again be a great manufacturing country, but by cultivating her connexion with England she may be one of the greatest agricultural countries in the world." [3] Let India grow and ship to England raw materials such as cotton, Grindlay and many others argued, and thereby obtain the wherewithal to pay for a swelling volume of finished goods from Britain. To carry through this program the merchants campaigned for rapid steamship service to the Indian ports and for railway lines throughout the vast hinterland. Added pressure was forthcoming from the textile manufacturers of Lancashire, uneasy over their complete dependence for raw cotton on the slave-owning American South, and the Midlands ironmasters who looked forward to providing at a handsome profit the rails, locomotives, and other heavy items. Considerations of high imperial policy brought support from generals and governors. The 1840s saw the introduction of modern steamers and the beginnings of two railways. Construction proceeded rapidly until India was covered with the most elaborate railway network in all of Asia.

The railways, the iron steamers, and the Suez Canal (the latter completed in 1869) effectively linked India's economy to the European and world markets. In practice, as mid-Victorian Britain was the world's foremost economic and industrial power, they bound India tightly to the British economy. The greatly increased shipments of raw materials for export—cotton, tea, jute, wheat, oilseeds, indigo—were consumed in Britain or were shipped via Britain. India's mounting imports were supplied primarily by British factories. The handling and financing of India's trade remained chiefly under the control of mercantile houses like the ones which had originally campaigned for steamers and railways.

India, in short, was converted into a subordinate and dependent part of the metropolitan economy of Great Britain. The profound repercussions of this transformation can be understood only if they

[3] Captain Robert Melville Grindlay, *A View of the Present State of the Question as to Steam-Communication with India* (London, 1837).

are seen against the background of the older structure of India's economy and society.

THE OLDER STRUCTURE OF INDIAN SOCIETY.—The outstanding feature of the traditional economy of pre-British India was the self-contained and self-perpetuating character of its typical unit, the village. Most of these communities came close to being little worlds unto themselves. The only outside authority which they acknowledged was that of some local princeling, who in turn might be subordinate to some distant overlord, whether Hindu Rajah or Muslim Nawab. The chief sign of submission to that authority was the payment each year of a share of the village crops, in some periods amounting to one-sixth or less, and in others to as much as one-third or even one-half. As a general rule the responsibility of making this payment, whether in produce or in money, was joint or collective, and rested upon the whole village considered as one single unit. Within the little world of the village, social and economic relations were governed by customary patterns and conventions of immemorial antiquity. The cattle were tended and the soil was tilled by peasants whose fathers had been cultivators and whose sons would take their places when they came of age. A kind of rough equality was maintained in some parts of India by a periodic throwing together of all the lands of the village, followed by a fresh redistribution of the land among the cultivating peasants. Cloth for the garments of the peasantry generally was spun and woven by families whose ancestors had been weavers long beyond the living memory of man. The other crafts were carried on by families which in effect were servants of the village. Their occupations passed on traditionally from father to son: blacksmith, potter, and carpenter who made and repaired the implements and utensils of the village; the silversmith, who made the village jewelry; and the oilseed-presser. For their services these craftsmen received a regular stipend out of the crops of the villagers. In some areas hereditary servants and slaves attached to peasant households performed both domestic and agricultural duties, and received from their masters food, lodging, and garments.

The pinnacle of authority within the village was either a headman or council of elders; associated with them were such officials as the village record-keeper, boundary man, supervisor of water courses for irrigation and the Brahman teacher-priest-astrologer. These hereditary and traditional divisions of occupation and function were confirmed and given the stamp of obligation by the caste system (the present character and sanctions of which are treated on pp 563–71).

The village itself consumed most of the foodstuffs and other raw materials it produced. Its needs in the way of handicrafts were satisfied by the families of craftsmen associated with the village. It was this tight union of agriculture and hand industry which made the village economically independent of the outside world, except for a few indispensables like salt and iron. The share of the village crops which went to the local magnate and moved on from him in a diminishing stream upward to the highest political over-lord sustained the structure of government and provided subsistence for the urban population. India's towns and cities generally were little more than headquarters for the top political overlords or im-perial courts. The industries which flourished in these cities provided mainly luxury goods for the upper classes or implements of war for the army. Economically the cities had a one-way relation with the countryside, taking foodstuffs as tribute but supplying no goods in return. A special class of cities included pilgrimage centers like Benares and Allahabad on the Ganges, Puri the home of Jagannath on the Bay of Bengal, and Trichinopoly in the far south. They lived off the great stream of pilgrims who flocked to them at the holiday seasons. In a number of cases these pilgrimage towns also served either as centers of administration or as commercial marts for the interchange of luxury articles from all over the country.

Relations between the imperial courts and their regional sub-ordinates varied widely. When the center was strong, the provincial governors and local potentates were compelled to forward regu-larly a prescribed share of the revenues they collected. If the cen-ter became weak, then revenue payments from distant areas soon fell off. Ambitious subordinates often attempted to set up empires for themselves.

The demand of local princelings upon the villages under their control also varied widely according to time and place. The chief demand, of course, was for a share of the principal crop, but gener-ally this was supplemented by a number of services and exactions. These might include forced labor on roads, customary "gifts" of food or animals upon specified occasions, and payments for permission to cut wood in forests, fish in lakes and rivers, or use mills for grinding grain.

This was the structure of Indian society which, with regional variations, the British found between 1750 and 1850 as they conquered one part of India after another. The basic land relations were rooted in century-old custom and usage rather than in any formal, elaborate set of statutes, legal cases, and court procedures about property. So long as the peasants turned over to the local potentate his customary tribute and rendered him the usual services, their right to till the soil and reap its fruits was taken for granted. Local rulers who repeatedly abused this right were considered oppressive; if they persisted, the peasantry fled to areas where the customs of the land were better respected. As land was still available for settlement and labor was not too cheap, local chiefs had to be careful lest they alienate the villagers.

The position of the chieftains and other intermediaries between the villages and the imperial courts or dynastic houses was less stable than that of the cultivating peasants. As empires and kingdoms rose and fell, and rival armies marched and fought, the local rulers had to line up and choose sides. Those who picked the winner were likely to be confirmed in their status or given a better place, while the unlucky ones who guessed wrong might be degraded or slaughtered. As the local chiefs often constituted a rallying point against the center, some emperors or kings tried to weaken their power by collecting the land revenue without their participation. Instead they created tax-farmers or operated through subordinate officials controlled by the imperial court.

There was nothing in India comparable to the highly developed forms of private property in land which were the rule in late eighteenth century England. There were no landlords and no tenants in the Western sense. The right to levy the land revenues was recognized to be the very essence of political power. In effect, the supreme political authority was the supreme landlord.

II. The Countryside

THE NEW LAND SYSTEMS.—By conquering India the British automatically became lords of the land. They then had to decide upon the kind of system they would employ to collect the land revenue due to them. In an epoch-making decision Lord Cornwallis in 1793 broke with previous tradition by introducing in Bengal a form of private property in land. He transformed the tax-farmers and revenue collectors of Bengal into private landlords, *zamindars*. His action sprang from two considerations: first, to find a dependable basis for obtaining annually the greatly enhanced payments demanded of the peasantry; and, second, to do this in such a way as to support "a regular gradation of ranks" which Lord Cornwallis believed was "nowhere more necessary than in this country for preserving order in civil society."

The new landlords were expected to collect a little more than £3,000,000 annually as rent from the peasantry, but they were required to pass on ten-elevenths to the British authorities; for themselves the landlords might keep the remaining one-eleventh. To persuade the former tax-farmers to accept this unusual kind of private property, Cornwallis offered them a special inducement. They were promised that the annual sum expected of them would remain permanently settled for all time to come at the level of the initial year, around £3,000,000. Thus Cornwallis renounced all claim of the state to any additional sums which in the course of time the new landlords might obtain out of their holdings, whether from rising crop prices, bringing of new lands into cultivation, or from sheer exaction. For the past century and a half, therefore, while the landlords of Bengal have continued to pay to the state £3,000,000 annually, the amounts taken from the peasantry have grown until in recent years they have ranged between £12,000,000 and £20,000,000 per year.

The hereditary rights of the cultivators to the soil they tilled were virtually ignored. As Lord Halifax [4] declared in 1861, Cornwallis's Settlement left the cultivating peasants to the mercy of the newly created landlords. The old body of custom was submerged by the formidable apparatus of law courts, fees, lawyers, and formal procedures. For with the introduction of private property in land the purchase and sale of *zamindars*' holdings were explicitly sanctioned by law. All of this was too much not only for the humble peasants

---

[4] Grandfather of the present holder of the title and from 1859 to 1866 Secretary of State for India in the British Cabinet.

but also for the new landlords. Most of the latter could not raise the heavy revenues inflexibly required by the government and soon defaulted or sold out to merchants, speculators, and other sophisticated persons from the cities. These new landlords by purchase were interested only in the rents they could squeeze from the land; often they delegated the collection to middlemen who contracted to pay high sums annually. The latter in turn sublet to still other classes of middlemen, so that before long the unfortunate peasantry of Bengal were supporting an impressive panoply or "gradation" of middlemen, speculators, and absentee landlords.

The new land system soon came into disfavor with the government. Partly this was because of its harmful effect upon the peasantry, but of equal or greater importance was the belated appreciation of the fact that it limited for all time the income which the state could expect from the principal source of revenue, the land. In applying the *zamindari* system to other parts of their territories the British shifted it over to a temporary rather than a permanent basis. Thus in making modified *zamindari* settlements in central and northern India in the beginning and middle of the nineteenth century, the British stipulated that the level of payments could be lowered or raised at the end of each twenty or thirty-year period.

An entirely different land system was devised for large parts of Bombay and Madras and later applied to areas in northeastern and northwestern India. Here the British generally by-passed and ignored the claims of tax collectors and other intermediaries. Instead they dealt directly with the peasantry on the land, for thereby they hoped to be able to obtain more revenue than under the landlord (*zamindari*) system. The peasant was recognized as holding the particular plot or plots he occupied, but his right to the land depended upon annual payment in full of a heavy money rent to the state. Because it dealt directly with the peasant or *ryot*, the new system was called the *ryotwari* settlement. As under the *zamindari* settlement, the first claim on the land was owed to the state, and this claim was so heavy as to resemble markedly that of a landlord. At the same time the new *ryotwari* system also introduced some features of private property in land. The holders were registered and empowered to sell, lease, mortgage, or transfer their right to the use of the land. In contrast to the more or less elastic demands upon the peasantry under previous indigeneous regimes, the British insisted rigorously upon prompt and complete payment of the stipulated sums. In cases of default livestock, household property, and personal effects might be attached, and the peasant might be evicted. The new land system thus made mobile both the land and the peasant, and opened the way for the growth in power of the moneylender and the absentee landlord.

SPREAD OF COMMERCIAL AGRICULTURE.—The middle and latter decades of the nineteenth century saw changes in India's agrarian economy even more drastic than those immediately entailed by the British land settlements. The key to this transformation was the striking increase in the production of crops for sale in distant towns or overseas markets, rather than for consumption within the village. This is of course the familiar process of commercialization of agriculture, a development which has taken place not only in India in the last century but in most other parts of the world. In recent years India has ranked as the world's largest producer and exporter of jute and burlap, tea and peanuts; it has been the world's second largest producer and exporter of cotton and cottonseed oil; and has been an important supplier of linseed oil, hides, and skins. Thus India came to fulfill the hopes of British merchant-bankers and manufacturers that it would serve as one of the world's premier sources of agricultural products. But of deeper significance is the fact that its new role has meant a new era of economic and social relations in the Indian countryside.

The speed of the peasants in expanding the cultivation of cash crops arose in great part from their search for ways of getting money to meet the steadily mounting demands upon them by the state and the new landlords. Industrial areas like Britain and western Europe offered good prices for raw materials which India could grow cheaply. Once the railways were opened it became possible for the inland areas to produce for the world market. Wheat poured out of the Punjab, cotton out of Bombay, and jute out of Bengal. As commercial agriculture and money economy spread, the older practices associated with a subsistence economy declined. In some districts the peasants shifted over completely to industrial crops and had to buy their foodstuffs from dealers. Villagers sent to market the cereal reserves traditionally kept for poor years. They became less prepared to meet poor harvests. Years of successive drought in the 1870s and 1890s led to great famines and agrarian unrest.

To produce crops for the market the peasant applied to the moneylender for credit to tide him through the long period of turnover. In pre-British times the local moneylender extended "casual credit" to meet occasional needs of the villagers, but he occupied a subordinate place in the subsistence economy of the countryside. In many areas the state refused to help moneylenders recover debts owed by the peasants, and in cases of default the legal system commonly did not permit them to take away the peasants' land. Usurers who tried to grind down the peasants laid themselves open to homemade rustic justice.

The new forms of land holding, land-revenue systems, legal procedures, and commercial agriculture of the nineteenth century opened up a golden age for the moneylender. The demand for his services became an integral part of economic life. The moneylender was encouraged to expand his activities by the fact that under the new order of things he could make a good and secure profit. If the peasants defaulted he could use the new legal procedures to attach their lands, livestock, and personal possessions. Furthermore, from the middle of the nineteenth century the price of land rose rapidly in value, thereby encouraging the moneylender to broaden his operations. He began to take over the peasants' land and rent it out. The moneylenders waxed fat and grew in number.

The same railroads which carried away the commercial crops brought back machine-made industrial products to the villages. Like the skilled urban artisans in the first half of the nineteenth century, the village weavers and traditional handicraft servants had to compete in the second half of the nineteenth century with products like Lancashire cloth, which was then overrunning the markets of the world. The village artisans no longer were sheltered by the friendly backwardness of the older village community. Furthermore the union of agriculture and hand industry which had been the basis of village life was disrupted. Under the impact of new forces the village could no longer remain the compact social and economic unit that it had been. The growing tendency was for each family to make ends meet as best as it could. Deep in the interior of central India and in other areas difficult of access, the handicrafts held on for a long time, and some still show strength today. But in the coastal zones and in the regions lying along the new railroads the ancient village handicrafts declined. The village potter, tanner, dyer, oilman, and jeweler all faced stiff competition from machine products, whether made in England or, since the close of the nineteenth century, made in the new industrial centers that sprang up in India. Over the course of the past hundred years a dwindling proportion of the village artisans have been able to subsist on what they have received for their services from the village. Millions of them have had to find other ways to gain a livelihood or to supplement their scanty earnings. In most cases the only avenue open to them has been agriculture and they have added steadily to the great pressure on the land which is one of the chief characteristics of contemporary Indian life.

As the villagers came to depend more and more on purchases from the outside of such daily necessities as oil for cooking and lighting, matches, cloth, farm tools, and cooking pots, the traditional weekly or semiweekly rural fairs were supplemented by the appearance of village shops. Soon the cloth and provision merchants were extending credit to their customers. Some found it profitable to take up agricultural moneylending as a side-line. Others developed into middlemen who bought up the village produce and marketed it in the cities. To a large extent the three functions of shopkeeping, moneylending, and marketing became combined in the same persons. The peasants, in disposing of their harvests, could not hope to bargain on equal terms with middlemen to whom they were already indebted and from whom they would soon again need advances of food grains or money for the next year's cultivation.

Commercialization of crops and the money economy which developed with it also affected the territories of the Indian princes. But they spread less rapidly than in British India and the transformation of the traditional village economy has proceeded at a correspondingly slower pace.

RURAL ECONOMY TODAY.—The power of the landlords and the moneylenders over the countryside shows itself today in the striking concentration of landed property. Less than a million great magnates, large landlords, and moneylenders own or control more than two-thirds of all the cultivated land. With few exceptions the members of these groups have no productive function. They do not farm their land with modern machinery nor do they apply fertilizers or worry about the latest techniques of scientific cultivation. Instead, they stay in the cities and lease out their property in tiny patches to peasants at all the traffic will bear. The income of the landlord and moneylender thus is drawn almost exclusively from rent and usury, and practically never from profits gained by growing crops on their own land. On a much smaller scale the same is true of some three million petty landlords and rent receivers, chiefly city dwellers, who obtain some income from their minor properties. There are few other ways of investing money, even for professional men such as teachers and lawyers.

Less than one-quarter of the land is owned by peasant proprietors who actually till the soil. Of this class only about one million peasants may be termed well off; many of these are half-landlord and half-peasant, for they rent out parts of their holdings to less fortunate cultivators. Except for this relatively small number of well-off peasants, India's working population on the land consists of poor proprietors, poor tenants, and propertyless agricultural laborers. These three groups, totaling more than 100,000,000 working men and women, exist *below* what by nineteenth century standards would have been the barest minimum considered satisfactory. The bulk of the cultivators during most of the nineteenth century had been either small holders or tenants. Except in several deeply impoverished regions a good percentage of them managed to remain free of debt, to run their affairs with a modest competence, and to hold on to enough livestock to work their fields and to carry their produce to market.

The contemporary position is shown by the evidence on the size of the holdings belonging to peasant proprietors or held by tenants with a recognized right to continue working the lands they till. Reporting on the Punjab, then the most prosperous part of India, the Royal Agricultural Commission of 1928 found that three-quarters of the holdings were less than ten acres. Moreover—at a time when the figure of five acres was commonly accepted as the absolute minimum to sustain a family—the Commission ascertained that three out of five holdings were less than this. While details were lacking for the rest of India, the Commission concluded that the picture was similar: the average holding "is small; and there are a very large number of such holdings under two or three acres." Many small proprietors, unable to make a living out of their *dwarf* holdings, cultivate additional land as tenants or hire themselves out as laborers. Some abandon independent cultivation altogether. They lease out their tiny plots and become indistinguishable from the landless laborers.

In the two decades since the 1928 Report of the Royal Agricultural Commission the concentration of property in the hands of landlords and moneylenders has increased and the size of the units held by petty proprietors and tenants has further shrunk. A prime factor in this process was the catastrophic world depression of 1929–1933, which also had the effect of more than doubling the total debts owed by the peasantry. The level of debt is such that it exceeds in amount the total annual income of the small proprietor and tenant. A government handbook put out a decade ago by retired British officials noted tersely that "indebtedness, often amounting to insolvency, is the normal condition of a majority of Indian farmers."

Poor peasant proprietors and tenants are becoming so dependent and tributary to the landlord-moneylender that their condition approaches that of servile or unfree labor. The clearest evidence for this is the character and extent of sharecropping. It already accounts for one-fifth of the sown area of Bengal and Bihar, one-fourth of the United Provinces, one-half of the Punjab, and for large parts of southern India, Sind, and the North West Frontier Province. Half or more of the gross produce goes to the landlord who often as not provides neither seeds nor implements nor work animals. The sharecropper hands over half the yield simply for the privilege of growing crops on the landlord's land.

At the bottom of the economic scale come thirty to fifty million totally landless laborers. During the last two generations they have increased more rapidly than any other significant part of India's population. The recruits to this class stem from peasants who have lost all their livestock and all their land to the landlord or moneylender. A striking number of them serve today in one or another form of unfree labor. Some are serfs who may be transferred from one master to another as the land changes hands. Others are bond servants who perform customary menial tasks for their masters.

Crop production methods remain the time-honored ones dating back to pre-British times and passed on from generation to generation. There is little or no impetus to technical progress, because the benefits of improvements generally are siphoned off by landlords, moneylenders, and middlemen. In most cases peasant holdings consist of a number of fragments scattered among the various grades and types of land in the village. The basic unit of field production on large and small holdings alike is a plot so tiny as to make modern methods irrelevant. In the Punjab two out of five peasants tilled units less than two and a half acres in size. In other provinces subdivision of holdings through inheritance has proceeded so far that units of cultivation as small as $\frac{1}{100}$ of an acre have been found. The only important example of large-scale cultivation is furnished by the tea, coffee, and rubber plantations, originally founded and managed by British capital. But even here cultivation has been by old-fashioned hand methods and the working force has consisted of indentured or semifree labor.

PATTERNS OF VILLAGE ORGANIZATION.—The typical Indian community remains the small village. Six out of seven Indians spend most of their lifetime within the circumference of one or another of the 650,000-odd peasant settlements which honeycomb the countryside. About one hundred to five hundred families with their houses, shops, wells, gardens, fields and places of worship, constitute a typical village. Scarcely visible from the road, the village proper is reached by a path through the crops. The actual dwellings, whether of brick, stone, baked mud, or straw matting, usually huddle together along narrow alleyways. Here and there an open space surrounds a well or threshing floor. In the craftsmen's quarters, front rooms, verandas and the unpaved footpaths serve as workshops and show windows. A clump of greenery may proclaim a temple or mosque or the residence of a more prosperous villager. Just beyond the knot of houses is almost always the "tank"—an artificial pond providing water for bathing, washing clothes, and for other household and agricultural purposes. Here stands perhaps another shrine, sacred tree, or domed tomb of a Muslim saint. Often separate smaller clumps of hovels shelter the families of the lowliest villagers. And on all sides stretch the fields. In regularly flooded lowlands like East Bengal, homesteads must be built on raised earth dikes. Here one family may occupy several tiny huts surrounding a courtyard or pool and the village site is more extensive. Similarly, in other areas such factors as vegetation, availability of water, type of crop, local tradition, number of people, nearness to the railroad, affect the arrangement of houses or compounds. But throughout India the predominant village pattern is the tight cluster of habitations within an irregular tract of cultivated soil.

Socially the village consists of one or more agricultural castes plus a handful of families from each of the artisan and specialized service castes. The craftsmen have lower status than the respectable peasant castes, but in turn outrank a substratum of agricultural labor castes and "unclean" menial castes.

Usually a small aristocratic segment of petty landlords, moneylenders and shopkeepers claims membership in the "twice-born" priestly (Brahman), warrior (Kshattriya), or mercantile (Vaisya or Bania) castes described in ancient Hindu literature. The cultivating castes are sometimes associated with the fourth and lowest of the traditional divisions, that of toilers (Sudras); more often this classification is reserved for artisans. Still further down the social and economic scale are the numerous and variously labeled outcastes, exterior castes, fifths (*panchamas*), untouchables, depressed classes, and so forth. These vaguely defined categories include two main types: (1) castes whose real or nominal occupation involves ritual impurities such as handling the skin of dead animals or laundering soiled linens; and (2) castes of (landless) laborers considered to be drawn from ethnic stock inferior to that of the dominant population of the area.

Each section of India has its own chief cultivating and herding tribes or castes, its own roster of necessary village servants, and its own submerged classes. Gangaikondan, a village of some 3,500 inhabitants near the southern tip of the Indian peninsula, is representative of the Tamil-nad districts of Madras.[5] An investigator in 1934 listed 19 groups in the "caste village" and 6 more in the near-by outcaste hamlets (*cheris*). He counted substantial numbers of Brahmans, Vellallas (the great cultivating caste of the Tamil-speaking south), Maravars (a clan of former robbers, marauders, and cattle

---

[5] B. Natarajan in *Some South Indian Villages,* ed. P. J. Thomas and K. C. Ramakrishnan (University of Madras, 1940).

thieves), Edayars (shepherds), betel-vine growers, and fishermen. In addition there were weavers, barbers, oilseed-pressers and sellers, carpenters, blacksmiths, goldsmiths, laundrymen, potters, lime-burners, musicians, and Muslim shopkeepers. The depressed classes included Pallars (both Hindu and Christian) by far the largest single caste in the village; Parayans (whence our word pariah); toddy-tappers, leather workers, a second weaver caste and broom-makers. The Brahmans formerly owned most of the village lands and although they did no actual field work, supervised the cultivation by hired Parayans and Pallars. By the time of the survey, Brahman holdings had decreased, their remaining lands were generally let out to tenants, and a goodly number of Brahman families had moved away to the cities in search of white-collar jobs to supplement their income from rents. The most prosperous Brahmans were those who engaged in moneylending as well as managing their properties. The Vellallas devoted themselves exclusively to agriculture, generally owning and cultivating their own plots. The only other caste with land-holdings of any size was the Maravars, who also cultivated fields on lease, and supplemented their earnings by such means as keeping poultry, tending sheep, and hiring out carts. In addition they extorted from the other villagers sums known as Swathanthriams —a kind of blackmail insuring protection of crops and cattlesheds, free passage for carts, and immunity of other worldly goods from banditry.

All the other castes of the village depended directly or indirectly upon the Brahmans, Vellallas, and Maravars. The specialized betel-growers had to rent land from them on which to plant their vines. Only a few of the Edayars actually owned sheep or even tended flocks. For the rest, the members of this caste were simultaneously very petty proprietors, tenants, and hired laborers. They supplied the household servants for the Brahmans. A considerable number had lost their holdings through debt and emigrated to the towns. Similarly the lowly Pallars worked on their own small fields, on rented fields, and on the fields of their high caste masters; many spent nine months of each year up in the hills as coolies on the plantations. None of the Gangaikondan Pallars or Parayans were *padiyals*, a kind of indentured farm servant common in Madras; but some families in the past had been slaves to particular Brahmans and continued to work for them. For the right to follow their traditional calling in the local tanks the fishermen paid rent to the Brahmans in the name of the village god. Carpenters, blacksmiths, and other artisans worked both for daily wages and for customary annual payments obligating them to repair implements or furnish other services upon demand throughout the year. The two castes of weavers were trying to eke out a living by farming on the side, but as was the case with the other skilled workers, several families had already left the village. Several small cloth provision and liquor shops were run by betelmen, oil-pressers, and toddy-tappers. The Muslims who kept stores in the outcaste settlements for the Pallars, also owned and tilled infinitesimal plots.

In many Madras villages there are few Brahmans or none at all, the higher cultivating castes providing both landlords and tenants, and even moneylenders. Perhaps the most marked feature of the whole southern region is the large number of depressed castes, who, whether as free or unfree laborers, tenants or small holders, form the basic agricultural labor force.

In Gujarat, where the west coast juts out into the Arabian Sea, the broad social division is into Kaliparaj ("black races") and Ujaliparaj (all other Hindu castes). The Gujarat village of Atgam in 1927 had a small elite in terms of prestige and standard of living consisting of five families of landowning Anavil Brahmans, one family of Modh Brahmans (the village priest), five families of Banias (moneylenders), three of Parsis (liquor dealers of Persian descent and Zoroastrian faith), and one Christian family employed by a mission.[6] Most of the other members of the Ujaliparaj were Kolis (farmers, tenants or farm laborers) formerly belonging to the Kaliparaj but now considered in the higher group because of their prosperity. The Kolis are numerous around Bombay and from their early services to European settlers comes our term "coolie." Together with Kolis in the village proper lived shepherds, fishermen, various artisan castes, untouchable Dheds (a caste of blanket-weavers here serving as policemen), and Bhangis (a depressed caste of latrine cleaners), all of whom worked on the land as well as following their hereditary trades.

In small hamlets on the outskirts dwelt the four castes of the

[6] G. C. Mukhtyar, *Life and Labour in a South Gujarat Village* (London, 1930).

Kaliparaj who together constituted two-thirds of the community. Even though some of the Kaliparaj owned their own plots of land, they were socially segregated from the other Hindus, carried out their separate religious rites involving wooden images of dead relatives, and celebrated their own holidays. Upon manhood they usually became halis or bond-slaves of the landlords. In consequence of modest loans which, in effect, they would never be able to repay, they were committed to serve one master the rest of their lives, receiving in return a homesite, certain gifts of clothing, and customary wages in kind. Their wives fetched water, washed cooking vessels, and removed dung from the cattleshed of the masters' houses, while the sons were engaged as cowherds.

The mid-Ganges Valley, the heart of northern India, has a substantial Muslim minority comprising a separate set of aristocratic, agricultural, and artisan castes. A study of 54 villages in the United Provinces yielded over 50 Hindu and Muslim castes, no one of which was found in more than 30 of the villages.[7] The largest land holdings (tenancies under absentee owners) were in the hands of the castes locally known as *sharif* (gentlefolk)—Hindu Brahmans and Thakores (claiming Kshattriya origin), and Muslim Syeds (descendants of the Prophet) and Sheikhs. Together they accounted for about one-seventh of the people. Five *razil* (lowborn) Hindu castes constituted the bulk of the working peasantry: Chamars (traditionally leather and hide workers of low standing but not here considered untouchable), Morais (vegetable gardeners), Kurmis (the most respectable peasant caste), Ahirs (cowherds), and Lodhas (another agricultural group). Specialized castes such as shepherds, Muslim weavers, barbers, laundrymen, and oil-pressers commonly cultivated small parcels of land in addition to practicing their hereditary occupations. Only six of the villages had one family each of untouchable Mehtars or Doms whose task was to clean the privies of the high caste houses.

Further to the north and west is the Punjab, where Muslims and Sikhs (a reformed Hindu sect) form the majority of the population. Here the common term of social identification is tribe as often as caste. In the central Punjab, the Jats—Hindu, Sikh, and Muslim—are the premier cultivators. Jat villages, in many of which the work-

[7] S. S. Nehru, *Caste and Credit in the Rural Area* (Calcutta, 1932).

ing peasants are themselves the holders of the land, tend to pay more respect to husbandry and wealth than to caste prerogatives. One such village is Kala Gaddi Thamman, colonized by settlers from northern Punjab about fifty years ago after the opening of a new canal.[8] In 1927 Sikh Jats were the sole landowners and provided most of their own labor. A handful of tenant farmers belonged to Muslim agricultural and artisan castes, while Christian sweepers (converted untouchable Bhangis) worked as field laborers. Of the three Brahman families, two lived on charity and religion while the head of the third was a shopkeeper and small trader. Far more esteemed in the village society were the prosperous Khatris and Aroras (two Hindu mercantile and moneylending castes). Those of the village artisans who worked at their trades ranked as menials in relation to the cultivating peasants and were paid largely in kind.

In the almost purely Muslim areas such as the west Punjab no caste distinctions separate landlords, tenants, and farm laborers. But the village servants are members of Muslim artisan castes and retain their caste designations even when they follow other occupations. The mullah (Muslim priest), ranks as a religious menial. Some larger villages have Qazis (Doctors of Muslim law)—of high social standing and economic position. And almost always a few families live on charity.

For India as a whole, a rapid survey such as the foregoing can do little more than indicate the regional diversity of patterns of village organization. Many important areas and castes have gone wholly without mention. The general picture that emerges, however, can be taken as roughly representative. High caste status is associated with land ownership or superior rights in the soil, higher living standards, and abstinence from manual labor. Similarly, the relative ranks of the various middle and low agricultural castes coincide, by and large, with their gradations from peasant proprietors or occupancy-right tenants to field laborers or serfs. The artisan castes, to the extent that they follow their trades, are regarded and treated as menials by the farming castes of good standing. The scale of their wages and customary receipts is such as to provide a level of existence as well as a social position just one notch above that of

[8] Randhir Singh, *An Economic Survey of Kala Gaddi Thamman* (Punjab Board of Economic Inquiry, Lahore, 1932).

the agricultural laborers. Merchant and moneylending castes enjoy both high prestige and high material rewards even where they own no land; petty shopkeepers are generally drawn from other and lower castes.

Only a minority of the village Brahmans actually serve as priests, temple-caretakers, teachers, or astrologers, and few of these obtain their whole livelihood from the fees and gifts due for these services. Where Brahman landlords and Brahmans who follow the traditional priestly vocations live in the same community, the priests are often looked down upon by the landed Brahmans, and treated as a separate subcaste. Many poor Brahmans in northern India are, in effect, religious mendicants, regarded contemptuously by the entire village community.

On the whole, the Muslims, while in practice adopting many features of the caste system, have exercised in the north of India something like a moderating influence. Social distances are observed most strictly in the south, which is also the home of the sharpest economic differentiation along caste lines. Rarely has the economic service of caste shown itself so starkly as in a list of 11 prohibitions for untouchables issued in 1931 by the Kallars, a prosperous cultivating caste of Ramnad district in the far south of Madras province. Sandwiched between enjoinders against the wearing of gold jewels or clothes below the knees, or the use of music or horses in wedding processions, are the following strictures:

1. Their men and women should work as slaves of the *Mirasdars* [landlords].
2. They must sell away their own lands to the *Mirasdars* of the village at very cheap rates.
3. They must work as coolies from 7 A.M. to 6 P.M. under the *Mirasdars* and their wages shall be for men Rs. 0–4–0 [$0.08] and for women Rs. 0–2–0 [$0.04] per day.[9]

FUNCTIONING OF CASTE.—Within the villages the ordinary cycle of daily life—going out to the fields and returning, fetching water from wells or tanks, arranging for repairs of ploughs and carts, buying and selling of foodstuffs, cloth and other necessaries, negotiating loans and land transfers, attending to births, marriages, funerals and festivals—inevitably brings all castes into frequent contact with

[9] J. H. Hutton, *Caste in India* (Cambridge, England, 1946), p. 179.

each other. But each caste preserves a separate tradition of common stock, common geographical or mythological origin, and common household regimen. It is this living complex of customs governing innumerable major and minor aspects of behavior that forms the distinctive inheritance of the caste and sets it apart from all other castes.

Thus the Brahmans in a particular community may eat only vegetables, take their water from the river, and marry off their daughters before puberty, while the main cultivating caste eats meat and fish, draws water from a well, and practices adult marriage. There is no rule that holds all over India for Brahmans or any other caste. Within a given locality, however, the practices of the various indigenous castes are fairly stable and a matter of general knowledge.

An individual, once born into the caste, remains in good standing so long as he abides by the appropriate regulations whether or not he follows the traditional caste occupation. The castes are self-perpetuating, for in almost all cases marriage is restricted to members of the same caste—and indeed to members of its particular local subdivision. Often fellow castemen of two near-by villages regularly provide each other's brides. Social precedence among castes is expressed in such terms as the number of other castes from whom any particular caste will accept water or cooked food, or with whom they will share the hookah (water-pipe). As a convenient exception, certain eatables such as confectionery can be taken from a wide variety of castes, including the customary vendors.

Each caste of any social pretensions whatsoever will also provide a list of one or more lower castes whose mere physical touch is considered contaminating and necessitates immediate ritual purification. The phenomenon of so-called "distance pollution"—outcastes whose approach within a certain number of feet defiles—although much discussed in the literature on the subject of caste has been actually reported only for Malabar, a small area on the southwest coast where matriarchal families and remnants of other atypical social forms have persisted to the present day. In practice, the concept of untouchability is honored not so much by the frequent ceremonial baths as by matter-of-fact avoidance of contact wherever possible. Exceptions in the case of river ferries and buses are common and illustrate precisely how flexible caste rules can be.

Real village problems arise over issues like common use of wells, temples, and schools.

Correct caste behavior is enforced largely by the weight of religious sanction. There are also, especially in the lower castes, councils of elders which may extend in authority over the castemen of a group of neighboring hamlets. These panchayats, often informal in character, are called upon to decide matters of ritual, relations with other castes, changes in caste rules and punishment for their infraction. They can impose fines, exact penances like giving a dinner for the caste or in extreme circumstances decree the complete outcaste-ing of the offender. Caste councils may also take up collections of money for religious observances.

Within his lifetime a Hindu remains a member of the caste into which he was born. But in the course of three or four generations a caste may rise or fall markedly in the social scale. In fact, there is a constant process of formation of new subcastes, amalgamation of old ones, and appearance of new claimants to the titles of Brahman and Kshattriya. Thus the Mochis (shoemakers), in North India, have separated from the Chamars (who skin dead animals and handle raw hides). The Kayasthas (scribes), who rank next to the Brahmans in most of Bengal, were not long ago simply on the "clean" artisan level. A steady small stream of tribespeople enters the Hindu fold each generation by settling on the edge of villages and accepting the most menial occupations. From making brooms or baskets they may proceed to hire out as day laborers and eventually to rent small plots of their own. Meanwhile if they are able to find a Brahman priest to minister to their worship, to give up eating the foods deemed impure in that locality, and often to reduce their marriage age, they are on their way toward forming a new agricultural or artisan caste or being accepted into an already established one. Another source of new castes arises from religious reform movements; even those specifically directed at the removal of caste have historically been perpetuated as castes or subcastes.

THE FAMILY.—Side by side along the village lanes inhabited by the various castes live old-fashioned joint families and the small families more typical of today. While households of twenty or more are still to be found, the average in most parts of India is five or six. Under the joint family system sons remain under the paternal roof together with their wives and children. The father functions as a domestic autocrat, expecting and receiving immediate obedience. Ownership of property is vested in the family as a whole; and every member of the household is entitled to maintenance from the common income. Upon the father's death the eldest son becomes head of the joint family. Should, however, a division of the inheritance be demanded, each son is entitled to an equal share. Or the brothers may set up separate households while still retaining common title to the family acres. If one or more brothers emigrate to the cities as industrial laborers, government employees, personal servants, or even professional men, the wives and children frequently stay in the family home. Although joint-family establishments are becoming fewer, very strong kinship ties are still felt to unite the individuals who would in the old days have resided together.

Marriage is a social duty incumbent on all villagers. The obligation of parents to find husbands for their daughters and wives for their sons has the whole weight of religion behind it. Matches are arranged by the heads of the families, who sometimes employ the services of a go-between. In the north, the village barbers typically carry on this occupation as a sideline. The most important criterion of suitability is proper caste status. Within many castes the matter of clans, branches, or kinship groups also arises. Astrologers may be consulted as to whether the union of a particular couple would be propitious. Dowry practices and customary gifts vary widely, and are often the subject of negotiation. The bride's family almost always pays for the expenses of the wedding, which can be quite substantial. The festivities prescribed by caste and local custom may take several days and involve most or all of the village.

Usually both parties to a marriage are in their teens—sometimes they are even younger. For almost two decades unions of boys under eighteen and girls under fourteen have been prohibited by law in British India, but there is little doubt that many such nuptials continued to be celebrated. Numerous castes, particularly the higher ones, hold to an interpretation of the Hindu scriptures requiring that girls be married before puberty. Where this rule is followed the wedding amounts to a betrothal ceremony: the child-wife does not go to live with her husband until she attains womanhood. Even in post-puberty marriages it is common for the youthful bride to visit back and forth between her new and old homes before she finally leaves her parents and settles down as a fully married woman; from that time on she is subordinate to her mother-in-law and must conform to her behests. Both Hindu and Muslim religions permit but do not encourage polygamy. A Hindu widow of high caste does not remarry. She stays in her husband's home, dependent upon his father or brothers. If she is the mother of sons, she retains full status within the household and will in time command the services of her daughter-in-law; otherwise she is often condemned to a lifetime as a domestic drudge. Should her husband have died before he grew old enough to consummate the union the "virgin widow" will remain under her father's roof. There are no barriers to the remarriage of high caste widowers. In fact social pressure is put on them to take new wives. Since the only available mates are young unmarried girls, great disparity in age often results. Muslim and Christian widows, and those of many low castes are permitted to and do marry a second time. Divorce is almost unknown in the village.

No matter how poor or low caste a villager may be, he expects and receives absolute deference from the womenfolk of his own household. The wife eats only after her husband has finished. A laborer's wife may toil with him in the fields, drudge as a servant in the home of his patron, or take daily employment with a road construction gang in addition to her home chores. In some castes of village servants the women have traditional functions: thus the barber's wife typically serves as a midwife. High caste women usually occupy themselves with domestic tasks. In some parts of northern and western India seclusion of the womenfolk is considered a mark of status and refinement among both Muslims and Hindus. Strict "purdah" women stay in so far as possible in the inner rooms and courtyards of their houses and pull their scarves or saris over their faces when the menfolk enter. In extreme cases they may deprive themselves completely of fresh air and contact with neighbors, female as well as male. Within the same community peasant women may go about freely while those of higher rank wait for dusk to quit their four walls. For outdoor protection, Punjabi Muslim women who observe purdah don tentlike "burkas" with tiny crocheted peepholes. South Indian women, with few exceptions, take full part in village social and economic affairs.

The most eagerly awaited event in the village family is the birth of the first son. According to Hindu practice, certain essential parts of the funeral rites can be performed only by a son. Childless or sonless women resort to the use of amulets, charms, and other forms of magic. They make special offerings to the local shrines and may even undertake long journeys to holy places of particular repute in this matter. If none of these methods suffice, a boy of the same caste may be adopted. Daughters are valued much lower than sons and, as a rule, less care and food is lavished upon them. Six is the average number of children born to a couple, but only half of these may be expected to live to maturity. High and low caste families alike have about the same number of babies.

Village mothers suckle their infants generously as long as two or three years. During these first years the baby is almost constantly in the arms of its mother, grandmother, or older sister. Young children have considerable freedom to run about in the village and in general are treated with great indulgence by men as well as women. As they turn eight or nine, boys of the working castes begin to herd animals or help in the care of the crops, while girls are taught to prepare and cook grains, pound spices, and so forth. Some villages have elementary schools attended by a number of the boys, mostly of the higher castes, and very few of the girls. Adulthood is commonly assumed to begin at about fifteen.

RELIGION.—Religion enters into all phases of Hindu life. Washing oneself in the morning, preparing and eating meals, sowing the fields take on sacramental quality when performed with the appropriate ritual. Religious merit is acquired as much by simple adherence to one's caste conventions and family obligations as by any special acts of worship. The pious village woman will offer grains to her household image before starting on her day's chores. On her way to the well she will stop to throw flower petals over a stone lingam in a wayside shrine. The most important deity to her is the local one she knows best—perhaps an evil spirit to be propitiated like Sitala, the goddess of smallpox, perhaps a genial character like Ayanar, the protector of the Madras Parayans. If there is a

Brahman priest in the neighborhood he will be called upon on special occasions to perform ceremonies in the homes of high caste families. The village temple, which may or may not have an attendant priest, is a place for individual homage and communion rather than congregational worship. Religious holidays are usually celebrated by the whole village; there may be processions with music, decorating of houses with lighted oil lamps, drawing fresh designs on white-washed mud walls, organizing dramatic performances of classic tales, special gifts for the deity whose day it is, or even animal sacrifices, although the Brahmans of the village would have nothing to do with this last. Holi, a widely observed festival in northern India, sanctions such jollities as masquerade, let-down of social barriers, and dousing with colored powder or water. At Ayuda Puja in the South each man worships his plough, chisel, blowpipe, pen or other tool of trade. Wandering holy men—Hindu *saddhus*, Muslim *fakirs*, Buddhist monks—are everywhere welcome to expound their own gods and gospels. But there is no real organizational hierarchy of priests. The relation between the village's own gods and the celestial figures in the populous Hindu pantheon is at best vague. Nonetheless, Siva, Vishnu, Brahma, and the other great names of Hindu mythology are generally known throughout the countryside. Hundreds of thousands of villagers from every corner of India annually stream to the holy cities such as Hardwar where the Ganges enters the plains, the cathedral shrines like the famous "Black Pagoda" of Konarak, and other pilgrimage points associated with the common Hindu tradition.

The great sectarian divisions of Hindudom, such as the distinction between Shaivites and Vaishnavites (respectively followers of Siva and Vishnu), are not matters of creed decided between an individual and his conscience but particulars of religious custom inherited in the same way as caste. Thus one group may mark the forehead with a sandal-paste trident, another with three parallel lines. But devotees of both would on occasion worship at the same shrine, and give a few coppers to its attendant priest. New cults arise frequently, usually among the lower castes, and after a few generations shake down into fairly stable sects or subcastes. The elaborate ethical and philosophical doctrines of the ancient Vedas

and Upanishads are kept alive mainly in Brahman families which
have been able to devote years to the study of the scriptures in the
original Sanskrit.

There is widespread belief in the rebirth of the soul in the form
of a man of higher or lower caste, or a heavenly being or even an animal or insect, according to the deeds of the individual. Associated
with this belief in transmigration is the injunction against injury to
any living creature. The cow is held in particular reverence. Cows
are not worshiped (the commonly seen bovine idol is Nandi, the bull
who is Siva's "vehicle"), but to the Hindus it is a great sin to kill
them. Persistence of fertility cults is indicated by widespread worship of representations of the male and female sexual organs (linga
and yoni). Other popular divinities such as the elephant-headed
Ganesh or the Serpent Gods of the south take animal or part animal
form.

Village Muslims, where they are few, are scarcely more distinguishable from their Hindu brothers in religion than in customs.
Though idolatry is strongly forbidden to followers of the Koran,
they may even worship at the shrines sacred to the local cults. Or
perhaps the grave of a Muslim pir (saint) will serve the same purpose. Characteristically Muslim areas preserve Islamic traditions
and customs better. Good Muslims do not eat pig or drink wine. Of
peasants who work in the fields, only a minority find time to perform the five lengthy daily prayers, each of which must be prefaced
by a thorough washing. They do attend mosque services on Fri-
days and observe the fasting month of Ramzan, during which no
food is taken from after dawn to before dusk. At harvest time,
the first deduction from the pile of threshed grain is a gift to the
mosque for "the soul of the Prophet." For holidays there are the
three Ids (on which goats or cattle are slaughtered), and particularly
the dramatic Mohurram when elaborate processions of tinseled paper caskets commemorate the death of the Prophet's grandsons.
Whoever can afford it attempts to make at least once the great pilgrimage to Mecca, and returns with red-dyed beard. Doctrinally,
the majority of Indian Muslims are Sunnis rather than Shias; but the
local mullahs are usually quite innocent of fine theological distinctions.

Village temples and mosques often have title to plots of land, sometimes at quitrents considerably lower than the general level of revenue assessment. A trustee or group of trustees manages the fields and expends the money realized from sale of produce on up-keep of the building, festivals, charities and so forth. These funds may also be used to pay a schoolmaster who holds his classes right in the temple or mosque.

Throughout the countryside, and more particularly in spots of great sanctity are to be found Hindu *maths*, Sikh *gurdwaras* (both similar to monasteries) and Muslim *wakfs* (foundations). Great merit can be achieved by willing property to these religious establish-ments, which house and support great numbers of priests, theologi-cal students, and holy men of all varieties. Villagers come from great distances to offer their small contributions and receive blessings. In recent years there have been charges of corruption against these institutions, but they are still highly venerated by most of the people.

The cultural life of the village is almost completely bound up with religious observances and religious holidays. Where small li-braries exist, most of the books are vernacular versions of the ancient religious epics. Traveling theatrical troupes perform the same classic tales. Young men sometimes organize hymn-singing sessions. A hand-ful of literate landholders or shopkeepers may subscribe to big city newspapers and perhaps read items aloud to their cronies. Radio receivers and phonographs are extremely rare. In different parts of the country various sports such as wrestling and cock-fighting have a long-standing popularity.

VILLAGE ADMINISTRATION.—The two indispensable village officials are the headman and the accountant. The headman is responsible chiefly for collecting the land revenue and secondly for assisting the authorities in matters of law and order and of the general wel-fare. As a sort of petty magistrate he is sometimes empowered to try minor cases, such as affairs of robbery or assault. The accountant keeps up to date the records of land holdings, rights, and transfers. According to its size a single village may have one or more head-men and accountants. In addition, there are usually a number of watchmen charged with protecting houses and crops, particularly at night. The watchmen also serve as news bearers and messengers.

Upon their oral report of births, deaths, and crimes the village vital and other statistics depend.

Although village functionaries today serve under and by appointment of the district officials, the posts are in many cases hereditary. They may carry with them grants of tax-free land as well as small stipends. The headman and accountant are almost invariably men of high caste. Since they act as intermediaries in all relations between the government and the other villagers, they are in a position to exercise great local power. The accountants, who can distort or falsify entries in the village register by a stroke of the pen, are subject, not surprisingly, to widespread corruption. Watchmen and streetsweepers, if there are any, are generally drawn from the lowest castes, often "untouchables." Since they are frequently servants of the village rather than of the state, they may depend for payment on customary daily donations of cooked food from certain households or on shares of grain distributed at the harvest. Only large villages have police stations with regularly employed full-time constables. Subordinated to district and subdistrict apparatus, the administration in most villages is a formal affair carried out without popular support, and with little reference to local needs.

The typical poor villager mistrusts and fears government in any of its manifestations. He considers the policeman a bully to be avoided, the accountant and headman as potential sources of harm to be conciliated or bribed. In many areas where the collection of land revenue and of rent is conducted as a single operation the villager draws no distinction between agents of the landlord and agents of the government.

Quarrels over land and attempts to recover unpaid debts often end up in long expensive legal suits. Once a villager becomes embroiled in litigation he has not only to pay his lawyer but also to take off time from his field work to attend court in the district town. Not infrequently both parties to a controversy are financially ruined by the time final judgment is handed down.

Some vital works, like the clearing of silted up irrigation ditches, are accomplished by cooperative effort of the villagers in accordance with ancient traditions. But where these old practices have broken down, or in fields where they never existed, the village simply goes without. Thus the vast majority of communities have no sewage arrangements, no safe drinking water, no public school, and no authority competent to settle civil disputes.

Village medical facilities generally consist of the services of the hereditary midwives, the *mantrams* (charms) of the priest, the patent preparations offered for sale by the shopkeeper, and the folklore preserved by the old men and women. Occasionally district surgeons or "lady health visitors" make tours, particularly for purposes of vaccination. A handful of villages enjoy a high quality of medical attention supplied by Christian missionaries or Indian social service organizations.

Fevers and stomach disorders such as dysentery are the villagers' most common complaints. Malaria is endemic in the river deltas and has spread to many other spots. It is believed to account for a million deaths per year, and makes a constant drain on the health and energy of vast numbers of people. Epidemics of cholera, plague, smallpox, and other diseases visit the villages every few years, taking a heavy toll each time.

In recent years attempts have been made to revive the *panchayat* (council of elders) form of village rule characteristic in most of India in the days when villages were self-sufficient political units. While the creation of panchayats has been authorized by numerous laws, they have actually been set up in perhaps one out of ten villages. Invariably the new *panches* (members of the council) whether appointed by the district officers, coopted by the headmen, or elected by some or all of the villagers have turned out to be the most prosperous members of the community—the high caste landlords, merchants, and moneylenders. A few of these panchayats have had considerable success in settling disputes and preventing recourse to superior courts. Other panchayats have concerned themselves with village improvements. Usually a single project such as the sinking of a well or the construction of a school building occupies the energies and resources of a panchayat for a year or more. Funds are derived from contributions by higher governmental bodies, from fines and fees, from sale of fishing rights or fruit picking concessions, and in some cases from small additional tax levies. Nowhere have they been sufficient to underwrite a consistent program of expansion of village services.

TRIBESPEOPLE.—All along India's mountainous northern border

and in the less accessible hilly and jungly tracts of the interior dwell tribes of people separated from the main body of plainsmen by custom and culture.

They number somewhere between twenty and thirty millions, according to the criteria of enumeration. Individual tribes vary in size from a few hundred (such as the primitive Toda dairymen of the Nilgiri Hills described in W. H. R. Rivers classic monograph) to more than two million. In the latter category are the Gonds of Central India, who include in their ranks shy foresters, ordinary villagers, and a few wealthy and educated rajahs.

The most backward tribes are to be found in the densest forests and jungles. Often naked or dressed only in leaves they flee at the approach of strangers. For food they gather nuts, berries, and roots or snare birds and small animals.

More numerous are the tribes who practice shifting or axe cultivation. They make small clearings in the woods by chopping and burning, and then sow seed in the ashes. After two or three years the plot is abandoned and a new clearing made in the same manner. In addition these tribes often weave baskets, hunt, and fish, or collect and sell such natural products as lac and resin.

Still other tribes are composed of settled agriculturists who keep poultry and cattle, cultivate with a plough, spin and weave cloth, and make pottery. Some of the Assam hill dwellers have developed an elaborate system of irrigation for their terraced hill-side rice fields. The tribes who inhabit the arid reaches of Baluchistan are sheep and cattle-herding nomads rather than farmers; they also engage in carpet-weaving.

The typical problem of tribal life at all levels of advancement is insufficiency and unreliability of the food supply. Although there are exceptional groups of outstanding health and physique, the great majority of tribespeople are undernourished and ravaged by disease.

To the extent that tribes have remained isolated they have preserved their characteristic institutions more or less intact. Hundreds of tribal languages still persist, although tribesmen today often speak the dominant Indian language of the area as well as their own tongue. Many of the tribes are subdivided into clans, each identifying itself with a particular animal or vegetable totem. Economically, the tribal villagers pool their labor and resources and share the fruits.

Where land is tilled it is either held in common or periodically redistributed. The high point of social life is communal dancing and singing, usually associated with ritual. Bachelor's halls, where the young men sleep under a common roof until marriage, are frequent, and sometimes there is a similar arrangement for the girls, sometimes for both together. Adult marriage is the general rule, together with ease of divorce and a considerable tolerance of extramarital sexual activities. Some tribal societies are matriarchal and on the whole the position of women is high.

In religious beliefs and practices there is a great deal in common between most tribal faiths and popular Hinduism. Tribal gods are many, but they are easily accommodated in the Hindu pantheon. Some characteristic features of tribal worship, however, are at variance with modern Hindu precept. Under this category come animal sacrifices or sacrificial hunts and the use of home-brewed liquor in religious rites.

It is among the tribes and particularly the more primitive ones that such curious religious practices as swinging from a pole by metal hooks passed through the flesh are occasionally found; all too often they have been given wide currency as though they were generally representative of Indian life.

The vigorous and warlike Pathans on the northwest frontier are, of course, militant Muslims, while Buddhist traditions are preserved by some of the peoples who nestle on the Himalayan slopes.

The period of British rule and modernization has speeded up the process of contact between the tribespeople and the Indians of the plains. New land systems have provided a basis upon which outside landlords and moneylenders established themselves in areas where tribespeople had hereditary cultivating rights. Eventually the original occupants were driven to the hills or reduced to depressed castes of agricultural laborers. At the same time, railway construction, tea and coffee plantations, and coal mines provided a demand for hard manual labor which attracted many tribespeople away from their ancient pursuits. Every advance in rail or road communications and every attempt to exploit mineral or forest resources brought more tribespeople face to face with the shops, ploughed fields, and caste rules of village India.

In the course of their incorporation into village life, the tribespeople have generally suffered a loss of their traditional culture without gaining much from their new neighbors. Songs, dances, languages, marriage, birth and death rites have progressively decayed. Only with great pains and, in some cases, with direct violation of deeply felt religious prohibitions have they learned to till the soil in accordance with approved local methods. Except in a very few instances tribespeople have been relegated to the lowliest caste status.

British administration of tribal areas was initially concerned with the stamping out of practices offensive to Western standards such as human sacrifice, head-hunting, blood-feuds, and trial by ordeal. Towards the end of the nineteenth century this process was well-nigh complete. Another problem of the early period was presented by the periodic raiding and looting forays of hillmen into the plains. For a long time official policy vacillated between sending punitive military expeditions and paying money subsidies as bribes for good behavior. The Santals of the east central uplands, after several bloody risings, were eventually pacified and settled on reserved lowland areas suitable for cultivation. On the northwest frontier, where no steps were taken toward providing an adequate livelihood for the tribespeople, there has been constant recurrence of trouble.

As in the rest of India, official land revenue settlements in the tribal districts tended to establish clear-cut ownership of land in the place of former customary relations. Thus many chiefs were transformed into landlords and the economic structure of the tribe was fundamentally altered.

At a later date the enactment and enforcement of forest protection laws, which were widely hailed by Indian opinion as essential for the preservation of a great natural resource, worked a great hardship on the forest-dwelling tribes. Forbidden from practicing axe-cultivation and to a large extent from hunting, many groups were totally at a loss for a means of subsistence. Similarly the excise laws which prohibited home production of liquor disrupted long-standing tribal customs. In large part indigenous brews were replaced by the stronger and more intoxicating distilled spirits obtainable in the villages, and drunkenness became a serious problem.

In recent years governmental agencies, missionary organizations,

individual reformers and nationalist leaders have all called attention
to the sad plight of the tribespeople and have asked special consideration of their economic and social needs. But there is great controversy as to how these needs can best be served, whether by cus-
todial isolation from the "tainting" influence of the modern world,
by religious or educational uplift, or by more intimate relations
with the rest of the Indian community.

III. The Cities

URBAN LIFE AND LIVELIHOOD.—One out of every seven Indians
lives in a town of 5,000 or more. The large cities which have sprung
up or expanded during the last century have functioned economically in a role supplementary to the needs of the British metropolitan economy. They have served as centers for handling the outward movement of raw materials and the inbound movement of
manufactured goods from abroad. The factories which have come
into existence in these cities are either devoted to consumers' goods
and other light industries, or they are workshops for maintaining
transport services, particularly the railroads. The only field of heavy
industry in which Indian enterprise has challenged Britain's supremacy has been the fabrication of steel; and even here Indian
production has been limited both in volume and range of output.

Indigenous enterprise made slow headway in the nineteenth
century partly because it came up against the power of the entrenched British mercantile houses. These houses had evolved a
form of business enterprise peculiar to India: the managing agency.
Under this system a single business organization runs the affairs of
a dozen or more concerns operating in a number of different fields.
The system arose when British merchant houses, like those which
had pressed in the 1840s for the opening up of India, themselves
later founded banks, opened coal mines, built jute factories, or
started tea plantations. Over the last century the great British concerns have played a predominant, almost quasi-monopolistic, role
in the economic life of India. In Great Britain, as well, these houses
occupied a vital economic position. From the ranks of their officials
were recruited more than one Governor of the Bank of England.
Lord Catto, who presently holds the post, was formerly the head

of Yule and Company, the largest of these concerns. Indians desiring to enter fields in which British managing agencies already operated came to find it sound or advisable to place themselves under the protection of one or another of these houses.

Throughout the nineteenth century would-be Indian captains of industry received little if any of the help from the state for "infant industries" which was so common in Europe and America. To British industrialists and shippers the idea of Indian protective tariffs behind which local industries might develop was anathema. The British had come to regard India as a country predestined for raw material production, a land in which machine industry would be an unnatural, misdirected, artificial development. When the Government of India after 1858 enacted low customs duties on textiles for the sole purpose of raising revenue, the Lancashire millowners forced the abolition of the imposts. When the desperate financial need of the Government of India later compelled their reenactment the Lancashire interests succeeded in having them matched by Indian excise taxes of nearly the same amount. The first tariffs of even a moderately protective nature did not come into operation until the 1920s. Up to that decade British managers of the railroads followed a policy of purchasing railway stores in England, even when Indian articles of the same quality were available at cheaper prices. And right down to 1947 the great British shipping lines were able to operate at will in the Indian coastwise trade.

The birth of modern industry in India has therefore been a prolonged and painful process. The total number of factory workers has never reached 1 percent of the population. And over the last fifty years the urban craftsmen and other hand workers have declined, not only in proportion to the rest of the population, but even in absolute numbers. It is the exceptionally slow rate of industrialization combined with the decline of the handicrafts which explains the otherwise astounding fact that over the last half-century India has become more and more of an agricultural country. Back in 1891 three persons out of five gained a living from the soil; in recent years the proportion is about three out of four.

There are two chief centers of modern industry in India: Calcutta with its environs, and Bombay and Ahmadabad in western India. Heavy industry is concentrated in the Calcutta area. At

Jamshedpur 150 miles west of Calcutta is the well-known Tata Iron and Steel Works, which produces annually about one million tons of finished steel. The coal mines which supply Tata's needs and most areas of India as well are concentrated around Asansol about 150 miles northwest of Calcutta. They employ some 250,000 miners who turn out each year about 25 million tons of coal. Near Asansol is the only other sizeable steel works in India, that of the Steel Corporation of Bengal (SCOB), with a capacity roughly one-third that of Tata.

The heart of industry in and around Calcutta itself is the spinning of jute and its weaving into burlap, in both of which fields Calcutta leads the world. More than 100 jute and burlap mills employing close to 300,000 workmen stretch along both banks of the Hooghly River, whose tortuous channel connects Calcutta with the Bay of Bengal. Calcutta's services as India's greatest rail hub and one of its two chief ports require a wide assortment of transport and engineering workshops and small metal-working plants, which together form the city's second largest industry. With these shops should be mentioned its three government ordnance plants; these are the best equipped in India and turn out rifles, machine guns, shells, and a few pieces of field artillery.

The mills and mines of the Calcutta-Asansol-Jamshedpur industrial complex employ altogether about a million workmen. The Tata works is the only important segment of industry which has always been owned by Indians and directed by an Indian managing agency house. Much of Calcutta's industry is still run by British managing agency houses and up to recent years most of it was British-owned. This is not surprising, since Calcutta is the oldest center of British influence in India and served as its capital up to 1911. Calcutta's Clive Street remains the home of the British banks in India, which to this day are the most powerful in the country.

The great industry of Bombay and of Ahmadabad 300 miles to the north is the spinning and weaving of cotton. Bombay has about 100 cotton mills employing more than 150,000 workmen, while Ahmadabad also has 100 mills but only about 100,000 workers. But whereas Ahmadabad economically is nothing but a cotton town, Bombay matches Calcutta as a great port and even exceeds it as a center of diversified light industries. These include the manufacture of tires, bicycles, and ordnance, and the assembly of automobiles.

In addition to the two great clusters of mills and shops in Calcutta (resting on jute grown in Bengal) and in Bombay (based on cotton of the Deccan), there are about a dozen smaller industrial centers of lesser rank. Cawnpore, an important transport point in the middle of the United Provinces, boasts a number of textile mills, railway shops, and leather tanneries, and some specialized metal-working plants. The State of Mysore, towards the southern tip of India, has sponsored a variety of industrial establishments, most of which are located in the rising town of Bangalore. Scattered widely in southern India are a number of textile mills, some old and antiquated, others relatively new and modern. In northern and eastern India are a number of sugar-refining factories, paper mills, and cement plants; most of these have come into being since the war of 1914–18.

India's industrial position on the eve of war in 1939 may be summed up with three facts: the value of capital invested in industrial plants was two billion dollars; net annual output was worth one billion dollars; and total factory employment was roughly two million workers.

For India as a whole, the great bulk of the town dwellers are not factory workers but either hand artisans, unskilled laborers, or domestic servants. Estimates of the size of these three groups vary, but the total number of persons employed in them is perhaps twenty millions. All three of these groups are employed by or otherwise subordinate to and dependent upon the urban middle classes. The hand artisans carry on their ancient crafts either at home or in tiny sweatshops. Most of them are at the mercy of the merchant-moneylenders who finance them. Often they have fallen into a relation of debt-slavery to their creditors, to whom they are bound to sell their products. Domestic servants work for money wages which are so small as to be little more than nominal. Custom prescribes, however, that the servant be provided lodging, clothing, maintenance during illness, and days off on certain holidays.

Large numbers of unskilled laborers are employed in construction and road building, and on docks and railways. Generally, they are not hired directly, but are engaged in groups through contractors. Since no regulations govern the terms of their employment or the conditions of work, the common laborers are completely dependent upon the contractors, many of whom have been known to pay their labor less than five cents a day.

The bulk of the middle classes in the cities consists of petty traders, shopkeepers, middlemen, sweatshop owners, and small absentee landlords. From their families come lawyers, schoolteachers, and the lower ranks of government employees, such as clerks. For the middle classes generally, the struggle to make ends meet is a hard one, and only a small percentage of them achieve a moderate degree of comfort. In this more fortunate group fall the larger merchants and the successful lawyers, whose main practice is taken up with corporation law and suits about land.

BIG BUSINESS.—Genuine economic power and influence in Indian hands is confined to a few thousand rather well-entrenched Indians who have successfully established themselves as industrialists. The degree of concentration is impressive: "500 important industrial companies are managed by 2,000 directors. 1,000 of these directorships are held by 70 men. At the apex of the pyramid stand 10 men holding 300 directorships. This oligarchy in industry is a closed preserve. The son succeeds the father." [10] Perhaps even more striking than the concentration of control is the narrowness of the three tiny social groups from which have come the leading Indian industrialists: the Parsis of Bombay, the Marwaris of Rajputana, and the Jains of Gujarat.

The Parsis are a community of 100,000 most of whom live in and around the city of Bombay. They are descendants of Zoroastrians who fled from Persia at the time of Islamic invasion by the Arabs more than a thousand years ago. For some centuries the Parsis were merchants and shipbuilders. They got on particularly well with the British after the latter conquered India. Prominent Parsis served as contractors for the British Government while others acted as brokers for British firms. With the economic opening up of India in the middle of the nineteenth century, a few of the most enterprising Parsis established factories on their own. They were among the first Indians to set up cotton textile mills both in Bombay and in cities of the interior. The most famous Parsi was J. N. Tata. Despite the doubts of his countrymen and the scoffs of British engineers and

[10] P. A. Wadia and K. T. Merchant, *Our Economic Problem* (2d ed., Bombay, New Book Company, 1945), p. 498.

government officials, he succeeded, relying almost solely on his own resources, in founding the great Tata Iron and Steel Company. Besides textiles, iron and steel, his descendants have large holdings in such key fields as electric utilities, chemicals, and machine tools. The Tata interests are today probably the strongest single Indian business group.

The chief Indian rivals of the Parsis in recent decades have been the Marwaris, who originally were a caste of moneylenders and merchants from Indian states in the deep interior of Rajputana. This relatively backward area was not opened up until the railway-building days of the 1870s. Thus until that late date the Marwari traders were sheltered against the competition of British houses, which had already overwhelmed many of the larger Indian mercantile firms in the chief ports. Quick to take advantage of the new economic opportunities presented in the late nineteenth century, the Marwaris moved south to Ahmadabad and Bombay, and east to Cawnpore and Calcutta. Along with the Parsis they were among the first to set up cotton textile mills. From these they have expanded into every conceivable type of business in India. The wealthiest of the Marwari houses, that of the Birla Brothers, has interests in cotton textiles, sugar mills, paper companies, cement plants, jute mills, insurance companies, newspapers, and weekly magazines. Birla Brothers is a power to be reckoned with in Bombay, Calcutta, and Delhi. Another Marwari outfit, the Singhania group, exercises the dominant business influence in Cawnpore.

The Jains of Gujarat profess a variant Hindu faith dating back to a religious teacher of the sixth century B.C. In pre-British times the Jain mercantile community consisted of small traders and petty local bankers. Like the Marwaris, the Jains began their rise to prominence in the second half of the last century. In the cotton textile industry their holdings are today second only to those of the Marwaris. The best-known Jain concern is the house of Dalmia, which controls cement plants, airlines, sugar refineries, vegetable oil mills, a daily paper and periodicals.

Wealthy Muslims, whose income was derived chiefly from the land, did not attempt until quite late in the day to enter the fields of industry and banking. Up to the present not a single Muslim concern has holdings of comparable dimensions with those of the

Parsis, Marwaris, or Jains. Of the substantial Muslim merchants and grain dealers, the best known is the Calcutta house of Ispahani, founded by immigrants from Persia.

The few great Indian houses have risen to their present position by dint of great effort in the face of many difficulties. Clashes among themselves and with their British competitors were frequent and severe in the opening decades of the twentieth century. In recent years, by which time the Indian houses may be considered to have "arrived," all the established firms have tended to work together to prevent outsiders from intruding into their domain. In organization and structure the dominant Indian firms of today resemble the old British managing agency houses.

The war years from 1939 to 1945 brought unprecedented profits to the large mercantile and industrial firms. Prolonged and intense shortages of food and cloth have prevailed since 1942. Many leading textile manufacturers and food merchants took advantage of this to extract fabulous prices from helpless consumers. Of some it is literally true that their normal sphere of operations became the black market. The Singhania mills of Cawnpore were heavily fined by Indian courts for flagrant activity in this direction. Connivance by government officials, both British and Indian, in such black market operations became widespread and led to a sharp decline of standards both of private business and of public conduct. By 1947 the Reserve Bank of India issued an open and official warning that if such practices were not checked, they would cause a breakdown of government and society.

A fifteen-year program for national economic development jointly authored by representatives of the leading Indian business houses was put forward early in 1944. The Bombay Plan, as it was called, proposed a great expansion of industry. It relied heavily for capital on the sterling balances India had accumulated in London during the war (Britain's unpaid five billion dollar debt to India for wartime purchases and services). It also envisaged a considerable degree of governmental regulation of economic enterprise. At the time, the Bombay proposals evoked much official and unofficial interest in planning. But developments since the end of the war have been little influenced by any blueprint. Pressure from industry has already forced the government to drop most of the wartime economic controls, let alone introducing any new ones. And the question of repayment of all or part of the sterling balances is still unsettled.

The postwar years have been marked by a series of understandings between the Indian houses and some of the largest manufacturing interests in Britain. Birlas, for example, has reached agreement with the leading British firm of Nuffields for the assembly and for the manufacture later on, of automobiles in India. Tata has concluded an arrangement with the British cartel, Imperial Chemical Industries, Ltd., for the manufacture of dyes for the Indian textile industry. There have also been negotiations in the direction of similar links with important American concerns.

URBAN SOCIETY.—The heart of every Indian city is a web of long narrow bazaar streets, thronged from morning to night with men, animals and carts of all descriptions. The wares offered for sale range from foodstuffs and cheap cooking vessels to tissue-thin fabrics and elaborately wrought jewelry. Back of the crowded rows of tiny shops stretch line upon line of squalid tenements. With one room and sometimes a veranda per dwelling, the low Calcutta *bustees* and the many-storied Bombay *chawls* resemble transplanted and closer-packed village huts. The *cheris* of Madras are nothing but clusters of mud and thatch hovels built by the occupants upon any available space.

In the midst of these mean alleys rise great temples and mosques, royal palaces—perhaps still inhabited—and stately mansions of former grandees. Often these buildings date back many centuries and are of great architectural beauty.

The "cantonment" area is usually quite separate from the rest of the town. Here, along broad tree-lined avenues, stand spacious houses each in its own sizable walled compound. Army barracks, law courts, clubs, Christian churches, colleges, and other public buildings of recent construction dominate the scene. Originally British enclaves, these residential suburbs have lately attracted more and more of the well-to-do Indians.

Only the chief cities have blocks of fairly commodious flats, large office buildings, and glass-windowed stores. Even in Calcutta, Bombay, and Madras, however, the modern downtown section is but a tiny enclave surrounded by several square miles of straggling bazaars, slums, mill districts, and better residential quarters.

The first Britons who came to India as traders, missionaries, and adventurers quickly found their own level in the variegated life of the great courts at Agra, Golconda, and Murshidabad. Later, as the East India Company expanded its operations, small colonies of English businessmen, soldiers, and administrators grew up in Bombay, Madras, and Calcutta. Except for occasional exchanges of courtesy visits with the Indian gentry, these little settlements kept very much to themselves. As a general rule relations between the English and Indians of rank were generally correct, often civil, seldom cordial. Some Britons distinguished themselves by their interest in Persian poetry, Hindu metaphysics, or ancient Indian epics. But on the whole the Indians were regarded as "men benighted." Any gestures in the direction of Western styles or customs excited mild amusement.

The expatriates, for their part, were quick to adopt such Indian ways as conduced to their own comfort. Above all they delighted in the cheap availability of great numbers of household servants, who were invariably docile and unused to such amenities as days off. As the number of Englishwomen in the various settlements increased, social life for the higher ranks of army officers and civilians became a gay whirl of balls, picnics, supper parties, and amateur theatricals. The Rebellion of 1857 put an end to this joyous era. The reported and imagined excesses of the mutinous troops and their supporters aroused the Anglo-Indians, as the British residents then called themselves, to cries for bloody revenge. For half a century thereafter the "white sahibs" tended to regard the "natives" with mingled fear and loathing. British soldiers and their officers became notorious for the vigor with which they insisted that Indians offer the customary signs of inferiority and submission. Self-conscious isolation was the order of the day for all ranks of the English, from the lesser officials of the upcountry "stations" to the Viceroy himself. In food, dress, household furnishings, and amusements the attempt was made to reproduce English existence as closely as possible, despite such difficulties as the Indian climate. English children were educated only with their own kind; usually they were sent home for their schooling.

In the most recent years the growing strength of nationalism and of the Indian business community has led to somewhat more equable social relations. English and Indian government officials, professional men and industrialists lunched together and occasionally dined in each other's homes. In rare cases, genuine friendships cut across the customary barrier. But right up to the actual withdrawal of British political rule, there remained in Bombay and Calcutta "European" clubs in whose premises no Indian could set foot except as a servant.

Marriages between the English and the Indians have not been tolerated by the social arbiters of either group. The descendants of these unions, typically of British Tommies or subordinate employees and Indian women, today constitute a separate community of about one hundred and forty thousand and are known as Anglo-Indians. Living in the large cities, they are employed almost entirely by the Government, particularly as policemen and on the railways. They generally wear Western clothes, profess Christianity, and try to identify themselves in so far as possible with the "European" population. In social contacts they are usually limited to British enlisted men and converted Indian Christians.

The one Indian group which has come nearest to being socially acceptable to the British is the older aristocracy of princely families and landed gentry. With their secure and in many cases vast incomes from rent, the rajahs, nawabs, and zamindars have been able to maintain palatial residences in the big towns as well as at their ancestral seats and to entertain on a lavish scale. The elements of Western culture which they have adopted most enthusiastically are horse-racing and motor cars. Although many have been willing to overstep caste or religious precepts in their personal lives (such as the prohibition against alcohol), they are generally supporters of orthodoxy and opponents of social progress.

It is the new Indian middle classes who have given the modern and modernized cities their characteristic flavor and culture. Drawn from many different castes and areas they have come to share common commercial or professional pursuits, common conditions of urban home life, and a common position in the national scene. As a group, the middle classes constitute the most Westernized element in Indian society. From them have come the pioneers of social change and religious reform. Culturally they have been torn between an intense admiration for Western forms and a deep desire to vindicate the traditions of their own country.

Internally the middle classes are divided into "communities" defined in terms of religion, caste, and language or region of origin. A few of these communities, such as the Parsis and the Marwaris, have already been mentioned. In the latter case the name is applied to the western branch (associated with the Marwar district in Rajputana) of the Agarwals, a trading caste of Vaisya rank. The Kashmiri Pandits are a Brahman group, originally from Kashmir, who held high posts under the Mogul emperors and produced many scholars of Persian, the official language of the Delhi court. Many of them have played important roles in current Indian politics, and the present premier of India, Jawaharlal Nehru, is numbered among them.

A wealthy and important Muslim trading community is the Khojas of Bombay, who as Shias of the Ismaili sect follow the religious leadership of the Agha Khan. The Nattukottai Chettis, a Madras caste of Tamil-speaking Shaivites, are well known as bankers in Burma, Ceylon, Malaya, Fiji, and South Africa, as well as India. The Rahri Brahmans of Bengal take their title from a riverside tract and include in their ranks the well-known literary family names of Bannerjee, Chatterjee, and Mukherjee (actually Anglicized versions of Bandopadhyaya, Chattopadhyaya, etc.). Bengali Brahmans were among the first Indians to write novels, poetry, and history in English. Mrs. Sarojini Naidu, the renowned poetess and nationalist, is a Chattopadhyaya by birth.

In the sphere of home life each one of these separate communities preserves to a considerable degree its own food and dress habits, its native language, its particular birth, death, and marriage customs, and often its tutelary gods and goddesses. Men returning from their daily work will doff their European clothes and put on the more comfortable *dhoti* (draped loincloth), pajama, or skirt of their native district. Outside the home there is greater uniformity and much less regard for caste restrictions. English and to a much larger extent Hindustani have become common languages. Men whose wives observe caste rules devoutly in their own kitchens sit down to luncheon in restaurants together with members of totally different communities. Even at home, tap water from municipal pipes which may have been touched by any number of low castes is not considered polluting. Castes which refused fish will take cod-liver oil on a doctor's prescription. Some "advanced" members of the present generation have gone so far as to contract intercaste marriages.

Reinterpretation of caste rules to conform with the needs of everyday urban existence and even occasional direct flouting of them do not mean that caste is unimportant to the city middle classes. In some ways modern developments have even served to strengthen caste influence and solidarity. Thus formerly caste councils or panchayats were restricted in scope to a small group of neighboring villages or a single town. During the course of the present century many new associations have sprung up uniting all members of a caste who speak the same language wherever they may be domiciled. The new organizations defend caste interests and status in such matters as the rank assigned by the decennial census. They often collect and administer funds which provide scholarships for needy young men of the caste. Sometimes they attempt to codify or change caste regulations on subjects such as age of marriage, amount of dowry, or widow remarriage. Other urban projects frequently restricted to members of a single community are cooperative apartment houses, student dormitories, credit societies, and even joint stock companies.

Indian movements for the purification of the Hindu religion and the reform of the caste system date back at least to the time of Buddha (about the fifth century B.C.). The distinguishing feature of the movements of this kind of the past hundred years is that they have been initiated by the urban middle classes and owe much of their stimulus to contact with both Christianity and Western rationalism. The great reformer Ram Mohan Roy founded in the early 1800s an organization known as the Brahmo Samaj, which campaigned for monotheism, the brotherhood of man, and widow remarriage. The Arya Samaj, started half a century later, took on the character of a fundamentalist Hindu revival and agitated for a consolidation of the numerous present-day castes into the four traditional *Varnas* (Qualities) mentioned in the old *Vedas*. Among the Muslims, Sir Syed Ahmed Khan in the early years and Sir Muhammad Iqbal in the present century were outstanding reinterpreters of Islam in a modern setting.

The field in which European influence has been welcomed most warmly by the middle classes is education. After several decades of sporadic and very limited assistance to Sanskrit or Persian academies conducted under religious auspices, the East India Company in 1835 announced an official education policy along the lines of Macaulay's famous "Minute" which condemned all Indian culture as barbarous. The object of instruction was declared to be the promotion of European literature and science, and the chief medium was to be English. Efforts were to be concentrated on the upper groups in Indian society with the prospect of a filtering down of benefits to the great masses. In the years that followed, government middle and higher schools, colleges, and eventually universities were established at the district and provincial headquarters. Government grants were extended to private Indian secondary schools and to the growing number of Christian mission schools.

The middle classes came to look upon English education as the stepping stone to social and economic advancement. Great sacrifices were made to enable sons to complete their college courses and thereby qualify for government service or the growing profession of law. Before the end of the century the number of graduates was already outstripping the possibilities of employment, but the stream of aspiring B.A.'s continued to grow. Thousands of young men went to England to continue their studies, and in particular to gain the coveted degree of Barrister-at-Law. The most recent years have brought a considerable widening of Indian collegiate curricula both in the addition of more scientific and technical subjects and in increased attention to India's own cultural heritage. There has also been a great increase in the number of girl students at every educational level. Of some eighteen universities of today, Calcutta is the oldest and best known. It confers degrees in Law, Medicine, Science, and Commerce, among others; possesses a cyclotron for research in theoretical physics, and has achieved international recognition for its statistical studies.

The daily and weekly press both in English and in the dozen-odd important Indian vernaculars is read with attention and wields considerable social as well as political influence. Circulation figures of individual newspapers range as high as 50,000. The cinema is perhaps the most popular form of urban amusement. Huge billboards advertising new screen plays abound in the large cities, and the Indian moving-picture industry—centered in Bombay—is one of the world's largest. The pictures usually deal with modern social problems or historical and mythological heroes, and are always embellished with music and dancing.

The working population of the cities is, to some extent, organized into its own communities according to caste and region like those of the middle classes. Hand artisans such as gold or silversmiths may be found following their traditional caste occupations. Certain castes from the Kathiawad peninsula provide the cooks in well-to-do Bombay Hindu homes. Large numbers of low caste Malayali-speaking workmen from the southwest coast are employed in the Bombay *bidi* (hand-rolled cigarette) sweatshops.

Most of the town laboring classes have been recruited within the past fifty years from the ranks of landless or almost landless peasants. As the Tamil proverb puts it, "After ruin, go to the city." At first, close ties are maintained with the home village. Frequently only one member of a family emigrates, or agricultural laborers may seek town work in slack seasons. Among the workers who stay in the cities and bring up their children in the shadow of the mills, there is a tendency in the course of a couple of generations to coalesce into a fairly homogeneous group.

Where the untouchables or depressed classes are involved, the assimilation is by no means complete. In the eating sheds outside factories separate arrangements for the mid-day meal are usually made for the castemen and the untouchables. But they frequently work on the same machines and they sit together at union meetings. Every large city has a sizable proportion of depressed classes: they form one-eighth of the population of Madras and about one-fifth of the industrial workers of Bombay.

Slum conditions place an almost unbearable strain on family life. It is by no means unusual for a tiny single-room dwelling to be shared by two or more households of adults and children, each with its own smoky little cookstove. Many working class families in Bombay sleep at night on the city streets, with a single length of cloth stretched protectively across the mother, father and youngsters. When the wives as well as the husbands work (during the war women accounted for 10 percent of the factory employees) the young children must be taken along on the job or left untended in the gloomy rooms. Maternal and infant mortality rates in the large cities are double and triple the rural figures; tuberculosis attacks ever increasing numbers of slum-dwellers. Prostitution and alcoholism have become prime social problems, particularly in the coal towns. In the balance, the economic opportunities offered by the city to the worker are purchased at the cost of extreme drabness and discomfort in daily life.

POPULATION.—In the present century India has begun to exhibit the rapid pace of population increase previously experienced by societies in process of modernization. Up to 1920 growth or stability varied sharply from decade to decade according to the occurrence and severity of famines and epidemics. Presence or absence of disaster raised or lowered the death rate while births held fairly stable.

Since 1920 there has been a sustained period of growth. This has been due not to any rise in the birth rate, which in fact has declined slightly, but to a consistent and significant fall in the death rate. This fall primarily reflects improved control over epidemic diseases, which, it must be emphasized, has been accomplished by public health techniques rather than by any rise in urban or rural living standards of the great bulk of the people. The extent to which the Bengal famine of 1943 and the recent communal disturbances have interrupted the trend is as yet unclear.

At the time of the 1941 census India's population was 388 millions; today the figure for India and Pakistan together must be near 420 millions. The recent rate of increase (about 1.2 percent per year) is by no means phenomenal. India is in fact growing no more rapidly today than did Britain in the last century or the United States as recently as the 1920s. But the base is so large that even this small percentage amounts to an addition of nearly five million persons per year.

The future prospects for the population trends in India and Pakistan depend directly on the speed and thoroughness with which the transformation of society and economy—now begun—is carried through. On the one hand the present death rate (about 30 per 1,000) is still so high that further advances in medicine and sanitation could bring it down considerably, thereby contributing still further to a population increase. On the other hand the high birth rate (about 45 per 1,000) cannot be expected to fall until and unless greater urbanization and higher standards of living for the vast majority of the people encourage voluntary reduction.

In the short run the question is not one of controlling fertility but of providing a better livelihood. There is no question that modern agricultural techniques could double the present food yield of India's cultivated soil. But this is not possible without a profound reorganization of peasant-landlord relations throughout the Indian countryside. India's "population problem," in effect, is a matter not only of social structure and economics, but of politics.

IV. The Rulers and the Unruly

THE GREAT REBELLION.—The first century of British power in India—from the Battle of Plassey in 1757 to the Rebellion of 1857 —saw the defeat of every Indian power or prince who would not bow to British rule. Three-fifths of India was taken over, divided into provinces, and ruled directly by British officials—this area was generally called "British India." The remaining two-fifths of the country was left in the hands of Indian potentates on condition that they acknowledge their submission to the "Paramount Power" of the British and consent to varying degrees of control or interference.

Rather slowly the British took a number of steps toward the modernization of India. Slavery, infanticide, and widow-burning (sati) were outlawed. Modern postal service, a free press, and a system of Western education were introduced. Indian reformers like the distinguished scholar and publicist Ram Mohan Roy and the great merchant Dwarkanath Tagore welcomed these measures and founded societies to organize popular support for them. Along with the opening up of the Indian economy in the 1840s and 1850s through railways, telegraphs, and steamships, the process of social and political change was greatly accelerated. One princely house after another was pushed aside and its territories absorbed into British India. Land settlements rode roughshod over the claims of conservative upper classes and dealt directly with the peasantry. Unlike the urban merchants and educated middle class, the older aristocracy resented British reforms and innovations. By the middle of the century they felt that their very existence was menaced, and more and more came to believe that the only way to preserve themselves was by driving the British out of India. In 1857 the drain of British troops from India for expeditions to Persia and China convinced them that a favorable moment had arrived. When mutiny broke out in the ranks of the East India Company's army over the issue of the famous greased cartridges, the dispossessed princes and by-passed nobility and former tax-farmers joined with the rebellious sepoys and called upon the people in city and country for support. The British struck back with all the force they could muster, but it took them more than a year to crush the revolt. During the hostilities rulers and ruled gave little quarter to each other.

STRUCTURE OF EMPIRE.—After the suppression of the rebellion Parliament ended the power of the old East India Company and vested complete control over India in a newly created department called the India Office. At its head was the Secretary of State for India, a full-fledged member of the British Cabinet. As Secretaries came and went with the turn of political tides in Britain, the substance of power was in practice exercised mainly by the seasoned advisers and permanent staff of the India Office, whose initial personnel had been taken over largely from the old East India Company. Typically, the Secretary was a man who had never set foot in India.

Within India the peak of authority was the Governor-General who was now also the Viceroy or representative of the Crown. Throughout the period since 1858 successive Governor-Generals waged a losing battle against the tendency for policy questions and other basic decisions to be made in London rather than India. Steamship, telegraph, radio, and airplane all combined to make the authorities in the field ever more subordinate to the hub of Empire. The Governor-General was left with prime responsibility for deciding how the policies set by London could best be carried out under the conditions existing in India. He was assisted by a number of appointed departmental chiefs, who sat together with him to form the supreme executive body of the Government of India.

Below the central government came the half-dozen-odd provinces into which British India had been divided for purposes of imperial administration. Their boundaries reflected the uneven course of imperial conquest rather than any genuine ethnic or historic divisions. For the bulk of British India, in fact, the provinces were little more than convenient clusters in which were grouped the 250 basic units of administration, the districts. Only at this level of authority did the ordinary Indian come into direct contact with the power of the British *Raj* (rule). The District Officer—whether officially known as "Collector" or "District Magistrate" or "Deputy Commissioner"—was all-powerful in his domain. He set and collected taxes, caught and tried offenders against the law, and did everything else that could be expected of a general factotum. The typical area over which he held sway was larger than Rhode Island but smaller than Massachusetts, with a population on the order of one million.

The several levels of administration were cemented together by the Indian Civil Service which provided the corps of highest permanent officials in the central government, the provinces, and the districts. Up to the War of 1914–18 the I. C. S. was staffed almost exclusively by Britons, many of them sons and grandsons of former East India Company officials. Upon retirement and return to their native country senior I. C. S. officers were often given key posts in the India Office in London, thereby contributing a further measure of continuity to the policy-making operations.

THE SEARCH FOR FRIENDS.—While the imperial administrative framework was being tightened and revamped, even more fundamental changes were taking place in imperial social policy. Despite the success of British arms in putting down the Rebellion, both Conservative and Liberal party leaders of Britain realized that India could not be held by the sword alone. It was essential to find and cultivate groups or classes who could serve as "steadfast friends and supporters" of British rule. After reflection upon the character of the Rebellion, especially in its storm center, the region of Oudh in the heart of the Ganges Valley, British statesmen rejected a policy of relying upon and seeking direct support among the peasantry. One scant year before the Rebellion, the dynastic ruler of Oudh had been deposed and his lands annexed by the British. A land revenue settlement had at once been made directly with the occupying peasants, while the claims of tax-collectors, local chiefs, and other would-be landlords had largely been ignored or rejected. The level of payments set by the new British administration, however, was substantially enhanced. When the Rebellion of 1857 broke out, the discontented chieftains of Oudh threw themselves behind it ardently. To the dismay of the British, who had hoped to be regarded as popular benefactors, the peasantry of Oudh followed the lead of their displaced masters and took sides with the rebels.

From this behavior the architects of imperial policy drew the conclusion that the people of Oudh and other provinces were deeply attached to their former masters and lacked the "independence of mind" to back the British against them. For the preservation of the Empire it was held to be "of greater consequence to secure the aid of the landed interest than even to deserve the gratitude of the masses of the people." Acting upon this conviction, the Government of India speedily restored the landed gentry of Oudh to their former station and confirmed their prerogatives. In the style of English country squires they were appointed magistrates, and Government officers were instructed to treat them courteously and to look upon them as "gentlemen of property and station, whose interests are identified with those of the Government, who are its natural born adherents, not opponents." [11] Throughout the rest of British India similar steps were taken. Ranking government officials advised the landed gentry and other privileged orders of the high esteem in which they were held, and sought their advice and aid in the tasks of local administration.

The ruling princes who had remained loyal to the British crown were rewarded with what amounted to security of status. Leading imperial authorities acknowledged that but for the fidelity of Hyderabad, Gwalior, and the Sikh states of the northwest, the Rebellion might have engulfed the entire country and swept away the British Raj. Queen Victoria issued a solemn proclamation of respect for "the Rights, Dignity, and Honour of Native Princes . . ." and a simultaneous repudiation of the policy under which Governor-General Dalhousie had from 1848 to 1856 taken over the territories of one princedom after another. The climax of the post-Rebellion

---

[11] Instructions of the Chief Commissioner of Oudh to district officers, cited by Lord Halifax (then Sir Charles Wood), July 25, 1861, in *Parliamentary Debates*, 3d ser., Vol. 164, p. 1521.

policy toward the princes was reached in 1881, when the fertile and populous state of Mysore in southern India was given back to its long-dispossessed ruling family after fifty years of direct British administration.

In a related phase of post-Rebellion policy the British abandoned their role as social reformers. In an effort to bind more closely to the Empire the conservative upper classes who embodied and benefited from social and religious orthodoxy, Britain formally promised respect and protection to all religious faiths and observances. Simultaneously, army reforms frankly took as their point of departure communal divisions and regional antagonisms. In analyzing the causes of disaffection, the military authorities concluded that sufficient precautions had not been taken by the East India Company to prevent the growth of fellowship, brotherhood, and common national sentiment among the different Indian elements in the old army. Soldiers of various origins had been mixed together rather heterogeneously so that "their corners and angles and feelings and prejudices" tended to get rubbed off. To guard against such developments in the future, a guiding principle of multiple segregation—communal, caste-wise, and territorial—was recommended and adopted. Hindus, Muslims, and Sikhs were not to be allowed to serve in the same companies. Furthermore, even members of the same community were to be allowed to serve only in units of the same territorial origin. The clearest statement of this basic principle of military policy was given in 1858 by the well-known administrator, Lord Elphinstone:

The safety of the great iron steamers, which are adding so much to our military power, and which are probably destined to add still more to our commercial superiority, is greatly increased by building them in compartments. I would ensure the safety of our Indian Empire by constructing our native army upon the same principle; for this purpose I would avail myself of those divisions of race and language which we find ready to our hands.[12]

STIRRINGS OF NATIONALISM.—By the 1870s vigilant British observers began to detect new signs of unrest. The world depression

[12] House of Commons, *Parliamentary Papers*, 1859, Vol. V., *Report of the Commissioners Appointed to Inquire into the Organization of the Indian Army*, Appendix, p. 146.

of that decade slashed the income of the peasants who produced crops for export, while tax collections were inexorably maintained. Successive droughts helped bring years of famine to large areas. In the cities the rising classes of educated professional men and merchants began to form local and regional organizations which petitioned for the appointment of Indians to responsible public offices. Alarmed by the outbreak of peasant riots, high government officials urged the moderate and loyal advocates of constitutional reform to unite their forces in order to guide public feeling along peaceful lines. Thus the celebrated Indian National Congress was formed in 1885 with the private blessing of the Viceroy. Its first president, Allan Octavian Hume, was an Englishman recently retired from the I. C. S.

The new organization caught on rapidly, and its provincial branches soon covered all of British India. Some of its early sessions attracted more than a thousand delegates, including Hindus, Muslims, Sikhs, Parsis, and a sprinkling of British merchants and retired officials. It put forward a program of mild administrative reforms intended to insure preservation of the Empire by reforming it in time *from within*. As a long-range goal the Congress aimed at parliamentary self-government within the imperial framework.

Governmental enthusiasm for the growing middle class forum soon cooled. Civil and military officials in up-country Oudh, Agra, and other areas where the 1857 Rebellion centered had never ceased to see mutiny around every corner. They hastened to attack the Congress as a seditious and dangerous body, a purveyor of grievances. The same Viceroy who had earlier encouraged the organization now publicly termed it a "microscopic minority." When Congress attempted at Christmas, 1888, to hold its annual session at Allahabad on the Ganges, it was refused the use of public meeting grounds. And the provincial Governor arranged to spend the holidays away from the city which was his capital. Only one year earlier, the Governor of Madras had welcomed and mixed with the delegates, and had sent his own brass band to entertain them.

Sir Syed Ahmed Khan, the outstanding Muslim political and intellectual figure of the day, was much disturbed at the rise of the Congress, and enjoined his co-religionists to shun it. He had been depressed by the harsh policy pursued toward the Muslims by the

British in the decade after 1858. The authorities blamed the rising primarily upon influential classes of the Muslim community, who had attempted, so the government believed, to reestablish the Mogul Empire. After the Rebellion, Muslims were regarded as unalterably hostile to the British Raj; as such they were generally excluded from government employment and were kept out of the professions. Sir Syed set himself very early the task of reconciling the Muslims and the British. In arguing with the British for a reversal of policy, he contended that the revolt was far more than a Muslim rising, that many Muslims had remained loyal, and that wise policy could reconcile even those who had been hostile. To his Muslim brethren, he pleaded for the ending of anti-British sentiment. In spite of fierce attacks by orthodox Muslims, he went ahead to found the Moham- medan Anglo-Oriental College (later called the Aligarh University, the most influential center of Muslim education in India), where advanced Western education and the religion of Islam were taught together. Convinced that British rule had brought "astonishing" advances to the people of India, Sir Syed saw the Congress as the fomenter of "unreasonable discontent." In this work Sir Syed received funds and support not only from the Muslim middle class and other advanced Muslim elements, weak as they were, but from government officials as well. Abandonment of the anti-Muslim policy was made plain to all when the Governor-General in 1877 laid the foundations of Sir Syed's college and later appointed Sir Syed to sit on his Legislative Council. To combat the Congress Sir Syed stated his views forcefully in a public address in 1887 at Lucknow. He fol- lowed this up by organizing the opponents of Congress into a "United Indian Patriotic Organization" with himself as secretary.

Sir Syed's pronouncement did not immediately deter Muslims from working with the Congress. During the succeeding two years, in fact, more Muslims than before appeared at its annual sessions, though they hailed mostly from the busy ports of Bombay and Cal- cutta rather than from the upper Ganges Valley where Sir Syed's influence centered. Even there a pronouncement was obtained from the outstanding Muslim divine of Lucknow that "it is not the Mus- lims but their official masters who were opposed to the Congress." Up to the eve of the war of 1914, however, Muslim participation was on a small scale. Sir Syed's opposition undoubtedly played a major part in this, as did the undeniable hostility of government to Congress. The weakness of the Muslim middle classes as a whole also had a retarding effect upon their entry into political affairs generally. In particular many Muslims kept away from Congress when strife developed within it between champions of militant Hinduism and the original leaders.

The Westernized founders of Congress admired the better aspects of parliamentary government and European society, and wished to reproduce them in India. The tediously deliberative pace of their political activity failed to satisfy the younger generation of middle class politicians born and raised after the Rebellion. They lost all patience in the world depression of the 1890s, which struck India even harder than the great crisis of the 1870s.

As prices of India's exports of foods and raw materials touched new lows, another series of droughts hit the hapless peasantry. The worst famines in all of Indian history followed, which, combined with epidemics of plague, resulted in a death toll officially estimated at more than 10,000,000. In those years talk of the benefits of Western civilization seemed ironic indeed; as elsewhere in the colonial and Oriental world, fresh efforts were made in India to vindicate the superiority of indigenous traditions against the much-vaunted rational and capitalistic culture of Europe. Restless Indian nationalists now carried forward the work of reviving and popularizing India's ancient heritage of Hinduism. Others developed the cult of such figures as Sivaji, the great Maratha hero of western India whose work contributed so largely to the breaking up of the Mogul Empire. Poona, the former capital of Sivaji's empire, and Calcutta became centers of rituals, festivals, and gymnastic societies, which were extremely popular among middle class students.

In response to repeated Indian pleas for increased participation in government, three successive Acts of Parliament in the first three decades after 1858 provided for the appointment of a few Indians to the Viceroy's Legislative Council, the election of some members of municipal and provincial councils, and later the indirect election of a minority of the Viceroy's Legislative Council. During the same years a few Indians were admitted to posts in the subordinate ranks of the Civil Service. This disappointing rate of progress led the younger nationalists, particularly in Bengal, to call for large-scale agitation and even the use of force. Alarmed at the militancy and potential dangers of the new doctrines, the government revived and utilized long-forgotten ordinances, including an authorization to deport any Indian from his own country without trial or public charges.

A drastic attempt to provide a counterpoise to the "rapidly growing strength of the educated Hindu community" was combined with a long overdue administrative reform when Lord Curzon in 1905 halved the huge, sprawling, province of Bengal. Rather than following any linguistic or historical principles, Curzon carried the dividing line right through the heart of the traditionally united Bengali-speaking area. The predominantly Muslim eastern and northern districts were proclaimed a separate province, whose first governor, Sir Bampfylde Fuller, jocularly remarked that he had two wives, one Hindu and one Muslim, and the latter was his favorite. Popular outrage against the partition of Bengal exploded into a six-year long campaign which proceeded rapidly from meetings and parades to boycotts of British goods and assassination of officials. Politically, the movement developed the most anti-imperial slogans since the Rebellion, with *swaraj* (self-rule) the chief demand. As unrest spread to the Punjab, the Government ordered further curtailment of the rights of press and assembly and sentenced opponents of partition to long-term deportation.

Right in the middle of the antipartition agitation, Lord Minto, Curzon's successor as Governor-General, received an impressive delegation of several dozen Muslim chieftains, landed magnates, and others of great wealth or station, headed by the Agha Khan. The deputation asked that the Muslims be considered a single bloc with interests distinct from the rest of the body politic. They prayed the Viceroy for the introduction of a separate Muslim electorate in the constitutional changes then contemplated. Lord Minto replied on the spot that he was "entirely in accord" with the general principle proposed and guaranteed that their rights would be respected. Capitalizing on the success of their mission, the members of the delegation founded at the close of 1906 the All-India Muslim League. Like that of the Congress two decades earlier their program began with a solemn profession of loyalty to the Empire. Their chief aim was to defend and further the political rights of Muslims, whose needs they undertook to voice to the Government "in temperate language." Alongside of this goal, but subordinate to it, they recorded a desire to promote friendly feelings between the Muslims and the other communities of India.

In London, where Parliament was discussing the proposed reforms, Lord Morley, the Secretary of State for India, explained the core of his policy to the House of Lords. In India, he stated, there were three classes of people to be considered: the small group of *irreconcilable* extremists who nursed "fantastic dreams that some day they will drive us out of India"; secondly, the advocates of self-government ("colonial autonomy") and lastly, the would-be cooperators who sought only a larger share in the administration of India. The essence of the reforms was to draw the second class, the advocates of "colonial autonomy, into the third class, who will be content with being admitted to a fair and full co-operation." In short, Morley announced he would march forward "with unfaltering repression on the one hand and vigour and good faith in reform on the other. . . ."

Finally passed in 1909, the Morley-Minto reforms permitted members of the provincial and central legislatures to criticize the British administrators of India more freely than in the past, but refrained from giving the legislatures any authority whatsoever to control the actions of those administrators. As Lord Minto had promised, the principle of communal representation was formally adopted. The number of members of the legislatures was greatly increased largely to provide places for special representatives of the Muslims, wealthy landlords, and the European and Indian mercantile communities. The disproportionate representation allotted to specially favored groups provoked the foremost organ of British opinion in India, the *Statesman* of Calcutta, to remark that the Reforms Scheme "amounts to little else than the provision for including in the Legislative Councils more landowners and more Mohammedans."

Meanwhile the antipartition movement led by the irreconcilables gained in strength. In 1911 on the accession of George V to the throne the reunion of Bengal was suddenly announced. Simultaneously, the capital of India and residence of the Viceroy were shifted from Calcutta to Delhi—a center of Muslim influence and the former headquarters of the Moguls.

Muslim reaction to the reunion of Bengal—a measure prepared in secret and sprung from them by surprise—was one of shock and indignation, which soon grew among the educated middle classes into a wave of anti-British sentiment. Prominent Muslim writers now undertook a passionate vindication of Islam (a *liberalized* Islam) against the supposed superiority of the West. They rushed to the defense of such Islamic countries as Persia and Turkey (the "sword of Islam") which were seen as the victims of Western imperialism. From criticism of Britain abroad, Muslim publicists moved swiftly to a nationalist position within India. The period of unqualified loyalty of the Muslim League to the British was brought to an end by the adoption in 1913 of a plank calling for "self-government" for India.

THE WARTIME HOPE.—The World War of 1914–18, with its severe economic drain upon India and its ideological identification of the Allies with democracy and national self-determination brought in its train vastly increased discontent at the continuance of India's subject status. Neither extensive application of punitive measures nor limited concessions in the direction of representative government succeeded in preventing the growth in the immediate post-war years of the first real challenge to imperial rule since the Rebellion. Little hint of the conflict to come was evidenced in 1914 when the onset of European hostilities evoked gratifying expressions of loyalty to Britain from the leaders of the Congress and the Muslim League, as well as from the princes and other conservative elements. After a quarter of a century of aloofness, British Governors in the early war years resumed the practice of honoring Congress sessions with their official presence. Official statements and private intimations encouraged the widespread belief that if India aided generously in the prosecution of the war she would be suitably rewarded with a new and favorable constitutional settlement.

The most influential Muslim writers and journalists held apart from the new wave of good feeling. When they persisted in expressing pro-Turkish sentiments and in poking fun at British war propaganda, their newspapers were shut down and they themselves interned. Other small groups of revolutionary-minded nationalists seized the opportunity to obtain aid and arms from Turkey and Germany with which they plotted to overthrow the British Raj. In the Punjab, which was the chief army recruiting ground, the heavy pressure exerted by government officials to raise "voluntary" loans and contributions and to enlist large drafts of "volunteer" soldiers occasioned deep resentment. A conspiratorial organization known as the *Ghadr* (Revolt) Party developed among the Punjab peasantry and received guidance and aid of emigrant Sikh communities in far-away California and British Columbia. The severe measures employed to break up the *Ghadr* and similar secret societies drew popular sympathy to the revolutionary currents. In their 1916 sessions at Lucknow, the Congress and the Muslim League reaffirmed their support for the British war effort, but called for an end to repression and release of political prisoners. At the same time the two organizations initialed an agreement on the form of government for India which they hoped the shortly anticipated constitutional revisions would provide.

The imperial authorities realized that the pace of reform would have to be speeded up or it would lag dangerously behind the march of popular sentiment in India. In August, 1917, the Secretary of State for India, Edwin S. Montagu, made his famous declaration to the House of Commons that Britain's aim in India was "the gradual development of self-governing institutions with a view to the progressive realization of responsible government in India as an integral part of the British Empire." Shortly thereafter Montagu departed for India to consult with the Viceroy, Lord Chelmsford; the outcome of their deliberations was the Montagu-Chelmsford Report of 1918 which became the basis for the Government of India Act of 1919.

Montagu's efforts to cultivate the support of the moderate nationalists met a cool reception from British officials in India, whose great concern was to nip in the bud what they conceived as a growing threat to the very structure of imperial rule. While Montagu was still in India, the Government appointed a special committee to advise on measures for dealing with "criminal conspiracies." This Rowlatt Committee in due course recommended a new set of sedition laws extending the Government's expiring wartime powers and providing for peacetime trial of political prisoners before special tribunals without benefit of counsel, jury, or the right to appeal even against sentences of death. The Rowlatt proposals were instantly denounced by all shades of nationalist opinion. Early in 1919, however, while Montagu's reforms were still undergoing the lengthy process of parliamentary consideration, the Government of India introduced and passed one of the Rowlatt Bills. The speedy enactment of this drastic and unpopular statute gave rise to suspicion that India's wartime contributions and aspirations were to be disregarded.

GANDHI'S DOCTRINE OF NON-VIOLENCE.—Initiative in organizing protests against the Rowlatt Bills was seized by a leader newly risen to prominence in nationalist circles, Mohandas K. Gandhi. Gandhi was a native of Gujarat in western India, where for several generations his forbears had served as chief ministers for some of the small Indian states on the Kathiawad peninsula. After completing his legal education in England, he set up practice in South Africa. There he attracted world-wide attention as the moving spirit of a series of campaigns by the resident Indians against political and economic discrimination. In the course of these struggles Gandhi developed his renowned technique of *satyagraha* or passive resistance.

Gandhi's basic aim was to make "religious use of politics" [13] for promoting an inner spiritual revival among men. His chief and abiding object of attack was not so much any one of the institutions of society as the evil embedded in men's minds. His method itself was only one part of a whole philosophy of non-violence, derived from Tolstoy's pacifism and Jesus' "Sermon on the Mount," as well as from the ancient Hindu concept of *ahimsa* (non-injury to living creatures).

Satyagraha (a coined word literally meaning "truth force" or "soul force") is the type of pressure which can be applied to remedy insufferable conditions without exceeding the bounds of non-violence. A Gandhian *satyagraha* campaign does not begin until a vow has been taken first by the leaders and then their followers to "adopt poverty, follow truth, cultivate fearlessness," and "observe perfect chastity." After this self-purification, the *satyagrahis* vow not to submit to the injustice against which they are protesting, and to endure cheerfully all penalties for their refusal. The success of

[13] "Religious use of politics" is Gandhi's own phrase. C. F. Andrews, *Mahatma Gandhi's Ideas* (New York, 1930), p. 110.

this extraordinary program in dramatizing the plight of the Indians in South Africa, and obtaining some concessions from the authorities was due in part to the strength of Gandhi's personal influence. An indefatigable worker, a gifted speaker and writer, he had limitless confidence in his own beliefs and unusual powers of persuasion.

Returning to India during the war years with his South African laurels, Gandhi gained new stature by conducting two important struggles among the peasantry. Under his leadership indigo-growing cultivators of Champaran, Bihar, won release from enforced planting of the dyestuff, and overtaxed peasants of the Kaira district of Gujarat gained a tax reduction. He also helped to persuade the Ahmadabad textile-mill owners to pay higher wages to their striking factory hands by resorting to a fast. Since he believed a large part of India's economic ills were due to the growth of a factory system, he introduced and popularized the idea of a revival of hand-spinning in peasant households. In the political field he soon took his place in the top ranks of the Congress.

The Rowlatt Bills shocked Gandhi to the core. He believed that inevitably they would so wither the projected Montagu-Chelmsford Reforms that the latter would emerge as a "whitened sepulchre." He called for passive resistance to the new law and designated April 6, 1919, as a day of protest and mourning, to be marked by a country-wide *hartal* (shutdown of shops and business—an ancient political technique in India). The *hartal* was an extraordinary success. At meetings in some of the most famous mosques, Hindu leaders were invited to appear and address the throngs. Hindus and Muslims publicly accepted water from each other. An atmosphere of high tension between the people and the authorities erupted here and there into rioting. Apprehensive officials, particularly in the Punjab, directed swift and severe pacification measures. Gandhi, upset at the outbreak of violence, concluded that he had erred in calling upon untrained people to offer *satyagraha*. A week after the *hartal* he called off the passive resistance. Meanwhile, the "restoration of order" in the Punjab proceeded. At Amritsar, the holy city of the Sikhs, soldiers were ordered to rake with gunfire a peaceful crowd of unarmed men, women, and children, in an enclosed square. The commanding officer later testified that he had aimed at "producing a sufficient moral effect" throughout the province.

THE SWELLING TIDE OF DISCONTENT.—Details of the Amritsar massacre and the other happenings in the Punjab seeped out through heavy censorship to rankle in the minds of all Indians. The substance of the Montagu-Chelmsford Reforms, finally promulgated from the throne at the close of 1919, fell far short of offering the "Home Rule" for which the nationalists had campaigned. While providing, for the first time, a partly elected central legislature the new constitution left the prerogatives of the Secretary of State and the Governor-General almost intact. In the provinces the governors and other officials were to continue to exercise full authority over land revenue, police, and other key departments. A separate sphere, including such subjects as health, education, and fisheries, was handed over to the provincial legislatures as a trial of "self-rule."

A belated government investigation of Amritsar eventuated in a report, issued in March, 1920, which Gandhi promptly characterized as "thinly disguised official whitewash." In May, feelings of indignation among India's Muslims rose to new heights because of the draft Treaty of Sèvres which the Allies, led by Britain, decided to impose upon Turkey. Under this treaty and related undertakings Turkey was to be partitioned among the powers and to lose all semblance of authority over Mecca, Medina, and Jerusalem. The Indians argued that these holy places should not be allowed to pass completely out of the hands of the Caliph (head) of Islam, a post claimed by the Sultan of Turkey. For India's Muslims the Khilafat agitation—as the activities on behalf of the Caliph came to be known —was both an attempt to soften the drastic penalties inflicted upon an Islamic nation and an expression of anti-imperial sentiments. To Gandhi, the Khilafat was not only a just demand in itself, but also an issue which furnished "an opportunity of uniting Hindus and Mohammedans as would not arise in a hundred years."

Parallel to the mounting political dissatisfaction in the Congress and in the Khilafat movement, there occurred unprecedented stirrings among the underprivileged populace in city and country. The first half of 1920 witnessed the birth of the labor movement in India, as shown in a wave of industrial strikes involving one and a half million workers. Simultaneously, in many parts of India the desperate peasantry formed local movements against oppressive landlords and sought counsel and support from nationalist leaders who had gained public fame. In Allahabad, Jawaharlal Nehru, young son of the wealthy nationalist lawyer who was then president of the Congress, was compelled by Oudh peasants to come and see for himself the inhuman condition to which they had been reduced by the landlords. He found "the whole countryside afire with enthusiasm and full of a strange excitement."

This high pitch of hope and excitement was part of and related to the immense tide of unrest and nationalism which swept the colonial world all the way from Morocco to China in the wake of World War I. The world economic crisis of 1920, which brought a ruinous fall in farm prices to the Indian countryside and a wave of unemployment in the cities, added to the reservoir of dissatisfaction.

Of the Indian leaders of the day, Gandhi was unique. The bulk of the nationalist leaders moved primarily among the middle classes and were familiar only with middle-class problems; in Nehru's words they were largely cut off from their own people and "lived and worked and agitated in a little world apart from them." To Gandhi it was clear that the people were moved by a deep fury at the unredressed wrongs done to the Punjab and at the savage treatment being meted out to Turkey; and it was also obvious to him how deep-seated was the unrest among the peasantry and urban labor. He knew that to organize a mass movement under such conditions was fraught with risk; but he felt that "the risk of supineness in the face of a grave issue is infinitely greater than the danger of violence ensuing from organizing non-cooperation. To do nothing is to invite violence for a certainty."

In the spring of 1920, therefore, Gandhi publicly pronounced British rule "a curse to India," and termed it "satanic." He called for a new satyagraha campaign in the form of "civil disobedience" or "noncooperation" with the Government. Satyagraha was to be applied in four successive stages. In the first, Indians were to give up all titles and renounce all honors they had received from their imperial masters. The second and main stage called for a triple boycott of the government legislatures, law courts, and schools. This was intended to make the functioning of the government collapse; at the same time, to strengthen India's home industries, foreign goods were to be boycotted, and hand-spinning and hand-weaving were to be encouraged. As a "distant" third stage, to be employed only if the first two were not efficacious, Gandhi listed withdrawal or resignation of soldiers and policemen from government service; fourth and lastly came the suspension of payment of taxes.

CIVIL DISOBEDIENCE, 1920–1922.—The main features of Gandhi's program were endorsed during the summer of 1920 by the Muslim leaders of the Khilafat agitation. The Khilafat leaders, steeped in the martial traditions of a militant faith, held no brief for nonviolence. They accepted Gandhi's doctrine as an expedient in order to get the immense non-Muslim backing for their cause which Gandhi held out to them. Later, and in the face of important opposition, Gandhi persuaded the Congress to adopt his plan of campaign. The worldly and wealthy leaders of the Congress prided themselves on their modernity and sophistication. Gandhi's asceticism and piety, his rejection of Western science and medicine as well as material progress, were alien to their temper. But they realized, partly because of Gandhi's own work, that India was passing through "a revolutionary period" and that their older techniques would not suffice. In part Congress accepted nonviolence as a policy necessitated by the fact that the masses of the Indian people had long been disarmed by law and were therefore unacquainted with the use of arms. The Muslim League followed the lead of the Congress, and both organizations changed their basic goals from self-government inside the Empire to "the attainment of *Swarajya* by the people of India by all legitimate and peaceful means."

Leadership of the noncooperation campaign was assumed by the Congress and exercised by a top committee of fifteen, the "Working Committee," headed by Gandhi. For the first time the Congress organized itself on a "grass-roots" basis, establishing branches or units right down to the village level. The Congress thus lost the overwhelmingly urban character which it had had during its previous thirty-five years of existence and became the first modern mass political party in India.

Gandhi scored his first important success in November, 1920, when the bulk of the electorate boycotted the first elections held under the Montagu-Chelmsford reforms. From this he moved on to the initial stages of the noncooperation campaign. Students all over the country responded eagerly to his call to boycott government schools and colleges. Professors deserted their posts and joined with nationalist leaders to found a number of new universities, notably the Benares Hindu University and the Jamia Millia Islamia near Delhi. Huge bonfires of imported British cloth dramatized the boycott of foreign goods. Large sums were quickly raised to support the campaign, women throwing in their jewelry for the cause. In the cities militant Congress and Khilafat youth organized themselves as volunteers; Hindus and Muslims together, they paraded through the streets in large numbers, shouting slogans and picketing shops selling foreign cloth. They paid scant attention to government declarations that this was illegal. By the end of 1921 most of the Congress and Khilafat leaders were in jail. Heavy pressure was brought to bear on Gandhi, particularly by the remaining Khilafat spokesmen, to intensify the campaign by moving on to mass civil disobedience, including nonpayment of taxes. Some even proposed abandonment of nonviolence, which irritated Gandhi intensely. At long last, he served an ultimatum on the Government in February, 1922, that unless the authorities released the volunteers from prison and desisted from "virulent" repression, he would begin mass civil disobedience in Bardoli, a small subdistrict of Gujarat.

Just at this critical moment word came from Chauri Chaura, a tiny hamlet in the United Provinces, that peasants who had been attacked by the police had turned upon the constables, forced them to withdraw to their station and set fire to the building, thereby burning or beating to death 22 persons. This news Gandhi treated as a direct, personal, divine warning, and he at once persuaded the Congress Working Committee to call off the entire noncooperation movement in all of India. The Government took advantage of the ensuing confusion to arrest Gandhi himself on charges of promoting "disaffection" and sentenced him to six years' imprisonment.

DECLINE AND REVIVAL OF THE CONGRESS.—The abrupt suspension of civil disobedience brought mass political activity to a sudden and staggering halt. Cooperation between the Congress and the Khilafat movement evaporated in an atmosphere of bitter mutual recriminations. Erstwhile lieutenants of Gandhi fanned the flames of communal rivalry and disorder by heading antagonistic Hindu and Muslim proselytizing organizations. The Khilafat campaign suffered a final and crowning humiliation in 1924 when the Turks under

Kemal Pasha abolished the office of Caliphate, and banished the deposed Caliph from the country. Within the Congress, which was greatly reduced in numbers, the influence of conservative and orthodox Hindu elements rose.

Gandhi, upon his release from jail in 1924, turned his attention to social rather than political problems. He campaigned against untouchability, for Hindu-Muslim unity, and for a program of rural reconstruction based on spinning and other village industries. During the same years Jawaharlal Nehru and Subhas Chandra Bose, a fiery Bengali, emerged as the leaders of a new left wing in the Congress. Nehru openly avowed himself a socialist, and he and Bose devoted much time to the student movement and to the problems of industrial labor.

The imperial authorities had been deeply perturbed by the scope of Congress influence shown in the civil disobedience movement. In the consequent readjustment of policy few opportunities were overlooked for creating or strengthening opponents of the Congress. The princes, who had been kept separate and disunited in the first half-century after the Rebellion of 1857, were now invited to join an advisory Chamber of Princes which met for the first time in 1921, right in the midst of the noncooperation campaign. In the same year the Government appointed special representatives of the "Untouchables" to seats in the provincial legislative councils. To dissuade the chief Indian business interests from backing the Congress with their prestige and funds, the Government for the first time levied protective tariffs on a number of commodities, beginning with steel. The principle of communal representation, under which the Muslims already received a disproportionate share of elective seats, was extended to the recruitment of personnel for the government services. In an attempt to relieve peasant discontent a number of officials undertook village reform activities emphasizing better farming methods and cooperative credit societies.

The 1919 Act embodying the Montagu-Chelmsford Reforms required that at the end of ten years a parliamentary committee reconsider the working of the reforms. In the fall of 1927 a review commission was appointed consisting exclusively of British members of Parliament and headed by Sir John Simon. Nationalist opinion in India was quickly beside itself with anger at the idea that a jury of seven Britons was to decide India's constitutional future. When the Simon Commission came to India, Congress took the lead in organizing a boycott against it. Outstanding Indian political figures, Moderates and Muslim Leaguers as well as Congressmen, refused invitations to confer with the visitors. Wherever the members of the Commission traveled, they were greeted by hartals and strikes, and their ears resounded with the cry shouted in unison by enormous crowds of demonstrators, "Simon go back! Simon go back!"

The reaction to the Simon Commission produced a great awakening. Already both agrarian and industrial unrest were on the upgrade. The political influence of the trade unions was reflected in a growing left-wing demand within the Congress for a new all-out struggle to win complete independence. At the Congress session in Calcutta at Christmas, 1928, there was a unique demonstration by mill hands. Tens of thousands of them marched in orderly and disciplined procession to the meeting grounds; took over the platform despite appeals from Bose and Jawaharlal Nehru; proclaimed their support of complete independence; and marched themselves off.

The following year a group of several dozen leading trade union officials—some of whom were well-known Communists—were arrested, whisked away to the small cantonment town of Meerut, and tried for "conspiring to deprive the King-Emperor of the sovereignty of British India." Independence-minded Congress leaders and the Indian public generally took offense at the flimsy nature of the case against the prisoners, who were in fact charged with no overt act, and contributed generously to their defense.

As the work of the Simon Commission drew to an end, it was announced that a Round Table Conference would be held in London. Various delegates from India and from the princely states would be invited by the British Government to sit together with representatives of the British political parties. When the Viceroy, Lord Halifax, was unable to give assurances that the conference would actually proceed to frame a Dominion Constitution for India, the Congress refused to participate.

The rapid onset in 1929 of the world economic depression strengthened and deepened the militant tide within the Congress. Reluctantly the older leaders came to the conclusion that another open struggle was unavoidable. Gandhi felt that there was a serious "danger of unbridled but secret violence breaking out in many parts of India owing to understandable and pardonable impatience on the part of many youths." Opposed to this "insignificant" but growing party of unorganized violence stood the might of the British Government in India, which Gandhi termed the party of "organized violence." The conviction grew upon him that "to sit still would be to give rein to both the forces above-mentioned."

CIVIL DISOBEDIENCE AGAIN, 1930–1934.—With Gandhi at the helm, the Congress annual session at the end of 1929 adopted "complete independence" as its creed and authorized the launching of a new civil disobedience movement in which nonpayment of taxes was explicitly sanctioned. To celebrate the new creed and arouse the people for the new struggle, the Congress Working Committee named January 26, 1930, as Indian Independence Day. At immense meetings held all over India on that date people pledged allegiance to a Declaration of Independence which began as follows:

We believe that it is the inalienable right of the Indian people, as of any other people, to have freedom and to enjoy the fruits of their toil and have the necessities of life, so that they may have full opportunities of growth. We believe also that if any Government deprives a people of these rights and oppresses them, the people have a further right to alter it or abolish it. The British Government in India has not only deprived the Indian people of their freedom but has based itself on the exploitation of the masses, and has ruined India economically, politically, culturally, and spiritually. We believe therefore that India must sever the British connection and attain *Purna Swaraj* or Complete Independence.

In its broadest outlines the second civil disobedience campaign of 1930–1934 employed the same devices as its predecessor of the previous decade. Chief emphasis was laid on the boycott of government offices and schools, and on the shunning of liquor, opium, and, especially, foreign cloth. Gandhi launched the campaign in March, 1930, by publicly defying the hated government monopoly of salt. At the head of a group of picked disciples from his retreat near Ahmadabad, he marched southward to the Arabian Sea. On its shores he flouted the law by making salt, using kettles to boil out the sea water. At this signal the various boycotts were put into operation all over the country with remarkable effectiveness. Popular action, however, went far beyond them. Without waiting for sanction from above, peasants in the United Provinces, Gujarat, and in western Bengal began "no-rent" movements; in eastern Bengal local terrorists raided the Chittagong armory and made off with much ammunition; and in the key city of Peshawar on the northwest frontier, the townspeople took complete control of the place and actually ran it for ten days.

The magnitude of the mass response to Gandhi's call for civil disobedience alarmed the authorities, who in May, 1930, placed Gandhi himself under arrest. This set off a fresh series of hartals and strikes, the most spectacular of which occurred in Sholapur, an important textile town in the southern part of Bombay province. There the aroused mill workers followed the example of Peshawar by setting up a regime of their own, yielding control only after martial law was proclaimed. Throughout the rest of 1930 the Government mobilized all its resources to break the Congress movement by repression: the order of the day included police charges on Congress Volunteers, firings on unarmed crowds, punitive expeditions into the countryside, confiscation of the property of Congress and Congressmen, and mass imprisonment. All these measures failed to bring the Congress to terms; its staying power was stiffened by the extraordinary activity of women of all classes, educated and illiterate. Casting off their traditional restraint, they threw themselves into the fight, many taking the lead of great processions. In sustained resistance to the authorities first place was taken by the city of Bombay. Its businessmen directed a comprehensive boycott of all British enterprise, while its militant industrial workers, the most highly organized labor group in India, again and again challenged the police and the military for command of the streets. In their parades the red flag of the trade unions flew side by side with the Congress flag, and in some places even overshadowed it.

While the civil disobedience movement was under way, the Round Table Conference opened in London. In view of the imposing strength which the Congress was demonstrating all over India, the British Government came to feel that the whole elaborate process of constitution-making inaugurated by the appointment of the Simon Commission three years earlier would be jeopardized, unless the Congress could be persuaded to take part. Accordingly, in January, 1931, the Viceroy set Gandhi free and negotiated a pact with him under which Gandhi called off temporarily the Civil Disobedience Movement and agreed to attend the next session of the Round Table Conference.

This suspension of civil disobedience had a frustrating and demoralizing effect in the Congress ranks quite similar to the termination of the previous campaign in 1922. These feelings were deepened during the course of Gandhi's fruitless visit to London as the sole representative of the Congress at the second Round Table Conference. Shortly after his return Gandhi himself was rearrested and Congress was declared an outlaw organization. The Government spared nothing in its efforts to break up the Congress; arrests far exceeded all previous records and were accompanied by systematic confiscation of Congress funds and property. While the Government was disappointed in its hopes of putting a speedy end to civil disobedience, it did disorganize the Congress further and keep it on the defensive. Eventually the immense mass civil disobedience of 1930 petered out into civil disobedience on a scattered, *individual* basis, and then was terminated altogether in 1934.

A MASS BASE FOR THE CONGRESS.—The debacle of the second civil disobedience campaign caused many nationalists to lose confidence in the older heads and policies of the Congress. A reassessment of values and objectives took place, accompanied by a search for a more effective program. This was particularly true among the left wing, where discontent and more than a trace of disillusionment had already been manifested in the twenties. The significant new development in the middle thirties was the determined effort of the left to found or strengthen independent organizations among the peasantry and urban labor and to make them the mass base for the Congress.

The trade union movement, which had been disrupted and disunited during the protracted Meerut conspiracy trial, emerged with renewed vigor in 1934. The strikes of that year in part reflected the appearance of a more militant leadership—in which, as in the twenties, a number of Communists were prominent—and in part were simply a phase of the wave of industrial unrest which swept many countries in the aftermath of the great depression.

National federation of all the union groups was achieved in part by 1938 under the aegis of the All-India Trade Union Congress, and completed in 1940.

Perhaps the most distinctive popular development of the thirties was the formation for the first time of a countrywide peasant league. Various local organizations of poor cultivators had already campaigned against evictions and for reduction of rents, debts, and taxes. Long-term goals, including the abolition of landlordism, the cancellation of peasant debts, and a revolutionary redistribution of land, were added when the local bodies joined together in 1936 to form the All-India *Kisan Sabha* (peasant league). Its founders (mostly left-wing Congressmen) stressed the need for close cooperation of the peasantry with the national movement for independence. They argued that, compared with the peasantry, the landlords and money-lenders were few and weak, and would not last long if they were not supported by the power of the British authorities in India. In enrolled membership the peasant league quickly outdistanced the organized urban workers; its annual conferences drew many thousands of delegates from all over the country.

Inside the Congress itself the left wing began to operate as a more coherent group than in the twenties. Militant Gandhians who had moved on from working for rural reform to an interest in socialism and a demand for thoroughgoing agrarian change cooperated with the new Congress Socialist Party. This was founded in 1934 by a group of students and intellectuals whose aim was to persuade the Congress to adopt as its goal not only the ending of empire but its replacement by a democratic and socialist society. With this objective the Congress Socialists restricted their membership to individuals already enrolled in the Congress, and made particular efforts to win over to their viewpoint prominent personages in the Congress hierarchy. At the same time the Congress Socialists participated actively in many of the organizing drives of the unions and the peasant leagues. The Communists, a party proclaimed illegal and hunted down by the Government, could not operate as a public group, but a small nucleus continued to carry on underground. Attacks upon them by the imperial authorities brought them a measure of popular sympathy. A number of Communists took part as individuals in *kisan* and labor activities, and a few held places in some of the councils of the National Congress.

In 1936 Jawaharlal Nehru, the foremost public exponent of the left trend, was elected Congress President. Nehru proposed the collective affiliation of the peasant leagues and trade unions to the Congress. The older Congress leaders, while believing in uplift of the poor, had mixed feelings about the growth of organizations of workers and peasants under independent leadership, and strongly opposed the idea of bringing them into the Congress as units. Nehru's proposal was defeated, but the Congress incorporated many of the basic demands of the peasant leagues and the trade unions in its own agrarian and labor program.

On this radical social platform the Congress contested the first elections held under the Act of 1935. The new constitution had in the main been written arbitrarily by the British authorities after the end of the Round Table Conferences; some changes were inserted while the measure ran the gauntlet of a Parliament in which Winston Churchill and other imperial diehards attacked the Cabinet for every modification proposed in the prevailing system of autocratic, centralized rule. The most important innovation was contained in provisions which extended the range of authority of the provincial legislatures and the popular ministries responsible to them. The provincial franchise was enlarged to include one quarter of the adult population. As a balance, the new scheme provided both at the center and in the provinces for a formidable extension of the system of communal and special electorates. A section slated to go into effect at a future date outlined means for joining British India and the princely states into a countrywide federation, in which the states would enjoy representation and power out of all proportion to their population, area, or resources.

Congress declared that its aim in seeking office was "not to submit to this Constitution or to cooperate with it, but to combat it, both inside and outside the legislatures, so as to end it." The polls held early in 1937 after the greatest electoral campaign in Indian history completely confirmed the Congress as the only major political force in the country. It received nearly 70 percent of the votes cast, while all other parties were restricted at best to purely local successes.

CONGRESS IN OFFICE.—The assumption of office in seven provinces by the Congress had an immense psychological effect upon the people. In the eyes of the peasantry, Nehru tell us,

Government was no longer an unknown and intangible monster, separated from him by innumerable layers of officials, whom he could not easily approach and much less influence, and who were bent on extracting as much out of him as possible. The seats of the mighty were now occupied by men he had often seen and heard and talked to. Sometimes they had been in prison together, and there was a feeling of comradeship between them.[14]

Feelings of exuberance spread, carrying along the message and raising the prestige of the Congress. Its membership grew amazingly, from some 600,000 late in 1936 to more than 3,000,000 early in 1938 and 5,000,000 in 1939. Of these, however, only one out of thirty were Muslims.

The initial measures of the new ministries were aimed at relief of the peasantry and improvement in the conditions of urban labor. But they were far from the "thorough change" promised in Congress electioneering. In part this was because the Act of 1935 had parceled out a limited share of authority to the provincial ministries, leaving the substance of power largely in the hands of the Governor-General, the provincial governors, and the civil services. In part, however, the gap between promise and performance resulted from the character of the Congress ministries. They were composed chiefly of seasoned figures in the Congress, who had been designated for their positions by the conservative and right-wing Congressmen who ran the party machinery. The latter had many ties of friendship and kinship with large landlords and wealthy industrialists, and tended to look with disfavor upon popular pressure for drastic restrictions on these vested interests.

The inevitable clash between the ministries and the mass organizations did not take long to develop. Some of the same ministers who upon entering office had received popular acclaim for releasing numerous political prisoners—a few of whom had been kept in jail since the first civil disobedience movement back in 1920—now shocked the country by using the kind of repressive measures which

[14] Jawaharlal Nehru, *The Discovery of India* (New York, 1946), p. 373.

they had previously condemned. In certain instances police were ordered to open fire on processions of peasants or used to help break strikes. Constructive measures, particularly in the field of education and health, did not awaken the enthusiasm that they might have, when they were accompanied in Bombay, for example, by a severe abridgment of labor's right to strike, and in Bihar, by a public pact between the ministry and the landlords.

The discontent within the Congress erupted in dramatic fashion in 1939. By that time the left wing groups had lost their patience with the provincial governments. They demanded an end to the policy of "drift," and called for a new mass struggle. In an unprecedented show of strength the Congress left reelected Subhas Bose to the presidency, defeating the candidate of the conservatives. But the old guard retained control of the Working Committee and were soon able to force Bose's resignation.

REBIRTH OF THE MUSLIM LEAGUE.—The abrupt termination of civil disobedience by Gandhi in 1922 had separated from the Congress the great body of Muslims organized in the Khilafat movement, and the gulf was widened by the communal controversies of the twenties. In contrast to the united effort of 1920–22, Muslim participation in the second civil disobedience campaign was on a significantly smaller scale. With the exception of certain groups in the North West Frontier Province and in the Punjab, Muslim organizations generally stood aside. In part this occurred because the Congress, despite its secular platform and noncommunal structure, could not escape the markedly Hindu flavor of Gandhi's leadership and doctrines. The latter inspired a distrust among Muslims which was not dispelled by the prominent place in the Congress hierarchy occupied by a number of staunch Muslim nationalists.

In its great election drive of 1936–1937, the Congress made no special effort to appeal to the Muslims. Of the hundreds of seats set aside for the Muslims under the system of special electorates, the Congress ran candidates for little more than one-tenth. Its neglect was soon capitalized by the Muslim League, which took a fresh lease on life during the election campaign.

The president of the League at this time and the guiding spirit of its transformation into a great popular party was a prominent Muslim politician, Mahomed Ali Jinnah. Like Gandhi a Gujarati in origin and a barrister by profession, Jinnah first entered public life in 1906 as private secretary to the then president of the Congress. Jinnah's marked abilities brought him rapid fame. Soon after 1912, when the League shifted its course to nationalism, Jinnah was invited to take part in its deliberations. As an outstanding figure in both the Congress and the League, he was dubbed "ambassador of Hindu-Muslim unity" in recognition of his efforts to bring the two organizations together. His election to the League presidency in 1916 led to the conclusion of a Congress-League "pact" to press for a common program of constitutional reform.

The climate of the Congress in 1920 under Gandhi's ascendancy became increasingly uncongenial to Jinnah. He argued that Gandhi's essentially spiritual program would not work, that nonviolence was suited more for saints than for ordinary mortals. Like other moderates of the day who did not seem fully to appreciate the depth of popular discontent, Jinnah believed that opposition to the Government should be expressed through proper legislative channels. When Congress proceeded with plans for mass action, he withdrew.

In the great upsurge of nationalist struggle that followed, the Muslim League was completely overshadowed by the Khilafat and noncooperation movements. When it emerged from obscurity in 1924, Jinnah was again called upon to serve as president. As in 1916 his principal object was to achieve an agreement on constitutional demands among the leaders of the various Indian parties and factions. But the effort came to nought and Jinnah retired to London, where he later attended the Round Table Conferences. Upon his return to India in 1934 he once again took up the reins of the League, and started it on the road to mass leadership of India's Muslims.

At the polls in 1937 the newly revived League did rather poorly, gathering only a small fraction of the total Muslim votes. But the aftermath of the elections brought Jinnah fresh and powerful allies. The immense popular support won by the Congress under a radical land and labor program astounded a wide range of conservative elements. Uneasiness affected landlords, title-holders and princes, Hindu as well as Muslim, and concern spread among the British business community, particularly in Calcutta. These various groups came to see in the League a possible alternative outlet for the energies of one section of the rapidly awakening peasants and townsfolk.

From the frightened vested interests the League was able to raise a war chest, with which it proceeded like the Congress in 1920 to establish a network of local chapters in villages, rural districts and urban areas. The program which the League presented to the Muslim public reflected both the Muslim desire for the freedom of India and the Muslim demand for reassurance on the status of their own community. It set as a goal the establishment in India of "a federation of free democratic states" in which full safeguards would be provided for all minorities. Socially, the League announced its concern for popular welfare and for "means of social, economic and political uplift of the Musalmans." Lest this plank seem too radical to the wealthy backers of the League, Jinnah simultaneously decried talk by Nehru and other Congressmen of India's hunger and poverty; this, said Jinnah, was "intended to lead the people toward socialistic and communistic ideas for which India is far from prepared."

The chief appeal of the League was to middle class Muslims in the towns. To them the League announced that rule by the Congress in the provinces meant Hindu rule. All lucrative positions, contracts, and special subsidies would go to Congress supporters, while Muslims would be relegated to menial posts. To Muslim peasants the League explained that the half-hearted and hesitant manner in which the Congress was putting into effect its radical agrarian program was due to a desire to protect Hindu landlords. No opportunity was lost to ascribe to the Congress a policy of keeping the Muslim community politically and economically backward. Islam was depicted as in immediate danger of destruction.

The League campaign proved highly effective in the United Provinces and Bihar, which contained eleven million Muslims, forming a small but important part of the population. Hundreds of new branches were opened and tens of thousands of new members enrolled. In the Muslim majority areas Jinnah succeeded in recruiting to the League banner the British-favored conservative Muslim premiers who headed the ministries of Bengal and the Punjab. With this added backing and prestige the League proclaimed itself in June, 1938, "the one and only authoritative and representative organization of the Indian Muslims."

Since the Congress professed to speak for India's nationalists of all creeds—Muslim as well as Hindu—the gulf between the two parties widened. Talks and exchanges of letters between the leading figures on both sides failed to achieve any degree of reconciliation. When the Congress in 1938 barred its ranks to members of "communal organizations" including the League, the break was complete.

For its part the Government proceeded to accept the League's self-estimate at face value. As the power and membership of the Congress, in the eyes of top Government of India functionaries, assumed dangerous proportions, imperial encouragement was increasingly extended to the League. Its officials were treated as though the League was the sole influential Muslim voice meriting serious recognition. In point of fact the League's control over the provincial ministries of the principal Muslim majority areas, the Punjab and Bengal, was by no means certain or dependable. In the two overwhelmingly Muslim provinces, Sind and the North West Frontier, the League had little influence at all. In short, by a policy of favoring the League, British officialdom was throwing its weight behind an anti-Congress organization whose main strength lay in the non-Muslim provinces where Congress ministries were in power. The importance assigned by the imperial authorities to the League helped it to obtain later on the unique status among the Muslims which it already claimed in 1938.

THE UNTOUCHABLES IN POLITICS.—Every part of India, as has already been noted, observes its own rules of social gradation and each section has its own roster of castes and tribes considered impure by the rest of the population. The constitution of the depressed classes into a political entity dates back to 1921 when British Provincial Governors appointed to the legislatures a number of individuals—some of high caste origin—to speak in the name of the untouchables. Until 1931, however, no official definition of the depressed classes had been established. The census of that year inaugurated an all-India category of Exterior Castes, but the criteria for inclusion varied from province to province and even according to the discretion of the individual enumerators. Thus arbitrarily delimited, the Exterior Castes were found to total over fifty millions or about one out of seven Indians.

After the census, the imperial authorities moved ahead with plans for detaching the depressed classes from the general body politic

by organizing them into a new communal electorate. The program
ran into formidable opposition from Gandhi. In his view Hinduism
neither included nor sanctioned discrimination against the *Harijans*
(Gandhi's own term for the degraded castes, literally "Children of
God"). He warned repeatedly that continued practice of such discrimination might lead to the breakup of the whole structure of
Hinduism. While still in South Africa he had introduced untouchables into his own household, and in the twenties he devoted much
of his time to campaigning for the right of *Harijans* to enter Hindu
temples. He rejected vehemently the idea of treating them as a
separate community, apart from the rest of the Hindus. In 1932
when the creation of special depressed-classes constituencies was announced, Gandhi secured their virtual abolition through one of his
most famous fasts. Subsequently he intensified his efforts on behalf
of the *Harijans*.

In the elections of 1937, Congress candidates won most of the
seats reserved for the depressed classes except in Bombay Province,
the home of the outstanding untouchable leader, the American-
educated Dr. B. R. Ambedkar. The Congress in office included a
small number of untouchables in its ministries. Some legislative
progress was registered in improving the civil and social status of
the depressed classes. On the whole the representatives of the untouchables were dissatisfied both with the number of positions
granted to them and the degree of social reform accomplished.

THE PEOPLE VS. THE PRINCES.—Around and between the provinces of British India spread in crazy quilt array the substantial
territories left in the hands of the princes. By the twentieth century
these enclaves, twenty large and 500-odd small, contained about a
quarter of India's population. The peculiar relation of the rulers to
the paramount power, which had chosen for reasons of high policy
to underwrite their position, allowed them to maintain absolute
autocracy without fear of popular revolt. To keep the princes in
line, the imperial authorities stationed in the states a corps of high-
ranking officials known as Residents, who were responsible directly
to the Viceroy. To these Residents was reserved the right to intervene in state affairs in cases of injustice, oppression or gross maladministration; but in practice the right was seldom exercised. On the
other hand the paramount power accepted full responsibility for

protecting the princes against insurrections, whether arising out of misgovernment or out of widespread demand for popular rule.

Thus buttressed by the whole weight of British power in India, the states with few exceptions preserved into the present the practice of arbitrary personal rule. Advisory councils, legislative bodies, or other representative institutions seldom existed to temper the sway of the rajahs, maharajahs, thakores, and nawabs who appointed and dismissed diwans (ministers) at will. No clear line separated state funds from the privy purse, which often accounted for a lion's share of the expenditures.

For the people the recognized varieties of forced unpaid labor included construction and repair of roads and palaces, cultivation of special crops as wild-animal bait for hunts, and carrying luggage of visitors. Special fees were collected for marriage, for the right to hire palanquins for wedding processions, and for other religious and social observances. Trade monopolies of such necessities as salt and kerosene were either auctioned off or assigned to royal favorites, and resulted in prices for such commodities much higher than those prevailing in British India. The land revenues and other taxes were augmented by additional exactions on the part of local officials.

Political activity in the states arose slowly and in the face of great obstacles. The All-India States' Peoples' Conference met for the first time in 1927, and persuaded the Congress session of that year to adopt a resolution calling on the princes to introduce responsible government. Subsequently, the Conference collected and publicized information on abuses and outstanding examples of misrule, and presented the case for the princely subjects to the Round Table Conferences in London. At the time of the second civil disobedience campaign, the ferment in British India was reflected in spontaneous movements for elementary political rights in a number of states, notably Kashmir in 1931. A series of local conferences in all parts of princely India in 1937 signalized a widespread popular awakening in the states. Demands were put forward for abolition of forced labor, for land reform, for civil liberties, and for government responsible to the people. Before long, full-scale conflict was raging between the rulers and their subjects from the Punjab States in the north to Hyderabad and Travancore in the south, and from

Rajkot in the west to the Orissa states on the east. The people dramatized their program by marching in processions, holding meetings, and by circularizing pamphlets and newspapers, and in some places by offering satyagraha. The princes and diwans struck back by outlawing the popular organizations; banning public assemblies; prohibiting entry into the states of books, journals, and political figures from British India; expelling local leaders or subjecting them to rigorous treatment in jails; and intimidating the populace by fines, seizure of property, police charges into unarmed parades, and, in a number of cases, firing by soldiers. But sustained popular pressure achieved a modest degree of reform in one princedom after another. Obligatory labor services were transmuted into taxes, commissions were appointed to investigate grievances, advisory legislative bodies were authorized, and in a very few instances, the principle of rule by law was incorporated into new constitutions.

Although the States' Peoples' Conference was never formally linked to the Congress, many individual Congressmen in the years after 1934 held posts as national officers of the Conference and assisted in its organizing drives. Within the Congress a sharp controversy developed as to whether the states' peoples' fight should be waged under the Congress banner. Gandhi held out firmly for noninterference by Congress in the affairs of the princes who, he insisted, could be persuaded to act in the best interests of their subjects. He argued that the states' peoples' movement would gain self-confidence by standing on its own feet. Gandhi intervened personally, however, by undertaking a fast when trouble broke out at Rajkot, where his own father had served as prime minister. The Congress Left attacked the princely system as a whole. They believed with Nehru that "The Indian Princes have hitched their wagon to the chariot of imperialism. They have both had their day and will go together." As 1939 drew to a close, the issue of noninterference or aid to the states' peoples was overshadowed by the outbreak of war, and, in fact, it was never resolved.

THE WARTIME CRISIS, 1939–1942.—The coming of war was no surprise to the Congress, whose focus of attention had been broadened, largely under Nehru's influence, to a deep concern with India's place in the shifting world scene. After Munich the Congress formally dissociated itself from British foreign policy, and recorded its opposition to "Imperialism and Fascism alike." India, the Congress declared, would not "permit her man-power and resources to be exploited in the interests of British Imperialism. Nor can India join any war without the express consent of her people."

When war actually broke out on September 1, 1939, the imperial authorities went ahead as if Congress did not exist and had never indicated any stand on foreign policy. The Governor-General automatically declared India a belligerent. Parliament in London passed in eleven minutes a bill permitting the suspension of the provincial autonomy sections of the 1935 Constitution under which the Congress ministries had been functioning. The Congress responded with a solemn reaffirmation of its position and a request that the authorities declare "in unequivocal terms what their war aims are in regard to democracy and imperialism and the new world order that is envisaged." Making its conditions clear, the Congress proclaimed that "if the war is to defend the *status quo*, imperialist possessions, colonies, vested interests, and privilege, then India can have nothing to do with it." On the other hand, "if Great Britain fights for the maintenance and extension of democracy, then she must necessarily end imperialism in her own possessions, establish full democracy in India, and the Indian people must have the right of self-determination by framing their own constitution through a Constituent Assembly without external interference." In any event, declared the Congress finally, the real test of any British declaration of aims would be its application to the present; the only way of convincing the people that a declaration is meant to be honored, is by giving "immediate effect to it to the largest possible extent."

The British reply was a vaguely phrased offer by the Viceroy to appoint an all-party advisory committee. Congress pronounced the Viceroy's statement "wholly unsatisfactory" and began to talk of a new noncooperation campaign. As a first step in this direction, the Congress ordered its ministries to resign, in all of the eight provinces it controlled. The trade unions and peasant leagues struck out on their own with antiwar strikes and demonstrations; the Government answered by seizing and imprisoning as many of their militant leaders as it could find.

When the Congress met for its annual session in March, 1940, it flatly concluded that "Great Britain is carrying on the war fundamentally for imperialist ends and for the preservation and strengthening of her Empire, which is based on the exploitation of the people of India, as well as of other Asiatic and African countries." Before Congress took any action to follow up this sharp declaration, the entire world was stunned by the sudden fall of France to the Nazi invaders. The Congress rushed to express its sympathy with a Britain whose forces had been driven off the European Continent and which stood exposed to the formidable menace of Nazi military power. Under such circumstances the Congress leaders felt Britain could not long refuse to come to terms. The Working Committee announced that the Congress was prepared to abandon nonviolence in the sphere of foreign affairs and cooperate with Britain in the defense of India. For its part of the bargain Britain would have to declare the complete independence of India, and as a first step in giving effect to such a declaration, allow a provisional National Government to assume office in Delhi. When this offer was rejected by the British Cabinet, then headed by Winston Churchill, the Congress fell back upon nonviolence and invited Gandhi to resume direction of its activities. Toward the end of 1940 Gandhi launched a limited form of satyagraha, which began on an individual basis, something like the closing phase of the unsuccessful 1930–1934 civil disobedience. The central aim of this campaign Gandhi declared, was not to hinder the war effort, but primarily to register India's moral resistance to its enforced and involuntary participation in the war. Following a procedure outlined by Gandhi, selected Congressmen made public declarations, often after giving advance notice to the police, that it was wrong to support the war. Thereupon they were arrested, and in May, 1941, as many as 14,000 were in prison for offenses of this character. In August, 1941, the issuance of the Atlantic Charter aroused hope in India of fresh political advance. The hope turned into bitterness and cynicism when Churchill announced that in his eyes the Atlantic Charter did not apply to India, but rather primarily to regions under the Nazi yoke.

The devastating Japanese attack upon Pearl Harbor in December, followed by the rapid conquest of Hong Kong, Singapore, and Rangoon, did serve to precipitate developments in India. Four days before the Japanese entry into the war, the British released Jawaharlal Nehru and other Congressmen who had courted arrest.

As after the fall of France in June, 1940, so again in January, 1942, during the humiliating fall to the Japanese of every bastion of the British Empire in Southeast Asia, the Congress indicated its willingness to take part in defending and governing India, provided that its basic demand for freedom was conceded and provided the interim government would be allocated real power. In anticipation of a fresh round of negotiations, Gandhi was again temporarily relieved of the Congress leadership. In March, 1942, as the Japanese outflanked Rangoon and approached the eastern land frontier of India itself, Prime Minister Churchill sent a special negotiator to India in the person of one of the members of his own War Cabinet, Sir Stafford Cripps. The latter had indicated for some years a considerable degree of sympathy for the demands of Indian nationalism and it was widely believed that he would be empowered to make significant concessions to the Congress viewpoint.

The Cripps Offer, as it came to be known, consisted of a short-term proposal for greater participation of Indians in the "counsels" of the Government, and a long-term perspective of India's emergence fairly soon after the end of the war as one or more Dominions. By raising the possibility of a division of India, the Cripps Offer shocked and angered the Congress leaders. In their eyes the "unity of India" was both an incontrovertible fact and a condition essential to preserve. Their main interest, however, was focused on the question of transfer of wartime governmental functions and powers to Indian hands, the extent of which the Congress had repeatedly termed the acid test of British intentions. Since the major share of India's budget was for the army, the Congress was particularly anxious to exercise civilian control over all aspects of military affairs apart from actual wartime strategy and operations. In this sphere the Cripps plan granted what appeared to Nehru and his colleagues no more than the right to run a Ministry for "Canteens and Stationery." Largely over this issue the negotiations broke down in April, 1942. Cripps flew back to London, and the political situation in India deteriorated rapidly.

ORIGINS OF PAKISTAN.—Although the Cripps Mission failed, the Churchill cabinet's post-war proposals which it brought to India had a far-reaching effect on subsequent political developments. The most novel provision of the Cripps Offer allowed to any province the right to stay out of the projected Indian Union. This new departure in imperial policy was a concession to the demand for *Pakistan*, raised barely two years before by the Muslim League.

From the outbreak of war the League had enjoyed renewed evidences of imperial favor. In refusing the Congress' request for an immediate and substantial transfer of power, the Viceroy cited as justification the "lack of prior agreement between the major communities." In effect, the Viceroy was treating the two parties, the League and the Congress, as though they spoke for socio-religious communities and not for political platforms—the League for the Muslims, the Congress for the Hindus. Furthermore, he was in practice treating them as equals, although in the sole election held throughout British India under the 1935 Constitution, the League had received little more than 300,000 votes, whereas the Congress had rolled up more than 10,000,000.

When the Congress ministries in the provinces resigned, the League was quick to capitalize on the situation. It called upon its members and friends to celebrate, in December, 1939, a "Deliverance Day" of thanksgiving and relief. On the day following this League celebration the Governor-General, in a long-delayed reply to a letter from Jinnah, assured him that the Government fully appreciated:

the importance of the contentment of the Muslim community to the stability and success of any constitutional developments in India. You need therefore have no fear that the weight which your community's position in India necessarily gives their views will be underrated.

As if to demonstrate unmistakably to the world that it was unsound to consider the Congress a communal or "Hindu" body, the Congress at its annual session in March, 1940, elected as president Maulana Abul Kalam Azad, one of the best-known Muslim political and intellectual figures in India. Congress supporters took care to point out that his election was no novelty, for over the half-century since Congress had been formed in 1885, half a dozen Muslims had preceded Azad in the highest office of the Congress.

Azad had scarcely finished delivering his presidential address to the Congress when the Muslim League began its momentous 1940 session at Lahore, and proceeded to pass Jinnah's resolution embodying the demand for "Pakistan." This term has a literary and religious origin, being based on the Persian word *Paki* (which carries many shades of meaning, including pure, sacred, and noble) and *stan*, country; *Pakistan* thus means land of the pure. The idea of grouping Muslims in a single political unit was adumbrated in the presidential address of Sir Muhammad Iqbal to the Muslim League in 1930. Iqbal, a great poet and probably the most influential Muslim writer since Sir Syed Ahmed Khan, suggested joining the Muslim-majority provinces of northwestern India into a single Muslim self-governing state, which would form one part of a loose all-India federation. This notion of setting up a "Muslim India" within a federal India was speedily welcomed and supported by British figures from the chief imperial security agencies, the Indian Army and the Imperial Police.[15]

Active political propaganda for the partition of India was first launched in 1933 by some Indian Muslim students at Cambridge University. In opposition to the scheme of federation then being discussed at the Round Table Conference, they advanced the slogan of a separate and distinct nation, based on the 35 million Muslims in "the five northern units of India, viz., Punjab, North West Frontier Province (Afghan Province), Kashmir, Sind and Baluchistan." Prominent Muslim League spokesmen at the Round Table Conference paid little heed to what they then considered as "only a student's scheme," impracticable and "chimerical." Even in 1937 Jinnah and the League were still thinking in terms of a united India based on a "federation of free, democratic states." As the breach between Congress and the League rapidly widened, the League became more and more attracted to the aim of gaining for itself a completely independent territorial entity.

In its 1940 resolution, the League did not specify clearly what areas were to be included in Pakistan, nor did it indicate what opportunity, if any, would be given to the inhabitants to indicate whether they actually wished to be separated from the rest of India. Even within the League itself it was by no means clear that the demand had been made for more than bargaining purposes. In the

[15] Cf. Col. M. L. Ferrar, co-author of *Whither Islam* (London, 1932), and John Coatman, who had served seventeen years in the Imperial Police and had been Information Officer for the Government of India during Lord Halifax's term as Governor-General, in his *Years of Destiny: India, 1926-32* (London, 1932).

chief provinces where the Muslims were a majority—Punjab, Bengal, and Sind—initial support for Pakistan was at best lukewarm. Active opposition was immediately forthcoming from the North West Frontier Province and from the Nationalist Muslims affiliated to the Congress. Their views that partition of India would be a national disaster were supported by several Muslim organizations in the Punjab and Sind, by the convocation of Muslim divines in India, and by some of the smaller Muslim sects. Representatives of all these groups shortly arranged a well-attended conference in Delhi, which formally condemned the idea of dividing India and repudiated the League's claim to be the sole, true voice of India's Muslims. In accordance with the common policy of ignoring the existence of non-League Muslims, the Governor-General took little notice either of this conference or of other indications of Muslim disapproval of Pakistan.

Carrying this support for the League one step further, the Cripps Proposals took Pakistan for the first time out of the realm of talk and dealt with it as a feasible political expedient. After the Cripps Mission the impression spread among the Muslims that the British considered Pakistan a reasonable claim and one that should be granted.

THE WARTIME CRISIS, 1942–1945.—The disappointment of the Congress with Cripps shortly gave way to a painful realization that Britain's wartime defeats in Europe and the Far East, instead of speeding concessions to nationalist opinion in India, were, if anything, leading to a hardening of British policy. Feelings of resentment quickly mounted and were intensified throughout the spring of 1942 by the apparent inability of British armed forces to defend India. Japanese divisions overran Burma, Japanese warships closed the Bay of Bengal, and Japanese airplanes bombed harbors and towns on India's east coast.

Within the Congress there was great confusion and conflict of opinion as to what course of action should be followed. Nehru urged the need for organizing guerillas and following a "scorched earth" policy in the event of a Japanese invasion. Gandhi proposed non-violent resistance to the Japanese but did not expect an attack. When impatient nationalists—particularly of the Congress Socialist persuasion—called for an immediate struggle to shake off British control, Nehru argued against harming the cause of China, the Soviet Union, and other victims of aggression. Gandhi, while "straining every nerve to avoid a conflict with British authority," summed up popular indignation by warning the British to "Quit India!" or face a new noncooperation movement. On the whole, the conservatives in the Congress felt that the threat of civil disobedience would be sufficient to wring new terms from the British.

After several months of intense discussion the All-India Congress Committee on August 8, 1942, issued its fateful "Quit India" ultimatum. The critical resolution reaffirmed the Congress willingness to organize India's defense, and its great reluctance "to jeopardize the defensive capacity of the United Nations." At the same time it authorized Gandhi to start, when he saw fit, "a mass struggle on nonviolent lines" to vindicate "India's inalienable right to freedom and independence." Gandhi indicated publicly that his first step would be to negotiate with the Viceroy.

But the Churchill Government was in no mood to negotiate. A few hours after the August 8 resolution was passed, and before Gandhi could even finish drafting his letter to the Viceroy, all of the Congress leaders and most of its key personnel were seized and imprisoned. Congress was at once declared an outlaw organization. At this moment a small group of Congress Socialists took the lead in calling on the people to cut telegraph wires, blow up railroad bridges, storm village police stations and so forth. These instructions were carried out in many places, and scattered local uprisings continued for several months in the face of the most extreme military and police measures. Official reports estimated that 940 Indians were killed in the course of the suppression. When the movement collapsed, the cup of nationalist frustration was full to overflowing. For the rest of the war, most important Congress leaders and large numbers of their supporters remained in jail and the Congress suffered an utter political eclipse.

The League meanwhile enjoyed from 1942 to 1945 the period of its most rapid growth and the greatest spread of its influence. This was particularly true in Bengal, Punjab, and Sind, where the League for the first time really caught on among the Muslim populace. In those three provinces alone the League claimed by the end of 1944 a membership of nearly 900,000. In this process the League shed its predominantly urban character and rolled up an impressive following among the peasantry. Simultaneously the League won over to its side or reduced the importance of the Muslim organizations and leaders who previously had worked with the Congress; the outstanding exception to this was in the North West Frontier Province, where pro-Congress sentiment remained strong.

As the popular following of the League grew, friction inevitably developed between younger middle class members, who identified themselves in part with the impoverished Muslim public, and the rich landlords and knighted gentry who controlled the League ministries. In the Punjab the clash over policy and discipline between the ministry and the younger groups became so acute that in 1944 the pro-British premier and most of the League representatives in the Punjab Assembly were expelled from the League. A similar conflict raged in Bengal. There the ministry, with the aid of the British members of the provincial legislature, managed to remain in office, despite the fact that its corrupt and incompetent regime was held in great part responsible for the disastrous Bengal famine of 1943. In Sind the clash was so bitter that in 1945 the Provincial League split into two hostile factions which were reunited only after heavy pressure from Jinnah.

The dissensions within the League resembled the conflicts which took place from 1937 to 1939 between the Congress left wing and the Congress ministries. One striking difference, however, is that the League issues were fought out on a provincial rather than on a country-wide basis. On the central level Jinnah and the right-wing landlords continued virtually unchallenged in their tight control of the League machinery. The League produced no one like Jawaharlal Nehru to speak on a national scale for its more progressive elements.

To a much smaller degree than for the League, the years after 1942 were boom times also for India's Communist Party. Its leadership of labor and peasant anti-war demonstrations had caused the imperial authorities to hunt it down with renewed vigor in the first two years of the war. On December 15, 1941, under the influence both of the Japanese onslaught in the Pacific, and of the earlier Nazi assault on the Soviet Union, the Communists had dropped their campaign against the war and raised the new slogan of "a people's role in the people's war." Six months later, as the Congress appeared to be moving in the direction of a new noncooperation campaign, the Central Government lifted the long-standing ban on the Communists and permitted them to function as a legal party.

When the "Quit India" resolution was under discussion in August, 1942, the handful of Communist members of the All-India Congress Committee voted against it, on the grounds that Congress should not run counter to the world-wide anti-Fascist war effort. With similar arguments they tried to dissuade the people from sabotage and other violent expressions of their anger at the arrest of the Congress leadership. During the years when the bulk of the Congress was immobilized they attracted public notice by their activities on behalf of price control, food rationing, and famine relief. This was considered reprehensible by many Congressmen, who held that any participation in official programs, even local food committees, was little short of treason to the martyred nationalist movement. But these same activities helped the Communists to extend their influence in the peasant leagues and trade unions, both of which advanced in strength and membership in the second half of the war.

A new development in Communist theory still further alienated the rest of the Congress. This was the declaration by the Communists in 1942 that India consisted of no less than 18 separate nationalities meriting recognition; and that the demand for Pakistan was a just one, in so far as it expressed the strivings toward statehood of the "Muslim nationalities." Although the Communists themselves proposed a single multi-national union of India modeled on the Soviet example, Congress spokesmen charged them with aiding and abetting the League campaign for "vivisecting" the motherland.

During the period that the Congress leadership was in jail several attempts were made by Congressmen who had opposed the August, 1942, resolution and had not been arrested to find a basis for agreement between the Congress and the League. Their purpose was to set up a popular government at the center, in order to deal more effectively with the wartime food and clothing shortages that were causing so much suffering. During these same years, while the Congress was paralyzed, the British authorities manifested a much less cordial attitude to the idea of Pakistan. The Viceroy, addressing the Central Legislature early in 1944, proclaimed the desire of His

Majesty's Government "to see India a . . . united country." In this setting, Jinnah and Gandhi, shortly after the latter's release on grounds of health, came together in September, 1944, for a series of talks. But they could not agree on a basis for Congress-League cooperation. In 1945, upon his return from an extended stay in London, the Viceroy announced his intention of forming a new Executive Council whose members would be drawn from the leaders of Indian political life. The Viceroy stated that he would select "equal proportions of Muslims and Caste Hindus." He called a conference at Simla to which he invited representatives of the various groups. Nehru and other imprisoned members of the Congress Working Committee were released so that they might attend.

As soon as the conference met, the issue arose whether the Congress could nominate a Muslim, such as Maulana Abul Kalam Azad, for a cabinet post. The League insisted that no other organization could nominate Muslims; the Congress rejected the implication that it spoke only for caste Hindus. The conference broke up and the Congress-League deadlock persisted.

THE ROAD TO PARTITION.—The two events of outstanding importance in India's most recent history have been its sudden partition into two separate countries and the simultaneous withdrawal of British political power. When the war ended in August, 1945, there was little outward indication that fundamental changes would occur rapidly. The first pronouncements by the new Labor Government headed by Clement Attlee were cautious affirmations of intent to resume negotiations for constitutional change upon the basis of the unsuccessful Cripps offer. The pace of development, however, was forced by a wave of mass unrest, which expressed the feelings of suffering, bitterness, and frustration repressed during the six long years of war. Unrest first showed itself in the fall and winter of 1945–1946. It has been sustained to the present not only by the political conflicts of the last three years, but by the daily hardships of postwar life. Prices have continually risen, black-marketing is widespread, industrial lay-offs are common, demobilization and reconversion have been thinly cushioned. Both urban labor and restless peasants have tried to prevent the stabilization of economic conditions at the unsatisfactory pre-war level. Factory owners and landlords, enriched by wartime profits, have had the resources to resist popular pressure. Clashes have been frequent and severe.

After the ban on it was lifted in June, 1945, the Congress throughout most of India again became the chief vehicle for voicing popular discontent. The Labor Government's failure in 1945 to make a new departure in British policy toward India was criticized unreservedly by the Congress, which termed Attlee's initial statement "vague, inadequate, and unsatisfactory." The influence of the Muslim League at the same time continued to spread. The League declared in ever more forthright terms its demand for a real transfer of power and its opposition to rule either by Britain or by the Congress (the latter invariably being labeled the "Hindu" Congress).

The most dramatic issue around which popular feeling crystallized in the closing months of 1945 was that of the "Indian National Army." This army has been associated primarily with the name of Subhas Chandra Bose. The latter became estranged from the Congress in 1939, was later placed under house arrest by the British authorities, and slipped out of India in the winter of 1940–1941. He turned up in Berlin, where he broadcast over the Axis radio that India's freedom could be achieved through the assistance of Britain's enemies, Germany and Japan. Bose became the leading spirit and "Marshal" of the Indian National Army (I.N.A.) with headquarters in Japanese-held Singapore. His troops were drawn largely from Indian soldiers who, along with their British officers, had surrendered to the Japanese early in 1942, at the capture of Singapore. In organization and leadership the I.N.A. disregarded communal lines. It played a part in 1944–1945 in several engagements against regular troops of the Indian Army. At the end of the war in August, 1945, the I.N.A. men fell into British hands and were moved to prison camps in India. Subhas Bose escaped by air from Rangoon, but later died in an airplane crash in Formosa.

When the Government of India announced its intention of trying some I.N.A. personnel for treason, the Congress at once came to the defense. It recorded the opinion that "it would be a tragedy if these officers, men and women were punished for the offense of having labored however mistakenly for the freedom of India." Protest meetings under Congress leadership were held all over India. By November, when the first trial of I.N.A. officers took place, public feeling had grown to the point where they were the outstanding heroes of the day. I.N.A. poems, portraits, photographs appeared everywhere. In the popular mind the key fact was that I.N.A. men had actually fought against the British authorities. By November the campaign seemed to be moving under its own momentum, gaining in breadth and intensity as it went. Police firings on pro-I.N.A. demonstrators led to some of the greatest processions Calcutta had ever seen, with Congress, League, and Communist flags flying together.

The unrest originally set off by the I.N.A. issue spread until it seriously affected even the ultimate safeguards of the empire, the armed services and the imperial police. In the regular Indian Army substantial funds were collected for the I.N.A. defendants. It became clear that Indian soldiers could not be relied upon, if they were sent into action to curb their fellow countrymen. Strikes by Britons in R.A.F. units, weary of years of service away from the British Isles, helped set off demonstrations by men of the Indian Air Force. In February, 1946, the core of the Indian Navy at headquarters in Bombay, struck *en masse*. The naval strikers appealed for public support which the Communists (whose headquarters are also at Bombay) hastened to organize. In the clashes which followed, hundreds of Indians were killed and wounded. Shortly thereafter, the police in the imperial capital, Delhi, went on strike. Their action was copied by the police of Patna, the capital of Bihar Province. Even Gurkha soldiers from Nepal, veritable Grenadier guards of empire, demonstrated in a cantonment of the Delhi zone against their British officers.

Prime Minister Attlee replied with a dramatic gesture in the direction of the Congress. Waiving his previous statements, he announced in February, 1946, that Sir Stafford Cripps and two other members of the British Cabinet would leave shortly for India. They would have full power to work out an immediate constitutional settlement, though of course Parliament would have to approve it. Amplifying this in March, Attlee declared that if India wanted independence she could have it. In words the Congress had long wanted to hear, he declared that no minority could be allowed to veto India's progress.

The response of the Congress to the announcement of the Cabinet Mission reflected the new line-up of forces within the Congress in the post-war period. After the release of the Congress heads from prison in the summer of 1945, the "Old Guard" assumed virtually complete control. Various sections of the left wing were either forced out of the Congress or threatened with expulsion if they did not keep in line. The first to go were the Communists. A list of charges was drawn up against them, centering on their opposition both to the August, 1942, resolution and the subsequent local risings, and on their qualified support of the League's Pakistan demand. In December, 1945, the Communists were barred from all elective positions in the Congress, but in anticipation of this the Communists had already resigned almost to a man from the Congress ranks. At the same time the right of parties like the Congress Socialists to function as organized groups within the Congress was sharply challenged. Open opposition was shown to the peasant league and the All-India Trade Union Congress, in both of which the Communists occupied key positions.

Nowhere was the dominance of the Old Guard shown more sharply than in the elections of 1945–1946. The imperial authorities had decided to hold elections both for the Central and the Provincial Legislatures. In formulating the Congress platform, the Old Guard sidestepped the economic and social issues which had proved so explosive in the 1930s. Instead they campaigned on a single slogan, a renewed call to the British to "Quit India!" Lest this seem too ominous, the Congress heads openly stated that they were willing to deal with the British either by noncooperation and direct action if necessary, or with negotiation where possible.

Attlee's announcement of an unprecedented Cabinet Mission was taken by the Congress heads as a clear indication that the method of negotiation would be both fruitful and desirable. Calls from the weakened and divided left wing for an immediate showdown with the British, or for "a final bid for power," were overridden or ignored. The popular response to the I.N.A. issue, far more intense and dramatic than the Congress had expected, had led the Working Committee in December, 1945, to reaffirm its basic belief in nonviolence. After the widespread disturbances connected with the Bombay naval mutiny, the Congress heads welcomed the opportunity presented by the Cabinet Mission to return to peaceful negotiations.

The proposals of the British Cabinet Mission of 1946 (led by Sir Stafford Cripps) constituted a striking reversal of the original Cripps scheme. At a period when India was menaced by Japan in the East and the Nazis in the West, the Cripps offer of 1942 had extended a broad right of secession to the component parts of India. In 1946, after hostilities had ceased, the Cabinet Mission proposed, on grounds of defense, that India remain a single united country; the center of such a union, it was provided, would be weak, while three distinct regions were made possible, virtually amounting to a "Hindustan" in the middle, flanked on the sides by an Eastern Pakistan and a Western Pakistan. Because the plan preserved a central government, the Congress leaders overcame their misgivings and accepted it. Thereby the Cabinet Mission was able to avoid a test of strength between the Congress and British power in India; by September, 1946, the Congress had taken office as an Interim Government, thus proving that the energies of India's most powerful political party had been turned away from the demonstrations and processions of the previous winter and into quieter constitutional channels.

In the process of conciliating the Congress, both the Cabinet Mission and later the Viceroy temporarily alienated the Muslim League. The latter body at first had been willing to go along with the Cabinet Mission plan. In July, 1946, because of allegedly undue concessions to the Congress, the League violently denounced the Viceroy and totally rejected the Cabinet Mission Plan. Speaking for the League, Jinnah said "goodby to constitutional methods and constitutionalism." The League designated August 16 as "Direct Action Day." The League Premier of Bengal declared the day a public holiday; his action helped to precipitate the worst communal riots that had ever taken place in Calcutta. From Calcutta they spread eastward in Bengal, and, later, westward up the Ganges Valley through Bihar, parts of the United Provinces and, eventually, into the Punjab. In the Deccan the area most affected was the city of Bombay. (The basis for this communal warfare is discussed below.)

After the Congress-headed Interim Government had definitely taken office, the Viceroy bent his efforts toward conciliating the League. By-passing the Congress, the Viceroy opened direct negotiations with the League and then on his own initiative placed League members in important posts in the Interim Government. The Congress was placated in part when the Viceroy announced that the League would participate in the forthcoming Constituent Assembly. This was the body which was to meet at the end of 1946 to draft India's constitution under the Cabinet Mission Plan. The League, however, denied that it had such an understanding with the Viceroy. When the Constituent Assembly met, the League held aloof. It declared that its only interest was in obtaining an entirely separate state of Pakistan, and it declared the proceedings of the Constituent Assembly, even though called by the Viceroy, would be "invalid and illegal." When, despite this declaration, the Viceroy retained the League members in the Interim Government, the Congress and other non-League groups exploded. Congress members prepared to resign, and threats of precipitate action were voiced.

At this critical moment Prime Minister Attlee intervened. Speaking in February, 1947—one year after his announcement of the Cabinet Mission—he proclaimed that by June, 1948, British power in India would be completely withdrawn. But on this occasion Attlee did not repeat his previous assurance that minorities would not be allowed to stand in the way of India's march to freedom. He rather declared that if by June, 1948, a constitution had not been drafted by a Constituent Assembly in which all important parties had worked together, then Britain would have to decide to whom power should be transferred: "whether as a whole to some form of central government, or in some areas to existing provincial governments," or in some other way.

The Attlee announcement in effect reversed the Cabinet Mission decision for a central, unified government and opened the door for the partition of India. The attention of the Congress and the League turned at once to the two provinces which had Muslim majorities but non-League governments: the Punjab and the North West Frontier Province. Whichever party retained control of those provinces down to June, 1948, would automatically retain power when the British quit India. Violent clashes quickly developed in both disputed areas. Large sections of the great Punjab cities of Lahore, Amritsar, and Rawalpindi were looted and burned to the ground. This violence and destructiveness made both the Congress and the League realize that if communal warfare of such a character continued, the country would be ruined for decades to come. By tactful negotiation, the new Viceroy, Lord Mountbatten, persuaded the League to accept a diminished version of Pakistan, and the Congress to accept reluctantly the partition of India. The Punjab and Bengal were to be split in two, with the Muslim majority districts going to Pakistan and the rest to India. The agreement was embodied in an Act passed by Parliament in July, 1947, providing for the creation of two new Dominions, India and Pakistan. The administrative departments, armed forces, funds, staff, and properties of the Government of India were promptly apportioned. As border commissions drew boundary lines across Bengal and the Punjab, there began in the latter province one of the greatest, bloodiest, and saddest migrations in all the history of mankind. On August 15, 1947—just one day short of the anniversary of the League's "Direct Action Day" of August 16, 1946—India was formally partitioned into the new Dominions of India and Pakistan.

V. Problems and Prospects

In the most recent past, communal antagonisms and communal warfare have occupied the center of the stage in India, and today they throw a dark shadow over the future of the country. The recent strife between Muslims on the one side and Hindus and Sikhs on the other has its roots deep in the history of India, particularly in the peculiar impact of imperial rule by a Western power upon the characteristic but disintegrating organization of the older Indian society. As has already been pointed out, religion in India has not been limited to or even chiefly concerned with beliefs about the nature of the deity or practices of worship. Rather Hinduism and Islam have performed a much wider range of social functions. They provide the sanction which gives each individual his place in society, his code of social relations, and his guide to personal behavior. Each religion gives a certain sense of solidarity to those born within its ranks and a sense of distinctness and difference from those outside the fold.

There were many wars between Hindu and Muslim powers in the centuries after the first invasion of India by Central Asian followers of Islam. In the course of the hostilities temples were desecrated and other savageries committed in the name of religion by both sides. But local Hindu-Muslim riots breaking out in city or country in the calm of peace were almost unknown in Mogul times. Rather, such conflicts are a modern phenomenon and appear to originate in the period of British rule in India, particularly in the decades of profound change after the Rebellion of 1857. The sweeping revolution of Indian economic, social, and political life has affected the Muslims at the various levels of society quite differently from the Hindus, and pitted important sections of each group against the other more sharply than ever before. The highest level among the Muslims consisted of descendants of landed magnates and nobles of Mogul days. These former grandees soon found themselves outstripped economically by Hindu and Parsi merchants who formed the new urban industrialist and banker class, and the well-known opposition of landlords and capitalists set in. With the rise of nationalism among the growing urban upper middle class, the gulf at the top widened, particularly after the British in the last quarter of the nineteenth century reversed their previous hostility to the Muslims and actively cultivated the loyalty of "the better classes of Mohammedans" as "a source of strength to us." [16]

Below the top level, there has also been a marked difference in development. Among the middle classes, the Hindus, generally speaking, have gained more advantageous positions in trade, in the professions, and in the government services. By contrast the share of the Muslims has been weak. Among the working population a larger proportion of Muslims than Hindus do the rougher sorts of common labor. Many Muslims work for Hindu factory owners and Sikh public works contractors, whereas the reverse is much less common.

In the countryside the great bulk of the peasantry are Hindus who are subject to Hindu landlords. But in three important areas a large part of the peasantry or agricultural laborers are Muslim while the landowners and moneylenders are Hindu or Hindu and Sikh: East Bengal, Malabar, and the Punjab. In the United Provinces, on the other hand, the peasantry are predominantly Hindu but there are a number of great Muslim landlords.

In the changed Indian scene Hindus and Muslims thus found themselves on opposite sides in such familiar struggles as those of

[16] Sir John Strachey, *India, Its Administration and Progress* (London, 1894), p. 241.

peasant vs. landlord; debtor vs. moneylender; factory worker vs. factory owner; hired laborer vs. sweatshop owner or contractor; landed aristocrat vs. industrialist. By the end of the nineteenth century these economic rivalries were already causing friction between India's two chief communities, but no other step did so much to poison their relations as the introduction of communal electorates under the Morley-Minto Reforms of 1909. For since then the various communities which were already set off from each other in "religion" and in many economic relations have been posed directly against each other as the basic entities of Indian politics.

The first important series of local riots broke out in Bombay and the United Provinces in the 1870s; and others have occurred in one or more parts of India in every succeeding decade. The immediate issues over which these riots and most subsequent ones erupted are numerous. Among the most important is the question of cow-killing. To orthodox Hindus the cow is a sacred animal whose life must not be taken, whereas to Muslims the cow is one of the animals suitable for sacrifice on high holidays. A second issue is that of the playing of music in the vicinity of mosques, an act which is considered by Muslims in India to be a sacrilege. Hindus, on the other hand, celebrate marriages and many holidays with colorful processions and joyous music. In periods of tension, gay processions passing mosques can easily set off serious clashes. These causes of provocation have existed for the whole of the nine hundred years that Muslims and Hindus have lived side by side in India. It is only in the last three-quarters of a century that they have become the focus of acute tension and dispute. The early communal riots were generally sporadic in character; and the areas most affected were regions of the severest agrarian distress. During the great wave of nationalist sentiment from 1919 to 1922, communal riots ceased almost altogether, except for the Malabar districts of southernmost India, where they were inseparable from a rising of Muslim peasants and laborers against their Hindu overlords. British Government of India officials in those years expressed their amazement at what they called the "unprecedented fraternization of Hindus and Muslims." By this time the imperial authorities had come to feel that Hindu-Muslim riots were the natural and expected thing, whereas Hindu-Muslim amity was unexpected and "unnatural."

After the staggering end of civil disobedience in 1922, the succeeding years were marked by a great many communal riots in the cities. Hindu and Muslim extremists by now had been formed into organized bodies headed by public figures who devoted full time to this work. The newspapers which they printed depended for their circulation upon the virulence of their accounts of the activities of the rival religious community. Each group retained its regular body of hooligans eager for a fight and for the looting that was certain to follow. The most serious urban riot of the 1920s occurred in Bombay in 1928. It originated after the British-owned Burma-Shell Oil Company hired Pathan (Muslim) strikebreakers against its Hindu workmen. The latter bore a double resentment against the Pathans, some of whom also functioned as petty moneylenders and usurers in the Bombay area.

During the second civil disobedience movement, and indeed during the whole of the 1930s, communal riots were fewer, with the notable exception of the very severe disturbances at Cawnpore in 1931. Here the local authorities were charged by British and Indian witnesses alike with culpable negligence both in their failure to try to prevent the outbreaks or to bring them to an early end. About this time responsible British writers openly charged the imperial authorities with occasionally resorting to the use of *agents provocateurs* to make trouble between Hindus and Muslims.

The revived Muslim League of the 1930s was an avowedly communal organization campaigning under the banner of anti-Hinduism. As the League grew, tension between Muslims and Hindus heightened. The political deadlock of the war years further embittered their relations. In the summer of 1946 the prospect of attaining actual political power over all or part of India dangled as a prize before India's two great parties. At this point, organized communal warfare became in effect a recognized tactic of important elements within both camps.

The leaders of the League had built up a large force of volunteers somewhat along the lines of the groups which had been organized by the Congress in the civil disobedience campaigns. The Congress, holding office in the Interim Government, continued to proclaim itself a non-communal organization; in any event, it was not prepared to measure its strength with the League because its village and district branches had been shattered in the sequel to the August, 1942, uprisings and never really rebuilt. Some of the most powerful right-wingers in the Congress preferred to work with extremist Hindu organizations in building up an independent counterforce to the League volunteers. Funds and leadership for this Rashtriya Swayam Sevak Sangh (National Voluntary Service Association) were obtained from Hindu industrialists, princes, and landlords. Similarly, as the division of the Punjab loomed, a Sikh striking force was rallied by the Sikh princes, landlords, and big contractors.

As these various semimilitary bodies came into open conflict with each other, the Indian police and the Indian units of the army, themselves recruited and organized on a communal basis, tended to divide on communal lines and the structure of civil administration began to break down. The imperial authorities seemed indifferent to the consequences of this process and hesitated to use British soldiers to preserve law and order. In the course of the official inquiry into the great Calcutta riots of August, 1946—the starting point of the recent phase of communal warfare in India—the British Commanding Officer in the area revealed that he had delayed the use of troops at the start of the disorders lest both side drop their quarrel and join against the military.

The creation of the two new dominions along communal lines required the drawing of a boundary line through the heart of the Punjab, where Muslims, Sikhs, and Hindus were closely intermingled. The minorities in the two parts of the divided Punjab were thus left open to the depredations of the most inflamed elements among the majorities. By train, boat, and on foot, harried millions of refugees poured across the new borders. At their inception the two new dominions were saddled with the responsibility of absorbing and resettling these hordes of ruined refugees. The tales told by the refugees gave fresh ammunition to the most virulent communalists in both countries, and deepened the atmosphere of suspicion and ill-will on both sides.

The recent past has been marked by increased activity of the princes in politics rather than by their eclipse. During the war the states' peoples' movement and the rulers clashed repeatedly, sometimes with considerable strife. At the end of the war in 1945 friendly overtures to the princes were made by both the League and the

Congress, with each organization striving to persuade as many of the princes as possible to cast their lot with it. For their part the imperial authorities continued their older policy of explicitly safeguarding the princes. The position of the princes became a sharp issue during the Cabinet Mission of the spring of 1946. In those negotiations the British unqualifiedly insisted that in any future Constituent Assembly the Indian states would be represented by their rulers rather than by popularly elected delegates.

At the time of the partition of India in 1947, the princes were given the option of becoming independent or of joining up with one or the other of the new dominions. Except for Hyderabad and Kashmir, most of them chose the latter alternative; their bargaining position was good, however, and in internal matters they retained virtual autonomy. In both India and Pakistan they have typically allied themselves with the most backward and undemocratic elements, and have figured prominently in the recent communal warfare.

While the British, the League, and the Congress were arguing out questions of transfer of power and partition, the peasantry and the impoverished classes of city dwellers were left to shift for themselves so far as their basic economic problems were concerned. But the latter were not so easily shelved. During the war there had always been the hope that peace would bring a better livelihood. The postwar frustration of these hopes has helped to fertilize the ground for the spread of communalism.

In the long run the governments of India and Pakistan cannot achieve stability until they overcome the heritage of communal antagonism. To do this they will have to deal successfully with the underlying problems of agrarian reform and healthy industrial development. It remains to be seen whether their governments as presently constituted are able or willing to grapple with problems of such formidable magnitude.

SUGGESTED READINGS

Anand, Mulk Raj. Coolie. [A novel.] London, 1941.
—— The Village. [A novel.] London, 1939.
Anstey, V. Economic Development of India. London, 1936.

Archer, J. C. The Sikhs. Princeton, N. J., 1946.

Chandrasekhar, S. India's Population. New York, 1946.

Coupland, Sir R. The Constitutional Problem of India. London, 1944.

Crookes, W. Religion and Folklore of Northern India, London, 1926.

Darling, M. L. The Punjab Peasantry in Prosperity and Debt. London, 1932.

Dutt, R. C. Economic History of India. 2 vols. London, 1906.

—— tr. The Ramayana and the Mahabharata. New York, 1944.

Dutt, R. Palme. India Today. Bombay, 1947.

Elwin, Verrier. The Baiga. London, 1939.

Emerson, Gertrude. Voiceless India. New York, 1944.

Forster, E. M. A Passage to India. [A novel.] New York, 1924.

Gadgil, D. R. Industrial Evolution of India. London, 1944.

Gandhi, M. K. His Own Story. Ed. C. F. Andrews. New York, 1930.

Ghose, Bimal C. Planning for India. New York, 1946.

Ghurye, G. S. Caste and Race in India. London, 1932.

Havell, E. B. A Handbook of Indian Art. New York, 1920.

Hodson, T. C. India: Census Ethnography, 1901–1931. New Delhi, 1937.

Hutton, J. H. Caste in India. Cambridge, England, 1946.

Iqbal, Sir M. Reconstruction of Religious Thought in Islam. Lahore, 1944.

Jinnah, M. A. Recent Speeches and Writings. Ed. M. Ashraf. Lahore, 1942.

Ketkar, S. V. An Essay on Hinduism. London, 1911.

—— The History of Caste in India. Ithaca, N.Y., 1909.

Macdonnell, A. A. India's Past. Oxford, 1927.

Mann, H. H., and N. V. Kanitkar. Land and Labour in a Deccan Village. Bombay, 1921.

Morrison, Cameron. A New Geography of the Indian Empire and Ceylon. London, 1926.

Mukerji, D. G. My Brother's Face. [A novel.] New York, 1924.

Mukhtyar, G. C. Life and Labour in a South Gujarat Village. Calcutta, 1930.

Nehru, Jawaharlal. The Discovery of India. New York, 1946.

—— Toward Freedom. New York, 1941.

Noman, M. Muslim India. Lahore, 1942.

Panchatantra, The. [Folk tales and proverbs.] Tr. Arthur W. Ryder. Chicago, 1925.

Punjab Board of Economic Inquiry. Punjab Village Surveys. Lahore, 1922—.

Radhakrishnan, Sir S. The Hindu View of Life. New York, 1927.

Rawlinson, H. G. India: a Short Cultural History. New York, 1938.

Rivers, W. H. R. The Todas. New York, 1906.

Shelvankar, K. S. The Problem of India. London, 1943.

Slater, Gilbert. Some South Indian Villages. Madras, 1918.

Smith, W. C. Modern Islam in India. London, 1946.

Stoll, D. G. The Dove Found No Rest. [A novel.] New York, 1947.

Tagore, R. Reminiscences. London, 1943.

Thomas, P. J., and K. C. Ramakrishnan. Some South Indian Villages, a Resurvey. Madras, 1940.

Thompson, E., and G. T. Garratt. Rise and Fulfilment of British Rule in India. New York, 1934.

Wadia, P. A., and K. T. Merchant. Our Economic Problem. Bombay, 1945.

Wernher, H. The Land and the Well. [A novel.] New York, 1946.

Weston, C. Indigo. [A novel.] New York, 1943.

Wiser, C. V., and W. H. Wiser. Behind Mud Walls. New York, 1930.

Raymond Kennedy

SOUTHEAST ASIA AND INDONESIA

THE SOUTHEASTERN EXTREMITY of Asia consists of a great peninsula and, off its tapering tip, a vast archipelago. The islands of the latter form an almost continuous mass of land, with only narrow straits separating its parts, so that the entire area—Southeast Asia and Indonesia combined—can be regarded as an interconnected unit from one end to the other. It is divided, however, into five distinct countries. Across the thick top of the peninsula, from west to east, lie Burma, Siam, and Indo-China, all almost equal in size. Malaya, much smaller in area than any of these, occupies the long, slender lower end of the peninsula. South of the narrow Malacca Strait begins the enormous island realm of Indonesia, which then stretches a full three thousand miles to the east, where it ends in New Guinea.

A great right angle drawn on the map, about three thousand miles long on each side, would approximately cradle the whole area. Its location on the equator, whence it extends thirty degrees north and ten degrees south, makes the climate very warm; and the only changes of season are between the wet and the dry times of the year, which follow the shifting of the monsoon winds. Although there is considerable local variation, the wet monsoon, bringing extremely heavy rainfall, generally prevails north of the equator during the months of the American summer and south of the line during the months of the American winter.

The Indian Ocean washes the southern shores of Indonesia, whose eastern border cuts across the middle of New Guinea; to the north lie the Philippine Islands. Much more important, however, are the land borders. Running along the western boundaries of Burma, and the northern limits of Burma, Siam, and Indo-China, rise some of the highest and most nearly impassable mountain ranges on earth. These have isolated Southeast Asia from the rest of Asia, and have made of it a region unto itself, remote and hard of access. These towering heights have had an even more significant function in terms of peoples and culture. Through them a few rivers have cut deep gorges from Tibet and China: the Chindwin, Irrawaddy, Salween, Menam, Mekong, and Red. These have let people in, but not very many at a time, so that Southeast Asia has never been subjected to mass invasions, but rather to small and slow infiltrations through the tortuous passages. Those who did get in have had a chance to retain and develop their own cultures, which have a character distinct from any others in Asia. China, for all its gigantic expansive power, has never been able to press conquest beyond the mountain barrier, except, for a period in the past, along the narrow coastal corridor leading into Tonkin and Annam on the shores of the China Sea. India, the other large neighboring country, to the west, has likewise been kept from any possibility of conquest here.

This is not to say that influences from these two great centers of civilization have not penetrated into the area. Chinese culture has made a strong impress upon Indo-China, particularly; and Indian civilization, far more than Chinese, spread steadily and strongly, for centuries, over almost the entire region. But the spread was gradual, coming in largely by way of the sea, and in driblets rather than mass inundations; so that the indigenous cultures absorbed Indian elements selectively, and never lost their own character. The history has been one of cultural diffusion from China and India, much more importantly the latter, and not one of cultural conquest. The consequence is that Southeast Asia and Indonesia have remained a true culture area, with a nature of its own, well marked off from all others.

The most striking characteristic of this isolated and protected culture area is its amazing diversity. There is no other part of the world where the range of racial and cultural differentiation is so wide. Many remote sections of swamp and mountain are still roamed by scattered bands of nomads, whose ways of life are as primitive as any found on earth. There are vast regions on the mainland and in the islands where tribes of thousands are now existing under conditions approximating those of prehistoric Europe. There are nations of millions who live in settled villages and towns, carrying on highly developed agriculture, skilled in intricate handicrafts, and following the beliefs and rituals of the world's great religions.

And there are scores of large cities with modern buildings and transportation systems, and populations who are in full contact with the currents of world affairs. In short, virtually the entire range of levels of human culture can be found in Southeast Asia and Indonesia, from the most primitive to the most advanced.

The racial diversity is also great. Pygmy Negritos and dwarf Australoid groups of the Veddoid type stand at one end of the scale, and the progression runs thence through dark-skinned Negroids, brown peoples of mixed Caucasoid and Mongoloid derivation, and lighter folk of virtually pure Mongoloid stock. All of the three main races of mankind—Negroid, Caucasoid, and Mongoloid—as well as the problematical archaic Australoid, are represented.

The size of the entire area, excluding the sea, is about 1,550,000 square miles, half that of the United States. Indonesia is the largest of the five countries, comprising nearly one-half of the combined area, slightly under 750,000 square miles. The three northern mainland countries compare closely in size, their approximate areas being as follows: Burma, 260,000; Siam, 200,000; and Indo-China, 280,-000 square miles. Malaya, the smallest, is only a little over 50,000 square miles. The division of population between Indonesia and the four mainland countries is also about equal; of the combined total of approximately 130,000,000—nearly that of the United States—the islands of the Indies have 70,000,000. The remainder is divided roughly as follows: Burma, 17,000,000; Siam, 16,000,000; Indo-China, 25,000,000; and Malaya, 5,500,000.

In any one of these countries except Malaya, the indigenous population—the "natives"—are the great majority. The only foreign groups who anywhere figure importantly are, as might be expected, Chinese and Indians. The Chinese are most numerous in Malaya, where they have actually come to outnumber the native Malays themselves, and comprise almost 2,500,000 of the total 5,500,000. Their next greatest concentration is in Siam, where, although the statistics are disputed, they are said to total around 2,000,000, or about 12 percent of the entire population. The Chinese in Indonesia number approximately 1,500,000 which, though a considerable figure, is only 2 percent of the total population. In Indo-China and Burma, the Chinese are numerically unimportant, with slightly over 400,000 and slightly under 200,000, respectively.

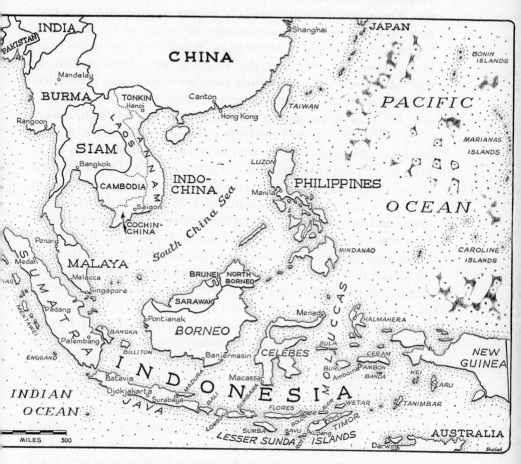

SOUTHEAST ASIA AND INDONESIA

The only countries where Indians are present in large numbers are Burma and Malaya. Burma has, or rather had before the recent war, a little more than 1,000,000 of them; while Malaya has around 750,-000. Europeans, though extremely important politically and economically, are very few numerically anywhere in the area. The largest concentration of them is, or was before the recent war, in Indonesia, but even here the total was only about 250,000. In each of the other countries their population was quite inconsiderable, the approximate pre-war figures being: for Indo-China, 40,-000; for Malaya, 30,000; for Burma, 30,000; and for Siam, a mere 2,000. Indonesia also had the largest number of Eurasians, who, it has been estimated, formed well over half, probably close to 70 percent, of the persons classed as Europeans. The Dutch, like the French, but unlike the British, gave these people of mixed European and native ancestry the legal status of Europeans; and social intercourse and intermarriage between Netherlanders and Eurasians were freer than such relations in other colonial countries. Burma and Malaya had 20,000 and 12,000 Eurasians, respectively, and the British placed them in a separate legal and political category, drawing strict lines of distinction in social relations as well. The number of Eurasians in Indo-China was not known precisely, but, since native women were frequently taken as concubines by the French, the mixed-blood population was probably quite large, at a rough estimate around 10,000. Here, as in Indonesia, the Eurasians, after 1928, had the legal status of Europeans; indeed, the French went further than the Dutch and gave them French citizenship; but social discrimination against them, though less marked than in the British possessions, was greater than in Indonesia. Siam had only a small number of Eurasians; but, as in most of Southeast Asia and Indonesia, intermarriage between Chinese and natives was very common.

With the native population so large and in such overwhelming majority as compared with the foreign elements, it might be expected that the former would be in political and economic control of their countries. But, except in Siam, this is not so. Indonesia, Indo-China, Burma,[1] and Malaya are colonial possessions, that is, lands subjected to conquest and consequent control and exploitation,

[1] Burma became an independent republic in January, 1948.

of both people and resources, by outside powers. Southeast Asia, with Indonesia, is one of the two great areas in the world where colonialism exists, the other being Africa. In both cases, the same combination of factors, historical and cultural, brought about the lowly status of these areas as compared with the independent countries of the earth. They were attractive economically; they were weak politically and militarily; they were culturally different from and, in the Western sense, more backward than the nations of the Occident; and their people were racially different from and, again in the estimation of the West, inferior to Europeans and Americans.

The first of the countries in the area to be subjugated by a Western power was the largest, Indonesia, or, as it was commonly known until recently, the East Indies. This island area, in the sixteenth century, when European explorers came upon it, was undergoing a tremendous revolution. Starting ten centuries before this, a great number of petty kingdoms had arisen in the archipelago, under the influence of adventurers from India. Before the Indians arrived, the native peoples had never evolved the centralized state form of political organization. These new kingdoms passed through countless wars with one another until finally, in the early fourteenth century, one of them, Modjopahit in Java, had performed the remarkable feat of achieving control over all of them. The organization of the empire was feudal, each of the princes of the subject states standing in the relationship of vassalage to the emperor. By the time the Europeans arrived, some two hundred years later, the great empire was in the last stages of breaking up, as a consequence of successive revolts of the vassal princes. Religion played a part in this catastrophe, for the common signal for revolt was the conversion of a prince to Islam; whereupon what might have been a mere uprising against the suzerain took on the guise of a holy war against the Hinduist center of Modjopahit.

The Europeans—first the Portuguese and Spanish and then the Dutch and British—thus encountered, not a unified empire, but a checkerboard of small states, newly independent and mutually hostile. It was not difficult to conduct a campaign of conquest under such conditions, for one prince could be played off against another, to the ultimate ruin of both. As a matter of fact, the various European powers spent more time and energy fighting with each other, or instigating the native states against rival Westerners, than in direct assaults upon the Indonesian principalities themselves. The Dutch were eventually victorious in this combination of guile and warfare, and by 1650 were well started on their way to ultimate control of the Indies. The complete conquest required over two hundred years, but throughout the whole history of it the superior statesmanship, military equipment, and technical ability of the Europeans easily outweighed the vastly preponderant numbers of the natives.

The organized conquest of the three other colonial countries in the area—Malaya, Burma, and Indo-China—did not begin until the nineteenth century, although seizure of small sections—trading ports and military posts—had gone on intermittently ever since earliest European contact. In Malaya the task of the British was easy, for this small country was split up into especially minute and weak sultanates; and a mere show of power usually produced whatever concessions the British desired. The annexation of Malaya was conducted in a leisurely, piecemeal fashion; and it was not until 1909 that the four final states, which happened to be under Siamese sovereignty, were taken over by the British.

Burma had had a troubled history of internal warfare lasting for centuries before the British started the process of annexation in the nineteenth century. Numerous small kingdoms, which had arisen under Indian influence, fought each other for supremacy, and one after another fell. Starting in the middle of the sixteenth century and continuing for two hundred years, various Burmese kings had carried on a series of remarkably savage and pointless wars with Siam. But an unusual degree of power and unity had been attained by the final Burmese kingdom in 1824, when the British undertook the first campaign into the country as a consequence of a dispute over the India-Burma boundary. The second war, in 1852, and the third, in 1885, completed the annexation of the country; and the last king, Thebaw, was banished to India.

The French took their turn, choosing Indo-China, very late. Persecution of French missionaries was given as the main reason for starting the conquest. Three of the sections of this country—Tonkin, Annam, and Cochin China—had long been under the rule of the Annamese emperor. This empire was unique in Southeast Asia in that its origin and organization had Chinese rather than Indian sources. A fourth section, Cambodia, was one of the oldest centers of Indian influence in all of Southeast Asia; and its past glories are signalized by the astounding ruins of the deserted capital of Angkor. The fifth section, Laos, largely primitive territory, was under the loose control of two small kingdoms, Luang Prabang and Vientiane. Into this picturesque and varied country the French moved in 1858, gaining a foothold in the port city of Saigon. Five years later, the emperor of Cambodia, whose realm was threatened with extinction by pressure from Siam on the one side and Annam on the other, voluntarily placed his domain under French protection. At the same time, the French forced the emperor of Annam to cede Cochin China; but the monarch achieved a sardonic revenge by removing all of his mandarins from Cochin China, leaving the French a country completely devoid of government. Abashed but undaunted in the quest for empire, the French entered upon a campaign against Tonkin; and, in 1883, the emperor of Annam accepted a protectorate over both Annam and its subsidiary, Tonkin. Finally, the French took over Laos in 1893, after some obscure border skirmishes with the Siamese, who had been raiding into the region, with, it is said, the encouragement of the British, who were concerned over French expansion in Southeast Asia.

The only independent country in the whole area is Siam, whose people are appropriately named Thai, "free men." There are other ethnic groups in the country, such as the primitive mountain tribes, but the Thai proper, the main stock, trace their origin back into the mountains of Yunnan in southern China, whence they came in successive waves many centuries ago, down the Mekong and Menam valleys. They brought with them a Chinese type of political and military organization, which was later blended with Indian-derived elements taken over from the Cambodians whom they conquered. They have always been not only a free, but an aggressive and conquering, as well as a remarkably adaptable, people. The last quality, especially, has served them well in their contacts with the rapacious Europeans, who have ranged the neighboring countries with conquest and subjugation always in mind. The Thai have remained free in recent times by learning to deal with Europeans in ways the latter understand, although it must be admitted that no small factor in their success has been British-French rivalry, because of which neither would allow the other to take over Siam. Relations with European powers began in the seventeenth century, but for a long period Siam stayed almost completely aloof from the white strangers.

Siam's bitterest enemy, Burma, with whom wars were fought for centuries, was effectively neutralized by the British early in the nineteenth century; and then Siam entered upon a hundred years of delicate maneuverings with Western powers, particularly France and Britain. Two remarkable kings, Mongkut and Chulalongkorn, whose combined reigns covered fifty years, conducted foreign affairs with such skill, while modernizing the country to a remarkable degree, that in 1896 a French-British agreement guaranteed Siam's freedom from encroachment by either power; although in the early years of the twentieth century France demanded and got certain cessions of land on the Indo-China border, and in 1909 Britain persuaded Siam to yield territory to Malaya.

What did the European powers—Britain in Burma and Malaya, France in Indo-China, and the Netherlands in Indonesia—obtain for their efforts, and what did they lose by not annexing Siam? Involved are the enormous areas of land listed above; but also involved are a remarkable array of peoples and cultures, and some extremely profitable resources.

To take the peoples and cultures first, there is, in every one of these countries, one majority tribal or national group. In Indonesia these are the Javanese, who, together with the closely related Sundanese and Madurese of the same island, comprise about 65 percent of the entire population of the Indies. In Burma the dominant group is the Burmese proper, who form over 60 percent of the total population. The Thai are by far the most numerous people in Siam, although their exact percentage of the total population has not been calculated. In Indo-China, the Annamese comprise fully 75 percent of all the people. Malaya is most peculiar, because the Chinese alone outnumber the indigenous peoples, and there are very large numbers of immigrant Indians and Indonesians. But among the true "native peoples" of Malaya, there are only a few thousand primitive and semiprimitive folk to compare with the more than 2,000,000 Malays proper.

Aside from the Javanese, Burmese, Thai, Annamese, and Malays, there are literally hundreds of other ethnic or tribal groups. Some of them are very large, numbering in the millions, and thence they range down all the way to minute tribes of only a few hundreds. In Indonesia, the Malays of Sumatra total well over 3,000,000; the Minangkabau about 2,000,000, the Batak 1,000,000; and the Achinese 750,000. The Makassarese and Buginese of Celebes have a population of close to 3,000,000; while the Balinese total over 1,000,000. In Burma, the Karen and Shan populations are over 1,000,000 each; the Kachin, about 400,000; and the Chin, around 300,000. The Malays of southern Siam run to approximately 400,000; and the Lao are also a very large group, although their population, as distinct from that of the Thai, has not been reckoned. In Indo-China, the Cambodians total about 3,000,000; the Cham, approximately 1,000,-000; and the Thai and Lao, around 800,000 and 600,000, respectively. These are the very large groups only; a complete listing of all the smaller ones in the area would fill pages. The mountain lands of Burma, Siam, and Indo-China are checkered with scores of more or less primitive tribes, such as the Naga, Wa, Palaung, Lolo, and Padaung in Burma; the Kamuk, Tin, Chao-Nam, Miao, Yao, Lahu, and Lawa in Siam; and the Muong, Man, Moi, Meo, and Kha in Indo-China. In Malaya, there are Semang, Sakai, and Jakun, all very primitive. For Indonesia, suffice it to say that there are well over 100 tribal groups—aside from the larger ones mentioned—in Java, Sumatra, Borneo, Celebes, the Lesser Sunda Islands, and the Moluccas. The number of tribes in Netherlands New Guinea, on the far eastern extremity of Indonesia, has never been estimated, although scores of names are known.

To an anthropologist reading the foregoing, the conclusion would be evident that in this part of the world the range and variety of peoples and cultures are truly unmatched anywhere else on earth. An attempt to describe the complete cultural situation, in all its diversity, cannot be made here; but the general outlines of the ways of life among the inhabitants of Southeast Asia and Indonesia can be indicated, for, despite extreme heterogeneity in details, there are many broad similarities.

The enormous array of languages spoken in the area can be generally classified in four great linguistic families or stocks. The languages of Burma, Siam, and Indo-China fall into three of these stocks:

Tibeto-Burmese, Siamese, and Annamese. The boundaries of these families do not coincide with the political borders of the three countries, but, reckoning by the number of speakers, it may be said that most of the people in Burma speak Tibeto-Burmese languages, which, as might be supposed from the name, run up into Tibet; that most of the inhabitants of Siam speak languages of the Siamese family, which extend north into Yunnan in China; and that the majority of the population in Indo-China speak languages of the Annamese stock, whose connections also are with China to the north. The fourth family of languages in the area is the Malayo-Polynesian, and languages of this group are spoken all the way from southern Siam down through Malaya and throughout all of Indonesia. In the extreme east of Indonesia, however, the linguistic picture shifts, and in New Guinea a new type of speech, which has scarcely begun to be investigated, begins. These non-Malayo-Polynesian languages of New Guinea have been tentatively placed in a classification called Papuan.

Although many of the languages in Southeast Asia and Indonesia have never been written, and while knowledge of writing is restricted to only a small proportion of the entire population, wherever script has been used for these languages four types have been employed. The Arabic script came into Indonesia and Malaya with the advent of Islam about seven centuries ago. Before this, alphabets derived from Indian writing were used, not only in Indonesia and Malaya, but in parts of Burma, Siam, and Indo-China as well. In Indo-China, and anciently in Siam, Chinese script was employed; and much of the Annamese writing is still in the Chinese style. Since the coming of Europeans, the Roman alphabet has been spreading rapidly, and it is likely that eventually it will replace the other kinds of writing.

In considering the economy of Southeast Asia and Indonesia, a peculiar situation, unfamiliar to people of the Western nations, must be mentioned at the start. These countries are in part merely a region of the world where the inhabitants, as they have throughout history, make their living from their own soil and by their own efforts, locally. But these countries are also spheres of exploitation for foreigners, who use them to produce goods to be exported to the outside world. One might put it thus: here are farms and hunting and fishing grounds geared to local needs, but here too are plantations and mines whose production has little relation to the requirements of the natives, except indirectly. The economy is, in short, dual. There is the native economy, and there is the export economy. The former is one of subsistence; the latter, one of exploitation for profit. The two touch upon each other at numerous points, but their organization and purposes are different.

The great majority of the native peoples of the area are farmers, and their main crop is rice. If one had to select a single symbol to express the life of the people of Southeast Asia and Indonesia, the best choice would be a picture of a farmer working his rice field with a plow and water buffalo. There are a few regions where other crops—such as maize or taro or sago—take precedence over rice, and a very few where rice is not even grown; but they are so rare as to warrant only passing mention, and rice cultivation is spreading to these areas. Also, in even fewer very remote jungle and swamp districts, there still roam small bands of extremely primitive nomads, who have no knowledge of any kind of agriculture, but live by hunting and collecting wild foods. As these groups gradually come in contact with more advanced neighbors, however, they tend to settle down and begin to grow rice. Although many of the people who live along the seacoast make their living by fishing, they sell their surplus fish for rice. And those relatively few millions of natives who have come to work for wages buy rice as their staple food. While rice is by all odds the main crop, other grains are grown, as well as a wide variety of vegetables and fruits. Maize is an important crop in many regions; a partial listing of other plant foods would include yams, taro, sweet potatoes, sago, cassava, sugarcane, and coconuts.

The principal non-vegetable article of diet throughout the entire area is fish; fishing is a very important occupation among the coastal peoples. Although various kinds of animals and fowls are raised and their flesh eaten, meat is unimportant in the diet of nearly all groups. The most common domesticated animals, water buffalo and cattle, are used mainly in agriculture or transportation. As a consequence of this aloofness from meat, animal breeding is a minor economic activity in all except a few regions of specialization.

The native economy, being based almost entirely upon subsistence farming, is very little involved with cash income. The people use money, of course, but they never have much of it. The farmer will sell his small stock of surplus rice for cash, but the payment will go at once for the purchase of salt, tobacco, cloth, and similar commodities which he does not produce for himself. The money income of the peoples of Southeast Asia and Indonesia is so small as to be astonishing to the European or American. In pre-war Java, for instance, the average income per person was about $15 a year; and in Siam in 1940 fewer than 3,000 individuals out of a population of 15,000,000 had annual incomes of over $1,100. The vast bulk of the peoples of this area are still living, then, in what might be termed a pre-capitalistic stage of economy. They pass their lives within the closed economic circle of their native villages, producing virtually all of their needs inside the self-sufficient unit and seldom coming into contact with the economy of the external world. Such contacts as they have are brief, and involve only a minimal use of the goods and monetary symbols of international commerce.

The other segment of the economy of Southeast Asia and Indonesia is, or rather was before the recent war, mainly the concern of foreigners who came to the area to develop it for financial gain. It was an economy of money, profit, exploitation, and export. As the years passed, an increasing number of natives became involved in it; and now in the postwar period the leaders of native revolutionary movements are demanding that their people actually take over control of this alien system. But before the war the Indonesians, Malayans, Burmese, Siamese, and Indo-Chinese had little to do with its direction; nor did they derive much direct profit from it, although many of them worked in it.

In this economy, the Europeans were the directors, and a certain proportion of the natives the laborers. Chinese and Indians participated in it also, both on the upper level of directors and on the lower level of workers. The activities involved in the pre-war profit economy of Southeast Asia and Indonesia were: (1) the production of commodities for export, (2) the importation of goods from abroad, (3) the distribution of these goods for sale, and (4) the transaction of the financial matters connected with export and import. In all of these activities except the first, the natives of the area had little share; and in the actual production they had a nonprofitable share. They were the laborers, a few of them the foremen. The Chinese and Indians, varying with the region, were concerned to some extent in all four activities. Some of them were laborers; others were owners and directors of production. Many of them, especially the Chinese, had to do with the importation of goods and selling; indeed, the Chinese were the main distributors throughout the area. And there were a few Chinese and Indian bankers and financial magnates; although their dealings in finance were concerned mainly with small-scale moneylending. But the Europeans were the ones who controlled most of the profit economy. Most of the great producing enterprises were European, as were most of the leading import firms and banking and financial houses. The only activities in which Europeans played almost no role were laboring and retail distribution.

The heart of the colonial profit economy was production for export. Plantations, mines, and oil wells supplied the commodities on which the system was based. And the factors which made the area extremely profitable were first, the fertility of the soil, augmented by little seasonal change and a wide range of crop possibilities; second, the mineral deposits in the subsoil; and third, the cheapness of the labor supply. Although vast sections were either undeveloped or unsuitable for cultivation, no other tropical area in the world had such great expanses of fertile land. The climate, continuously warm, and varying only between wet and dry seasons, with annual rainfall plentiful in most sections, made of Southeast Asia and Indonesia a kind of natural hothouse, producing throughout the year. And the variations in altitude, soil conditions, and, to some extent, in climate allowed for an extensive diversity of crops. Moreover, this agricultural wonderland had beneath its fertile soil other sources of wealth in the form of a great variety of minerals, chief among which were petroleum and tin. Finally, to work this treasury of natural resources, there was available, either within the area itself or in the teeming countries of China and India near by, an ample supply of extremely cheap human labor, unorganized and easy to handle. It was indeed a colonist's paradise.

As might be expected from its vast size, Indonesia was the richest prize of all. The six main export commodities of the Indies were, in order of value, rubber, petroleum, vegetable oils, tin, sugar, and tea. But these were only the start of a much longer list of extremely profitable export goods, which included, to mention the chief ones only, quinine, spices, kapok, coffee, sisal, tobacco, and tapioca. And ready at hand, in the heart of the islands, was one of the world's greatest reservoirs of cheap labor, Java. The human fecundity of this island matched the fertility of Indonesia's soil. With a population already standing at almost 50,000,000 just before the war, the Javanese were increasing at the rate of over a half million a year. This constantly expanding supply of potential labor could be tapped at will for workers to be employed in any part of the archipelago, and the wages offered need be only a few cents a day. Java was a veritable treasurehouse of human capital. With all these factors in its favor, it is not surprising that Indonesia was the most profitable colonial possession in the world excepting India. It is also not surprising that the Netherlands should now be so desperately anxious to retain this tremendously rich possession.

Malaya, despite its small size, was also a remarkably profitable dependency. Its wealth, from the British standpoint, was owing almost entirely to three factors: rubber, tin, and the commerce of Singapore. Viewed from the perspective of the profit economy, it was a vast rubber plantation and tin mine combined, with the greatest port city in the Orient attached. In its case, a local labor supply was not available, for the native population, consisting of only about 2,000,000 Malays and a small number of primitive tribesmen, was both inadequate in size and disinclined to work for wages. But the problem was solved by the mass importation of Chinese and Indian laborers. The Indians usually left Malaya after their period of service was over; but as time went on more and more of the Chinese stayed after their term of labor, so that by 1940 their numbers had increased to the point where they were the majority group in the colony, surpassing even the native Malays. Tin and rubber had made Malaya, in human terms, a Chinese country.

The three northern nations of Southeast Asia—Burma, Siam, and Indo-China—presented one striking contrast with Malaya and Indonesia, and this was that their main export commodity was rice. Indonesia had barely enough rice for its own needs, and Malaya had to import supplies of the grain; but the three northern countries produced surpluses of it. Each of the three has a great river flow-ing down its middle—the Irrawaddy in Burma, the Menam in Siam, and the Mekong in Indo-China—and each of these three valleys is a vast complex of rice fields. A century ago, rice was grown here by the natives for themselves. But during the past hundred years rice production became commercialized, and the economy shifted from one of subsistence to one of profit. The great difficulty was that the natives, except to some degree in Siam and Indo-China, did not get the profit. They worked the land, along with Chinese and Indian laborers, but the monetary yields went largely to European, Chinese, and Indian buyers and exporters. Moreover, because of the shift from a subsistence to a money economy, and the natives' lack of familiarity with the new system, they lost ownership of their land steadily to credit merchants and money lenders. The new landlords were Chinese, Indians, and, in Siam and Indo-China, a small number of natives as well. In Indo-China alone did Europeans become landlords of rice plantations. In the other two countries their role in the commercialized rice economy was played at a higher level, in the actual exportation of the finished product; although in this enterprise the Chinese and Indians had a large part also.

Burma before the war was the greatest rice-exporting country in the world. Rice accounted for almost half the total value of all exports. Most of the rice went to India, which was the main market for Burmese exports in general. The two other leading export products were petroleum and timber, principally teak; and India took almost all of these. Although Burma's share of the total oil exports of the world was small, the business was extremely profitable, and was operated by the Burmah Oil Company, a British concern. Burma was also the world's largest producer of tungsten, and led the Orient in the export of silver and lead. Other products of the country which entered into world trade included zinc, copper, nickel, gems, and small amounts of tin and rubber. The transportation system centered on the Irrawaddy River, where a British firm, the Irrawaddy Flotilla Company, had a virtual monopoly on shipping, which yielded rich returns.

By comparison with the other colonial countries of Southeast Asia, Indo-China was a poor possession. In terms of the profit economy, it was kept jealously by France as a reserved area for French trade and investment; and the French never developed it as the British and Dutch did their colonies. Rice was by all odds the main commercial product, accounting for about 70 percent of the total value of all exports. Half of it went to China, and nearly all the rest to France or the French African possessions. The rice for export was grown almost entirely in the southern regions of Cochin China and Cambodia, in the Mekong valley. The small farmers here were heavily in debt to Chinese and Annamese usurers; but most of the rice was produced on plantations, largely French-owned. This was the only place in Southeast Asia where Europeans dominated the actual growing of rice. Other landlords were Chinese and, also a very unusual feature, natives. In the northern regions of Tonkin and Annam, even though the Red River valley of Tonkin was a rich rice-producing area, not enough of the grain was grown to supply the extremely dense population. Here was the source of labor for the southern rice plantations, and thousands of Annamese and Tonkinese went seasonally down to the Mekong valley to work in the rice fields. It was a situation similar to that in Indonesia, where overpopulated Java supplied the manpower to operate the plantations of Sumatra and the other sparsely settled outer islands. Indo-China was the only important coal-producing country in Southeast Asia. The mines of Tonkin yielded a fairly large supply of coal, most of it anthracite. Zinc and tin were the other leading mineral exports. Indochina has good potentialities for rubber-growing, and the French were expanding this enterprise before the war. But in general the development of Indo-China lagged; profits from it were poor; and as a late venture in imperialistic enterprise its record was one of disappointment and failure.

Although Siam was the only independent country in Southeast Asia, its profit economy had a colonial character. The main export commodity was rice, which accounted for almost half the value of all exports, but the farmers who produced it were largely either tenants or so deeply in debt to moneylenders that they were little better off than sharecroppers. The landlords and usurers who dominated the rice growers of the great Menam plain were Chinese and native Siamese. The production of tin, the second most valuable export, was carried on by Chinese and Europeans. Rubber, third on the list, was controlled by Chinese and Malays. And teak, the only other significant export, was exploited by Europeans and Chi-

nese. The Siamese government, in recent years extremely nationalistic, grappled vigorously with the problem of getting the profit
economy of the country into the hands of the Siamese people. But
not only did the native inhabitants show no talent for or inclination toward business enterprise; they would not even accept wage-
earning jobs. The government experimented with nationalization of
commercial enterprises, but foreign capital was needed; European
firms were subjected to discriminatory regulations, but for its development the country required skills and techniques not possessed
by its own people; obstacles were put in the way of Chinese immigration, but 90 percent of the wage-workers in Siam were Chinese,
and the Siamese refused to labor for pay. Here was a remarkable
case of an independent nation surrounded by colonies and trying to
get rid of the elements of colonialism that infected its own economy.

In the foregoing discussion of the economy of Southeast Asia
and Indonesia, industry has not been mentioned. The reason is that
there was almost no industrial development in the entire area. These
countries, to the extent that they were incorporated at all in the
industrial economy of the Western world, were utilized as producing hinterlands for the factories of Europe and America. This
was not entirely a matter of deliberate design, for these areas lack
the two natural resources on which modern manufacturing is based,
namely, iron and coal. Every one of them has some iron and some
coal, but both are present only in poor supply and quality. What factories there were had been built to provide for local needs only
(textile and rice mills) or to perform the preparatory processing
of raw materials for export (tin smelters and rubber and sugar and
lumber mills). Indeed, the only large manufacturing installations
which produced a finished product were the oil refineries of Indonesia and Burma.

Throughout Southeast Asia and Indonesia, then, the economic organization was constructed upon a stratified plan. At the bottom
were the natives, the overwhelming majority of them still living on
a subsistence level within the closed economy of their local communities; and a few millions of them absorbed into the profit economy, but almost entirely in the capacity of laborers. In the middle
were the Chinese and Indians, who operated small-scale businesses
for the most part, and controlled almost completely the retail trade;

although many of them were laborers. At the top were the Europeans, who really managed the export economy, including its associated large-scale importing aspects, and drew the high salaries and the big profits.

The crucial question is why the natives, who were in so great a majority and whose homelands were involved, never made a better showing in the profit economy which extracted such enormous monetary yields from their own countries. Partly, no doubt, it was because the colonial powers saw only danger to themselves in providing natives with education and training which might fit them to take over management of the rich enterprises; but this would not explain the case of Siam. It would be a more general truth to say that the natives simply had not learned the ways of the profit economy. Throughout ages of history they had been engrossed in their own self-sufficient, nonprofit, subsistence economy; and they had not had time to adjust themselves to the completely unfamiliar system in such a way that they might participate profitably in it. They also lacked capital, so that they could neither buy nor invest in profit-making enterprises. The Europeans had the capital and the technical skills; and the Chinese and Indians, with their organizing ability and commercial acumen, took over what was left in the profit economy. In Burma, Siam, and Indo-China, the natives were so unused to the new pecuniary system that, when rice production was commercialized, they even lost their land to sharp-dealing Indians and Chinese; and the same thing would probably have happened in Indonesia if the Dutch had not wisely forbidden the alienation of native land under any circumstances. This question has more than academic interest, for the native revolutionary governments which have arisen since the war are insisting upon a termination of alien control of the profit economy. Their problem is how, in such an event, they could provide the requisite technical skill, capital, and commercial ability to keep the systems operating properly. This point will be considered more fully later, in connection with the current nationalistic developments.

Closely linked with the problem of the low economic status of the native peoples, and also vitally important in relation to their past and future political development, is the matter of education. The fact is that the educational level of the masses and their rate of literacy are lower in Southeast Asia and Indonesia than in most other areas of the world. The colonial governments until recently were either indifferent or opposed to native educational improvement. The Europeans were in the area to make profits and to control the local populations for this purpose; and they were unwilling to spend money for native schooling and to run the risk of sowing seeds of revolt by educating potential leaders in the knowledge, skills, and human ideals of the Western world. The monopoly on economy and government, to be preserved, had to be accompanied by a monopoly on education, for ideas are the stuff of revolution. Although progress has been made in all of the countries, the figures on literacy show how slow it has been. After three centuries of Dutch rule, the Indonesians were still 93 percent illiterate in 1930, and it is very unlikely that this proportion had dropped below 90 percent by 1941. Precise statistics for Indo-China are not available, but the rate of literacy there was probably not over 15 percent. In Malaya, literacy was much more widespread, with 24 percent of the total population able to read and write in 1931. The large numbers of Chinese in Malaya, however, raised the average literacy considerably, and the rate among the native Malays was much lower. Siam and Burma had the best records. Although the Siamese were said to have been 95 percent illiterate in 1934, the census of 1937 claimed that literacy had risen to 31 percent. According to the constitutional provisions of 1932, completely representative government was to be withheld until over 50 percent of the population were literate; and since in 1946 the legislature was elected entirely by popular vote, this evidently indicates that the goal has been reached. Burma, thanks mainly to its system of native Buddhist parochial schools, had a literacy rate of 37 percent in 1940.

The school systems were remarkably complicated, including private establishments operated by missionaries, the natives themselves, and various alien groups such as the Chinese, and, besides these, government schools for Europeans, natives, and alien Asiatics, as well as mixed institutions, using different languages and methods. In Burma, Siam, and Cambodia and Laos in Indo-China, most of the elementary schools were Buddhist monastic establishments, with monks as teachers. The Siamese and Indo-Chinese "pagoda" schools were being absorbed steadily into the government systems; just before the war in Siam, for instance, 70 percent of all the state schools were located in monasteries. In Malaya and Indonesia, every large village had a Koran school, run by Moslem priests. Above the elementary level, however, nearly all of the education was carried on in government or mission establishments, and European languages were used for instruction. In all of the countries except Siam fees were charged for schooling, rising with the grades; and in none of them except Siam was education of any kind compulsory. The proportion of the government budget devoted to education was very small in every instance. Taking the Philippine school expenditure of about 25 percent of the total budget as a standard, the comparable allotments for the countries of Southeast Asia and Indonesia were approximately as follows: Siam, 12 percent; Indo-China, 10 percent; Burma, 10 percent; Indonesia, 8 percent; and Malaya, 5 percent.

Another basis of comparison is the proportion of the children of school age and the proportion of the total population attending school. In Indonesia, with 2,700,000 students, 20 percent of those of school age were receiving instruction, or 4 percent of the population. In Indo-China, 570,000, or 17 percent, of the children were in school, representing 2.5 percent of the population. Malaya had 270,000 students, 33 percent of those of school age, and 7 percent of the population. In Burma, the comparable figures were, respectively, 850,000; 37 percent; and 5 percent. Siam, in proportion to population, was far in the lead, with 1,570,000 children, or 78 percent of those of school age, in school, representing 11 percent of the total population.

All of the countries had numerous specialized schools for training in technical subjects such as manual trades, agriculture, and forestry; and each of them had a university. But the number of students enrolled in the universities was very small. Raffles College in Malaya had only about 700 students; Rangoon University in Burma, approximately 2,500; and the University of Batavia in Indonesia, Hanoi University in Indo-China, and Chulalongkorn University in Siam each had about 1,000 students. Thus in the entire area, with a combined population of 130,000,000, the number of university students was less than 7,000. A few hundred students from Southeast Asia and Indonesia went abroad each year for higher education, and, though inconsiderable in numbers, these men became very important in the developing nationalist movements. But in the whole area there was a great scarcity of natives with advanced education. The entire school enrollment, very small in proportion to population, was concentrated in the primary level, 98 percent of all students attending the local elementary schools, most of which had courses of only three or four grades. All of the nationalist programs placed better educational facilities high on their lists of demands for reform, and the new revolutionary governments have included in their programs vastly expanded provisions for schooling.

While these countries have thus accepted wholeheartedly the educational ideals of the Western world, they have resisted the intrusions of European religions. There are two great religious divisions in the area. One, in which Buddhism is the main religion, includes Burma, Siam, and the two western sections of Indo-China: Cambodia and Laos. The other, where the majority of the people are Mohammedans, comprises Malaya and Indonesia. There is still a third, smaller region, including the three other parts of Indo-China —Annam, Cochin China, and Tonkin—where a peculiar combination of Confucianism, Taoism, and Buddhism prevails. In Siam, 95 percent of the population is Buddhist, and about 4 percent Moslem, the latter being nearly all Malays in the southern districts. In Burma, Buddhism is professed by 85 percent of the people, the minor groups being pagan, with 5 percent; Mohammedan, with 4 percent; Hinduist, with 4 percent; and Christian, with 2 percent. The pagans are all members of the primitive hill tribes; the Moslems, mostly the Arakanese of the Coastal districts bordering on India; the Hinduists, almost entirely Indian immigrants; and the Christians, for the most part formerly pagan Karens.

The Annamese, forming 75 percent of the total population of Indo-China, might be classed as Confucianists, but Taoist, Buddhist, and surviving pagan elements give their religion a unique and amorphous character. The remarkably synthetic disposition of Annamese religion has been demonstrated recently in the birth and rapid growth of a sect called Caodaism, which originated in Cochin China in 1926 and now claims a million adherents. It is a blend of Buddhism, Taoism, Christian Catholicism, Confucianism, and spirit worship. Its guiding spirit is a deity named Cao-Dai, whose spokesman on

earth is a kind of pope who holds court in a Cochin China pagoda.
It has an element of nationalism in it also, for Cao-Dai is expected
to restore independence to Indochina, whose loss of freedom, it is
believed, was a punishment for sins. The Buddhist peoples of Indo-
China—mostly concentrated in Cambodia and Laos—comprise
slightly over 20 percent of the total population of the country. The
remainder of the Indo-Chinese are mostly pagans of primitive culture
inhabiting the isolated mountain sections.

The Malays of Malaya are nearly all Moslems. There are a few
thousand primitive pagans; and the Chinese and Indians of the country have their own religions. The Indonesians are over 90 percent
Mohammedans; the Javanese, who constitute two-thirds of the entire population, being the largest Moslem group. Indonesia is the
second greatest Islamic country in the world in terms of population,
following the new state of Pakistan in India, which has a population of about 90,000,000. Approximately 5 percent of the Indonesians are pagans, and 3 percent are Christians, the latter mostly
former pagans who have been recently converted. The Balinese
have retained the Hinduist religion which once prevailed in Java
and Sumatra, and they constitute about 2 percent of Indonesia's
population.

The Buddhists and the Moslems of the area, the majority religious
groups, observe most of the outward forms of these faiths. The
Javanese and Malays, for instance, attend the mosque, regard the
Koran as sacred scripture, and follow the precepts of Islam. The
Burmese, Siamese, and Cambodians profess the Buddhist ethical and
theological beliefs, and carry on the associated rituals. Every youth
among these Buddhist peoples spends some period of his life as a
novice in a monastery school. Siam, for example, has over 18,000
temple monasteries, served by 150,000 monks and 100,000 attendants, in which at any time nearly 100,000 novices and 5,000 advanced scholars of the sacred Pali literature may be studying; and
Burma has 120,000 Buddhist monks and 20,000 monastery schools.
The form of Buddhism practiced in these countries, and in Cambodia and Laos also, is the so-called Hinayana, which survives elsewhere only in Ceylon. The kind of Buddhism which has influenced
the Annamese, however, is that prevalent in China, known as Mahayana. But adherence to these great world religions is in large part

merely nominal among the masses of the people in Southeast Asia and Indonesia. In every case, the successive adoptions of Hinduism (now virtually vanished except in Bali), Buddhism, and Mohammedanism—and Confucianism and Taoism in Annam—have represented only superimpositions upon the old pagan cults and beliefs. Underneath the varied veneer of these later faiths the ancient religions survive, and the new systems are fitted into the old.

Throughout the whole area, the basic substratum of paganism is made up of three main elements: beliefs and practices associated with magic, spirits, and the ghosts of ancestors. The pristine form of this pagan pattern still survives among the primitive tribes, but it shines through the covering fabric of Islam, Buddhism, and other more recent faiths in every country. Thus the Burmese hold strong beliefs in spirits of various kinds, called "nats." Every rural village has, along with its Buddhist monastery, shrines of these spirits. The Siamese have numerous beliefs and practices connected with magic and spirits; and their worship of the king, though formally derived from Hinduism, is related to the pagan concept of personal magical power which prevails throughout the area—appearing, for instance, in the highly charged persons of the Annamese and Cambodian monarchs—and which extends across the Pacific into Oceania, where the Polynesians call it "mana." The Annamese have taken over the magical elements in Taoism, and merged them with their own traditional sorcery; and Confucianism has been selectively adapted to the extremely vital ancestor cult. Indeed, the official religion of the Annamese empire, although it included Confucianist, Buddhist, and Taoist elements all intermixed, was essentially a combination of magic, animism, and ancestor worship. Every Annamese family has an altar on which the named tablets of deceased relatives are kept, to which offerings are made regularly; and after five generations the tablets are stored in clan shrines. The empire itself was founded upon a harmonious synthesis of Confucian ideals of filial piety and the vastly ramified ancestor cult, which culminated in the worship of the ancestors of the royal clan. Just as there never was a mass cultural invasion and transformation in Southeast Asia and Indonesia, but rather a slow infiltration of alien influences, leaving the native cultural core largely intact, so also religious pressures from the outside have seeped rather than swept in, being gradually and smoothly added to and fitted in with the traditional beliefs and practices of the native peoples. There has never been a religious revolution in the area; the record is one of evolutionary synthesis of old and new.

Of all the great world religions which have been introduced into the area, Christianity, despite centuries of missionary effort, has been least successful. Christian missionaries have made most of their converts among the pagan tribes, but have been unable to turn Moslems and Buddhists to the new faith. In Burma, there are about 250,-000 Christians, the great majority of them being of the Karen and other predominantly pagan tribes. The Baptists are the largest group, with 225,000, followed by the Catholics, with 90,000, and the Anglicans, with 10,000. The remaining 10,000 are of several minor denominations. Siam has about 50,000 Christians, including 35,000 Catholics, 10,000 Presbyterians, and 5,000 others. Statistics are not available for Indo-China, but the number of Christians there, mostly Catholic, is probably not large. In Malaya, hardly any of the Moslem Malays have turned to Christianity, and most of the work of missionaries has been among the resident Chinese. After hundreds of years of missionary effort, Indonesia has only about 2,500,000 Christians, who, as elsewhere in the area, are mostly converts from paganism. The proportions of Catholics and Protestants there are approximately equal. Thus the influence of Christian missions in Southeast Asia and Indonesia has not been primarily religious, for relatively few converts have been made; far greater have been the missionary contributions to education and medical care. Here can be noticed the readiness of the native peoples to receive the learning and the material aids of the West; but they have refused, usually calmly and without fanaticism, to accept the religious doctrines of Christianity.

Perhaps the inconsistency between the preachings of Christ and the imperialistic activities of the Christian Europeans has been a factor impeding the spread of the new faith. But another handicap of the Christian churches has been their intolerance toward existing religions. They permit no compromise with paganism, Islam, Buddhism, or any other system; these must be condemned as totally false, and the Christian beliefs upheld as totally true. The Mohammedan proselytizers, and the Hinduist and Buddhist bringers of foreign religions before them, came in as individuals, subject to no doctrinal orthodoxy. They were not dominated by church hierarchies which dictated what they must preach. They were able, therefore, to adapt their teachings to the prevalent beliefs of the natives, to grant the possible power of a local god or the efficacy of a traditional ritual while indicating the additional virtues of their own faiths. This, in large part, is why Islam and Buddhism have succeeded so well, and also why the texture of religion in the area is so varied and mixed. The new has been overlaid on and intermeshed with the old all through the centuries, and the result is a composite of paganism and later religions. The Christian missionaries have refused to let their faith be woven into the pattern, demanding that the latter be destroyed and replaced by Christianity alone. They have refused to compromise; but compromise and synthesis are the very essence of the religions of the area, and consequently Christianity has failed.

Just as the natives of Southeast Asia and Indonesia had no control over the profit economy of their countries, so also they had little voice in the higher levels of government, except, of course, in free Siam. As a matter of fact, there was a striking parallel between the economic and the political situations. The native peoples in their countless village communities lived under the rule of their own chiefs and councils, with little interference from the colonial administrations. As long as they kept the peace and paid their taxes, the central government let them alone. Like the profit economy, it was a thing separate from their traditional life.

Long before the coming of the Europeans, however, higher forms of social and political organization than the village commune had evolved in Southeast Asia and Indonesia. The Siamese brought in with them from southern China a centralized form of government, and the Annamese had an elaborate monarchical system also derived from China. Many of the other peoples—including the Burmese, Cambodians, Malays, and Javanese—developed centralized state governments as a consequence of contact with Indians; and most of the empires, kingdoms, sultanates, rajadoms, and the like which the Europeans encountered were constructed on the Indian pattern. But throughout the centuries these states actually represented a mere superimposition on the basic form of social and political organization, the village commune. This was true even in highly centralized Siam, where theoretically the entire nation was arranged in an hierarchical continuum from lowliest serf to king, but where in reality the communities were the functioning units of society for the great majority of the common people.

When the European conquerors entered, they abolished many of the native states, usually in cases where the monarch refused to submit; as the British did in Burma and the Dutch in various parts of the Indies. But they generally sought to keep these traditional rulers in power, in order to use them as puppets through whom they might govern under the guise of noninterference with existing local regimes. This technique of colonial administration, widely practiced in Africa also, is known as indirect rule, and it took various forms in different parts of Southeast Asia and Indonesia. Actually, whatever form the relationship between the colonial administration and the native states assumed, real control passed into the hands of the Europeans. In every instance, the native prince had beside him an official of the European administration, who told him what to do: a French Resident Superior with the Emperor of Annam, a British Adviser with the Sultan of Kelantan, a Dutch Governor with the Susuhunan of Surakarta, and so on. Large areas of Southeast Asia and Indonesia, however, were not within the domain of any native monarch when the Europeans came in. Here there were only village communities, or, in remote sections, nomadic or seminomadic primitive tribes. Such districts were put under direct rule, with no intermediary political organization between the colonial government and the chiefs and elders of the villages and bands.

The foregoing discussion of the economic, social, and political situation in Southeast Asia and Indonesia has been written in the past tense. In the case of the economy, this could hardly be otherwise, for information on economic conditions since the end of the war is extremely scanty and unreliable, and, more important, the area has not yet recovered from the catastrophic effects of the war. To present anything like a picture of the normal economic situation —the organization of the economy, the actualities and potentialities of production, internal and external trade, and the like—it is necessary to go back before the Japanese invasion. The present conditions are abnormal, extremely so, and when peace and stability return to the area the economy will pick up at the point where it was before the war; although, as will be seen when current political developments are described, some very great changes will probably be introduced, especially as regards the role of the natives in economic affairs. But the basic pattern of organization, production, and distribution will be about the same as in pre-war days, not as it is under present conditions of excessive disturbance.

In the case of social and political organization, it is simply impossible to discuss this area except in two phases: pre-war and postwar. The recent war brought a revolutionary change in Southeast Asia and Indonesia, and the conditions before and those after the conflict are so completely different as to constitute an epochal transformation. Nevertheless, as in the case of the economy, there will be no clean break with the past. No one could understand the present revolutionary movements and events, or venture anything like a valid prediction as to the future, without knowledge of the situation as it existed before the war. Therefore, in the following detailed discussion of social and political conditions in each of the five countries of Southeast Asia and Indonesia, the state of affairs just preceding the war will first be described, and then the developments since the end of the war, with some intermediate consideration of happenings during the Japanese occupation.

Burma has been the fourth largest unit in the British Empire in terms of population, coming after India, the United Kingdom, and Nigeria. Before the war, it had the greatest measure of self-government of any tropical dependency except the Philippines. Its status within the Empire, after separation from India in 1937, was never exactly defined. It was usually characterized in rather vague terms as a self-governing unit of the British Commonwealth. Actually, it had many of the features of a true dominion, but it also bore certain marks of colonial status. Thus, although it had a cabinet of native ministers responsible to an elected legislature, the control of foreign affairs, finance, and defense was in the hands of a British Governor. The Governor appointed the cabinet from among the members of the legislature. The legislature consisted of two houses, a Senate and a House of Representatives. The latter was entirely elected, but half of the members of the Senate were appointed by the Governor and the other half were elected by the lower House.

While the legislature enacted laws for the country, its powers were restricted in some very important respects. It had no authority over foreign affairs, defense, and monetary policy; and large sections of Burma were not under its jurisdiction. The Governor had charge of these matters, which were known as "reserved" subjects. The absence of native control over the first three is invariably a characteristic of colonialism everywhere, and in these respects Burma had mere colonial status. The fourth reserved subject, the Governor's power over the so-called "excluded areas," had reference to the parts of Burma which were mostly non-Burmese in population and culture. Although they covered over 40 percent of the land area, they included only about 15 percent of the total population. The most important of these regions were the Federated Shan States on the eastern border. Great parts of the mountain territory on the north and west were also in the excluded areas, over which the legislature had no authority whatever. The peoples of these sections were ruled by their own chiefs, with whom the Governor dealt through the Burma Frontier Service.

The powers of the Governor went much farther than his control over the "reserved" affairs. He could issue ordinances at any time, although these were subject to legislative review. The legislature could introduce bills on certain matters only with his consent. A most remarkable provision was one enabling the Governor to enact laws, which could not be contravened by any subsequent acts of the legislature. Such laws, however, had to go at once to the Secretary of State in London, to be laid before Parliament. By declaring a state emergency, the Governor could even assume unto himself "all or any of the powers vested in or exercised by any body or authority in Burma." In justice it must be said that these various extraordinary privileges of the Governor were seldom exercised, although just prior to the war mass arrests and peremptory censorship of the press occurred under the Governor's emergency authority. In the ordinary course of events, the greatest single power vested in the Governor was that of absolute veto over acts of the legislature. This right of absolute executive veto is another characteristic feature of colonialism.

Elections to the legislature were conducted on a communal basis, that is, Burmese voted for Burmese candidates, Karen for Karen,

Indian for Indian, and so on. Several seats were reserved for special interests, such as commerce, industry, and labor. But the system insured a majority of about ninety for the Burmese proper, on whose side the 12 Karen delegates usually voted. According to the terms of the franchise, the voting privilege was open to about a quarter of the population, but the rules had some amazing features. A man, to vote, had to possess real estate of at least $30 value (approximate equivalent in American currency), or have paid municipal taxes, or have rented a building for at least three months which had a monthly rental of about $1.20; or he might vote, regardless of these financial requirements, if he had served for a year in the army or the police force. He had also to be eighteen years old or over. Women could vote if they were twenty years old or over, fulfilled the financial requirements, and were literate. Remarkably enough, men did not have to be literate to vote. To become a senator (half of the number were elected by the House of Representatives), a person had to receive an income of at least $3,600 a year, or pay a minimum land tax of $300. The British were evidently determined that only solid citizens, by monetary standards, voted.

Nevertheless, with all these restrictions on true popular government, by comparison with other dependent countries Burma had obviously progressed far along the way to real self-rule. This fact appears also in the large proportion of Burmese who were employed in government service. For purposes of administration, the country was divided into seven parts, called divisions, each under a commissioner, and these in turn were subdivided into districts, under deputy commissioners. Of the 37 deputy commissioners in pre-war Burma, over half were native Burmese. Full control over the civil service, however, was vested in the Governor; the legislature had no control over appointments, dismissals, promotions, or pay. More than half of the magistrates in the country were natives, and the police force was largely Burmese. In the government service as a whole, Burmese outnumbered Europeans by 15 or 20 to 1.

The rank and file of the administration of Burma consisted of the headmen of villages. Each village headman had to be approved by the deputy commissioner of his district; but the selection of the headman was dependent upon the will of his people. These village chiefs were petty judges, police officers, and tax collectors combined; and they received as salary a share of the tax receipts. They were assisted by an elected committee in each locality. The fundamental democracy of Burmese society is shown also by the fact that there is no native aristocracy, and the caste system of India does not exist. Under the old kingdom, of course, there was a royal family; but, although little is known of social organization during the monarchy, apparently there was either no real class of nobility, or at best a very weak one. Certainly fewer traces of it survive now, only sixty years after the abolition of the kingdom.

Another common feature of social organization in Southeast Asia and Indonesia appears also in Burma, namely, the high status of women. In this part of the world the complete subordination of females which is found in India has no place. In Burma, although men are clearly the heads of the households, women suffer no restrictions on their freedom, and they have equality of rights with males in respect to property ownership, inheritance, and divorce. They carry on a good deal of the small trading in the local markets, as elsewhere in Southeast Asia, and several of them have actually become chiefs of villages. There is no concubinage; polygamy is rare; and widows may remarry without hindrance. And, very unusual for the Orient, almost 20 percent of the women are literate.

Despite the unusual degree of autonomy in administration and native participation in government, Burma in the pre-war period had a strong and constantly growing nationalist movement, agitating strenuously for an even greater measure of self-rule. The nationalists wanted not only political rights, however; almost an equal part of their demands were for economic improvement, including protection of native land-ownership and more opportunities for Burmese in the better kinds of wage-earning occupations. Because the Indian immigrants were the main source of trouble in the land question, and also competed for jobs with natives, Burmese nationalism always had an anti-Indian as well as an anti-government aspect. Consequently, many of the violent manifestations of nationalism took the form of riots directed against the Indians; and the worst of them occurred in the period of the 1930s, just when political nationalism was running at full flood. The Indians thus became the targets of economic nationalism, while the British bore the attacks of political nationalism.

In its early stages, Burmese nationalism was linked with Buddhism; in fact, its roots run back to the Young Men's Buddhist Association, founded in 1908, which later merged with other similar organizations in the General Council of Buddhist Associations. But, as in the case of most of the Islamic parties of Indonesia, religion was really only a symbol of the national spirit, and played a role similar to that of Catholicism in the Irish struggle for freedom. Moreover, Buddhist monks became constantly more active in political agitation, taking part in riots and inciting strikes, although the more conservative Burmese had doubts about the propriety of such behavior by priests. As the size and powers of the legislature increased, and as more and more natives participated in government, the pace of political activity accelerated and the number of parties multiplied. Indeed, one of the greatest defects of Burmese politics was the proliferation of parties, which led not only to disunity and confusion in the whole nationalistic movement, but also to personal politics of an undignified character. The separation of Burma from India in 1937 brought political activity to a high point, but, despite internal dissension, the common political goal of most parties became dominion status for Burma within the British Empire. So insistent did the pressure become that in 1940 the Governor declared officially that the eventual aim of British policy was to transform Burma into a self-governing dominion.

During the Japanese occupation of Burma, a puppet government headed by Ba Maw, a former premier who had been imprisoned by the British for sedition in 1939, was established. But although many Burmese worked with the Japanese and even fought on their side, a sizable proportion of the nationalist elements formed a guerrilla army, the Burmese National Army, and harassed the occupying forces in the late stages of the war. The ranks of this army were drawn mainly from the young members of the Thakin Association, a group of nationalist youths who, to symbolize their rebelliousness against the colonial system, had adopted the custom of addressing each other as Britishers were usually addressed, by the term "thakin," meaning "master." In the closing period of the war the Burmese guerrillas and British soldiers fought together against the Japanese, but cooperation ended with the war.

For the first time, the Burmese formed almost solidly behind a single party, which was the postwar descendant of the former Thakin organization and took the name of the Anti-Fascist Peoples Freedom League. The freedom movements in Southeast Asia and Indonesia since the war have produced a remarkable group of native leaders, and in Burma the revolutionary genius was a young man, only a little over thirty years old, named Aung San. He had been a guerrilla leader during the war. As head of the League, he brought Burma to the threshold of freedom in less than two years. The British tried in every possible way to induce the Burmese irregulars to give up their arms after the Japanese surrendered, but the natives refused. The British could easily have taken any point in Burma they desired, but guerrilla operations would have made such positions virtually useless. It became quickly apparent that Burma was not tenable without the consent of the natives.

Military means failing, the British began negotiations for a peaceable settlement. In May, 1945, they offered a plan whereby, after an interim period of reconstruction and preparation, Burma might revert to its pre-war status, and then decide freely what form of government it desired. During the interim period, which would last for a maximum of three years, the country was to be ruled by an executive council under the chairmanship of a British Governor, who would appoint the members of the council. The Burmese claimed that the 1945 plan offered almost no fundamental change from the pre-war situation, and, although the interim government was set up, with the Governor at its head, the Freedom League refused to participate unless it were given all the seats on the executive council. Instead, the League was offered three places out of fourteen, and, declining these, the only important party in Burma began a campaign of terrific resistance. The Council was filled with old-line politicians, and it had no prestige whatever. For over a year the interim government staggered along, resisted by the Burmese at every turn, and finally, in September, 1946, after a countrywide police strike followed by a general strike, the council resigned. The British had to yield, and they granted six seats out of eleven on a new council to the League. Aung San was made deputy chairman and minister of both defense and external affairs, two posts which the British had formerly insisted upon retaining. The strikes were called off, but Aung San warned that unless a suitable settlement were made quickly, the Freedom League would withdraw from the government.

In December, 1946, the British government announced that Burma was to be granted either complete independence or dominion status, whichever it desired, by the quickest means possible. A delegation, headed by Aung San, went to London, and in January, 1947, an agreement was signed by the Burmese and the British, which marked complete surrender on the part of the latter. Burma was to elect a constituent assembly in April, which was to draw up a constitution under which the country might become either a dominion or an independent nation. The frontier regions, formerly "excluded areas," were to decide whether or not they wished to join the rest of Burma or remain in a special relationship to the British Empire. The April elections resulted in a sweeping victory for the Freedom League, and the new assembly set about drafting a constitution. Within two months the tentative decision was made that Burma was not to become a British dominion, but rather an independent sovereign republic, to be called the Union of Burma.

The revolutionary governments of Southeast Asia and Indonesia which have arisen since the war all show, on the economic side, a strong tendency toward state socialism. In Burma, an announcement by Aung San clearly foreshadowed this development. He stated that although the country is not yet ready for socialism, and capitalist economy must continue, the latter is to be regarded as purely transitional. In the eventual socialization of the national economy, all subsoil resources, forests, national power sources, and railways, ports, and other public utilities are to become national property; and, while private ownership of such properties, as well as land, may persist for a while, the goal will be state ownership.

The executive council was at work on these plans, both political and economic, when suddenly, in July, 1947, a catastrophic event threw Burma into confusion. A band of men armed with automatic weapons invaded the council chamber and assassinated eight of the nine ministers, including Aung San. This mass murder was evidence of the chronic disease of Burmese politics, extreme divisiveness aggravated by resort to violence. Political assassinations were frequent in the history of the old kingdom, down to the days of Thebaw, the last monarch, who had scores of members of the royalty murdered; and bloody riots were common occurrences in recent years. But the astounding massacre of the council reflects an even more pervasive pattern of violence which prevails in the country. During the period of attachment to India, Burma had the highest crime rate of any Indian province; and since then there has been no improvement. In 1942, one out of every 215 Burmese was in jail, many of them for political offenses, but an astonishingly large proportion for murder and banditry. Gangsterism, known in Burma as "decoity," flourished as in no other Southeast Asian country. Various reasons have been given for this peculiarity—among them the great number of transient laborers, the constant shifting of people owing to the troubles of farm tenantry, and the relative lack of public condemnation of crime—but the problem as a whole has never been satisfactorily accounted for. Perhaps the general air of political and economic frustration which one senses in Burma is the covering reason for the widespread crime; but there may be some element in Burmese culture itself giving rise to an abnormal manifestation of aggression and violent action.

For all the turbulence of the postwar period, culminating in the assassination of the great leader, Aung San, and his ministers, the record is one of steady progress toward the liquidation of colonialism, which is the dominant trend in the current history of Southeast Asia and Indonesia. There is no question that, so far as Burma is concerned, the British grant of autonomy to India set the specific pattern for this neighboring dependency of the Empire. But the occupation of the country by the Japanese for over three years, which broke the hold of imperial authority and gave the natives not only experience in administration—both as puppets of the Japanese and as independent leaders of resistance organizations—but also possession of large supplies of military weapons, was the catalyzing factor which precipitated the successful revolution. Burma was well on the way, before the war, to actual self-government and probably dominion status; but the war itself hastened the process and carried it much farther than the British expected it to go. Burma has actually been given the choice of complete independence, something which would never have been predicted for this early date by even the most pessimistic of British imperial authorities.[2]

[2] An independent republic since January, 1948, Burma became a member of the United Nations in April of the same year.

Siam has emerged from the war with no fundamental change in its internal social and political organization; but before the outbreak of war the country had passed through a decade of revolutionary transformation. The revolution of 1932, which was virtually bloodless, brought to an end the absolute monarchy which had ruled Siam for five centuries. Under the remarkable king, Chulalongkorn, who reigned from 1873 to 1910, and his successors, the ground had been laid for the eventual revolution by numerous departures from traditional Siamese political organization. The great king continued the practice, begun by his father Mongkut, of hiring European advisors to assist in the modernization of Siam. He centralized the administration of the country, rendering obsolete the feudal rule of hereditary nobles by subordinating them to royal officials in the provinces, who in turn were placed under the jurisdiction of a Minister of the Interior. The monarch also established a cabinet on the Western model, although it was purely advisory except for administrative duties. He regularized and codified the system of law, instituted a national budget, and encouraged education, which in 1921, under his successor, became compulsory on the elementary level for all children.

The centralization of government and the suppression of feudalism caused a great expansion in the civil service, whose members were drawn from the constantly increasing class of educated Siamese, more and more of whom went abroad to study. Until 1932, the high posts in the government, however, were held by the nobility; and one of the principal reasons for the revolution was the resentment of the growing body of officials against this aristocratic monopoly. Moreover, the ruler at this time, Prajadhipok, had cut the civil service sharply in the interests of economy, and hundreds of dismissed employees were antagonistic toward the government. But the main reason for the revolution was the spread of democratic ideas among the educated Siamese. The internal revolution in Siam had its source in the emerging intelligentsia, just as the nationalist movements against colonialism elsewhere in Southeast Asia and Indonesia arose as a consequence of the slow but steady diffusion of education among the native peoples. In its inception, however, the revolution was not a mass movement, and at the time of its occurrence scarcely 5 percent of the Siamese were literate. It was the achievement of a small elite of government officials and military officers. Indeed, the new Siam has still to rise above the oligarchic kind of government which replaced the absolute monarchy in 1932. Nevertheless, the constitution adopted after the revolution made provisions for the eventual development of true democracy.

The 1932 revolution was carried out by a combination of civilian and military officials. The two groups were incompatible from the start, however, and this split still persists in Siamese politics. The leader of the civilian contingent was a young lawyer educated in Paris, Luang Pradit, and the military group was led by two colonels trained in Germany, Phya Bahol and Phya Song. Siam's first constitution provided for a legislative assembly with full law-making powers. Half of the members of the assembly were elected by popular vote (all citizens over the age of twenty were eligible for the franchise) and the other half were appointed by the king. After ten years, or before then if the literacy of the population rose to 50 percent, all of the delegates were to be elected. The executive branch of the government was in the hands of a state council, whose president was also prime minister. The king appointed the members of the council, which was responsible to the assembly and could be dismissed on a vote of no confidence.

The king could propose legislation to the assembly and could veto acts passed by it, but his veto could be overridden by a simple majority vote. The king could also dissolve the assembly, but then new elections had to be held within three months. Thus, although the powers of the ruler were still considerable, there was a profound change from the absolute royal authority of the former regime. Moreover, the grip of the nobility on the higher administrative offices was broken completely, as princes of royal blood were not allowed to hold positions in either the legislature or the ministry; they were eligible only to advisory or diplomatic posts. By curbing the nobility, however, Siam did not become a democracy. The record of the government since the revolution has been one of rule by small cliques, and during most of the time the group in power has been dominated by military officers or by conservative civilians whose dictatorial tendencies have been supported by the military. The royalists, however, lost virtually every vestige of control after an abortive attempt at counterrevolution in 1933. The king, in protest against the infringement of his powers, abdicated in 1935 while living in England, and was succeeded by his nephew, Ananda.

The struggle for power between the different groups of the ruling oligarchy caused numerous conspiracies and a few actual civil revolts during the 1930s. The most liberal among the leaders was Pradit, but, although he was a member of the state council throughout most of the period, he never was able to gain firm control of the government. Indeed, he was continually branded by rivals as a dangerous radical, and on one occasion he was sent out of the country to France, in virtual banishment. Nevertheless, his socialistic ideas were shared by the more conservative politicians who dominated the administration. The reason for this was not that these men were radical in any ideological sense, but rather that another factor, extreme nationalism, led them, because of the peculiarities of Siam's economic situation, into a program of state socialism. Since native capital was lacking and most of the profit economy—ownership, management, and wage-earning jobs—was in the hands of foreign elements, principally British and Chinese, the only means of gaining native control over the commercial life of the country seemed to be through state enterprise. Consequently, economic nationalism became the keynote of Siam's pre-war history. The government entered many businesses, including shipping, the distribution of oil and tobacco, and the buying, milling, and selling of rice. The development of local industries was encouraged by subsidization, and expansion of agricultural enterprise in rubber, sugar, and rice was fostered by state research and aid. Food production for domestic needs was required of the people, in order to make the country self-sufficient; and the state set up agricultural credit facilities and sponsored the promotion of cooperative societies. The program also had an aggressive aspect, in the form of laws discriminating against foreign business, both European and Chinese, and restricting immigration, which was almost wholly Chinese. The entire scheme had little success, because of the lack of native capital and specialized skills, as well as the stubborn reluctance of the Siamese to enter commercial occupations or even to accept wage-earning positions in the profit economy. The underlying reason for the latter difficulty was the almost complete lack of training for and experience in such work.

The program of economic nationalism is of interest mainly, how-ever, because it is a part of a general trend in Southeast Asia and Indonesia toward state socialism. The leaders of the revolutionary movements in Indonesia, Indo-China, and Burma all see socialism as the appropriate and desirable economic system for their countries when self-government is won. Siam, although it is an independent nation, has had a colonial type of economy, and here also state con-trol of the means of production has been the goal of the government, no matter what may be the purely political opinions of the differ-ent groups in power. Perhaps the reason for this is that any colonial country, in order to gain control of the profit economy, must, in the absence of native private capital for local investment and de-velopment, yield the direction of capital and production to the gov-ernment in a system of state socialism.

Along with economic nationalism, Siam developed a strong tend-ency toward political nationalism before the war. The Siamese have always been a militaristic, aggressive nation; and the revolution seemed to intensify this characteristic. In 1939, the name of the country was changed to Thailand, to symbolize the nationalistic sig-nificance of the tribal name, Thai. Military expenditures rose to al-most a quarter of the total budget, and the army and navy, though small, were continually modernized. Siam, in proportion to its size, was carrying on a more intensive rearmament program than any other nation; and while the Siamese gave little direct military aid to the Japanese, they did take advantage of France's defeat in 1940 to seize by force of arms a sizable slice of territory in Indo-China. The military emphasis was strengthened by the influence of army officers in the government. Nevertheless, the virtual dictator-ship of the 1930s was increasingly challenged by the rising power of the assembly, and particularly by the growing confidence and prestige of the elected members. When the Japanese moved into Southeast Asia in 1942, however, the government became virtually an outright dictatorship, and the nationalistic spirit was manifested by declaration of war upon Great Britain and the United States.

Although Siam was until 1932 an absolute monarchy, and after the revolution a virtual dictatorship, the underlying democracy of Southeast Asian and Indonesian society always survived at the level of the masses of the common people. The king was regarded as semidivine, a Boddhisattva or reincarnation of Buddha, and until the revolution the immensely ramified nobility had an important share in the government of the country. A peculiar feature of the royalty, however, was that membership in it was not hereditary, except temporarily. After five generations of descending status, any branch of the royal family reverted to the level of commoners. The monarchy itself passed from father to son, the eldest son of the premier queen being the heir apparent. In the absence of male offspring, the throne went to a brother of the king. So strong was the idea of the king's divinity that royal incest was frequently resorted to for the sake of preserving the sacredness of the sovereign lineage, and four of the five kings in the nineteenth century were the children of half-brothers and half-sisters. The royal family was extremely large, as the monarch had an enormous harem; Chulalongkorn, for instance, died leaving 600 widows and 370 children. Slavery, officially abolished in 1905, was widespread until late in the nineteenth century; and it is said that a hundred years ago over a quarter of the population were slaves. Moreover, every freeman had a patron in the upper class; so that a form of feudalism permeated the country.

The vast majority of the people live in small villages of ten to eighty families. The members of the village, both men and women, elect their chief. The villages are grouped in subdistricts, and the chiefs of the communities elect the presiding chief of the subdistrict. Above this level, the administrative officials are appointed by the central government. The subdistricts are grouped in 406 districts, these in 70 provinces, and these finally in 10 "circles"; the three categories of territory being under administrators of progressively higher rank. But the important strata from the standpoint of the masses of the people, the village and the subdistrict, are democratic in organization and functioning. Siam never accepted the Indian caste system; and the status of women, as elsewhere in the area, is unusually high for the Orient. Men are the family heads, but women may own and inherit property, and, except in the old harem system of the upper class, they have always had complete personal freedom. Over 15 percent of them are literate, and under the constitution they may vote in national elections.

The extension of democratic government provided for under the 1932 constitution has progressed steadily in spite of the oligarchic character of the central administration. Provincial councils were soon set up; as in the national legislature, half of the members were to be elected and half appointed until the rise in educational level warranted a completely elective membership. King Ananda, just before his death—apparently by assassination in the old royal tradition of Siam—signed a new constitution in 1946, which provided for a senate and a house of representatives, elected by popular vote; and the first entirely elected Siamese legislature was opened in June of that year.

During the war, Siam gave little substantial aid to the Japanese, but the dictatorial government made a bad error in declaring war on the United States and Great Britain. The British were disposed to force a punitive peace settlement on Siam, and only the intervention of the United States softened their terms. Britain had replied to Siam's declaration of war in 1942 with a counter-declaration, but America refused to acknowledge a state of hostilities with Siam, on the ground that the declaration of war was made by a puppet dictatorship of the Japanese; and in 1945 the king ruled both declarations void. During the war, moreover, the Siamese underground, in which several high officials of the government secretly participated, cooperated effectively with the Allies, especially the Americans. Without consulting the United States, the British first tried to impose a treaty which would have given Britain military bases on Siamese soil and have reestablished trade monopolies. Under American protest these demands were withdrawn, and the final settlement of early 1946 required that Siam give up its new name of Thailand, reverting to the traditional designation; that it return to Malaya the four northern states and to Burma the two Shan states which had been ceded to Siam by Japan during the war; and that disposition of the territory in Cambodia and Laos which Siam had seized from Indo-China be decided by international arbitration.

Siam thus reverted to its pre-war condition, with little change in status or in social and political organization. The disturbed economic conditions of the rest of the world have affected the country adversely, however, and it has yet to return to normal production and trade. The government is still dominated by cliques, as it has been since the revolution of 1932; and graft and corruption have increased to a degree previously unknown. But for a technically enemy nation situated in a colonial area, large parts of which are undergoing bloody revolution, Siam is holding its balance surprisingly well in the turbulent aftermath of war. It is preserving its remarkable record of maintaining independence in a part of the world where it alone has been a free country.

Malaya, a British dependency, has also come through the war relatively unchanged, especially by contrast with Burma, Indo-China, and Indonesia, the three other colonial countries of the area. In all of Southeast Asia and Indonesia, Malaya before the war had the smallest measure of self-government; and it promises to continue to have less than any of its neighbors. One reason for this is its small size, and another is its inconsiderable population, about 5,500,000. Furthermore, even this number is divided among three markedly different ethnic groups: the Chinese being in a majority with close to 2,500,000, the Indians totaling almost 800,000, and the native Malays constituting the remainder of about 1,250,000. The two large alien groups, forming 60 percent of the entire population, were in Malaya as immigrants, and were therefore in a poor position to make political demands; while the Malays were not only few but also politically unsophisticated.

It has been remarked that Malaya was an ideal colony from the imperial standpoint, being richer in profit yield than all of the other British colonies combined, and having a consistently favorable budget, no income tax, and no politics. The actual control of the British administration was more complete than that of any other colonial government in the area. Nevertheless, the fiction of native self-rule by Malay sultans was elaborately maintained, each of the nine states being theoretically independent, and all of them functioning as protectorates of Britain under treaty arrangements. The truth was that the British Resident or Adviser of each state ruled it, under direct supervision of the High Commissioner in Singapore. Since there were no elections, there were no political parties and no organized political activity. The Malays, with no experience or hope of experience in government, had almost no national political consciousness. The British cited the preponderance of Chinese as a major justification for continuing their strict control over the gov-

ernment, representing themselves as protectors of the indigenous
Malays against possible encroachments of the politically and economically more experienced Chinese.

The administrative organization almost surpassed belief in its
fantastic intricacy. There were three separate governmental juris-
dictions in the small area: the Straits Settlements, a crown colony
consisting principally of the three port cities of Singapore, Penang,
and Malacca; the Federated Malay States, a loose union of four of
the pseudo-independent protected sultanates; and the Unfederated
Malay States, an unamalgamated checkerboard of the other five
protected states, each independent of the others; and there were
seven tariff boundaries. The British justified perpetuation of this
plan of utter confusion on the ground of protecting the integrity
of the native states. An obvious advantage to the British was that
such minute subdivision of the territory insured against any possibility of Malayan unity. It was, on a small scale, the ultimate application of the principle of divide and rule. The complex system was
held together by lines of control that came together in the hands
of the Governor of the Straits Settlements and the High Commissioner of the Malay States, the same man, in Singapore. The main
channels of administration within the country ran down through
the Malayan Civil Service, an almost completely British corps directed by the Governor-High Commissioner. All of the sultans,
chiefs, and other official and semiofficial organs of administration
were actually nonfunctional embroidery on the pattern of concentrated control.

The Malays had no political program; most of the Chinese were
so strongly oriented toward their home country that they felt only
a tangential interest in Malayan affairs; and the Indians, almost all
of them transient coolies, were even less concerned with Malaya
than the Chinese. The British apparently felt little apprehension
about the possibility of political action on the part of the Malays.
They rightly saw no reason for being concerned about the Indians.
The Chinese alone seemed to be a potential source of trouble in
the future, for among them were two elements—the one communistically inclined and the other Kuomintang nationalists—who
might some day attempt a movement toward either Malayan independence or a closer link with China. The police and the secret

service spent more effort on surveillance of the Chinese than of any other group.

The British, like the native and alien peoples of Malaya, had no plans for changing the *status quo*, except that from time to time they reviewed the possibility of federal unification of the nine states and the Straits Settlements under a single administration. After the war, during which Malaya was occupied by the Japanese for almost four years, the British came forth with a plan for federalization. It proposed a new name for the country, the Malayan Union and Singapore. The Union would include all nine of the native states, as well as Penang and Malacca, which were formerly parts of the Straits Settlements. Singapore would become a separate colony, although it would be linked in certain ways with the Union; and both would be under the supreme authority of a single Governor. The entire plan represented a departure from the past only in the greater centralization of government; in the lowered dignity of the nine sultans, who henceforth would recognize the British Crown as sovereign in their states, rather than as a foreign protecting power under treaty relationships with them; and in the proposed establishment of a Malayan Union citizenship, to be granted to all persons, of whatever origin, born anywhere in Malaya. The last provision, which, if carried out, would give to Malaya-born Chinese citizenship rights equal to those of the native Malays, is potentially important, but, since the scheme made no mention whatever of voting or popular government of any kind, the question of citizenship is only of latent significance. The plan had nothing at all to do with democratic progress, although this was mentioned in the preamble; it was designed merely to simplify the governmental structure and thus render more efficient the colonial administration of Malaya.

During the Japanese occupation, numerous resistance groups conducted guerrilla warfare against the occupying forces. Most of the members of these groups were Chinese, but many Malays and some Indians were also in the guerrilla bands. Out of these wartime resistance units, and also from the more educated segment of the general population, there emerged, after the Japanese surrender, types of organization which had never existed in Malaya before, or at best had been very weak, namely, political parties and labor unions. One party, which claims the support of the majority of

Malays, is the United Malay Nationalist Organization, dominated
by aristocrats and native governmental officials. It is conservative,
asking mainly for a return to pre-war conditions, with the sultans'
sovereignty intact, the states mutually independent, and requirements for citizenship strict. The Malayan Communist Party, largely
Chinese, which, though small in numbers, is strong in the labor
union movement, calls for the independence of Malaya, democratic
government, and universal suffrage. The most interesting political
organization is the Malayan Nationalist Party, claiming a membership of 60,000. It desires a republic, in which the majority of the
legislature would be elected and half of the delegates would be
Malays. The Governor would have no power of absolute veto over
acts of the legislature, to which the executive council or cabinet
would be responsible. This republic would eventually join the Indonesian Republic in a general Malaysian federation. All of these organizations were opposed to the original British plan for the Malayan
Union on one ground or another. The Chinese parties, of which
there are several, liked the citizenship provisions, but many Chinese
criticized the absence of any provision for a democratic form of
government. The Indians, while rather active in the new labor unions, are politically apathetic, being much more interested in the
political developments in their own country than in Malaya. The
British, taken aback at the opposition to the original Union plan,
formed a political working committee in 1946 to consider revisions.

   In the meantime, the government of the colony of Singapore was
established, with a constitution which for the first time in Malaya's
history provided for popular elections. But the innovation was a
cautious one, for only six members of the legislative council would
be elected—by vote of all British subjects over 21 years of age,
regardless of sex and literacy—while nine of them would be appointed by the Governor and three others would represent the
European, Chinese, and Indian chambers of commerce. In a British
White Paper issued in July, 1947, a revision of the Union plan
was offered. Instead of a Union, it proposed a Malayan Federation,
to include the nine states and the British settlements of Penang and
Malacca. Whether Singapore would eventually become a part of
the Federation was left an open question. The traditional sovereign
rights of the sultans would be preserved under the Federation, but

their exact status was not defined. There would be a strong central government headed by a British High Commissioner, who would be assisted by an executive council. A legislative council would have jurisdiction over domestic affairs, but the High Commissioner would have "reserved powers" to enforce any bill which the council did not pass in a reasonable time and which he considered to be necessary. A form of federal citizenship would be established, but no specifications concerning this were given; and there was no mention at all of voting or popular government. The entire plan actually offered little change from the former system of administration in the country, and provided for no progress whatever toward democratic self-government.

Thus, although Malaya remains largely in its pre-war status, even here, in the politically most backward country of the area, there are stirrings of protest against the traditional colonial system. But the division of the population into different ethnic groups, particularly the Malays and Chinese, as well as the generally conservative tendencies of the Malay intelligentsia, many of whom are officials in the native states and therefore have a personal stake in the preservation of the old organization, impede any united movement toward genuine self-rule and popular government. Even in its orientation toward neighboring countries Malaya is divided. The Indians, a majority of whom are transient laborers on rubber plantations, have interest only in India. The native Malays, mostly farmers and fishermen living in small villages in the rural regions, are culturally and linguistically related to the Indonesians, especially to their brother Malays of Sumatra. And the Chinese, like their kinsmen elsewhere in the world, do not forget their ties with the homeland. They are principally urban dwellers or businessmen in the small towns; and in the future development of Malaya will probably constitute the bulk of an emerging middle class whose point of view on foreign relations will be strongly influenced by their ancestral link with China.

Indo-China, with the largest population and the greatest potential wealth of all the French colonies, was, immediately upon the close of the war, the scene of a revolution whose violence has been surpassed only by the Indonesian struggle for freedom. The French had held Indo-China for less than a century, many parts of it for less than that, when the Japanese invaded the country in 1940, more than a year before they moved into the other parts of Southeast Asia and Indonesia. Indo-China is actually a complex of five small countries, and before the war the form of administration differed for each. Annam and Cambodia were protectorates, under the nominal rule, respectively, of an emperor and a king. Laos also was technically a protectorate, but it was ruled directly by the French except for the territory of the king of Luang Prabang. Tonkin, formerly a part of the empire of Annam, was classed as a protectorate too, but since there was no native ruler the protectorate was one in name only, and the region was actually under direct French administration. Cochin China, also a territory of the Annamese empire in the past, became a colony, directly ruled, under the French. The French official in charge of each of the four protectorates had the title of Resident Superior, while the colony, Cochin China, had a Governor. The supreme head of government was the Governor-General.

The administration of the entire country, despite superficial variations, was actually in the hands of the civil service, headed by the Governor-General, and supreme authority was thoroughly centralized. Decisions as to policy and even relatively minor matters were made by the Governor-General, or, very commonly, by the Ministry of Colonies in Paris. In fact, most of the direction of government in Indo-China was by executive decrees. There were numerous kinds of councils in the country, but, although the members of many of them were chosen by a very restricted electorate, their functions were purely advisory; and, of course, the native rulers whom the French kept in office were mere puppets of the colonial regime as represented by the Residents Superior of the protectorates. French administration was handicapped not only by the fact that, unlike the British and Dutch officials in neighboring dependencies, the French civil service personnel was largely incompetent, but also by the constant shifting of the higher officers of government. In forty years there were 52 different Governors-General; and in the same period Cochin China had 38 Governors, while Annam, Tonkin, and Cambodia had 32, 31, and 22 different Residents Superior. Laos, in thirty years, had 17 Residents Superior. The instability of French politics was reflected in the colony, for, unfortunately, the colonial administration, which in the British and Dutch dependencies was mainly staffed by career experts, was used in Indo-China, especially on the important upper levels, as a device for rewarding party politicians with lucrative jobs. The civil service was also, by comparison with neighboring countries, overstaffed, and it had a higher proportion of European personnel in proportion to native employees—one French official to every five native—than elsewhere in the area. As a consequence of all of these factors, the administration was very expensive, absorbing fully a half of the total government budget.

Annam, nominally ruled by its emperor, had an assembly whose members were elected by a small elite consisting of mandarins, native "notables" of the upper class, licensed merchants, and graduates of French-Annamese schools. But it was merely a consultative body, and the actual administration was carried on, under the direction of the Resident Superior, by the French and native civil service. The latter was staffed by mandarins, who represented a survival from the days of the independent empire. In former times they were chosen by competitive examinations which concentrated on the Chinese classics, as in old China, and even the poorest youth could aspire, through learning, to the mandarinate. After 1915, the French abolished the examination system, and merely appointed the mandarins. Although there was a class of nobility under the independent empire, noble titles were granted by the emperor, and in each generation a noble family dropped down one level until, after five generations, it merged with the common folk, a procedure identical with that prevalent in Siam. The nobility, as such, had no part in government, but many mandarins had noble status. The emperor was an absolute monarch, semidivine, and the representative of Heaven on earth. All authority was concentrated in him, and from him it was diffused downward through the mandarins, the chiefs of villages, and the heads of families. The entire concept of government was paternalistic, and Confucian ideas of filial piety permeated not only the administration, but also the highly developed ancestor cult and the system of clans, of which there were many, grouped under the dozen or so great family names of Annam. The organization was thus strikingly similar to that of the Chinese empire; whence indeed it was derived. The emperor kept a large measure of his prestige under French rule, although he lost nearly all of his actual power.

Tonkin, which until 1897 had been ruled by a viceroy of the Annamese emperor, was directly administered by the French, even though it was classed as a protectorate. The mandarinate was retained, and, with the European bureaucracy, represented the central government in the various parts of the country. Tonkin had a protectorate council, whose relationship to the Resident Superior was purely consultative, and whose members were chosen by a small electorate of French citizens and prominent natives. Cochin China had no mandarins, for they were all withdrawn by the Annamese emperor when he ceded the country to the French in 1862, and the administration was carried on, under the Governor, by French and native officials. The Colonial Council here had more power than any other council in Indochina, and yet its authority was restricted to questions of taxation. This body elected one delegate to the French Chamber of Deputies, the sole representative of Indo-China in the central government of the French empire. Laos, nominally a protectorate, was ruled directly, except for the kingdom of Luang Prabang, whose monarch was retained in office. The administrative officials, under the Resident Superior, were native mandarins and functionaries of the French civil service. The advisory council was similar in composition and functions to the protectorate council of Tonkin.

Cambodia also had a consultative assembly, elected by a limited elite, but here the traditional native system of government was preserved more nearly intact than anywhere else in Indo-China. French civil officials and a corps of about a thousand mandarins carried on the administration, but the king had more prestige and the forms of monarchy were observed more fully than in the other two royal domains of Annam and Laos. The king of Cambodia, heir to the glories of Angkor, was, like his counterpart in Siam, a divine ruler, and a feudal kind of hierarchical social organization, again similar to that of Siam, placed every individual in a status of either client or patron to persons above and below himself. This system was extended to include not only nobles and freemen, but serfs and slaves as well. All persons in positions of authority were graded by the number of "dignities" they possessed; thus the highest nobles had 10,000 dignities and an ordinary village chief about 500. Even the marriage rules were affected by status, the king being allowed eleven wives and a commoner only three. The number of concubines, however, was not limited, and the king's harem regularly had a personnel of between three and four hundred. One further feature of the Cambodian kingdom had a parallel in Siam, and in Annam as well. Any branch of the royalty, after five generations of descending rank, dropped to the level of the common people; the only difference from the other two countries was that in Cambodia such *déclassé* aristocrats became members of a special ex-royal clan. Evidently the function of this remarkable scheme in all three countries was to keep the noble class fluid and thus reduce the possibility of any challenge to the supremacy of the royal line.

It might be supposed that with the strong development of monarchical systems in Indo-China, followed by the autocratic French colonial regime, democratic forms of social and political organization could hardly survive. And yet here, as elsewhere in Southeast Asia and Indonesia, beneath the elaborate cover of despotism, the basic units of society, the village communities, have always functioned as small republics, autonomous in local affairs and mutually independent. In Annam alone was there any tendency toward oligarchic government on the village level. Here the communities have been dominated by so-called "notables," the prominent men and taxpayers. But, since Annamese villages are virtually clan units, the ruling elders have been relatives of everyone else in the community. Their status as heads of the village has been merely analogous to that of heads of families within the locality. These village notables formerly elected the chiefs of cantons, or subdistricts, in Annam; but the French violated this democratic practice by making canton heads appointive. Throughout Indo-China, except in a few primitive matrilineal tribes, men are dominant in the families, but the status of women is high by Oriental standards. And even in Cambodia and Laos, where Indian influence was strongest, there are no caste divisions.

The French colonial policy had as its ideal goal assimilation of the Indo-Chinese to French culture, so that eventually the country might form an association with the mother country on a basis of equality. The first part of this program was realized among only a very small proportion of the natives. This elite group received French education, and even, in a relatively few cases, secured French citizenship. Naturalization usually required that a person be able to speak the French language and have performed some distinguished service; but, remarkably enough, citizenship so obtained was not inherited. Not only was citizenship sparingly granted; the whole scheme of concentrating upon production of a native elite impeded the development of mass education. Moreover, confidence in the ideal of assimilation itself was shaken when it was noticed that a very large proportion of the Indochinese who were given the advantages of a Western type of education became fervent national-ists, leading their less enlightened brethren in movements against French colonial rule. The basic difficulty was that the political and economic emancipation of native society did not keep pace with even the slow educational advances; and the young Indo-Chinese instructed in the notions of liberty, equality, and free competition found themselves with no voice in the government of their home-land, unable to rise above a subordinate level in the administrative services, and facing a barrier of discrimination which shut them out from employment in the better wage-paying jobs of the profit economy. In the second part of the French ideal program, equal association of the mother country and the overseas territory, no progress was made at all; but the principle is important because it became the basis for concessions offered to Indo-China by France after the recent war.

Perhaps the quickest way to summarize the defects of French colonial administration in Indo-China is to outline the program of reforms demanded by the Viet Minh, the dominant and most rep-resentative party of the country today. The party drew up its mani-festo in 1942, and it is a striking commentary on French rule. The party demanded a democratic constitution, providing for a popu-larly elected legislature. It insisted that the rights of personal liberty be guaranteed, and that irregular police methods, including torture, brutality, and peremptory trials, be abolished. It called for an end to arbitrary censorship of the press and to restrictions on the right of free assembly, as well as for freedom of movement from one part of the country to another. It asked for a reform in the tariff system, so that foreign trade need not always be sacrificed for the protection of French interests. The other demands were for an opening of government employment, on all levels, to Indo-Chinese, without discrimination; a program of industrial development; attention to modernization of agriculture, and with more emphasis on native food crops and less on export products; improved provisions for public health and medical care; vastly expanded educational facilities, from elementary school to university; abolition of forced labor on public works in lieu of tax payment; labor legislation to improve particularly the poor conditions among the coal miners of Tonkin and the contract laborers on the Cochin China rice and rubber plantations; and finally a reform in the system of taxation, especially abolition of the hated government monopolies on the sale of opium, alcohol, and salt.

In all of these respects, French administration was implicitly condemned by the Viet Minh, and the record justified the indictment. Two striking examples from the list of grievances above are the matter of police methods and the government monopolies. Indo-China, particularly in the period around 1930, was little better than a police state. There were mass arrests, followed by brutality, torture, and star-chamber trials under a special Criminal Commission. Thousands of Indo-Chinese, most of them for mere political agitation, were imprisoned in jails where conditions were appalling even for the Orient or were banished to the sinister concentration camp on the island of Pulo Condore, and hundreds were executed. As for the monopolies, since almost half of government revenue came from their sale, the use of opium and alcohol was encouraged and almost forced upon the people, and salt was sold only by the government at a high price.

In the face of such conditions, and with improvement stubbornly refused by the French, it is not surprising that nationalistic movements struck constantly against the colonial regime. The earliest organized uprising occurred in 1908, just at the time when the first signs of nationalism were appearing in Burma and Indonesia. There is no doubt that the victory of the Japanese over the Russians in 1905, symbolizing for the masses of Orientals the revolt of Asia against European domination, had an important part in the rise of colonial nationalism. The uprising of 1908 was suppressed, and hundreds of natives were banished to the prison on Pulo Condore.

Here among the prisoners was established the first recorded party, called the Restoration of Annam. During the first World War, over 100,000 Annamese were conscripted for labor service in France, and these men returned home with rebellious ideas which grew out of their comparison of conditions in free Europe with those in their own country. They formed the base of two parties which arose in 1925 and 1927, the Revolutionary Party of New Annam and the Annamese Nationalist Party. The former group split apart because of dissension between its communist and more conservative elements, and the communists established a party of their own. The Nationalist Party attempted an assassination of the Governor-General in 1929; in the bloody suppression following this incident it was discovered that half of its members were government employees. This party also instigated a mutiny of a garrison of native troops in Tonkin in 1930, and the French reprisals were so violent that the organization was forced into clandestine activity only.

The Indo-Chinese Communist Party reached its high point in 1931, when it claimed 100,000 followers. Its leader, a remarkably able and courageous Annamese, is the greatest figure in modern Indo-Chinese history and the president of the Republic of Viet Nam. Originally named Nguyen Ai Quoc, he changed his name to Ho Chi-Minh. The son of a mandarin, he traveled around the world as a sailor, and spent several years in Europe, where he became prominent in communist circles. He spent most of his time outside Indo-China in the 1930s, for the French were desperately eager to seize him; but he was constantly at work organizing nationalist groups, on the pattern of communist cells, within the country. His main base of operations was Canton, which became the center of Annamese plotting, but he also lived for brief periods in Russia and Siam. Before the British arrested him in Hongkong, he had been able, with the aid of small funds provided by Moscow, to set up a network of village cells, canton sections, and a central committee in Indo-China. The French hunted down the communist-nationalists ruthlessly, mobilizing, in a campaign of terror, the police, a large corps of spies, and the Foreign Legion. Mass arrests, torture, and arbitrary trials were employed to break the movement, and thousands of natives were imprisoned or executed. By 1933, the communists had been forced underground, but the organization built up by Nguyen Ai Quoc survived.

When the Japanese occupied Indo-China in 1940, the nationalists saw their chance to emerge. Just before this, in 1939, a new party, largely communist in composition, had been established. It took the name of Viet Minh, or the League for the Independence of Viet Nam (the ancient name of Indo-China, meaning Land of the South). In 1942, all of the other important parties—including the Annamese Nationalists, the Association of Revolutionary Annamese Youth, and the Indo-Chinese Communist Party—merged with the Viet Minh, which was pledged to fight against both Japanese and French for the cause of independence and democracy. The French administration was kept in office by the Japanese and cooperated with them in attempting to suppress the Viet Minh, which conducted effective guerrilla warfare all during the war, especially in the final months, when it aided Allied operations, mostly American, behind the Japanese lines. Nguyen Ai Quoc, now Ho Chi-Minh, was the organizing genius and leader of the resistance forces, which numbered fully 50,000 and functioned through a central national committee and local people's liberation committees. In 1945, the Japanese, foreseeing the probability of defeat, took over administration from the French, who had collaborated wholeheartedly with them during the occupation, both in combating the Viet Minh resistance and in supplying material, especially food, for the Japanese forces. In place of the French, the Japanese set up a puppet government, nominally independent, in Annam, Tonkin, and Cochin China, under the Annamese emperor Bao Dai; Cambodia and Laos were declared independent under their kings.

The Viet Minh would have nothing to do with these Japanese puppet governments, and, when the surrender came, they proclaimed the establishment of the Republic of Viet Nam. Meanwhile, France, now liberated, broadcast a plan for the future administration of Indo-China which was intended to counteract the Japanese declaration of independence for the country. Indo-China was to become a semiautonomous dominion in a French Union, with full control over domestic affairs. But the Governor-General would still be supreme, and a new state council, composed of appointed members, would have only advisory functions. In other words, Indo-China would have domestic autonomy, but this would be under French direction. Furthermore, defense and foreign affairs would be managed by France. The new Republic saw no progress in this plan and refused to consider it. Because French forces were not available, the British were given the task of disarming the Japanese and restoring order in the southern part of Indo-China, while the Chinese were assigned the same duties in the northern half of the country.

The British refused to recognize the authority of the new Republic, and, after they had armed the French who were in Saigon, incidents occurred which led to open warfare. As in Java, the British used Japanese troops to reinforce their own largely Indian contingents; and they stayed in southern Indo-China long enough to protect the entry of some 50,000 French troops. The Chinese in the north dealt openly with the Republic, recognizing it as the *de facto* government of the area, but in early 1946 they withdrew and French forces moved into Tonkin. The French, using mostly American weapons and other equipment, as well as Foreign Legionnaires largely recruited in Germany, started the full-scale warfare which was still going on in 1948.

It appeared for a while in the middle of 1946 that a negotiated peace would put an end to the French-Viet Nam war, for in March of that year France recognized the Republic "as a free state with its own government, parliament, army, and finances, forming a part of the Indochinese Federation and the French Union." Cambodia and Laos had meanwhile been established, under the nominal rule of their kings, as autonomous states within the Federation. The French promised that the question of inclusion of Tonkin and Cochin China with Annam in the Viet Nam Republic would be decided by referendum; but they quickly broke this pledge by setting up a separate republic in Cochin China. This country is the richest part of Indo-China, and without it the Republic would be economically crippled. Moreover, there is little doubt that its population, mostly Annamese, would vote to join Viet Nam if a referendum were held. Tonkin would certainly declare for Viet Nam. The president, Ho Chi-Minh, made a fruitless trip to France to try to settle the issue of Cochin China, arrange for a cessation of hostilities, and initiate concrete action in establishing the Republic, the

Federation, and the French Union, as provided for in the March agreement. The French continued to behave as though the agreement never existed, and, in a country raging with open warfare, the Viet Nam Republic set about organizing a government while the French poured in troops and tried constantly to spread out from their bases in the principal cities and in Cambodia and Laos. Even in the latter two countries, however, they were resisted by native forces sympathetic to Viet Nam, the Free Cambodia and Free Laos groups. In Annam, Tonkin, and Cochin China, the French organized a party of native politicians who were at odds, for one reason or another, with the Viet Minh. It was called the National Union Front, but its name was utterly inappropriate, as it had almost no support among the populace.

The Viet Nam Republic held its first elections, for a provisional government, in early 1946, openly in Annam and Tonkin, and clandestinely in French-occupied Cochin China. The result was a complete victory for the Viet Minh, and Ho Chi-Minh became president of the Republic, while his party won nearly all of the seats in the provisional assembly, among whose members was the ex-emperor Bao Dai, who had abdicated, giving the Republic his blessing, at the time of the Japanese surrender. He was elected, under his nonimperial name, from his ancestral province. In November, 1946, the assembly adopted a constitution providing for a unicameral legislature to be elected triennially by popular vote of all persons over the age of 18, a president elected by a two-thirds vote of the legislature, and a prime minister and cabinet responsible to the legislature.

The plan of local government owed something to both the traditional village autonomy of Indo-China and the Soviet system, which Ho Chi-Minh learned well in his days as a communist. Each village would have an elected council, and elected councils would also govern the provinces. The local councils would elect the executive committees of the prefectures, districts, and the three great divisions or "bos" of Viet Nam: Annam, Tonkin, and Cochin China. Although the pattern of organization had a relationship to the Russian plan, and while Ho Chi-Minh himself was once an avowed communist, the Republic is not communistic. There is not even a communist party in Viet Nam, the old one having been disbanded in 1945.

The Republic is nationalistic and opposed to imperialism, but is not anticapitalist. President Ho, when asked about communism, has stated that it may come some day to Indo-China, but that now the goals are freedom and democracy. Still, while the Republic has declared that foreign capital will be welcome and its rights thoroughly protected, it is probably inevitable that the new regime, once firmly established, would lean toward moderate state socialism in the economic sphere. The constitution makes careful provision for the safeguarding of civil liberties and the rights of minority peoples, and stipulates that education on the primary level will be compulsory and free. Already a great literacy campaign is under way and has produced surprisingly successful results.

The French have never had true peace in Indo-China, except in the backlands of Cambodia and Laos. The Annamese have resisted colonial rule constantly, and even when lying low under military suppression they have busily organized for the uprisings which burst forth every few years. The French reaction has been, not compromise and concession, but force and still more force against the aspirations of the natives. French colonial policy, ideally directed toward assimilation of the subject peoples to French culture, with eventual equal association as the final goal, has not succeeded because it was never tried. The assimilationist program was carried out for a very small proportion of natives, but their education in Western ideas of freedom and democracy made them revolutionaries against the colonial system, which in its very essence is opposed to self-government, personal liberty, and equality.

The Dutch, although they never had the principle of cultural assimilation of their colonial subjects in their imperial policy, have nevertheless faced the same problem in Indonesia as the French have in Indo-China. Here, in the largest and richest of all the countries of the Southeast Asian area, a violent revolution, led by men who have received the Western kind of education and absorbed the social and political ideals of Europe, started immediately after the Japanese surrender. As in Indo-China, the groundwork had been laid over a period of many years, during which nationalist movements gained steadily in strength despite the frantic efforts of the Dutch to suppress them.

Indonesia before the recent war was not designated by the Dutch as a colony, although it had most of the characteristics of a colonial dependency. It formed, according to the Dutch definition, a "part" of the Kingdom of the Netherlands, and had its own governmental system to handle domestic affairs; but the islands were under the "guidance" of Holland and the latter controlled relations with foreign states. Thus legislation concerning matters of broad and fundamental import for Indonesia went through the parliament in Holland, while strictly domestic affairs were directed by the Governor-General and the organs of government in Batavia. These included the Council of the Indies, a five-man advisory board of the Governor-General; a cabinet; and a legislative assembly. Defense, as usual in colonial countries, was in the hands of the controlling power.

The legislature, known as the Volksraad or People's Council, was potentially the most important organ of the Indonesian government. Created in 1916, at first it could not initiate legislation, but might merely call upon officials for explanation and defense of their policies and activities. Later, however, members could introduce bills on their own initiative, and the Volksraad could amend bills presented to it by the Governor-General, who was required to lay every legal measure he advocated before the Volksraad for a vote. If the legislature and the Governor-General could not reach agreement on a bill, the deadlock was resolved by vote of the Netherlands parliament or by royal decree. The Volksraad was therefore almost, but not quite, a true legislative body. The partially democratic nature of this assembly was also demonstrated by the way in which its members were chosen. There were 61 delegates, the chairman being appointed by the Crown, and the remaining 60 members being partly elected and partly appointed by the Governor-General in the following manner: of the 30 Indonesian delegates, 20 were elected and 10 appointed; of the 25 European members, 15 were elected and 10 appointed; and of the 5 "alien Asiatic"—Chinese and Arab—deputies, 3 were elected and 2 appointed. Thus the elected members totaled 38 and the appointed ones 22. The method of election was indirect, and only members of the local councils—provincial, regency, and municipal—voted for delegates. These council members in turn were partly chosen by the people themselves, in most places voting as village units, and partly appointed by the head of the local civil service administration. The whole process was remarkably complex and cumbersome.

There were three kinds of administration outside the central government. The civil service, headed by extremely well-trained Netherlanders but including a large proportion of natives on the lower levels, acted as the representatives of the central government in the various districts. The native rulers, mostly hereditary, were survivors from the feudalistic system which was traditional in many parts of the Indies before the arrival of the Europeans. And third, there were councils of the different kinds of administrative districts: provinces, regencies, municipalities, and communities.

The civil service included several grades of officials, virtually all of the superior posts being manned by Netherlanders. About 7 percent of the area of Java was nominally governed by four native sultans, while fully 60 percent of the territory outside Java was technically controlled by local rulers. In Java, although 90 percent of the island lay outside the domain of the native states, "semi-indirect" rule operated throughout these sections, which were divided into 70 regencies, each under the nominal authority of a native prince of noble ancestry. The lowest unit in the hierarchy of native regional government was the village community. Each village throughout the Indies had a chief, in some cases elected and in others hereditary. He was assisted by certain other officials, usually including a secretary, a messenger, a bailiff, and a priest. Each village had an assembly, to which generally all adult males in good standing were eligible. Many of the communities had, in addition, a council of elders, a kind of senate drawn from the members of the assembly. These village republics, as elsewhere in Southeast Asia, were the real centers of native social and political life. Only a small proportion of the common folk ever had anything to do directly with the higher native officials or the Dutch administrators. The third type of regional government was a recent innovation, although the groundwork for it had been laid as far back as 1903, by the first "decentralization" law. The plan was to develop in each section of Indonesia a complete local government to handle internal affairs. Each of the major administrative divisions was to have not only its civil service officials representing the central government, and its native rulers, but also a council, partly appointed and partly elected. The scheme would eventually have worked out into a system similar to the American federal type of government, with its state legislatures and city councils, and by 1941 the plan was in partial operation.

The development of Dutch rule was thus one of cautious and gradual liberalization of the governmental system, slowly tending toward the goal of native self-rule under European supervision. Still, the Indonesians had little share in the central administration, for above the village level the Dutch held firm control in their own hands. The only important governmental body in which natives were even fairly well represented was the Volksraad; but the thirty native delegates formed only half of the legislature, and ten of them were appointed by the Governor-General, while the method of election of the other twenty was so indirect that true popular representation was not achieved. Moreover, acts of the Volksraad were subject to initial veto by the Governor-General, and to final veto by the Netherlands government. The highest officials of the central administration included only two Indonesians in 1941; and the civil service above the rank of *controleur* was almost entirely staffed by Netherlanders.

The nationalist movement began in Indonesia in 1908, but it was not until the first World War and the establishment of the Volksraad that the Dutch began to be alarmed at the growing radicalism, from the Netherlands standpoint, of the Indonesian nationalists. Some of the parties and leaders of the 1920s were undoubtedly communistically inclined, but the great majority of them were purely anti-imperialist nationalists who agitated for improvements in the political and economic status of their people. The Dutch reaction to the growth of nationalism was to make the laws regarding native political activity ever stricter, and to impose heavy penalties on all who violated them. The worst period was in 1926 and 1927, when armed rebellions broke out in Java and Sumatra. The Dutch reprisals were swift and ruthless, and thousands of natives were imprisoned, a large proportion of them being banished either to a political concentration camp deep in the wild interior of New Guinea or to remote towns in the outer islands. Hundreds of prisoners were still in the Boven Digoel camp in New Guinea when the Japanese invaded the Indies, and many of them had been there ever since the uprisings of the late 1920s. Nearly every one of the leaders of the postwar revolution had spent years in Dutch prisons for political activity. The censorship laws were so strict that almost any statement of the nationalist position was punishable, while public advocacy of Indonesian independence was legally classed as sedition. It was forbidden to use the term "Indonesia" in print or on the radio, because, the Dutch said, it had become a symbol of independence; and singing of the nationalist anthem "Indonesia Raya" was prohibited by law.

After the harsh repression of the uprisings in the late 1920s and the imprisonment or banishment of many of the nationalist leaders, the movement subsided, and from 1930 until the Japanese invasion there was a period of wary political action, marked mainly by efforts to organize a united front of nationalist parties so as to strike strongly when the time was ripe. The most powerful of the parties were the Parindra, a federation of moderate nationalist groups; the Gerindo, a coalition of more radical organizations; the PPBB, a party of native government officials; and the MIAI, a federation of Moslem societies. Both the Parindra and the Gerindo were composed mainly of Western-educated Indonesians, supported by local patriots of lesser learning; the PPBB membership, drawn from the ranks of government employees, was well educated; while the MIAI followers were, for the most part, comparatively uneducated and unsophisticated Mohammedan enthusiasts. Until 1939 these parties, and a score of minor ones, had never achieved any appreciable unity of goals or action. In that year the relatively conservative Parindra and the relatively radical Gerindo both shifted ground sufficiently toward a compromise central position to render amalgamation possible. They were joined by several minor parties, and the result was a coalition under the name of GAPI, *Gaboengan Politiek Indonesia*, or Indonesian Political Union. The PPBB and the MIAI did not join the Union, but in policy and voting in the Volksraad they followed the GAPI line. This coalition thus became a unified and powerful pace-setter for Indonesian political action, the first effective Indonesian political federation.

The GAPI policy represented a strategic compromise with necescity. All through the 1920s and 1930s, whenever an Indonesian party showed signs of real power and embarked upon a program of spirited agitation, the government, on one pretext or another, jailed its leaders or sent them off into exile. Since, under the East Indian laws, advocating independence could be construed as treasonable activity, any "secessionistic" nationalist was liable to legal prosecution. The native political leaders came to realize that the government was determined to enforce this type of law, and therefore, in self-defense and for the sake of staying free, most of them came around to the point where they carefully refrained from preaching complete independence for Indonesia, and instead advocated self-government within the Netherlands Kingdom. The GAPI epitomized this device of strategic retreat. Just before the Japanese invasion, the Dutch made a final gesture which left a bitter memory of them among the Indonesians: they rounded up and imprisoned scores of native political leaders.

The Japanese occupation of Indonesia for nearly four years changed the political situation in the islands tremendously. The revolution really started during this period, although the nationalist movement had prepared the way for it. The quick conquest indicated to the Indonesians the fact that the European masters, who had seemed invincible for three centuries, could be beaten. Moreover, it suited the purposes of the Japanese to indoctrinate the masses of the natives with anti-imperialist propaganda, and they did this on a much wider scale than the nationalist leaders, constantly suppressed by the Dutch, were ever able to. Unfortunately for the Dutch, there was enough of truth in what the Japanese said about imperialism to make a profound impression upon the Indonesians. At the same time, they never accepted the idea of Japanese domination. They responded enthusiastically to the Japanese slogan "Asia for the Asiatics," which they interpreted as meaning "Indonesia for the Indonesians." But the Japanese did certain other things which had greater practical significance. They removed all of the Dutch from their positions of control in government and economic affairs, replacing them with Indonesians who had previously been subordinates of the Europeans. Thus native politicians and government employees, who had been either excluded from political life or given minor administrative posts by the Dutch, were placed in the highest levels of authority. Indonesian foremen and technical assistants became managers and superintendents of plantations, mines, and petroleum installations. And for the first time in modern Indonesian history, large numbers of natives were recruited, trained, and armed for military service; and former petty officers in the small Dutch forces were elevated to positions of command in the new militia. For almost four years the Indonesians, under Japanese direction, actually ran their country, and they gained skill, experience, and confidence in themselves. The Japanese occupation was thus the training period for the postwar Indonesian Republic. Finally, when the Japanese saw that their defeat was certain, as a last gesture, since they had nothing to lose by it, they set up an Independence Preparatory Committee, which, immediately upon the surrender, proclaimed the Republic of Indonesia.

Unlike Burma, the Philippines, and most other liberated areas, the Indies were not entered by the Allied forces at once. The suddenness of the Japanese surrender left the islands "unoccupied," which gave the Republic time to get organized and to mobilize its troops for defense of the new independence. When the Dutch heard about what was occurring, they frantically broadcast from Australia an order placing Indonesia under interim control of the Japanese, and condemned the Republic as a Japanese-inspired creation of native radicals and collaborators. The natives were shocked that the Dutch preferred to designate the enemy as their deputies rather than allow Indonesians to take charge. Since the Dutch had only small forces available, the British were given the responsibility, as in Indo-China, of entering the Indies to receive the surrender of the Japanese there, disarm them, and release Allied prisoners of war and internees. The Republican leaders assured the British that they would cooperate fully with them in their mission, but warned that if they brought any Dutch troops or officials with them this would be resisted by armed force. The British landed, accompanied by Dutch contingents, and warfare began.

As in Indo-China, the British did two things which went contrary to their mandate from the Allied command. In the first place, instead of disarming the Japanese troops at once, they used them to police sections of Indonesia, and also, to the scandalized amazement of the natives, they even employed them in actual combat against the Indonesians. The second British breach of contract came when they announced that they intended to insure the safe entry of Dutch forces before leaving Indonesia. They decided this entirely on their own; it was not in the terms of the mandate defining their mission in the Indies. Dutch forces poured in steadily under cover of British protection, until by the end of 1946 they numbered about 100,000. They had mostly American weapons and equipment, just as the French troops did in Indo-China. The best Dutch units were marines who had been trained and fully fitted out in the United States. In both places, British protection and American supplies paved the way for attempts at colonial reconquest.

During a full year of warfare, partially interrupted from time to time by temporary truces, the Dutch and Indonesians carried on negotiations with the mediation of the British. The Dutch first offered a settlement based upon a statement of policy which had been made by the Queen in 1942. According to this, Indonesia would in the future become an equal, autonomous partner of Holland in a Netherlands commonwealth. In the early months of 1946, the Dutch elaborated this vague plan, and, although commonwealth partnership was still the basis of the new offer, they outlined a procedure whereby, after an indeterminate period, Indonesia might, if it so desired, become completely independent. In the meantime, a democratic, representative legislature would be established in the Indies, with a majority of native members. The Governor-General, however, would retain extensive special powers to guarantee "fundamental rights, efficient administration and sound financial management." Whether he and the Netherlands government would still have veto power over acts of the Indonesian legislature was not stated. There would be a central government of the whole Kingdom, including a commonwealth cabinet and administrative bureaus, and composed of members drawn from all parts of the union, but no central legislature. Probably the question of the latter was deliberately avoided, because the problem of proportional representation would have surely arisen, and Indonesia, with over 70,000,000 population, might logically have claimed a larger representation than Holland, with only 9,000,000 population. The Dutch promised that if Indonesia accepted the proposal, Holland would sponsor Indonesian membership in the United Nations.

The Netherlands plan just outlined became the basis for protracted negotiations between the Dutch and the Republican officials.

Finally, in November, 1946, just as the British were withdrawing their forces from Indonesia, the Netherlands and Republican representatives initialed an agreement which was then submitted to their respective governments for approval. In the meantime, the Dutch, who had succeeded in occupying most of the islands outside Java and Sumatra, called a conference of delegates from these outer territories at Malino in Celebes in July, 1946, and resolutions were passed which, among less important provisions, favored the establishment of a United States of Indonesia, to consist of four federated states: Java, Sumatra, Borneo, and the Great East (eastern Indonesia). Another conference, held in Bali in December, 1946, formally established the state of Eastern Indonesia; but this event attracted little attention because of the far greater significance of the initialing of the general Dutch-Indonesian agreement in November.

This agreement, which was approved by the Netherlands and Republican governments and formally signed in Batavia by their respective delegates in March, 1947, was named the Linggadjati or Cheribon Agreement, after the place near the city of Cheribon in Java where it was drawn up in November. It consists of eighteen articles, the main points of which may be summarized as follows.

The Netherlands Government recognized the Republic of Indonesia as the *de facto* government of Java and Sumatra. The Malino resolutions had favored separating Java and Sumatra, and the Dutch showed a disinclination to allow incorporation of the latter island in Republican territory; but this first provision of the Cheribon Agreement represented a crucial victory for the Republic, which thus would gain control over the two most important islands of the Indies and more than 80 percent of the total population. Areas in Java and Sumatra occupied by Dutch troops were to be gradually yielded to the Republic, at the latest by January 1, 1949. The Netherlands Government and the Republic were to cooperate in the rapid formation of a sovereign democratic state, a federation to be known as the United States of Indonesia. This would include all of the Indies, in three divisions, namely, the Republic of Indonesia (Java and Sumatra) and the States of Borneo and the Great East. If the population of any territory should decide, by democratic process, that they were unwilling to join the United States of Indonesia or preferred a special relationship to the federation or to the Netherlands, their desires would be recognized. The constitution of the United States of Indonesia was to be drawn up by a constituent assembly composed of the democratically nominated representatives of the Republic and the other divisions of the federation. The method of participation in the constituent assembly would be determined by consultation between the Republic and the Netherlands.

The Netherlands and the Republic would cooperate in establishing a Netherlands-Indonesia Union, with the Netherlands Queen (or King) as head, this Union to be composed of two equal partners: the Kingdom of the Netherlands (Netherlands, Surinam, and Curaçao) and the United States of Indonesia. There would be joint organs of government in the Union, formed by both partners, to handle matters of mutual interest, such as foreign relations and defense, and, if necessary, fiscal, economic, and cultural affairs. One very important article dealt with economic adjustments, and provided that the Republic would recognize the claims of all non-Indonesians to the restoration of their rights and the restitution of their goods within the territory of the Republic. A joint commission was to be set up to effect such restoration and restitution.

The date by which the Netherlands and the Republic would endeavor to have established the United States of Indonesia and the Netherlands-Indonesia Union was set at January 1, 1949. In the meantime, the Netherlands would take steps to obtain the admission of the United States of Indonesia as a member of the United Nations immediately after the formation of the Netherlands-Indonesia Union. In the period of preparation, an interim organization would be established, consisting of delegates of the Netherlands and the Republic and of a joint secretariat. If the delegates were unable to settle a point of dispute, it would be submitted to arbitration, with a chairman of another nationality presiding and having the deciding vote. If the two delegations could not agree upon a chairman for arbitration, the chairman would be appointed by the President of the International Court of Justice.

The question of the "right to secede" was avoided, which marked an important difference in status between the projected United States of Indonesia and the British Dominions. Indonesia was to become a sovereign state, but her bond with the Netherlands would be indissoluble, a remarkable and unique innovation in international relations. In the Cheribon Agreement, moreover, no mention was made of the office of Governor-General, which would evidently be abolished; and High Commissioners would be exchanged by the Netherlands and the United States of Indonesia. This arrangement was in keeping with the basic principle of equality of partnership. The Cheribon Agreement was vague concerning details of the central government of the Netherlands-Indonesia Union; even the central cabinet envisaged in the early 1946 Dutch proposal was not mentioned.

After the signing of the Cheribon Agreement, movement toward implementation of its provisions was slow, and numerous instances of differences in interpretation occurred. Considering the lack of precision in its terms, this was not surprising. Basically, the questions of interpretation centered upon the degree of sovereignty the Republic was to exercise in the interim period before the United States of Indonesia and the Netherlands-Indonesia Union became finally established. The very nature of the interim government of Indonesia (the joint "organization" mentioned in the Agreement) was not precisely defined. Among other vexing questions were the right of the Republic to trade directly with other nations, the diplomatic and consular representation of the Republic in foreign countries, the rate and manner of reduction of the armed forces of both parties, and the interim status of the territories outside Java and Sumatra. The police organization of the Republican area was also a subject of dispute, the Netherlands favoring a joint Dutch-Indonesian gendarmerie, the Republic insisting that this would be an infringement upon its sovereignty.

For several months these questions were the subject of tense parleying. Meanwhile, the Republican government in the interior of Java, headed by three former Dutch political exiles—President Soekarno, Vice-President Hatta, and Premier Sjahrir—proceeded with its own organization. A national convention—representing various parties, regions, and minority, religious, and occupational groups—drew up a draft constitution, providing for a president, vice-president, premier, and cabinet; a Peoples Congress; and a Council of Representatives. The Peoples Congress, elected by universal suffrage, would meet at least every five years, and would have the power of electing the president and vice-president and of amending the constitution. The Council of Representatives, selected by the Congress from among its own members, would meet at least once a year to carry on legislation; and the premier and cabinet would be responsible to it. The national economy would be organized on a basis of partial state socialism, with emphasis on the establishment of cooperative enterprises. Branches of production vital to national welfare would be controlled by the state, as would natural resources, which would be exploited in the public interest. At the same time, however, foreign capital and enterprise would be welcomed and subjected to no discrimination.

Starting in June, 1947, the Dutch began to issue ultimatums to the Republican government. They demanded that native forces be withdrawn from the perimeter of Netherlands-held territory in Java and Sumatra, claiming that the Republican troops were violating the lines which had been agreed upon; yet there were probably as many Dutch attacks across the Indonesian lines as the reverse, and the Dutch showed no intention of removing their forces from Java and Sumatra, which they already recognized as under the *de facto* authority of the Republic. The Netherlands insisted that Republican territory be opened to trade, complaining that the Republic refused to allow rice shipments to the Dutch-occupied districts, where the shortage of food was starving the people; yet the Dutch blockaded all Republican ports, and forbade any trade between the Republic and other countries. The Netherlands demanded that a joint Dutch-Indonesian gendarmerie be set up to police Republican territory; yet the Dutch were unwilling to give the Republic time to consider this bold intrusion upon its sovereignty. Finally, in direct violation of the article in the Cheribon Agreement which stipulated that points of dispute which could not be settled by negotiation would be submitted to international arbitration, the Dutch suddenly and without warning, in July, 1947, struck with full military force against the Republic, just as the Germans had attacked Holland seven years before. The Netherlands pleaded due provocation, but the suspicion cannot be avoided that during all of the negotiations of 1946 and early 1947, plans were being laid for eventual resort to armed force, and that the conciliatory gestures of the Dutch were made in order to play for sufficient time to bring in troops and deploy them for the ultimate attack.[3]

Thus, while Burma has moved steadily toward national self-determination, with a British guarantee of either dominion status or independence; [4] while Siam has resumed its independent progress toward democratic government; and while Malaya, relatively undisturbed, has subsided into a close approximation of its pre-war condition; the two largest countries in the area, Indonesia and Indo-China, have been ravaged by bitter revolutionary wars against Dutch and French forces determined to reimpose their domination against the will of the people. These continuing conflicts have kept the whole area in a state of extreme disturbance, which is damaging to world economy and potentially threatens world peace; for, as the preliminary skirmishes of the second World War proved, war anywhere is a danger to peace everywhere. Aside from the importance to international economy of the products of Southeast Asia and Indonesia, the immense population of the area makes it impossible to regard these countries as insignificant territories of perpetual backwardness, especially since the population is increasing at an amazing rate. Death rates are high all over the area, but within the last century the cessation of native warfare and the introduction of modern medical facilities—both benefits brought in by the Western powers—have remarkably reduced the incidence of mortality, infant and adult. The extremely high birth rates have continued, however, and the net population gains have been enormous.

Indonesia's population, probably not much over 8,000,000 at the beginning of the nineteenth century, totaled in 1940 over 70,000,-000. The increase of 20,000,000 since 1920 represents an average annual growth of a million a year, or between 15 and 20 percent per decade, about twice the rate of increase in the United States. Java, the center of population, is the most densely inhabited region of its size in the world, with about 1,000 persons per square mile.

---

[3] A United Nations Security Council Committee of Good Offices succeeded in terminating the warfare by a truce agreement in January, 1948. Certain basic principles were accepted by both sides, and negotiations for peaceful settlement were again undertaken.

[4] Independence was chosen, and was secured in January, 1948.

By 1940 the Javanese had increased to nearly 50,000,000 from 35,-000,000 in 1920. Burma at the beginning of the present century had a population of approximately 11,000,000, but now it has approximately 17,000,000, which represents an increase of about 12 percent per decade. Malaya, with only 3,000,000 in 1910, almost doubled its population in thirty years, having a total of 5,500,-000 in 1940; but this increase was owing largely to immigration, most of it Chinese. Indo-China's population rose from 23,000,-000 in 1936 to almost 25,000,000 in 1941, representing an increase per decade of about 14 percent. Finally, Siam reported a total population of 14,500,000 in 1937, the date of the last census, but it had only 11,500,000 in 1929; these figures, if correct, would mean an increase of between 25 to 30 percent in eight years, which is almost incredible. The truth is probably that the last census was more complete than the previous one; but even if this is granted, the rate of increase of Siam's population is tremendous. In the whole area, the population increased by around 20,000,000 between 1930 and 1940, or between 15 and 20 percent in a decade, a rate exceeded in few other parts of the world. If it were to continue, by the year 2000 the combined population of Southeast Asia and Indonesia would be over 300,000,000.

But, in the current and future fate of the area, more is involved than natural resources and great masses of human beings. There is also the question of the relations of man to man, and nation to nation. Is one kind of social order, based upon the ideals of national liberty and democracy, to prevail in the Western world, while a different one, involving racial discrimination and economic and political subordination of whole populations to foreign states, continues in vast areas of the Orient and Africa? Can what is right in Europe and America be wrong in Asia and Africa? Is there a double standard of democracy? It happens that these are exactly the questions being raised by the leaders of colonial revolutionary movements in Southeast Asia and Indonesia. They are pointing out the embarrassing paradox of colonialism: that countries like Britain and France and the Netherlands, which base their entire national code upon the principles and ideals of democracy, have suppressed the rise of democracy in their own dependencies. In broader perspective, the colonial revolutions raise the question of whether it is possible to have a peaceful international order without international democracy.

It would seem that the things the colonial peoples are demanding —native self-government, freedom from outside domination in economic affairs and international relations, equality of representation in international organizations, military self-defense, and removal of arbitrary lines of social and educational discrimination based upon race, creed, and nationality—are indissolubly linked with the problem of international peace, for it is denial of these rights, or aggressive attempts by nations to deprive other nations of them, which cause wars. It was because Germany threatened other nations in these particulars that the last war was fought. There will never be peace in the dependent areas until these rights are won. There is proof of this in Southeast Asia and Indonesia now, and Africa and the other colonial regions will follow the same course sooner or later. The dependent areas are, then, zones of actual or potential revolution. As such, they constitute a threat to world peace, for wars tend to spread. Colonial revolts are different from civil wars, because, much as the statesmen of Britain, France, and the Netherlands may insist that warfare in their overseas dependencies is a matter of domestic concern only, actually such conflicts are international, involving two different and quite separate entities: the ruling nation and the subject nation. The Indonesians are not Netherlanders, the Burmese are not Britishers, the Indo-Chinese are not Frenchmen. Indeed, the whole institution of colonialism emphasized the differences, and was constructed upon the basic premise of the racial and national divergence between the ruling group and the subject group.

The very possession of subject territories affects the international policies of imperial states so that, while supporting democratic principles and ideals for themselves and other independent nations of the Western world, they tend to follow undemocratic lines in their policies toward the dependent peoples of the Eastern world. Thus there grows up among the imperial powers a colonial bloc in the councils of the nations, which stands together against the rest of the world. It would be difficult, perhaps impossible, to create an international order based upon democratic principles if the imperial states were to continue to form a bloc opposing the development of independence and democracy in the colonial half of the world. This is not to say that the independence of such countries of Indonesia and Indo-China might not create new problems of peace and international accord. But a retrogression toward the traditional system of half-a-world ruled by the other half would be sure to bring constant and increasing unrest and revolutionary wars in the dependent areas; and probably suspicion, mutual resentment, and a breakdown of democratic morale among the independent powers. It might even produce such exacerbation of racial feelings —for the colonial line happens to be a racial line—that eventually the world would be split apart by interracial warfare. Remarkably enough, however, the imperial states—including, despite their recent military activities in Indo-China and Indonesia, the French and the Dutch—accept as a worldwide ideal the principle of national self-determination and self-government. Not one of the major imperial powers now takes a stand against colonial nationalism as such. They justify their policies by insisting that these are planned to develop national maturity in their dependencies, however long this may take. The Dutch, for example, deny that they are fighting against nationalism or nationalist Indonesians in the Indies; they say that they are fighting "extremists." The French claim that they are not attempting to suppress the cause of nationalism in Indo-China; they are fighting "communists." And yet the vast majority of the natives in both countries are either enrolled in the ranks or supporting the struggle of the alleged extremists and communists.

The most disillusioning experience of the Indonesians and Indo-Chinese since the war has been their failure to gain the support of the United States in their bids for freedom. The Americans, or rather American statesmen, have either taken no stand on the colonial issue or have sided with the imperial powers. America, the great hope of world democracy, has virtually become a member of the colonial bloc. There are several reasons for this. One is the provincialism and ignorance of the American public and American statesmen regarding the Far East. The issues involved are not appreciated because the true significance of Oriental resurgence to our own future and that of the world is not understood. We have retreated therefore into a policy of hands-off, with a hope that all will turn out well. Related to this are the prevalent American preconceptions about race. White people and white nations are, Americans think, superior by nature, and should stay in control over the darker races. We have also been reluctant to take any action which might embarrass our recent allies, Britain, France, and the Netherlands. Moreover, on the economic side, American business concerns with interests in colonial countries generally distrust native independence movements; and their opinions carry great weight in the governing circles of the United States. They prefer to carry on in the old ways, fearing the possibly adverse effects of colonial nationalism on their properties and profits.

But above all, the policy of the United States toward the colonial revolutions has been determined by the morally devastating fear of Russia which infects the American government and military command—which, in foreign affairs, are becoming one and the same thing. We refuse to take any action which might weaken the Western European powers, because they would be our allies in a war against Russia. Since they include the three most important imperial states—Britain, France, and the Netherlands—we therefore support the preservation of the colonial system. Because the Navy, looking toward Russia, insisted upon unrestricted control of island bases in the Pacific, our statesmen renounced the principle of international trusteeship for all colonial areas. And there has crept into the parlance of American international statesmanship a pernicious doctrine of "power vacuums." With regard to colonies, the idea is that if imperial control is withdrawn this will create a power vacuum, into which some new system might creep. This could be communism, and therefore we must be careful to avoid the possible development of power vacuums in the dependent areas. The fact, however, is that, while the revolutionary regimes in Southeast Asia and Indonesia favor a form of state socialism no more extreme than that of Britain or of Sweden, they have shown no tendencies toward communism, nor have they made any overtures to Russia at all. There are communist parties in Burma, Malaya, and Indonesia, but in each case they are in a small minority. The Indo-Chinese Communist Party was dissolved in 1945.

As for the possibility that fascistic dictatorships might arise in these countries, there have been no signs of this whatever. The revolutionary governments have operated on the basis of proportional representation of all parties and ethnic groups; and the leaders—Aung San in Burma, Ho Chi-Minh in Indo-China, and Soekarno in Indonesia—have made no gestures toward dictatorship. Perhaps one reason for this is that the nationalist movements developed in protest against dictatorships of the colonial variety. But another reason is that in none of these countries was there a native upper class or even middle class of any importance. The governing officials of the nationalist regimes are common men, distinguished from the masses only by their better education; the revolutions are people's movements, led by men of the people. In fact, a peculiar service of the colonial administrations was to stifle the development of a native elite, who, with independence, might seize power and set up oligarchies. This is what happened in the Latin American revolutions of the nineteenth century, when mestizo cliques replaced the Spanish and Portuguese imperialists. It has also occurred in the Philippines, where the government, as before the war, is run by an inside circle of Filipino politicians and capitalists, largely mestizos. Following the same pattern, the French have tried to establish counter-revolutionary puppet governments in Indo-China under docile members of the gallicized elite, but these efforts have failed.

The real danger is not that the native peoples of Southeast Asia and Indonesia will turn of their own accord to either communism or fascism, but rather that the policies of the Western democracies will force them into an alignment opposed to the imperial powers and their supporters, particularly the United States. There never was much friendly feeling toward Britain, France, and the Netherlands in the area, but America has had a tremendous "reservoir of good-will" among the colonial peoples, largely because of its record in the Philippines. This was demonstrated rather pathetically when the Indonesians, expecting American troops to enter Java after the Japanese surrender, painted the walls of Batavia with phrases from the American Declaration of Independence and Constitution. But faith and hope in the United States are dwindling in Southeast Asia and Indonesia, while the Dutch and the French are taking the surest means to alienation of the colonial peoples from the imperial powers and perhaps from the Western democracies. A great opportunity to establish firm outposts of loyalty to democracy in Southeast Asia and Indonesia is being lost.

This discussion of the implications of the colonial revolutions in the area has concentrated upon political developments, but something should be said about the economic aspects. With regard to the national economies of the area, the question is whether, if natives took control of the government, they could also handle the economy. Concerning the basic subsistence economy there can be no doubt, for this has always been in native hands. In the profit economy, however, they have had little share except as laborers and petty bosses. Still, much of the export production of the area is carried on by uncomplicated agricultural methods, easily learned. This is true of rubber, quinine, tea, coffee, sugar, copra, fibers, and indeed just about all of the export crops. Rubber perhaps requires the most skill, not in the growing, but in processing before shipping; and yet the procedure is relatively simple. Two native foremen, one experienced in the growing and the other in the processing, could direct the operations of a rubber plantation; indeed, this was quite usual in Indonesia during the Japanese occupation. The operation of mines, oil installations, and factories on a large scale by natives would require far more training than they have had. Likewise, there have been few natives who could handle the intricacies of foreign trade and banking, or even any commercial enterprise of appreciable size. But all of the revolutionary governments insist that, despite their socialistic tendencies, they would welcome foreign capital and talent; and, given educational and training facilities, the natives could learn. Even if some of the foreign concerns abandoned their activities in the area and local governments had to take them over as state enterprises, foreign experts could be hired to run them until natives had time to learn, as happened in Russia after the 1919 revolution. Siam, also, has regularly employed foreigners as government advisors. Education is the solution to native deficiencies in the economic as in the political sphere, and this is the reason why the programs of all of the revolutionary governments give such heavy emphasis to its rapid and intensive development.

In the perspective of international economic interests, the covering question is whether the economic good of the dependent peoples and the rest of the world would be better served if the former colonies became self-governing than if they were to continue as subject countries. It is virtually certain that if the dependent peoples were to gain self-government they would quickly set about raising taxes, always remarkably low in colonies, thus reducing the profits on foreign investments. They would turn the funds so obtained into an expansion of public services, especially education. They would certainly try to raise wages too, which would cut further into profits. The indications are that they would also attempt to develop a more self-sufficient, balanced domestic economy, with local industries to supplement the raw-material production which has been almost the sole economic function of colonies in the past. All of this would amount to an effort to raise the standard of living of the colonial masses. One school of thought in international economics, with which the present writer agrees, claims that the net result would be a great expansion of markets for imports in the colonial areas, which would redound to the advantage of the Western countries as well as to that of the native peoples themselves. The phenomenally low purchasing power and depressed standard of living in the colonies have kept these potential markets of millions from anything except a small volume of consumption. In Southeast Asia and Indonesia imports have been only a small fraction of exports, and hardly a quarter of the imports have been for native consumption, because of mass poverty.

This point of view might be called the argument for the economic indivisibility of the world, and it claims that poverty and low standards anywhere reduce potential prosperity everywhere. It matches the other argument here advanced, that of the political indivisibility of the modern world, which claims that an absence of national freedom anywhere is a threat to peace and freedom everywhere.

SUGGESTED READINGS

NOTE: Items marked with an asterisk are standard comprehensive references for each of the five countries of Southeast Asia and Indonesia.

Bell, H. H. J. Foreign Colonial Administration in the Far East. London, 1928.
Brodrick, A. H. Little China: the Annamese Lands. London, New York, 1942.
Callis, H. G. Foreign Capital in Southeast Asia. New York, 1942.
*Christian, J. L. Modern Burma. Berkeley, Calif., 1942.
—— Burma and the Japanese Invader. Bombay, 1945.
Cole, F.-C. The Peoples of Malaysia. New York, 1945.
Collis, M. S. The Burmese Scene. London, 1943.

Crosby, J. Siam: the Crossroads. London, 1945.

Deignan, H. G. Burma. Washington, 1943.

—— Siam. Washington, 1943.

Eerde, J. C., van. De Volken van Nederlandsch-Indië. 2 vols. Amsterdam, 1921.

*Emerson, R. Malaysia. New York, 1937.

Emerson, R., L. A. Mills, and V. Thompson. Government and Nationalism in Southeast Asia. New York, 1942.

Ennis, T. E. French Policy and Developments in Indochina. Chicago, 1936.

Enriquez, C. M. D. Malaya. London, 1927.

Furnivall, J. S. Educational Progress in Southeast Asia. New York, 1943.

—— Netherlands India. Cambridge, England, 1939.

—— Progress and Welfare in Southeast Asia. New York, 1941.

Furnivall, J. S., and J. A. Andrus. An Introduction to the Political Economy of Burma. Rangoon, 1938.

Heine-Geldern, R. "Südostasien." In Vol. II of *Illustrierte Völkerkunde*, ed. G. Buschan, Stuttgart, 1923.

Janse, O. R. T. The Peoples of French Indochina. Washington, 1944.

Kaberry, P. British Colonial Policy in Southeast Asia and the Development of Self-Government in Malaya. New York, 1945.

*Kennedy, R. The Ageless Indies. New York, 1942.

—— Islands and Peoples of the Indies. Washington, 1943.

Klerck, E. S., de. History of the Netherlands East Indies. 2 vols. Rotterdam, 1938.

Landon, K. P. Siam in Transition. Shanghai and Chicago, 1939.

Lasker, B. Peoples of Southeast Asia. New York, 1944.

Levy, R., G. Lacam, and A. Roth. French Interests and Policies in the Far East. New York, 1941.

Pelzer, K. J. Pioneer Settlement in the Asiatic Tropics. New York, 1945.

Robequain, C. The Economic Development of French Indo-China. New York, 1944.

Scott, J. G. The Burman. London, 1910.

*Thompson, V. French Indo-China. New York, 1937.

—— Postmortem on Malaya. New York, 1943.

*—— Thailand: the New Siam. New York, 1941.

Vandenbosch, A. The Dutch East Indies. Berkeley, Calif., 1942.

Vlekke, B. H. M. Nusantara: a History of the East Indian Archipelago. Cambridge, Mass., 1943.

Wickizer, V. D., and M. K. Bennett. The Rice Economy of Monsoon Asia. Palo Alto, Calif., 1941.

Francis L. K. Hsu

CHINA [1]

Natural Resources

LAND AND CLIMATE.—The most important of China's natural resources is her agricultural land, since about 70 to 80 percent of her population are farmers. Buck [1] and his associates (*Land Utilization in China*, Chicago, 1937, pp. 161–74) divide China proper into eight agricultural areas totaling 1,320,000 square miles or about 41 percent of the land area of the country. These eight areas are divided into two groups. The three northern areas, covering roughly Kansu, Shensi, Shansi, Hopei, Shantung, Hunan, northern Kiangsu and Anhwei and much of Inner Mongolia, comprise the wheat region. The five southern areas, covering roughly Szechwan, Hupei, Anhwei, Kiangsu, and all provinces south of the Yangtze, comprise the rice region. Twenty-five to 27 percent of the gross area of the two regions is cultivated and about 90 percent of the cultivated land is devoted to crops, in contrast to the United States where 42 percent of the cultivated land is in crops and 47 percent in pasture.

Taking China as a whole, rice and wheat together occupy more than half the total crop area, being grown in almost equal proportions. Other crops include millet, maize, *kaoliang*, soy beans, barley, and cotton. Manchuria raises little wheat or rice but produces about 40 percent of China's *kaoliang*, 37 percent of her soy beans and 20 percent of her corn.

Only Manchuria and two of the four Inner Mongolian provinces produce a surplus of goods above the local requirements. The largest net imports for China as a whole have for many years been rice, wheat, and sugar. In 1932 these items in American dollars amounted to $34,800,000, $24,955,000, and $14,235,000, respectively. With

[1] The author is greatly indebted to Dr. M. J. Herskovits, who has read the galleys and given many valuable suggestions.

the return of Formosa to China, China's position with relation to sugar will be materially improved.

Tea was the eighth largest item among Chinese exports in 1936, but the importance of China as a producer of tea for the world market has declined steadily during the last fifty years, due to the development of tea plantations in India and Ceylon. Hunan is the largest tea-producing province in China.

Silk is another product for which China was once famous, but its production and export has also suffered severe setbacks due to foreign competition. In 1860, China supplied half of the world's export silk; by 1936 most of the market had been captured by Japan. The Yangtze delta produces the most silk, with Canton second.

Cotton is grown in the basins of the Yangtze and Yellow Rivers, the former accounting for about two thirds of the commercial crop. China is the third largest producer of cotton, ranking next to the United States and India.

The climate of China is one of extremes. The winter monsoons begin early in December, bringing a period of cold dry winds and radiant skies. The ports are ice bound and the earth is frozen to a considerable depth. In the summer, with low pressure areas spreading out from Indo-China, the southern monsoons bring a period of moist wind, heat, and heavy rains. Spring in China is a short season of blustering wind, rain and persistent fog along the coasts. Autumn is the most delightful season, clear and windless with gentle sunshine.

MINERAL RESOURCES.—Coal is China's most abundant mineral resource, almost every province having some deposits. The greatest concentrations occur in Shansi, Shensi, Kansu, and Honan, these four provinces together accounting for over 80 percent of the total reserves in China proper. Unfortunately, much of the coal is of poor quality and some of the deposits are too deep or in too thin veins to make mining profitable. The only fields likely to supply coal suitable for metallurgical purposes are those of Hopei, Liaoning, and Shansi in the north and Kwangsi in the south.

Actual coal production in the various provinces is not proportional to the reserves and has been much influenced by the use of modernized equipment. Thus in 1940 the Kailan Mining Company in Hopei, a Sino-British concern, produced over 60 percent of the

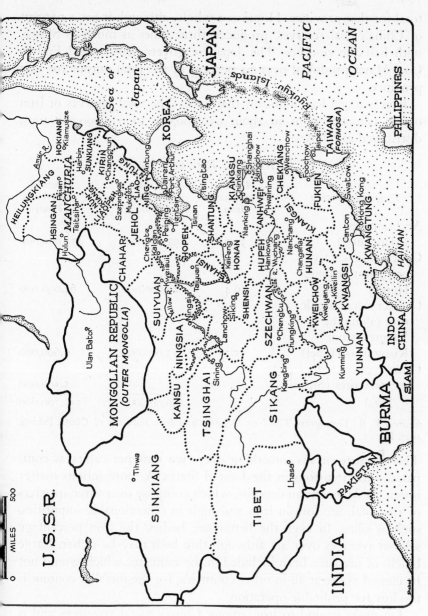

CHINA

total for that province. Hopei as a whole, although it has 1 percent of China's coal reserves, produced three times as much as Shansi, with half the reserves. The Japanese government-owned Fushan Coal Mines produced over 74 percent of the total output for that province.

Iron is much less abundant than coal. The total reserves of iron ore in China amount to about 950,000,000 to 970,000,000 tons, distributed as follows:

Iron-Ore Reserves

Type and Location	Average Percentage Iron	Actual Reserves (Tons)	Potential Reserves (Tons)
Archean ores (Manchuria)	34.9	295,000,000	477,000,000
Oolitic sedimentary ores (near Peiping, Hopei)	50.4	28,000,000	64,000,000
Contact metamorphic ores (along the Yangtze, from Kiangsu to Hupei)	55.3	73,000,000	9,600,000
Other types of sedimentary ores (Shansi)	..		5,100,000
Total		396,000,000	555,700,000

Source: F. R. Tegengren, *The Iron Deposits and Iron Industry of China*, Peking, 1920, II, 293.

This estimate means something over two tons per capita as compared with 37.9 tons in the United States. A more serious matter is that the Manchurian deposits, which comprise over three-quarters of the total, are low in iron and high in objectionable impurities, such as silica. In even the better ore bodies, the iron percentage seldom averages over 40, although thin beds may be richer. Large blocks of ore have been included in the estimates which would not be classed as ore at all in other countries, for the metallic content is too low for profitable operation.

Tin ranks second to iron among China's metal resources and is more important commercially. No estimates of tin reserves are available, but for a long time China has been the fourth largest world producer. Most of the tin comes from the Kochiu district in southeastern Yunnan. This province accounts for about 90 percent of the total output, but small amounts are produced in Kwangsi, Hunan, Kwangtung, Kiangsi, and Manchuria.

Copper has been mined in China since the seventh century B.C., but never in large quantities. The peak of production was in the time of Emperor Chien Lung (1775), when the output reached 7,000 tons a year, mostly from Yunnan. Since that time it has steadily diminished until, in the 1930s, it amounted to only 500 tons for the whole of China. Small reserves are to be found in many provinces.

Antimony is a unique feature of China's mineral wealth, 70–80 percent of the entire world output coming from there. About 90 percent of the deposits are in Hunan province, with estimated reserves of 1,500,000 to 2,000,000 tons.

Tungsten is an important metal in Chinese economy. China produces at least 40 percent of the total supply and dominates the world market. Reserves, estimated at 949,000 metric tons are found in three provinces: Kiangsi, Hunan, and Kwangtung.

Manganese has been mined in five provinces: Kiangsi, Hunan, Kwangtung, Kwangsi, and Liaoning. The total reserve is estimated at 22,000,000 tons of ore with an average metallic content of forty-five percent.

Lead and zinc are found in many parts of China, but few known deposits are large enough to justify modern operations. The principal producing provinces are Hunan (by far the most important), Szechwan, Yunnan, and Kweichow.

Silver has been used in China for many centuries but her reserves of this metal are insignificant. Gold deposits are found mainly in Northern Manchuria, near the Siberian border, Outer Mongolia, and the Tibetan borderland. The two former areas produce the bulk of China's gold, about 180,000 to 190,000 Chinese ounces per annum.

Salt is China's most important nonmetallic mineral. She produces between 3,500,000 and 4,000,000 metric tons yearly, 80 percent of this by the evaporation of sea water. This industry is carried on mainly in Liaoning, Hopei, and northern Kiangsu provinces.

Cement, made from limestone and clay, has become economically important in recent years. China proper produced 15,247,000 barrels in 1940; Manchuria, 19,000,000 barrels in 1937 (last date for which figures are available). Ninety-five percent of all production was in less than a dozen modern cement works.

China clay is important as the raw material for porcelain. It occurs mainly in Anhwei and Fukien provinces, but is also found in several other areas, including Manchuria. Between 1932 and 1934 China produced 860,000 to 965,000 metric tons, or about 25 percent of the world's total.

Mercury, sulphur, gypsum, and alum are produced in small quantities and a 340,000 ton deposit of nickel ore is reported from the eastern border of Tibet.[2]

Bauxite, chief commercial source of aluminum, has recently been discovered in China. A recent estimate of reserves give 271,000,000 tons for Shantung and 461,000,000 for Manchuria and Kansu.[3] These estimates may be too optimistic.

Magnesite may assume commercial importance in the future. Deposits of over 1,000,000 tons are reported from Manchuria.

Oil is not an important resource, China's known oil reserves being only about 206,000,000 metric tons of petroleum and 314,750,000 metric tons of shale oil.[4] The total, including the inferior shale oil reserve, is equivalent to 3,645,000,000 barrels.[5] At the present per capita rate of consumption in the United States, this would meet the needs of China's population for less than two years. Most of the existing oil fields are difficult of access and the amount of oil produced has been very small. In 1936 it was only 677,481 barrels, of which China proper produced 2,613 and Manchuria 674,868.[6]

Water power is confined to native water mills and is thus negligible. However, there is considerable potential. The latest estimate puts the total for the whole of China, excluding Tibet, Sinkiang, and Outer Mongolia, at 64,696,000 kilowatts a year at the minimum flow.[7] About 60 percent of this power potential comes from the Yangtze Valley and about 20 percent from the Southwest International waterways. Admittedly there are serious difficulties in the way of utilizing this potential, but the Yangtze gorges alone would

[2] China Handbook, 1937–43, p. 491.   [3] Ibid.
[4] U.S. Bureau of Mines, "Mineral Resources, Production, and Trade of China," Foreign Minerals Quarterly, October, 1941, p. 112.
[5] Chuh-Yueh Li, Mining and National Defense (Shanghai, 1944), p. 45.
[6] Chinese Yearbook, 1944–45, p. 655.   [7] Ibid., p. 645.

have a generating capacity of 10,600,000 kilowatts a year, twice as much as the present total of TVA, Grand Coulee, Boulder and Bonneville.

Population

No exact figure on China's total population is available. Ta Chen regards the gross figure of 400,000,000 as being substantially correct.[8] Buck thinks it is somewhere between 400,000,000 and 600,000,000.[9] On the other hand, a great deal about the essential features of the population has been uncovered through a number of surveys and regional studies in recent years, although the resultant findings do not entirely agree.

BIRTH AND DEATH RATES.—Ta Chen's estimate of the country's crude birth rate is 38.0 per 1,000 population. Compared with the United States (18.9, 1940), Italy (23.5, 1939) the Philippine Islands (32.7, 1937), and even the British Provinces of India (34.1, 1938), China's birth rate is high, but compared with such countries as Egypt (43.4, 1938), Mexico (43.2, 1940), and the U.S.S.R. (44.1, 1921–25), its figures are low.

Ta Chen's estimate of China's crude death rate is 33.0 per 1,000 population. This differs considerably from the figure obtained by Buck, which is 27.1. According to Buck's estimate the natural increase is 11.2 per 1,000 and it will take less than sixty-five years for the population to double itself; according to Ta Chen the natural increase is 5.0 per 1,000, and it will take 139 years for the population to double itself.

By either estimate, the rate of natural increase is not among the highest of the world. It is higher than France (–0.8, 1938), England and Wales (0.3, 1940), Belgium (1.9, 1939), Sweden (3.6, 1940) and a few other European countries; but much lower than Mexico (22.0, 1940), U.S.S.R. (20.0, 1921–25). Ta Chen's figure puts it below Japan (9.3, 1938) and the United States (8.1, 1940), and about the same as Norway (5.6, 1940), Hungary (5.2, 1940), and Latvia (4.6, 1939). This remarkably low rate of natural population increase is certainly not consistent with popular conception, but

---

[8] See the discussion by Ta Chen, *Population in Modern China* (Chicago, 1946), pp. 4–5.

[9] J. L. Buck, *Land Utilization in China*, p. 363.

it is not a matter for optimism. In most countries where the rate of natural increase is low, the result has been achieved by low birth and death rates; China on the other hand has achieved it by high birth and death rates.

Infant mortality is high in China. According to Ta Chen's estimate it is 275 per 1,000 live births and is the highest in the literate world. According to Buck's survey the figure is much lower (156), but still puts China the sixth highest and way above the United States and Scandinavia. Buck's survey reveals a slight difference in infant mortality rate between North China where it is 155, and South China where it is 157. But in Ta Chen's tabulation of returns from different parts of China it appears to vary widely, between 122.6 (Nanking) and 555 (Canton). Chen is of the opinion that the variation is caused by inaccuracy in the sources. However, some such real differences may actually exist, although the exact extent is hard to determine. For example, infanticide is generally assumed to prevail in all China, but this is highly doubtful. The practice is far from being universal. Nor is it resorted to the same degree in different localities or within the same locality at different times. It is also generally known that during a famine infanticide tends to increase.

LIFE EXPECTANCY.—The life expectancy of Chinese is among the lowest of the literate world. According to Buck's study [10] it is as follows:

Expectation of Life in Years ($^{0}e_x$) at Selected Ages

SEX	AGE					
	0	*5*	*10*	*20*	*50*	*60*
Male	34.85	47.58	47.05	40.74	26.84	14.19
Female	34.63	46.95	46.00	40.08	28.05	15.22

With good reason Buck regards this as an overoptimistic picture because,

It must be re-emphasized that the experience on which the life tables were constructed is simply that of a population observed during the enumeration years. The life tables organize this experience as if it were the experience of a generation passing through life. Actually the two are the same only if the risk of death remains unchanged for nearly 100 years. In

[10] *Land Utilization in China*, p. 389.

China any generation must be exposed in some degree to the risks of famine, war, flood and epidemic.[11]

For this reason the data obtained under Ta Chen are particularly valuable. Chen has separated his data for Cheng Kung, Yunnan, into two groups: those which exclude deaths due to cholera epidemic in one of the enumeration years and those which include them. The two life-tables are as follows:

Expectation of Life in Years (^0e_x) at Selected Ages for Cheng Kung, Yunnan (Excluding Deaths Due to Cholera Epidemic of 1942)

SEX	AGE					
	0	*5*	*10*	*20*	*40*	*60*
Male	33.8	47.3	46.7	38.2	25.5	12.4
Female	38.0	53.5	51.8	43.8	28.6	13.4

Source: Abridged from Ta Chen, *Population in Modern China*, pp. 105–7.

Expectation of Life in Years (^0e_x) at Selected Ages for Cheng Kung, Yunnan (Including Deaths Due to Cholera Epidemic of 1942)

SEX	AGE					
	0	*5*	*10*	*20*	*40*	*60*
Male	31.9	44.4	42.9	35.6	23.5	11.5
Female	34.2	48.6	47.1	39.3	25.6	12.1

It will be clear that, excluding the deaths due to the cholera epidemic, the average expectation of life as a whole becomes higher than Buck's figures, but including those deaths, it becomes lower.

The Chinese picture of life expectation is interesting when compared with certain other countries. At birth the expectation of life in China is only higher than that in India (26.91 *m.* and 26.56 *f.*) but lower than other literate countries including Japan (42.06 *m.* and 43.20 *f.*) and the United States (59.31 *m.* and 62.83 *f.*). As the individual grows, the life expectation in China increases, until at ten years it is only about ten years shorter than in Britain and the United States. At twenty the difference is less than ten years.

HEALTH.—There are two things which bear most on the matter of health: nutrition and disease. The figure generally used in the West for the standard minimum intake of energy value for an adult male is 3,000 calories. After considering such factors as smaller size

[11] *Ibid.*, p. 390.

and lower metabolism of the Chinese, Buck regards the figure 2,800 calories as a more reasonable substitute.

After computing the individuals as adult-male units according to Atwater's factors, Buck has found about 29 percent of the localities studied to be below the minimum standard.

Daily Intake of Calories per Adult-Male Unit

Regions	No. of Localities	No. of Adult-Male Units	Intake of Calories	No. of Localities with Intake Below Standard
China	136	13,341	3,295	39
Wheat region	67	6,996	3,186	27
Rice region	69	6,375	3,400	12

Source: J. L. Buck, *Land Utilization in China*, p. 407.

The Rice Region is much better off than the Wheat Region because a smaller percentage of its localities are below the minimum standard. Within the Rice Region the Rice-tea, the southwestern Rice, and the Yangtze Rice-wheat areas enjoy the highest intakes of calories. On the other hand, "in the Winter Wheat-kaoliang Area, where the average intake is 20 percent above the modest standard of 2,800 calories, consumption in eleven of the thirty-three localities where studies were made falls below it; and the average calorie intake in one of these is only one-half of the minimum." The extreme variation often reflects the "actual state of relative prosperity" because, as a whole, the Chinese peasant is so close to the poverty line.

In comparison with the diet of 224 urban and rural families in the United States, Buck has found the Chinese diet extraordinarily high in seed products (91.8 percent: 38.2 percent) and very low in animal products (2.3 percent: 39.2 percent). It is also very low in milk, eggs, green leafy vegetables and fruits, which are classed as protective foods.

Other features of Chinese diet are a fat supply of 9 percent of the total calories consumed, which is below the recommended minimum; seventeen out of 136 localities are deficient in protein; and there is, in general, a markedly low calcium intake, averaging little more than one-half the standard of 0.8 gram. But the intake of phosphorous and iron is generally above the standard.

There is no doubt that many diseases are prevalent in China but the exact extent of their incidence is hard to determine. Buck has shown that the five most important were smallpox, dysentery, typhoid, tuberculosis, and cholera, in the order named. An epidemic is liable to take a large toll of the population. In Cheng Kung, Yunnan, a community of about 70,000, Ta Chen counted 1,002 dead in a cholera outbreak in 1942. The percentage of the toll in West Town, Yunnan, was even higher, when the same epidemic took over 150 lives out of about 8,000.

Figures of the National Health Administration on the 1936 incidence of certain diseases in three large cities may be profitably reproduced here:

Disease	Nanking (pop. 1,000,000)	Shanghai (pop. 3,400,000)	Peiping (pop. 1,500,000)
Dysentery	387	1,095	1,119
Typhoid	347	1,115	152
Diphtheria	308	946	292
Smallpox	287	362	247
Cerebrospinal meningitis	145	188	44
Scarlet fever	..	593	609

Source: Chinese Yearbook, 1937, p. 1142.

It is not clear whether the figures represented the numbers of cases treated in institutions or whether they were the total incidence of such diseases in those cities. Whatever the case may be, one has to observe that a much smaller incidence of such diseases in the modern West would have made the headlines in all newspapers but that they practically passed unnoticed in national China.

There is even less information concerning the incidence of less fatal or slowly fatal diseases. For example, the Kala-Azar Research Station at Tsingkiangpu, northern Kiangsu, recorded in 1936 an attendance of 28,938; but it is estimated that more than 200,000 people in that one endemic area were suffering from the disease. Malaria sometimes breaks out in the form of an epidemic. In the fall of 1936 the Central Field Health Station examined 1,008 children under twelve years of age in nine hsien (districts) in Anhwei and Kiangsu for spleen and parasite indices and found 78 percent of them suffering from subtertian infection. Tracoma is observed everywhere.

CURRENT POPULATION TRENDS.—As observed earlier, there is a

considerable difference between two leading writers as regards the
natural rate of increase of China's population, but neither estimate
puts China among the highest group. After a survey in 1929–1933
of 137 districts in 21 provinces, Buck found that, for all areas as a
whole, 58 percent showed an increase, 35 percent, a decrease, and
7 percent, no change. The increase is greater in the Rice Region,
being 60 percent of the localities surveyed as compared with 53
percent in the Wheat Region. Thirty-one percent of the localities
in the Rice Region showed a decrease and nine percent showed no
change while 42 percent of the localities in the Wheat Region
showed a decrease and five percent showed no change.

The most important evidence given for the increase was "increase of buildings," and the most important evidence given for
the decrease was bandit trouble, disease, famine, or drought. Immigration has been given as evidence for increase in 18 percent of the
localities but actually other data show that it often resulted in net
loss of population for the rural areas. In a sample study of 206,274
residents and nonresidents Buck found that, if those migrants who
made round trips during the survey year were disregarded, there
was a net loss of 48 males per 10,000 total male population. In this
the North showed a net loss of 133 per 10,000 whereas the South
indicated a net gain of 32.

In 1936 a survey was conducted by the National Agricultural
Research Bureau on the outward movements of entire families in
1,001 hsien of 22 provinces. It was found that for a three-year period
5 percent of the families emigrated. Among emigrant families those
with three acres or less (that is to say, below the median size of 3.3
acres in the survey) represented 86 percent.

The recent war with Japan saw much heavier movements of
the population. Between July, 1937 and September, 1944, after the
fall of Changsha, it was estimated that 3,400,000 persons emigrated
from 24 leading cities and 10,750,000 persons from seventeen occupied provinces (including Manchuria). The former represented
about 25 percent of the total population in those cities and the latter
represented about 5 percent of the total working population of the
occupied provinces excluding the cities referred to above.[12]

Whether the wartime migration will stimulate the permanent

[12] Ta Chen, op. cit., pp. 60–62.

rate of migration is not possible to tell. Nor do we have any concise information as to whether it has meant an increase or decrease of the birth rate. No exact figures on war casualties are available. From the start of the war to June, 1944, roughly 9.7 percent of the total males of the district of Cheng Kung, Yunnan, was drafted. If we use this percentage as a rough measure and apply it to the entire male population of the country we shall have something like 20,000,000 draftees. Supposing half of them to be casualties, we shall see that they can be made up by five to ten years of peace.

Economics

The vast majority of Chinese are still rural farm people. Buck's study of 168 *hsien*, 173 localities in 19 provinces in 1929–1933 has shown that, for all regions, 79 percent of the population are in farm villages and hamlets. An additional 11 percent of the population investigated are inhabitants of market towns, some of whom are also tillers of the soil. Manufacturing holds a relatively unimportant place in China's economy. The total labor force engaged in modern manufacturing was, by 1933, about 1,200,000. Even granting further increases by 1937 it could not have been more than 2,000,000.[13] This is about 0.5 of the total population, compared with 10,000,000 wage earners engaged in manufacturing or 7.6 percent of the population in the United States in 1939.

LAND UTILIZATION.—The most important factor in agriculture is land. As pointed out before, the proportion of cultivated land to the total area of China excluding Outer Mongolia and Farther Tibet is about 11 percent but in agricultural China the figure is raised to about 25 to 27 percent. This compares favorably with Russia (12 percent, 1928), Japan proper (17.2 percent), Great Britain (22.5 percent) and the U.S. (22.6 percent), but unfavorably with Germany (43.8 percent), Italy (44.6 percent), and British India (46.3 percent). (The data for this section is based largely on Buck's *Land Utilization in China*.)

Within this agricultural China the agricultural resources are intensively utilized relative to the population. The average density of

---

[13] See Ta Chen, "Labour," *China Yearbook*, 1933, p. 6; and see *Nankai Weekly Statistical Service*, Vol. IV (1931), Nos. 7, 9.

population per square mile of gross area of the country is 504, but that per square mile of crop area is 1,485. The density is much higher in the Rice Region (1,746) than the Wheat Region (1,128), and varies between the low (858) of the unfavorable Spring Wheat Area to the high (2,636) of the very favorable Southwestern Rice Area.

As a result of this high density of population the average size of the farm is very small. The usual size of a farm is 4.0 acres; the mean crop area, 3.76 acres. In the Wheat Region the usual size is about 5.5 acres; in the Rice Region, about 3.0 acres. The extremes are 0.6 acres to 18.6 in the south and 1.2 acres to 78.0 acres in the north. For the entire country the extreme averages are 0.8 to 42.7 acres.

The mean size of China's crop area compares favorably with only that of Japan (2.67 acres); it falls far below that of Germany (21.59 acres, 1933), Denmark (39.74 acres, 1919), England and Wales (63.18, 1924), and the United States (156.85 acres, 1930).

The picture of agricultural production presents some interesting features. After converting all products into grain-equivalent and after computing twelve months of man labor each year on a farm into man-equivalent, it is found that the average production of grain-equivalent per man-equivalent for all localities is 1,400 kilograms. The figure is lower in the Wheat Region and higher in the Rice Region. In the Spring Wheat area the figure is only 787 kilograms while in the Southwestern Rice area it is 1,830 kilograms.

Considering the fact that farms are smaller in the south than in the north, these production figures are a contradiction, because the production of grain-equivalent per man-equivalent is lowest on the small farms (833 kilograms) and highest on the very large farms (2,073 kilograms). But this is not hard to explaian. We have merely to realize that land productivity is higher in the Rice Region than in the Wheat Region. For when we compare the production of grain-equivalent per capita in the different regions according to the size of farms we find that within the *same* size group the figures are progressively higher as we move from the Spring Wheat Area through Winter Wheat-millet, Winter Wheat-kaoliang, Yangtze Rice-wheat, Rice-tea, Szechwan Rice, Double-Cropping Rice to Southwestern Rice areas. Farms of the most economic-size are, then, the very large group in which the farm business is on the scale of 13.02 acres per farm. Only a little over 7 percent of the 16,786 farms studied were found to be in this group.

Compared with an estimate for the United States of 20,000 kilograms of grain-equivalent per man-equivalent, the Chinese figure of 1,400 kilograms is pathetic. Yet a fact not commonly appreciated is that, in terms of *per acre* crop yields, China has a place higher than the United States in both rice and wheat. Compared with the United States, China produces an average of 67 bushels of rice per acre as against 47 bushels, and an average of 16 bushels of wheat per acre, as against 14 bushels. In both yields China is below Italy and Japan; in wheat, both Japan and Italy are below Great Britain and Germany. The United States has a higher yield per acre than China in barley, corn, Irish potatoes, and cotton, although in the last-named crop China's per-acre yield compares favorably with that of the United States (168 bushels as compared with 177 bushels).

It is thus obvious that, while there is some room for improvement, China's agriculture has reached a very intensive state. This has been achieved at the expense of a tremendous amount of cheap labor which works the land not only by double-cropping but also by the modification of physical conditions through private and government irrigation, drainage, terracing, and to a smaller extent, through fertilization.

The pressure of population is equally obvious in the picture of land ownership. In this respect, however, there is some popular misunderstanding. Over 93 percent of China's farm land is privately owned. According to current belief, China is a country where most land is owned by a small number of landlords who do not till their soil but suck the blood of their tenants. But Buck's survey [14] has shown that, in the Wheat Region, which includes most of north China, only 12.7 percent of the farms are rented. In the Rice Region, which includes most of south and central China, the corresponding figure is 40.3 percent. The tenancy figure for all China

[14] Buck, *Land Utilization in China*, p. 192. In certain local areas of Kwangtung clans own up to 40 percent of all farm land in the area (see H. S. Chen, *Landlord and Peasant in China*, New York, 1936). H. T. Fei is of the opinion that Yunnan province is also characterized by clan ownership of land, but the proportions shown by his examples of clan owned land never came to as high as those in Kwangtung (see his *Earthbound China*, Chicago, 1945).

is 28.7 percent. In some local areas the percentage of farms rented is sometimes as high as 99 percent and in others as low as zero. Buck estimates that over one-half of the farmers are owners, less than one-third part-owners, and 17 percent are tenants.

It does not follow, of course, that we should dismiss the problem by saying that the extent of farm tenancy in China is no greater than in many other countries. For when we look at the comparative picture the size of farms by ownership changes drastically. For all China, the size of farms of owners is 4.22 acres, of part-owners, 4.25 acres and of tenants, 3.56 acres. The difference in size of farms privately owned and those rented is less than one-fifth. This difference is smaller in the Wheat Region than in the Rice Region. In the Winter Wheat-kaoliang area the owners and part-owners even have smaller sized farms than tenants, the three figures being 5.34, 5.73 and 6.40 acres.

The landlord may be a working farmer who resides in the same village as his tenant, he may be a nonfarming resident in the same village as his tenant, or he may be a resident in some town or city. In the last case he is known as an absentee landlord and his only relation with his tenant is the collection of rent. He or his family members may collect the rent, or he may, if he is a big owner, appoint an agent to do so.

In proportion to the tenants' returns the amount of rent is high. There are three generally recognized types of rent systems: (a) the landlord and the tenant share the risk and the former receives a fixed proportion of the crop; (b) the landlord receives a payment in kind stipulated in advance; and (c) the landlord receives a payment in cash stipulated in advance. Buck has found in his studies that roughly one-half of all lease contracts run on the cash payment system, one-quarter on crop payment system and somewhat less than one-quarter on a share-the-risk system. Two percent of the tenants are croppers.

Generally the rent is between 40 to 50 percent of what the tenant produces in the case of share and crop systems and between 6 and 11 percent of the price of the land in the case of the cash system. Wide variation exists. Thus in Kwangtung province, on very poor land the landlord receives about 20 percent of the produce but on better lands his share may amount to over 70 percent. In one local area the rent to the landlord amounts to 50 percent of the income from crop but an extralegal and yet customary extortion from a bandit organization called "black ticket fee" took up another 30 percent of the income, thus reducing the tenant's net return to about one-fifth of what he produced. In some villages of Yunnan the rent apparently varies around the figure of 50 percent of the produce.

The relation between tenants and landlords is uneven throughout the country. In certain areas, especially southern Kiangsu, Fukien, and Kwangtung provinces a system prevails whereby the surface (namely, the right to cultivation) and subsoil (namely, the title to the land) are dissociated. Where this system is found, the tenant, who is the surface holder, has permanent claim over the land he cultivates and cannot be ousted by the landlord. In Kwangtung province there are also hereditary tenants known as Sia-Wu (servile family) who pay rent and perform additional free services to the landlord in exchange for the security of the land. Where there is collectively owned land, the tenant may be in a very favorable position. In one Yunnan village, Fei has found that those who rent land owned by their own clan are particularly fortunate. "The poorer households have a traditional right of occupancy and, though they are theoretically bound to pay fees fixed for that privilege, the treasurer will have difficulty in dispossessing them for delinquency in payment or for any other reason." [15] Fei seems to think that it is virtually impossible to dispossess members of the owning group, but this is not true elsewhere. In Kwangtung clan-owned lands are usually in the hands of an "oligarchy" composed of a few powerful families, who were as hard on the member tenants as on anybody else and shared the spoils among themselves.[16] When agricultural conditions are unfavorable and competition for land is not keen, or in places where there are only a few absentee landlords controlling large areas, the relationship between landlord and tenants tends to be amiable. The reverse is true in areas where land is exceptionally fertile and there is much absentee control.

In spite of the heavy rent, some renters of land are no worse off than some owners. Fei (p. 77) says that, in some Yunnan villages,

[15] Fei, *Earthbound China*, p. 78.
[16] H. S. Chen, *op. cit.*, pp. 37-41.

"the holders of relatively large properties, as well as the landless and small owners, tend to expand the amount of land under their management through renting rather than through purchasing land . . . those who rent land are not necessarily landless or even poor people; the rich rent land too. This is because tenants can enjoy a profit even if they operate their rented land by hired labor." This is contrary to the generally accepted impression that tenants are worse off than part-owners; the latter, worse off than owners.

HANDICRAFT INDUSTRIES.—These are present, in one form or another, in practically every village, market town, district city, or metropolitan center of the country. These industries are carried on by farmers and their families as a side line during the slack season, and by itinerant craftsmen who may stay in a village for a couple of weeks or for several months. The work is done in the shops which line the streets of many towns and cities and which also serve as the dwelling places of the workers. It is also carried on under an arrangement known as the merchant-employer system. Under this system the merchant supplies the raw materials and, if necessary, the tools, but the people work in their own homes at a piece wage. Often the home workers finish only a part of the product, which is completed at the establishment of the merchant-entrepreneur. Nankai Institute of Economics studied this system in Kaoyang, Hopei. I have seen it operating in many parts of China, including South Manchuria and Western Yunnan.

The extent of handicraft manufacturing is problematic since few figures have ever been collected. On the basis of the data collected by Buck and Chen, I have estimated the total number of home-industry workers in agricultural China to be 22,660,000, and in urban China, 2,000,000, including 1,000,000 apprentices. If we add to this about 1,200,000 laborers in native mines, we then come to a grand total of about 25,000,000, or roughly 5 percent of the total population.[17]

Before contact with the West, handicrafts were the only industries in the country. They are still important. A strong evidence for

[17] Enormous local variations exist. In their study of Kaoyang, Hopei, H. D. Fong and his associates estimated that there were about 110,400 hand loom workers among a population of 154,130 which brought the percentage of handicraft or non-farm subsidiary workers to 71. (H. D. Fong, *Rural Industries in China*, Tientsin, 1933, pp. 49, 58). Kaoyang probably topped the entire country in this respect.

this is the place occupied by them in the export trade. In 1931 sixteen groups of products of principal rural industries made up 12.6 percent of the country's total export.[18] In 1932 both the total export and the export of rural handicraft products declined, but the percentage occupied by the latter was increased to 16.6 percent.

The list referred to above does not include cotton goods and mining. In 1930 handlooms accounted for 78.5 percent of cotton yarn consumed in China.[19] Native collieries were estimated to produce about 25 percent of China's annual coal output before the war,[20] and native mines of all kinds were said to produce 30 to 40 percent of the total mineral products in China.

MODERN INDUSTRIES.—Modern manufacturing has been present in the country for only the last hundred years. Unlike Western Europe, which has produced machinery for its growing industries, China has to import her machinery from abroad. For this reason the net import has been used as an index of the extent of the country's industrialization. Between 1887 and 1890, the imports of machinery amounted to 382,000 Hk. taels. Between 1906 and 1910 the figure rose to 6,406,000 and between 1931 and 1934 it rose to 35,376,000.[21] An analysis of the principal categories gives a clearer picture: in 1918 these were prime movers valued at 646,000 Hk. taels and textile machinery, at 1,650,000 Hk. taels. Both imports rose tremendously in the years which followed and by 1934 the two figures were, respectively, 5,274,000, or about nine times the 1918 figure, and 9,118,000 or about six times the 1918 figure. In both cases there were heights and slumps during the intervening years but the main trend is clear. The importation of electrical machinery has shown a continuous upward trend, rising, between 1924

[18] The figures on rural products were taken from *ibid.*, p. 19.

[19] H. D. Fong, *Rural Industries in China*, p. 20.

[20] In 1934, 24 principal mines in China Proper and 11 principal mines in Manchuria and Jehol produced 26,862,701 short tons of coal. Subtracting this amount from the total coal output for the year—namely 36,079,147 short tons (*League of Nations Statistical Yearbook, 1941*, p. 132), we get 9,216,446 short tons; that is, about 25.5 percent of the total output of the year. This percentage agrees with that given by D. N. Rowe, "20 to 25 percent" (*China among the Powers*, p. 58) and also in a broad way with an estimate given in *Chinese Economic Journal and Bulletin*, Vol. X, No. 4 (April, 1937). In the last-named source about six to seven million tons of coal were attributed to native collieries.

[21] H. D. Fong, *Industrial Organization in China*, p. 27.

and 1934, by four and a half times. The increase in the import of machine tools was also more or less continuous.

In terms of an average for the period of 1928–34, textile machinery accounted for 46.8 percent of the total import of industrial machinery; prime movers, 20.9 percent; electrical machinery, 14.9 percent; and machine tools, 4.5 percent. Textile machinery and power machinery (prime movers and electrical machinery) made up 83.0 percent of the total import of industrial machinery.

This shows the increased use of power in Chinese industries and the great importance of the textiles in China's industrialization. This fact is borne out by a classification of workers. The 1930 government survey covered a total of 1,204,317 industrial workers in 29 cities of 9 provinces. Of these, textile workers composed 47 percent. Foods came second with 14.7 percent. Shanghai, which before the last war had about 40 percent of the country's industries, presented the same picture. In 1931 the textile industry occupied 60.1 percent of all workers in Chinese factories and the food industry occupied 10.9 percent of the workers. In 1933 the two figures, respectively, were 56.0 and 12.7.[22]

Some interesting facts were compiled by R. H. Tawney in the early thirties on the growth of China's industrialization. The accompanying tabulation is abridged from one of his tables:

Indices of Industrial Development in China, by Certain Years

Factor	1896	1913	1929	1930
Cotton mills	12	28	..	127
Cotton spindles (thousands)	417	1,210	..	4,223
Cotton looms (hundreds)	21	..	..	293
Flour mills	..	57	193	..
Factories in Shanghai (over 30 workers)	..	70	648	837
Factories in China (over 30 workers, not including Tientsin)	..	245	1,747	1,975
Iron ore production (thousand tons approx.)	..	959	2,003 [a]	..
Pig iron production (thousand tons approx.)	..	150	433 [a]	..

[22] D. K. Lieu, *The Growth and Industrialization of Shanghai* (Shanghai, 1936), pp. 347–48.

Factor	1896	1913	1929	1930
Coal production (million tons)	..	14.0	25.1	..
Motor vehicles in China (thousands, approx.)	..	..	30	35
Imports of gasoline (million gallons)	..	0.5	35.0	..
Miles of motor roads (thousands, approx.)	250	5,400	9,500	..
Exports (quantity index)	..	100	165.5 [a]	..
Imports (quantity index)	..	100	188.3 [a]	..
Steamers entered and cleared (tons)	..	89,614	150,203	..
Imports of raw cotton (thousand piculs)	..	134	2,515	..
Exports of bean cake (million piculs)	..	11.8	18.7	..
Exports of bean oil (million piculs)	..	492	1,115	..
Imports of machinery (million Hk. taels)	..	4.6	29.9	..

Source: R. H. Tawney, op. cit., p. 196.
[a] 1928 figures.

Another study was made by Nankai Institute of Economics, Tientsin, on the geographic distribution of industrialization. By about 1930 industrialization was concentrated in the six provinces of Kiangsu, Liaoning (South Manchuria), Hopei, Kwangtung, Shantung, and Hupeh, which have 10 percent of the total area and 36.3 percent of the total population. These six provinces had 55 percent of the country's mining (value of output, 1927), 65 percent of coal (quantity of output, 1928), 64 percent of iron (quantity of ore output, 1928), 93 percent of cotton manufacture (spindles, 1930), 92.6 percent of silk (export value of filature, 1929), 86 percent of beans (export value of oil and cake, 1929), and 87.6 percent of electricity (power capacity, 1929). These six provinces also accounted for 92.5 percent of China's foreign trade and 84.0 percent of all trade (1929). Lastly, they also had 54.4 percent of China's railways, (length, 1924), 42 percent of the motor roads (length, 1930) and 42 percent of the telegraphs (length, 1928).[23]

There is no doubt that modern industries have been gaining im-

[23] H. D. Fong, China's Industrialization (1931), p. 30.

portance. In addition there is also evidence for the increasing size of factories. In Shanghai alone, between 1931 and 1933 the number of factories which employed more than 30 workers and used motive power was increased from 710 to 1,186.[24]

However, when we look at the size of the country and its potentialities and compare it with some nations of the West, we have to admit that modern industries are, except in certain fields, comparatively insignificant. First, with regard to the entire population, the proportion of workers engaged in modern industries is small. A 1930 government survey numbered them at 1,204,317, of which 92.6 percent were in four provinces: Kiangsu, Kwangtung, Hupeh, and Shantung. Since these four provinces had, according to Nankai Institute's compilation, 82.9 percent of the country's cotton manufacture, 55.2 percent of its electricity, 55.8 percent of all trade and 62.0 percent of foreign trade, it is obvious that the total labor force in modern manufacturing could not be more than double 1,200,000 of the 1930 survey; in fact there is good reason to think that by 1933 the total was still under 2,000,000.[25]

Even if we add to the industrial workers the estimated 800,000 workers in modern mines, the 22,660,000 in rural industries and the 2,000,000 in urban handicraft shops, the number engaged in manufacturing and mining comes to about 27,460,000 or about 6 percent of the total population. This compares with about 12 percent in the United States (1930), 18 percent in United Kingdom, (1931), 17 percent in France (1931), 20 percent in Germany (1933), 9 percent in Japan (1930), and 4 percent in India (1931).[26]

Secondly, although import of machinery has increased enormously since 1887, the total amount during each year was insignificant as far as country of origin was concerned. In 1928 the United States accounted for 34.8 percent of the total world export,

---

[24] D. K. Lieu, op. cit., p. 112.

[25] In the 1933 survey D. K. Lieu estimated that the total labor force engaged in all types of manufacturing in Shanghai, including factories of all sizes, was about 350,000 (op. cit., pp. 112-3). If we use the proportion obtained through the 1930 government survey, namely, Shanghai had about 30 percent of the total number of industrial laborers, then the grand total for all China would be still about 1,200,000 in 1933.

[26] These percentages are only approximate. They are computed from data given by Ryoichi Ishii, Population Pressure and Economic Life in Japan, p. 83, Table XXIX; Japan, Cabinet, Statistical Bureau, Statistical Yearbook, no. 58, 1939, p. 411. Quoted in Ta Chen, Population in Modern China (Chicago, 1946), p. 117.

Great Britain 21.1 percent, Germany 24.1 percent, and France 5.2 percent. China took in only 3.1 percent of the total.[27]

Thirdly, China's industrialization does not represent Chinese interest alone. Before the war the total capital in modern industries was about U.S. $1,300,000,000, of which only one-quarter was Chinese owned.

Fourthly, while foreign investments occupy a major place in Chinese industries, they are comparatively insignificant with regard to the total foreign investments of most of the creditor countries involved. Britain's Chinese investments represent about 5.9 percent of her foreign investments, and those of the United States, 1.3 percent. Only in the case of Japan and the U.S.S.R. do Chinese investments represent a high proportion of the total foreign holdings.

Lastly, in spite of its growing industrialization, China continued to suffer from a seriously unbalanced trade. In 1935 the net import was about 75 percent over net export. In 1936 exports picked up slightly, so that the net import was about 30 percent over the net export.[28] Between 1930 and 1936 the export of raw materials and semimanufactured articles had increased by 7 percent and that of manufactured goods by only 2 percent; but while the import of raw materials and semimanufactured articles decreased by 2 percent, that of manufactured articles rose by 11 percent.[29]

LIVELIHOOD.—In his 1930 work Buck found the average value of family earnings to be about Chinese $291.[30] In his 1937 work, Buck has raised this figure to about Chinese $367.[31] This, in the pre-war ratio of about 3.6 to 1, would be U.S. $100. However, we must look at the actual purchasing power rather than figures in currencies. Earlier reports on Chinese rural livelihood were especially pessimistic. In the words of R. H. Tawney, "more than half the families in eastern villages then examined, and more than four-fifths in the northern, had an income below the minimum required

---

[27] Siao-Mei Djang, *The Position of China as a Producer of Raw Materials and a Consumer of Manufactured Products* (China Institute of Pacific Relations, Shanghai, 1933), p. 26.

[28] *Chinese Yearbook, 1937*, pp. 671–72.

[29] Between 1931 and 1936 the import of machinery (which would be in the manufactured category) decreased rather than increased.

[30] *Chinese Farm Economy* (Chicago, 1930), p. 87. The actual income in money was lower. See Summary by R. H. Tawney, *Land and Labor in China*, pp. 69–70.

[31] Buck, *Land Utilization in China*, p. 468.

to support life." [32] Buck's larger and later findings were slightly more optimistic. In connection with nutrition, as we have seen, over two-thirds of the families studied have at least an adequate intake of calories for bodily efficiency. But there is no doubt that the lot of the average farmer in China is far from enviable. The bulk of his expenditures goes for food. Cotton is used for most of the clothing. "Even in the Double Cropping Rice Area, where silk is most used, only one out of every nine dress garments is of silk." Nearly half the farm buildings in China have walls of tamped earth or earth brick; brick walls, representing the best type of construction, are found in 17 percent of all farm buildings in the Wheat Region and 23 percent in the Rice Region; in more than 75 units out of 100, the floors are of earth. [33]

The number of rooms per person is 1.3, in both of the agricultural regions studied by Buck, with the small farms having fewest rooms, doors, and windows per person. But most of the farms in China "use one or more of their rooms both for family and for farm use." Bedrooms often serve for storing farm equipment and also as barns. On 12 percent of the farms livestock were kept in bedrooms, and in the kitchen in 3 percent of the large farms but in 7 percent of the small farms.

Only one-fifth of the farmers reported savings of any kind. While this is possibly underreporting, the real situation could not have been very much better if judged by the use of credit. At the average high rate of interest of 32 percent per year, about 39 percent of the farms studied obtained credit. Roughly only one-quarter was for productive purposes and the other three-quarters for food and special occasions like birthdays, weddings, and funerals. The average cost of a wedding ran to about four months' net family income, while the customary funeral cost about three months' net family income. A dowry meant almost an equally burdensome outlay. In every area, the cost of a wedding exceeded the total value of a laborer's yearly earnings. [34]

Among both rural and city workers overpopulation keeps incomes low. Generally speaking, a high proportion of urban workers have come directly from villages. According to Olga Lang's find-

[32] Tawney, *Land and Labour in China*, p. 70.
[33] See Buck, *Land Utilization in China*, pp. 439, 443, 445.
[34] *Ibid.*, pp. 449, 461–3, 466, 468.

ings, at least 30 percent of the workers in craft and other old-fashioned shops in Peiping were unable to support their families even on the bare subsistence level.[35]

The industrial workers were somewhat better situated. According to Lang's research, the compilation of Nankai Institute of Economics, and the more extensive and intensive investigation under D. K. Lieu, the wages of industrial workers often were two or three times more than those in craft or other old-fashioned shops, but showed a much wider range of variation.

However, here again, cash figures tell us little with reference to the actual business of living. To what extent are the industrial workers better off than craft and other city workers? In this connection two summaries of various inquiries are very useful.

Percentage Distribution of Living Expenses

Type of Studies	No. of Studies	Food	Clothing	Rent	Fuel and Light	Miscellaneous
(PART A)						
Rickshaw or largely rickshaw families	2	73	6	8	10	3
Village families	5	67	15	5	5	8
Miscellaneous city workers families	4	60.5	8	7	7	17.5
Buck's farm families	13	58.9	7.3	5.3	12.3	16.2
Industrial workers' families	8	56	9	10	8	17
Unweighted average		63.1	9.1	7.1	8.4	12.3
(PART B)						
Urban working families	48	55.7	8.0	8.8	8.8	16.7
Rural miscellaneous families	5	66.8	18.7	4.7	3.6	6.2
Farm families	16	59.0	8.2	4.7	12.3	15.8

Source: Part A was constructed by H. D. Lamson, "People's Livelihood as Revealed by Family Budget Studies," *Chinese Economic Journal* (May, 1931), p. 452; Part B was based upon L. K. Tao, *Standard of Living among Chinese Workers,* 1931, pp. 35-37. Both are quoted in Ta Chen, "Labour," *China Yearbook* (1933), p. 384.

It is generally accepted that the lower the standard of living, the higher the percentage expended on food and the less on the other

[35] Lang, *Chinese Family and Society,* p. 83.

items, especially "miscellanies." Between the two parts of the tabulation there is a general area of agreement. The figures for farm families under the column "food" practically correspond, as do those under "miscellaneous." If we take the "Urban working families" in Part B to mean the same thing as "Industrial workers' families," in Part A, we find a similar correspondence in the same items. The "rural miscellaneous families" were probably non-farm families; their expenditures were higher for food indicating a lower standard of living than the two classes of people mentioned above. The lowest on the scale are the rickshaw families which expended the highest percentage on food. According to the evidence here presented it would seem probable that modern industrial workers in cities enjoy a somewhat higher standard of living than rural, or other urban, workers.

However, the one outstanding thing is the low standard of living of all workers. In none of the Chinese groups studied was the proportion of income expended for food less than 55 percent. But in Japan, manual workers spent 35.1 percent for food; in the United States farm families spend 39.5 percent for food and semiskilled workers in San Francisco spend 38 percent for food.[36]

It has often been remarked that Chinese workers have low wages because their rate of efficiency, compared, for example, with American workers is so low that a higher wage would be economically unjustifiable. That the relative efficiency of the Chinese workers is lower than American workers, under present circumstances, cannot be denied. We have seen the low per man-equivalent production of the Chinese farmer compared with the American farmer. In such an urban industry as cotton spinning, it can also be easily shown that the output per worker in China is, even allowing for longer hours, only about a quarter to a third of that in the United States.[37]

[36] In the Japan figure drinks and tobacco were moved from food to miscellaneous. These figures were quoted by S. D. Gamble, *How Chinese Families Live in Peiping* (New York, 1933), p. 323.

[37] The Chinese productivity data are based on H. D. Fong, *Cotton Industry and Trade in China* (Tientsin, 1932) and the U.S. data are based upon *Census of Manufactures*, 1935, 1937. I am indebted to Allan D. Searle and C. J. Sterling for their computations.

However, an analysis of labor's share in industrial profit shows a different picture. In 1932 the total output of Shanghai factories covered by Lieu's survey was valued at Chinese $557,690,754;[38] 60 percent of it represented raw material,[39] which reduced the net value to Chinese $223,076,300. The wage earners' share in that year was Chinese $37,787,625 or 16.9 percent.[40]

In 1931 the output of United States manufactures was valued at U.S. $39,829,888,000. After subtracting the cost of materials, containers, fuel, and purchased energy from the above figure the net value is reduced to U.S. $18,600,532,000. The wage earners' share in that year was U.S. $6,688,541,000, or 35.9 percent. In subsequent years United States wage earners have taken about the same percentage of the nation's industrial receipts.[41]

Chinese industrial workers, though enjoying a slightly higher standard of living than farm workers, have not been receiving a fair share of the results of their labor. For this reason the per unit labor cost of Chinese products is lower than that of American products, despite the much lower productivity per worker. In 1931 the labor cost of a short ton of coal in the U.S. was 60 cents (U.S.), while the corresponding figure was 29 cents (U.S.) for China. In textile mills the cost in the U.S. per 100 square yards was $2.84 (U.S.), while the corresponding figure was $1.49 (U.S.) for China.[42]

We have then to ask, who has collected the bulk of the industrial profits of China? Not the salaried workers. In the United States they took 16.25 percent of the net income from manufactures in 1931; in Shanghai the figure was only 3.3 percent in 1932.[43] But after deducting salaries and wages from the total net value of the

[38] Lieu, *op. cit.*, pp. 325-35.   [39] *Ibid.*, p. 109.

[40] D. K. Lieu specified that 60 percent of the value of total output had to be accounted for by *raw material*. He did not say whether that would cover also such other items as containers, fuel, and purchased energy, which the American figures generally took into consideration.

[41] Computed from *Statistical Abstracts of the U.S.* (Washington, D.C., 1946), p. 809. The percentages quoted in H. B. Parkes, *Recent America* (New York, 1941) are somewhat different because the total income (net) quoted was smaller.

[42] I am indebted to Searle and Sterling for this information.

[43] The value of total output figure pertained to 1932, but the numbers of salaried workers as well as wage earners pertained to 1933. In 1933 there were, in the factories covered by the survey, 17,592 salaried workers and 214,736 wage earners. The salary payments of 1932 amounted to Chinese $7,275,753.00 and wage payments of 1932, $37,787,625.00.

output we find 80 percent of it went to capital.[44] The Shanghai
factories included in the 1933 survey, it will be remembered, were
ones employing 30 or more workers and some motive power, and
were all Chinese owned. In those days they did not even pay any
income tax. Such economic inequality cannot but produce some
spectacular contrasts in a country hard pressed by overpopulation.
No wonder the manner of life of some privileged Chinese is so extravagant as to amaze not only Chinese but Americans.

There have been, and are, only two classes of any importance in
China. On the one hand there are those who may be designated as
the literati-gentry-bureaucrat; the other group comprises illiterate
peasants and city workers, who are the majority. The former group
is the master; the latter, the subject or servant. During a period of
civil disturbance, such as between two dynasties or the one through
which China is now passing, military power becomes supreme; then
the administering group must be designated literati-gentry-bureaucrat-militarist.

Apart from other considerations, the contrasts in salaries and other
income between these two groups are tremendous. Since about 1700
the salary of the magistrate, at the lowest level of the bureaucracy,
had been between 600 to 1,200 taels a year.[45] But by 1900 the wage
of the average skilled worker in Peking (Peiping), which for many
centuries had been the national capital, was only about Chinese $68
to $102 a year (counting 200 to 300 days of work per year), or
less than one-tenth of a magistrate's salary.[46] Since the coming of
the Republic the difference has widened. By about 1930 the vast
majority of the industrial and other workers were earning less than
Chinese $20 a month, as we have seen earlier, while the magis-
trate's regular salary ranged between Chinese $300 to $500 per
month. In other words the difference became from 15 to 25 times
greater.

Needless to say, the bureaucrats on higher levels than the magis-

[44] After deducting the salary and wage payments from the 1932 value of total
output minus "raw materials," we have Chinese $178,012,922.00 or 79.7 percent of the
value of the total *net* receipts.

[45] Kung-lu Chen, *Chinese History in Modern Times* (Shanghai, 1934; in Chinese),
I, 8.

[46] T. P. Meng and S. D. Gamble, "Prices, Wages and the Standard of Living in
Peking, 1900–1924," *Chinese Social and Political Science Review* (Special supp., July,
1926), pp. 96, 99. One tael equaled about Chinese $1.30.

trate enjoyed much higher salaries. In the Ch'ing Dynasty, the salary of a provincial governor was 10 to 15 times that of a magistrate; and that of a viceroy 15 to 20 times.[47] That however, is not the entire story. While the wages of the workers were all that they received, the actual income of the official was far above the quoted figures.

To be sure, not every one in the literati-gentry-bureaucrat-militarist group commanded these high incomes. Many were less fortunate, and many scholars achieved imperial degrees or school or university certificates but did not succeed in entering bureaucracy. However, even the less fortunate bureaucrats earned more than manual workers and all of them had higher aspirations. That is why at one end of the social scale people starve to death or exist like animals, knowing no recreation, no human dignity, while at the other end of the social scale the comparative few have so much comfort that they live like uncrowned kings, without work and without limit to their excesses.

Thus, in terms of reward we must say that there were no lucrative industries in China except the bureaucratic industry. In premodern China this "industry" was based upon land tax and upon other legal or illegal levies and graft. In recent decades two more important sources have been added: customs revenues and, secondarily, industrial returns. Since the economic margin of the people as a whole was so low and since the bureaucrats were the only ones with any sizable accumulation of wealth it was only natural for them to enter the new industries when the latter began to appear.

This bureaucratic "industry" determined the character of Chinese urban economy and largely doomed her rural economy. First, in spite of the smallness and the uneconomic nature of the average farms, Chinese peasants pay a higher rate of land tax than American farmers. In Buck's inquiry, covering 1907 and 1933, it was found that the average tax on all farm land in China was one to three times higher than in the United States.

The extent of levies and graft are hard to determine. However, it was estimated that in 1929 no less than Chinese $300,000,000 worth of property was held by various retired ministers, politicians, and warlords in the treaty port of Tientsin alone, while in June, 1931, the total amount of capital of the 1,100 registered corpora-

[47] Kung-lu Chen, *loc. cit.*

tions in the whole of China was only Chinese $556,000,000.[48]

Thirdly, the most revealing evidence of the importance of the bureaucratic industry is found in the nature of Chinese urban centers. According to a compilation in 1937, in 21 of 26 provinces the capital was the largest city, far outnumbering the other cities of the same administrative unit. In two provinces the capital was matched in population by one other city. In three provinces it was smaller than one or more other cities of the province; in one of these three it was outranked by Peking, the former national capital; in another of the same group it was outranked by Nanking, the present national capital, and by Shanghai. Taking the country as a whole, Peking outranked all cities except Shanghai; and Nanking, all cities except Shanghai, Peking, and Canton.[49]

The concentration of urban population in the United States is practically the reverse. According to the 1947 *World Almanac*, there are fifteen states of the Union in which the state capital is the biggest city of the state; in 9 states it was outranked by one other city of the state; in 5 states, by two others; in 4 states, by three; in 3 states, by four; in 6 states, by five; in one state, by six; in two states, by eight; and in three states, by ten other cities. Even in twelve of the fifteen states in which the capital is the largest city, it is at least matched in size by one to five others. Washington, D.C., the national capital, ranks eleventh among the cities in the United States.

After over half a century of industrialization the bureaucratically dominated nature of Chinese urban economy is still clear. In the United States the cities are both consumption and producing centers. In China they are, in terms of the national economy as a whole, chiefly consumption centers. The chief consumers were and are members of the literati-gentry-bureaucrat (and lately, militarist) group. They are the employers of those peasants and their children who have been crowded out of their villages, and who have no alternative but to accept practically subhuman wages.

It may be asked how such a system managed to work for so many centuries. The answer is that the evils of the system were not as

[48] F. L. K. Hsu, "A Closer View of China's Problems," in *Far Eastern Quarterly* (November, 1946), p. 55.

[49] Based upon Ju-Sheng Shen, "The Distribution of Cities in China," *Journal of the Geographic Society of China*, IV (Nanking, 1937), 929-35.

serious as they seem at first sight until they were aggravated by huge
war indemnities and an unfavorable trade balance. Before China's
door was opened wide, the ruling classes collected from the people,
but they also spent among the people. Through their urban ex-
travagance they at least disseminated a part of their riches among
thousands and millions of other Chinese. But a different picture resulted with the beginning of the unfavorable trade balance. Since
imported goods have always been more in line with the consumption habits of the upper classes, it is easy to see that, together with
huge indemnities, the same corruption has become much more
vicious than before. The same wealth held by officials which used
to be redistributed among the people through spending now left
the country by way of foreign concessions, treaty ports, and pleasure tours in Europe and America.

Social and Political Organization

The basic unit of the society is the family. The Chinese family
is a patrilineal, patrilocal, and, in large measure, patriarchal organization. But a number of features distinguish it from many other
patrilineally, patrilocally, and patriarchally organized families.
These features may be summarized under one principle: glorification of the father-son relationship at the expense of the husband-
wife relationship. This principle is built on a number of secondary
patterns, the first on the list being filial piety.

THE FAMILY PATTERN.—Filial piety means that a son owes everything to his parents. While they are alive it is his duty to support
and obey them; after their death it is his duty to continue the support
and obedience in the form of elaborate funerals and ancestor worship. The second pattern is, for lack of a better term, estrangement
between the sexes. Marriages are arranged, with no freedom of
choice whatever. A married couple, by custom, must live under the
same roof with the husband's parents. Their duty to his parents are
primary, and to each other secondary. Furthermore, there must
be practically complete suppression of the expression of erotic life
between man and wife. Husbands and wives are seen together only
on special occasions. Since the society is patrilineally organized, the
second pattern also means inequality between the sexes, accompanied

by high premium on virginity, male dominated rules of divorce, and concubinage.

However, contrary to popular assumption, filial piety is not a one-sided affair in favor of the parents. Father and son are both links in the infinite continuity of the family lineage. Their roles are complementary; thus the sons owe all they have to their progenitor, but whatever the father has also automatically belongs to the sons. The rule of inheritance is, therefore, that sons receive as a matter of course the father's entire property. The older man cannot make a will in favor of any one else. For the same reason, while the son's status in early life is determined by the achievement of his father, the father's status in his later life is determined by that of his sons. It is such features which have prompted me to speak of, elsewhere, father-son identification.[50]

This father-son identification leads to the fourth and last secondary pattern of Chinese family, namely, the big family ideal. All men desire to have more than one son. In order to maintain the father-son identification it is necessary to protect the family from the defection of some sons who break away. For this reason a high degree of solidarity among brothers is important. The logical extension of this pattern is the ideal of solidarity among all members of the kinship group.

This being the essential manner of its organization, it is clear that the Chinese family is potentially larger than any modern European or American family which emphasizes the husband-wife relationship, which, according to the Chinese pattern, is suppressed in the interest of the father-son relationship and joint family unity.

IDEAL AND REALITY.—However, a complete glorification of the father-son relationship and suppression of the husband-wife relationship is not practical. Within the Chinese family there is always the duel between these two basic sets of relationships. If the former predominates, the family will be large. If the latter, the conjugal relationship triumphs and the family will be smaller.

Investigations made in recent years have demonstrated conclusively that there is a close and direct correlation between the size of farm and size of household. Buck's surveys show that on "small" farms the average number of persons per household is 4.4. Extend-

[50] See F. L. K. Hsu, *Under the Ancestors' Shadow* (New York, 1948).

ing this to "medium," "medium large," "large," and "very large" farms, the respective numbers of persons per household are, respectively, 5.5, 6.9, 8.3, and 10.1.[51] This has been thought to indicate a close adjustment between land and population.

But if we consider this in terms of population pressure only, we shall be unable to explain why, in spite of the fact that the majority of Chinese farm holdings are below the most economic unit for production, families keep dividing. According to my own observation it is precisely the poorer families which most frequently divide.

Some years ago I ventured to explain it by the differential adherence, between the rich and the poor, to the social ideal of elevating father-son relationship at the expense of husband-wife relationship. In the poorer family the difficulty of living together is easily aggravated by poverty and often overshadows even the wisdom for continued economic cooperation, so that the more a family needs to keep the land intact the more it tends to divide. Here the husband-wife relationship has greater weight for the additional reason that, for the poor man, to secure a wife is a matter of immediate economic consequence. On the other hand, with the wealthier families the social ideal has more weight because that is the mark of prestige and status. Furthermore, men in such families have also no worries about remarriage. When brothers do not actively side with their wives, the family can hold out together longer and the father-son relationship has a greater chance.[52]

We now have some additional evidence in support of this thesis, namely the preponderance of conjugal families among the poorer groups. In her inquiries, Olga Lang confirms the observation that family size increases with social and economic status. But she also gives the following data on 485 north China village families: among farm laborers the percentage of conjugal families was 54; next came peasants, 41 percent; then came middle peasants with 27; among well-to-do peasants the percentage was 17, while among landlords it was 12.[53] Thus the wealthier the family the greater its tendency toward the Chinese large-family ideal, which is not only measured by the number of children born and brought up, but also by the

[51] Buck, *Land Utilization in China*, p. 278.
[52] F. L. K. Hsu, "The Myth of Chinese Family Size," *American Journal of Sociology*, XLVIII, No. 5 (March, 1943), 555–62.
[53] Olga Lang, *Chinese Family and Society*, p. 138.

number of collateral kinsmen and their wives and children under the same roof. Lang's data on city families points in the same direction.[54] This is why, though China has been regarded as a country of large families, the actual mean or average size is only about 5.2 or even less.[55]

The extension of the family group is the clan. Contrary to popular assumption, clan in China is not strong. It is better organized in some areas, notably the extreme south and central China. It has form but not substance in the southwest. The external manifestations of a clan are as follows: (a) surname, which all members of the same clan share, but which no longer corresponds to clan; (b) exogamy, which is sometimes ignored; (c) common territory (some villages are composed solely of members of the same clan); (d) clan land, which is common in the extreme south, less common in the southwest, and very rare in the north; and (e) clan temples where common worship of the same ancestors takes place on appointed dates every year. Clan temples are absent in many parts of the country.

For reasons which we have no space to dwell on here, the areas in which the clan organization has been described as strong, and where clans possess large pieces of common land, are often areas where the clan is under the dictatorship of an oligarchy, who appropriate the common goods for furthering their own immediate family interests.[56]

THE CHANGING FAMILY.—The picture of the Chinese family today is one of contrasts. For the vast majority of peasants there has been little change. As far as they are concerned the family picture as described above is still essentially true. On the other hand, changes have occurred increasingly among the urban population. Some changes are noticeable even in small interior towns.

Changes are taking place faster in some aspects of the traditional pattern than in others. Lang found the vast majority of the students investigated desired freedom of choice in marriage whether for

[54] *Ibid.*, p. 142.

[55] C. M. Ch'iao, *Rural Population and Vital Statistics for Selected Areas of China, 1929-31* (Shanghai, 1934); L. S. C. Smythe, "The Composition of the Chinese Family," *Nanking University Journal*, II (1935), 371–93, and J. L. Buck, *Land Utilization in China*, p. 368. In fact, Ta Chen put the figure at 4.84 persons (*op. cit.*, p. 23).

[56] See H. S. Chen, *Landlord and Peasant in China*, and F. L. K. Hsu, *Under the Ancestors' Shadow*.

themselves or for their children.[57] She also found evidence for a change in the husband-wife pattern not only among educated groups, but also among industrial workers in cities. The wife in the new situation has more influence over her husband and children and in many cases controls the family purse. Lastly, there is evidence for change in the father-children relationship. Although the presentation of her material is peculiar, Lang's data do indicate, for example, that college-educated fathers have developed more intimate relationship with their children of both sexes.

These three lines of change will naturally increase the demand for the conjugal rather than the joint type of family. It is not surprising, therefore, to find that in the modernized groups, the percentage of conjugal families is higher among the upper class than among the middle class families—a tendency exactly in reverse to that found among the tradition-bound groups.

Certain other aspects of the family pattern have shown little change even in urban areas. The vast majority of students questioned considered the support of parents to be important. Many students and others feel that if they do not live under the same roof with their parents they ought at least to offer them material assistance. In my own observation I have found, on more than one occasion, university professors being severely criticized for neglecting their parents.

The other aspects of the family pattern which have changed but slowly may all be put under the heading: lack of organic relationship between custom and the written law. Modern Chinese law is comparable to the most advanced laws of the West. In the case of family behavior it provides, for example, that parents are free to leave their property to anyone they like by a will. In reality, such property goes to the sons as a matter of course in the vast majority of cases. The law also says that betrothal is not binding unless both parties assent and that men and women have the same right to divorce on specific grounds. Finally, concubinage has been outlawed since 1936. The reality is that only among the highly educated is the view concerning betrothal commensurate with the spirit of the law; the vast majority of women do not go to law even when mis-

[57] Lang, *op. cit.*, pp. 303-4; see also pp. 142, 203, 222, 300.

treated; and concubinage goes on as before, even though it is highly disliked among educators and some others.[58]

THE VILLAGE AND THE DISTRICT.—Next to family and clan is village. Chinese villages are rarely walled and are usually irregular in shape. The larger ones have a population exceeding 200 households, or approximately 1,000 individuals. The smaller ones may have only 20 households, or fewer than 100 persons. In some villages all or the majority of the families belong to the same clan. More often a village is dominated by members of a few clans.

Most villages are organized in two ways: an organization originated from the central government and a second originated from the indigenous community. The organization of government origin is more or less uniform throughout the country: Every ten families near each other form one *Chia;* every ten *Chia* form one *Pao;* every ten or more *Pao* form one *Hsiang,* while a district may be divided into three or more *Hsiang.*

In each of the subdivisions within the district there is an elected headman (sometimes two), empowered to handle certain affairs of the unit. According to government regulations the functions of this organization are numerous, but only a few definite ones are ever discharged. First, organizing for public safety against bandits and communists; secondly, the keeping of census (which during and since the war has been used by the government for conscription and forced labor); thirdly, the headmen of the *Pao* and *Hsiang* often passing judgment on simple cases or serving as mediators in intra-village disputes; fourthly, managing the local schools; and lastly, serving as intermediaries between the people and the government functionaries. In the last function their maneuvers are often to the advantage of the villagers.

The organization originating from the village also covers the entire community. In this matter, villages in different parts of China vary so much that it is better to take a specific example. This example is taken from a community in Yunnan, southwest China, which will be referred to as South Village. The observations pertain to the years 1941 to 1944.

The village is divided into two parts known as Upper and Lower

[58] In this connection see "Some Problems of Chinese Law in Operation Today," *Far Eastern Quarterly* (May, 1944), pp. 211-21.

South Village. The Upper section is divided into six equal units of the local organization and the lower section is divided into five. Each unit has a head and consists of an approximately equal number of households. Broadly speaking, the chief function of this organization is the handling of matters of public worship. Public ancestor worship takes place once every year in the seventh Moon according to the lunar calendar. The birthday of Buddha is celebrated on the eighth day of the fourth Moon. The women's "blood lake" meeting, to insure safety in child delivery, takes place in the middle of the fifth Moon. The "green crop" meeting, for the purpose of insuring freedom from locusts and other insects, takes place on the first day of the sixth Moon. Besides these, one or more nonperiodic affairs always take place in any given year. These prayer meetings are organized when worms or locusts appear in the fields; when fire has burned down some houses in the village; and when some unexpected or unexplainable accident occurs. In many villages, Japanese bombing was included in the last category.

All these meetings, though different in purpose, are similar in procedure. One or more priests are hired, either locally or from a distant village. They recite the scriptures and perform special rites on a platform in the temple yard. The entire proceeding may last from one to three days and nights. The costs of these events, which are often considerable, come partly from rent collected from communal land and partly by voluntary donation from practically every household in the community. Although occasioned often by a communal crisis, these meetings, with their rituals, music, and feasting, are also an important source of recreation. They also bring the community together so that its solidarity is reaffirmed.

Theoretically these meetings have nothing to do with the *Pao* organization. Actually they are sometimes sponsored by present or past *Pao* headmen, who are also chief participants. The personnel in the government organization take part for the simple reason that the local functions are events through which the place of the individual in the community is seen. An individual who has no place in these functions has also no place in the community.

In addition to the local organization we must mention the gentry group as an organized force. There is no local name for this organization and, if questioned, villagers deny its existence. But there is no question about its reality and its power. In brief, this is a prestige group. To qualify as a member, one must have, in addition to land, some connection with bureaucracy or the army. If a man is neither a bureaucrat nor a militarist, he must have some close relative, a brother, a son, or a son-in-law, who works in the provincial government or the central government, or on higher levels of the army. Members of the gentry associate freely with the magistrate as equals. The latter must profess that he would be guided by their opinion in all important decisions concerning the community.

South Village had no recognized member of the gentry at the time of investigation, but when occasion demanded it, the village was guided by some who were inhabitants in another village and in the district seat. The gentry did a number of things. It was largely responsible for the creation of a lower middle school with about 150 students, a primary school with about 200 students, and a district clinic with one doctor and two nurses trained in the Western style. At the time I left the village, the magistrate, with the help of the members of the gentry, had just paid Chinese $1,000,000 for a house to be made into a hospital with 20 beds.

Whenever there is any friction between the magistrate's office and some visiting military or civil group the gentry will be the mediator, by virtue of its connection with higher offices. On the other hand, the group certainly perpetuates its own vested interests and, in doing so, maintains a symbiotic relationship with the magistrate. It can influence land taxes or "fix" law suits, or protect anyone from the local law more or less as it pleases.

The district is the basic political unit in the country and is supposedly administered by a magistrate of the people's choice. But the highest so-called elected official so far is the headman of the *Hsiang* or subdistrict. Even in this case, popular election is often a myth. In size, districts very considerably; they are classified according to size and importance.

A district always comprises a walled city, where the magistrate resides and administers, and a large number of villages. It may also comprise one or more market towns.

Within the district the magistrate has final authority on all important issues, except that he has to play ball, so to speak, with the local gentry. He has charge of the public safety, collection of land tax, public works, conscription, and opium suppression. If there is no law court he is the district judge as well.

CLASS STRUCTURE.—Before discussing the wider political organization it will be necessary to understand the class structure. In pre-modern China the popularly conceived classes and their ratings were first, scholars; second, farmers; third, craftsmen and laborers; and fourth, merchants.

The actual picture was considerably different. At the very top sat the aristocracy. Then came the literati-bureaucrats on whom rested the practical business of administration and who were literally servants of the imperial households. Then came the gentry, who, in more ways than one, could be classed with the literati-bureaucrats, and heads of important clans. Then came small owners, farmers, and merchants. City and farm laborers came at the bottom of the class scale. Military personnel rated in a number of ways: common soldiers were considered outcasts, but as they rose in rank they became comparable to members of the various classes; at the upper end there would be some military officials enjoying the same social esteem as high bureaucrats, not usually so much because they were great soldiers, but because they were learned and were known as "scholar-generals."

It is, of course, to be understood that the interclass demarcations of this structure were ambiguous at many points. The term bureaucrat could be applied to anyone including the prime minister, the clerk in a district magistrate's office, and anyone with a title. Needless to say, their prestige and privileges varied tremendously. Many heads of clans and members of the gentry would also have bureaucratic connections. Rich merchants would climb socially by purchasing a title, by winning a title through charity or generous contributions to government causes in an emergency, or by sending their sons to tutor schools and halls of examinations. Finally, everyone of any means owned some land or other forms of real estate. In this sense every class above the group termed small-owner farmers would also be landlords.

Following contact with the West, and particularly after the downfall of the imperial dynasty, some changes have occurred. There is no longer an aristocracy recognized as such, but its position is occupied by a much larger group claiming the right to rule the country and backed, as was the old aristocracy, by armed forces. An equally important change has occurred in the gentry-literati-bureaucrat group. It has broadened to comprise a larger number of persons and more diverse elements, including the literati-reformers, among whom are also professionals such as doctors, architects, accountants, writers, and teachers. There are also bureaucrats who have invested their gains in industries and commerce, and industrialists and modernized merchants who derived their capital by purely business means. Finally there are the compradores, who rose by being agents between Western firms and Chinese customers, or raw material suppliers, and others who rose through missionary assistance.

The outstanding things about this new gentry-literati-bureaucrat group are that, on the one hand, literary achievement per se is no longer as important as before, and on the other, bureaucratic positions share the attraction with industry and commerce. There are even scholars whose chief concern is the improvement of society. But in the minds of the lower classes and even among the literati, all such persons are closely associated with the bureaucracy. Even many professionals, industrialists, and merchants were forced to play ball with bureaucrats on the higher levels to get what they want or to get into the most advantageous business positions.

Some changes have also occurred among the other classes. For example, a new class of technical workers in factories has risen from the ranks of the common industrial workers. These technical workers often commanded, before the war, a wage six to eight times that of the other workers. But the most spectacular change has been in the position of the military. Common soldiers are still without class, but the highest military men have risen above their station in the imperial days. Instead of being the instruments of the rulers, as bureaucrats were, they are now the top rulers.

The most unusual thing about the class structure is that, in its functioning, it exhibits the polar characters of fluidity and rigidity. There is no doubt that social mobility has been considerable in the long run. Nevertheless, at any given point of time the class structure is extremely rigid. The most basic divisions as we see them today are the ruling military and bureaucracy on the one hand and the vast majority of the people on the other. The gulf between them is great, not only in remuneration but also in power and prestige. As a matter of fact the structure is so rigid that differences between separate classes are often translated into differences between separate statuses within the classes. That is to say, there is not only sharp inequality between the two large divisions, but also among the high military and the bureaucrats.

GOVERNMENT BY BUREAUCRACY.—The magistrate is not responsible to the people but to the head of the provinces. Before the Republic, there were a number of intermediary divisions between the province and the district. These have since been abolished. At the head of the province is the governor, appointed by the central government, who administers with the aid of the boards of civil affairs, of education and of finance, and a number of bureaus, including that of police.

Like the magistrate in his district, the governor is the supreme authority in his province. Just before the war with Japan there were twenty-eight provinces in China. Since the war, Manchuria, which formerly comprised three provinces, became nine provinces, thus bringing the total number to 34. But unlike the magistrate the provincial governor is never vested with judiciary function. There is a higher court in every province.

The governor is responsible to the Executive Yuan (Council), the leading organ of the central government. Also responsible to it are the various ministries and bureaus and special municipalities, such as Shanghai, Canton, Hankow, Peiping and Tientsin.

There are four other *Yuan:* legislative, judiciary, examination and control. The Control Yuan investigates all officials and government organizations and indicts those which fail in integrity or efficiency. The Judiciary Yuan, through the Ministry of Justice, has charge of the Supreme Court in the national capital, the Higher Courts in the provinces, and the District Courts. The chief of the Executive Yuan is the premier, who is nominally responsible to the President of the Republic.

The command of the armed forces is a separate system. Under the Executive Yuan there are Ministries of War and Navy. But the supreme command of the armed forces is vested not in the Executive Yuan, but in a separate organ called the Military Council. When the chairman of the latter council happens to be also the Premier or the

President of the Republic, then the separate powers merge, but not otherwise. For many years one man served as the figurehead Chairman of the Republic, another man served as the chief of the Executive Yuan, while General Chiang Kai-shek was Chairman of the Military Council. Upon the death of the former chairman, Lin Sen, Chiang took over the chairmanship while concurrently holding the reins of the armed forces. For a short period Chiang was also the Chief of the Executive Yuan.

The Kuomintang, or Nationalist Party, has been the power behind the government since 1929, when the armies of General Chiang, chief of the party, unified China. Like the administration, the party is organized on three levels: in the national capital, the provincial capitals, and the district seats.

To understand the functioning of the Chinese government one must, however, look into the basis of its power. Before the Revolution of 1911 the power of the ruling group rested upon three things: military force, execution of certain essential functions, and tradition. For the founder of a dynasty the first weapon was military prowess. When a man emerged victorious after successfully eliminating all contending factions in an interdynastic chaos, then tradition would have it that he possessed the Mandate of Heaven. He and his lineal descendants would then be entitled to rule the country until they lost the Mandate.

However, to remain in power the ruling group, with the help of the bureaucracy, had to fulfill certain essential functions. These functions included maintenance of law and order and water control chiefly for purposes of irrigation. Some students insisted that it was *primarily* water control which gave the ruling group its power and determined the characteristics of the entire Chinese society.[59] There is no doubt that the government's part in water control was economically important at least to a part of the agricultural population.

As long as the ruling group kept the people happy, it possessed the Mandate. When civil disturbances failed to be pacified and famines and other catastrophes became too serious and too frequent,

[59] See K. A. Wittfogel. "Economic and Political Features in China's Social Heritage" (mimeographed lecture originally given at the Princeton University Bicentennial Conference on Far Eastern Culture and Society, April 1–3, 1947).

it was a sign that the Mandate of Heaven was withdrawn. Then the people had the right to revolt and a new ruler with the Mandate would sooner or later emerge.

The foundation of power of the bureaucracy rested in the execution of certain essential functions, and also in tradition but particularly in favor. As servants of the ruling group they helped to execute the essential functions expected of that group. Tradition was behind them because literacy carried with it a very high premium. The vast majority of the bureaucrats were members of the literati, who were regarded as traditionally entitled to rule. However, the power of the bureaucracy cannot be properly understood without reference to favor. With relation to the people under their administration, the bureaucrats literally had absolute power. The magistrates, who were really at the tail end of the bureaucracy, were called "father-mother officials," expressing definite meaning that while they might love the people like parents they also, like them, held the power of life and death over their children.

This great power lasted only as long as the bureaucrats remained in favor with their superiors and the latter in the favor of the emperor, who had the final authority. While in favor they had all the power they wanted and abused it; but this power might be taken away suddenly, so that overnight a bureaucrat would become the most miserable wretch in the streets. Throughout the dynastic histories of China ups and downs of individual bureaucrats, and even of eunuchs, occurred time and again.

Under this pattern of political power, economics had little to do with the case. Money came after power, not vice versa. After attaining power one was in a position to acquire wealth and to protect it after he got it. A millionaire who had no bureaucratic connection was an easy prey to the man with political power. That is why the wealth of individual merchants and others could be confiscated by imperial decree with comparatively little provocation.

After the fall of the imperial dynasty new factors came into play. For the upper level of the ruling group, military force has become increasingly prominent. But while formerly this was an indigenous matter, now much of it depends upon the place of foreign powers in China. The traditional Mandate of Heaven idea has been severely shaken because, while the peasants have the right to revolt, with the introduction of modern weapons it is hard for unarmed and unorganized common peasants to revolt.

However, the introduction of modern weapons has also brought about other consequences, one of which is the growing influence of financiers. For the first time most implements of war have to be obtained from abroad. Yet extraterritoriality and foreign concessions have enabled Chinese wealth to escape the reach of the government; to get at this wealth, coercion no longer suffices; the money has to be attracted. For this reason wealth has come into the picture as a source of political power for practically the first time in Chinese history. The fact has become particularly manifest since the Kuomintang completed its northern expedition under Chiang in 1929.

But this influence has been partially destroyed by the war in two ways: A sky-rocketing inflation and physical devastation have wiped out large fortunes, and war has also seen the end of extraterritoriality and foreign concessions. Chinese fortunes are no longer secure from the grasp of the government. The many decrees during and since the war, whether enforced or not, aiming at fixing prices, government budget, and even confiscation of Chinese savings, whatever their origin, in American banks are good evidence that, once again, in the minds of the ruling group, private wealth is subordinate to political power backed by military force. This political power can determine everything. It can give favors, but it can also deprive anyone of whatever he has by a simple stroke of the pen. To insure this absolute power it has been necessary to initiate, in recent times, along with the sharpening of military power, party activities among and supervision on all levels of the bureaucracy and among such groups as the workers in cities and towns.

The essential key to the functioning of the political organization in recent decades is, therefore, armed force on the part of the top ruling group and favor on the part of the bureaucracy. In the struggle to hold on to their privileges the bureaucrats look to their superiors and their superiors, in turn, to the armed forces, for guidance. The people, instead of being the masters, are actually regarded as servants and inferiors.

The struggle for existence in Chinese bureaucracy has produced two results. On the one hand, the struggle for favored positions has made the capital outlay of being a bureaucrat very large. Just before the war's end a commissioner of the Board of Civil Affairs of a certain province celebrated his mother's birthday. Every one of the magistrates of the province sent him a gold ingot. Not that all of the magistrates wanted or could afford to do so, but if most of them did, the few who did not would evidently be out of favor. The least one could do was follow the crowd. But if one of the magistrates wanted a larger favor it would be logical for him to express his wishes in the form of a larger ingot.

Because the remuneration of being a bureaucrat is much higher than other lines of work in a country hard pressed by poverty, the struggle for entry into the group, as well as for a higher place in it, is very intense. On the other hand, by virtue of this very intensity, the salaries, however large, tend to fall short of expenses. Corruption is the natural answer. As the favors increase, the expenses rise, demanding greater corruption and still larger favors to make sure that the corruption will go unpunished. Thus the cycle perpetuates itself.

The second result is an apathy on the part of the people towards bureaucracy and matters of government concern. Centuries of practical living as servants and inferiors to the administrators have taught the people how to behave, unless too hard pressed.

During and since the war years there has been an awakening among the masses. Looking over the Chinese newspapers for the last three years I have been amazed at the amount of criticism of the government and the bureaucracy. Even some peasants and laborers write (or ask some literate person to write on their behalf) letters to the editors voicing complaints of one sort or another. Literacy has gradually lost some of the aura which used to be its distinction. Undoubtedly a much larger proportion of the population can now read and write, or at least read. However, as long as the age-old master-servant pattern persists it will be a long time before any political party may be said to enjoy popular support in the true sense of the term.

Religion

The so-called "Three Religions" are: Buddhism, Taoism, and Confucianism. Considerable amounts of ink have been spilled over the question whether Confucianism is a religion or not. Such arguments are irrelevant, for religion in China does not follow the pattern of interreligious exclusiveness of the West; there is no sharp dividing line as between Christians and Mohammedans, or even between Presbyterians and Baptists. Theological formulations may be argued by a few learned monks or priests but the average man has no interest in such things.

CHINESE RELIGION IN ACTION.—The first thing that impresses one on entering a Chinese village or market town is the size and number of the temples. This is particularly true in South and Southwest China. The temples are usually built of bricks, in contrast to the family houses which have earthen walls and thatched roofs. The temples are brightly colored, whereas ordinary houses are gray or brown.

The temples are dedicated to a variety of gods, as the Dragon God or the Goddess of Mercy, or they may be dedicated to dignitaries of the community who were deified after death. In market towns, and particularly in district cities there are usually three other kinds of temples: to the God of Wealth, to Confucius, and to the district patron god, who is equivalent in the spiritual hierarchy to the district magistrate in the political one. Several others may also be found, including temples to the Goddess of Measles and to San Kuan (literally three gods—heaven, earth, and man). In some villages there is one to the Sun God, and in the towns and cities there are frequently temples to the God of Agriculture, or the God of War, or to Lu Chu, the most important Taoist god.

Most of the temples, however, house more than one god. Usually they are sacred to a number of gods or goddesses who are from what would be, judged by Western standards, different religions. Consider, for example, a modest temple of the patron of a town in southwest China. The main altar is dedicated to the patron god, who in this case has the appearance of a warrior. The two side altars are occupied, respectively, by the Goddess of Measles and of Mercy. In front of the main altar are some tablets and images dedicated to the third son of the Dragon God and some lesser spirits. This is by no means atypical. There is one temple in this region which houses Confucius, Laotze (Founder of Taoism), and Buddha. This mixture is seen everywhere in China and indicates that the idea of monotheism is alien to the Chinese.

Identical patterns prevail in ritualistic observance as well as in the concrete arrangements of the temple. To quote from one of my notebooks:

Today saw old Mrs. Y with two grandsons and three granddaughters in the Pan Chu Temple. Offerings are made to every god (that is, every image) in the whole temple, both inside and outside the main hall. Even the dragons winding around the two main pillars of the main entrance received a share of the "food and money."

The old lady first burns paper money in front of the three main shrines. There are four gods occupying these shrines. As the paper burns she kneels down to koutou fifty times to each of the four gods.

When these gestures of homage are over, she takes some of the food offered at the main shrines and puts it in a tray. She takes this tray and offers it in front of every other image in the temple one by one. In front of each image the procedure is as follows: She offers the tray by lifting it up with both hands to a position over her head. She lays it on the table. She burns some paper money. She kneels to koutou eight or more times. She prays only to the first four gods as she kneels to koutou.[60]

The old lady on this occasion went to express gratitude and to report to the local patron god on the third day after the birth of a grandson; but she wanted to make sure that all gods, whether directly concerned or not, would be pleased, so that the child would grow up well. The yearly cycle of offerings to the gods observed in the same community reflects a similar attitude. Altogether thirty-four days of ritual observances are recorded for the community throughout the year. These include the birthdays of thirty different gods and one occasion on which all gods are worshiped.

During any emergency, such as an epidemic, drought, earthquake, or even after Japanese air raids, prayer meetings take place at which numerous gods and spirits are invoked. At one of these meetings, one scripture contained 608 gods with specific titles, including Jesus Christ and Mohammed, who are called sages and are subordinate to the Jade Emperor, the supreme ruler of heaven. Then the scripture goes on as follows:

In addition to the above the following gods are hereby invoked: Gods of ten directions; all fairies and sages; all fairy warriors and soldiers; ten extreme god kings; gods of sun, moon and nine principal stars; three officers and four sages; the stars of five directions; gods guarding four

[60] F. L. K. Hsu, *Under the Ancestors' Shadow*, Chapter VIII.

heavenly gates; thirty-six thunder gods guarding the entire heaven; twenty-eight principal stars of the Zodiac; gods for subjugating evil ghosts; god king of flying heaven; great long life Buddha; gods of Tien Kan and Ti Tze; great sages of Trigrams and Nine Stars; secondary officials of five directions; secondary officials of ten directions; gate gods and kitchen gods; godly generals in charge of year, month, day and hour; gods and spirits in charge of four seas, nine rivers, five mountains, four corners; of hills, woods, all rivers and lakes, wells and springs, ditches and creeks, twelve river sources; every and all gods; Cheng Hwangs and their inferiors; local patron gods; minor local officials; gods of roads and bridges; of trees and lumber; spiritual officers and soldiers under the command of priests; all spirits in charge of protecting the taboos, commands, scriptures and the right way of religion.

THE SPIRIT WORLD.—The Chinese believe strongly in an afterlife; in their conception, the spirits of the dead are closely bound to earth and interested in human affairs. The spirit world is divided into three parts: first, the Upper Heaven, ruled over by an Emperor with an extensive hierarchy of gods beneath him; second, the Western Heaven, where Buddha is the supreme ruler, also with a large group of high gods all of whom are subordinate to the ruler of the Upper Heaven (the exact relationship of the Upper and Western Heavens is never clearly expressed); and third, the Lower Spirit World where the spirits of the dead enter and are processed according to their record on earth. A ruler with ten judges working under him goes over the records. Those who have led good lives are rewarded with titles, leisure, and comfort. Exemplary characters may become gods in the Upper or Western Heavens or may be reincarnated into another existence on earth in which they attain honor and luxury. The wicked are punished by severe tortures, such as being sawed in half or boiled in oil. They may be banished permanently into hell, or reincarnated into another life beset with poverty and degradation, or they may be reincarnated as worms or rats or other lowly animals.

In broad outline this concept is not far different from that held by Western society, but the spirit world and the world of humans is more closely allied in China than in the West. The Emperor of China was known as the Son of Heaven and many heroes of history and legend are considered to be gods reincarnated. New gods are continually being created and many return to earth in reincarnation. The emperor, as son of Heaven, has power over both humans and spirits. Even high bureaucrats, by virtue of the power vested in them by the emperor or because they are gods incarnate, have power over the lesser spirits.

Most important in this link between heaven and earth is the belief that spiritual reward or punishment may come in one's lifetime as well as after death, or may be visited upon one's children. Thus death by lightning, sudden and violent illnesses, as cholera, and serious accidents are generally regarded as punishments originating from the spirit world. For this reason the most important measure against epidemics and accidents is the prayer meeting, in which hired priests invoke the mercy of the superior deities who are presumed to have ordered the disasters as punishment against the community. Conversely, wealth and good fortune are usually held to be rewards originating from the spirit world in payment for good deeds performed by the recipient or his ancestors.

Thus the spirit world and the human world are counterparts of one another. The spirit world is based upon and functional to the existence of the world of humans, and the human world is in turn supervised and guided by the spirit world. They exchange personnel. They endorse the same virtues and condemn similar evils; they express mutual approval or disapproval. In the popular mind the spiritual hierarchy is a part of the social order just as much as the bureaucratic and political hierarchy is. That is why it is irrelevant or even erroneous to speak of different religions in China. To the Chinese there is only a spiritual order which stands as firm as the social order. As there is no question of a community living under two social orders, so it is inconceivable that there should be two spiritual orders. If two religions are both true, they must find their place in the existing hierarchy. A creed for which this adjustment cannot be made is destined to be disregarded or forgotten.

ANCESTOR WORSHIP.—The basic religion of China is ancestor worship, but, here again, it is a mistake to regard the cult as a separate religion, for it is part of the larger, all-inclusive structure. The ancestors have gone through the life-death routine of all human beings; that is, they have died and been processed by the Lower Spirit World. They may have been so good that they were received directly into one of the Heavens. Also they may have been reincarnated. If they were evildoers they may have been doomed to eternal punishment; but ancestors, in so far as their descendants are concerned, are different from all other human beings. To their descendants, ancestors are all great men and women with a glorious past and an exalted status. One may believe that someone else's ancestors are in hell, or reincarnated in some base animal form, but no true descendant believes that such a thing could happen to his own ancestors.

To understand this cult one must take cognizance of the family organization and the father-son identification. Between the father and the son there is not only a complete community of interest, but complete social identification. The son not only inherits all his father possesses but he is judged by his father's achievements. Conversely, the father not only has complete rights over his son's wealth but, when the son has reached maturity, the father is evaluated by his son's abilities. When we realize that a particular father and son are but a link in the infinity of many generations in any given family line, we see how the father-son identification becomes the foundation for the religious cult of ancestor worship.

The basic assumptions of the cult are threefold. First, the living owe everything to the departed ancestors, who are, therefore, regarded as persons of great magnitude. Since death only puts the relationship on a somewhat different level, and since the dead have the same needs as the living (namely, food, money, housing, and so on) it is necessary for the descendants to provide for them as if they were alive. Secondly, while the ancestors have already made their imprint on the fate of the descendants, their actions in the spirit world continue to affect the living. Conversely, the actions of the living descendants have bearing on the spirits of the ancestors. Thirdly, the interest of the ancestors is confined to their own descendants, particularly lineal ones. They concern themselves not only with ceremonial occasions—weddings, division of the family, birth of sons—but also take action in emergencies, as when a deserving descendant is about to be flunked by the chief reader in an imperial examination. On ceremonial occasions the presence of the ancestors is recognized by offerings of incense and food. But on occasions of emergency, the ancestors intervene in the form of apparitions.

Thus the ancestral cult shows the same close interrelation between spirit world and human world and the same close correlation between religious structure and social organization. As the family is the foundation of the wider society, so ancestor worship forms the link between the individual and the supernatural.

MISSIONARY ACTIVITIES.—It is easy to understand why Western missionaries encounter difficulties in their efforts to win China for Christianity. First, the Chinese fail to see any reason why one religious cult should be adhered to, exclusive of all others. "All religions are for the good of mankind," say the Chinese, who are practical about their gods. They have gods for measles, eye disease, safe birth, fertility, agriculture, fire, and for astronomical phenomena such as the sun and the moon. It is hard for them to conceive of an all-inclusive god who alone is omnipresent and supersedes all others. Furthermore, the Chinese deduce their belief in spirits from their experience in life and living and shun theological arguments.

I have no intention of showing that conversion of Chinese to Christianity is impossible, nor do I wish to belittle what Christianity has done for China. Schools for women and coeducation were first started by missionaries. Abolishment of foot-binding has long been one of their objects of reform. Many orphanages and institutions for the mentally and physically defective have been sponsored and supported by missions. Missionary universities, such as that at Nanking, have contributed greatly to the study of China's agricultural economy; West China Union contributed most to dentistry in China; Yenching, the ace of missionary universities, has made large contributions to China's sociological sciences; and Peiping Union Medical, one of the finest institutions of its kind, was started as a missionary adventure and later supported by the Rockefeller Foundation.

But after a thousand years of intermittent efforts and over three centuries of concentrated efforts, the results today cannot be described as impressive. No one knows the exact number of church members in China, but even the most zealous have never put the figure higher than 3,000,000, or less than one percent of the total population. Of this number, about 400,000 are Protestants and the rest Catholics.[61]

[61] It has also been pointed out that the figures represent great increases for both

It may, of course be argued that quantitative strength is not as important as qualitative. Numerous stories of conversion, some of which are practically like legends and miracles, have been published and told by missionaries, but the following excerpt presents the other side of the picture, ignored in missionary chronicles.

In the first half of 1947, *Hsin Min Pao*, a small newspaper in Peiping, opened a column under the general title "How to Conquer Poverty" and asked for public contributions. A large number of responses were subsequently published. One of these expressed a deep feeling of depression, which struck the sympathy of another reader, who wrote the editor as follows:

DEAR SIR: I have read a number of your letters in this column and am particularly in sympathy with Mr. X. Now if we look at the conditions of society today, we may ask, who can really conquer poverty? Although Mr. X has a job and his monthly salary is a little over Chinese $90,000, this is only adequate to take care of himself. If he desires to support his family and accumulate some savings, it will be impossible. Nowadays rice and fuel are as expensive as pearls. Relatives cannot help continuously; nor are friends in a position to help each other. The proverb has it that incidental help can lift one from an emergency, but it cannot take him out of poverty. If we placed ourselves in Mr. X's position we shall agree that it will be impossible for Mrs. X to take up some job, since they have some small children and a nursing baby. But of course human beings cannot wait for their own extinction and not do anything about it.

I would like to offer my own experience. About twenty years ago I was in a similar plight. Then the times were good. I had a big family, I had no selling skill and I was unfit as a heavy manual laborer. So a friend introduced me to the Catholic Church at XX, where they have a Woman's Home. I took my wife and three small children to this Home to pledge their faith in the creed (Catholicism). I did some work elsewhere. I visited them once a week, bringing them some gift every time. In that enclosure they studied characters, learned the scriptures, and worshipped God. At first they were very unhappy and bored. But after some days they felt all right. They got to know other inmates and fell in with the routine. The children no longer craved for home, since they had lots of other children to play with. They had three meals a day. The food was not too good, but having lived in poverty, that was not unbearable. Those women who had nursing babies usually got a little more food. Generally

---

Roman Catholics and Protestants since 1911. See K. S. Latourette, "Christianity," in H. F. McNair (ed.): *China*, (Berkeley, Calif., 1946), p. 306.

a person could be eligible for Baptism three months after admission. There was no male in the Home. All teachers and other officers were nuns. Boys over six years of age were not admitted. Those admitted did not have to bring their own bedding.

In short, this is one of the ways of meeting an emergency. Just get Baptized and don't worry about the rest. The proverb says: "The Supreme Ruler of Upper Heaven will not starve even a blind sparrow." He will certainly not starve us, who are human beings, the most exalted of all living creatures, and are all children of God.

Don't be afraid of poverty. Ask God's mercy. If, three or four months after Mrs. X and her children have been admitted to the Home, Mr. X has found a good job, then he can at once ask for their discharge. If Mr. X cannot get a good job, then just let them stay there and save worries. Don't you think this is wise?

Another way out is to go to the Relief Department of the Bureau of Social Welfare of the Municipal Government. Mr. X should write to them describing his present plight. In conclusion I would say that Mr. X's difficulties are temporary. Just struggle hard for another few years and the children will grow up. They will then become independent members of society and understand the hardship of the times. Then they will surely be filial to their parents and support them.

My last word is, Mr. X, more patience for the sake of your children.

P. L. Liu
No. 68, Chiang Yank Fang,
Fifth Police Area, Peiping [62]

No one can fairly say what percentage of the less than 3,000,000 Chinese Christians are such rice-bowl converts. I can point out a number of great churchmen, both native and missionary, whose sacrifice, achievement, and integrity are unquestionable. The reverse side of the picture, however, has not been sufficiently emphasized, nor squarely examined, and it applies, not only to poor people like Liu, quoted above, but also to some men and women who have had the opportunities of advanced education.

So far as my personal experience and observation go, missionary activities in China today suffer from the following defects. First, there is no doubt that there has been a drop in the quality of missionaries going out to China in the last century. Many missionary teachers have been underqualified, particularly at the university

[62] Translated from *Hsin Min Pao*, Peiping, April 6, 1947. The original letter was written in not very fluent Chinese.

level. With a few outstanding exceptions, those who can find suitable posts in their home countries do not go abroad. For others, missionary positions have become merely job opportunities. Since these people do not have the dedication to the work of the earlier missionaries, desire for comfort and power tend to overshadow their real mission.

Recently I studied some of the missionary chronicles of recent years and also those of the period of the Boxer Uprising in 1900, when a large number of Western missionaries were massacred by rebels. In the older pictures, the missionaries, men, wives and children, dressed in Chinese costume. In the more recent pictures, the missionaries might have been in London, Paris, or New York. Dress, however, is only one of the indications of the way in which the manners and standard of living of the modern missionaries differ from those of the vast population whom they are trying to win for Christ.

A second defect lies in the fact that, being without deep conviction, many of the missionary personnel become overly susceptible to the backward environment in which they work. Thus a British lady missionary, coming from the home of liberty, slapped the face of an old servant who talked back to her.[63] I, personally, have seen similar happenings in Western Yunnan and my wife has seen them in central China. Because it is common for Chinese masters to manhandle their servants, and bureaucrats their subjects, the modern missionary feels that there is no reason why he should not do likewise.

A third defect is most fundamental. This is the failure to recognize the priority in importance of society over individual religion, especially in China. The traditional missionary line is that the key to conversion is the heart and, therefore, all hope is focused on individual salvation. But there can be no real happiness for the few when the vast majority are in misery. Particularly with the Chinese, who are accustomed to deduce the existence of the spirit world from the existence of the world of humans, such missionary tactics cannot but fall wide of the mark.

As a result, in all missionary institutions that I know, particularly on the higher education level, the tendency is for missionaries to cling to a few selected Chinese converts with whom they have

[63] Ida Pruitt, *A Daughter of Han* (New Haven, Conn., 1945).

achieved mutual understanding and who can provide them with a perfect sense of mastery and justification of their missionary work. But the student converts who are favorite protégés of missionaries are often extremely unpopular with the other students. They are regarded as opportunists who frequently turn out to be lesser Christians than others who have never been baptized.

The final defect is that missionary work in China has never achieved Chinese financial support and has, so far, been largely dependent on donations from abroad. It has been said that this was due to the poverty of the country. However, when the people are eager and interested, they do spend often excessive amounts on things which are quite removed from matters of bare existence. Poor Chinese families spend several months' income on a funeral or a wedding. They also spend large sums on matters of communal concern. In one southwestern community, for example, during the cholera epidemic of 1942, the total publicly announced expenditure for nineteen prayer meetings for combating the epidemic was Chinese $43,175. This amount represented, according to my calculation, the average monthly expenditure of 39 to 79 average families of the locality. The most significant thing about it was that, although the locality contains some outstandingly wealthy families, two of which spent, respectively, Chinese $300,000 and Chinese $1,000,000 for funerals, the largest single donation in these prayer meetings was Chinese $2,400. This means that the major part of the fund came, not from a few top families, but from many humble contributors. Moreover, the entire fund was collected and expended within a month.

To bear permanent fruit in China, the missionary movement must have indigenous financial support. Only when people contribute to a cause of their own accord will they take such a cause seriously. And this brings us back to the earlier observation that society comes before religion. It would be fantastic to expect missionaries to shoulder the entire burden of social reform in China. But it will perhaps be wise for missionaries and those in charge of missionary policy-making to think less in terms of individual conversion and more in terms of the social forces which shape the individual converts.

Education

Until recent times the illiteracy rate in China has been extraordinarily high. The number of persons who could read and write probably did not exceed one percent of the population. School education was a privilege of the few. The idea of universal literacy, if such a thing had ever been suggested, would have sounded absurd, not only to the administrators but to the people as well.

Because of this situation, scholarship has been accorded a more exalted position in China than anywhere else in the world. The written word has been regarded as sacred. All over the country there were Societies for the Protection of Lettered Papers, whose members collected every piece of waste paper on which there were written characters. These papers were burned to prevent the sacred words from being defiled.

The state provided no public education of any sort, except on the highest level for scholars who had already advanced themselves through the imperial examinations. Education was a private affair. A wealthy family in a village might decide to hire a tutor for the benefit of its own sons but allow sons of relatives and neighbors to attend the lessons also. Sometimes several families in the community would form a tutor-school jointly, sharing the salary and keep of the teacher. Or an elderly scholar might decide to form a tutor-school of his own, charging the parents of his students a sum for tuition each year. In all cases the ability to pay determined the amount of tuition.

The Chinese tutor-school was a very simple affair. It usually consisted of two rooms: the outer one, where a tablet in honor of Confucius was located, served as the entrance and also held some desks and benches; in the inner room was the teacher's desk and chair, set upon a platform facing rows of desks and benches for the pupils.

Upon entering the school, each pupil would make a gesture of homage to the tablet of Confucius, repeating the gesture every time he left or entered the school. Then the children would read or write from seven or eight in the morning to four or five in the afternoon. Sometimes there were night classes, lasting two or three hours. There were no intermissions, except for lunch and supper. The main part of the schoolwork consisted in reading, reciting from memory, practicing caligraphy, composing couplets, poetry, and essays. The textbooks were the standard classics, beginning with the Three Character Classics and going on to the Confucian Analects. Free thought was not encouraged. The pupils not only had to learn the texts by heart, but also the comments and interpretations of well-known and accepted scholars through the centuries. None of the books was indexed. When a pupil became sufficiently advanced to compose an essay of his own, he was expected to be able to draw freely, from memory, texts, comment, and interpretations from various books. A scholar who had to refer to a book before he could quote from it was a very poor scholar indeed.

This was a tremendous feat, for by the middle of the last imperial dynasty the known number of Chinese classics came to hundreds of thousands of volumes. No scholar amounted to anything unless he knew by heart at least several hundred. One of the usual sayings in praise of a genius was that he could "recite from memory what he had read only once." This sort of education encouraged great perseverance but it killed most, if not all, initiative. Instead of looking to the future, scholars looked to the past for inspiration and material for thought. If scholars and reformers wanted their new ideas accepted, they had to pretend that those ideas were taken from some great men of the past.

Parents had close supervision over the schools. Teachers could not teach anything or use any method that was not agreeable to the parents. It would be no exaggeration to say that the school was a mere appendage to the family. Those fathers who were literate would make secret tests of the literary qualifications of the teachers from time to time. Teachers were sometimes dismissed for not being strict and severe enough with their pupils.

School and state met only in the imperial examinations. These were arranged at four levels: the first was held every year in every district; the second, every three years at every provincial capital; the third, every three years in the national capital; and the last, at the Emperor's Court immediately after the completion of the third examination. These were not civil service examinations in the Western sense, because they were based on the classical training, and successful candidates were not automatically entitled to any definite or permanent position. On the other hand, success was an almost absolute prerequisite to most offices, especially the higher ones, such as ministers or their assistants. Successful candidates in the third and fourth examinations were more sure of being appointed to offices.

Competition became progressively stiffer. No exact data on the relative numbers of successful and unsuccessful candidates for the district examinations are known. At the provincial examinations the ratio was often something like 10,000 unsuccessful candidates to 100 successful ones. Since each province had a fixed quota according to its total population, but not according to the number of scholars, the competition was less keen in border provinces, such as Yunnan where there were comparatively few scholars, than in central and coastal provinces, such as Kiangsu and Hupei, where the degree of literacy was higher. Less than 300 candidates emerged triumphant from the third examination at the national capital. The fourth examination at the Emperor's Court was a sort of check-up on the third.

Thus Chinese education in former times was not merely for the few; it was also an education for bureaucracy, not for citizenship. In view of the fact that bureaucracy always benefits from and supports the existing order, there is little wonder that this education was focused on the past and uninterested in the future.

MODERN SCHOOLS.—Modern education had its start shortly after the Opium War of 1841. The first schools were founded by the Manchu government either for the purpose of training interpreters or for producing a few men with some knowledge of the technical sciences. In 1872 the first group of thirty chosen boys was sent to study in the United States. Later other students were sent to England, France, and Germany to study naval subjects and ship construction.

After the Sino-Japanese War of 1892 more schools were established along military and technical lines. The main idea then was to form a blend in education in which "Chinese learning would be the foundation and Western learning would be the use." None of these institutions and movements, which were under fire from bureaucrats and aristocrats alike, bore much fruit. Nevertheless, they blazed the trail, for in 1905 the old examination system was definitely abolished by an imperial edict and a Ministry of Education was set up which had the function of promoting and administering institutions of higher learning. A commissioner of education in each province carried on the same function with relation to other schools.

At this time there were five notable colleges, among them St. John's and Nanking, both missionary schools. Schools of the modern type appeared in the provinces much earlier. Through the promotion of Kang Yu-wei, Liang Chi-chao, and their associates, 19 schools, 24 associations, and 8 newspapers were founded in 1897–99 in Hunan, Kiangsu, Kwangtung, Peiping, Kwangsi, Shen, Hupei, Chekiang, and Fukien. Although the purpose of some of the associations was nonscholarly and the scale of the schools and newspapers was small, their establishment represented a spontaneous movement. During the same period, the number of Chinese students abroad increased tremendously. By the end of 1905 there were 8,000 or more Chinese students in Japan. By 1907, according to official statistics, there were 1,500,000 students in the secondary and primary schools in the provinces. Many of the schools were private; and, of these, the most notable were founded and financed by Yeh Ch'eng-chung, who began life as a peddler on the Whangpoo River, and Yang Sze-sheng, a mason. In 1899 Yeh contributed over 200,000 taels and a large estate for a middle school which bore his name. In 1904–5 Yang gave considerably more than this for the founding of a primary school and a middle school called Pootung.

Of course, modern education could not be developed merely by increasing the enrollment and the number of schools. In the first place, the spirit of most government schools was far from modern. Their main emphasis was loyalty to the emperor, reverence to Confucius, as well as endeavor in industrial pursuits. Secondly, most students and their parents regarded the new school education in the traditional light, as merely another gateway to bureaucracy.

More important changes came with the Revolution of 1911. These included alteration of the curriculum, enlargement of school facilities, increase of emphasis on handicraft and physical exercise, introduction of coeducation in primary schools, and the elimination of the classics from the lower schools. With the Second Revolution of 1926–27 and the assumption of power by the Kuomintang, came further changes, including greater compulsory registration of all schools (indicating greater centralization), an emphasis on vocational education, and a popular movement against illiteracy.

Before the outbreak of hostilities between China and Japan in 1937, there were about 2,000 straight middle schools, 800 normal schools for the training of teachers and 400 vocational schools. As the war progressed over one-third of these schools were in Japanese-occupied territory. But the number of secondary schools in Free China was increased to 2,483 in 1940–41 and to 2,819 in the following year. The total enrollment of secondary schools was 622,800 in the year 1939–40, of which over 80 percent were in middle schools and a little over 10 percent in normal schools.[64]

There were many more elementary schools and pupils. In the year 1937–38 there were 229,911 such schools with over 12,800,000 pupils. About 2,500,000 pupils graduated from the elementary schools during that year. The figures for 1938–39 and for 1939–40 were similar, showing slight decreases rather than increases. What percentage of the pre-war elementary schools were in occupied territory and whether the 1937–38 figures represented large increases or decreases over the pre-war years is unknown.

The number of Christian middle schools was 255 by 1942. One hundred thirty-three of them maintained also elementary schools. The total enrollment was 53,673 by the same year. No figures on Catholic schools for the same year were available; it is reported that in 1935 there were 480 with an enrollment of 26,263 students, of which about 17,000 were males and 9,000 females. It is presumed that a number of these were elementary schools. The number of pupils in missionary establishments thus came to less than 0.5 percent of the total enrollment of all secondary and elementary schools in the country.

The subjects taught in Chinese schools were similar to those in American schools. The classes in civics were based upon Dr. Sun Yat-sen's *Three People's Principles*. Many Chinese graduates who entered American universities suffered no particular disadvantages because of their educational foundation. In terms of the ratio between the numbers of teachers and students, the Chinese schools

[64] These and other similar figures are extracted from *China Yearbook*, 1938, and *China Handbook*, 1937–43.

appear to be better off than those in the United States. However, Chinese schools in different parts of the country vary tremendously in quality. The average graduate from such famous schools as Nankai in Tientsin is apt to be far ahead of the average graduate from schools in interior provinces like Szechuan and Yunnan.

MASS EDUCATION MOVEMENT.—One of the greatest needs in Chinese education is a change in the attitudes toward schooling, so that the people will feel that education, instead of being a privilege for the few, designed solely for the purpose of admitting the scholar to the bureaucracy, is the right of every human being and designed to further good citizenship for all. This attitude is far from being established, but a beginning has been made. First, the much larger number of people who have attended schools than ever before cannot but have a broadening effect. Scarcity always creates the illusion of special value for the scarce item. Secondly, mass education is something unique in Chinese history.

The exact figures on literacy in contemporary China are not obtainable. Sample studies in selected communities show that among men, this varies between 5 and 30 percent, and that among women it varies between 0.5 and 4 percent. In 1938 over 80 percent of the total population were illiterate.[65] The mass education movement concerns itself with about 40 percent of the population between the ages of 15 and 45.

Both government and private initiative attacked the problem, the government through a system of *Hsiang* (subdistrict) nucleus schools, and *Pao* (one-tenth of a subdistrict) people's schools. It claims that, since 1938 and up to 1942, roughly one-fourth of those between the ages of 15 and 45 have been educated. Having seen a number of *Hsiang* and *Pao* schools, I wonder how realistic it was to deduct from the total number of illiterates those between the ages of 6 and 15. All such schools which I have seen in various parts of the country included children within this latter age bracket. The exactness of the government figures is dubious, but there is no doubt that some effort and some progress have been made.

[65] *China Handbook*, 1937–43. Ta Chen's findings in Kunming city, the capital of Yunnan province, shows that 41.1 percent of the males and 75.4 percent of the females have never been to school. For Kunming *hsien*, which included some semi-rural areas, the percentage of illiteracy for males and for females was raised to 49.6 and 79.8 respectively. These figures pertain to 1942, five years after the area had been subject to a tremendous outside influence due to the war against Japan. See Ta Chen, *Population in Modern China*, p. 123.

The private efforts were represented at their best by James Yen in Ting Hsien, Hopei province. Other efforts were made in Kiangsu, Shantung, Chekiang, and Kiangsi provinces, but Yen's work was on a larger scale, developing the most elaborate theoretical ground-work and receiving the most attention at home and abroad. It is based on Yen's diagnosis that the masses of China suffer from four "illnesses": poverty, ignorance, poor health, and selfishness. To combat these, he designed a fourfold program: (1) improvement of livelihood, (2) literacy, (3) public hygiene and (4) civics. The main technique is demonstration and participation. For improve-ment of livelihood the peasants are taught how to use better seeds, fertilizers, and marketing techniques, and how to breed farm ani-mals, including pigs and poultry. Peasants are taught to use the Thousand Characters, a simplified written Chinese corresponding more or less to Basic English. For betterment of public health, a system of public medical centers are organized which not only treat illnesses but also teach peasants the fundamentals of hygiene and child care. For training in civics Ting Hsien became an experiment in popular self-government.[66] Yen employed a large number of assistants, many of whom had received their training in philosophy, political science, sociology and economics, agriculture, entomology and other technical subjects, either in China or abroad. The entire project was financed largely by voluntary contributions from abroad.

There is no doubt that Yen's work provided numerous benefits for the people of Ting Hsien, and that it brought about an awaken-ing among the intelligentsia to the existence of rural problems. But two general criticisms can be made. First, from our analysis of the economic situation, it is clear that rural livelihood is closely linked with urban development. The lack of industrial development, to-gether with age-old tradition, is at the root of a corrupt bureaucracy which keeps the country in a feudal state and effectively prevents it from solving its problems. A nation is not a mere conglomeration of districts and villages. Even if Yen has found a complete solution of all the recognized problems of the village, as indicated in the program, we are still far from being in sight of a solution of those

[66] For an eloquent exposition of the program and its ideals, see P. S. Buck, *Tell the People* (New York, 1945).

problems for the nation as a whole. What is more, without taking into account the organic nature of the nation as a whole there can be no real or permanent solution of the village or district problems as separate entities.

The second criticism is a financial one. Yen's results in Ting Hsien were achieved with the assistance of a large number of graduates from Chinese and Western universities and with large amounts of money donated chiefly from abroad. One must ask how other districts in China can afford such a program. Of course, Yen's idea was that his work was an experiment, the results of which would serve as a model for others. However, most rural work stations have so far claimed to be similar but smaller experiments. Because this costly attack on the observed problems has no roots among the masses of the people, there grew up an untenable slogan that college students should go back to villages to improve the life and conditions of the peasants, while, at the same time, they would have to live under more or less the same conditions that they were trying to improve. The basic trouble is that without some strong incentive, such as religious fanaticism or the exigencies of war, it is impossible to persuade people to give up comforts and a high standard of living in favor of a life of hardship and discomfort. The Mass Education Movement, as James Yen founded and promoted it, has reached a dead end. For results commensurate with the expenses and personnel involved, it will have to find new definitions of methodology and a new approach.

HIGHER LEARNING.—Shortly after the turn of the century a national university was established in Peking by imperial decree, but it was an institution of higher learning in form only. By 1905 there were five institutions of higher learning of some standing, two of which were missionary establishments. The three indigenous ones were Peiyang University, Tientsin; Nanyang College, Shanghai; and Shansi College, Taiyuan. By 1928 there were 74 institutions of higher learning in the country. By 1936 the number went up to 108. The war against Japan destroyed some of them, but by 1944 the total number was 137, nearly 40 percent more than in 1936. The total enrollment is unknown. The Ministry of Education's figures on those who have passed the entrance examinations each year is indicative of the actual enrollment. Between 1931 and

1936 the number was about 40,000 per year; in 1940 the figure was raised to 50,000.[67]

If we compare the available enrollment figures for the primary schools, middle schools, and institutions of higher learning, we get a rough ratio of 12:6:0.5. That is to say, there are twelve primary school pupils to six middle school pupils, and six middle school pupils to half a college student. The comparable ratio in the United States in 1930–40, the last normal years before the war, is 21:7:1.4.[68] Thus there are more middle school pupils in China than in the United States, compared with the enrollment of the other levels. Also it is not true, as generally believed, that, in proportion to elementary and secondary enrollments, the college and university enrollments are larger in China than in the United States.

But when we consider the expenditure per pupil in the three types of institutions in China and in the West, the Chinese figure becomes fantastic. According to reports of the League of Nations for the year 1930, the expenditure per pupil in the lower primary schools was $3.50 to $4; in higher primary schools, $17; in middle schools, $60; in normal and professional schools, $120, but in colleges and universities, $600 to $800 (all Chinese national dollars). The ratio between the lower primary schools and the colleges is about 1:200. In the West the ratio is about 1:10.

The same picture is seen in the relative salaries of university professors and schoolteachers in the two countries. In China the earnings of the former group are generally 15 or more times those of the primary schoolteachers. In the United States the corresponding figures are generally 3:1 to 5:1.

The same overemphasis on higher education is shown in the expenses of Chinese students abroad. In 1930 there were 50,032 Chinese students attending institutions of higher learning in Japan, Europe, and the United States. If we take the low estimate of Chinese $3,000 to $4,000 (corresponding to U.S. $900 to $1,200, before World War II) as being the annual expenditure per student, we have a total of about Chinese $15,000,000 to $20,000,000. In 1934 the total budgeted expenditure of all registered universities and colleges in China was only Chinese $24,480,000.

[67] Figures taken from *China Handbook*, 1937–43, and "Facts on China," mimeograph publication by United Service to China.
[68] John Kieran, ed., *Information Please Almanac* (New York, 1947), p. 209.

In view of the country's traditional background this emphasis on higher education is natural. For centuries bureaucracy offered the only really lucrative opportunities, as the attainment of a higher degree through the system of imperial examinations was essential to a place in this bureaucracy. Since the abolition of the examination system, and since, in the popular mind, degrees in the modern educational scale are equated with the old imperial honors, the emphasis on higher education is easily understood.

This situation has resulted in serious unemployment among modern educated Chinese. Unemployment was most acute in the years between 1925 and the outbreak of hostilities with Japan and is present again since the Japanese surrender. Unemployment among the literati is nothing new to China. Before 1911, Peking and the various provincial capitals had thousands of scholars waiting for official posts. The situation became worse after the 1911 Revolution, for at this time the bureaucratic offices were reduced while the number of students on all levels increased. There are few opportunities in industry or commerce to absorb the increasing numbers of new literati who crowd government departments, bureaus, and, as second choice, institutions of higher learning. The government and its bureaucracy being what they are, the talents of those fortunate enough to receive appointments are not put to good use.

Those who are eager to help China have wondered why students with excellent academic records and general worthiness should either become corrupt and inefficient officials or disillusioned money-makers; they have concluded that the younger generation is selfish and irresponsible. The difficulty lies, however, in the training and conflicting standards to which the young Chinese is exposed. He has been brought up in a background which stresses success and the need to bring honor and wealth to his family and his relatives. He was educated in schools modeled on those of Western industrialism, in which the economic incentive is dominant. He may have spent several years in Europe or America where financial gain is the measure of the worth of the individual. However, when he returns to China to take up his life work, he is expected to discard the monetary incentive and assume a totally different attitude toward life and success. Nothing can be so absurd. This folly has already been demonstrated in connection with the college-student-return-to-village movement.

It has been generally believed that the government's effort at centralization of schools and especially of institutions of higher learning was opposed by educators and students, and that the national universities were unpopular. As a matter of fact, faculties and students in the government-controlled universities are for the most part better qualified than those in private universities. Moreover, more teachers and students apply, respectively, for posts and for enrollment in government universities than in private ones. Even more significant is the fact that some private universities are eager to be taken over by the government. In 1946 students and faculty members of China University in Peiping organized into an Association for Nationalization of China University. Early in 1947, Ta Hsia University in Shanghai organized a similar campaign.

The key to this situation is the lack of permanent financial backing in all Chinese universities. With the exception of National Tsing Hua University in Peiping, which was built and founded on the American share of the Boxer Indemnity Funds, no university in China has an endowment which makes it self-sufficient. In fact, because of the poverty of the country, the vast majority of colleges have no endowment whatsoever. The government universities, because they are founded and operated on taxes, have greater security and can therefore afford better equipment and attract more highly qualified teachers.

In spite of the political and civil disturbances, bureaucratic corruption, poverty, lack of facilities, student strikes, and the devastations of war, the institutions of higher learning have, in the last two or three decades, made tremendous progress, particularly when one considers the tradition-bound background from which they are slowly emerging. Chinese researches have achieved international reputation in geology, linguistics, philology, history, certain branches of medicine, agricultural economy, and social anthropology. There are even Chinese inventions in certain applied branches of the natural sciences.

Acculturation

Many centuries before its first contact with the West, acculturation had been going on in China in varying degrees of rapidity at
different stages of its history. For the most part, it was a one-
sided operation, as the various ethnic groups, even when they came
as conquerors, took on the culture of the Chinese rather than imposing their own ways upon the conquered people. In spite of the
great climatic differences and regional diversities of the country,
there is a very real and fundamental unity in its culture. To be
sure the dialects may vary so greatly that people from some districts
will not be able to understand those from others, but there are
certain things, such as the kinship nomenclature, which are universal in their basic pattern. In spite of local differences in customs and manners there are many things, particularly the uni-
versality of the family structure, which give unity to the culture.
Unless we make the unlikely assumption that all inhabitants of
China are descendants of the original Chinese, we must assume that
the present basic unity is a result of sinification, or absorption of
other ethnic groups through acculturation.

Not all the ethnic groups which have inhabited the country and
maintained more or less close contact with the Chinese have become completely sinified, however. When the Republic was
founded after the Revolution of 1911, the popular slogan was, "A
Republic of Five Races," meaning the Hans (original Chinese,
who arrived in early times from Central Asia), Manchus, Mongols,
Mohammedans, and Tibetans. In recent years a sixth group, the
Miaos, has been added. The actual number of ethnic groups to be
found in China is, of course, much larger than this.

Most of the original inhabitants were absorbed in early times by
the invading Hans from Central Asia; a few in the border regions
to the east and south have remained more or less culturally independent. The Min Chias of the southwest and the Manchus of the
northeast are examples of nearly complete sinification. The former
group has only its language to show its non-Chinese origin. The
Manchus came into contact with the Chinese as conquerors. While
their political power lasted, they constructed rigid barriers against

assimilation in an attempt to maintain their cultural identity. Even so, they accepted so much of Chinese culture that, as soon as their political domination ended, the Manchus differed little from the Chinese except in such matters as their head-dresses and the unbound feet of the older women. On the other hand, the Chinese took over from the Manchus a number of cultural items, including gowns for males and females.

Other conquerors have ended their rule differently. The Ch'itans (founders of the Liao dynasty), the Jurchens (founders of the Chin dynasty), and the Mongols (founders of the great Yuan dynasty) all more or less effectively resisted sinification. When their political power was lost, the majority of them left the mainland and joined their tribal brothers. Culturally they remained apart from the Chinese, even though there was mutual influence.[69]

A number of other ethnic groups have remained outside the orbit of Chinese culture, but most of them maintain some kind of contact with the Chinese in the form of trading, or tenant-landlord relationship, or management of the local government. In this group may be mentioned many branches of the Miaos, Lo Los, Yis, Tibetans, Nakhis, Shans, Mohammedans, Yus, and others. The relationship between the majority of these ethnic groups and the Chinese is cordial, and there is little question of racial superiority or inferiority. The Mohammedans, who have a religion as aggressive and monotheistic as Christianity, are the only group which has had constant friction with the Chinese.

CONTACT WITH THE WEST.—With the coming of Europeans and Americans to the East, the situation changed. Westerners visited China long before the middle of the nineteenth century, but it was after the Opium War with Britain in 1841 that the government and people were forced to give serious attention to the problem of the West. Hitherto Chinese culture had not been seriously challenged. Even under Mongol or Manchu domination, there was no thought, by Chinese scholars and others, of any change in the age-old Confucian traditions; such an idea would have been considered monstrous. But with military defeats and definite signs of economic and

[69] See K. A. Wittfogel, "General Introduction" to "History of Chinese Society, Liao (907-1125)," *Transactions of the American Philosophical Society*, XXXVI, pp. 3-14.

technological superiority in the West, some Chinese leaders and thinkers became alarmed. Many of those who initiated and promoted the cause of radical economic and political Westernization did so at the risk of their lives and the lives of their families and relatives. The preaching of such changes was regarded as treason.

The arguments of those who opposed the new ways were simple enough. Chinese culture and sagely teaching were the best in the world. Historically, the Chinese had always civilized their barbarian invaders; the reverse had never happened. To adopt the ways of barbarians was contrary to the Confucian teaching and would also affect the ancestral laws. Thus such changes would be both illogical and unfilial.

To break down the opposition, the reformers had to resort to half-measures and propose changes only in technological fields, but not in customs, government and traditional ways and ideals of life. Some tried to crush the opposition by pointing out that their arguments were absurd even in the light of the Confucian classics. However, the opponents to change had the political vested interests and the apathy of the people to support them. Seldom have autocratic groups given up even a little of their power without a bloody struggle. Many intellectuals were arrested and executed as a result of their somewhat radical ideas. But nevertheless the West came closer and closer. More missionaries came to China, more gunboats, more soldiers, more merchants, more diplomats, more commodities, including British opium. There were more military setbacks and concessions. All these things led to the Boxer Uprising of 1900, which the Manchu rulers endorsed but which marked the end of all serious thought of shutting the West out of China.

It is impossible to deal with all the phases of acculturation in modern China in this short space. I can only go into a few selected aspects of the problem.

LAW.—China had written law long before any of the European countries, but in the nineteenth century the progressive countries of the West were far in the lead in matters of legal administration. The most important principles of the Chinese system of justice at that time were: (1) the principle of restitution and even revenge, in which an eye was actually taken for an eye; (2) the emphasis on family and political status. For example, if a father killed a son, even with malicious intent, the punishment was often light. If a son killed a father, even accidentally, the punishment was extreme, often death by being cut into minute pieces. The same principle held with regard to crimes involving superiors and inferiors in the bureaucracy.

The judges had too much power; in all litigation, all plaintiffs, defendants and witnesses had to prostrate themselves before them. Torture was not only resorted to as punishment but openly as a means of extorting confessions. Prisons were literally hells on earth and the prisoners were at the complete mercy of greedy jailers.

By the end of the first quarter of the twentieth century, Chinese written law had been transformed into one of the most up-to-date bodies of law in the world, based upon the best of Western law. It is now practically free of the old familial influence, so that the son is no longer a lesser human being than his father. It is free of the domination of the Christian Church, so that the sexes are completely equal, with equal rights to property and divorce. The principle of divorce by mutual consent, which has been fought in England and the United States, was arrived at in Chinese written law with the greatest of ease. In addition, prisons have been remodeled, corporal punishment and koutouing in courtrooms have been abolished.

However the law in practice is closely dependent upon the people who come before it, express opinion about it, and operate it. The people are guided by customs and traditions, which do not change as quickly as written law. For this reason there has been a sharp conflict between the spirit of the law and of the customs. Elsewhere I have described the situation as follows:

The spirit of modern law demands uniformity; the customs of a land so vast and so poorly equipped with means of communication as China are bound to be highly diversified. The law has abolished concubinage; the people regard concubinage as a matter of course. The law prescribes that no private individual has the right to commit violence against any other individual; the people regard beating of slave-girls and apprentices as something inherent in the order of nature. The law says that social or sexual intercourse between unmarried adults is entirely their own concern; the people look upon any unrelated two persons of different sex having close contact with each other as having committed adultery and are to be treated as such, with violence sometimes endangering the lives of both parties.[70]

Some of these contrasts and conflicts are particularly sharp in the interior provinces and villages. The operators of the law resolve the conflicts, not by strict enforcement of the law, but by a spirit of compromise. For example, mistreatment is admitted as grounds for divorce, but this is interpreted in the light of the education and standing of the parties involved. If both parties are college educated, a slap on the face might be admitted as constituting maltreatment; if they are not highly educated, and particularly if the wife is illiterate, divorce is not granted even if the mistreatment is of a more violent nature. The judges, in applying such a sliding scale, are actually protecting the ignorant women to whom divorce usually means ruin.

Nevertheless, much hardship, misery, and violence have resulted from the difference between the written law, on the one hand, and custom and tradition on the other. The written law has scored victories because the legislative bodies so far do not represent the entire population. Some of the legislators are progressive elements of the intelligentsia. Some of them are bureaucrats, whose main interest in this instance has been to make China look as modern as any advanced Western country. But there is no doubt that the law, once written and enacted, serves as a guiding influence. There is equally no doubt that its sphere of operation is daily widening in many respects.

STUDENTS.—The break from the past has been more complete in education than in law, as has been shown previously, and the change is largely of Western origin. The old schools offered training for bureaucracy; the new institutions provide training for citizenship and for the various trades and professions. Because of the undeveloped state of the country's industry and commerce, and because the attraction of bureaucracy is still great, the products of the new schools, like those of the old ones, are still largely unable to fulfill their functions adequately.

However, the trend toward practical subjects has been noticeable, especially in recent years. This point is demonstrated by a

[70] F. L. K. Hsu, "Some Problems of Chinese Law in Operation Today," *Far Eastern Quarterly* (May, 1944), pp. 212–13.

study of university enrollments. Before 1937 the most popular subject in the curricula were arts and literature. By 1940, this had changed. According to Quenten Pan, dean of National Tsing Hua University, Peiping, the vast majority of the science students enrolled in the practical branches of the sciences.

Comparing students of today with their predecessors, one cannot help but be impressed with the broadening interests of the newer generation. The old-fashioned scholars lived in a narrow world of Confucian classics and bureaucratic wire-pulling. Modern students have active views on Christianity and other religious matters, on the liberation of the Chinese people, the future of the peasant, humanity, politics, romantic love, and on the different schools of philosophy and literature and their bearing on life. According to the inquiries of Olga Lang, the most popular authors in contemporary China are those who write social fiction, including Lu Hsun, Pa Chin and Lao She. Only one of the popular authors, Ping Hsin, is nonpolitical and writes of things not dealing with existing social conditions. Among foreign authors the Russians Maxim Gorki and Anton Chekhov are the most popular, with Goethe, Dumas, and Bernard Shaw coming next. In the nonfiction field Marx's *Das Kapital*, Engel's *Origin of the Family*, and H. G. Wells's *Outline of History* have been found to be the most popular. Lang further found that of the 350 Christian informants questioned only 32 mentioned the Bible.[71]

Since the revolution, students have not merely expressed their views by reading and writing; they have participated in rural reform projects, beaten up traitors and other officials who were believed to have unnecessarily given away the country's rights, marched to the capital in protest against the nonresistance policy toward Japan, organized newspapers, established factories, led strikes, staged lectures to the masses, helped the universities to move equipment and men over several thousand miles on a land route to the southwest when the Japanese occupied the coastal provinces, and joined the guerilla forces against the enemy. Of course, in their youthful enthusiasm they made many blunders. Sometimes their actions were instigated by ambitious politicians. Some became reactionary bureaucrats. But as a whole, the students

[71] Olga Lang, *Chinese Family and Society*, pp. 275–77.

are that portion of the people who have taken an active part in counterbalancing the autocratic power of the government. They have helped to waken the masses to their rights and responsibilities.

FAMILY RELATIONSHIP.—The outstanding changes in family relationship pertain to parental authority and the position of women. As is the case in other fields, the changes in the cultural patterns concern the comparatively few educated people more than they do the vast majority, but the new patterns have come to stay and will propagate themselves.

The relationship between father and son is fast becoming a matter of companionship rather than the old one of authority and submission. However, there does not seem to be any outright desire to override the persons of the older generation. Olga Lang has found a hesitation among students to criticize their parents, although there is dissatisfaction, couched in comparatively mild terms, against the older man's conservative political views, his gambling and frequenting of brothels, and so on. As far as can be observed, the age-old father-son identification remains an important factor in shaping the personality of the young.

The rise in the position of the educated women has been spectacular. Within less than three decades the women who were once confined to their homes, illiterate and with bound feet, are now enjoying a position comparable to that of their American sisters. They are admitted to most professions, including medicine, law, and the diplomatic service. They have equal rights with men before the law. Although the percentage of women in schools was only about 10.3 of the total enrollment by 1932, the vast majority of the educational institutions are coeducational.

RECREATION AND SPORT.—Fifty years ago there was no Sunday in China. Regular holidays only occurred on festivals, New Years, and the end of the harvest. Now a Sunday holiday is a matter of course for a large number of city workers in modernized establishments and for all government offices. The moving picture has become so popular that Hollywood now makes films with subtitles in Chinese script. In contemporary China there are movie fans, sport fans, stamp collectors, match cover collectors, and a host of other hobbyists of recent Western origin. Drama in Western style has become very popular. Student amateur groups produce well-known plays like *Lady Windermere's Fan*, and *Salome* in Chinese. Chinese opera in the traditional style has remained popular, but even here, operatic themes, sequences, and stage production have been drastically changed as a result of Western influence, much to the distaste of the orthodox connoisseurs. After a tour in the United States, where he received an honorary degree from one of the colleges, Mei Lan-fang, the first actor of China, tried to introduce the violin on the Chinese stage, without much success, however.

Social dancing has become popular, not only among the intelligentsia but also among merchants and other city dwellers. Dancing is not unusual in the best circles, but, on the other hand, some dance halls are little more than brothels. Because of the old patterns of segregation of the sexes, and because dancing has been taken up in the brothels, there have been violent attacks upon the pastime on moral grounds; some city authorities have attempted to put a ban on it, but without much success.

Through social dancing and the movies, jazz music has gained a foothold. The tunes which are popular in the United States are soon taken up by the younger people in the Chinese cities also. But in areas where the movies and dancing have not penetrated, the foot-pedaled organ is a popular instrument. I sang to the tune of such an organ 29 years ago under the leadership of a teacher in a market-town school in Manchuria. During the long war years numerous Chinese composers have arisen who combine Chinese themes with Western technique. Group singing has become widespread.

Music of the classical Western type is less appreciated. There is one conservatory at Nanking staffed largely by Europeans. Before the war there was a symphony orchestra in Shanghai and another in Hankow. In addition, a number of the missionary universities have college orchestras which are able to perform such numbers as the *William Tell Overture* or *Poet and Peasant*. They also have glee clubs and choirs. For several years Handel's *Messiah* was performed by a choir of over a hundred singers under the auspices of Yenching University at Peiping. However, because of the lack of cultural background for the appreciation of classical European music, this type of music has little appeal even for the intelligentsia of China. Kreisler, Moiseiwitsch, Heifetz, Elman, Szigeti, Feuermann, and many other fine soloists who have toured China have drawn only limited audiences.

In sports the change has been revolutionary. There was no organized sport in old China. Ice-skating and wrestling were done by professionals for the entertainment of the officials and aristocracy. The gentleman scholar was one who did not have to move a finger. Since the coming of modern schools, the pursuit of sports has become part of the regular curriculum. Tennis, basketball, football (soccer), volley ball, track, baseball, skating are all usual in Chinese schools and colleges. There are also interschool, intercollege, and intercity sports events. There are North China Meetings, South China Meetings, as well as National Meetings. In addition, China has been a regular participant in the Far Eastern Olympics, competing with athletes from Japan, the Philippine Islands, and recently, India. Since the early thirties China has also participated in the World Olympics.

LIVING CONDITIONS.—Agriculture, on which the livelihood of over 90 percent of the population directly depends, has remained very much in its traditional state. This means that, for the vast majority, there has been little or no change in the basic means of subsistence.

For the educated and most of the urban population, on the other hand, there has been considerable change in living conditions. They are becoming conscious of the importance of vitamins and a balanced diet. The drinking of coffee and milk is a developing urban habit. Although most European and American foods are not generally popular and are only eaten occasionally as a novelty, ice cream and dairy products have become widely accepted in the cities.

For the educated and some of the urban population there has also been a change in manner of dress. Twenty years ago European costume was rare, even in big cities like Tientsin and Mukden. Today European dress is seen even in the small market town of the interior. Leather shoes have become common and are worn even in the small villages by both adults and children. Even the students who still wear the Chinese costume are shod in leather shoes of Western style. Chinese women, however, have not taken to European dress as the Japanese women have. The Chinese woman who wears European dress is rare in China on any level of society and, even when living abroad, Chinese women usually prefer to wear their traditional costume. In the United States only the second generation wear American clothes.

In the matter of housing there has again been little change for the majority of the people. In the cities and among the wealthier families, Western-style houses are becoming prevalent. In the five or six years just before the outbreak of hostilities with Japan, Nanking witnessed the biggest boom in house construction in the history of any Chinese city. I toured all sections of the capital in 1937 and saw not one house of the traditional type which had been built within the last six or seven years.

Of course, many of the public buildings which have been built in recent decades exhibit a sort of marriage between Chinese and Western styles. This is true of Peiping Union Medical College, financed by the Rockefeller Foundation; and also of the National Library and Yenching University in Peiping; and of Wuhan University in Hankow and Ginling's Women's College in Nanking. These buildings have the appearance of traditional Chinese palaces, but are equipped with the most up-to-date conveniences of the West. In most of the wealthier homes there are modern toilets, bathrooms, electric lights, running water, wooden floors, and soft chairs and sofas, in contrast to the old-style hard, square wooden chairs and benches.

Considering the size of the country and the population, improvements in transportation have been negligible. But these changes affect a much larger number of people than the previous items discussed. Up to 1945 there were about 80,000 miles of highways of all sorts and about 13,000 miles of railways, mainly concentrated in North and East China and Manchuria. The Yangtze River is navigable up to Hankow for ocean vessels (595 miles from the sea) and up to Chungking by special steamers (1,427 miles from the sea). Before the war a large number of British, Japanese, and Chinese modern vessels plied this river, as well as traveling between the coastal ports of Dairen, Tientsin, Tsingtao, Shanghai, Foochow, Swatow, and Hong Kong. Air service is comparatively recent but now connects all the large cities.

Nowadays not only the wealthy and educated, the bureaucrat, merchant, and militarist travel by these modern means, but also thousands and thousands of peasants and laborers, in search of more fertile soil, better work, or more profitable markets for their wares. However, China's unfortunate masses travel only for necessity, rarely for pleasure, except occasionally for important visits to relatives. Travel is not a pleasure but a dreary toil, as anyone can understand who has visited the steerage class of any steamer plying Oriental waters or seen the teeming coaches of Chinese railways. Yet even the poor peasant who sees no reason for sending his daughter to school now takes a train or a bus if he must travel instead of making the laborious trip by foot or by donkey as in the old days.

I could mention hundreds of other items which have come to be accepted in the context of Chinese culture; among them, fountain pens, eye glasses, watches, and rubber and leather shoes have come to be in general use in the most remote parts of the country.

COMMUNISM.—The 1911 Revolution changed the form of government and made China into a "republic" with a president and a cabinet who were supposed to be responsible to the two houses of parliament which were supposed to represent the people. But this was a change in form rather than in substance. The old bureaucratic machinery remained, with much the same personnel. The members of parliament, instead of being elected by the people, won their chance to sit in the impressive buildings at the capital by licking the boots of the president and his immediate satellites. The masses continued in the old subservient way, not only at the mercy of the president but also of minor bureaucrats and military commanders.

The second revolution of 1927, led by the Kuomintang, which has since become the dominant party, unified the country politically and militarily more successfully than any government since the downfall of the Manchu dynasty, but the same bureaucratic machinery still administered the country. While the Manchu dynasty lasted, all power was vested, ultimately, in the emperor or the dowager. When the dynasty fell, the final source of power rested with a number of independent militarists who were practically little kings within their own domains. After the ascension of the Kuomintang the source of power was broadened. It was vested in a much larger number of members of the party who form the inner cliques and who are backed by armed forces.

Some new names came to high places in the government, only to reappear year after year, if not in the same high places, in different places of equal importance. The bureaucrat who is powerful in the party or who has the favor of the supreme leader is assured of his place in the sun. Even the Control Yuan, which supposedly has the power of impeachment, has been repeatedly shown to be power-less in the case of erring but favored high officials. Not even the office of the magistrate has ever been filled by popular election. The people are organized, ruled, told what to do and how much to pay. They do not know why they must do certain things or to what use the taxes which they pay are put.

The outstanding phenomenon of political acculturation in mod-ern China is the growth and rise of the Communist Party. It is popularly believed that China is a country in which most of the land belongs to landlords who suck the blood of the tenants who do the actual tilling of the soil. Dr. Sun, founder of the Republic, who was born in Kwangtung province where the tenancy figure is ex-ceedingly high, was under this impression when he specified in his program that "all tillers must own their own soil." Both American and British journalists merely repeated this popular belief.[72]

From my own analysis of land tenancy, I believe that it is not a main cause for complaint and dissatisfaction among the peasants and not the main reform instituted by Communists. In spite of some reported abuses, the real strength of Chinese Communists lies in their drastic reduction of state expenditures, of the cost of keeping armed forces, and particularly in the fact that their administrative personnel, regardless of rank, maintain much the same standard of living as the common citizen. By reducing expenditures the Reds have been able to lower taxation and curb corruption, thus reduc-ing the financial burden of the peasants. By lowering their own standard of living to meet that of the common people, they give the peasants for the first time a feeling that there is some affinity between themselves and the functionaries of the government. The

[72] See Edgar Snow, *Red Star over China* (New York, 1938), p. 216; and Harrison Forman, *Report from Red China* (New York, 1945), p. 178.

curtailing of luxuries among officials does away with the usual incentives for corruption, makes crushing taxation unnecessary, and destroys the conditions which have enabled usury to flourish. Thus with the people gaining an increasing share in the local government, the Reds, in spite of the obviously autocratic nature of their administration on the higher levels, have made possible a degree of social and economic equality which has never before been known in Chinese history but which is also the universal demand of our times. This demand is the most outstanding contribution of the West to culture, not only in China, but in most of the world.

Conclusions

In a short survey on so vast and complex a country as China it is impossible to do more than touch upon a few salient features. The most urgent problems of today are the alleviation of poverty and the establishment of a truly democratic government.

Poverty is partly a result of overpopulation. Birth control has been suggested as a solution, but until the standard of living is raised (it is now among the lowest of all literate countries), it will be almost impossible to introduce methods of birth control among the Chinese peasants. The standard of living can only be raised through industrialization.

Many factors have hampered industrialization in China. Before 1942, unfair foreign competition through such things as a nominal tariff on foreign goods under the "Unequal Treaties" was one of them. But since the abolition of the "Unequal Treaties," two indigenous factors have remained paramount. First, the family, with its emphasis on father-son identification and the big family ideal, was essentially conservative and frowned upon new ideas and ways which could not be fitted into the traditional scheme of things. It bred an intense feeling for the family as an in-group and prevented wider cooperation.

A second and more basic factor is the nature of the government structure. The bureaucracy, on which the government rests, enjoys such liberal economic privileges that it absorbs whatever small margin the people achieve in the economy. In addition it had dug deep into the flesh of the people by pseudo-legal practices and outright corruption. It would not be far wide of the mark to say that no public funds are safe.

Furthermore, since it is upheld primarily by military force, the government has never provided the people with any framework of security. In this respect the members of the bureaucracy are no more secure than the rest. The success or failure, whether social or economic, of the individual has always depended upon the pleasure or displeasure of the top rulers. Under the emperor a favored eunuch could amass a tremendous fortune, but anyone who incurred the displeasure of the emperor could find even his hard-earned and well-deserved wealth confiscated overnight.

The Revolution of 1911, which changed the form of the government, did not change its substance. The Revolution of 1927 achieved a little more, but only under the necessity of buttressing the military power of the government. When such necessity is removed, because inflation and war have destroyed the importance of accumulated wealth, the same traditional government attitude toward the people returns.

Because of the backwardness of the economic conditions and because of an indifferent bureaucracy and government, the drastic changes which have occurred in education have produced a tragic dilemma. On the one hand, the various lines of constructive work in the country urgently need competent men. On the other hand, a large number of graduates of modern institutions find that their training has failed to prepare them to fit into the existing order. Some go into various organizations for reform; some are fortunate enough to find opportunities in professional work or research; but many become disillusioned moneymakers. Others fall in line with things as they are and become the most corrupt of bureaucrats. Finally many drift about in confusion and despair because they find no compromise between what they want to do and what they have to do.

The situation in China today, with its contrast between the poverty and misery of the people and the corruption and indifference of the bureaucracy and government, can be truly described as a conflict of cultures. For in a genuine sense, the changes and the resulting chaos which have come about in China since the middle of the last century have come through contact with the West. In premodern China there was a sort of internal equilibrium, in spite of the despotic and extravagant ruling group and a corrupt and as a whole irresponsible bureaucracy. The bureaucracy taxed the people harshly but they also spent almost exclusively among the people. The emperors and his descendants had absolute power, but if they went too far the people had the strength and the occasion to revolt and overthrow them. In modern China this old equilibrium has been destroyed and a new one has not yet been worked out. In this situation the bureaucracy, including the military, have extorted from the people more harshly than before, but the sums thus collected have largely left the country by way of treaty ports, concessions, unfavorable trade balance, and all kinds of official tours to Europe and America for pleasure or propaganda, and for technical investigation by politicians who often have not had an elementary course in the subject they are dispatched to investigate. The top ruling group has also more absolute power than before over the people, because they have learned some methods of organization and secret police from the West; at the same time, ordinary citizens have less strength and occasion to exert their rights, because the introduction of modern weapons has made it impractical for unarmed masses to defend themselves against organized tyranny. Under these circumstances, the undoubted integrity, brilliance, and patriotic endeavors of some individual officials and others, high and low, inside and outside the party, have meant little.

Of course, it is a mistake to equate the present government with any previous one. Even two indigenous dynasties in pre-modern times were never exactly alike. The Kuomintang, notwithstanding its manifold shortcomings, has been instrumental in creating a degree of political consciousness among the people and has brought about advances in urban industrialization, in public finance, and in international relations within the very short period of 1929 to 1937. And Chinese society today is a very different entity from before. Never before has Chinese culture been so seriously challenged. Never before have the Chinese people even conceived of any modification in their age-old traditions and institutions. Today there is no longer the question of change or no change, but of how much change. The outstanding battle among the intelligentsia and others has been waged between those who urge complete Westernization and those who preach a surgical combination of the best of the East and of the West. In the light of the science of culture either position must be modified to be tenable. Cultures achieve permanent change by processes of integration but not by complete dropping of what has existed for centuries or by a fortuitous combination of the new and the old.

The present conflict between the government and the Communist forces is but an outward expression of the culture crisis. But so also are the runaway inflation, the unemployment of the educated, the increasing number of suicides, which has alarmed some newspapers, the student strikes and demonstrations, the impoverished universities, the clashes between police and soldiers, and the numerous difficulties and crimes arising out of differences between customs and the written law. The armed conflicts between government and Communist forces overshadows the others because it is of more immediate and decisive importance.

The duration of the culture crisis and the shape of the integration to come will depend, on the one hand, upon how truly the contending factions understand the importance of the active interest and support of the masses as a basic key to success and consolidation of their powers, and on the other, upon how wisely the United States will exercise her responsibilities as the most important stabilizing factor in the world today.

SUGGESTED READINGS

Buck, J. L. Chinese Farm Economy. Chicago, 1930.
—— Land Utilization in China. Chicago, 1937.
Chang, C. C. China's Food Problem. Shanghai, 1931.
Chen, H. S. Landlord and Peasant in China. New York, 1936.
Chen, Ta. Population in Modern China. Chicago, 1946.
Cressey, G. B. China's Geographic Foundations. New York, 1934.
Fei, H. T. Earthbound China. Chicago, 1945.
—— Peasant Life in China. London, 1939.
Fong, H. D., and Chih Wu. Rural Industries in China. Tientsin and New York, 1933.
Forman, Harrison. Report from Red China. New York, 1945.
Gamble, S. D. How Chinese Families Live in Peiping. New York, 1933.
—— Peking, a Social Survey. New York, 1921.

Hsu, F. L. K. Under the Ancestors' Shadow. New York, 1948.

Kulp, Daniel H. Country Life in South China. New York, 1925.

Latourette, K. S. The Chinese, Their History and Culture. New York, 1946. Vol. II.

Lattimore, Owen. Solution in Asia. Boston, 1945.

Lattimore, Owen, and Eleanor Lattimore. China, a Short History. New York, 1947.

Lin, Yu-tang. My Country and My People. New York, 1934.

Lowe, Chuan-Hwa. Facing Labor Issues in China. London, 1933.

MacNair, H. F., ed. China. University of California, United Nations Series. Berkeley, Calif., 1946.

Mitchell, Kate G. Industrialization of the Western Pacific. New York, 1942.

Pruitt, Ida. A Daughter of Han. New Haven, Conn., 1945.

—— The Flight of an Empress. New Haven, Conn., 1936.

Rosinger, Lawrence. China in Ferment. New York, 1943.

Rowe, D. N. China among the Powers. New York, 1945.

Snow, Edgar. Red Star over China. New York, 1938.

Stein, Gunther. Challenge of Red China. New York, 1945.

Tao, L. D. The Standard of Living among Chinese Workers. Shanghai, 1931.

Tawney, R. H. Land and Labour in China. London, 1932.

Taylor, George E. Changing China. Institute of Pacific Relations, New York, 1942.

Vinacke, Harold M. A History of the Far East in Modern Times. New York, 1941.

Yang, Martin C. A Chinese Village. New York, 1945.

Douglas G. Haring

JAPAN AND THE JAPANESE
1868–1945

THE UNITED STATES is ruling Japan. To the Japanese that is a fantastically preposterous outcome of their long-planned war to dominate. To Americans it still seems unreal. The traditional American aversion to conquest has not been reconciled with the cold fact that Tōkyō takes orders from Washington. The American habit of ignoring Asia facilitates an unconscious evasion of this contradiction. Inadvertency of dominion, however, does not abate the urgency of certain inevitable problems.

No one can rule wisely unless he understands those whom he rules. In this case, pink-tea concepts of international understanding lead to blundering and muddle; the need is for insight. Achievement of insight into Japanese society is the more imperative because of American ideals of responsible democratic government. Wise, blind, or indifferent, American voters ultimately determine the limits and trends of their nation's policy; the consequences of their decisions are inescapable for Americans and Japanese alike. Ordering the Japanese to abjure militarism and adopt democracy may be no more effective than commands shouted at the moon.

Before wise policies can be carried out over a term of years such policies must be understood. No one can command the American public to understand the Japanese. If such understanding is achieved there is a fighting chance that wisdom may prevail. If not, the wisest policy is defenseless against demagogues and popular ignorance. The penalty for bungling is automatic, inherent in the situation.

There have been many Japans. Intriguing glimpses of the earliest Japan, about the time of Christ, are possible through archaeology and history. Chinese documents refer to the curious customs of Wa, island country in the eastern ocean. There dwelt numerous tribes whose cultural practices hint at South Sea contacts as well as affiliation with the Asiatic mainland. The Wa regarded their chieftains—some of whom were women—as supernaturally potent. The dominant tribes were immigrants who were driving back the aboriginal Ainu toward the northeast.

The next Japan, revealed in the earliest native records, was in process of consolidation under the rule of the tribe of Yamato. Stubborn chieftains who balked at subordination to the Yamato king received short shrift. Customs and folk tales of Asiatic origin were blending inextricably with beliefs and customs from the South Seas. The Yamato chieftain was beginning to assert prerogatives resembling those of the Chinese emperors, but his divine kingship was attested by hereditary sanctity like that of kings in the South Sea Islands. To record his mythical divine genealogy he employed Korean scribes who introduced Chinese writing and thereby inaugurated Japanese history.

The Japan of the eighth, ninth, and tenth centuries A.D. reflected the glory of China's preeminent T'ang dynasty. To absorb this brilliant civilization at its fountainhead, Japanese students flocked to China. In Nippon the court, capital, patterns of governance, Buddhist religion, art, literature, and architecture acknowledged Chinese models. Particularly in art and architecture the Chinese originals often were equaled, now and then surpassed. Women still retained a time-honored freedom and set the pace in literature, in contrast to the situation in China. The common folk experienced the new civilization only through relentless taxation and *corvée;* the Imperial court constituted an unreal world of aestheticism and pseudo-Chinese sophistication. Along the adventurous northeastern frontier, far from the brilliant courts of Nara and Kyōto, semi-independent feudal lords pressed back the still vigorous Ainu.

As the glory that was T'ang faded, Chinese influence waned. By the eleventh century another Japan was emerging. Power shifted to aggressive war lords bred on the Ainu frontier, whose bases lay remote from Kyōto and its sacrosanct, politically impotent emperor. Distinctively Japanese rituals of social intercourse and patterns of governance took shape. Subtly, however, Confucian and Buddhist influence abetted the fighting men in achieving the subordination of women. Buddhism, already decadent, underwent rejuvenation as new sects arose with uniquely Japanese characteristics.

The political struggle, ever more ruthless and disorganized, culminated in the devastating Ōnin war (1467–1477).

A strikingly different Japan was created when three great dictators enforced peace about 1600 A.D. Thenceforward for nearly three centuries political power resided in the House of Tokugawa. The emperor retained his sanctity but the Tokugawas kept the power. Closing Japan to the outer world symbolized a determination to petrify a feudal *status quo* by sumptuary legislation and rigorous policing. Successive dictators strove to control ideas and to congeal the arts; Confucian doctrine was reinterpreted to justify the Tokugawa regime. Continued peace, however, nurtured new arts and a new money economy. The ban upon change may have invited the internal collapse that neared its climax as Perry's famous expedition arrived to "open" Japan.

"Modern" Japan dates arbitrarily from 1868 when the Emperor was restored as nominal head of the state. Whether the surrender to American arms in 1945 will initiate another metamorphosis cannot be foreseen. Whatever may impend, the brief "modern" period from 1868 to 1945 affords a key to current happenings. This essay accordingly focuses upon the thought-ways, customs, and societal patterns of that brief epoch, and the historical present tense refers to the "modern" period, not to the MacArthur episode.

Despite the plural historical Japans, certain persisting themes impart unity to Nippon's history. Assiduously as the Japanese copied China, they nevertheless boast that they remained Japanese, distinct in personal character, tastes, and feelings. Japanese cultural patterns subtly belie their superficially Chinese aspects. In contrast to China, warriors held ascendancy over literati, priests, peasants, and craftsmen. Confucianism, Buddhism, familism, and even Chinese writing underwent adaptation, not mere adoption. The Japanese learned avidly from foreign sources while aggressively asserting their superiority to all foreign barbarians. In patterns of governance, despite radical outward transformations, a consistent theme persists: like the South Sea peoples they have maintained a sanctified supreme magical ruler shorn of political power. The real controls have operated quietly from behind the scenes. Even *shōguns*, ostensibly masters of the Emperor, often answered in their turn to hidden manipulators. In religion, durable folkways of ancient belief in intangible supernatural power called *kami* infused a characteristic content into both Buddhist and Confucian forms. Another age-old theme is enforcement of Spartan frugality upon the many to provide luxury for the few. Recurrently such luxury outruns the slender resources of islands whose bounty cannot multiply as do the inhabitants. When this economic strain attains a climax, transition to another Japan impends.

The consequences of Japan's geographic position with respect to several great cultural areas are basic, though difficult to analyze. Ethnographically Japan is marginal to China, to the Malayo-Polynesian region, and to Siberia with its Central Asiatic affiliations. From all of these regions—by channels often lost to history—customs, ideas, beliefs, and technologies were carried to Japan. Since the middle of the nineteenth century still another alien influence has overshadowed these older contacts: the expanding Euro-American civilization. Attempts to allocate specific Japanese cultural forms among these sources are beyond the scope of the present discussion. The historical record, however, is clear with respect to Chinese and Euro-American influence. The record also shows that the Japanese have not received passively whatever foreigners chose to bring them; rather they have aggressively sought out what they desired from China, Europe, and America.

Deliberate exclusion of all the historical Japans save that of the "modern" period simplifies the present description. A related difficulty, however, persists.

Today, also, there are many Japans. The fundamental, enduring Japan is that of the forty-odd million peasants. Persistent also is that of the Imperial Court, now allegedly waning. The omnipresent bureaucracy is another world of its own. Still another Japan includes the families who dominate commerce, finance, and industry; while that of the military hierarchy has usually dominated the historical records. Then there are intelligentsia, technicians, the "fleeting world" of demimonde and *geisha*, soldiers, sailors, factory hands, advocates of modernization, nostalgic but violent reactionaries who inhabit a feudalistic dreamland—to say nothing of fascists, Rotarians, cinema addicts, communists, Salvation Army, labor bosses, political gangsters, Buddhist monks, and politicians. Perhaps still another Japan—MacArthur's Japan—may hold sway for a moment, to disappear like Cinderella's coach when the time comes.

If the Japanese people act so much alike that their homogeneity is impressive, it must be acknowledged that this homogeneity is highly diversified. How then can Japan be described briefly? It is possible to outline the limits of conduct within which diversities occur. Beyond these limits an individual promptly is stigmatized as un-Japanese. Within them one may be queer, his ideas may be suspect, but he remains a queer Japanese, a radical Japanese, or a maladjusted Japanese. The usual patterns of conduct may be conceived after the analogy of statistical modes—the most frequent among many varying occurrences—in order to describe them broadly. The range of variation so patent to anyone who learns to feel at home in Japan is omitted deliberately.

GEOGRAPHY.—Japan is not a Japanese word. The Chinese name *Jih-pen* means "Source of the Sun" and derives obviously from the position of the islands off China's eastern coast. Anglicized it became "Japan." The Japanese pronounce these Chinese ideographs *Nihon* or *Nippon;* by these words they denote both their country and its people.

Japan Proper comprises thousands of small islands and four large ones: Honshū, Kyūshū, Shikoku, and Hokkaidō. From northeast to southwest for fifteen hundred miles they dangle in a long festoon off the Asiatic coast. Hokkaidō, the northernmost, resembles Maine or Newfoundland in climate. Southward, mean temperatures increase until in Kyūshū the climate compares with that of Florida. Bathed by the warm Black Current, the southern tips of Kyūshū, Shikoku, and the Kii peninsula of Honshū enjoy subtropical climate. Across the islands on the shores of the Japan Sea deep snows persist throughout the winter from Toyama north.

In land area, Japan Proper approximates California. Rugged mountains, however, permit cultivation of but 15 percent of the area. This compares favorably with California, where about 7 percent is cultivated and not more than 13 percent is regarded as arable. Wherever feasible, Japanese hillsides are terraced to utilize every inch of tillable land.

The mountains, geologically young, are unusually precipitous. Several peaks exceed 10,000 feet in height; the sacred Mount Fuji

NORTH

HOKKAIDŌ

HONSHŪ

TŌKYŌ
YOKOHAMA
NAGOYA
KYŌTO
KOBE
ŌSAKA

Ancient Yamato

Ancient Izumo

SHIKOKU

KYŪSHŪ

JAPAN PROPER, SHOWING DENSITY OF POPULATION AND SIX LARGEST CITIES

Density of population indicates approximate distribution of level land. Urban areas, shown in black, indicate also the industrial concentrations. Non-urban areas of high density indicate approximate distribution of rice land. Areas of low population density generally are rugged and mountainous.

By courtesy of the Department of Geography, Syracuse University. Drawn by Lillian Johnson, after Trewartha.

reaches over 12,000 feet. In dry weather the short steep rivers dwindle to mere creeks and river navigation is negligible. In the rare level districts, canals carry much heavy freight. Abundant rainfall keeps the mountains green to their very summits. Slight earthquakes occur frequently in most districts and every few years a major earthquake strikes. Many of the severe shocks occur in rural areas; when one wrecks a city as in 1923 in Tōkyō and Yokohama, the disaster transcends description.

The few plains, which include most of the arable land, adjoin the sea. Proximity of the ocean and scarcity of agricultural land foster general dependence on marine products. Fish, crustaceans, mussels, and seaweed afford protein in the popular diet. For more than fifteen centuries Japanese fishermen have exploited the rich coastal waters and have ventured far into the Pacific.

Mineral resources, area for area, compare favorably with the Asiatic mainland. The fairly abundant bituminous coal is costly to mine because the seams are badly folded and broken. Sulphur is plentiful. Copper and gold occur in moderate quantities. Tin, silver, lead, magnesium, mercury, antimony, and petroleum exist in small deposits. Iron ore is plentiful but of a grade thus far unsuited for industrial use. Japan's mineral resources do not suffice for intensive industrialization although they would be adequate for a smaller population. They might supply the present population under an economy oriented to consumer needs rather than to war and heavy industry.

Apart from the ocean the outstanding natural resource is the forests. On most of the mountainous 80 percent of the land, conservation and reforestation have maintained forest cover. The forests could provide the building materials, fuel, synthetic fibers, and chemicals necessary to a standard of living like that of 1930. Such a wood-based economy differs from an iron-based economy, but permits a scale of living better than current Chinese or Indian standards. Wooden or concrete instead of steel-framed buildings; wooden railway cars and bus bodies; wooden utensils; charcoal fuel; rayon textiles; wood-derived chemicals such as industrial alcohol; wooden ships—this specialized economy can be maintained and extended with proper forest management.

Steep gradients and heavy rainfall encourage hydroelectric development. Inadequate space for storage reservoirs, however, drastically reduces the supply of electricity in the dry months. The heavy industries of the 1930 decade obtained supplementary power from high-cost steam plants. Even so, the per capita consumption of electricity exceeds that of European countries; short distances mean short transmission lines, and almost every peasant's hut has electric lights.

The islands are famed for scenery. The green-clad, precipitous mountains often descend directly to the islet-studded ocean; the white surf, dashing streams, and almost vertical mist-shrouded peaks of Japanese paintings have their counterpart in reality. From sea level to mountain heights climate and vegetation pass through a wide gamut of change; for this reason most of the attempts to map clear-cut climatic zones are misleading. Nature, however, seems capricious; not only earthquakes and volcanic eruptions, but typhoons, floods, landslides, and other frequent disasters contribute to the psychological tension so often cited as a Japanese characteristic.

BODILY APPEARANCE.—Racially the Japanese derive from a Mongoloid-Malayan mixture. There is considerable diversity in physical characteristics, but practically any type of Japanese can be matched in some part of China. From bodily features alone it is impossible to be certain that an individual is a Japanese and not a Chinese. Improved habits of diet and living during the modern period have been accompanied by an increase in average stature. There are many tall Japanese, but they are outnumbered by the short ones; the average male stature is five feet three inches.

Black hair, straight, abundant, and rather coarse in texture, is practically universal. Beards are scanty and body hair sparse. Noses are generally rather flat and bridgeless. The combination of high cheekbones and slight pads of fat on the cheeks often imparts a well-fed look even when the rest of the body is emaciated. Skulls are broad, rarely long. The "Mongoloid eyelid"—a fold of the upper lid that covers the inner junction of the two lids—occurs more frequently than among Europeans, less often than in many parts of China. Skin color ranges from near-white to brown; the light brown incorrectly called yellow predominates. Japanese are rarely fat; most of them are small-boned and slender. In bodily contour females differ but slightly from males.

For many centuries intermarriage has occurred without regard for differences in bodily type. Consequently diverse features and proportions occur in combinations that interest a trained observer. For example, a slender wiry "Malayan" type of body will carry a broad head with Mongol features. Some persons, allegedly of noble ancestry, are tall and slender, with narrow aquiline noses and high sloping foreheads. A few round-headed, light-skinned, hairy individuals recall the Ainu aborigines who inhabited the islands prior to the coming of the Japanese.

POPULATION.—In 1940 an official census yielded a total of 73,-114,308 persons in Japan Proper.[1] There were almost exactly 100 males per 100 females; the war that followed probably reduced this ratio slightly. The population had increased 13.4 percent between 1930 and 1940, but the rate of increase has been declining.

In the feudal period before 1864 the population had hovered between twenty and thirty millions. The total number of Japanese has nearly tripled in less than a century—a fact that underlies many of Japan's acute social problems.

Conspicuous trends of population change in the modern period include: (1) rapid increase in total numbers; (2) lowered death rates and increasing average length of life; (3) mounting birth rates, changing later to a decline; (4) stability of rural population; (5) tremendous expansion of large cities that absorbed practically all the increase in general population. Apart from reliable census data for 1920, 1930, and 1940, Japanese published statistics often are misleading—a fact notable in discussion of birth and death rates.

The swift increase in total population attained a peak in 1930–1935 with a net five-year increase of 4.8 millions or 7.5 percent. From 1935 to 1940, net increase fell to 3.9 million or 5.6 percent. Comparisons of data of expectation of life with similar data from Occidental countries are vitiated in part by special conditions in Japan. The increase, however, is significant:

[1] The sources for Japan's 1940 census in English are: Kōjima Reikichi, "The Population of the Prefectures and Cities of Japan in Most Recent Times," tr. Edwin G. Beal, Jr., *Far Eastern Quarterly*, III (August 1944), 313–361; U.S. Department of State, *Administrative Subdivisions of Japan* (Washington, 1946).

Expectation of Life at Birth [a]

	1921–1925	1935–1936
Males	42.06 years	46.92 years
Females	43.2 years	49.63 years

[a] Data from Irene B. Taeuber and Edwin G. Beal, "The Dynamics of Population in Japan," *Demographic Studies of Selected Areas of Rapid Growth* (New York, 1944).

The average Japanese lives about twenty years less than does the average American. The rapid increase in Japan's population therefore means that many more babies are born than would suffice to attain a similar increase in the Occident. As recently as 1935–1936 the death rate of male infants stood at 114 per 1,000. Death rates at all ages, despite improvements, exceed those of Western nations and are reminiscent of Occidental death rates in the early years of the industrial revolution. Such rates differ widely in various parts of Japan, both rural and urban.

Crude birth rates apparently increased from 28.5 in 1886–1890 to 34.6 in 1921–1925; perhaps the increase merely indicates more complete registration. From 1925 to 1938 the figure fell gradually to 27. The following figures indicate a change in marriage customs, probably later marriages of women:

Children Born per 1,000 Younger Women [a]

Age of Mother	In 1925	In 1937
15–19	21	fewer than 10
20–24	112	87

[a] Data from Taeuber and Beal, "The Dynamics of Population in Japan."

Births to women aged 30 or above also have declined somewhat; women of ages 25–29, however, maintained earlier birth rates unchanged. In the large cities fertility was lower than in the rest of the country.

It is difficult to compare statistics of urban and rural populations in Japan with those of Occidental countries. For example, the U.S. Census classes as rural all places of less than 2,500 population. Japanese peasants, however, live in villages, and isolated farmsteads are unknown. Statistical practice selects 10,000 as the maximum rural population and classes larger places as urban. Officially, however, Japanese deem a population urban only if the place has been incorporated as a *shi* or city. This latter standard loses meaning because incorporation practices differ widely.

Whatever the basis of classification, it is apparent that the rural population remained almost constant in numbers until the bombing of cities in 1943–1945. In 1920, places of 10,000 or less included 37.9 million people and in 1935, 37.5 million. Taking the unincorporated places as a basis, the rural population continued close to 45 millions in both years. Obviously the larger cities absorbed the total increase in population; peasant Japan was saturated. The bulk of the increase went to a few supercities. In 1920, 12.1 percent of the total population resided in cities of 100,000 or more; by 1940 this percentage had increased to 29.1. The six largest cities grew most rapidly, especially during the war preparations of 1935–1943. Dissipated by wartime bombings, the people gradually returned after fighting ceased.

Population of Japan's Six Largest Cities [a]

CITY	POPULATION	
	1940	*Nov. 1945*
Tōkyō	6,778,804	2,675,203
Ōsaka	3,252,340	1,102,959
Nagoya	1,328,084	597,941
Kyōto [b]	1,089,726	866,153
Yokohama	968,091	624,994
Kōbe	967,234	376,166

[a] Data from *Administrative Subdivisions of Japan*, Dept. of State, Washington, 1946.

[b] Not bombed. Apparently the people expected bombings.

Note: Tōkyō and Yokohama included extensive suburban areas that suffered less destruction, otherwise population might have dropped further.

The urban picture gains force from data of rural populations. City dwellers fled to the security of rural homesteads, and rural areas were overcrowded severely during and just after the war. A small but representative sample of rural *gun*—roughly equivalent to American counties—appears on the following page.

These drastic wartime shifts of population are typical. Manifestly the results of a regular census in 1950 would be difficult to predict. Whether the urban areas recover completely and rural districts continue overcrowded depends upon the extent to which industry and foreign trade are rehabilitated.

Population of Sample Rural Gun [a]

DISTRICT (GUN)	POPULATION		REMARKS
	1940	Nov. 1945	
Higashi Iwai, Iwate Prefecture:	86,451	100,797	Remote North Japan area
Naka, Ibaraki Prefecture:	131,553	173,761	Within 75 mi. of Tōkyō
Kamo, Shizuoka Prefecture:	84,244	99,856	Between Tōkyō and Nagoya
Fugeshi, Ishikawa Prefecture:	84,341	102,151	Remote; shore Japan Sea
Saeki, Hiroshima Prefecture:	110,410	149,860	Near Hiroshima City
Kuma, Kumamoto Prefecture:	80,700	92,853	On Kyūshū; includes Suye [b]

[a] Data from *Administrative Subdivisions of Japan* (Dept. of State, Washington, 1946).

[b] Village made famous by J. F. Embree's book, *Suye Mura* (Chicago, 1939).

Despite the mixture of races in early Japan the modern population includes very few immigrants. Until recently, most of the peasants never had seen a foreigner, and until the last two decades no foreigner could own land under any circumstances. In the 1930 decade less than 25,000 Chinese, approximately a million Korean laborers, and nearly 10,000 Europeans and Americans who were concentrated in the larger cities provided the sole foreign elements. Isolation from personal acquaintance with foreigners enabled the militarists to instill fear of other nations and insure support for an aggressive war.

LAND AND PEOPLE.—Until the modern period Japan maintained relatively few large cities. That cities existed at all was due to the excellent keeping qualities, small bulk, and ready transportability of a single grain—rice.[2] Also important were the analogous qualities of a standardizable, easily handled fuel, namely, charcoal. Rice, a swamp grass that grows in standing water, requires back-breaking hand labor but yields abundantly. A toiling peasant family can raise about double the amount required for their own sustenance. As soon as early societal organization provided for the transportation and

[2] Urban populations are possible only when the economy achieves a surplus beyond the needs of the farmers, fisherfolk, or herders who wrest livelihood directly from nature. So-called primitive societies remain primitive because they are unable to produce such a surplus. If and when their harvests or food animals are abundant the surplus is consumed in feasting or is wasted for lack of means of preservation. Urban or civilized societies depend upon foods that can be transported and kept. With such foods available, civilization is possible if patterns of societal organization are adequate to the complexities of exchange, storage, and transportation. In this sense the words *civilized* and *primitive* denote differing societal patterns without invidious comparisons. Cf. the present author, "God Kings and Cosmic Government," in *Christian Leadership in a World Society, Essays in Honor of C. H. Moehlman*, ed. J. W. Nixon and W. S. Hudson (Rochester, N.Y., 1945).

storage of surplus rice and charcoal, it became possible for part of the population to live in towns. Even today Japan's economy is based on rice.

Prior to the modern period, agents of the feudal lords collected the rice harvest from the peasants who were bound inseparably to the estates. The overlord stored the rice in his castle and doled it out in annual stipends to his retainers and craftsmen—sometimes to the peasants as well. To his superiors he paid tribute in rice. Incomes were reckoned in *koku* (5.119 U.S. bushels) of rice; a minor fief was appraised in thousands of *koku*, and the great feudatories in millions. The landless *samurai* warriors afford one of the many contrasts with feudal Europe: the European horseman fed himself from his own land holdings and was in a position to rebel against undue exactions. Japanese overlords controlled the food of warrior and peasant alike. To rebel was to starve. When mouths outran the available food, famine, war, abortion, infanticide, and killing of the aged restored a balance.

If in feudal times famines were frequent and peasants often rioted in protest at their miserable lot, how then could modern Japan manage with more than double the former population? By careful estimate there were 3,067 persons per square mile of crop land in 1938, in contrast to 2,365 in Great Britain and Northern Ireland and a mere 238 in the United States. Statistics of Japan's national wealth seem to support the facile answer that industrialization and importation of food did the trick. In 1904 Japan's foreign trade amounted to *Yen* 14.63 per capita; in 1937, despite greatly augmented population, it came to *Yen* 95.41 per capita. On the whole, however, the gains from foreign trade did not go to importation of food; they paid for armament, capital goods, raw materials for industry, and luxuries for the few. The food supply was increased by better fertilizers, adequate storage and distribution, and some increase in cultivated area. Korea and Formosa, and later Manchuria, supplied large amounts of food. The Japanese increased Formosan sugar production manyfold, for example, while drastically reducing the per capita consumption of sugar by the Formosan population. The bulk of the sugar went to Japan and, as domestic trade, did not enter into foreign trade statistics. By such devices food imports from outside the Japanese Empire were held to a minimum.

The fact of more than 3,000 persons per square mile of crop land requires analysis. From Tōkyō south, many farms yield two crops annually and the subtropical peninsulas permit three crops. Hence one square mile of Japan's crop land is equivalent roughly to two square miles of Euro-American farmland. Edward Ackerman has estimated the number of square miles of Occidental crop land required to equal Japanese production, assuming one crop yearly. On this basis he computes the density of Japan's population "per square mile of equivalent food producing capacity" as about 1,200 persons —a figure well below the population density in the British Isles.[3] The British, however, import much food; the Japanese home islands, prior to World War II, produced 85 percent of the food consumed. Strenuous efforts reduced the amount of imports. Every weed was analyzed and methods of utilizing its food values were worked out. Until the warlords were assured that food imports could be cut below 10 percent, and that this 10 percent could be eliminated by strict frugality, the attack on Pearl Harbor was held in leash.

In the 1930s the average daily food allotment per person was 2,500 calories, in comparison with the American standard for good health of 2,200–3,000 calories. The small-bodied Japanese require less and maintained health on the 2,500 average. Of course the rich got more and the poor less; in northern Honshū famine has been endemic since 1930. Most of the protein came from marine sources; Japan not only consumed much fish but led the world in fish exports.

It has been asserted that *if* all land were fertilized and cultivated scientifically, *if* only high-grade seed were used, and *if* distribution were efficient and rationing equitable, the pre-war daily allotment of 2,500 calories could be restored without importing more than 10 percent of the nation's food. Whether this is feasible in the face of postwar chaos, and whether in the long run the people would endure the regimen, remain to be seen. The Japanese were willing to fight to improve their lot, especially after seeing fabulous wealth portrayed in American films. At the moment this discussion would seem academic to the undernourished masses.

Japanese cuisine affords an outstanding instance of aesthetic em-

[3] Data of population density in relation to crop land from E. A. Ackerman, "Japan: Have or Have-not Nation?" in *Japan's Prospect,* ed. D. G. Haring (Cambridge, Mass., 1946).

bellishment of a few simple, scanty foodstuffs. Formal etiquette governs the serving and eating of meals. Food is served daintily on individual trays. Exquisite artistry characterizes the lacquered bowls, porcelain cups, and other dishes. Soups and other liquids are drunk from bowls or teacups without handles. Cut in small pieces in the kitchen, solid food is conveyed to the mouth by chopsticks. Boiled rice, many kinds of raw or cooked fish, pickled radishes, a slice of lotus root, half a pickled potato and a few peas, and other side dishes are arranged skillfully to appeal to the eye. Traditionally the culinary skill of the hostess was acknowledged by a hearty belch after each course, and tea, like soup, was ingested with appropriate sound effects. All this is effective in giving the impression of a bounteous feast despite the limited resources at the disposal of the kitchen.

Rice, the foundation of every meal, is cooked in most homes but once a day and is served cold at two of the three meals—an economy dictated by the scarcity of fuel. In good times a normal meal involves from five to seven small bowls of rice per person. Polished rice is preferred to brown rice, despite its lack of vitamins; consequently *beri-beri*, a disease of malnutrition, connotes prosperity since the poor cannot afford white rice. In wartime other grains or beans were mixed with the rice and sometimes replaced it. Every child is taught at home and in school never to waste even one grain of rice; formerly etiquette required that one rinse his rice bowl with tea and drink the tea at the end of a meal, lest grains of rice be wasted.

Varied and tasty soups are concocted with slender resources. Other dishes include shrimp, octopus, seaweed; noodles and noodle soups; relishes of greens, pickled vegetables; beans, bean pastes, bean curds, and bean confections; eggs, shreds of meat, rice wafers, and fruits. Clear green tea, the universal beverage, is served to guests, to customers in stores, in business offices, in schools between classes, at roadside stands, and is sold at railway stations to passengers on trains. Within the past fifty years milk has gained acceptance; sold in small long-necked bottles, boiled rather than pasteurized, and stoppered with a cork, it is used chiefly to dilute whiskey. Cows are kept in pens in the towns and food is brought to them. *Sake*, the favorite intoxicant, is brewed from rice and is served hot. *Shōchū*, a potent distillate of low-grade *sake*, takes the place of the more expensive brew in some villages. A nutritious, aromatic sauce brewed from soy beans provides salt, flavor, and vitamins for many cooked foods. In pre-war days bread, cake, caramels, chocolate bars, and other foreign-style edibles were stocked in many stores. Bread (called *pan*) and sponge cake (called *kasutera*—"Castile") have been used for three centuries; as the names indicate, they reached Japan via early Portuguese adventurers.

Many of these foods are luxuries scarcely known to the peasants, who subsist on brown rice in good times, otherwise on millet, beans, and miscellaneous vegetables. Meat is a luxury for all classes; to Buddhists it is taboo, though that rule now is honored in the breach. Poultry butchers frugally dissect a chicken and sell the meat in tiny slices; the bones are sold separately for soup. A popular seasoning is a rock-hard sun-dried bonito; a little of this is grated into soups or other dishes.

Perhaps the restricted diet and the widespread complaint of "indigestion" help to explain the universal passion for pills: vitamins, patent medicines, breath purifiers, liver pills, alleged aphrodisiacs, muscle builders, female pills, and just pills. The demand seems insatiable. Tobacco is used sparingly. A cigaret is extinguished after a puff or two and placed economically over an ear until time to re-light it for another puff. The Japanese pipe, dear to the old folks, has a long bamboo stem and a tiny brass bowl that holds but a pinch —whence its name of *ippuku*, "one puff." From time immemorial Emperors and officials have preached frugality to the people.

HOUSES.—Although Occidental-style houses have appeared in the cities since the 1920s, most Japanese live in traditional types of dwelling. These are of wood with mortised joints to economize on nails—scarcely any metal is used. There is no cellar; the house is raised two or three feet on posts that rest on foundation stones driven into the earth. Town houses are roofed with tiles laid without nails in mud spread on thin shingles; in the hamlets, roofs are thatched. Roof profiles copy that of Mount Fuji.

Long porches three feet wide, with polished unpainted floors, extend along the sunny exposures. They serve as hallways into which all rooms open. At either end, boxes hold sliding wooden doors to enclose the porch in bad weather and at night. Between each room and the porch are other sliding doors—light wooden-latticed frames over which translucent white paper is stretched to admit light. These paper-covered frames explain the Occidental myth that Japanese houses are made of paper. Glass replaces this paper in the better homes. Broad thin clapboards cover the outer walls. Solid partitions are of plaster on bamboo lath, but usually the principal rooms are separated by sliding doors covered with decorated opaque paper. In hot weather all doors may be removed to allow the breeze to circulate. Ideal in hot weather, the Japanese house affords slight protection against the cold. The fact that houses so well adapted to the tropics characterize even the cold northern districts has been advanced in support of the theory that the Japanese originally came from the tropics.

Neither the exterior nor the interior of these houses is painted, except as lacquer may be used on interior woodwork in the lacquer-producing areas. Ceilings are of thin wood. At one end of a room, ornamental posts and shelves mark off a slightly elevated "place of honor," before which guests are invited to sit. Unpainted wood is preferred for beauty of grain in conformance with the canon of severe simplicity embodied in Japanese aesthetic ideals. A single vase with one flower and some twigs, or a scroll painting, in the place of honor may constitute the sole decoration of a room.

Straw mats called *tatami,* two inches thick, cover the floors. *Tatami* consist in a tightly woven core of coarse straw faced with finely woven matting. Narrow black cloth strips bind the long edges. *Tatami* always measure six feet by three feet and provide a convenient unit of area. Thus a house may be advertised as follows: "House with garden, 3,6,6,8,2; Kotobuki *chō* 194." This means that the house contains one 3-mat room, two 6-mat rooms, one of 8 mats, and one of two. The remaining words give the address. The immaculate cleanliness of well-kept *tatami* and their ready disintegration under hard shoes explain the inexorable rule that shoes must be removed before entering a Japanese house.

Furniture is limited strictly in quantity. People kneel on square cushions and require no chairs. Consequently tables are about a foot high. A chest of drawers, one or two low tables, a diminutive dressing cabinet with a mirror—such is the furniture of the average home. Beds are piles of padded quilts spread on the floor. A sleeper inserts himself midway in the pile and places his head on a hard round pillow. By day the folded quilts repose in large closets.

Kitchens traditionally lack windows and often have dirt floors. Rice is cooked over a quick fire of twigs and straw in a large clay firebox. In cities gas and electricity often replace the old arrangements. A well with a sweep or with a pair of buckets on a rope hung from a pulley provides water. In cities with a modern water supply the less prosperous homes share a faucet on a street corner. Except in the largest cities drainage flows in open gutters beside unpaved streets.

Scarcity of fuel explains meager heating facilities. In winter one may toast his chilblains at a tiny charcoal fire half-buried in ashes in a large wooden or earthenware container. In colder districts a concrete firepot three feet square is built in the floor of one room, to be concealed beneath half a *tatami* in warm weather. When in use it may be covered with a light wooden framework a foot high, over which quilts are spread. A table top may be placed on top of the quilts and people sit around the firebox with their feet in the warm space under the quilts while they use the table top for reading, eating, writing, or a friendly card game. At night the quilts are spread around the firebox like spokes of a wheel and each sleeper may poke his feet into the pleasant warmth of the central tent.

Attached to the house is a privy. A large earthenware jar beneath a hole in the floor provides a storage place for Japan's principal fertilizer. A long-handled wooden dipper serves to transfer the accumulated contents of the jar to wooden tubs with lids. Cartloads of these "honey buckets" impart a pungent aroma to city and countryside on Sunday mornings when farmers come to town for their fertilizer.

In the towns—except for some districts along the Japan Sea— high wooden or stone walls separate houses and gardens from the streets. Usually the kitchen is nearest the street to facilitate delivery of supplies. Living rooms face southward when possible toward an ornamental garden. Low projecting eaves admit the slanting winter sunshine but shut out summer heat. With space at a premium, houses are packed together, and since their construction is light fires spread explosively. The labyrinthine older streets are six to ten feet wide and have no sidewalks. House numbers often follow no consecutive or rational order; thus 842 may adjoin 76 and 328, since the numbers are scattered at random over a *chō* or ward. Seeking an address, one asks people on the street for help, and anyone who knows the place escorts the stranger with unfailing politeness.

Peasants enjoy few of the comforts available to townsfolk. Mud-walled and thatched, their one-story huts contain few rooms. Some of these have only the earth for flooring. Relatively prosperous landowners enclose their quarters with fences.

Occidental-type houses gained popularity in the cities in the 1920s. The Tōkyō-Yokohama earthquake of 1923 demonstrated the superiority of steel and concrete buildings and the lesson was not forgotten. Offices, stores, and public buildings tend to follow Western architectural ideas.

CLOTHING.—Occidental styles of clothing gain or lose popularity with changing fashions, especially in the cities. In their homes, however, most Japanese prefer the traditional garb: the *kimono* of cotton, silk, or wool. Extra layers of *kimono* are donned in cold weather. Men and women wear contrasting styles and colors but both sexes lap the left front over the right as American men button their coats. For men the preferred colors are brown, dark blue, grey, or black; decorative patterns are limited to small figures or inconspicuous stripes, and the sash is narrow and tied in a small knot. Boys wear cotton *kimono* decorated with small white figures on a blue or grey background, made usually by printing or tie-dyeing the yarn; the size of the figures is larger for small boys, smaller for older boys. Except in infancy and after the age of sixty, men never wear red.

Women of good repute wear subdued colors in public but like under-*kimono* gay with color. Young girls appear in bright colors and large decorative designs; the patterns grow smaller and the colors less vivid as the wearer approaches womanhood. The *obi* or sash is broad, colorful, and tied in an elaborate vertical bow at the rear. Brocades and other expensive materials are preferred in the *obi*. Pregnant women wear a special *obi* and are treated with consideration. *Geisha* and prostitutes wear bright colors with large designs; only licensed prostitutes tie the *obi* in front.

On ceremonial occasions a plain outer *kimono* coat of subdued color, known as a *haori*, is worn over the *kimono*. The decorations consist in subtle patterns in the weave of the cloth, and of family crests. Some occasions call for the *hakama*, a loose divided skirt worn over the *kimono* and tied at the waist; this served as standard garb for many students until the textile shortage of the late thirties drove it into eclipse along with the *kimono*. Women factory workers took to wearing *mompei*, trousers of coarse cotton cloth like the traditional *momohiki* worn by laborers. The two legs of *momohiki* are separate garments, tied about the waist by narrow strips of cloth. Skin-tight, they may be accompanied by rolled puttees. The appropriate upper garment, a blue-jean jacket, has wide sleeves and *kimono*-type collar adorned on lapels and in the back with ideographs that symbolize the trade guild or employer of the wearer. In hot weather the jacket is discarded and white *momohiki* remind an American of a suit of old-fashioned long underwear. As the temperature rises men discard garments until only a loin cloth remains. In the steaming rainy season laborers of both sexes may toil in the fields quite divested of clothing. If necessitated by weather or hard work, nudity involves neither immodesty nor offense, although uncovering of the body for display is disapproved strongly. To the old-time Japanese, the cheesecake leg art of American newsstands would have meant nothing; his interest focused on the nape of a woman's neck as the critical test of beauty. Ladies planned their costumes to display a pretty neck to advantage.

Traditional Japanese clothing is laundered by the housewife in a wooden tub with wooden scrubbing board. First she pulls all threads from the seams and then washes each strip of cloth separately. Instead of ironing a garment, each piece of cloth is spread smoothly on a board and dried in the sun. The process is completed by sewing the garments together again.

While city-dwellers favor Occidental-style leather shoes, rubber shoes, or rubber boots, peasants and some city folk cling to the traditional sandals of straw, wood, or rubber. Outdoor sandals are wooden clogs, called *geta*, with flat tops and two transverse cleats beneath to raise the foot above the mud. When it rains *geta* with cleats as high as three inches are worn. Thongs passing between the great toe and its neighbor hold them precariously in place; hence the ankle-length socks have a separate space for the big toe. Laborers and peasants wear straw sandals secured by straw rope, or cotton socks with heavy canvas or rubber soles and a separate compartment for the big toe, or more recently, *chika-tabi*, the rubber-soled shoes known in America as "sneakers."

Despite incongruity with Japanese costume, men universally wear European types of hats. In feudal times men wore long hair and favored elaborate coiffures; now, except for professional wrestlers, close-cropped heads are the rule. Except when clad in European-style dress women never wear hats; together with details of *kimono* and *obi*, the hair-do indicates a lady's age and marital status.

A JAPANESE NEIGHBORHOOD.—Renting or buying a house generally is complicated by the fact that land and building usually belong to different owners. One who moves into a neighborhood is expected immediately to call upon each adjoining household. A present such as a cotton towel, box of matches, mosquito incense, tickets good for a bowl of noodle soup at a local restaurant, or some other small gift is offered each neighbor; the servants ascertain via the gossip grapevine the local custom in such gifts. During the brief formal visit one does not enter beyond the *genkan* or paved entrance hall; he introduces himself, presents the gift, apologizes for intruding into the neighborhood, and requests the gracious favor of his new neighbors.

Enterprising tradesmen soon appear to solicit the trade of the new household. Representatives of neighborhood organizations call to inform the newcomer of the extent to which his support is expected. Such include the *seinenkai* or young men's club, which generally holds forth at a nearby shrine and collects funds for festivals, patriotic celebrations, and parties. Perhaps a *fujinkai* or women's club solicits the support of the lady of the house. The most active body, the *tonari gumi* or neighborhood association, includes from five to a dozen adjoining households and is a subdivision of the *chōnaikai* or ward association. The *chōnaikai* employs collectors of garbage and sewage, night watchmen, and other district functionaries. Charitable activities formerly were conducted in part through the *chōnaikai*. These groups are effective agencies of government, for their chain of command extends all the way to Tōkyō.

About 1927 the growing movement for social welfare began to utilize the *tonari gumi*. Subsequently the Welfare Ministry named in each *tonari gumi* a sort of local welfare officer who was required to keep in touch with every family, to promote mutual aid, and to report cases in need of public relief. With the onset of war, military leaders turned the *tonari gumi* into a tool of psychological mobilization, an agency for detection of "dangerous thoughts," air-raid defense, rationing, and like functions. Official orders are attached to a small board called "mobile bulletin board," entrusted to the head of the *tonari gumi*, and circulated among the members. Each head of a household must affix his seal in acknowledgment of the order.[4]

Other frequent visitors include Buddhist priests and monks, solicitors for nondescript and often fraudulent charities, book agents, gadget vendors, craft guild members soliciting funds, and even an occasional gangster conducting a disguised house-to-house shakedown.

The policeman calls as soon as the new family moves in. He obtains a meticulous record of every member of the household, whether or not the official family register (*koseki*) is transferred from the former residence. Should a visitor stay overlong or a house guest arrive with baggage, the policeman turns up to probe the visitor's past. He calls regularly for a routine check, and cultivation of his good will pays dividends. From his street-corner police box he keeps tabs on everyone; should a crime occur, he will know who was away from home at the time. When a fire starts, someone runs to the police-box, bows, and requests that the fire brigade be called. A stiff fine awaits the household where the fire started; on such occasions, unless an unpopular family was at fault, neighborhoods may stand together solidly—no one can guess where the fire might have started.

Two or three times annually the policeman informs each householder of the day set for compulsory housecleaning. All furniture, including *tatami*, must be removed from the house and cleaned. Before anything may be put back the policeman inspects the house to make sure that all is spotless.

Neighborhood gregariousness focuses in the bathhouse. Distinguished by a tall smokestack, a commercial or municipal bathhouse provides facilities used by all except the prosperous. In the cool of the evening one dons a *yukata* (cotton bathrobe), takes a brass basin, toothbrush, soap, and towel and, with an occasional pause to chat, saunters to the bathhouse. In an anteroom he scrubs himself with

[4] The occupying American army failed to understand and utilize these neighborhood groups. Thus was forfeited an opportunity to explain democracy to every last Japanese. The way was left open for other interests to use these *tonari gumi*, perhaps for subsequent reassertion of Japanese aggressiveness.

rich lather; courtesy requires young people to wash the backs of their elders. In the next room he slips into a tank perhaps fifteen feet square where neighbors soak contentedly and gossip for an hour or so. In cold weather the main purpose of a bath is warmth, for the water is heated to 110° F. Men, women, and children bathe together with no self-consciousness or embarrassment. During the 1920s the shocked protests of foreign missionaries and tourists evoked a government order requiring separation of the sexes in public baths. Puzzled by the strange requirement, many establishments complied by stretching a rope across the tank and posting signs, "Gentlemen" on one side, "Ladies" on the other. While the gossip may wax lively, the gathering in the steaming tank is completely decorous.

Apart from bathhouses, men congregate in a neighborhood teahouse or beer hall where flippant waitresses join in the repartee and earn pin money as unlicensed prostitutes. Rarely does anyone drop in casually at a neighbor's house for a chat. Houses are small and calls are formal.

Children play in the streets and in the roomy grounds of shrines or temples. Nursemaids accompany the smaller children and carry babies on their backs. A *chōnaikai* or *tonari gumi* may collect funds to pay attendants and provide equipment for play. More recently, playgrounds adjoining public schools have come into vogue.

Festivals at Buddhist temples or Shintō shrines are neighborhood affairs. Children stage playlets and perform folk dances, priests mime the legendary gods, women compete in exhibitions of flower arrangement, and vendors of food, drink, and trinkets set up booths in the shrine yard. The youths of the *seinenkai* bear the *mikoshi* or god-car through the streets and get drunk on *sake* offered to the god. Tradition insists that the god guides his bearers; if so, the gods have long memories for the heavy *mikoshi* is wont to smash the fence of someone who has not contributed to the shrine.

A general air of excited anticipation develops in the neighborhood as the grand festival of the New Year approaches. Since all debts must be paid before the old year passes, many a burglary is committed to enable some conscientious debtor to meet the deadline. Before every house, at either side of the gate, decorations of pine for longevity, plum for vigor, and bamboo for fertility are erected

—save for the licensed prostitute quarters which logically omit the bamboo.

The New Year breakfast, a gift from servants to the family, consists of a soup made of fermented beans, and a doughlike pounded rice called *mochi*. Since a person's age is reckoned by counting the calendar years in which he has lived, New Year's is a universal birthday and the usual greeting is congratulatory. A baby born on December 31 is two years old the next day—has he not seen two years?

Following breakfast the man of the house dons his best *kimono* and *haori*. If he favors European dress, striped trousers, morning coat, and top hat are *de rigueur*. Thus adorned he sallies forth to visit neighbors, business associates, and relatives. The calls are brief. He enters the *genkan*, exchanges congratulations with the hostess, and invites favorable treatment in the new year. Each hostess serves him a cup of *sake* and the ceremonies proceed in an increasing alcoholic haze. The next few days are given over to family recreation, calls by the ladies on their friends, and other festivities.

In rural Japan, neighborhood relations involve participation in various cooperative groups. Some work in rotation on roads and bridges, others provide mutual aid in rice planting or harvest. The purpose of other groups is monetary or other material assistance; called *kō*, these vary widely in purpose and membership. The general scheme runs as follows: a man in need of cash, say to meet a mortgage, organizes a *kō* by inviting acquaintances who have the funds. The *kō* meets monthly, semi-annually, or annually and meetings rotate from house to house. At the first meeting every member except the organizer brings a predetermined sum of money—say twenty *yen*. If there are twenty members, the first meeting pays the organizer 380 *yen*. At each of the twenty subsequent meetings, the organizer, like his fellow members, contributes his twenty *yen* and the winner of the current total is selected by lot from those who have not yet won. The pattern varies endlessly; in some schemes those who win pay a higher amount thereafter while others pay less. Toward the end of the cycle the few remaining potential winners may sell their chances for a good sum. In some *kō* the members are villages, not individuals; the annual prize finances the people of the winning village in a mass pilgrimage or a sumptuous picnic.

Craft guilds may be active in neighborhood affairs. Carpenters, gardeners, cooks, potters, stonecutters, makers of sandals, fire brigades, collectors of sewage—even pickpockets and burglars—maintain guilds whose ancient pattern is challenged by modern labor unions. The guild limits apprentices, maintains prices, provides mutual aid, and holds parties. At the New Year for example, the carpenters may don costumes and escort a fearsomely masked synthetic beast from house to house to scare evil spirits—for a fee.

The more prosperous families learn how to deal with the servants' guilds. Custom sanctions the petty graft that diverts about 20 percent of the family purchases to the servants. Tradesmen render one bill to the employer and another to the servant, who gets his rebate even if the master pays his own bills. Servants, however, regard themselves as loyal defenders of their employer; their 20 percent graft may be a moderate price for the vigilance with which they defend the household from the exactions of gardeners, carpenters, sewage and garbage collectors, and tradesmen generally. Servants participate freely in most family activities.

Like any employee, a servant receives pay monthly, neatly enclosed in an envelope to maintain the fiction that money is beneath notice. Dismissal is effected by deducting a few *sen* from the usual wage. The dismissed person counts his pay, waits discreetly for a few days, then resigns formally on the pretext that his grandmother has died, his father is ill, or his brother needs help in his business. If the employee's face has been protected, his departure synchronizes with the arrival of a new worker supplied by the guild.

All employees expect bonuses, especially at the end of the year. No worker may be released without a dismissal allowance equal to at least a month's wages, perhaps several years' salary in case of long service. A vigilant guild sees to it that a niggardly employer cannot replace the departed worker.

Leases, receipts, contracts, and other legal papers are validated by imprint of the personal seal (*han* or *hankō*) of the householder or businessman. If revenue stamps are required on the document—as on all receipts for more than a trifling sum—they must be affixed and canceled by the *han* so that part of the imprint overlaps on the document. A contract may be torn in two, each party retaining half. Matching of the halves verifies the agreement. Use of the seal corresponds exactly to Occidental use of signatures. Contracts, however, are not regarded as sacred; if either party stands to lose heavily by fulfilling the contract no great disgrace attends refusal to honor it.

Every neighborhood has its shopping street. Small shops are tended by a wife or daughter while the head of the house works elsewhere. Specialization is minute; thus beef, pork, chicken and eggs, beans, fruits, and confections are sold in separate shops. *Sake*, soy sauce, and pickles occur in the same malodorous shop. Restaurants, also highly specialized, not only serve food on the premises but deliver meals when ordered. A boy on a bicycle delivers such meals, kept hot in lacquered wooden boxes, to home or office. Other neighborhood shops offer hardware, cloth and thread, notions and incense, porcelain, household supplies, *geta* and sandals, medicines, electrical goods, soft drinks, shaved ice with syrup, and so on. Barber shops, hairdressers, and beauty parlors are near at hand; the beauty parlors usually adjoin a shrine of Benten, goddess of beauty. A tiny branch postoffice provides not only postage and revenue stamps, money orders, parcel post and registry, but also accepts telegrams, handles postal savings and life insurance, and operates a postal transfer system that takes the place of bank checks in most transactions.

When a shop displays price tags, no haggling is tolerated. But some shopkeepers—fish dealers in particular—relish a good argument over prices. Buyer and seller may grasp hands under the protective cover of *kimono* sleeves and bicker silently by a code of hand pressures, leaving the way open for a fresh start with the next customer.

Unless a purchase be trivial a shopkeeper serves tea, and conversation wanders lightly before approaching business. The fiction is that gentlemen never count money. One pays furtively by depositing money in a tray; the shopkeeper computes price and change on an abacus since mental calculation is neither attempted nor accepted. Pretending not to count the change, the customer removes it almost surreptitiously from the tray.

Once or twice a month a *yomise* or night-show is set up somewhere in the *chō* (ward). Under glaring electric bulbs, sidewalk stands display everything from dwarf trees to German cameras. The entire neighborhood turns out; people stroll about eyeing the displays, flirting, gossiping, haggling over an occasional purchase. *Caveat emptor* is the law of the *yomise*.

From early dawn till long after dark the streets are busy. Motor trucks and taxis have not eliminated ox-carts, two-wheeled hand-carts, bicycle trailers, horses, and human burden-bearers. In villages and the older parts of cities where streets are too narrow and tor-tuous for taxicabs, the *jinrikisha* survives. Before dawn vendors of *tōfu* (bean curd) blow a soft-toned horn as they trot from house to house. At either end of a shoulder-pole an itinerant fish seller carries tubs that contain live fish swimming in water and cleaned fish packed in ice. Pushing a tiny cart that holds his tools, a mender of *geta* beats a small drum shaped like an hourglass. Drowning the lesser noises, a large steam whistle announces the cleaner of tobacco pipes, whose cart carries a copper boiler to provide high-pressure steam. The colorful wares of a flower seller—who probably is an *eta* or one-time outcaste—hang in baskets from his shoulder-pole. The milkman, the knife grinder, the street sprinkler, delivery boys, and many others pull or push two-wheeled carts. Others carry their stock in trade on their shoulders. At night a plaintive bamboo flute announces the blind masseur; at a street corner the cooking facili-ties of a two-wheeled cart provide midnight lunches. Throughout the wee hours the clapping together of two sticks reminds house-holders that all is well as the watchman makes his rounds to scare burglars and look out for fires. Until the advent of American troops holdups were rare and even at night in slum districts a woman never was accosted on the streets.

ETIQUETTE.—Designed to create and maintain "face," etiquette is fundamental and its rules are complex and onerous. All schools teach etiquette, especially to girls. Classical etiquette followed vari-ous patterns, but the school founded by Ogasawara Shinano-no-kami (14th century A.D.) prevailed and his descendants still teach the subject.

The introduction of strangers occurs even less frequently than in England. One introduces two persons only if both actively desire it. The bow, equivalent to the Occidental handshake, is a universal greeting, except that men who pass on the street may merely nod and lift their hats. Men do not raise their hats to women. The kiss and the embrace are deemed grossly sexual and are taboo in public.

A standing person bows slowly from the hips at a 90° angle, sliding his hands down his thighs as he bows. When seated on the floor both hands are extended froglike, palms down, and the forehead is touched to the floor. The bow is a formal courtesy, not a token of servility.

A call is a courtesy extended by the guest to the host. One apologizes for failure to call, prolongs his stay to show politeness, and apologizes profusely for leaving. Close friends minimize these formalities. The same principle leads a guest to refrain from eating all of the confections set before him. He always carries paper in which to wrap some of the dainties and stow them in his *kimono* sleeve. In effect this says to the host, "These cakes are so delicious that I wish my household to taste them."

Formal calls are expected when a man occupies a newly built house, on occasions of congratulation such as the birth of a son or at the New Year, and after a conflagration. The occupant of a burned house holds forth amid the embers to serve tea and cakes to callers offering sympathy.

Rigid etiquette governs formal exchanges of gifts. In general the recipient of a gift returns a present of equal value. If the return gift be delayed the value of the return increases by a sort of compound interest. Such postponement of the return carries subtle implications; it may hint that the original gift was inadequate or even that it was excessive. When calling, small gifts to children of the household are appreciated and involve no complex etiquette of exchange. On returning from a journey gifts are presented to business associates and employees, even to neighbors; such giving is friendly, informal and with no savor of competition. The preferred traveler's gift is a local product of some place visited, since every locality fosters a distinctive handicraft.

Men precede women in boarding trams, entering doors, and in receiving service of any kind. A lady offers her seat to a gentleman, and a wife always walks two or three steps behind her husband. Hotel maids and sleeping car porters aid a gentleman to undress and fold his clothing neatly. One tips an innkeeper, not his employees; the amount generally equals one third of the bill. A symbolic present of a towel or matches acknowledges a satisfactory tip.

Both sexes carry and use fans in hot weather. A closed paper umbrella is carried by the tip, handle down. In sun or rain, one's open umbrella must be folded if one talks with a person who does not carry one; a gentleman shares his companion's discomfort rather than humiliate him by offering what he lacks.

One deprecates his own household and possessions; he refers to his wife as a fool or a poor cook, to the food that he serves as filthy stuff, to his children as ignorant brats. A guest counters by exaggerated praise of the food, the skill of the cook, the beauty and breeding of the children. Japanese, however, rarely invite anyone to a meal in their homes, in accordance with the principle that a visit honors the host. Friends are feasted at a restaurant or tea house and professionals, usually *geisha*, provide the entertainment.

Even within the family, etiquette restrains expression. Children approach their parents with bows and formal phrases. Husband and wife avoid public display of affection. In the service of etiquette the language has attained esoteric complexity, with separate vocabularies for the sexes, for differing degrees of social status, and for formal occasions; elaborate circumlocutions protect the polite fictions of "face." In this brief sketch, the meticulous etiquette of eating, of tea, of drinking, of women's activities, of formal conduct in relation to house and garden and special occasions, must be omitted.

MORALITY.—By insensible stages etiquette blends into morals. If morality be defined as habitual conformance to *mores*, the Japanese are highly moral. If in contrast to morality, ethics be defined as the discriminating refusal to use another person as means to an end that he does not acknowledge freely, ethical conduct is rare in Japan. From infancy to death individuals move in an intricate network of duties and obligations. The most important obligation, *on*, denotes those infinite debts incurred by being alive and which no amount of effort can repay. Foremost is the debt to the divine Emperor by whose grace one exists and whose air one breathes; death in battle is but a token appreciation of the Imperial mercy. *On* inheres in the parental gifts of life and nurture, and in feudal times the overlord's *on* was regarded as modern Japanese regard the Emperor's *on*. Teachers confer "the life of the spirit" and thus create *on* that cannot be repaid. A person is said to "wear the *on*" of these benefactors.

Obligations capable of exact repayment are called *giri*. Favors received and insults that rankle alike involve *giri*. *Giri* prompts the meticulous calculation of equivalence in gift exchanges and dictates the repayment of favors and other reciprocal transactions. *Giri* is felt as a burden; *giri* also implies a sense of honor. Of an ill-bred person it is said, "He does not know *giri*." The touchy sense of honor of the medieval European knight is analogous to one aspect of *giri*.

Underlying this morality are teachings of Confucius, of Buddhism, and of the native cults of Shintō and Bushidō. The Confucian doctrines of duty, loyalty, and filial piety deal with attempted repayment or acknowledgment of *on* received from ruler or parent. The good ruler sets an example to his subjects, the good father to his household. Duty and filial piety consist in following these examples and in loyal devotion. The Japanese rejected the Chinese distinction between good and bad rulers or parents. Good or bad, rulers and parents impose on their subjects or children all the obligations implied by *on*. Confucius' insistence on propriety is implicit in Japanese *giri*, which requires formal balance and harmony; the social order is preserved by action precisely appropriate to every situation. Note Confucius' negative Golden Rule: "Do nothing to others that you do not wish done to you." Japan's meticulous etiquette serves Confucian propriety, even though formal rules of conduct may nullify ethical considerations. Through etiquette one preserves his own face and protects the face of others. At whatever cost one maintains face for his family.

The customs that center in maintenance of face impress an Occidental observer as extravagant and neurotic. *Yūmei mujitsu*—the polite fiction—is carried unblinkingly far past the point where anyone could be deceived. Loss of face results from denying the reality of fictions or from being found out, not from the actual conduct that fictions camouflage.

Coequal with the polite fiction and the maintenance of face is the Buddhist doctrine of noninterference. The belief in fate determined by events of a previous existence implies that whatever happens to one is part of the rewards and retributions due him for conduct in a former life. Anyone so reckless as to interfere in this order of the universe deserves whatever unpleasantness he may incur. For example, if a maniac attacks an innocent person, bystanders make no move to protect the victim or to apprehend the attacker. The seemingly innocent victim must be reaping the penalty for crime in a past incarnation; anyone interfering invites trouble. If a boy falls in the canal, let him drown; who knows that in a previous life he did not drown a sweetheart in a jealous rage?

One automatically aids others in protecting their face for he may need their cooperation in maintaining his own face. For example, after twenty-five years of faithful service the dean of a girls' college belatedly sows his wild oats. Parents complain that the dean has seduced their daughters. In interminable secret sessions the trustees examine the evidence and decide to oust the dean. There is no hint of scandal. On the contrary the school stages a celebration of the twenty-fifth anniversary of the dean's appointment to the faculty and invites notables from near and far. At the climax of the speech-making the dean, overcome by emotion, resigns dramatically. In recognition of his splendid record the trustees present him with a house and lot. Some of the outraged parents may have contributed to the fund that purchased the house and lot, but no matter how many persons know the real situation, never a word or a glance betrays their awareness.

A year later the former dean is appointed to the faculty of a college for boys, whose president also serves as trustee of the girls' college. A puzzled American ventures to ask, "Why do you appoint X to your faculty when you know all about the mess at the girls' college?" Smiling, the president replies, "X knows who saved his face. He dare not let me down; he will be the most dependable man on my faculty!" For twenty years the erstwhile dean leads an exemplary life, and the respect paid at his grave is genuine. He wore the president's *on* with dignity and devotion.[5]

Martial bravery and ruthless direct action are the tokens of manly excellence. The much-uttered word *makoto* ("sincerity"?) denotes direct action, or single-minded pursuit of a goal, no matter how devious the means. Weakness is despised. Unscrupulous exploitation of the weak is expected from a strong brave man. Protection of the weak is limited strictly to one's household, clansmen, or faction. Suicide, not a cowardly retreat from living, incontrovertibly demonstrates sincerity; it establishes one's integrity when motives are questioned, offers honorable death to political opponents, defeated

---

[5] For a clear if somewhat tight account of Japanese morality, see Ruth Benedict's excellent book, *The Chrysanthemum and the Sword* (Boston, 1946).

dignitaries and prisoners of battle, and cuts the Gordian knot of dilemmas in which *on* and *giri* conflict with each other or with "human feelings." Successful violence may achieve like ends.

Self-control is a cardinal virtue. Despite highly emotional temperament the Japanese despise any show of emotion, except at funerals, in the theater, or in aesthetic enthusiasms. Sorrow is camouflaged by a smile; one who inflicts his troubles upon others *"does not know giri."*

Early in the modern period the codes of *on* and *giri*, embroidered with stirring tales of the bravado of the *samurai* of old, were revived and deliberately elaborated into the patriotic cult of *Bushidō*—"The Way of Knights." The word *Bushidō* was new, and so were the startling applications of its principles to Japan's international relations. From the moment that the 1924 immigration law passed the United States Congress, the Japanese deemed themselves insulted. National *giri* and *taimen* (face) required Pearl Harbor.

Morality and etiquette do not apply to foreigners, who disregard *on* and "do not know *giri*." Since Japanese, unlike foreigners, are descended from the gods, no Japanese could be wrong in a difference with a foreigner. Children are indoctrinated with the idea that the god-descended Japanese are incapable of sin. Habitual practice of *on* and *giri* leaves them helpless when facing persons whose *mores* are alien; baffled and infuriated by the incomprehensible ways of foreigners, Japanese seize any opportunity to humiliate hapless aliens who fall into their power. When the alien holds the power, the Japanese is outwardly obsequious and complaisant; he holds his temper and bides his time. Japanese *mores* and Euro-American *mores* are incommensurable. The Occident proclaims the value of the individual; Japanese tradition explicitly denies the value of the individual.

FAMILY CUSTOMS.—The Japanese family supplies the model for all social and political behavior. In theory, individuals exist solely for the family and the nation. Each person is responsible to and controlled by his family; in turn the family bears responsibility for its members; the success of one is the boast of all, the disgrace of one loses face for all. Individual earnings belong to the family, and the aged and disabled receive care from the same source. Individuals rarely execute a last will and testament; the family continues to hold all property regardless of individual deaths.

Descent is traced in the male line. The surname is spoken and written before the given name. Family names were not permitted to commoners until the official ending of feudalism, when the new government ordered all families to assume surnames. The illiterate commoners appealed to local scholars to provide them with names. If the putative scholar really knew the ideographs, his clients acquired original surnames written with beautiful and suitably obscure characters; more often it happened that no one in the village knew more than a few ideographs and every local family took the same name without regard to kinship.

Normally a household includes grandparents, their eldest son, his wife and children, other immature sons, unmarried daughters, and unattached collateral relatives in the male line. Lacking food and space for so many people, urban households average smaller. At maturity younger sons leave the household to enter the army, business, or a profession, or to work in factories. Consequently urban populations include disproportionate numbers of younger sons and unmarried women who still belong legally to rural households. When ill or unemployed these persons return to the old home. The family register or *koseki* usually is maintained in the police station nearest the ancestral residence. Persons who "have no *koseki*" are hounded as suspicious characters by the police; it is assumed that only the vilest of men would be repudiated by their families.

When there is no son, or if the sons are incompetent, a man adopts as heir a youth who normally marries a daughter of the house and assumes the family name. Such adopted sons may be younger sons of relatives.[6] Rich men commonly adopt poor, honest youths who through hard work know the value of money. The position of adopted son sometimes carries an aroma of subservience that provokes trouble; many adopted children, however, are indistinguishable socially from natural children.

[6] At a reception a distinguished Japanese educator introduced the writer to his wife, a son, and two daughters. Later the professor introduced his brother with a wife and two children; his sister, her husband and two children; and finally another brother with wife and three children. With keen appreciation of the humor of the situation, the educator confided, "You are interested in kinship customs. All ten children originally were ours. Now you see how our adoption customs enable a professor to live on his salary—if his relatives happen to be childless!"

The father rules the household. A distinction between "external matters" and "internal matters" allots responsibility for the domestic establishment to the wife. Except for the purchase of supplies, all business outside the house is handled by the husband. Whatever the situation, respect and obedience are due to the husband and father. From infancy, females learn that the word, even the whim, of a male is law; even in the modern period girls must memorize the "Great Learning for Women," a classical formulation of this doctrine.[7] At the age of sixty a man retires from the family headship in favor of his eldest son. In unusual situations a younger son, or even the wife, may succeed to the headship of the family. The domestic hierarchy is based on age and sex; specific kinship terms differentiate elder brother, younger brother, elder sister, and younger sister.

Extraordinary events convene the family council, which includes the head of the house, his brothers, father's brothers, and for certain types of inter-family problem, males of the wife's family. Family councils deliberate the marriage of a son or daughter, adoption of a son, release of a son for adoption elsewhere, new ventures and business crises, education of a child, employment of a daughter, erection of a house, land transactions, or a crime or offense involving the family. Supposedly no woman except a grandmother sits in the council; in practice capable women may participate and wield much influence. The composition of the council varies with the occasion; the more serious the problem, the wider the representation of related households. Joint councils of two or more families are not uncommon.

All sorts of meetings—committees, village councils, boards of directors, teachers' meetings—are patterned on the family council. The word *sōdan* denotes an interminable solemn discussion in which everyone talks, beginning with those low in status and working up to the head of the senior branch of a family or the chairman of a committee. No vote is taken and the weight of a majority is not calculated formally. The head discovers what the others think, learns what he can "get away with" and delivers his decision. Failure of

[7] The article "Woman" in B. H. Chamberlain's *Things Japanese* (London and Kōbe, 1927) contains a full translation of *Onna Daigaku* (The Great Learning for Women).

the others to acquiesce is rare. The process may occupy several days and most of the intervening nights.

Families of wealth or distinguished lineage often maintain codes of household law inherited from feudal days. These rules bind every member to conduct uncompromisingly oriented to family interests. The extreme penalty for violations is ousting from the family and removal of the offender's name from the *koseki* in the police station. No employer hires a person until he is assured that the candidate is backed by his family. Even a family head, if he disgraces the house or wastes its resources, may be deposed by the council; if his sons are too young for responsibility and his wife is capable, she may be named as head.

Family obligations extend beyond the grave. Confucian, Buddhist, and Shintō practices converge in emphasis upon the duty of ancestor worship. Neglected spirits may become "hungry ghosts" that harm everyone. One's highest duty, after loyalty to the Emperor, is to rear sons to continue the family and the ancestral rites— hence the imperative necessity for adoption if no son grows up. National and family ancestors are honored with libations of *sake* and prayers at a "god-shelf" in every home. To the ancestors, formal announcement is made of the birth or naming of a child, business successes and reverses, school graduations, marriages, entrance of a son into the army, journeys, adoptions, and deaths. At *O-bon*, midsummer festival of all souls, ancestors are believed to return and share a feast with the family. In addition to the Shintō god-shelf, Buddhist households maintain a family altar.

Marriages traditionally are arranged by the family council, in which the prospective groom or bride does not participate. A married person—perhaps a relative—of the same sex as the candidate is selected as go-between. Aided by the gossip grapevine, go-betweens who represent different families meet and discuss possible matings. The go-between investigates the family of each proposed mate, inquires into social and financial status, education, personal tastes and habits, health, and probable compatibility of each candidate, and further ascertains the record of the family with respect to leprosy, syphilis, tuberculosis, and crime. Supposedly the bride and groom meet for the first time at their wedding, but in late years they manage to get acquainted earlier. Often the go-betweens sign the wedding invitations; always they permanently sponsor the new household. Should a marriage fail because of circumstances overlooked by a careless go-between, the offending negotiator suffers rigorous social ostracism.

The wedding ceremony traditionally occurs at the groom's home. Some modern weddings, in imitation of Christian custom, are held at a Shintō shrine or Buddhist temple. Shintō shrines are preferred because of the association with fertility magic and life; death is abhorrent to Shintō, while Buddhist temples are associated with death and funerals. The go-betweens have already negotiated the value of the presents contributed by the two families; the bride's family practically provide a dowry. At her home the bride dresses in white, the color of mourning, to symbolize death to the family of her origin. She enters the groom's house clad in garments symbolic of birth into the new family; these, however, are exchanged for other garments that indicate acceptance of household duties. The ceremony consists in three exchanges of three cups of *sake* between groom and bride—called *san-san kudō* or "three-three nine times."

Families who cannot afford the expense and formality of gobetweens turn to relatives for aid in finding mates for their children. Marriage brokers who guarantee a properly investigated mate for a fee exploit the poor by fostering matches that are sure to break up and thus occasion another fee; the Ministry of Welfare has established public marriage bureaus to combat such abuses.

To legalize a marriage the bride's father obtains her registration card at the police station and sends it to the groom's father, who deposits it in his own *koseki* at the appropriate police station. Thenceforth the bride is a legal member of the groom's household, subject to control by its head. Her parents-in-law may divorce her, even against her husband's wishes, if she displeases them. Among the middle classes, concern for a daughter's welfare motivates many a trial marriage; the ceremony is performed and the couple live together, but the girl's father shrewdly postpones the transfer of registration until he is assured of the success of the marriage. The poor often dispense with all formality and a couple simply live together at their pleasure; in rural hamlets a number of studies reveal almost universal coincidence of the date of registry of a marriage with the date of birth of the first child.

An adulterous wife is punished severely while her male partner goes scot free. Except in districts where local reforms have outlawed the traffic in girls, commercialized prostitution has status like that of a public utility. Published indices of economic conditions usually include statistics of the patronage at houses of prostitution. Since a father has full power to contract for the labor of his daughters, many girls lose their freedom for terms of years as factory hands or as involuntary prostitutes. Industrial employment of women, therefore, does not connote their social and economic freedom. Girls often volunteer as prostitutes to pay their fathers' debts, to finance a brother's education, or to defray family medical expenses. Newspapers feature these "noble sacrifices" in pictures and sentimental write-ups. Modern women, however, repudiate the dual standard of morality. They no longer respect the dutiful wife who rises in the wee hours to serve tea to the girl her husband has brought home from a party, and who later serves breakfast to the pair.[8]

Geisha are trained entertainers who dance, sing, play musical instruments and amuse guests with salacious repartee. A *geisha* establishment provides girls at fixed hourly rates for dinners and parties. Novels portray the struggles of an innocent *geisha* to retain her virtue; such an achievement probably would deserve a novel. Despite glamorous apologia for the system no *geisha* may look upon a member of the Imperial family. An army general once allowed a *geisha* to smoke a cigarette from an Imperial gift package and the ensuing scandal ended in the general's suicide. In contrast to the relative freedom and education of *geisha*, *jōrō* or licensed prostitutes are guarded day and night lest they escape. Thousands of these girls have burned to death in locked quarters during the frequent city-wide conflagrations. Every *jōrō* fills out a daily report to the police on her clients—a system that locates many a criminal.

The Japanese idolize infants. Babies ride on someone's back by day and sleep with the mother at night. After a year or so of indulgence they encounter increasingly rigorous training in etiquette and family *mores*. No longer fondled, their relations with parents assume formalized patterns. A child does not rush home from play

---

[8] For data of registration of marriages and births, see, for example, T. E. Jones, *Mountain Folk of Japan* (New York, 1926). The American Occupation ordered the closing of all licensed prostitution districts. Another order enforced woman suffrage. Apparently these orders won popular support.

or school to mother's welcoming arms. He bows gravely at the threshold and announces "Now I return!" The parent returns the bow with the phrase, "Come back!" He learns the momentous details of family face; many adolescents and even younger children have committed suicide to atone for the disgrace of failure at school or in games. Girls discover that their role in life centers in obedience: in childhood to father, brothers, and mother; in maturity to husband, parents-in-law, and sons; in old age, to a son. A boy must obey his father and elder brothers but may lord it over his mother and sisters. Adolescent emotional conflicts often focus in the dilemma of enforced submission to the family versus the dismal reality of internal quarrels and jealousies that preclude psychological security in the family situation. Repressed conflicts of this sort have been adduced in explanation of the sadistic brutality of soldiers and police.[9]

The superficially idyllic calm of a Japanese community conceals underlying tension and insecurity. Every family is pitted against every other family in the struggle for face and prestige. No individual dares relax his guard psychologically, and even within families emotional security is relatively scarce. Emotional habits nurtured in such a milieu inevitably transfer to foreigners and foreign nations. Every Japanese expects hostility toward his nation just as he expects other families to search out the vulnerable aspects of his family face, or as his forebears expected other duchies to maintain unrelenting espionage. Always in the presence of foreigners he is self-consciously a Japanese making face for Japan and hypersensitive to slights, insults, and fancied plots.

Family behavior sets norms for all social relations. Subtly and irretrievably institutional forms copied from the Occident are differentiated from their prototypes; a school or business corporation looks Occidental but the pattern of interpersonal dealings is that of the Japanese family. The go-between is indispensable to all negotiations that involve relations with persons not of one's family. Before a new employee meets his employer a go-between threshes out all details and achieves an understanding. When the principals finally meet they observe a polite fiction and smoothly work out

[9] For additional details, consult Benedict, *The Chrysanthemum and the Sword;* LaBarre, "Some Observations on Character Structure in the Orient: the Japanese," *Psychiatry,* Vol. VIII, No. 3 (August, 1945), and Haring, "Aspects of Personal Character in Japan," *Far Eastern Quarterly,* November, 1946.

the prior agreement as if neither had heard of it previously. Commercial contracts, educational appointments, the building of a house, the choice of a physician—every human contact is prepared in advance. Often the go-between is more than a negotiator; he acts as sponsor of one of the parties in dealing with the other. A responsible sponsor or *hoshōnin* stands back of every important transaction and guarantees the integrity of social relations of all sorts.

A man is bound to his *hoshōnin* by obligations of both *on* and *giri*, depending upon the favors he has received or expects. If a position be obtained through the good offices of a *hoshōnin* of standing in the community the dependability of the employee is assured. Bond may be required and if so the *hoshōnin* sees to that; but posting a bond means less in practice than the fact that the new employee wears the *on* of his guarantor. Such patron-protégé relationships may obtain between families and thus outlast the lifetime of an individual. The employer in his turn protects the face of his employees; no one, from school teacher to bureau chief, from peasant to *Zaibatsu*, ever publicly reproves or corrects another person. A schoolchild, corrected by the teacher before the class for slight mistakes, may commit suicide to restore lost face, and when that occurs the teacher's position is not enviable. If a pupil, an employee, or an associate requires correction or merits rebuke, the go-between or the *hoshōnin* is informed of the situation and improvement is instantaneous. *Yūmei mujitsu*, the polite fiction, has preserved everyone's face. In such a society the American businessman who prides himself on "dealing straight from the shoulder" encounters repeated frustrations. At the very least he runs the risk of getting uniformly untrustworthy employees and clients whom no one will sponsor, and he complains bitterly of the dishonesty and unreliability of the Japanese.

One who achieves face cannot retain it passively. Others feel out the crevices in his armor of propriety. Cleverly devised situations test his resourcefulness and ability "to take it." Loss of temper practically obliterates the angry man's face; ability to meet extreme provocation smiling and unruffled multiplies face tenfold. There arise, however, exceedingly rare crucial situations in which one may fly into a violent rage and thereby enhance his face.

In discussing the relation of Emperor and subject, patron and protégé, employer and employee, or the organization of associations, corporations, and political parties, the Japanese cite the analogy of the family *ad nauseam*. The hierarchical pattern of human relations is the only one they really understand, for they absorb it in the family before they learn to frame a sentence.

SOCIAL CLASSES.—Feudal Japan was composed of small duchies, each ruled by a *daimyō* backed by armed retainers called *samurai*. The *shōgun*, a super-*daimyō*, governed at Edo (now Tōkyō); the Emperor was secluded at Kyōto. Commoners and *samurai* owed supreme loyalty to their local *daimyō*; even modern peasants turn for advice to his legally powerless descendant, the *tono sama*. Thus tightly organized on the pattern of the extended family, subjects of the same *daimyō* held suspect any person from another duchy. Such local loyalties persist; a Japanese refers to his native district as his *kuni* (country). Almost the first question after an introduction is "What is your *kuni*?" The tradition of solidarity among subjects of a *daimyō* continues in the *mores*; fellow "countrymen" aid each other, mutually protect face, stop at each other's homes when traveling, contribute funds to educate boys of their *kuni*, and intermarry. Many an Ōsaka or Tōkyō factory is manned by employees who all come from one *kuni*.

The English word "clan" has been applied to these compact *kuni* groups although they do not resemble the clans or sibs of primitive tribes; they neither practice exogamy nor believe in common descent. Ostensibly *kuni* were abolished at the opening of the modern period; actually they continue real though intangible. The elusive factions in army, navy, bureaucracy, and financial oligarchies often stem from *kuni* loyalties and rivalries.

Feudal law defined social classes rigidly. Styles of dress, dwellings, food, districts of residence, even toys, were prescribed for each class. Broadly summarized the feudal classes were: (1) nobility, *samurai*, and priests; (2) farmers, honored in theory as food producers but actually oppressed; (3) artisans; (4) merchants, despised as economic parasites; and (5) outcastes.

Four classes were recognized legally in the modern period: *nobility, gentry, commoners*, and *"new commoners"*—a euphemism for erstwhile outcastes. None of these constituted a "landed class"; even the gentry were descendants of *samurai* and nobles, not landholders as such. The suffrage was extended to males of all classes in 1925. Even at that late date, however, it was necessary to record one's social class when filling out blanks such as school registration forms, police reports, and the like. Particularly in the cities these obsolescent legal class categories no longer fit the real situation. Socially significant groupings based in observable cultural differences yield categories such as: (1) Imperial family and titled nobility; (2) *Zaibatsu*, families who control finance and big business; (3) civil servants, teachers, professional people; (4) a small middle class of businessmen, industrial executives, supervisors, technicians, retail merchants, white-collar workers; (5) army, navy, and police officers; (6) farmers and peasants; (7) laborers and factory hands; (8) vagrants and descendants of outcastes. These groups vary in cohesiveness and class-consciousness; Japanese, however, act habitually in terms of class and "respect their betters." Money income plays little or no part in winning the respect of others; save as the first two groups constitute a plutocracy, wealth and income do not parallel the foregoing divisions.

Incomes as represented in *yen* have risen throughout the modern period, especially at the close of the first World War. The attempt to represent Japanese incomes in U.S. dollars involves innumerable difficulties; wage levels may be indicated roughly, however, by counting two *yen* as equal to one dollar. On this basis, taking 1930 as indicative of the latter decades of the modern period, some approximations may be stated. The net *annual* income of a peasant family, after charging minimum subsistence to expenses, ranged from a profit of about four dollars to a loss of about $60, with losses predominating. Day labor and spare time piecework helped recoup the losses. A farm laborer could earn a daily wage of slightly less than fifty cents. Apprentice boys received board, space to sleep, and simple clothing, plus from twenty-five to fifty cents a month in cash. A girl factory worker living in a company dormitory, with strict economy could net from twenty-five cents to a dollar a month after paying room and board. Skilled laborers received from fifty cents to two dollars a day; these workers, like office workers and salesmen, might expect a year-end bonus that in prosperous times ran from a month's pay to a full year's pay. Schoolteachers and civil servants generally counted prestige above money, and their pay ran from $15 per month upward. Upper-level civil service men, engineers, and other experts received salaries of $1,500–$2,500 annually. A foreign ambassador's salary stood at about $3,000 plus expense allowances for foreign duty. In the port cities in the 1920s, room and board, native style and first-class, could be had for about $15 per month. All of these figures are completely meaningless in the whirl of postwar inflation, black markets, and near-starvation that marks the 1940s.

The stability of class alignments has decreased steadily during the modern period. Numerically perhaps, most people continue in the class of their origin. Industrialization and urban growth, however, enable younger sons to move to cities, forge out their own careers, and shift to new classes. Sons of peasants have been able to advance in the army, and hence peasants generally have regarded the army as their champion against landlords, bureaucracy, and the *Zaibatsu.* Until the last two decades the higher civil service ranks were open only to graduates of the Imperial Universities; other universities, however, now share that privilege. Because of the expense, this road to advancement was more feasible for sons of the well-to-do, but an extended family could pool their resources to see an unusually bright boy through college. Advancement via commerce and finance involves handicaps; partly because businessmen inherit the stigma that a feudal society attached to merchants, partly because feudal-mindedness allots top positions to members of the *Zaibatsu* plutocracy. Of course minor civil service officials usually have risen in life—railways, post office, telephone and telegraph services and the like have discovered that prestige is a cheaper and more welcome reward than high salaries.

Between 1927 and 1945, the advocates of totalitarianism agitated for elimination of class differences. Their goal was not democracy, but rather the egalitarianism beloved of dictators who govern on the principle, "If a head rises above the others, cut it off."

RELIGIONS.—Before the introduction of Buddhism from Korea in the sixth century A.D., the amorphous indigenous cult of *kami* held sway. Generally mistranslated as "god," *kami* resembles the South Sea concept of *mana*, which denotes mysterious impersonal supernatural power resident in places, persons, material objects, or ghosts. In the habits of the common folk the *kami* cult persisted and permeated Buddhist practice. To distinguish this cult from Buddhism, it received the name of Shintō (The way of kami).[10] Japanese, however, rarely accept one religious sect to the exclusion of others; almost everyone considers himself at once a Buddhist, a Shintōist, and a Confucian. Individuals worship any deity regardless of cult affiliations, as indicated by the oft-repeated sentiment, "You never know; there might be something in it!" The Shintō pantheon nominally includes eight hundred times ten thousand kami —a statement that should be interpreted in the light of the use of the numbers eight, 80, 800, 8,000, 80,000 and "eight hundred times ten thousand" to indicate vague quantities of the appropriate orders. The supreme kami, though not the creator, is Amaterasu ō-mikami, sun goddess and heavenly ancestress of the Yamato dynasty. She originally sent a heavenly host to take over Japan and to rule the world—a myth that accounts for the belief in the divine descent of all Japanese, and for their aggressions on foreign countries. This myth is indoctrinated thoroughly in all schools. Other kami include popular food and fox deities, "clan" kami, local kami, and historical personages whose superior powers indicated possession of kami. At the Grand Shrine of Ise, dedicated to Amaterasu ō-mikami and the Imperial Ancestors, the Emperor or his deputy reports to the heavenly kami all important national events.

The peasant pays his respects to the kami at shrines presided over by sundry local deities, at his family god-shelf, and at national shrines dedicated to the Heavenly kami and the Imperial Ancestors. His reverence extends to sacred mountains, waterfalls, trees, phallic symbols, and scenes of natural beauty. He has gone on pilgrimage to one or more well-known national shrines and has good-luck charms from each of them. His infant sons are dedicated at a shrine of Hachiman, the war kami. Planting and harvest he times by the ancient festival calendar whose events provide his recreation. At school he learned to bow toward Tōkyō in reverence to the Emperor, but the doings of earthly kami in high places are as mysterious to him as are the Heavenly kami. When one of his household dies he summons a Buddhist priest and tries to pay for masses on behalf of the departed. His grandmother fingers a Buddhist rosary

[10] Chapter VII of Japan's Prospect (ed. Haring) sketches the history of the kami cult.

and repeats the *Nembutsu;* he omits these rituals and remains innocent of theology. Christianity he knows by rumor as a subversive
alien cult that neglects ancestors. He visits Shintō shrines and Buddhist temples when he feels so inclined; they hold no regular congregational services.

In early times the prestige of Buddhism was enhanced by association with Chinese culture. Its magnificent temples, centers of
learning and the arts, often succumbed to corruption. Nearly all
Japanese Buddhist sects followed the Mahayana tradition which
emphasizes Bodhisattvas (saints) at the expense of the teachings of
Sakyamuni, the Hindu founder of Buddhism. Amida, a Bodhisattva
who is believed to maintain a glorious western paradise for those
who accept his grace, occupies a prominent position in the more
popular sects. The bliss of paradise and the tortures of hell intrigued
the Japanese, but, habitually incapable of a sense of guilt, they at-
tenuated the fear of hell to an aesthetic melancholy. Sakyamuni's
metaphysics they leave to priests and monks. In the twelfth and
thirteenth centuries A.D., several new sects emerged in a Buddhist
Reformation that strikingly parallels the European Protestant Reformation. Against the magic and superstition of the former hierarchy
the reformers pitted the free grace of Amida; some of them abolished
monasteries and instituted marriage for the clergy. One of these reformed Buddhist sects, Shinshū, continues the largest and most vigorous in Japan. Nichiren, an anti-Amida reformer, founded in the
thirteenth century Japan's sole belligerent Buddhist sect and won a
large following of peasants, fisherfolk, and minor gentry. Nichiren's
devotion to the Emperor stimulated later movements toward restoration of the Imperial power, and his chauvinistic followers helped to
set the stage for the second World War.

Long dormant under Buddhist ascendancy, Shintō was revived in
the eighteenth and nineteenth centuries by scholars who discovered
ancient documents indicating that Buddhism was a foreign cult and
that Japanese are descendants of the *kami.* They proclaimed Japan's
mission to extend the blessings of divine rule to all the world. An
upsurge of patriotic devotion to the Emperor swept away the
Shōgun and effected the Restoration of 1868. When foreign powers
demanded freedom of religion in Japan, the fiery patriots solved the
dilemma by creating two Shintō cults. The first, State Shintō or

Shrine Shintō, was declared to be nonreligious and purely patriotic, hence obligatory upon all Japanese and State-supported. The second, Sectarian Shintō, was declared to be a voluntary religious system. Thus was laid the basis of that national indoctrination which prepared the people for the successive wars of conquest that came to a climax in 1941. Public officials were required to conduct State Shintō rituals; Shintō mythology was taught in all schools; all children were compelled to worship at Shintō shrines in the schools and to bow in reverence at frequent readings of Imperial Rescripts. In practice, State Shintō was a genuine religious cult which the common people did not differentiate from sectarian Shintō.[11]

Current magical practices include divination, exorcism, fertility rites, rain-making, and superstitions regarding foxes and badgers; both Buddhist temples and Shintō shrines are involved in magical activities. A few old-time shamans, called *yamabushi*, still follow their vocation.

Only one tenth of one percent of the population is reported as Christian, although the Bible has been a best-seller during most of the modern period. Christianity is strongest in urban districts.[12]

WRITING AND BOOKS.—Although spoken Japanese differs from Chinese quite radically, Chinese ideographs form the basis of the written language. Some 2,200 ideographs are essential to reading a newspaper; six to eight thousand give access to more sophisticated publications. Each ideograph represents an idea and has several pronunciations. A purely Japanese phonetic syllabary enables the common people to write with only fifty characters; however, thus far no phonetic writing system has given satisfaction because the language contains thousands of homophones, and only ideographs differentiate these words. Although few people remember the more complex ideographs after they finish school, they understand most of what appears in the newspapers because the phonetic syllables are printed beside each ideograph.

Paper-making was learned from the Chinese centuries before Europeans knew the secret. Small handicraft establishments turn out hundreds of kinds of papers for as many special uses, and modern

---

[11] In 1945 the American occupation suppressed State Shintō; Sectarian Shintō was not disturbed.

[12] Post-surrender developments are indicated by R. J. D. Braibanti, "Religious Freedom in Japan," *Christian Century*, July 9, 1947.

factories produce newsprint and other Occidental types of paper. Printing also reached Japan from China long ago; in fact, the oldest extant printed matter was made in Japan. Multiple color printing of pictures was accomplished successfully more than a century before Europeans learned the trick. Although most modern books are printed and bound in the Occidental manner, the true Japanese book is printed on one side of a long strip of thin paper. This is folded accordion-fashion, stitched at one edge, and the pages left uncut to give the effect of a single sheet printed on both sides.

The traditional writing materials comprise a brush and India ink, with long rolls of soft paper. Writing begins at the upper right and runs in vertical columns, down the page and from right to left. Consequently the beginning of a book corresponds to the "Finis" of a European book. Letters are written on long strips of paper with the columns at right angles to the long dimension of the strip. The completed letter is rolled, beginning at the end, then pressed flat, to be unrolled as read. In writing an address, the narrow envelope is held with the long dimension vertical. The postage stamp occupies the upper left corner; when the envelope is turned to bring the long edge in a horizontal position the stamp appears in the place expected by international usage.

The Japanese read omnivorously. Prior to the war the number of books in print was very great; [13] and while public libraries do not play an important role, books circulate rapidly via second-hand bookstores. Once read, a book is resold for a few *sen* less than it cost. As for newspapers, Japanese dailies lead the world in circulation. Rigid censorship of all printed matter is traditional and the public has read only such ideas and information as the government approved. In the cities, English-language books circulate briskly, since all graduates of middle schools know a smattering of English.

FINE ARTS.—Calligraphy underlies all drawing and painting and stands first among the fine arts; the same word means "to write" and "to draw." Specimens of the brush-writing of noted artists or of famous men become collectors' items at high prices. Painting, sculpture, architecture, poetry, drama, ceramics and lacquer work, prose

---

[13] Dr. A. K. Ch'iu of Harvard-Yenching Institute reported to the American Oriental Society in 1946 that 1,918,870 different Japanese-published books were extant in 1940.

literature, pantomime dancing, ritual and ceremony, and formal garden design all attain high aesthetic standards in terms of cultural patterns quite different from those of Occidental arts. The sophisticated simplicity that characterizes most Japanese arts appears also in the rituals of old-time etiquette. Uninformed reporters from the Occident sometimes mistake this conscious striving after simplicity for inability to devise complex forms. For example, the Japanese have been alleged to lack inventiveness because their houses contain so little furniture; actually they achieve aesthetic perfection by meticulous elimination of every superfluous object and every needless detail.

Uniquely original Japanese arts are not numerous; they include the pantomime drama called *Nō*, the seventeen-syllable poetic epigrams known as *haiku*, the designing of gardens by specialized artists, who have been known to sign a finished garden as a Western artist signs a canvas, and the formal tea-ritual. In all of these arts the keynote is sophisticated simplicity. For example, both the *Nō* and the *haiku* aim to suggest a compact emotional situation with utmost economy of material. Consequently the tiny *haiku* are well-nigh untranslatable, as one instance shows:

Asagao ni	By morning-glory
Tsurube torarete	Well-bucket captured
Morai midzu	Borrow water!

The picture is that of a sentimental maid, impressed by the morning-glory that has entwined the ropes of the well-bucket overnight, going to a neighbor's well rather than disturb the plant. Only in the Japanese original can the economy of words be appreciated.

Appreciation of the native arts, like sensitivity to natural scenic beauty, has been cultivated most successfully among the people. Coolie laborers bow reverently before a beautiful natural panorama and quote poetry unembarrassed. The feeling for art is akin to the mystery that surrounds *kami* and no clear line can be drawn between Japanese aestheticism and religion.

Occidental arts are imitated eagerly but without the facility that marked the adoption of Western technology. Thanks to the phonograph, however, European music has attained not only popularity but appreciation—and this among a people who a quarter-century ago seemed incapable of hearing intervals other than those of the ancient five-toned scale.

EDUCATION.—Although universal compulsory education was advocated in Japan centuries ago, temple schools cared for most of those who desired formal education. The Imperial University of Kyōto, founded in the eighth century A.D., admitted only sons of the nobility. With the return of students sent abroad to acquire Occidental knowledge at the time of the Restoration, public schools were organized on Western models.

Primary school is compulsory for both sexes through six grades. A five-year middle school provides for those who pass rigorous entrance examinations, with alternative courses in technical, commercial, agricultural, and normal schools. Part-time continuation schools for employed young people were organized in the 1930s. Higher education of women was restricted narrowly and despite recent expansion it falls far short of that provided for men. Middle-school graduates are eligible for the entrance examinations of the three-year junior colleges that prepare for the five-year university courses. During the first two thirds of the modern period, private universities suffered disabilities in competing with the Imperial universities. Most private schools, however, were profit-making commercial ventures; the quality of their work can be estimated. The government finally forced the more mercenary schools out of business.

In the modern period, Japanese society has been a planned society, with education as a major tool of social control. Schools are designed to produce specific kinds of persons in predetermined numbers for the service of the state. Policies actually maintained, whether stated explicitly or followed implicitly, include the following:

(1) Conformity, not freedom, is the goal. The despised ideal of academic freedom, attributed to disorderly, unplanned democracies, is shunned and feared by administrators and teachers alike. Only an inarticulate minority desire freedom.

(2) Education aims to maintain class domination and further expansion of the Empire for the benefit of the oligarchy, not for the general good.

(3) The masses of the people are to be literate but docile and un-

thinking. Small numbers of highly trained experts are to be prepared for service in the government and the great corporations. An excess of intellectuals might breed revolution. Hence the types of schools, the numbers of each type, and the numbers of students admitted to each, are limited strictly.

(4) Indoctrination of the orthodox ideology is attained by an official monopoly of textbooks, curriculum, visual education, and radio. All textbook copyrights automatically become the property of the Ministry of Education.

(5) Control of normal schools and strict inspection of classroom teaching in all schools insures conformity on the part of the teaching staff.

(6) A Bureau of Educational Reform is charged specifically with "thought control" of students.

(7) To insure that women shall become the mothers of good soldiers, higher education of girls is limited. Girls' textbooks are simpler than those of boys for the corresponding grade; hence women are usually unable to pass the entrance examinations to the universities.

(8) Teachers receive low pay but high status in local communities. Their loyalty to the central administration is insured by frequent shifts of locality.

(9) Foreign geography and history are practically inaccessible to primary school students. Shintō myths are taught as history. In even the higher schools the social sciences, except for economics and biased history, are taboo. *Shūshin*, euphoniously translated "Morals" but really a blend of platitudes and superpatriotism, is a required subject at all levels.

(10) Physical exercise is stressed, and military training is compulsory for boys. In the 1930s a determined effort was made to suppress baseball, soccer, basketball and other Occidental games that inculcate sportsmanship; they were replaced by *jūdō* (which Americans call jiu jitsu), *kendō* (fencing), and other ancient sports that fostered the mentality of win or die.[14]

---

[14] A vivid, accurate account appears in Lamott's *Nippon: the Crime and Punishment of Japan* (New York, 1944). Under the American occupation the teaching of "militarism" has been banned, *Shūshin* courses were banned temporarily, athletics have been remodeled, and textbooks are being rewritten. Despite purges of teachers, it is not demonstrated that the old teachers can learn new ways.

The educational system manifests both American and German influences, but the indoctrinated ideology is *Kōdō*, the Shintō doctrine of the Imperial Way. German influence appears in the limitation of higher education and in university organization. The strenuous curriculum of the higher schools combines Chinese classical studies with Occidental languages and science. This heavy curriculum, together with the intense struggle to gain admission, places a heavy strain on students, as indicated by a high number of student suicides. Since families sacrifice rigorously to keep a boy in school, failure means loss of face for the family, to be atoned in suicide. Despite relentless efforts of the army and the police to purge all liberal teachers, a number of forward-looking minds manage to survive. At best, however, the schools offer a warped view of the world.

Local governmental units are responsible for financing the schools, but have no voice in policy or in the selection of teachers.

POLITICS AND GOVERNANCE.—Traditionally, government has two aspects: one, the mundane business of political control, the other a system of cosmic magic centered in the Emperor as embodiment of the *kami* of the sun goddess and the Imperial Ancestors. Secluded from the vulgar eye and remote from the storms of politics, the Emperor as super-father maintains the psychological unity of the great nation-household. In China's history an emperor was regarded as a representative of heaven; Japan's Emperor, however, is *kami*, indistinguishable from heaven and the gods. No mere vehicle of greater powers, his person sums up and incorporates the mystical *kami* of the nation.

The mystical status of the Emperor has served the interests both of *status quo* and revolution. Standpatters have been able to point with pride to the fact of Imperial sanction of an existing order; revolutionaries, on occasion, have been able to revolt on behalf of the Emperor, that he might be freed from the sinister machinations of evil leaders of a decadent clique. Japanese never have revolted against their Emperor; the purpose of civil war has been capture of the Emperor, whose sanction gave effect to whatever measures the victorious faction desired. From behind the scenes the dominant clique ordered the Emperor to issue certain commands, edicts, and rescripts; when he divinely uttered these as sacred words, the clique fell on their faces before him and joined all his loyal subjects in obedience.

The political, judicial, legislative, executive, and military activities of government have been patterned after Occidental forms in the modern period. Many observers have insisted that all this Western façade is merely another polite fiction. Partisan allegiance, for example, is not what it seems to be. Both *on* and *giri* channel loyalty and allegiance of every sort toward individuals or families; loyalty to principle or to an abstract ideal is inconceivable and nothing in Japanese history offers precedent therefor. Political parties mouth all the Occidental slogans of their craft without the slightest interest in what the slogans mean. Parties dissolve and regroup regardless of previous platforms; the key to this confusion is study of the personal and *kuni* loyalties of individuals. Japanese government is magically sanctioned government by men, never impersonal government by laws. "Written law is not something permanent, guaranteeing rights and privileges to individuals, but simply a temporary expression of the opinion of the official class as to what is good for the nation at a given moment." [15] Deep down in his feelings the commoner prefers to leave such matters to "those above," just as in family council he has his say and then accepts without question the decision of the head. To him obedience is the right and proper habit of life.

The Constitution of 1881, copied from the Prussian constitution, aimed to perpetuate the ruling oligarchy, not to insure popular rights. Although it established a bicameral Diet, the Emperor alone could give effect to legislation and his decisions were controlled by an unofficial, often anonymous, inner circle. Of the cabinet, only the ministers of army and navy had the right of direct access to the Emperor. Since these two always were officers on active duty and subject to orders from headquarters, a premier who questioned army and navy policies speedily faced a cabinet crisis. Should an unruly Diet fail to pass a budget bill, that of the preceding year automatically took effect—a fact that enabled the military to maintain effective though camouflaged control of government.

The "police state" has been a reality in Japan for more than three centuries. Secret police, uniformed police, "thought police," mili-

[15] R. K. Reischauer, *Japan: Government—Politics* (New York, 1939), p. 30.

tary police, and gendarmes practiced all the ancient feudal devices of control plus whatever could be learned from Europe's tyrants. Their objectives were to care for the people, to maintain the Imperial dignity, and to control ideas. For example, they boasted of their skill in opening mail without leaving a trace. In the 1920s they opened and copied the first six letters of any correspondence with a foreigner. If these were adjudged innocuous, subsequent correspondence was sampled.[16] Later this censorship became total; for example, publishers who had always hired dummy editors to serve jail terms came to the end of their resources. While the efficient apprehension of nonpolitical criminals accounts for most of the orderliness of the Japanese population, the real efforts of the police centered on persons suspected of "dangerous thoughts"—democratic leanings, socialist or communist sympathies, liberalism, labor union activities, or interest in the social sciences.

Courts assume that the Emperor's police make no mistakes. Accused persons are deemed guilty unless they prove their innocence. On occasion legal counsel and jury trial may be permitted, but one may be arrested, held in jail for months, tried, and convicted without knowing the charge or hearing of the trial until informed of the sentence. A prisoner who asks on what charge he is convicted may receive a blow on the mouth as his sole reply. The rare defendant who proves his innocence receives a suspended sentence lest the police lose face.

Procurators and judges, who are not lawyers, are appointed from civil service panels of specially trained men, and the status of lawyers is low. Legal technicalities and citation of precedents rarely delay court procedure; the intent of the law and the state of official opinion outweigh the text. It suffices that the general public know only that certain acts and ideas are forbidden. The fact of arrest prejudices the case against the prisoner; presumably the police know what they are doing. To a considerable extent this is true, for the police are a nation-wide organization relatively free from local corruption and influence.

---

[16] Statement to the author at the time by a ranking police officer. Is it cricket to smile when Military Government imports American police officers to "bring the Japanese police up to date"? Of course Japan now has a new "democratic police" force who carry clubs instead of swords and who wear uniforms not too different from those of G.I. Joe.

While the centralized organization of the modern government utilizes older elements developed in the feudal regime, one of the first steps in the Restoration of 1868 was designed to end the divisive *kuni* system and to reorient local loyalties toward the Emperor. *Kuni* boundaries supposedly vanished when the feudatories returned their fiefs to the Emperor in 1868. New administrative divisions called *ken* (prefectures) were established with studied disregard of the former *kuni*. Each *ken* is administered by a governor nominated by the Premier from the high ranking civil service personnel; he administers the *ken* as a branch of the central government, not as a locally self-governing unit. Since all governors are under the Home Ministry, this portfolio stands preeminent in the Cabinet. While the minor divisions have been changed repeatedly, in general *ken* are subdivided into *shi* (cities) and *gun* (counties). In turn the *gun* are divided into *machi* or towns and *mura* or townships. The English translations imply a misleading parallel with American civil divisions. The metropolitan areas (Tōkyō, Ōsaka, Kyōto) are not *ken* but fall into a separate category. Each *ken* has its assembly and council, which are consultative, not administrative or legislative. *Shi* and other local units are administered by mayors or other officials chosen by local bodies with approval of "those higher-up"; these men are not civil service appointees but usually are selected from the list of retired officials because of their experience and contacts. Most of the national bureaus—for example, the police and the Ministry of Agriculture—operate directly in local communities through their own personnel or through *ken* governors and mayors.

In every branch and at every level of government a highly trained civil service bureaucracy consistently wields power. The bureaucratic tradition goes back to feudal days when the *Shōguns* developed a bureaucracy to control the local feudal lords. In the 1930s the modern bureaucracy numbered about 450,000. The planned scarcity of highly educated men enhances the prestige of the civil service; the educational system channels the very ablest men into the bureaucracy. There are carefully graded ranks of civil service with appropriate titles, court privileges, and other perquisites of office. Outside of the *Zaibatsu* financial groups, the civil service includes nearly all of the men with administrative experience and "know-how"; consequently, whoever governs Japan is forced to rely on this reservoir of specialized knowledge. When in the 1920s business seemed to be gaining control of the government, the militarists inveighed against the bureaucracy. But when the army regained the upper hand they were powerless to govern without the bureaucrats. Consequently they used the civil service for their own ends.[17]

The scope of bureaucratic influence is very wide, for the government owns and operates not only schools and postal services, but also telephones, telegraphs, banks, railways, insurance and savings banking for the common people, arsenals, various public utilities, and monopolies of tobacco, camphor, and salt. Heavy industries and water transportation, while not operated directly by the government, are subsidized and officials make the final decisions. This makes jobs for bureaucrats. On the whole, civil service morale is high and rewards are in prestige rather than in high salaries; the bureaucrats regard themselves as arms of the Emperor just as do the army and navy. A detail sometimes unnoticed by foreign political scientists is the power that prestige-rewards for public service place in the hands of officials who control the granting of decorations, court honors, and similar coveted tokens of recognition.

ECONOMIC ORGANIZATION.—In the long course of the Tokugawa regime (c.1600–1864) Japan's economy moved from a feudal to a monetary basis. Despite frantic efforts to retain feudal usages, both *daimyō* and *samurai* found rice incomes inadequate to new urban luxuries. Irresistibly the control of wealth shifted from the feudal aristocracy to merchants, bankers, and money lenders who shrewdly backed the antifeudal Restoration of 1864–1868. The new government rewarded its supporters by giving them control of the nascent Occidental-type industrial and financial enterprises. Thus arose the *Zaibatsu*, the fewer than fifteen families who own 80 percent of Japan's fluid capital and most of the producing equipment.[18] Each of these great houses tended to develop a separate industrial empire consisting of mines, factories, steamship lines, banks, warehouses,

[17] Apparently the American military government did the same thing. They also needed men with "the know-how" and turned to the bureaucrats.

[18] *Zaibatsu*, a slang term, means "money crowd." The four largest are: Mitsui, Iwasaki (corporate name, Mitsubishi), Sumitomo, and Yasuda. The Imperial Household really was the largest *Zaibatsu;* the public did not know this and never dubbed it a *Zaibatsu*. General MacArthur ordered the *Zaibatsu* to dissolve. How this will work out is not clear at the time of writing.

textile mills, land, and other enterprises. Most of these industries receive fat governmental subsidies, either directly or through tax reductions, tariffs, and similar devices.

Zaibatsu patterns of organization show little originality. They modified and adopted the general scheme of the defunct feudal duchies—the only pattern of organization that Japanese know how to operate. In the new far-flung commercial and industrial enterprises the workers are analogous to feudal serfs, the technicians and salesmen occupy a position comparable to the *samurai*, and administrative officers even bear the same name as did their feudal counterparts, the *bantō*. With an autocratic head and a semisecret house law, each owning family parallels the proud house of a *daimyō;* this resemblance is not accidental but asserts a positive claim to social prestige.

Modern Japanese industry exhibits four main patterns of organization: (1) Small household industries produce consumption goods for the domestic market, goods made and sold by one family. (2) Handicraft and household industries that accept apprentices to live in the household supply traditional consumer goods that require special raw materials and skills, such as pottery, paper, and lacquer ware. An entrepreneur called a *toiya* supplies raw materials, finances production, and buys the finished product. (3) Workshops with five to fifty employees produce both native and Occidental-type consumer goods such as hosiery, bicycles, pencils, rayon cloth, or electrical gadgets for both domestic and export markets, and these also work through a *toiya*. (4) Factories with 100 or more workers produce Western-type capital goods and producers' goods for both foreign and domestic markets. In equipment and technology these companies approach Occidental standards.

The first two types of enterprises are basically household industries, dependent on hand labor and traditional tools. In the second type, occasional machines of modern design speed production—for example, rice mills and lathes. The workers, often relatives, may live together or in adjoining houses. The third type clings to the household pattern despite its unwieldiness for the larger unit. Power machines are used; a single machine tool installed in a frame building may provide the nucleus of such a shop. Many of these concerns operate in rural towns, financed by a peasant who has risen to a small landowner or money lender and thus accumulated a little capital. Operations may be seasonal to take advantage of part-time labor from peasant families. The fourth type has outgrown the household model and patterns after the West, but retains the paternal-feudal ideology. Workers may or may not be related but often hail from the same *kuni*. Some companies provide dormitories and welfare services; more frequently perhaps, workers shift for themselves and live under the worst of slum conditions. Labor unions first attained importance in the early 1920s, but police and legal restrictions combined to prevent the survival of any union whose activities went beyond mutual aid, hospitalization, or similar benefits.

Against the psychological heritage of feudalism the idea of free enterprise and the idea of communism are deemed equally radical. Both are strange, foreign, and subversive of *Kōdō*, Japan's "Imperial Way." Management and workers alike hesitate to act except on orders "from above."

Since the military controlled the government throughout the modern period, economic development was oriented toward war. Those heavy industries essential to war but normally unprofitable under Japanese conditions—such as arsenals, steel, heavy chemicals, motor vehicles, airplanes, light metals—were developed by the government or with governmental subsidy and protection. In free competition with foreign enterprise such industries might not survive. Private industries on a sound and profitable basis have included cotton textiles, paper, foodstuffs, and other consumer-oriented production.

In general, industries producing native-type consumer goods follow traditional patterns of technology, finance, labor utilization, and distribution. Industries oriented to the export market and to services copied from the Occident tend to follow the Occident in mechanization. Financing and control represent a combination of Western ideas with traditional practices, and something like this applies to labor utilization. Distribution in such industries follows Western models, especially German practices, since the bulk of the product must be sold competitively in foreign markets. Government support plays a large part in nearly all such industries, and here too German influence can be traced. Government influence also restricts popular consumption habits to traditional goods; except in the larger cities, imported goods rarely appear in stores, and many

Japanese-made products for the export market are unobtainable in Japan. Nevertheless the Japanese urban standard of living has risen well above standards prevailing in China, India, and other Asiatic countries—a situation evident not only in better nutrition, but in general use of bicycles, electric lights, cameras, beer and liquors, transportation and communication, Occidental-style clothing, education, the press and printed matter, public water supplies, public health and medicine.

The peasant, however, fares but little better than under feudalism. The tremendous increase in statistical tokens of national wealth makes slight difference in his lot. His household numbers six or seven persons and they cultivate less than two and a half acres of land. A field approximates an American city lot in area, and consequently he measures land in *tsubo*, a unit equal to 35½ square feet. Often his fields are scattered at varying distances from the hamlet where he lives. Rent and taxes take from 40 to 65 percent of all that he produces; even with the help of sidelines such as silk cocoons and part-time factory labor by his family he faces mounting debts on which usurers collect interest averaging 15 percent per annum. Illness or other crisis may force the sacrifice of a daughter, who is contracted out to a factory or to a brothel under terms that amount to sale.[19]

Like peasants everywhere he wants more land. More than two thirds of Japan's peasants rent all or part of their fields. Those who own land hold an undue proportion of hillside fields; the good rice land is beyond their means. Only since the Restoration has anyone been able to obtain a title to land. At that time the government attempted to establish titles in the interest of the actual cultivators. The legal work, however, was done hurriedly and considerable injustice resulted—as for example, when village lands were allotted to the family that happened to be cultivating them at the moment, although by ancient custom they had been worked in rotation by different families. A money tax replaced the feudal exactions in kind; the proceeds of this tax were used to retire the government bonds issued to former *daimyō* in compensation for their expropriated fiefs. A few ex-*daimyō*, by devious political maneuverings, managed

---

[19] Currency inflation in the 1940s probably helped the peasants; black markets paid them high prices for food. Presumably debts were paid off. Such prosperity, however, is ephemeral; there still are too many mouths for the land.

to emerge from the Restoration as large-scale landlords; but most of the fiefs passed into peasant hands. The new land tenure did not solve the problem of farm tenancy; it merely shifted the fortunes of individuals.

The Japanese landlord, however, must be understood against a Japanese background and not interpreted in terms of Europe's great landholders. All holdings, by Occidental standards, are infinitesimal. Ownership of more than 25 acres puts one among the top $94/100$ of one percent of Japan's landholders. Only six hundredths of one percent of all landowners hold more than 122 acres. Fifty percent of all landowners hold less than 1.236 acres; 75 percent of all landowners have less than 2.43 acres. Superficially, this looks like a wide distribution of ownership. If a large holder be defined in Japanese terms, however, the picture is more sinister, for a few "large" owners control the livelihood of a good many peasants.

In 1935, 3,415 individuals each owned 50 *chō* (122.5 acres) or more of cultivated land. Although they constituted but sixty-six thousandths of one percent of the landowning class, they owned a total of 980,000 acres—about 7 percent of all the cultivated land—on which lived more than 620,000 tenant farmer households. Thus about 16 percent of all tenant families were under the control of this small number of landlords.[20]

About 20 percent of all landowners do not cultivate their holdings. They work for salaries, operate shops, lend money, run hotels and brothels, and some are officials. Their holdings average under seven acres each. They represent a step above the peasant, a step achieved by careful management of a small inheritance, by working at some nonfarm occupation and renting out their land, with purchase of more land as savings accrue. Many of them illustrate the Occidental saying that "the poorer a landlord, the meaner he is." Much of the misery of the tenant farmer is explainable by the fact that he rents from a landlord who aims to support two households —his own and the tenant's—on land where one household formerly lived precariously.

The tenant ekes out a livelihood by making straw rope or mats in

[20] Quoted from S. Wakukawa, "The Japanese Farm Tenancy System," in *Japan's Prospect* (ed. Haring), p. 137. Data of land ownership and tenancy from the same source.

the wintertime, working for other farmers, earning wages in small near-by factories, or by such occupations as he can find in addition to farming. For example, a tenant farmer also sold fish in a small *mura* near the Japan Sea and his daily schedule was recorded.[21] Four days a week in summer he worked his land; three days weekly he dealt in fish. On "fish days" he arose long before dawn, dug 75–100 pounds of snow from a hillside cave where he had stored it in winter, wrapped the snow in straw mats, loaded it on his back and trotted eighteen miles to the Japan Sea where he traded most of the snow for fish. Packing the fish in the remaining snow, he set off at a dog trot for home, 2,200 feet above the sea. Going the rounds of the *mura* he sold most of the fish, and when his daily profits exceeded a *yen* (50¢) he was a happy man. Toward the end of July his snow would give out; thereafter he would rise at 3 A.M. and climb a mountain 8,000 feet up, where he could get snow from the high gullies. With this added distance, he would cover a total of over fifty miles before he finished at night.

Income taxes bear heavily on the rural population. A tenant who takes in 300 *yen* pays 35 percent in income tax; an urban merchant pays 12.5 percent and a manufacturer pays but 1.1 percent. The great corporations usually receive subsidies instead of paying taxes. The peasant has borne the burden of modernization and militarism; he has paid for the battleships, airplanes, and propaganda brochures, for the factories and luxuries of the *Zaibatsu;* his sons die for the glory of the Emperor. Even if adequate medical and dental care were accessible he could not afford them. His children, however, attend school and usually he has one electric lamp in his cottage. The government provides agricultural advisors and experiment stations help him select seed and reforest his hillside land. His younger sons make their way in the army or in the cities, but they bring their families to crowd his household and attenuate his meals when times are bad. Otherwise modern Japan has passed him by.

The Japanese nation, throughout the modern period, has been in transition from a feudal order to a new alignment whose pattern is not yet clear. By militaristic regimentation under an efficient bureaucracy Occidental patterns of governance and industrial economy have been enforced without changing the basic patterns of

[21] By the present author, from daily observation and interviews, 1924–25.

personality and social organization. By indoctrination and liberal use of professional thugs—a process that became increasingly galling—the people were led to sacrifice for a dream of world conquest. Despite opposition from a minority, popular support for that war was inevitable in a society that inculcated obedience and backed precept with force.

The old order symbolized by *Kōdō* doctrine, however, failed to afford psychological security. In his deepest feelings many a Japanese anticipated—and dreaded—a catastrophe whose outcome he could not foresee. The fear of democracy was based in a repressed intimation that inevitably a new order would come. The apparent acceptance of the American occupation involved a feeling of relief —relief that the catastrophe had come and that a spectacular father-figure had accepted responsibility for instituting the new order. Whether such a new order can endure unless numerous able administrators are trained in the ways of democratic action is debatable. It will be much easier for the Japanese to undergo conversion to any ready-made system of ideas that promises security. Democracy is not a ready-made system, and the art of personal autonomy and individual social responsibility is hard to learn.

SUGGESTED READINGS

Allen, C. G. Japan the Hungry Guest. New York, 1938. On economic life.

Anesaki Masaharu. History of Japanese Religion. London, 1930.

Ballou, Robert O. Shinto the Unconquered Enemy. New York, 1945. Source documents.

Benedict, Ruth F. The Chrysanthemum and the Sword. Boston, 1946. Moral code.

Borton, Hugh. Japan Since 1931. New York, 1940.

Byas, Hugh. Government by Assassination. New York, 1942. Pre-war politics.

Chamberlain, Basil H. Things Japanese. 5th ed. 1905. London and Kōbe, 1927.

Colegrove, Kenneth. Militarism in Japan. Boston, 1936. Government.

Cressey, George B. Asia's Lands and Peoples. New York, 1944. Geography.

Eliot, Sir Charles N. E. Japanese Buddhism. London, 1935.

Embree, John F. The Japanese Nation. New York, 1945. Systematic general account.

—— Suye Mura, a Japanese Village. Chicago, 1939. Village study.

Fennollosa, E. F. Epochs of Chinese and Japanese Art. 2 vols., New York, 1912.

Harada Jirō. The Gardens of Japan. London, 1928. Photographs of garden art.

Haring, D. G. "Aspects of Personal Character in Japan," *Far Eastern Quarterly*, November, 1946.

—— Blood on the Rising Sun. Philadelphia, 1943. War preparations.

—— ed., Japan's Prospect. Cambridge, Mass., 1946. Symposium by nine writers. Extensive classified bibliography. Appendices.

Henderson, H. G. The Bamboo Broom. Boston, 1934. Poetry, especially *haiku*.

Holtom, D. C. Modern Japan and Shinto Nationalism. Chicago, 1943; rev. ed., 1947.

—— The National Faith of Japan. London, 1938. Shintō; authoritative.

Jones, Thomas E. Mountain Folk of Japan. New York, 1926. Village survey.

Keenleyside, H. and A. F. Thomas, A History of Japanese Education and Present Educational System. Tōkyō, 1937.

Kuck, Loraine E. The Art of Japanese Gardens. New York, 1941.

LaBarre, Weston. "Some Observations on Character Structure in the Orient: the Japanese," *Psychiatry: Journal of the Biology and Pathology of Interpersonal Relations*, Vol. VIII, No. 3 (August, 1945).

Lamott, W. C. Nippon: the Crime and Punishment of Japan. New York, 1944. General; excellent section on education.

Maki, John M. Japanese Militarism, Its Cause and Cure. New York, 1945.

Murdoch, James. A History of Japan. 3 vols. London, 1910; Tōkyō, 1926; New York, 1927.

Norman, E. H. Japan's Emergence as a Modern State. New York, 1940, 1947. Transition from feudalism.

Redesdale, Lord. Tales of Old Japan. London, 1871. Folklore and popular tales.

Reischauer, R. K. Japan: Government—Politics. New York, 1939.

Robertson-Scott, J. W. The Foundations of Japan. New York, 1922. Rural Life.

Sansom, Sir George. Japan, a Short Cultural History. New York, 1931, 1943. The best general history; up to the modern period.

Schumpeter, E. C. G. Allen, M. Gordon, and W. Penrose. The Industrialization of Japan and Manchukuo, 1930–1940. New York, 1940. Economic analysis.

Sugimoto, E. I. A Daughter of the Samurai. Garden City, N.Y., 1925. Autobiography.

Trewartha, G. T. Japan: a Physical, Cultural and Regional Geography. Madison, Wis., 1945.

Uyeda Teijirō. The Small Industries of Japan. New York, 1938.

Waley, Arthur, tr. The Nō Plays of Japan. New York, 1922. Text and description.

Warner, Langdon. The Craft of the Japanese Sculptor. New York, 1936.

Schumpeter, E. C., Allen, M. Gordon, and W. Penrose. The Industrialization of Japan and Manchukuo, 1930-1940. New York, 1940. Economic analysis.

Sugimoto, E. I. A Daughter of the Samurai. Garden City, N.Y., 1925.
Autobiography.

Trewartha, G. T. Japan: a Physical, Cultural and Regional Geography.
Madison, Wis., 1945.

Ueda, Teijiro. The Small Industries of Japan. New York, 1938.

Waley, Arthur, tr. The Nō Plays of Japan. New York, 1922. Text and
description.

Warner, Langdon. The Craft of the Japanese Sculptor. New York, 1936.

INDEX

Abbas, Farhat, 453

Abdeljalil, Omar, 455

Abdul Aziz Ibn Saud of Arabia, 478, 489, 494, 520; achievements, 501, 502, 503; an autocratic monarch: government of, 505; aids to education, 506

Abdullah, ruler of Transjordan, 520, 522

Abraham, prophet, 525

Acculturation, North Africa: events that aid evaluation of, 440 ff.; China, 797-809; *see also* Culture

Achimota (Prince of Wales College), 371, 373, 400

Ackerman, Edward, 827

Acre, Territory of, Brazil, 212, 228; *map*, 227

Aden, Arabia, 485, 495

Administration, *see* Political structure, administration

"Administrations, Native": Africa, 357, 364

Aesthetics, Japan, 830, 860

Affaires Indigènes, Bureau des, 452

Afghanistan, 24, 25, 70, 464, 466, 484

Africa, uneasy peace of colonial empires: their importance, 13, 71; mineral raw materials in South America and, *map*, 69; sections with primitive peoples unable to attain independence, 71; white man's country, 76; similarity between South America and, 80; their differences, 81; population trends, 99, 100, 132; *tables*, 101, 104, 148; changes taking place in, compared with effects of Industrial Revolution, 312; one of two great colonial areas: causes of lowly status, 659; will follow course of revolution against outside domination, 724; *see also names of countries and colonies, e.g.*, Algeria; Niger Colony; Rhodesia; *etc.*

—— east and south: region, people, resources, economy, 70 f., 75-80, 91, 271 ff.; race relations and policies, 271-

330; clash of cultures, 271-74; white colonization, 271; and distribution, 272; *map*, 273; physical background: topography, climate, 274-76; affinities between the Central Protectorates and countries of the East African plateau: area, population, land reserves, *with table*, 276-82; location of native reserves, 276, 277, 280; agriculture, 282-300; hazards of farming, 282-86; government and legal protection of Europeans, 283, 284, 289; prospects of settlers in tropics, 286-88; the African peasant, 288-90; depressed areas, 291-94; deplorable health and housing conditions among natives, 293 f., 316-21; education: race discriminations, 293, 312-15; migrant workers, 294-97; program for rehabilitation, 297-300; mines and industries, 300-304; labor market: industrial color bar, 304-12; legal discrimination against non-Europeans, 305 ff., 308, 319, 325, 326; extent of their participation in labor welfare legislation, 321-23; past and possible future of race relations, summarized and appraised, 323-29; racial policy of imperial government in British dependencies, 327, 328; resistance to segregation: political education and action of Indians and Colored, 327 f.; forces that may weaken structure of caste society, 328 f.

—— north: region, people, resources, economy, 14-21, 25, 405-60; role in world affairs, 25; the six political provinces and their European overlords, 405; agriculture, 406, 411, 441 f.; *map*, 407; people and their history in historic times: invasions and infiltrations, 408 ff.; regional isolation, always an extension of some other culture area, 408; languages, 409, 410, 446, 449; economic and governmental system of

ABOUT THE AUTHORS

WILLIAM R. BASCOM, Assistant Professor, Department of Anthropology, Northwestern University, has spent four years in West and Central Africa, three on missions for the United States Government and one as an anthropologist. He is the author of numerous articles on African folklore, divination, art, and social organization.

JAMES BATAL, newspaperman and editor, spent eighteen months in the Near East with the Office of War Information doing special research for the United States Government on the subject of Western (particularly American) impacts on the Arab world. As Nieman Fellow at Harvard University, 1945–46, he specialized in studies on the Near East.

CARLETON STEVENS COON served as Professor of Anthropology at Harvard University from 1925 to 1948, when he became Professor of Anthropology at the University of Pennsylvania and Curator of Ethnology, The University Museum. He has explored various parts of Morocco, Albania, Ethiopia, and Arabia, and has written extensively on the anthropology and ethnology of the Near East.

JOHN P. GILLIN is Professor of Anthropology and Research Professor in the Institute for Research in Social Science, University of North Carolina. He has led several expeditions specializing in the cultural anthropology of Latin America, and his contributions to the literature are many.

DOUGLAS GILBERT HARING, Professor of Anthropology, Maxwell Graduate School of Citizenship and Public Affairs, Syracuse University, has lived, worked, and studied in Japan, and has written and lectured on its history and culture. His publications include *The Land of Gods and Earthquakes*, 1929, *Blood on the Rising Sun*, 1943, and (ed.) *Japan's Prospect*, 1946.

FRANCIS L. K. HSU, formerly Professor of Anthropology, National Yunnan University, Kunming, is now Assistant Professor of Anthropology at Northwestern University. Besides contributions to anthropological and sociological journals, he has written *Magic and Science in Western Yunnan*, 1943, and *Under the Ancestors' Shadow*, 1948, and (ed.) *Labor and Labor Relations in the New Industries of Southwest China*, 1943.

RAYMOND KENNEDY, Professor of Sociology, Yale University, lived for four years in the Philippine Islands and the East Indies. He is a member of the Board of Directors of the Far Eastern Association and of the Board of Trustees of the Institute of Pacific Relations. Besides numerous articles, his publications include *Islands and Peoples of the Indies*, 1943, *Islands and Peoples of the South Seas*, 1945, and *Bibliography of Indonesian Peoples and Cultures*, 1945.

RALPH LINTON, Sterling Professor of Anthropology, Yale University, is editor of the Viking Fund Publications in Anthropology, and author of numerous books and monographs.

HOWARD A. MEYERHOFF is Professor of Geology and Geography at Smith College. He has worked and traveled widely in Latin America, and for many years has taught courses on the world's natural resources, applying this knowledge as consultant to foreign and domestic government agencies and private corporations.

STEPHEN W. REED served with the carrier task forces in the Pacific, 1942–45, and in 1946 returned to Yale University as Assistant Professor of Sociology. He has written papers and articles on the Kwoma People of Northeast New Guinea, and, with John W. M. Whiting, *The Making of Modern New Guinea*, 1943.

F. L. W. RICHARDSON, JR., spent several years studying the geography and anthropology of the Near East at Harvard and Princeton. For two years he was engaged in geographical exploration and anthropological field work in Iraq, and in World War II he returned to the Near East as a Lend Lease representative in Egypt and as an independent observer in Palestine, where he witnessed the never-ending conflict between power groups. Now at the Yale Labor and Management Center, he is continuing the study of conflicts, his present field being our own industrial society.

H. J. SIMONS is a lecturer in Native Law and Administration at the University of Capetown, and the author of numerous contributions to journals and periodicals.

ALICE THORNER was wartime Analyst on India for the Foreign Broadcast Intelligence Service of the Federal Communications Commission and contributed to the symposium edited by Gardner Murphy on *Human Nature and Enduring Peace.*

DANIEL THORNER is Research Assistant Professor of Economic History and is on the faculty of the Program of Regional Studies on South Asia at the University of Pennsylvania. As a member of the Walter Hines Page School of International Relations, Johns Hopkins University, he contributed to a symposium on *Chinese Central Asia.*

CHARLES R. WAGLEY, Assistant Professor, Department of Anthropology, Columbia University, has spent many years in Brazilian research. In addition to studies of the Tapirape and other aboriginal groups of the Amazonian region, he has made extensive studies of modern Brazilian rural communities. He is the author of numerous articles and monographs.